POLYTOPES:
Abstract, Convex and Computational

NATO ASI Series

Advanced Science Institutes Series

A Series presenting the results of activities sponsored by the NATO Science Committee, which aims at the dissemination of advanced scientific and technological knowledge, with a view to strengthening links between scientific communities.

The Series is published by an international board of publishers in conjunction with the NATO Scientific Affairs Division

A Life Sciences **B Physics**	Plenum Publishing Corporation London and New York
C Mathematical and Physical Sciences **D Behavioural and Social Sciences** **E Applied Sciences**	Kluwer Academic Publishers Dordrecht, Boston and London
F Computer and Systems Sciences **G Ecological Sciences** **H Cell Biology** **I Global Environmental Change**	Springer-Verlag Berlin, Heidelberg, New York, London, Paris and Tokyo

NATO-PCO-DATA BASE

The electronic index to the NATO ASI Series provides full bibliographical references (with keywords and/or abstracts) to more than 30000 contributions from international scientists published in all sections of the NATO ASI Series.
Access to the NATO-PCO-DATA BASE is possible in two ways:

– via online FILE 128 (NATO-PCO-DATA BASE) hosted by ESRIN, Via Galileo Galilei, I-00044 Frascati, Italy.

– via CD-ROM "NATO-PCO-DATA BASE" with user-friendly retrieval software in English, French and German (© WTV GmbH and DATAWARE Technologies Inc. 1989).

The CD-ROM can be ordered through any member of the Board of Publishers or through NATO-PCO, Overijse, Belgium.

Series C: Mathematical and Physical Sciences - Vol. 440

POLYTOPES: Abstract, Convex and Computational

edited by

T. Bisztriczky

Department of Mathematics and Statistics,
The University of Calgary,
Calgary, Alberta, Canada

P. McMullen

Department of Mathematics,
University College,
London, U.K.

R. Schneider

Mathematisches Institut,
Universität Freiburg,
Freiburg, Germany

and

A. Ivić Weiss

Department of Mathematics and Statistics,
York University,
North York, Ontario, Canada

Kluwer Academic Publishers

Dordrecht / Boston / London

Published in cooperation with NATO Scientific Affairs Division

Proceedings of the NATO Advanced Study Institute on
POLYTOPES: Abstract, Convex and Computational
Scarborough, Ontario, Canada
August 20 – September 3, 1993

A C.I.P. Catalogue record for this book is available from the Library of Congress.

ISBN 0-7923-3016-1

Published by Kluwer Academic Publishers,
P.O. Box 17, 3300 AA Dordrecht, The Netherlands.

Kluwer Academic Publishers incorporates the publishing programmes of
D. Reidel, Martinus Nijhoff, Dr W. Junk and MTP Press.

Sold and distributed in the U.S.A. and Canada
by Kluwer Academic Publishers,
101 Philip Drive, Norwell, MA 02061, U.S.A.

In all other countries, sold and distributed
by Kluwer Academic Publishers Group,
P.O. Box 322, 3300 AH Dordrecht, The Netherlands.

Printed on acid-free paper

Printed in the Netherlands

TABLE OF CONTENTS

PREFACE

The NATO Advanced Study Institute devoted to *Polytopes: Abstract, Convex and Computational* took place in Scarborough, Ontario, Canada, from 20 August until 3 September 1993. There were fourteen principal lecturers (one of whom was unfortunately absent because of ill health), and around sixty other participants, several of whom gave talks in sessions additional to the main lectures. There were also conducted several workshops on various topics, as well as a mathematical film show put on by Tom Banchoff. The contributions of the principal lecturers are printed in these *Proceedings*, together with a number of other submitted (and refereed) articles. The principal contributions are briefly discussed in the sections which follow.

Polygons and polyhedra appear to have been among the very earliest objects of systematic mathematical investigation, and there is a nearly continuous history of their study. Indeed, two of the three strands of the meeting carry on traditional areas of Greek mathematics. It has sometimes been claimed that Euclid's *Elements* is intended to be a treatise on the five regular (or Platonic) solids. Whether or not that is genuinely the case, certainly *Book XIII* is devoted to a classification of the five regular convex polyhedra. Moreover, the two strands of convexity and regularity remained closely entwined until at least the last century. The third strand, that of the computational aspects of polytope theory, has grown naturally out of convexity, so that it still shares much with it. Thus, although at first sight it might have seemed a little unnatural to yoke together such disparate subjects that now come under the heading of polytope theory, in practice the participants at the meeting in different areas found much common ground. This tends to foster the belief that it is likely that the three areas will continue to cross-fertilize each other for much time to come.

1. Abstract Polytopes

The theory of abstract polytopes, particularly of those that are regular, involves a mixture of algebra and geometry. As we have already mentioned, regular polytopes in two and three dimensions (polygons and polyhedra) were among the earliest objects to be treated mathematically, in the sense that we would now conceive of mathematics. As symmetric objects, they received much attention considerably earlier, for example, as cubical and even dodecahedral dice. Already in the first mathematical studies in the Pythagorean school, the non-convex regular pentagram occurs, though perhaps more than a little mystically, as a symbol of good luck. Nevertheless, some beginning of the parting of the ways can already be sensed.

Regular polytopes in general cannot seriously be investigated without the use of group theory, although those that are convex can be treated synthetically. Naturally, the symmetry groups of the regular polygons and convex polyhedra were among the earliest groups to be looked at. However, the systematic classification of the groups of regular polytopes and their close relatives undoubtedly owes most to Donald Coxeter, although with his typical generosity, in his article *The evolution of the Coxeter-Dynkin diagram* he gives much credit to earlier researchers. It is no accident, though, that the groups which lie at the basis of regular polytope theory bear his name. Coxeter describes the genesis of an important family of finite reflexion groups (their Coxeter diagrams are certain unmarked trees, and include E_n for $n = 6, 7, 8$), and their connexion with del Pezzo surfaces . We were fortunate in one of his lectures to be given a new and strikingly simple proof of the finiteness criterion for this class of groups.

Arjeh Cohen's article *Recent results on Coxeter groups* describes some of the latest developments in the theory. We now know much more about Coxeter groups than we did a few years ago, particularly about their subgroups structure (for example, about which of their parabolic subgroups are conjugate). Among other topics considered were the (solved) word problem for Coxeter groups, and their automaticity. An important tool here is the Moussang complex, which sits in the chamber system obtained from a Coxeter group. It is interesting here how explicitly computable bounds play a vital part in the theory.

Two of the remaining articles under this heading are closely linked. The first, *Modern developments in regular polytopes* by Peter McMullen, surveys how the subject has evolved from earliest times, into what appears to be the main thrust of research at present. An increasingly important rôle is now played by group theory, as was already foreshadowed by Coxeter's work on the groups of regular maps of various kinds, and abstract regular polytopes can now be defined in such a way that their study becomes a branch of pure group theory. This basic theory is outlined, together with many standard constructions to obtain new regular polytopes from previously existing ones. Finally, the theory of realizations brings the abstract back to the geometric.

Describing all the regular polytopes is a somewhat hopeless task. But if attention is restricted to certain special classes, rather more can be said. In his article *Classification of locally toroidal regular polytopes*, Egon Schulte surveys what progress has been made in one particular area, where the basic building blocks for constructing regular polytopes from those of lower rank are allowed to be toroidal (that is, among the simplest which are not of spherical type, corresponding to the convex regular polytopes), as well as spherical. A fairly complete description is now available in this special case, although some gaps remain to be filled even when the rank is small.

Branko Grünbaum's article *Polyhedra with hollow faces* (it is to be regretted that, through illness, these lectures were not delivered in person) takes a new look at (regular) polyhedra in ordinary space. Though he adopts a slightly different viewpoint from the previous two authors, his new results can be translated into the language of that theory. They provide examples of constructing new polyhedra as covers of old ones, and give continuously deformable families of isomorphic polyhedra joining (for instance) dual regular ones.

2. Convex Polytopes

Convex polytopes are of increasing utility in areas of computation mathematics, as we shall see in the next section. But they are still intensively studied for their own sake, and a rich theory now exists, which has almost changed out of all recognition over the last thirty years or so. However, the older traditions are also being maintained, and we shall begin with these.

For many reasons, polytopes are often easier to treat than general convex bodies, and so on practical as well as theoretical grounds, how closely a convex body can be approximated by a polytope with given properties (such as the number of its vertices) is of great importance. Peter Gruber's article *Approximation by convex polytopes* describes what is known about such problems. Naturally, more is known about the planar case, but even in higher dimensions there are many good asymptotic estimates available. It is interesting to note that the algorithmic aspect of the problems provides a connexion with the computational aspects.

A classical topic is discussed in *Polytopes and Brunn-Minkowski theory* by Rolf Schneider. Minkowski summation is a fundamental operation applying to convex bodies, and many open questions remain concerning mixed volume inequalities which result from it (for instance, that of Aleksandrov-Fenchel), particularly such as when equality occurs, as well as those about the closely related surface area measures. For polytopes, problems of this kind may be more approachable; this article describes several new results in this direction.

An extremely lively area of the theory concerns the combinatorics of polytopes. The exact conditions characterizing the f-vectors (that is, vectors of numbers of faces of various dimensions) of simple (or simplicial) d-polytopes were conjectured in 1970 by McMullen, and established around 1979–80 by Billera & Lee (sufficiency) and Stanley (necessity). However, only more recently has it been possible to substitute the original deep proof of Stanley by one which works strictly within convexity itself, and a few related questions still remain open. In his article *Generalized stress and motions*, Carl Lee outlines the background to another possible approach to these problems, derived from notions of stress and rigidity. He demonstrates the interesting and unifying rôle played by the volume form of a simple polytope.

Similar questions can be asked about general polytopes. For example, one might demand a description of their f-vectors, or even more ambitiously, their flag-vectors, which enumerate the incidences between their faces of various dimensions. In her article *Face numbers and subdivisions of convex polytopes*, Margaret Bayer discusses such problems, and how subdivisions, and especially triangulations, affect the h-vector, which encodes a part of this information which is important for the connexions with toric varieties which provided Stanley's original proof mentioned above.

Richard Stanley's article *A survey of Eulerian posets* delves further into some of the more purely combinatorial aspects of polytope theory. The Eulerian posets of the title generalize the face lattices of polytopes; their utility comes from the fact that the combinatorial formula which calculates the h-vector of a polytope (through intersection homology theory) extends naturally to them. An important property to be retained here is that suitably associated rings be Cohen-Macaulay. Connexions with various kinds of subdivision (again) and the Ehrhart formula for the lattice-point enumerator are explored here.

Gil Kalai discusses *Some aspect of the combinatorial theory of convex polytopes* in his article. In fact, the article is centred around three results on the facial structure of polytopes. The first and third are due to Perles, and are as yet unpublished; they concern the number of (combinatorially) distinct k-skeleta of d-polytopes with $d+b$ vertices for fixed k and b, and how the facial structure of a simplicial d-polytope is determined by its $\lfloor d/2 \rfloor$-skeleton. The second gives an elementary proof of the theorem of Blind & Mani, showing that the graph of a simple polytope determines its combinatorial type.

3. Computation

Two of the lecture series (and three of the lecturers) fall into this section. First, there is the long article *Recent advances in computational convexity* (which is divided into two parts) by Peter Gritzmann and Victor Klee. The topic of concern here is the difficulty, in the precise sense of finding bounds, of computing even such quantities as the volume of a polytope. A polytope may be presented as the convex hull of its vertices, or as the intersection of the half-spaces determined by its facets; which presentation is given crucially affects the consequent difficulty of the appropriate computation. A nice feature in this is the notion of an oracle, which lives up to its name in only providing restricted information about bodies when it is appealed to. Some of the problems which naturally occur here turn out to be NP-hard, or even NP-complete (or variants involving actual numbers), in terms of their input data.

The article *Graphs of polytopes: theory and applications* by Peter Kleinschmidt could have been listed in the previous section, on account of the theoretical nature of a large part of it. However, although some of the problems discussed here, such as the Hirsch conjecture about path lengths in the edge-graph of a polytope, are severely classical, the tenor of the article is computational. Other problems here involve the dual transportation polyhedron, and its applications to the classification of human chromosomes.

4. Submitted Articles

We shall not comment on the submitted articles individually; it is worth repeating that they have been refereed before acceptance. Although the primary purpose of the conference was instructional, a number of the "students" (as the non-principal participants seem officially to have been called) also contributed talks, and these articles provide a record of some of them. One at least amounts to a substantial survey of a lively area of computational convexity.

London, 9 May 1994 — Peter McMullen

ACKNOWLEDGEMENTS

Firstly, I wish to express my gratitude to the Scientific and Environmental Affairs Division of N.A.T.O., the Junta Nacional de Investigação Científica e Tecnológica of Portugal, the Scientific and Technical Research Council of Turkey, the Department of Mathematics and Statistics of the University of Calgary, and the Department of Mathematics and Statistics of York University for the financial support provided for this Advanced Study Institute.

To P. McMullen, R. Schneider and A. Ivić Weiss, my fellow organizers and editors, I am particularly grateful for their help and support, which made it possible for me to proceed with this meeting and to prepare this volume.

I am most appreciate of all the effort expended by the participants to make the Institute a productive and enjoyable one. In particular, special thanks to R. Erdahl, J. Fine and N.W. Johnson for organizing the workshops, J. Schaer for assisting with the finances, and W. Whiteley for looking after the transportation on the day of arrival.

At this point, I wish to acknowledge my debt of gratitude to Joanne Mellard for assuming, and fulfilling exceptionally well, the duties of administrative assistant and technical editor. For their help in the preparation of these proceedings, my thanks also to J. Kissinger and T. Morrison.

Finally, to my wife Margaret and daughter Deborah, my heartfelt thanks for helping me get through these past two years.

Calgary, 10 May 1994 Tibor Bisztriczky

NATO ADVANCED STUDY INSTITUTE

on

POLYTOPES: Abstract, Convex and Computational

The Guild Inn, Scarborough, Canada
August 23 – September 4, 1993

PARTICIPANTS

ALTINEL, Kuban – Istanbul, Turkey
ANDERSON, Laura – Cambridge, U.S.A.
BANCHOFF, Thomas – Providence, U.S.A.
BAYER, Margaret – Lawrence, U.S.A.
BISZTRICZKY, Tibor – Calgary, Canada
BOKOWSKI, Jürgen – Darmstadt, Germany
BORGWARDT, Karl – Augsburg, Germany
BÖRÖCZKY, Károly jr. –Budapest, Hungary
BRASS, Peter – Greifswald, Germany
BREDA, Ana Maria – Coimbra, Portugal
BREHM, Ulrich – Dresden, Germany
BURGIEL, Heidi – Seattle, U.S.A.
COHEN, Arjeh – Eindhoven, The Netherlands
COXETER, Harold Scott – Toronto, Canada
CROWE, Donald – Madison, U.S.A.
DAWSON, Robert – Halifax, Canada
DEZA, Michel – Paris, France
DINH, Thi – Calgary, Canada
ERDAHL, Robert – Kingston, Canada
ERDOGAN, Ismael Haki – Instanbul, Turkey
FINE, Jonathan – Cambridge, United Kingdom
FISHER, Chris – Regina, Canada
GOODEY, Paul – Norman, U.S.A.
GRITZMANN, Peter – Trier, Germany
GRUBER, Peter – Vienna, Austria
HENK, Martin – Siegen, Germany
HUBER, Birkett – Ithaca, U.S.A.
IM HOF, Hans–Christoff – Basel, Switzerland
JOCKUSCH, William – Ann Arbor, U.S.A
JOHNSON, Norman – Norton, U.S.A.
KALAI, Gil – Jerusalem, Israel
KÁROLYI, Gyula – Bonn, Germany
KELLERHALS, Ruth – Bonn, Germany
KLEE, Victor – Seattle, U.S.A.
KLEINSCHMIDT, Peter – Passau, Germany
KÜHNEL, Wolfgang – Duisburg, Germany
KUPERBERG, Wlodzimierz – Auburn, U.S.A.
LASSAK, Marek – Bydgoszcz, Poland

LEE, Carl – Lexington, U.S.A.
LEYTEM, Charles – Chruchten, Luxembourg
LINDERBERG, Jan – Aarhus, Denmark
LUDWIG, Monika – Vienna, Austria
MAHJOUB, Ridha – Brest, France
MARTINI, Horst – Chemnitz, Germany
MCMULLEN, Peter – London, United Kingdom
MONSON, Barry – Fredericton, Canada
MOODY, Robert – Edmonton, Canada
MUHLHERR, Bernhard – Tübingen, Germany
NOSTRAND, Barbara – Boston, U.S.A.
PATERA, Jiri – Montréal, Canada
POCCHIOLA, Michel – Paris, France
POPOV, Antony – Sofia, Bulgaria
RIVAL, Ivan – Ottawa, Canada
ROSE, Lauren – Wellesley, U.S.A.
ROTE, Günter – Graz, Austria
RUSH, Jason – Seattle, U.S.A.
SCHAER, Jonathan – Calgary, Canada
SCHMITT, Peter – Vienna, Austria
SCHNEIDER, Rolf – Freiburg, Germany
SCHULTE, Egon – Boston, U.S.A.
STANLEY, Richard – Cambridge, U.S.A.
THOMEIER, Siegfried – St. John's, Canada
THOMPSON, Anthony – Halifax, Canada
TSINTSIFA, Lydia – Thessaloniki, Greece
URRUTIA, Jorge – Ottawa, Canada
WEISS, Asia Ivić – North York, Canada
WHITE, Neil – Gainsville, U.S.A.
WHITELEY, Walter – North York, Canada
WILKER, John – Toronto, Canada
YU, Jietai – Notre Dame, U.S.A.
ZIEGLER, Günter – Berlin, Germany
ZVENGROWSKI, Peter – Calgary, Canada

PROGRAMME

MONDAY, August 23

Chairman:	**P. McMullen**
9:10–10:00	**H.S.M. Coxeter** – The del Pezzo surfaces
10:30–11:20	**R. Schneider** – Polytopes and the Brunn–Minkowski Theory
11:30–12:00	APPLICATIONS **I. Rival** – Order, drawing and surfaces
14:00–15:00	WORKSHOP: General convex polytopes (Organizer: **J. Fine**)
Chairman:	**T. Banchoff**
15:40–16:40	**P. Gritzmann** and **V. Klee** – Recent advances in computational convexity
16:50–17:40	**P. Gruber** – Approximation by convex polytopes
17:50–18:10	SPECIAL POLYTOPES **C. Leytem** – Hidden symmetries of the snub dodecahedron

TUESDAY, August 24

Chairman:	**A. Ivić Weiss**
9:10–10:00	**P. McMullen** – Modern developments in regular polytopes
10:30–11:20	**R.P. Stanley** – A survey of Eulerian posets
11:30–12:00	POLYTOPES AND RIGIDITY **W. Whiteley** – Stresses, liftings and parallel drawings
14:00–15:00	WORKSHOP: General convex polytopes (Organizer: **J. Fine**)
Chairman:	**A. Ivić Weiss**
15:40–16:30	**A.M. Cohen** – Recent results on Coxeter groups

WEDNESDAY, August 25

Chairman:	**D. Crowe**
9:10–10:00	**H.S.M. Coxeter** – Uniform polytopes
10:30–11:20	**R. Schneider** – Polytopes and the Brunn–Minkowski Theory
11:30–12:00	ALGORITHMS **K.H. Borgwardt** – Probabilistic analysis of an algorithm for determining the convex hull in *d*-dimensional spaces
14:00–15:00	WORKSHOP: Polytopes – regular, uniform and symmetric (Organizer: **N.W. Johnson**)

Chairman:	**R. Kellerhals**
15:40–16:30	**V. Klee** – Recent advances in computational convexity
16:40–17:30	**P. Gruber** – Approximation by convex polytopes
17:40–18:40	APPLICATIONS
	S. Thomeier – Applications of polytopes in coding theory
	J. Linderberg – Finite element methods and partial waves: discrete quantum mechanics on the {3,3,5} polytope

THURSDAY, August 26

Chairman:	**M. Bayer**
9:10–10:00	**P. McMullen** – Modern developments in regular polytopes
10:30–11:20	**R.P. Stanley** – A survey of Eulerian posets
11:30–12:00	APPLICATIONS
	P. Brass – On some Erdős distance problems in Euclidean and normed spaces
14:00–15:00	WORKSHOP: Polytopes – regular, uniform and symmetric (Organizer: **N.W. Johnson**)
Chairman:	**N. White**
15:40–16:30	**A.M. Cohen** – Recent results on Coxeter groups
16:40–17:40	PACKINGS AND COVERINGS
	R. Dawson – Tileorders
	M. Henk – The sausage conjecture for higher dimensions

FRIDAY, August 27

Chairman:	**C. Fisher**
9:10–10:00	**H.S.M. Coxeter** – Kaleidoscopes
10:30–11:20	**R. Schneider** – Polytopes and the Brunn–Minkowski Theory
11:30–12:00	APPLICATIONS
	W. Kühnel – Manifolds as subcomplexes of polytopes
14:00–15:00	WORKSHOP: General convex polytopes (Organizer: **J. Fine**)
Chairman:	**M. Deza**
15:40–16:30	**P. Gritzmann** – Recent advances in computational convexity
16:40–17:30	**P. Gruber** – Approximation by convex polytopes
17:40–18:10	MEASURES AND POLYTOPES
	A.C. Thompson – Zonotopes as solutions to the isoperimetric problem

SATURDAY, August 28

Chairman:	**B. Monson**
9:10–10:00	**P. McMullen** – Modern developments in regular polytopes
10:30–11:20	**A.M. Cohen** – Recent results on Coxeter groups
11:30–12:15	PROBLEM SESSION

MONDAY, August 30

Chairman:	**R.P. Stanley**
9:10–10:00	**E. Schulte** – Classification of locally toroidal regular polytopes
10:30–11:20	**M. Bayer** – Face numbers and subdivisions of convex polytopes
11:30–12:00	ALGEBRA AND POLYTOPES
	J. Fine – Some reflections on convex polytopes and Betti numbers
14:00–15:00	WORKSHOP: Polytopes – regular, uniform and symmetric (Organizer: **N.W. Johnson**)
Chairman:	**A.C. Thompson**
15:40–16:30	**G. Kalai** – Some aspects of the combinatorial theory of polytopes
16:40–17:40	MEASURES AND POLYTOPES
	R. Kellerhals – On volumes of non–Euclidean polytopes
	M. Lassak – On parallelotopes of possibly large volume contained in a convex body

TUESDAY, August 31

Chairman:	**P. Gruber**
9:10–10:00	**C.W. Lee** – Generalized stress and rigidity
10:30–11:20	**P. Kleinschmidt** – Graphs of polytopes: theory and applications
11:30–12:00	APPLICATIONS
	H.I. Erdogan –A study on webs and quasigroups
14:00–15:00	WORKSHOP: L-polytopes (Organizer: **R. Erdahl**)
Chairman:	**J. Wilker**
15:40–16:30	**G. Kalai** – Some aspects of the combinatorial theory of polytopes
16:40–18:10	ALGEBRA AND POLYTOPES
	P. Goodey – Translation invariant valuations
	B. Muhlherr – An alternative proof of the classification of the spherical Coxeter diagrams
	G.M. Ziegler – Coxeter associahedra

21:00–22:00	VIDEO AND SLIDE PRESENTATION **T. Banchoff** – Interactive computer graphics and four–dimensional geometry

WEDNESDAY, September 1

Chairman:	**H.C. ImHof**
9:10–10:00	**E. Schulte** – Classification of locally toroidal regular polytopes
10:30–11:20	**M. Bayer** – Face numbers and subdivisions of convex polytopes
11:30–12:30	LATTICE POLYTOPES **M. Deza** – The ridge graph of the metric polytope and related graphs
14:00–15:00	WORKSHOP: General convex polytopes (Organizer: **J. Fine**)
Chairman:	**W. Kuperberg**
15:40–16:30	**P. Kleinschmidt** – Graphs of polytopes: theory and applications
16:40–18:10	MEASURES AND POLYTOPES **G. Károlyi** – Irregularities of point distributions and convex polytopes **H.C. ImHof** – Complex volumes of orthoschemes **U. Brehm** – Polyhedral surfaces
21:00–22:00	PROBLEM SESSION

THURSDAY, September 2

Chairman:	**W. Whiteley**
9:10–10:00	**E. Schulte** – Classification of locally toroidal regular polytopes
10:30–11:20	**C.W. Lee** – Generalized stress and rigidity
11:30–12:20	**G. Kalai** – Some aspects of the combinatorial theory of polytopes
14:00–15:00	WORKSHOP: General convex polytopes (Organizer: **J. Fine**)
Chairman:	**T. Bisztriczky**
15:40–16:30	**P. Gritzmann** and **V. Klee** – Recent advances in computational convexity
16:40–17:40	SPECIAL POLYTOPES **A. Mahjoub** – On 2–connected subgraph polyhedra **P. Zvengrowski** – Regular polyhedra and Seifert fibrings of 3–manifolds

FRIDAY, September 3

Chairman:	**A.M. Cohen**
9:10–10:00	**C.W. Lee** – Generalized stress and rigidity
10:30–11:20	**P. Kleinschmidt** – Graphs of polytopes: theory and applications
11:30–12:00	REGULAR POLYTOPES
	H. Martini – Regular polytopes and their generalizations from a metrical point of view
Chairman:	**R. Schneider**
15:40–16:30	**M. Bayer** – Face numbers and subdivisions of convex polytopes
16:40–18:10	ABSTRACT POLYTOPES
	N.W. Johnson – Regular compounds
	B. Nostrand – Chiral polytopes
	J. Bokowski – On methods of proving abstract polytopes to be concrete

RECENT RESULTS ON COXETER GROUPS

ARJEH MARCEL COHEN
Fac. Wisk. en Inf.
Technische Universiteit Eindhoven
5600 MB Eindhoven
The Netherlands

Abstract. The last few years have been good for the knowledge of Coxeter groups: the conjugacy problem has been solved, Coxeter groups have been shown to be automatic, and the structure of subgroups has been further exploited. In these notes, we survey some of these results, thus providing a sequel to three earlier ASI lectures on Coxeter groups.

1. Basic definitions

Although this paper updates Cohen [1991] we facilitate its reading by repeating some of the basic definitions. The "classical" reference for Coxeter groups is Bourbaki [1968]. Besides that, the recent introductions in Humphreys [1990] and Scharlau [1993] are recommended.

1.1. Coxeter groups

A *Coxeter matrix* of rank n is an $n \times n$ matrix $M = (m_{i,j})_{1 \le i,j \le n}$ with $m_{i,i} = 1$ and $m_{i,j} = m_{j,i} > 1$ (possibly ∞) for all $i, j \in \{1, \ldots, n\}$ with $i \neq j$. The *Coxeter group* associated with the Coxeter matrix M is the group generated by a set R of elements ρ_i $(i = 1, \ldots, n)$ subject only to the relations

$$(\rho_i \rho_j)^{m_{i,j}} = 1.$$

It is denoted by $W(M)$ or just W. Since $\rho_i \neq \rho_j$ for $i \neq j$, the two sets I and R will often be identified by means of the map $i \mapsto \rho_i$. The pair (W, R) is called a *Coxeter system* of type M. The number n is called the *rank* of the system. The rank will be assumed finite throughout these notes.

It is common practice to provide a pictorial presentation of M by means of the labeled graph (I, M) with vertex set I; the pair $\{i, j\}$ is an edge whenever $m_{i,j} > 2$ and this edge is labeled $m_{i,j}$. If $m_{i,j} = 3$, the label is often omitted. This picture is usually called the Coxeter diagram.

1.2. Chamber systems

Let (W, R) be a Coxeter system of type M. The free monoid on the alphabet R with unit is denoted by R^*, and $\rho : R^* \to W(R)$ stands for the monoid morphism determined by $\rho(r) = r$ $(r \in R)$. A typical element $\mathbf{r} \in R^*$ of length q will be written $[r_1, \ldots, r_q]$ to distinguish it from its image $\rho(\mathbf{r}) = r_1 \cdots r_q$ in W. We shall write $\ell(\mathbf{r})$ to indicate the length q of $\mathbf{r}$. In particular, $[]$ denotes the unit of R^*.

T. Bisztriczky et al. (eds.), POLYTOPES: Abstract, Convex and Computational, 1–19.

The *length* of an element $w \in W$, denoted by $\ell_R(w)$, or just $\ell(w)$ if no confusion is imminent, is $\min\{\ell(\mathbf{r}) \mid \rho(\mathbf{r}) = w\}$. Each element $\mathbf{r} \in R^*$ with $\rho(\mathbf{r}) = w$ is called an *expression* for w; if $\ell(\mathbf{r}) = \ell(w)$, the expression is called *reduced*.

The *chamber system associated with* (W, R), denoted by C, is the labeled graph whose vertex set is W and in which the edges labeled r (for $r \in R$) are all $\{w, wr\}$ for $w \in W$. Note that the edges are undirected as $r^2 = 1$ for all $r \in R$. The (label-preserving) automorphism group of C contains W via left multiplication, and is actually equal to W. The graph-theoretic distance between the 'chambers' $w, w' \in W$ of C, denoted by $d_{\mathsf{C}}(w, w')$, equals $\ell_R(w^{-1}w')$.

Given $\mathbf{r} = [r_1, \ldots, r_q] \in R^*$ and $c \in \mathsf{C}$, there is a unique path in the graph starting at c along edges with subsequent labels $r_1, \ldots, r_q$. It runs through cr_1, cr_1r_2, $\ldots$ and ends at $c\rho(\mathbf{r})$; it is denoted by $c \cdot \mathbf{r}$. Thus, if $w = \rho(\mathbf{r})$, the chamber wc is the image of $c \in \mathsf{C}$ under $w \in W$ in the isometric regular action of W on C from the left, and the image cw of c under w (in the regular action on W from the right) is the end point of the path $c \cdot \mathbf{r}$ in C.

The chamber system can be given the structure of a combinatorial cell complex by letting its cells be the empty set and all subsets of the form $c\langle J\rangle := \{cw \mid w \in \langle J\rangle\}$, where J runs through all subsets of R and $c \in \mathsf{C}$. Instead of $\langle J\rangle$, we also write W_J. Observe that cW_J is a coset of W_J in W.

The cell complex construction works because of some beautiful properties of Coxeter groups, the most remarkable of which are

- for each subset J of R, the pair $(\langle J\rangle, J)$ is again a Coxeter system; its type is $M\,|_{J\times J}$, and $\ell_J(w) = \ell_R(w)$ for each $w \in \langle J\rangle$;
- for every pair J, K of subsets of R and every pair x, y of elements of W, the intersection $xW_J \cap yW_K$ is either empty or of the form $zW_{J\cap K}$ for some $z \in W$.

By leaving out all cells with infinitely many vertices, one arrives at a topological description of the metric cell complex that will be described in §2. Obviously, the cell cW_J has finitely many vertices if and only if $\langle J\rangle$ is a finite group.

1.3. Examples

For the time being, let M be a Coxeter matrix of rank n, and let (W, R) be a Coxeter system of type M, so that $n = |R|$.

(o) If $n = 0$, then W is generated by the empty set, and so W is the trivial group.

(i) If $n = 1$ then $M = (1)$ and $W = \{1\} \cup R \cong \mathbf{Z}/(2)$, the group of order 2.

(ii) If $n = 2$, then

$$M = \begin{pmatrix} 1 & m \\ m & 1 \end{pmatrix} =$$
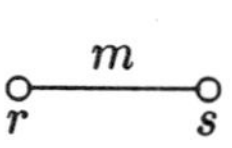

for some $m \in \mathbf{N} \cup \{\infty\}$,

and $W(M) = \langle r, s \mid r^2 = s^2 = (r\,s)^m = 1\rangle$ is the dihedral group of order $2m$, where, if $m = \infty$, the relation $(r\,s)^m = 1$ is void. If $m = 2h$ with $h \in \mathbf{N}$ odd, then

$$M' =$$
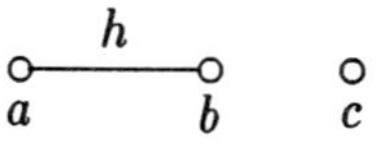

satisfies $W(M') \cong W(M)$, so that the Coxeter diagram of a Coxeter group is not uniquely determined by the group.

(iii) If $M = A_n$, where

$$A_n = \begin{pmatrix} 1 & 3 & 2 & \cdots & \cdots & 2 \\ 3 & 1 & 3 & 2 & \cdots & 2 \\ 2 & 3 & 1 & 3 & \cdots & 2 \\ \vdots & 2 & \ddots & \ddots & \ddots & \vdots \\ \vdots & \ddots & \vdots & \ddots & 1 & 3 \\ 2 & \cdots & \cdots & 2 & 3 & 1 \end{pmatrix}$$

or, in diagram form

$A_n =$ 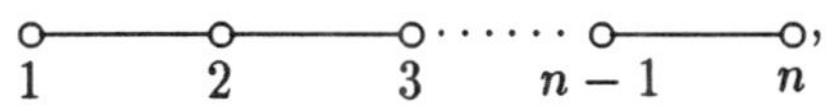

then $W(M)$ is the symmetric group Sym_{n+1} on $n+1$ letters. The evident morphism $W(A_n) \to \mathrm{Sym}_{n+1}$ sending ρ_i to $(i, i+1)$ for each $i \in I$ is in fact an isomorphism.

(iv) Take

$$\rho_1 = \pm\begin{pmatrix} 0 & 1 \\ 1 & 0 \end{pmatrix}, \quad \rho_2 = \pm\begin{pmatrix} -1 & 1 \\ 0 & 1 \end{pmatrix}, \quad \rho_3 = \pm\begin{pmatrix} 1 & 0 \\ 0 & -1 \end{pmatrix} \quad \text{in} \quad PGL(2, \mathbf{Z}).$$

Then, putting $R = \{\rho_1, \rho_2, \rho_3\}$ and $W = \langle R \rangle$, we obtain a Coxeter system (W, R) of type

$$M = \begin{pmatrix} 1 & 3 & 2 \\ 3 & 1 & \infty \\ 2 & \infty & 1 \end{pmatrix} =$$

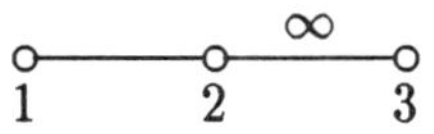

so that $W(M) \cong PGL(2, \mathbf{Z})$.

(v) If $R = R_1 \cup R_2$ is a partition of (R, M) into disjoint graphs (here disjoint means: $m_{i,j} = 2$ whenever $i \in R_1$ and $j \in R_2$), then $W(M) = W(M_1) \times W(M_2)$, where M_k is the restriction of M to $R_k \times R_k$ $(k = 1, 2)$. This explains why, in addressing many questions concerning Coxeter groups, we can restrict ourselves to the case where M is connected.

1.4. The Reflection Representation

Consider a Coxeter system (W, R) of type $M = (m_{rs})_{r,s \in R}$. Let V be a real vector space with basis $(e_r)_{r \in R}$. There is a real linear representation $\sigma : W \to GL(V)$ such that, for $r \in R$, the transformation $\sigma(r)$ is a reflection. We shall give a more specific description of this important representation.

Denote by B the symmetric bilinear form on V defined by

$$B(e_r, e_s) = -2\cos(\pi/m_{rs}) \quad (r, s \in R)$$

with the understanding that $B(e_r, e_s) = -2$ if $m_{rs} = \infty$. We call B *the symmetric bilinear form associated with* M. For each $r \in R$, consider the linear transformation σ_r of V defined by

$$\sigma_r(x) = x - B(x, e_r)e_r \quad (x \in V)$$

This defines a reflection of $GL(V)$ in the hyperplane $e_r^{\perp}$ with *root* (i.e., eigenvector with eigenvalue -1) e_r.

A representation is called faithful if it is an injective morphism of groups. Faithfulness of the reflection representation is due to Tits, cf. Bourbaki [1968]

Proposition 1.5 *Let B be the symmetric bilinear form associated with the Coxeter matrix M, and let σ_r $(r \in R)$ be as above. Then the mapping σ on R given by $\sigma(r) = \sigma_r$ extends to a faithful orthogonal representation $\sigma : W \to O(V, B)$.*

1.6. The Tits cone

Tits' proof that the above representation σ is faithful makes use of the representation of W on V^*, the dual of V. The action σ^* of W on V^* is induced from its action on V. Explicitly, for $w \in W$, the image $\sigma^*(w)f$ of $f \in V^*$ is given by

$$(\sigma^*(w)f)x = f(\sigma(w^{-1})x) \qquad \text{for } x \in V.$$

Let $f_1, \ldots, f_n$ be the basis of V^* that is dual to $e_1, \ldots, e_n$. Then

$$A = \mathbf{R}_{\geq 0} f_1 + \cdots + \mathbf{R}_{\geq 0} f_n$$

is called the *fundamental chamber* of (W, R). The stabiliser in W of a vector $a = \sum_{i=1}^{n} a_i f_i$ in A is W_J, where $J = \{1 \leq i \leq n \mid a_i = 0\}$. This shows that there is some control over the part $T = \bigcup_{w \in W} wA$ of V that is covered by images of A under W. This set is a cone and is called the *Tits cone*.

1.7. Finite Coxeter groups

A Coxeter group W is finite if and only if the quadratic form $x \mapsto B(x, x)$ on V for V and B as in Proposition 1.4 is positive definite. This, combined with Example 1.3(v), gives a straightforward method of classifying the finite Coxeter groups.

For instance, a Coxeter system (W, R) of type

$$\begin{pmatrix} 1 & p & q \\ p & 1 & r \\ q & r & 1 \end{pmatrix}$$

is finite if and only if the quadratic form with matrix

$$\begin{pmatrix} 2 & -2\cos\pi/p & -2\cos\pi/q \\ -2\cos\pi/p & 2 & -2\cos\pi/r \\ -2\cos\pi/q & -2\cos\pi/r & 2 \end{pmatrix}$$

is positive definite, which in turn is straightforwardly checked to be equivalent to

$$\frac{1}{p} + \frac{1}{q} + \frac{1}{r} > 1.$$

Let T be the Tits cone as above. In the finite case, $T = V^*$ and V^* can be identified with V as B (being positive definite) is non-degenerate. Conversely, if $-v$ lies in T for some v in the interior of A, then W is finite .

1.8. Connection with diagram geometry

The chamber system C of a Coxeter system is *residually connected.* That means that, for any collection of pairwise non-disjoint cells $c_1\langle J_1\rangle, \ldots, c_t\langle J_t\rangle$ of C, the intersection is a cell of the form cW_J where $J = J_1 \cap \cdots \cap J_t$. This condition, which is a direct consequence of what has been said at the end of §1.2, gives rise to the construction of a residually connected geometry from the chamber system (cf. Tits [1981]), which, in the case where M is a string diagram, is an abstract polytope.

2. The conjugacy problem

The conjugacy problem for Coxeter groups is the question of finding an algorithm that, with input a Coxeter matrix M of rank n and two words $\mathbf{v}, \mathbf{w} \in R^*$, decides if $\rho(\mathbf{v})$ and $\rho(\mathbf{w})$ are conjugate in $W = W(M)$. The solution we present finds numbers $K, L \in \mathbf{N}$ such that, whenever $v = \rho(\mathbf{v})$ and $w = \rho(\mathbf{w})$ are conjugate in W, they are conjugate by an element $g \in W$ of length at most $n^{K \max(\ell(v), \ell(w))+L}$. Thus, the solution is far from efficient, and leaves much in the direction of implementation to be desired. Yet, the setting of the proof, the Moussong complex to be defined below, gives hope for improvements.

We describe the construction of a metric complex that is based on Moussong [1988]. It will lead to a solution of the conjugacy problem for Coxeter groups. In fact, in his PhD thesis, Moussong does almost all of the work leading to the solution, but does not draw the conclusion. I gratefully acknowledge the help of Michael Shapiro and Daan Krammer, to whom I owe much of the content of this section.

Throughout this section, we take M to be a fixed Coxeter diagram and (W, R) a Coxeter system of type M.

2.1. Cells

A *Euclidean cell* is a metric space (X, d_X) which is isometric to a compact subset of Euclidean m-space E^m (for some $m < \infty$) which is the intersection of finitely many closed halfspaces. Here, the Euclidean metric induces the metric d_X on X. Without loss of generality we may assume that $X \subseteq E^m$ (with induced metric) and that X spans E^m. A *face* of a Euclidean cell X is, by definition, either X or $X \cap H$ for H a hyperplane of E^m which is disjoint from the interior of X. Thus faces, with the induced metric, are again Euclidean cells. The faces do not depend on the isometric embedding of X in E^m.

We recall that a Coxeter matrix M is called *spherical* if W is finite. The spherical Coxeter matrices have been classified. A subset J of R is said to be *spherical* if $M|_{J\times J}$ is spherical. If W is finite and M is connected, the reflection representation in $\mathbf{R}^n$ is irreducible and leaves invariant a positive definite form $(\cdot,\cdot)$, which is a positive scalar multiple of B as in §1.4. Thus, the reflection representation can be seen as an isometric action on the Euclidean space $E^n = E(\mathbf{R}^n)$, with Euclidean distance $|x-y| = \sqrt{(x-y, x-y)}$ $(x, y \in \mathbf{R}^n)$. The origin of the linear reflection representation is recovered from the metric action as the unique fixed point.

Proposition 2.2 *Let M be a spherical Coxeter matrix of rank n, and (W,R) a Coxeter system of type M. Let W act on $E^n = E(\mathbf{R}^n)$ via the reflection representation on $\mathbf{R}^n$. Then there is a W-equivariant injective map $\kappa : W \to E^n$ such that:*
(i) $|\kappa(w) - \kappa(wr)| = 1$ *for all* $w \in W$, $r \in R$;
(ii) *the metric space $E(W)$ consisting of the convex hull of $\kappa(W)$, with metric induced from E^n, is a Euclidean cell;*
(iii) *the nonempty faces of $E(W)$ are precisely the convex hulls of $\kappa(xW_J)$ for $x \in W$ and $J \subseteq R$.*

The cell $E(W)$ is unique in $E(\mathbf{R}^n)$, the map κ is unique up to W-isometries.

PROOF. Let ω_i $(i \in I)$ be the fundamental weights in the reflection representation space $\mathbf{R}^n$, in the sense that they form a dual basis to the normalised fundamental roots with respect to the W-invariant positive-definite form. Take $v \in \mathbf{R}^n$ to be half the sum of all ω_i for $i \in R$, and put $\kappa(w) = w \cdot v$. Then $|\rho_i(v) - v| = 1$ for all i. Since the stabiliser of v in W is trivial, κ embeds the chamber system $\mathsf{C}(\mathsf{W},\mathsf{R})$ into Euclidean space $E(\mathbf{R}^n)$. Moreover, vertices corresponding to adjacent chambers have distance 1, and, for any 2-set $\{i,j\} \subseteq J$, the subsystem $\mathsf{C}(\mathsf{W}_{\{i,j\}},\{i,j\})$ is a regular $2m_{ij}$-gon whose angles at the vertices are $\pi - \pi/m_{ij}$. Letting $E(W)$ be the convex hull of $\kappa(W)$, (i) and (ii) follow.

For $J \subseteq R$, let v_J be half the sum of all ω_i for $i \in R \setminus J$, so that $v_\emptyset = v$. Then $(wv - v, v_J) \leq 0$ with equality only if $w \in W_J$. This yields that the convex hull of $\kappa(W_J)$ is the face $v_J^\perp \cap E(W)$ of $E(W)$. A little more effort gives that each face on $\kappa(1)$ is of this form for some $J \subseteq R$. The rest of the proof follows from transitivity of W on the vertex set $\kappa(W)$. □

Corollary 2.3 *Each face of the Euclidean cell $E(W)$ is of the form $xE(W_J)$ for some Coxeter subsystem (W_J, J) of (W,R). Two such faces $xE(W_J)$ and $yE(W_K)$ meet if and only if xW_J and yW_K do, in which case there is $z \in W$ with $xW_J \cap yW_K = zW_{J\cap K}$. Such an element z also satisfies $xE(W_J) \cap yE(W_K) = zE(W_{J\cap K})$.*

PROOF. By (i) and (iii) of the proposition, the convex hull of $\kappa(xW_J)$ equals $xE(W_J)$, and multiplication by x is an isometry between $E(W_J)$ and $xE(W_J)$. □

2.4. COMPLEXES

A *Euclidean complex* is a pair $\mathsf{M} = (U, \mathsf{X})$ consisting of a set U and a collection X of metric spaces (X, d_X), where X is a subset of U, called *cells* of M, such that
(i) $U = \bigcup_{X \in \mathsf{X}} X$;
(ii) each member of X is a Euclidean cell;
(iii) if Y is a face of a cell (X, d_X), then Y is again a cell, and $d_Y = d_X|_{Y\times Y}$;
(iv) if (X, d_X) and (Y, d_Y) are cells, then $X \cap Y$ is a face of either cell.
Observe that, because of (iii), each cell has a unique metric. A Euclidean complex $\mathsf{M} = (\mathsf{U}, \mathsf{X})$ has a natural topology in which N is open if and only if $N \cap X$ is open for each cell $X \in \mathsf{X}$.

The chamber system of a Coxeter system can be turned into a Euclidean complex by viewing its edges as Euclidean line segments of length 1.

2.5. The Moussong complex

We now construct a Euclidean complex from the Coxeter system (W, R) of type M. Let $\widehat{U}$ be the collection of all triples (x, J, u) with $x \in W$, J a spherical subset of R, and $u \in E(W_J)$. On $\widehat{U}$, define the relation $\sim$ by

$$(x, J, u) \sim (y, K, v) \Leftrightarrow \exists z \in W \quad \begin{cases} xW_J \cap yW_K = zW_{J\cap K} & \text{and} \\ (z^{-1}x)u = (z^{-1}y)v. \end{cases}$$

Observe that, whenever $xW_J \cap yW_K \neq \emptyset$, the point $(z^{-1}x)u$ lies in $E(W_{J\cap K})$ and does not depend on the choice of z. Using the fact that $(x, J, u) \sim (y, K, v)$ implies that

$$(x, J, u) \sim (z, J \cap K, (z^{-1}x)u) = (z, J \cap K, z^{-1}yv) \sim (y, K, v),$$

one can show that $\sim$ is an equivalence relation. Define U to be the quotient of $\widetilde{U}$ by this relation, and denote the $\sim$-class of (x, J, u) by $[x, J, u]$.

In order to define Euclidean cells on U, we let X be the collection of subsets of U consisting of the empty set and all

$$xE_J := \{[x, J, u] \mid u \in E(W_J)\}$$

for $x \in W$ and spherical $J \subseteq R$. Then, as a set, xE_J bijectively corresponds to $E(W_J)$, and $xE_J \cap yE_K$ is either empty, or of the form $zE_{J\cap K}$ for some $z \in W$. Using this correspondence, each member of X can be given the structure of a Euclidean cell. If $K \subset J$ and $y \in xW_J$ then the embedding of yW_K into xW_J corresponds to an embedding of the cell yE_K into xE_J via the mapping $[y, K, v] \mapsto [x, J, (x^{-1}y)v]$ $(v \in E(W_K))$.

We shall refer to the pair (U, X) as the *Moussong complex of type M* or the *Moussong complex of* (W, R), and denote it by M. This definition is justified by the following claim.

Lemma 2.6 *The Moussong complex $\mathsf{M} = (\mathsf{U}, \mathsf{X})$ is a locally finite Euclidean complex. There are finitely many W-orbits on X and W acts discretely as a group of isometries on M with compact quotient and finite point stabilisers. The 1-skeleton of M coincides with the Euclidean complex defined by the chamber system $\mathsf{C(W, R)}$.*

Proof. Straightforward. For example, the quotient of U by W is the union of the cells E_J for $J \subseteq R$, J spherical, and hence is compact. □

The complex has been first constructed in Moussong [1988], whence its name. Prior to this, a non-metric version was given by Davis [1983].

2.7. Paths

Let N be a Euclidean complex. A *path* in N is a continuous map $\alpha : T \to U$ defined on a real interval T with the property that, for any $t \in T$, there is $\epsilon > 0$ such that $\alpha|_{T\cap[t,t+\epsilon]}$ and $\alpha|_{T\cap[t-\epsilon,t]}$ are differentiable paths with continuous derivatives, each contained in a cell. (In fact, it suffices for our purposes to restrict attention to paths that are piecewise linear.) Thus, if T is bounded and closed, the *speed* of α, that is, the absolute value of the derivative of α, is defined at all but a finite number of points. If $T = [0,1]$ and the speed of α is constant, then α is said to be *normalised*. The *length* of the path $\alpha : T \to U$ is the integral of the speed of α over T. We shall write $\ell_{\mathsf{N}}(\alpha)$ for the length of α.

Proposition 2.8 *Let* $\mathsf{N} = (\mathsf{U}, \mathsf{X})$ *be a locally finite connected Euclidean complex having only a finite number of isomorphism types of cells. Then the function* $d : U \times U \to \mathbf{R}$ *determined by*

$$d(x,y) = \inf \ell_{\mathsf{N}}(\gamma) \mid \gamma \text{ a path of } \mathsf{N} \text{ from } x \text{ to } y\}$$

is a metric on U.

2.9. GEODESICS

A *geodesic* in a Euclidean complex N is a path $\alpha : T \to \mathsf{N}$ such that

$$\ell_{\mathsf{N}}(\alpha|_{[t,u]}) = d(\alpha(t), \alpha(u)) \quad \text{for all } t, u \in T.$$

If N is as in Proposition 2.8, then geodesics exist between any two points in the same connected component.

Now let M be the Moussong complex of the Coxeter system (W, R), and let C be the chamber system of (W, R). Then, by Lemma 2.6 and Proposition 2.8, both are metric spaces. In C, the distance $d(x, y)$ between two chambers x, y is an integer coinciding with the graph-theoretic distance $d_{\mathsf{C}}(x, y)$. The path $x \cdot \mathbf{r}$, where $\mathbf{r} \in R^*$ has length q, can be viewed as a path $\alpha : [0, q] \to W$ of the Euclidean complex C by extending it to a path with constant speed on the edges. It occurs in the Moussong complex as its 1-skeleton, so, for any two elements $x, y \in W$,

$$d_{\mathsf{M}}(x,y) \le d_{\mathsf{C}}(x,y) = \ell_R(x^{-1}y).$$

Since W acts discretely on M with compact quotient, an argument of Milnor's gives a bound in the other direction; see below. In the metric space on C, geodesics exist, but need not be unique. For instance, if $M = A_2$, the paths $1 \cdot [r, s, r]$ and $1 \cdot [s, r, s]$ are distinct geodesics, both starting at 1 and ending at rsr. In the Euclidean cell $E(W(A_2))$, there is a straight line from 1 to $rsr = srs$; it is the unique geodesic between the two chambers. This illustrates one reason why the metric space on the Moussong complex is useful.

The following result shows that geodesics can be "approximated" by paths in the chamber system.

Proposition 2.10 *Let* M *be the Moussong complex of type* M*, so that the chamber system* C *of* (W, R) *embeds in* M*. Then the following hold.*

(i) *Milnor's inequality* (Milnor [1968]) *holds, i.e., there are non-negative constants* K*,* L*, such that, for every pair of chambers* x, y *of* C*, we have*

$$d_{\mathsf{C}}(x,y) \le K d_{\mathsf{M}}(x,y) + L.$$

(ii) *There exists a constant* A *such that, for each normalised geodesic* γ *with end points* $c = \gamma(0)$*,* $d = \gamma(1)$ *in* C*, a path* $c \cdot \mathbf{r}$ *of length* $q \in \mathbf{N}$ *and a strictly monotonous function* $f : [0,1] \to [0,q]$ *can be found satisfying, for all* $t \in [0,1]$*,*

$$d_{\mathsf{M}}(c \cdot \mathbf{r}(f(t)), \gamma(t)) \le A.$$

2.11. Links

Instead of Euclidean cells, other kinds of cells could have been employed. For instance, if "Euclidean" is replaced by "spherical" in the above sections 2.1, 2.4, 2.7, 2.8 and 2.9, most constructions and properties continue to hold. We shall need the notion of spherical complex to cope with a metric space capturing the local structure around a point of M. Here, as usual, the *star* of a point (or a face) is the union of all cells containing it.

Let x be a point of a Euclidean cell X of M. By the definition of Euclidean cell, we may think of X as a convex part of a Euclidean space E. Then the *link of* x in X, denoted $Lk(x, X)$, is the spherical cell obtained by taking all points y of the unit sphere around x in X such that xy contains a segment of X of positive length. If, for example, X is the regular m-gon with sides of length 1, the distance of the two vertices of the spherical cell is $\pi - \pi/m$. The link of a Euclidean complex M at a point x is the spherical complex whose cells are obtained by taking the union of all links $Lk(x, X)$ for X in the star of x, with proper identifications.

A *closed geodesic* is the image of an isometric embedding of a circle of strictly positive length. We have angular distances in spherical complexes. A Euclidean complex M is said to satisfy the *link condition* if there are no closed geodesics of length less than 2π in links.

Theorem 2.12 Moussong's theorem. *For each Coxeter system (W, R), the Euclidean complex M is simply connected and satisfies the link condition.*

Idea of proof Tits' rewrite rules:

$$[\rho_i, \rho_i] \Rightarrow []$$
$$\underbrace{[\rho_i, \rho_j, \rho_i, \ldots]}_{m_{ij}} \Rightarrow \underbrace{[\rho_j, \rho_i, \rho_j, \ldots]}_{m_{ij}}$$

are known to solve the word problem. Consequently, any closed path of C starting and ending at 1 is homotopic to the trivial path 1 in the 2-skeleton of M. Since any closed path in M is homotopic to a path in C, this gives simple connectedness. For the link condition, the main idea is that, in the link of a point, each closed geodesic goes via the 1-skeleton, where it can easily be seen to have length at least 2π. Here an inductive argument on links is needed. See the examples below for an impression of what happens in the low-dimensional cases. □

2.13. Examples

(i). Let $n = 2$, say $R = \{r, s\}$. If $m_{rs} < \infty$, then M has a single 2-cell: the metric regular $2m_{rs}$-gon. The link at a point of M is a circle if the point is in the interior, an arc of length $\pi - \pi/m$ if the point is a chamber, and arc of length π if the point is on an edge but not a chamber. If $m_{rs} = \infty$, then M has no 2-cells, and, as a complex, coincides with the chamber system, which is an infinite tree of valency 2.

(ii). For arbitrary n, with $m_{rs} = \infty$ whenever r, s are distinct, the complex M is the infinite tree of valency n.

(iii). Let $n = 3$, say $R = \{i, j, k\}$. The sum of the lengths of the spherical arcs in the link of 1 in the three polygons on the chamber $1 \in W$ amounts to

$$\pi\left(3 - \left(\frac{1}{m_{ij}} + \frac{1}{m_{jk}} + \frac{1}{m_{ik}}\right)\right).$$

Now, as stated in §1.7, W is finite if and only if

$$\frac{1}{m_{ij}} + \frac{1}{m_{jk}} + \frac{1}{m_{ik}} > 1.$$

Thus, if the sum of the arc lengths is less than 2π, the three angles border a 2-cell. This is at the heart of the argument that M satisfies the link condition. For example, for M the Moussong complex of the Coxeter group of type

$M =$ 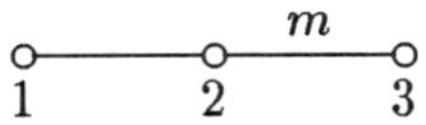

the link at a chamber is

- a spherical 2-cell if $m \leq 5$ (because then M is spherical),
- a concatenation of two arcs, one of length $\pi/2$ and one of length $2\pi/3$, if $m = \infty$, and
- a triangle with edges of length $\pi/2$, $2\pi/3$ and $(m-1)\pi/m$ if $5 < m < \infty$.

The latter triangle is a closed path of length at least 2π. The case $m = 6$ shows that equality may occur.

2.14. Unique geodesics and convex metric

We say that a metric space has *unique geodesics* if any two of its points are connected by a unique normalised geodesic. A metric space (U, d_U) is said to have a *convex metric* if, for any two normalised geodesics α, β,

$$d(\alpha(t), \beta(t)) \leq (1-t)d(\alpha(0), \beta(0)) + td(\alpha(1), \beta(1)) \quad \text{for all } t \in [0, 1].$$

A metric space with a convex metric in which geodesics exist is easily seen to have unique geodesics.

The following result expresses the major step of this section; it is of a "local-to-gobal" nature.

Theorem 2.15 *Suppose that* N *is a simply connected Euclidean complex satisfying the link condition. Then* N *has unique geodesics and convex metric.*

Proof. Krammer [1993] has given a fully elementary proof. More general results, using heavier machinery, are derived in Bridson [1991]. □

One of the consequences of the theorem is that local geodesics are geodesics. Hence, for each cell X of N, we have $d\,|_{X \times X} = d_X$.

For $v, w \in W$, set $N_W(v, w) := \{g \in W \mid gvg^{-1} = w\}$. Clearly, an algorithm to determine whether or not $N_W(v, w)$ is empty solves the conjugacy problem. The result below shows that an exhaustive search for an element of $N_W(v,w)$ among all elements of length exponentially bounded by $\max(\ell(v),\ell(w))$ provides such an algorithm.

2.16. Solution to the conjugacy problem. *Let M be a Coxeter matrix and (W,R) a Coxeter system of type M. Then there are constants K, $L \in \mathbf{N}$ such that, for all $v, w \in W$ with $N_W(v,w) \neq \emptyset$, there exists $g \in N_W(v,w)$ with*

$$\ell(g) \leq n^{K\max(\ell(v),\ell(w))+L}.$$

PROOF. Without loss of generality, we may take $\ell(w) \geq \ell(v)$. Suppose that $N_W(v,w) \neq \emptyset$. Take $g \in N_W(v,w)$ of minimal length, denote by γ the normalised geodesic from 1 to g in the Moussong complex M of (W,R) guaranteed by Theorem 2.15, and choose a presentation $\mathbf{r} = [r_1,\ldots,r_q]$ of g such that the path $1\cdot\mathbf{r}$ approximates γ in the sense of Proposition 2.10(ii) and such that q is minimal with respect to these requirements. Thus there are $A \in \mathbf{N}$ and $f : [0,1] \to [0,q]$ independent of g, v, w such that, for all $t \in [0,1]$,

$$d_{\mathsf{M}}(\mathbf{r}(f(t)),\gamma(t)) \leq A.$$

By the convex metric property applied to the geodesics γ from 1 to g and $w\gamma$ from w to $wg = gv$ and using the triangle inequality, we see that, for any time $t \in [0,1]$,

$$\begin{aligned} d_{\mathsf{M}}(1\cdot\mathbf{r}(f(t)), w\cdot\mathbf{r}(f(t))) &\leq d_{\mathsf{M}}(\gamma(t), w\gamma(t)) + 2A \\ &\leq (1-t)d_{\mathsf{M}}(1,w) + t d_{\mathsf{M}}(g,wg) + 2A \qquad (1)\\ &\leq \ell_R(w) + 2A \end{aligned}$$

The last inequality follows from $wg = gv$ and the fact that g acts from the left as an isometry, so that $d_{\mathsf{M}}(g,wg) = d_{\mathsf{M}}(1,v) \leq \ell_R(v)$. Using Proposition 2.10(i) for times t at which $\mathbf{r}(f(t))$ represents an element of C, we obtain the existence of constants K, L' such that, for all $k \in \{0,1,\ldots,q\}$,

$$d_{\mathsf{C}}(r_1\cdots r_k, wr_1\cdots r_k) \leq K d_{\mathsf{M}}(r_1\cdots r_k, wr_1\cdots r_k) + L'.$$

Together with (1), putting $L = L' + 2AK$, we find

$$d_{\mathsf{C}}(r_1\cdots r_k, wr_1\cdots r_k) \leq K\ell_R(w) + L.$$

For each k, let P_k denote the label of a path in C of length at most $K\ell(w)+L$ connecting $r_1\cdots r_k$ and $wr_1\cdots r_k$. If $q > n^{K\ell(w)+L}$, there are k, k' such that $k < k'$ and $P_k = P_{k'}$. But then $g' = r_1\cdots r_{k-1}r_{k'+1}\cdots r_q$, is an element of W with $g' \in N(v,w)$ and $\ell(g') \leq q - (k'-k) < q$, and so is a shorter expression than $\mathbf{r}$ for an element of $N(v,w)$, a contradiction. Hence $\ell(g) \leq q \leq n^{K\ell(w)+L}$, as required. □

2.17. REMARKS

Corollary 4.5 of Alonso and Bridson [1993] states that, if W acts properly and cocompactly by isometries on a space of non-positive curvature, then W is "semihyperbolic". Theorem 5.2 of Alonso and Bridson [1993] states that, if W is semihyperbolic, then it has a solvable conjugacy problem. These results can be applied to give a proof that Coxeter groups have a solvable conjugacy problem. The method presented above follows similar lines (namely, those set out by Gromov, cf. Gersten and Short [1990], Gromov [1987]), but is simpler and we hope leads to a more efficient solution.

A first step towards a better algorithm might be found by use of reductions of w of the form

$$w \mapsto sws \quad \text{whenever } \ell(sws) \leq \ell(w). \tag{2}$$

We shall call w *conjugacy-reduced* if each series of reductions as in (2) starting with w leads to an element w' of W with $\ell(w') = \ell(w)$.

Conjecture 2.18 *Let C be a conjugacy class of W and put $\ell_C = \min\{\ell(w) \mid w \in C\}$. Then, for any $w \in C$, we have $\ell(w) = \ell_C$ if and only if w is conjugacy-reduced.*

By Geck and Pfeiffer [1992], the conjecture holds for Weyl groups. The authors use the result for Hecke algebra representations.

2.19. Conjugacy of parabolic subgroups

Let (W, R) be a Coxeter system. Deodhar [1982] describes how, for two subsets I, J of R, it can be decided if the corresponding subgroups W_I and W_J of W are conjugate. Subgroups of the form W_I are called *standard parabolic*; their conjugates are called *parabolic*. Thus, the algorithm helps to find out how many distinct conjugacy classes of parabolic subgroups W has.

Let $I \subset R$ and $s \in R \setminus I$. Write K for the connected component of the subgraph of M induced on $I \cup \{s\}$ that contains s. Suppose now that K is spherical. Then the longest elements w_K of W_K and $w_{K\setminus\{s\}}$ of $W_{K\setminus\{s\}}$ exist (they are involutions). In this situation, we set

$$\nu(I, s) = w_{K\setminus\{s\}} w_K.$$

This element of W has the property that, for some $s_0 \in K$,

$$I^{\nu(I,s)} = (I \cup \{s\}) \setminus \{s_0\}.$$

Thus, there is a subset J of R such that conjugation by $\nu(I, s)$ not only maps W_I to W_J, but even gives a bijection from I to J.

Now consider the directed graph $\mathcal{K}$ whose vertices are the subsets of R, and in which two vertices I and J are connected by an edge pointing to J labelled s if the connected component of $I \cup \{s\}$ containing s is spherical (so that $\nu(I, s)$ exists) and $I^{\nu(I,s)} = J$.

For each $s \in I$, denote by e_s the fundamental root corresponding to s.

Theorem 2.20 *Let I and J be subsets of R. If $w \in W$ satisfies*

$$\{e_s \mid s \in I\} = \{we_t \mid t \in J\},$$

then there is a directed path $I = I_0 \xrightarrow{s_0} I_1 \xrightarrow{s_1} \ldots \xrightarrow{s_{t-1}} I_t = J$ in $\mathcal{K}$ such that

$$w = \nu(I_0, s_0) \cdot \nu(I_1, s_1) \cdots \nu(I_{t-1}, s_{t-1})$$

and $\ell(w) = \sum_{i=0}^{t-1} \ell(\nu(I_i, s_i))$.

PROOF. See Proposition 5.5 of Deodhar [1982]. □

By arguments similar to those below, this leads to an effective method for deciding if two subsets of R generate conjugate parabolic subgroups of W. Rather than giving details, we shall elaborate on the determination of the normaliser of a parabolic subgroup.

Corollary 2.21 *Suppose that I is a subset of R. Let $\mathcal{T}$ be a spanning tree of the connected component of $\mathcal{K}$ containing I. For each vertex J of $\mathcal{T}$, denote by $\mu(J)$ the element of W corresponding to the unique path in $\mathcal{T}$ from I to J. Then the normaliser of W_I in W is generated by I and all products of the form*

$$\mu(J)\nu(J,s)\mu(K)^{-1}$$

for edges $J \xrightarrow{s} K$ of the connected component of $\mathcal{K}$ containing I.

PROOF. If $w \in W$ normalises W_I, then, up to left multiplication by elements of W_I, it can be assumed to satisfy $\ell(sw) \geq \ell(w)$ for all $s \in I$. It follows that $\{w^{-1}e_s \mid s \in I\}$ is a fundamental system contained in Φ^+ of the group $\langle w^{-1}Iw\rangle = I$, whence it coincides with $\{e_t \mid t \in I\}$. Consequently, w satisfies the conditions of Theorem 2.20 with $J = I$. By the theorem, w is a product corresponding to a closed path in $\mathcal{K}$. The result follows as the fundamental group of $\mathcal{K}$ with base point I is generated by the paths $I \xrightarrow{\mathcal{T}} J \xrightarrow{s} K \xrightarrow{\mathcal{T}} I$, for all vertices J, K of $\mathcal{T}$ such that $J \xrightarrow{s} K$ is an edge of $\mathcal{K}$ but not of $\mathcal{T}$. Here, $\xrightarrow{\mathcal{T}}$ stands for a path in $\mathcal{T}$. □

2.22. IMPLEMENTED ALGORITHMS

We note that the description of the finite generating set for the normalizer of W_I is effective. An algorithm to this effect is implemented in a new version of LiE, cf. Leeuwen, Cohen and Lisser [1992].

For Weyl groups, Carter [1972] has given very explicit methods to resolve the conjugacy problem. Recently, they have been implemented in LiE by Pasqualucci [1992]. In the software package GAP, cf. Geck and Pfeiffer [1992], algorithms solving the conjugacy problem for finite Coxeter groups have been implemented by use of the permutation group algorithms available in the package.

3. The word problem

The word problem for Coxeter groups is the question of finding an algorithm that, with input a Coxeter matrix M of rank n and a word $\mathbf{w} \in R^*$, decides if $\rho(\mathbf{w}) = 1$ in $W(M)$. Solutions to the word problem have been discussed in Cohen [1991]. One of them was used in §2.12.

Another solution is based on the reflection representation. For, take $a = f_1 + \cdots + f_n$ with notation as in §1.6. Then, by §1.6, $\sigma^*(w)a = a$ if and only if $w = 1$, so that we can test whether $\rho(\mathbf{w}) = 1$ for $\mathbf{w} = [r_1, \ldots, r_q] \in R^*$ by computing

$$\sigma_{r_1}(\sigma_{r_2}(\cdots\sigma_{r_q}a))$$

and verifying if this vector coincides with a.

This algorithm is implemented for Weyl groups in LiE, cf. Leeuwen, Cohen and Lisser [1992]. To implement it on computer with exact arithmetic for arbitrary Coxeter groups, one would need to handle an extension field of $\mathbf{Q}$ containing the algebraic integers $2\cos(\pi/m)$ for each entry m of M, or an equivalent thereof.

3.1. A COMBINATORIAL SOLUTION TO THE WORD PROBLEM

In du Cloux [1990], a fast algorithm is presented that, given M, merely uses operations on words from R^* to rewrite an input word $[r_1, \ldots, r_q]$ (where $r_i \in R$) to the lexicographically minimal reduced word in $\rho^{-1}(r_1 \cdots r_q)$. Here a total ordering on R is used, e.g., $\rho_1 < \rho_2 < \cdots < \rho_n$.

For $r \in R$, let X_r denote the set of distinguished left coset representatives of the subgroup $\langle s \in R \mid s < r\rangle$ of $\langle s \in R \mid s \leq r\rangle$. Thus, if $r = \rho_j$,

$$X_r = \{w \in W_{\{1,\ldots,j\}} \mid \ell(sw) > \ell(w) \text{ for all } s \in R,\ s < r\}.$$

Then the multiplication map $X_1 \times X_2 \times \ldots \times X_n \to W$ is a bijection. The lexicographically minimal reduced word for any element of W respects this decomposition. Du Cloux describes

- how to obtain the lexicographically minimal reduced words for each X_r;
- how multiplication on the right by an element t from R affects an element $x \in X_r$: it either produces the element of X_r again or it is of the form sw with $s \in R$, $w \in X_r$, and $\ell(w) = \ell(x)$.

These facts suffice to process any expression for an element of W into the form $X_1 X_2 \cdots X_n$.

3.2. REGULAR LANGUAGES

The set of all lexicographically minimal reduced words turns out be remarkably well behaved, as the theorem below exhibits. If U and V are subsets of R^*, their *product* is the subset

$$UV := \{[u, v] \mid u \in U, v \in V\}$$

of all products of elements from U with elements from V. The *Kleene closure* of U is the subset

$$U^* := \{[u_1, \ldots, u_q] \mid q \in \mathbf{N}, \quad \text{each } u_i \in U\}.$$

(Note that this notation is consistent with the earlier definition of R^*.) A subset of R^* is called a *regular language* if it can be obtained from the subsets $\{[\,]\}$ and $\{r\}$ where $r \in R$, by a finite number of applications of the operations union, product and Kleene closure.

An alternative definition of a regular language uses the notion of a finite state automaton: it is the set of words accepted by such an automaton. For every Coxeter group, Brink and Howlett provide a finite state automaton and construct additional multiplier automata, one for each $r \in R$. This proves that Coxeter groups are automatic (cf. Epstein et al. [1992]). To be somewhat more precise, we define the notions automatic and bi-automatic.

3.3. Automatic structures

Extend R with a padding symbol $\$ \notin R$, and set

$$R_\$ = ((R \cup \{\$\}) \times (R \cup \{\$\})) \setminus \{(\$, \$)\}.$$

For $\mathbf{r} = [r_1, \ldots, r_t]$, $\mathbf{q} = [q_1, \ldots, q_s] \in R^*$, put

$$(\mathbf{r}, \mathbf{q}) = \begin{cases} [(r_1, q_1), \ldots, (r_s, q_s)] \text{ where } r_j = \$ & \text{if } t < j \leq s, \\ [(r_1, q_1), \ldots, (r_t, q_t)] \text{ where } q_j = \$ & \text{if } s < j \leq t. \end{cases}$$

As the notation suggests, this embeds $R^* \times R^*$ into $R_\$^*$. The triple (W, R, L) is said to be an *automatic structure* if L is a regular language in R^* with $\rho(L) = W$ such that

$$\{(\mathbf{r}, \mathbf{q}) \in R_\$^* \mid \mathbf{r}, \mathbf{q} \in L, \quad \rho(\mathbf{r}) = \rho(\mathbf{q})\}$$

and, for each $s \in R$,

$$\{(\mathbf{r}, \mathbf{q}) \in R_\$^* \mid \mathbf{r}, \mathbf{q} \in L, \quad \rho(\mathbf{r}) = \rho(\mathbf{q}s)\}$$

are regular languages in $R_\$^*$. If, in addition, for each $s \in R$, the language

$$\{(\mathbf{r}, \mathbf{q}) \in R_\$^* \mid \mathbf{r}, \mathbf{q} \in L, \quad \rho(\mathbf{r}) = \rho(s\mathbf{q})\}$$

is regular in $R_\$^*$, then the triple (W, R, L) is called a *bi-automatic structure*. A group is called *automatic* if it has an automatic structure, and similarly for *bi-automatic*.

Theorem 3.4 (Brink and Howlett [1993]) *For each $w \in W$, let $\sigma(w) \in R^*$ be the lexicographically first reduced expression for w. Then*

$$L = \{\sigma(w) \in R^* \mid w \in W\}$$

is a regular language, and (W, R, L) is an automatic structure.

Theorem 3.5 (Brink and Howlett [1993]) *Coxeter groups (with finite diagram) are automatic.*

The proof heavily uses the reflection representation and the corresponding root system. If W is finite, the number of accepting states of the automaton constructed for L is close to the number of subsets of the set of positive roots. So, for the Coxeter group of type E_8 there are more than 10^{36} states.

If a group is bi-automatic, it has a solvable conjugacy problem. Thus one may ask:

3.6. Question. *Are Coxeter groups bi-automatic?*

Continuing the work in Le-Chenadec [1986] for some finite Coxeter groups, Hermiller [1992] has given finite complete sets of rewrite rules for all Coxeter groups of rank at most 3 and for all Coxeter groups whose types M have the property that all of its subdiagrams of the form

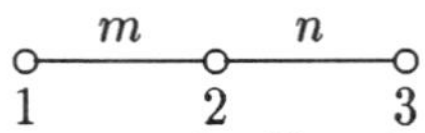

satisfy either $\infty \in \{m, n\}$ or $m = n = 2$. If such a group has a complete rewrite system, the set of canonical forms with respect to the rewrite system is a regular language. Thus, for the class of Coxeter groups W covered by Hermiller's work, this result also gives a regular language L in R^*, together with a computable section $\sigma : W \to R^*$ such that $\sigma(W) = L$.

4. Subgroup structure

In order to construct quasi-crystals, Moody and Patera [1993] have distinguished certain subgroups of Coxeter groups, which are Coxeter groups in their own right. Independently, Mühlherr [1993] has indicated similar subgroups in a more general approach, which we follow here. Both papers are inspired by Scherbak [1988]. The subgroups found in Dyer [1990] are usually different from those Mühlherr pointed out, for the simple reason that the former are generated by reflections whereas the latter may not contain reflections at all. Earlier work in this direction, but from a polytopal point of view, can be found in Monson [1987].

For the remainder of this section, let M be a Coxeter matrix and (W, R) a Coxeter system of type M. Fix a partition Π of R whose parts are spherical (cf. §2.1), and let R^Π be the set of all longest elements w_J of W_J for $J \in \Pi$. Thus, $|R^\Pi| = |\Pi|$.

Theorem 4.1 (Mühlherr [1993]) *Suppose that Π is a partition of R whose parts are spherical with the property that, for all $w \in \langle R^\Pi \rangle$ and $J \in \Pi$, either $\ell(wr) = \ell(w) - 1$ for each $r \in J$ or $\ell(wr) = \ell(w) + 1$ for each $r \in J$. Then $(\langle R^\Pi \rangle, R^\Pi)$ is a Coxeter system.*

A further result in Mühlherr [1993] states that it suffices to verify the length condition for Coxeter subsystems based on the union of two parts from Π. More specifically, the length condition is equivalent to the requirement that, for each pair $J, K \in \Pi \times \Pi$ and each $w \in \langle w_J, w_K \rangle$, either $\ell(wr) = \ell(w) - 1$ for each $r \in J$ or $\ell(wr) = \ell(w) + 1$ for each $r \in J$.

4.2. Applications

If α is an automorphism of W which leaves R invariant, then the orbits of α on R provide examples of partitions satisfying the hypotheses. With this method, for instance, the Coxeter group of classical type B_n can be seen to embed in the symmetric group $W(A_{2n-1})$. More generally, any Coxeter group can be shown to occur as a subgroup in the guise of the theorem of a Coxeter group whose type has labels at most 3 only. Other applications lead to a new proof of the classification of spherical Coxeter matrices.

4.3. Example

Consider the diagram

$E_8 =$ 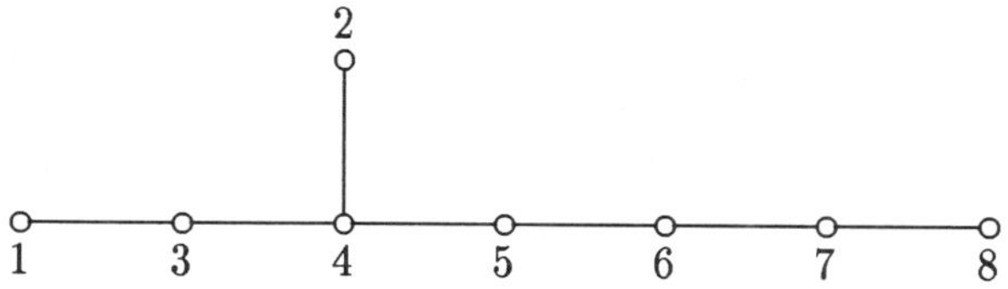

.

The partition $\{\{1,8\},\{2,5\},\{3,7\},\{4,6\}\}$ satisfies the hypotheses of the theorem, so there exists a subgroup of $W(E_8)$ isomorphic to $W(H_4)$ as follows:

.

Here the nodes are labeled by the longest elements of the group W_J for J running over the parts of Π, and the entry (J,K) of the corresponding Coxeter matrix equals the order of the element $w_J w_K$. For instance, for $J = \{2,5\}$ and $K = \{4,6\}$, the element $w_J w_K = 2546$ is a Coxeter element in the Coxeter group $\langle J \cup K\rangle$ of type A_4, and hence is of order 5.

This example occurs in both Mühlherr [1993] and Moody and Patera [1993]. The latter paper exhibits beautiful configurations arising from projections of lattice points using the ring of "icosians".

4.4. Maximal finite subgroups

Using the Tits cone of the Coxeter system (W,R), it is easy to prove that every finite subgroup of W is contained in a spherical parabolic subgroup. Thus the conjugacy classes of maximal finite subgroups of Coxeter groups coincide with the conjugacy classes of maximal spherical subsets of R, a problem that can be algorithmically solved, cf. §2.19.

For example, from §1.3(iv) it is immediate that $PGL(2,\mathbf{Z})$ has precisely two conjugacy classes of finite maximal subgroups, one isomorphism type being the Klein four group (corresponding to the spherical subset $\{1,3\}$ of R) and the other isomorphism type being the dihedral group of order 6 (corresponding to the spherical subset $\{1,2\}$ of R).

Conversely, the determination of conjugacy classes of finite maximal subgroups of $PGL(4,\mathbf{Z})$ can be employed to establish that the latter group is not a Coxeter group. For, the quotient of $W(F_4)$ by its center is a maximal finite subgroup of $PGL(4,\mathbf{Z})$ which is not a Coxeter group (this can be verified by a check of finite Coxeter groups). Since, by the above, maximal finite subgroups of Coxeter groups are Coxeter groups, it follows that $PGL(4,\mathbf{Z})$ is not a Coxeter group.

References

Alonso, J.M., and Bridson, M.R. [1993], "Semihyperbolic groups", preprint Princeton University.

Bourbaki, N. [1968], *Groupes et algèbres de Lie, Chap. IV, V, VI,* Hermann, Paris.

Bridson, M.R. [1991], "Geodesics and curvature in metric simplicial complexes", in *Group theory from a geometrical viewpoint,* Ghys et al., ed., World Scientific, Singapore.

Brink, B., and Howlett, R. [1993], "A finiteness property and an automatic structure for Coxeter groups", *Math. Ann.*296, 179-190.

du Cloux, F. [1990], "Un algorithme de forme normale pour les groupes de Coxeter", Report Ecole Polytechnique, Palaiseau.

Cohen A.M. [1991], "Coxeter Groups and three Related Topics", in *Generators and Relations in Groups and Geometries, Castelvecchio Pascoli, Italy (April 1990),* A. Barlotti, E.W. Ellers, P. Plaumann, and K. Strambach, eds., NATO ASI Series C: Math. and Phys. Sciences #333, Kluwer Acad. Publ., Dordrecht, 235-278.

Davis, M.W. [1983], "Groups generated by reflections and aspherical manifolds not covered by Eulidean space", *Annals of Math.* 117, 293-324.

Deodhar, V. [1982], "On the root system of a Coxeter group", *Comm. Algebra* 10, 611-630.

Dyer, M [1990], "Relection Subgroups of Coxeter Systems", *Journal of Algebra* 135, 57-73.

Epstein, D.B., Cannon, J.W., Holt, D.F., Levy, S.V.F. Paterson, M.S. and Thurston, W.P. [1992], "Word Processing in Groups", Jones and Bartlett, Boston.

Geck, M., and Pfeiffer, G. [1992], "On the Irreducible Characters of Hecke Algebras", preprint, RWTH, Aachen.

Gersten, S.M. and Short, H.B. [1990], "Inventiones Mathematicae", 102 (1990), 305-334.

Gromov, M. [1987], "Hyperbolic groups", in *Essays in Group Theory,* S.M. Gersten, ed., MSRI series, No. 8, Springer-Verlag.

Hermiller, S. [1992], "Rewriting Systems for Coxeter groups", preprint, Cornell University, May.

Humphreys, J.E. [1990], *Reflection groups and Coxeter groups,* Cambridge University Press, Cambridge, UK.

Krammer, D. [1993], "The Moussong complex and the conjugacy problem for Coxeter groups", Preliminary notes, Univ. of Utrecht.

Le-Chenadec, P. [1986], *Canonical forms in finitely presented algebras,* Pitman, London.

van Leeuwen, M.A.A., Cohen, A.M., and Lisser, B. [1992], *LiE, A package for Lie group computations* #ISBN 90-74116-02-7, CAN, Amsterdam.

Milnor, J. [1968], "A note on the curvature and fundamental group", *J. Diff. Geom.* 2, 1-7.

Monson, B. [1987], "A family of uniform polytopes with symmetric shadows", *Geom. Dedicata* 23, 355-363.

Moody, R.V., and Patera, J. [1992], "Quasicrystals and Icosians", J. Physics A: Math. Gen. 26, 2829-2853.

Moussong, G. [1988], "Hyperbolic Coxeter groups", *Ph.D. thesis*, Ohio State University.

Mühlherr, B. [1993], "Coxeter groups in Coxeter groups", preprint, Math. Univ. Tübingen.

Pasqualucci, R. [1992], "The conjugacy classes in the Weyl groups", Math. Degree Thesis, La Sapienza University, Roma.

Scharlau, R. [1993], "Buildings", preprint 93-016, to appear in Handbook of incidence geometry, ed. F. Buekenhout, Bielefeld.

Scherbak, O.P. [1988], "Wavefronts and reflection groups", *Russ. Math. Surveys* 43, 149-194.

Tits, J. [1981], "A local approach to buildings", in *The Geometric Vein (the Coxeter Festschrift)*, C. Davis, B. Grünbaum and F.A. Sherk, eds., Spring-Verlag, Berlin, 519-547.

THE EVOLUTION OF COXETER-DYNKIN DIAGRAMS

H.S.M. COXETER
Department of Mathematics
University of Toronto
Toronto, Ontario M5S 1A1
Canada

Abstract. 'Dynkin diagrams' or 'Coxeter-Dynkin diagrams' are graphs whose vertices and edges represent generators and relations (respectively) for so-called *Coxeter groups*. The generators have period two, and each of the relations specifies the period of the product of two generators. For instance, the group of automorphisms of the 27 lines on the general cubic surface (the Weyl group E_6) is generated by six involutions, each of which interchanges the two rows of a 'double six'. An early precursor of the appropriate diagram (see page 4) was used in 1904 by C. Rodenberg to indicate how six double-sixes are related. The same diagrams arise in various connections; for instance, the vertices may represent the mirrors of a kaleidoscope while the edges indicate the angles between pairs of mirrors.

1. Introduction

The use of trees as diagrams for groups was anticipated in 1904, when C. Rodenberg was commenting on a set of models of cubic surfaces. He was analyzing the various rational double points that can occur on such a surface. In 1931, I used these diagrams in my enumeration of kaleidoscopes, where the dots represent mirrors. E.B. Dynkin re-invented the diagrams in 1946 for the classification of simple Lie algebras. For instance, the graphs

are related to Wilhelm Killing's algebras

$$A_4, \qquad D_5, \qquad E_6, \qquad E_7, \qquad E_8.$$

Many other applications have arisen since that time (see, for instance, Hazewinkel, Hesselink, Siersma and Veldkamp 1977).

Since Rodenberg made use of double-sixes, it seems appropriate to begin this history with a summary of the pioneering work of Arthur Cayley, George Salmon, Jacob Steiner, and Ludwig Schläfli.

2. The Non-Singular Cubic Surface

Since a straight line usually intersects a cubic surface in three points, a line that contains four points of the surface must lie entirely on the surface. In 1849, such considerations persuaded Cayley that, since the whole projective 3-space contains

T. Bisztriczky et al. (eds.), POLYTOPES: Abstract, Convex and Computational, 21–42.

∞^4 lines, the general cubic surface should contain a finite number of them. He communicated this idea to Salmon, who replied that this finite number is 27. Five years later there was an equally fruitful correspondence in Switzerland between Steiner (1857) and Schläfli (1858). Steiner showed that a certain set of 9 among the 27 lines can be regarded as the intersections of two trihedra, that is, two sets of three planes. Since each of the 27 lines lies in 5 such planes, there are altogether 45 of them. Moreover, there are 120 of his pairs of 'conjugate' trihedra.

In his reply, Schläfli described his famous *double-six*

$$\begin{bmatrix} a_1 & a_2 & a_3 & a_4 & a_5 & a_6 \\ b_1 & b_2 & b_3 & b_4 & b_5 & b_6 \end{bmatrix} \tag{2.1}$$

such that two of these twelve lines intersect if and only if their symbols occur neither in the same row nor in the same column; thus a_1 intersects b_2 but is *skew* to a_2 and b_1. The planes a_1b_2 and a_2b_1 intersect in a line $c_{12}(= c_{21})$ which, containing four points of the surface, lies entirely on it. By considering the possible intersections of other lines with the plane a_1b_2, Schläfli easily verified Salmon's number 27: the 2×6 lines a_ν, b_ν, and the $\binom{6}{2}$ lines $c_{\mu\nu}(\mu < \nu)$.

Any two of the 27 lines are either intersecting or skew. The intersecting pairs are

$$a_1b_2,\ a_1c_{12},\ b_1c_{12},\ c_{12}c_{34},\ etc.,$$

while the skew pairs are

$$a_1b_1,\ a_1a_2,\ b_1b_2,\ a_1c_{23},\ b_1c_{23},\ c_{12}c_{13},\ etc.$$

(Schläfli 1858, p. 213). In particular, two lines of the form $c_{\mu\nu}$ intersect if their subscripts have no common digit, but are skew if they have one common digit.

3. The Weyl Group E_6 of order 51840

The group of automorphisms of this configuration of 27 lines was investigated by Jordan (1870, p. 317), Maschke (1889, p. 320), Burkhardt (1891, p. 317) and Dickson (1916, p. 348). Although these men made successively simpler choices of generators and relations, it seems that none of them saw how to generate this group by six involutions, suitably chosen from among the 36 involutions observed by Burkhardt (1891, pp. 324-326). Each of these 36 involutions interchanges the two rows of a double-size while leaving invariant the remaining fifteen lines. Those authors apparently failed to notice the simple structure of the set of 36 double-sixes, namely, that any two of them contain either 4 or 6 common lines: 4 forming a "double-two" (two skew pairs of intersecting lines) or 6 forming a "grid" (two intersecting triads of skew lines). When there are 4 common lines, the corresponding involutions *commute* ($AB = BA$, or $A \leftrightarrows B$, or $A \overset{\leftrightarrow}{2} B$); but when there are 6, the involutions are *braided* ($ABA = BAB$, or $B^A = A^B$, or $A \overset{\leftrightarrow}{3} B$). For instance, the double-six (2.1) shares with

$$\begin{bmatrix} a_1 & b_1 & c_{23} & c_{24} & c_{25} & c_{26} \\ a_2 & b_2 & c_{13} & c_{14} & c_{15} & c_{16} \end{bmatrix}$$

the 4 lines a_1, a_2, b_1, b_2, and with

$$\begin{bmatrix} a_1 & a_2 & a_3 & c_{56} & c_{46} & c_{45} \\ c_{23} & c_{13} & c_{12} & b_4 & b_5 & b_6 \end{bmatrix}$$

the 6 lines a_1, a_2, a_3, b_4, b_5, b_6. The corresponding involutions are the permutations

$$\begin{aligned} N &= (a_1b_1)(a_2b_2)(a_3b_3)(a_4b_4)(a_5b_5)(a_6b_6), \\ N_{12} &= (a_1a_2)(b_1b_2)(c_{13}c_{23})(c_{14}c_{24})(c_{15}c_{25})(c_{16}c_{26}), \\ N_{123} &= (a_1c_{23})(a_2c_{13})(a_3c_{12})(b_4c_{56})(b_5c_{46})(b_6c_{45}) \end{aligned}$$

(Gruber and Wills, 1983, p. 112). Cartan (1894, p. 331) called them T, S_{12}, S_{123}. One might say briefly

$$N = (a\ b), \qquad N_{12} = (1\ 2),$$

and observe that N transforms N_{123} into N_{456}. Any three involutions A, B, C may conveniently be said to form a *braided triad* if

$$ABA = BAB = C;$$

for then also $BCB = CBC = A$ and $CAC = ACA = B$. Among the 36 involutions arising from the double-sixes, typical commutative pairs are

$$NN_{12}, \qquad N_{12}N_{123}, \qquad N_{12}N_{34}, \qquad N_{12}N_{345}, \qquad N_{123}N_{145},$$

and typical braided triads are

$$N_{12}N_{13}N_{23}, \qquad NN_{123}N_{456}, \qquad N_{12}N_{134}N_{234}.$$

In particular, two involutions of the form $N_{\lambda\mu\nu}$ are commutative or braided according as their subscripts have an odd or even number of common digits.

4. Singularities

Rodenberg (1904, pp. 5, 32) used the notation $\lambda\mu\nu$ for the double-six corresponding to $N_{\lambda\mu\nu}$, and invented a diagram in which symbols for two double-sixes are linked when they share 6 lines (so that the involutions are braided) but not linked when they share only 4 lines (so that the involutions commute). He was investigating the possible singularities of cubic surfaces. In this project he essentially anticipated the discovery by Du Val (1934) of a connection between double points

$$B_3, \quad B_4, \quad B_5, \quad B_6, \quad U_6, \quad U_7, \quad U_8, \quad U_9$$

(*Biplanar* or *Uniplanar*) and reflection groups

$$A_2, \quad A_3, \quad A_4, \quad A_5, \quad D_4, \quad D_5, \quad E_6, \quad \tilde{E}_6$$

(see also Arnold 1974, pp. 21, 24; Fischer 1986b, p. 13). For instance, Rodenberg described the uniplanar double point U_8 by a diagram

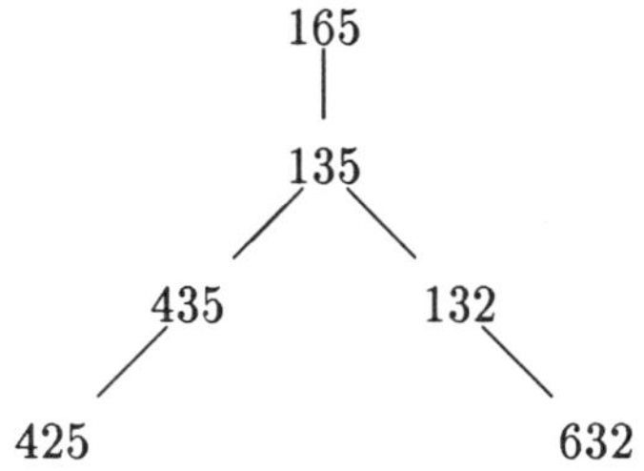

and mentioned that U_9 could be derived by appending 164, just above 165, so as to make a "triquetra" (Coxeter 1988, pp. 23, 26) which would be symmetrical by the permutation

$$(1\ 3\ 5)(2\ 4\ 6).$$

His seemingly unnatural ordering of digits (165 rather than 156, and so on) was determined by his insistence that, when two neighbouring triads have two common digits, these should occur in the same two positions.

In a trigonally symmetrical model of a cubic surface with 27 real lines (Fischer 1986a, Figures 10, 11, 12) those seven double-sixes appear as seven "holes" or "passages" (Durchgänge), any two of which are visibly adjacent or non-adjacent according as the corresponding double-sixes share 6 or 4 lines.

Rodenberg's diagram for U_8 exhibits six involutions which suffice to generate the Weyl group E_6. He happened to choose six which are all of the form $N_{\lambda\mu\nu}$. But instead of

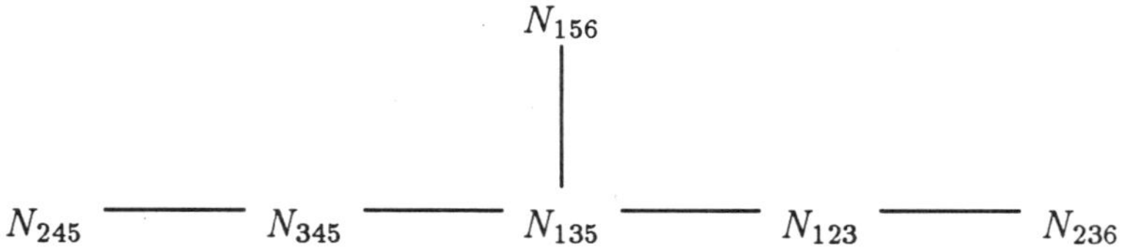

he could just as well have chosen

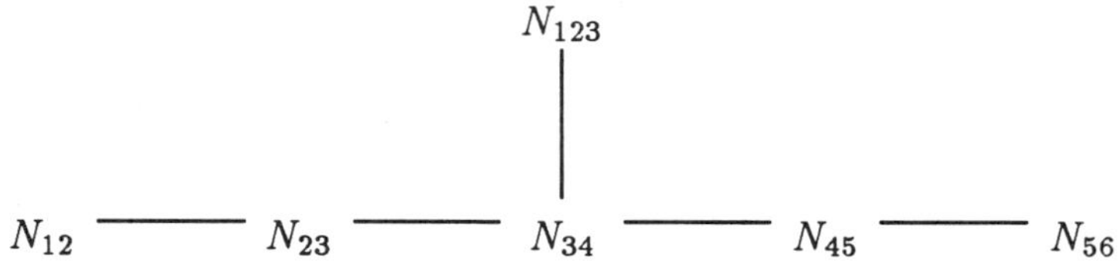

so as to reveal Schläfli's symmetric subgroup of degree 6, generated by the transpositions

$$(1\ 2),\qquad (2\ 3),\qquad (3\ 4),\qquad (4\ 5),\qquad (5\ 6).$$

The same diagram, with the six generators differently named, appeared independently in one of my early works (Coxeter 1932, p. 164). While that paper was in press, I realized that, instead of naming the generators, we could simply use dots. In fact, the graph

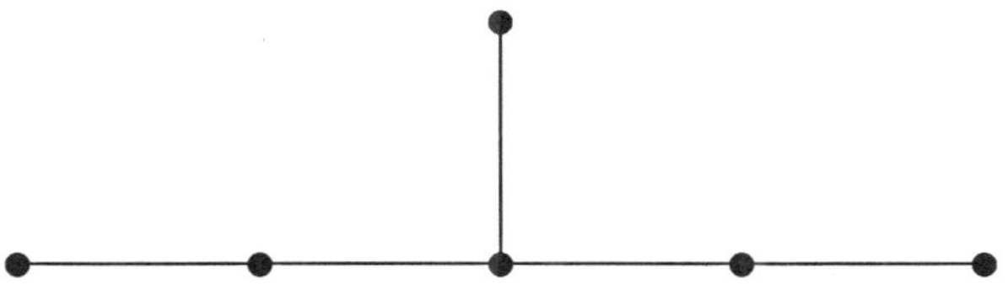

symbolizes a presentation for the group $[3^{2,2,1}] \cong E_6$ (Coxeter 1931, p. 133; 1934, p. 618; 1935, p. 25): the dots represent involutory generators, and two of them are linked or not linked according as the corresponding involutions are braided or commutative. By removing generators successively from the left, we can similarly present subgroups D_5, A_4 and $A_2 \times A_1$ (Rodenberg 1904, pp. 31, 27).

5. The del Pezzo Surfaces

P. del Pezzo (1887, pp. 243-253) regarded the cubic surface in 3-space as the member F_2^3 of a family of surfaces F_2^n of order n in n-space for $n \leq 9$. This surprising inequality arises from del Pezzo's ingenious argument, developed synthetically (without any direct appeal to coordinates). He first observed that an algebraic surface, being a 2-manifold, must be either a *cone* or another kind of *ruled* surface or a *non-ruled* surface, and that these three categories are preserved by projection from a point into a hyperplane. The *order* of the surface (its maximal number of intersections with a general $(n-2)$-flat) also is preserved, except when the centre of projection lies on the original surface, in which case the order is diminished by 1. In particular, a surface F_2^n is projected, from a point on itself, onto a surface F_2^{n-1} in $(n-1)$-space. Projecting this again, and continuing, we eventually reach a cubic surface F_2^3 in 3-space. When F_2^n is projected onto F_2^{n-1}, a set of m mutually skew lines on the former yields a set of $m+1$ such lines on the latter, the extra one arising from the neighbourhood of the centre of projection, or from the tangent plane there. Since F_2^3 has a maximal set of *six* skew lines (such as $a_1, ..., a_6$), F_2^4 has a maximal set of five, F_2^5 of four, and so on. Thus F_2^n contains no set of $10-n$ skew lines $(3 \leq n \leq 9)$. In particular, F_2^9 contains no line. Finally, a surface of order 10 in 10-space must be ruled (or a cone) because a non-ruled F_2^{10} would yield a line on F_2^9. This shows that non-ruled surfaces F_2^n occur only when $n \leq 9$.

Since the tangent plane to F_2^n at the centre of projection yields a line on F_2^{n-1}, all the lines on F_2^n have the same relations of incidence as those lines on F_2^{n-1} which are skew to one. Among the 27 lines on F_2^3 , each is skew to 16 others; for instance, b_6 is skew to

$$\begin{array}{ccccccccc} & & & & a_6 & & & & \\ b_1 & & b_2 & & b_3 & & b_4 & & b_5 \\ & c_{12} & & c_{23} & & c_{34} & & c_{45} & \\ & & c_{13} & & c_{24} & & c_{35} & & \\ & & & c_{14} & & c_{25} & & & \\ & & & & c_{15} & & & & \end{array}$$

and we can use these same symbols for the 16 lines on the quartic surface F_2^4 (which is the intersection of two quadric 3-folds). Similarly, the quintic surface F_2^5 contains 10 lines having the same incidences as those lines on F_2^4 which are skew to one, say a_6, namely

$$c_{\mu\nu} \qquad (\mu < \nu < 6),$$

these being the 10 in the last four lines of the above list. The sextic surface F_2^6 contains 6 lines having the same incidences as those on F_2^5 which are skew to (say) c_{45}, namely the "double-three"

$$\begin{array}{ccc} c_{14} & c_{24} & c_{34} \\ c_{15} & c_{25} & c_{35} \end{array}$$

which may be regarded as a skew hexagon $c_{14}\ c_{35}\ c_{24}\ c_{15}\ c_{34}\ c_{25}$. The septimic surface F_2^7 contains 3 lines having the same incidences as those on F_2^6 which are skew to (say) c_{25}, namely

$$\begin{array}{ccc} & c_{24} & \\ c_{15} & & c_{35} \end{array} \ .$$

Here a new situation arises because c_{24} intersects both c_{15} and c_{35}, which are skew. Accordingly, there are two distinct octavic surfaces F_2^8: one containing just one line, c_{15} or c_{35}, and the other none. Finally, the unique nonic surface F_2^9 again contains no line.

6. Uniform Polytopes

A few years before his work on the cubic surface, Schläfli (1852, pp. 215, 224) discovered the regular polytopes, including the *simplex* α_n, the *cross polytope* β_n and the $n-$*cube* γ_n. For instance, α_1 is a line-segment, α_2 is an equilateral triangle, $\beta_2 = \gamma_2 = \alpha_1 \times \alpha_1$ is a square, which may be regarded as the Cartesian product of two line-segments; α_3 is a regular tetrahedron, β_3 is a regular octahedron, γ_3 is a cube, and $\alpha_2 \times \alpha_1$ is a triangular prism with square side-faces. This prism, having faces of two kinds (α_2 and β_2), is not regular but 'uniform'.

A polygon is said to be *uniform* if it is regular, and any polytope is said to be uniform if all its facets are uniform and its symmetry group is transitive on the vertices. The farther ends of the edges at one vertex of an n-dimensional polytope Π_n are the vertices of a polytope Π_{n-1} called the *vertex figure*: for instance, the vertex figure of α_n or γ_n is α_{n-1}, the vertex figure of β_n is β_{n-1}, and the vertex figure of $\alpha_2 \times \alpha_1$ is an isosceles triangle.

My own contributions to the subject began about 1922, when my school friend, John Flinders Petrie (who recognized the so-called 'Petrie polygon') helped me to appreciate the five Platonic solids, the fourteen Archimedean solids and the uniform prisms and antiprisms. In particular, we rediscovered two sequences of polytopes such that each member of the sequence is the vertex figure of the next.

One of my most satisfying experiences (about 1924, while I was sitting under a tree in Savernake Forest, Wiltshire) was my realization that there is a third such sequence, beginning with the triangular prism $\alpha_2 \times \alpha_1$ (which has 6 vertices, 2 triangular faces and 3 square faces), the truncated 5-cell $t_1\alpha_4$ (whose 10 vertices are the midpoints of the edges of the 5-cell α_4, while its facets consist of 5 tetrahedra α_3 and 5 octahedra β_3), the 5-dimensional *hemi-cube* $h\gamma_5$ (whose 16 vertices are alternate vertices of the 5-cube γ_5, while its facets consist of 16 simplices α_4 and 10 cross-polytopes β_4), and ending with an infinite honeycomb filling the whole 8-dimensional space with $\alpha_8 s$ and $\beta_8 s$, $17280 + 2160$ surrounding each vertex. I was, of course, blissfully unaware that the same figures had been discovered by Thorold Gossett (1900) and that the vertices of the honeycomb would later be identified with the lattice of centres of balls in the densest 8-dimensional packing (Conway and Sloane 1988, p. 185).

A still more memorable experience (about 1928, while I was cycling along Hills Road, Cambridge) was my realization that, if Gosset's polytopes and honeycomb are named

$$(-1)_{21},\quad 0_{21},\quad 1_{21},\quad 2_{21},\quad 3_{21},\quad 4_{21},\quad 5_{21},$$

there is a generalization $p_{qr}(= p_{rq})$ in $p+q+r+1$ dimensions, such that the Cartesian product

$$(-1)_{qr} = \alpha_q \times \alpha_r$$

is the vertex figure of 0_{qr} and, for all $p \geq 0$, $(p-1)_{qr}$ is the vertex figure of p_{qr}. Moreover, the facets of p_{qr} are of two kinds,

$$p_{(q-1)r} \text{ and } p_{q(r-1)}$$

(which may still be distinguished if $q = r$, say by being differently coloured); and the three polytopes

$$p_{qr},\ q_{rp},\ r_{pq} \tag{6.1}$$

are so related that the facets of each contain the vertices of the other two (Coxeter 1930, pp. 372, 375).

At first, because of the familiar relationship between reciprocal *regular* polytopes, I was inclined to call this three-way relationship 'semi-reciprocation'; but a better term is *triality* (by analogy with *duality*).

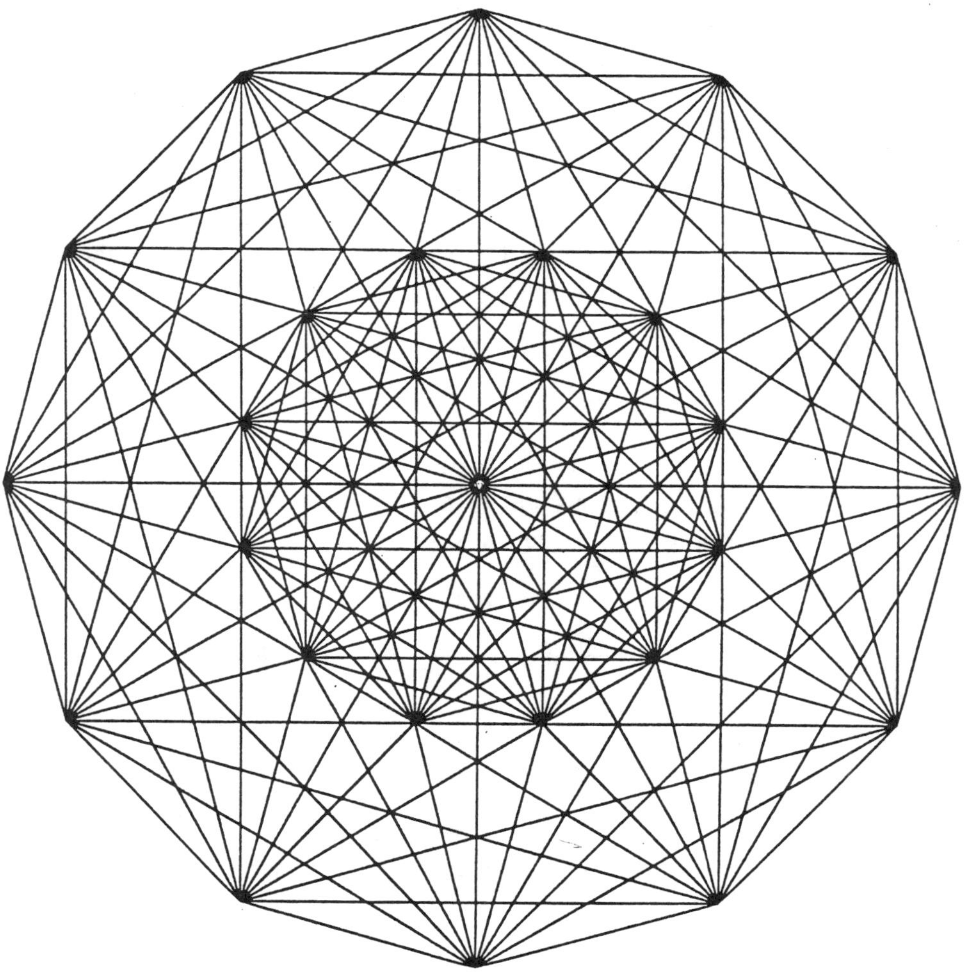

Fig. 1. The 6-dimensional polytope 2_{21} (27 vertices, 216 edges).

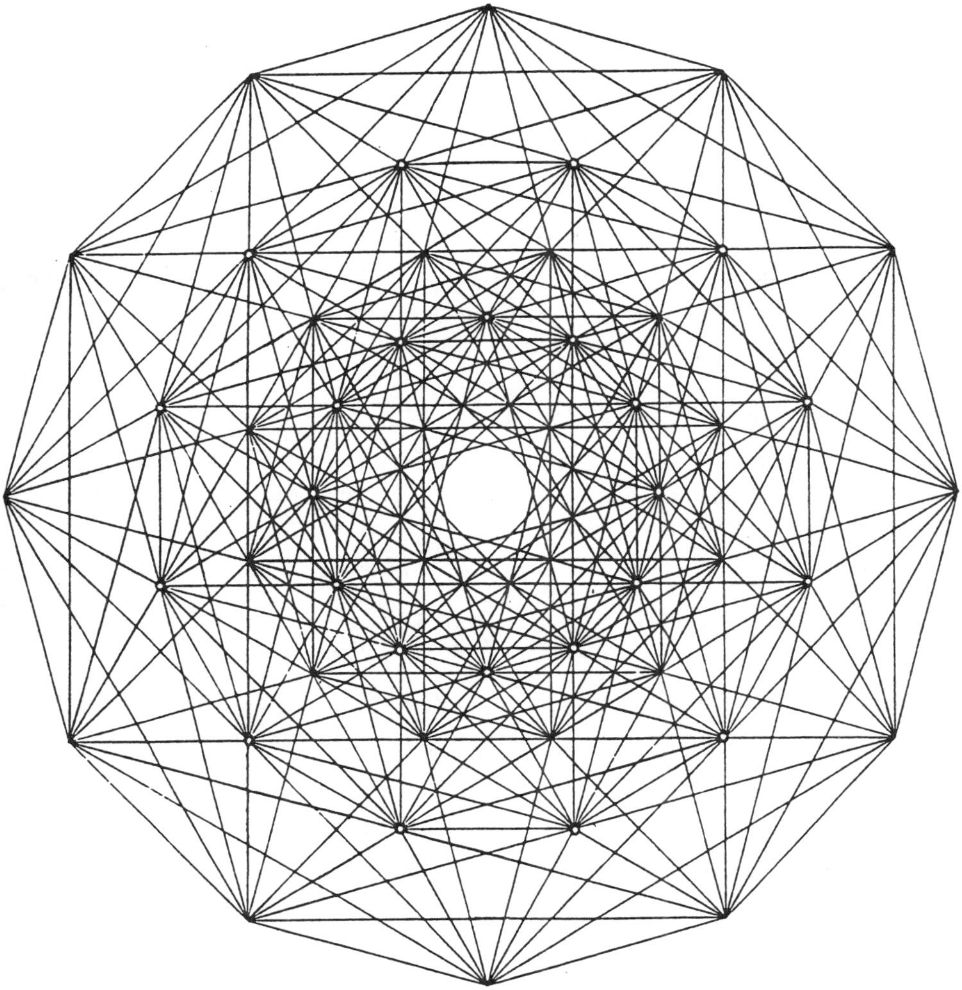

Fig. 2. The 6-dimensional polytope 1_{22} (72 vertices, 720 edges).

The p-faces of a regular n-polytope are tangent to an $(n-1)$-sphere which touches the $(n-p-1)$-faces of the reciprocal polytope at the same points of contact (Coxeter 1930, p. 354). Analogously in triality (*ibid.* p. 372) the three uniform polytopes (6.1) can be so placed that the centres of their faces

$$\alpha_p, \quad \alpha_q, \quad \alpha_r$$

(respectively) all coincide. These points are, in fact, the vertices of a fourth uniform polytope

$$0_{pqr}$$

(*ibid.* p. 410) whose vertex figure is the triple Cartesian product

$$\alpha_p \times \alpha_q \times \alpha_r.$$

(When $p = 0$ or 1, the 'face' α_p of p_{qr} is, of course, a vertex or an edge.)

The simplest example arises when $p = q = r = 1$, so that triality relates the three regular 16-cells

$$1_{11} = \beta_4 = \{3,\ 3,\ 4\}$$

which can be inscribed in the regular 24-cell $\{3,\ 4,\ 3\}$ (Coxeter 1990, p. 32). The midpoints of the 24 edges of each 16-cell coincide with the 24 vertices of another (reciprocal) 24-cell 0_{111}, whose vertex figure is the cube

$$\alpha_1 \times \alpha_1 \times \alpha_1 = \{4,\ 3\}.$$

Each vertex of this last 24-cell is the common midpoint of three edges of the respective 16-cells, crossing one another at right angles. These edges are the diameters of the 24 octahedral facets of the first 24-cell.

Other instances are

$$q_{r0} = \alpha_{q+r+1}, \quad 0_{qr} = t_q\alpha_{q+r+1} = t_r\alpha_{q+r+1},$$
$$p_{11} = \beta_{p+3}, \quad 1_{p1} = h\gamma_{p+3}, \quad 0_{p11} = t_2\gamma_{p+3}$$

(Coxeter 1930, pp. 372, 410), where $p,\ q,\ r$ may be any non-negative integers.There are also sporadic polytopes such as 2_{21} (Figure 1), 1_{22} (Figure 2), and honeycombs such as 2_{22} and 0_{222}, determined by the inequality

$$\frac{1}{p+1} + \frac{1}{q+1} + \frac{1}{r+1} \geq 1.$$

This restriction is most easily established by comparing the circumradius ρ of 0_{pqr} with the circumradius σ of its vertex figure

$$\alpha_p \times \alpha_q \times \alpha_r.$$

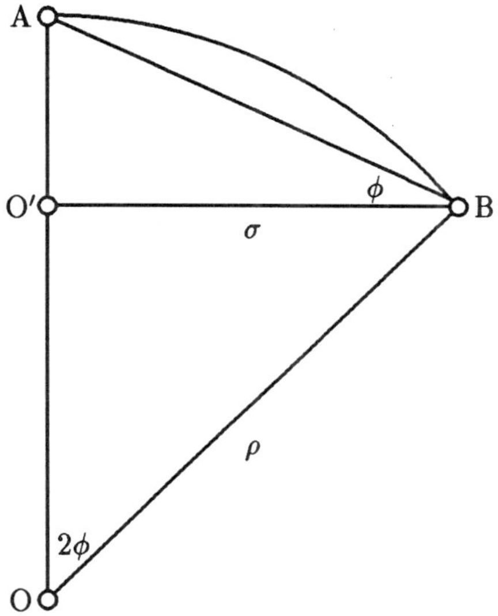

Fig. 3. The circumradii $OA = \rho$ and $O'B = \sigma$ of a uniform polytope and its vertex figure.

In Figure 3, AB is an edge (of length 1) of a uniform polytope with centre O. The vertex figure at A has a vertex B and centre O'. Let $\rho = OA = OB$ and $\sigma = O'B$ be the circumradii. By similar triangles,

$$\sqrt{1-\sigma^2}\ /1 = \tfrac{1}{2}/\rho = 1/2\rho,$$

so

$$\frac{1}{2\rho^2} = 2 - 2\sigma^2.$$

Since $\frac{n+1}{n} = 2 - \frac{n-1}{n}$, it follows by induction that the circumradius ρ of the regular simplex α_n is given by

$$2\rho^2 = \frac{n}{n+1}\ .$$

Since the square of the circumradius of a Cartesian product is equal to the sum of the squares of the circumradii of the constituents (Coxeter 1930, p. 351), the circumradius σ of $\alpha_p \times \alpha_q \times \alpha_r$ is given by

$$2\sigma^2 = \frac{p}{p+1} + \frac{q}{q+1} + \frac{r}{r+1} = 3 - \frac{1}{p+1} - \frac{1}{q+1} - \frac{1}{r+1}$$

and the circumradius ρ of 0_{pqr} is given by

$$\frac{1}{2\rho^2} = 2 - 2\sigma^2 = \frac{1}{p+1} + \frac{1}{q+1} + \frac{1}{r+1} - 1.$$

It follows that 0_{pqr}, and so also p_{qr}, q_{rp}, r_{pq}, are finite polytopes or infinite honeycombs according as the number

$$\frac{1}{p+1}+\frac{1}{q+1}+\frac{1}{r+1}$$

is greater than 1 or equal to 1 (Coxeter 1930, pp. 369-371). This criterion was obtained independently by Nathan Jacobson (1962, p. 133) whose p, q, r are my $p+1$, $q+1$, $r+1$.

Thus the 'triquetral' family includes the eleven sporadic polytopes

$$2_{21},\quad 1_{22},\quad 0_{221},\quad 3_{21},\quad 2_{31},\quad 1_{32},\quad 0_{321},\quad 4_{21},\quad 2_{41},\quad 1_{42},\quad 0_{421},$$

related to the spherical reflection groups (or 'Weyl groups') E_6, E_7, E_8; and the nine sporadic honeycombs

$$2_{22},\quad 0_{222},\quad 3_{31},\quad 1_{33},\quad 0_{331},\quad 5_{21},\quad 2_{51},\quad 1_{52},\quad 0_{521},$$

related to the infinite Euclidean groups (or 'affine groups') $\tilde{E}_6$, $\tilde{E}_7$, $\tilde{E}_8$.

Various formulae have been found for the *orders* of the spherical groups (Coxeter 1988, pp. 11, 19). One of the most remarkable was given by Peter McMullen (1991, p. 124).

7. A Surprising Analogy

One Saturday afternoon about 1928, I was invited to give a talk at one of Professor H.F. Baker's tea-parties in the Arts School of the University of Cambridge, attended by about a dozen of his student and colleagues. I spoke about Gosset's sequence of polytopes

$$(-2)_{21},\ (-1)_{21},\ 0_{21},\ 1_{21},\ 2_{21},\ 3_{21},\ 4_{21},\ 5_{21}$$

and wrote on the blackboard their numbers of vertices:

$$3,\quad 6,\quad 10,\quad 16,\quad 27,\quad 56,\quad 240,\quad \infty.$$

One member of the audience, being well acquainted with algebraic geometry, excitedly remarked that the first six are precisely the numbers of lines on the del Pezzo surfaces F_2^{5-p}.

If, instead of British geometers, my audience had consisted of French analysts, someone would just as eagerly have remarked that the same numbers appeared in the work of Élie Cartan (1894, p. 343) as factors of the order

$$240 \times 56 \times 27 \times 16 \times 10 \times 6 \times 2$$

of the Weyl group E_8, which was already implicit in "The greatest mathematical paper of all time" (Coleman 1989, p. 35; Killing 1889, pp. 23, 28-29; Weyl 1935, Appendix).

It was during my visit to Princeton in 1932 that I thought of representing a kaleidoscope by a graph of dots and links indicating the mirrors and their mutual inclinations. Two of the dots are linked or not linked according as the angle between them is 60^0 or 90^0. Later, Ernst Witt (1941) extended this notation by letting two dots be joined by $q-2$ links when the angle is π/q $(q \geq 2)$. The same notation was rediscovered, quite independently, by E.B. Dynkin, with the slight difference that the number of links joining his pairs of dots is not $q-2$ but $4\cos^2\pi/q$. (This difference causes confusion only when $q > 4$.) In a letter to me, dated April 3, 1984, Dynkin wrote: 'It is striking that my notations turned out to be so similar to yours. This probably shows how natural these notations are'.

In particular, the common symmetry group $[3^{p,q,r}]$ of the polytopes (6.1) is generated by reflections in hyperplanes which can be neatly represented by the $p+q+r+1$ dots in the *triquetra*

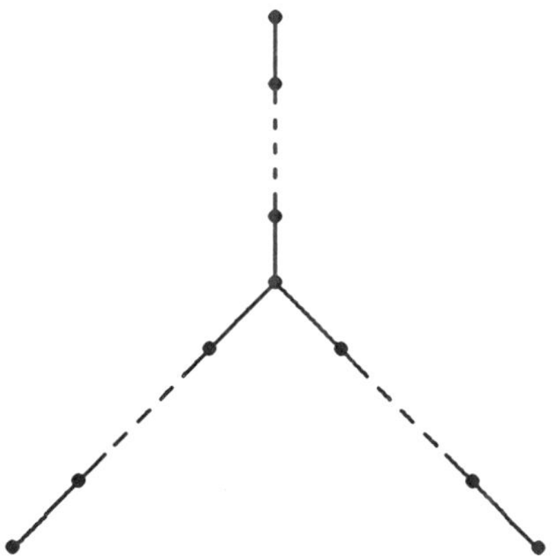

having legs of length p, q, r. The polytope or honeycomb p_{qr}, derived by inserting a ring round the first dot, has for its set of vertices the orbit of a point lying on all the mirrors exept the one indicated by the ring.

More generally, for any Coxeter-Dynkin diagram with rings drawn round one or more of the n dots, Wythoff's construction yields a uniform polytope Π_n whose typical vertex A lies on the intersection of all the 'un-ringed' mirrors, and is reflected into a neighbouring vertex B by each ringed mirror. In order to yield edges AB of length 1, the point A must be at the same distance $\frac{1}{2}$ from all the ringed mirrors. For instance, the kaleidoscope indicated by three disjoint dots consists of three mutually perpendicular mirrors in ordinary space; and when all the dots have rings, the polytope is a cube γ_3 of edge 1.

When only one dot has a ring, the corresponding mirror perpendicularly bisects the edge AB of the polytope Π_n. The remaining $n-1$ mirrors form an $(n-1)$-dimensional kaleidoscope which operates on AB to yield all the edges through the vertex A. More precisely, the mirrors indicated by dots joined to the first dot reflect AB into other edges through A, while the remaining mirrors, being perpendicular to the first, all contain the line AB. The orbit of B in this simpler kaleidoscope is precisely the set of vertices of the vertex figure of Π_n. It follows that this vertex figure is indicated in the diagram by deleting the single ringed dot and drawing rings round the neighbouring dots (which may be one or more). In particular, we see again that the vertex figure of p_{qr} is $(p-1)_{qr}$.

This recipe for the vertex figure of a 'one-ring' polytope shows that the same triquetra, with just the central dot ringed, describes the polytope 0_{pqr} whose vertex figure is $\alpha_p \times \alpha_q \times \alpha_r$ (Coxeter 1930, p. 416).

8. Projective n-Space and Euclidean $(9-n)$-Space

Gosset's polytopes had already been rediscovered in 1910 by Elte(1912). Schoute (1910) recognized one of them (Elte's V_{27}, my 2_{21}) as representing the cubic surface in such a way that the 27 vertices, 216 edges and 135 diagonals of the polytope correspond to the 27 lines, 216 skew pairs and 135 intersecting pairs on the surface. This remarkable correspondence was rigorously explained by Todd(1932). Figure 1 (drawn by Peter McMullen) shows the most symmetrical two-dimensional projection of the six-dimensional polytope. Its 27 vertices appear as 12+12+3: two concentric dodecagons with the remaining three vertices all projected into the centre (Coxeter 1940, pp. 461-463; see also Edge 1970, p. 757).

Since *skew* lines on the cubic surface F_2^3 correspond to *adjacent* vertices of 2_{21}, the lines on F_2^3 that are skew to one of them correspond to those vertices of 2_{21} which are adjacent to one vertex. In other words, the lines on F_2^4 correspond to the vertices of the vertex figure of 2_{21}, which is the *hemi-cube* $1_{21} = h\gamma_5$ (Elte's HM_5). Continuing, we see that the lines on F_2^n correspond to the vertices of $(5-n)_{21}$, for $n < 8$.

In detail, 1_{21} has the 16 vertices

$$(0,0,0,0,0),$$

$$(0,1,1,1,1),\ (1,0,1,1,1),\ \cdots,\ (1,1,1,1,0),$$

$$(1,1,0,0,0,),\ (1,0,1,0,0),\ \cdots,\ (0,0,0,1,1)$$

(with an even number of ones), and these represent the $1+5+10$ lines

$$a_6,$$

$$b_1,\ b_2,\ \cdots,\ b_5$$

$$c_{12},\ c_{13},\ \cdots,\ c_{45}$$

on the quartic surface F_2^4. Similarly, the vertex figure of 1_{21} is the truncated 5-cell

$$0_{21} = t_1\alpha_4$$

(Elte's tC_5), whose 10 vertices

$$(1,1,0,0,0),\ (1,0,1,0,0),\ \cdots,\ (0,0,0,1,1),$$

being the midpoints of the edges of the four-dimensional simplex

$$(2,0,0,0,0)\ (0,2,0,0,0)\ \cdots\ (0,0,0,0,2),$$

represent the 10 lines

$$c_{12},\qquad c_{13},\qquad \cdots,\qquad c_{45}$$

on the quintic surface F_2^5. And the vertex figure of 0_{21} is the triangular prism

$$(-1)_{21} = \alpha_2 \times \alpha_1$$

(Du Val 1933, p. 33), whose 6 vertices

$$(1,0,0,1,0),\ (0,1,0,1,0),\ (0,0,1,1,0)$$

$$(1,0,0,0,1),\ (0,1,0,0,1),\ (0,0,1,0,1)$$

represent the lines

$$c_{14},\ c_{24},\ c_{34}$$

$$c_{15},\ c_{25},\ c_{35}$$

on the sextic surface F_2^6. The vertex figure of $(-1)_{21}$ is the isosceles triangle

$$(0,1,0,1,0)$$

$$(1,0,0,0,1) \qquad\qquad (0,0,1,0,1)$$

whose 3 vertices represent the lines

$$c_{24}$$

$$c_{15} \qquad c_{35}$$

on F_2^7. Finally, at distance $\sqrt{2}$ from $(1,0,0,0,1)$ we have only $(0,0,1,0,1,)$, and from $(0,1,0,1,0)$ we have none, agreeing with del Pezzo's discovery of two distinct kinds of F_2^8.

Proceeding in the opposite direction, we would like to utilize the fact that 2_{21} is the vertex figure of the seven-dimensional polytope 3_{21} (Coxeter 1988, p. 32) and hence to describe the 56 lines on some kind of two-dimensional 'del Pezzo surface' F_2^2. C.F. Geiser (1869, p. 129; Dickson 1916, p. 351) observed that the enveloping cone from an arbitrary point P on the cubic surface intersects an arbitrary plane in a quartic curve whose 28 bitangents arise from the 27 lines along with the tangent plane at P. Patrick Du Val (1933, p. 57) interpreted this result as meaning that the cubic surface F_2^3 is projected from P onto a "Geiser surface" F_2^2 which consists of a repeated plane, branching along the quartic curve whose 28 bitangents count as $\binom{8}{2}$ line-pairs

$$c_{12}C_{12},\ c_{13}C_{13}, \cdots,\ c_{78}C_{78}.$$

The line-pair $c_{\mu\nu}C_{\mu\nu}(\mu < \nu < 7)$ arises from the line $c_{\mu\nu}$ on F_2^3, $c_{\nu 7}C_{\nu 7}$ from a_ν, $c_{\nu 8}C_{\nu 8}$ from b_ν, and $c_{78}C_{78}$ from the tangent plane at P. The two coincident lines $c_{\mu\nu}$ and $C_{\mu\nu}$ are regarded as intersecting each other twice: at the two points of contact of this bitangent with the quartic curve. Pairs that intersect just once are $c_{12}C_{13}$, $c_{12}c_{34}$, $C_{12}C_{34}$, etc., while skew pairs are $c_{12}c_{13}$, $C_{12}C_{13}$, $c_{12}C_{34}$, etc. In particular, the lines skew to C_{78} are $c_{\mu\nu}(\mu < \nu < 7)$, $C_{\nu 7}$, $C_{\nu 8}$; these are projections of the lines $c_{\mu\nu}$, a_ν, b_ν on F_2^3.

We see from the coordinates (Coxeter 1988, pp. 8, 25, 28) that, in terms of the edge-length as unit, the polytope p_{21} for $p \leq 3$ has diagonals of length $\sqrt{2}$, while 3_{21} (Coxeter 1928) has also *diameters* of length $\sqrt{3}$. Thus, in every case with $2 \leq n \leq 7$ and $1 \leq m \leq 3$,

Two vertices of the polytope $(5-n)_{21}$ *at distance* $\sqrt{m}$ *represent two lines on the surface* F_2^n *having* $m-1$ *points of intersection.*

In particular, as we have seen, any two of the 27 vertices of 2_{21} belong either to an edge ($m = 1$) or to a diagonal ($m = 2$), and any two of the 27 lines on F_2^3 are either skew or intersecting.

Taking successive vertex figures, we see that, since $E_6 = [3^{2,2,1}]$ is both the symmetry group of 2_{21} and the automorphism group of the lines on F_2^3,

$[3^{5-n,2,1}]$ *is the symmetry group of* $(5-n)_{21}$
and the automorphism group of the lines on F_2^n.

In detail (Coxeter 1988, p. 17), the groups

$$E_7, \qquad E_6, \qquad E_5 = D_5, \qquad E_4 = A_4, \qquad E_3 = A_2 \times A_1$$

(mentioned at the end of §4) are the symmetry groups of the polytopes

$$3_{21}, \qquad 2_{21}, \qquad 1_{21} = h\gamma_5, \qquad 0_{21} = t_1\alpha_4, \qquad (-1)_{21} = \alpha_2 \times \alpha_1$$

and the automorphism groups of the lines on the del Pezzo surfaces

$$F_2^2, \qquad F_2^3, \qquad F_2^4, \qquad F_2^5, \qquad F_2^6 \quad .$$

9. Diagrams Containing Circuits

The complete list of irreducible reflection groups, finite and Euclidean (Coxeter 1934, p. 619), has been quoted so often that there is no need to repeat it here. (Some confusion is caused by many authors' unfortunate habit of calling the *Euclidean* groups "affine".) One notices that the Coxeter-Dynkin diagram is a *tree* in every case except

$$[3^{[n]}] = \tilde{A}_{n-1},$$

where it is an n-gon (Coxeter 1988, p.5). In this case, if we work in the hyperplane $\sum u_\nu = 0$ of Cartesian n-space, the n generators R_ν are reflections in the hyperplanes

$$u_1 - u_2 = 0, \quad u_2 - u_3 = 0, \cdots, \quad u_{n-1} - u_n = 0, \quad u_n - u_1 + 1 = 0 :$$

each $R_\nu (\nu < n)$ transposes the coordinates u_ν and $u_{\nu+1}$, while R_n is the transformation

$$u_1' = u_1 + 1, \quad u_2' = u_2, \cdots, \quad u_{n-1}' = u_{n-1}, \quad u_n' = u_1 - 1..$$

This infinite reflection group has an infinite normal subgroup generated by translations such as T^p, where

$$T = R_1 R_2 \cdots R_n R_{n-1} R_{n-2} \cdots R_2;$$

that is, T is the translation

$$u'_1 = u_1 + 1, \quad u'_2 = u_2 - 1, \quad u'_3 = u_3, \cdots, \quad u'_n = u_n.$$

To symbolize its quotient group $G(p,p,n)$ of order $p^{n-1}n!$ (Shephard 1953, p. 379; Shephard and Todd 1954, p. 277; Coxeter 1957, pp. 244, 248, 251), we decorate the Coxeter-Dynkin diagram by inserting the number p inside the n-gon. Just as in the case of $\tilde{A}_{n-1}$ itself, any two of the n generating involutions $R_1, \cdots, R_n$ are braided if their subscripts are consecutive *in cyclic order* and are commutative otherwise, and we have also the extra relation

$$T^p = 1.$$

Since T^p is the translation

$$u'_1 = u_1 + p, \quad u'_2 = u_2 - p, \quad u'_\nu = u_\nu \quad (\nu > 2),$$

we may now regard the coordinates as residues modulo $2p$. Thus when $p = 2$, the group is D_n, permuting the 2^{n-1} vertices of the hemi-cube $h\gamma_n$ (Coxeter 1988, p. 5), namely the points

$$(\pm 1, \pm 1, \cdots, \pm 1)$$

with an even number of minus signs. We may equally well take the coordinates to be homogeneous and replace R_n by

$$u'_1 = -u_n, \quad u'_\nu = u_\nu \quad (1 < \nu < n), \qquad u'_n = -u_1.$$

In other words, the Y-shaped diagram for D_n can be replaced by an n-gon with the number 2 written inside.

When $n = 3$, so that $T = R_1R_2R_3R_2$, the diagram

for the octahedral group $D_3 = A_3 \cong \mathfrak{S}_4$, means that the generators R_1, R_2, R_3 satisfy the presentation

$$R_\nu^2 = 1, \quad R_1R_2R_1 = R_2R_1R_2, \quad R_1R_3 = R_3R_1, \quad R_2R_3R_2 = R_3R_2R_3.$$

Replacing R_3 by its conjugate $\bar{R}_3 = R_2R_3R_2$, we have

$$\bar{R}_3R_2\bar{R}_3 = R_2R_3R_2R_3R_2 = R_2\bar{R}_3R_2,$$
$$\bar{R}_3R_1\bar{R}_3 = R_3R_2R_1R_2R_3 = R_1R_3R_2R_3R_1 = R_1\bar{R}_3R_1$$

and the period of $R_1R_2\bar{R}_3R_2$ is 2. Hence

In terms of fundamental regions (Coxeter 1990, pp. 16, 20), the spherical triangle (3 3 2) is combined with one of its neighbours to form the larger triangle $(3\ 3\ \frac{3}{2})$, as in Figure 4 (see Davis, Grünbaum and Sherk 1981, p. 163, where B should be D).

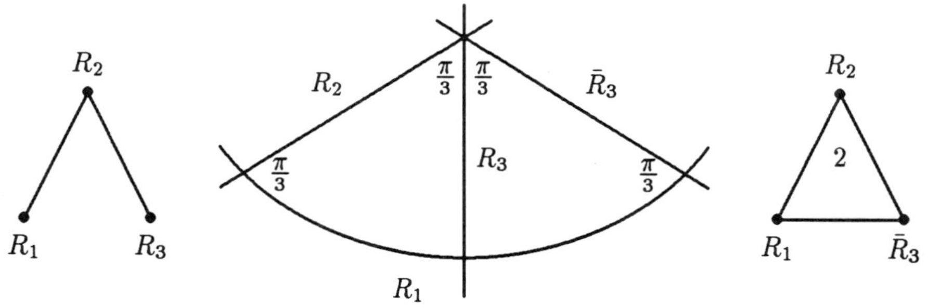

Fig. 4. The triangle $(3\ 3\ \frac{3}{2})$ dissected into two triangles $(3\ 3\ 2)$

10. Exotic Presentations

In the same spirit, whenever two generating involutions A and B are braided, we can replace one of them by $ABA = (= BAB)$ so as to obtain a possibly different presentation for the same group. In particular, we can add a tail (or two or three tails) of any length, to a triangle marked 2, and the group will still be symmetric (Coxeter 1957, p. 250). Similarly, D_n can be presented as an n-gon marked 2 or, if $n > 4$, as an m-gon marked 2 with $m < n$ and one tail of length $n - m$. In particular, D_4 can be presented as a square (marked 2); not a triangle with a tail (which is A_4) but instead, a pair of triangles both marked 2, with a common side.

For E_6 we obtain, in succession, the presentations.

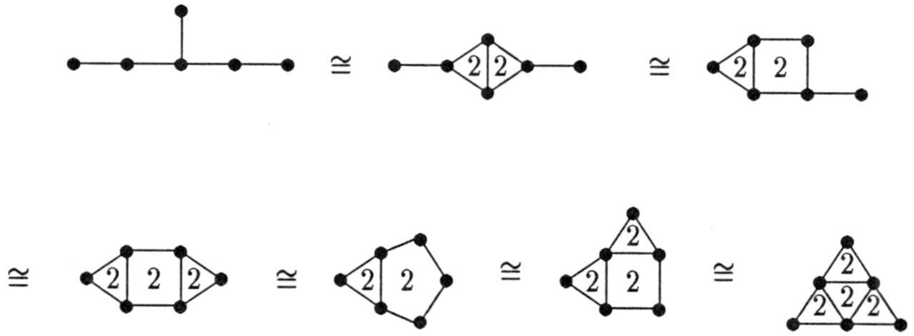

In the last case, the generating reflections may be taken to be

N_{156}

N_{12} N_{13}

N_{246} N_{23} N_{345}

A still more symmetrical presentation for E_6, suggested by Tsaranov (1989), is

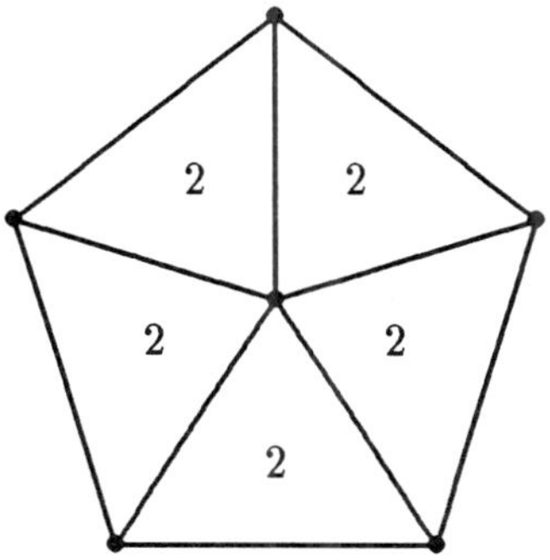

with generators

$$N_{126}$$

$$N_{236} \qquad N \qquad N_{156}$$

$$N_{346} \qquad N_{456}$$

The group generated by Tsaranov is not E_6 itself but its 'even' subgroup E_6^+, the simple group of order 25920, generated by the five rotations

$$x_1 = NN_{126}\ ,\quad x_2 = NN_{346}\ ,\quad x_3 = NN_{156}\ ,\quad x_4 = NN_{236}\ ,\quad x_5 = NN_{456}$$

which are easily seen to satisfy his presentation

$$x_i^3 = 1, \quad (x_i^{-1}x_{i+1})^2 = 1, \quad (x_i x_{i+2})^2 = 1,$$

$(i = 1, \ldots, 5;\ x_6 = x_1;\ x_7 = x_2)$. For E_6, E_7, E_8, Peter McMullen found many other exotic presentations, including

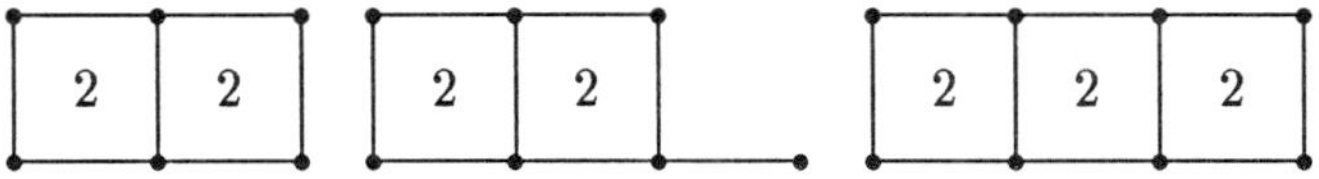

which were almost anticipated by Carter (1972, pp. 10, 11: see also Schellekens and Warner 1988).

Finally, extended diagrams for hyperbolic reflection groups have been used by Du Val (1933, p. 73), Vinberg (1971, p. 1084), Conway and Sloane (1988, pp. 529, 570) and many others. It was Arnold (1974, p. 20) who remarked that the "classification of more complex singularities provides new wonderful coincidences, where Lobachevsky traingles and automorphic functions take part."

References

1. Arnold, V.I., 1974: Critical points of smooth functions. *Proc. Internat. Congr. Mathematicians, Vancouver*, I, 19-39.
2. Burkhardt, H., 1891: Untersuchungen aus dem Gebiete hyperelliptischen Modulfunctionen. III. *Math. Ann.* **41**, 313-343.
3. Cartan, E., 1984: Sur la réduction à sa forme canonique de la structure d'un groupe de transformations fini et continu. *Œuvres complètes*, I, 293-353.
4. Carter, R.W., 1972: Conjugacy classes in the Weyl group. *Compositio Math.***25**, 1-59.
5. Coleman, A.J., 1989: The greatest mathematical paper of all time. *Math. Intelligencer* 11.3, 29-38.
6. Conway, J.H. and Sloane, N.J.A., 1988: *Sphere packings, lattices and groups.* Springer, New York.
7. Coxeter, 1928: The pure Archimedean polytopes in six and seven dimensions. *Proc. Camb. Phil. Soc.* **24**, 1-9.

   ____1930: The polytopes with regular-prismatic vertex figures. *Phil. Trans. Roy. Soc. London*, A 229, pp. 329-425.

   ____1931: Groups whose fundamental regions are simplexes. *J. London Math. Soc.***6**, 132-136.

   ____1932: The polytopes with regular-prismatic vertex figures II. *Proc. London Math. Soc.*(2) **36**, 126-189.

   ____1934: Discrete groups generated by reflections. *Annals of Math.* **35**, 588-621.

   ____1935: The complete enumeration of finite groups of the form $R_i^2 = (R_iR_j)^{k_{ij}} = 1$. *J. London Math. Soc.***10**, 21-25.

   ____1940: The polytope 2_{21}, whose 27 vertices correspond to the lines on the general cubic surface. *Amer. J. Math.* **62**, 457-486.

   ____1957: Groups generated by unitary reflections of period two. *Canad. J. Math.* **9**, 243-272.

   ____1978: Polytopes in the Netherlands. *Nieuw Archief voor Wiskunde* (3), **26**, 116-141.

   ____1988: Regular and semi-regular polytopes. III. *Math. Z.* **200**, 3-45.

   ____1990: *Regular complex polytopes* (2nd ed.) Cambridge University Press.
8. Davis, C., Grünbaum, B. and Sherk, F.A., 1980: *The geometric vein.* Springer, New York.
9. del Pezzo, P., 1887: Sulle superficie dell' n^{mo} ordine immerse nello spazio di n dimensione. *Rend. Circolo Mat. di Palermo* **1**, 241-271.
10. Dickson, L.E., 1916: Applications of finite groups. Part III of Miller, Blichfelft and Dickson, *Finite groups,* Wiley, New York.
11. Du Val, P., 1933: On the directrices of a set of points in a plane. *Proc. London Math. Soc.*(2), **35**, 23-74.

    ____1934: On isolated singularities of surfaces which do not affect the conditions of adjunction. *Proc. Camb. Phil. Soc.* **30**, 453-491.
12. Edge, W.L., 1970: Permutation representations of a group of order 9, 196, 830, 720. *J. London Math. Soc.* (2), **2**, 752-762.

13. Elte, E.L., 1912: *The semiregular polytopes of the hyperspaces.* Groningen.
14. Fischer, G., 1986a: *Mathematische Modelle.* Vieweg, Braunschweig.
____1886b: *Mathematische Modelle, Kommentarband.* Vieweg, Braunschweig.
15. Geiser, C.F., 1869: Ueber die Doppeltangenten einer ebenen Curve vierten Grades. *Math. Ann.* **1**, 129-138.
16. Gosset, T., 1900: On the regular and semi-regular figures in space of n dimensions. *Messenger of Mathematics* **29**, 43-48.
17. Gruber, P. and Wills, J.M., 1983: *Convexity and its applications.* Birkhäuser, Basel.
18. Hazewinkel, M., Hesselink, W., Siersma, D. and Veldkamp, F.D., 1977: The ubiquity of Coxeter-Dynkin diagrams (an introduction to the A-D-E problem). *Nieuw Arch. Wisk.* **25**, 257-307.
19. Jacobson, N., 1962: *Lie algebras.* Interscience, New York.
20. Jordan, C., 1870: *Traité des substitutions et des équations algébriques.* Paris.
21. Killing, W., 1889: Die Zusammensetzung der stetigen endlichen Transformationsgruppen. *Math. Ann.* **33**, 1-48.
22. Maschke, H., 1889: Aufstellung des vollen Formensystems einer quaternären Gruppe von 51840 linearen Substitutionen. *Math. Ann.* **33**, 317-344.
23. McMullen, P., 1991: The order of a finite Coxeter group. *Elemente der Mathematik* **46**, 121-152.
24. Rodenberg, C. 1904: Modelle von Flächen dritter Ordnung, in *Mathematische Abhandlungen aus dem Verlage Mathematischer Modelle von Martin Schilling.* Halle a. S.
25. Schellekens, A.N. and Warner, N.P., 1988: Weyl groups, supercurrents and covariant lattices. *Nuclear Physics* B, **308**, 397-435.
26. Schläfli, L., 1852: Lehre von den linearen Kontinuen, *Gesammelte Mathematische Abhandlungen* I, 177-226, Birkhäuser, Basel, 1950.
____1858: An attempt to determine the twenty-seven lines upon a surface of the third order..., *Gesammelte Mathematische Abhandlungen* II, 198-218, Birkhäuser, Basel, 1953.
27. Schoute, P.H., 1910: On the relation between the vertices of a definite six-dimensional polytope and the lines of a cubic surface. *K. Akad. Wetens. Amsterdam, Proc. Sec. Sci.* **13**, 375-383.
28. Shephard, G.C., 1953: Unitary groups generated by reflections. *Canada. J. Math.* **5**, 364-383.
29. Shephard, G.C. and Todd, J.A., 1954: Finite unitary reflection groups. *Canad. J. Math.* **6**, 274-304.
30. Steiner, J., 1857: Über die Flächen dritten Grades. *J. für r. und a. Math.* **53**, 133-141.
31. Todd, J.A., 1932: Polytopes associated with the general cubic surface. *J. London Math. Soc.* **7**, 200-205.
32. Tsaranov, S.V., 1989: On a generalization of Coxeter groups. *Algebras, Groups and Geometries* **3**, 281-318.
33. Vinberg, E.B., 1971: Discrete linear groups generated by reflections. *Izv. Akad. Nauk SSSR, Serv. Mat.,* Tom **35** = *Math. USSR Izvestija* **5.5**, 1083-1119.

34. Weyl, H., 1935: The structure and representation of continuous groups. *Institute for Advanced Study, Princeton; Notes.*
35. Witt, E., 1941: Spiegelungsgruppen und Aufzählung halbeinfacher Liescher Ringe. *Abh. Math. Sem. Hansischen Univ.* **14**, 289-322.

POLYHEDRA WITH HOLLOW FACES

BRANKO GRÜNBAUM*
Department of Mathematics
University of Washington GN-50
Seattle, WA 98195, USA

1. Introduction

The Original Sin in the theory of polyhedra goes back to Euclid, and through Kepler, Poinsot, Cauchy and many others continues to afflict all the work on this topic (including that of the present author). It arises from the fact that the traditional usage of the term "regular polyhedra" was, and is, contrary to syntax and to logic: the words seem to imply that we are dealing, among the objects we call "polyhedra", with those special ones that deserve to be called "regular". But at each stage—Euclid, Kepler, Poinsot, Hess, Brückner, ...—the writers failed to define what are the "polyhedra" among which they are finding the "regular" ones. True, we now know what are the convex polyhedra, which we think are the polyhedra Euclid had in mind; hence there is no stigma attached to the use of a term like "regular convex polyhedron". But where in the literature do we find acceptable definitions of polyhedra that could be specialized to give the "regular Kepler-Poinsot polyhedra"? For these, a better expression would be to say that they are "regularpolyhedra"—a distinct kind of objects, constructed according to more or less explicit procedures, and without any connection to what the separate parts of that ungainly word may mean.

One aim of the present paper is to provide a usable and intrinsically consistent foundation for a geometric theory of polyhedra which need not be convex; in fact, as defined here, they can be very general. An effort was made to formulate the relevant concepts in such a way as to "legitimize" not only the regular and uniform nonconvex polyhedra, but also the various interesting kinds of polyhedra introduced by Hess and Brückner, as well as other polyhedra described below and believed to be new.

Another aim of this paper is to atone for a cardinal sin committed by Poinsot and all later writers on regular polyhedra (again including the present author): Poinsot's stated definitions of "polygon" and "regular polygon" do not tally with his actual usage of the terms. In the theory of polyhedra this inconsistency causes needless restrictions and loss of contact with other mathematical disciplines, thus leading to the widespread impression that the theory of (nonconvex) polyhedra is an obscure

* Research supported in part by NSF grants DMS-9008813 and DMS-9300657. Comments by Heidi Burgiel on an earlier version of this paper are acknowledged with thanks.

T. Bisztriczky et al. (eds.), POLYTOPES: Abstract, Convex and Computational, 43–70.

and arcane deadwater. As will become clear in the sequel, this impression is unjustified; in fact, one may imagine that an understanding of these general polyhedra may well inspire and contribute to new results in graph theory, the study of maps on manifolds, and topology in general.

The present paper is meant to be an introduction to a topic for which no satisfactory exposition is available, but on which much that is inaccurate and misleading has been written. Hence I hope that the reader will excuse the details, and the slow pace. A great effort was devoted to making the exposition precise and clear, without being pedantic. Before starting on the chosen topic, a few general considerations seem appropriate.

In the study of convex polyhedra it makes no essential difference whether a polyhedron is understood as a "solid" or as the collection of its faces. The combinatorial aspects are most often related to the facial structure, while other properties are conveniently studied in the general setting of convex bodies; since the transition between the two points of view is easy, in each specific problem the most appropriate one may be used without causing confusion or complications. For nonconvex polyhedra, however, even the most cursory examination shows that the choice between these points of view is of critical importance. For better or worse, throughout this paper we are guided by the idea that a polyhedron is an object formed and determined by the polygons that are its faces, and we relegate to another occasion the study of nonconvex polyhedral solids; for some steps in the latter direction see [21].

With this understanding, it is convenient to distinguish three main levels of generality of nonconvex polyhedra. The smallest departure from convexity occurs among **acoptic** polyhedra (from the Greek $\kappa o \pi \tau \omega$, to cut); these are polyhedra in which the faces are planar, simple polygonal regions, and different faces intersect only along common edges or vertices. Next in generality are polyhedra in which faces may be selfintersecting planar polygons (such as the pentagrams in some of Kepler's polyhedra), and different faces may intersect and cross in various ways. Finally, there are polyhedra which may have nonplanar faces. There exists no systematic presentation of a theory of polyhedra at any of these levels of generality, and it was originally our intention to discuss here all three levels. However, constraints of time, space and energy led us to restrict the attention to the middle level, and to polyhedra in Euclidean 3-space. We hope that the reader will consult some of [2], [17], [18], [20], [30], [31] in order to see at least a few of the results concerning acoptic polyhedra, and the references mentioned in Section 7 for polyhedra whose faces need not be planar. Each of these directions presents ample opportunities for further research.

We start by clarifying (in Section 2) the definition of polygons which we shall use in later sections as building blocks of polyhedra. In Section 3 we define a class of nonconvex polyhedra, which may be called "hollow-faced polyhedra", and illustrate the definition by several examples. This concept of polyhedron is then used to discuss isogonal polyhedra (Section 4), regular polyhedra (Section 5), and polyhedra we call "noble" (Section 6). (A polyhedron is called noble if it is both isogonal and isohedral; these polyhedra were last studied in a very long paper by Max Brückner in 1906.) Some general remarks and problems are collected in Section 7.

Part of this material was presented in courses given by the author at the University of Washington in 1990 and 1992. Acoptic polyhedra have been considered in joint work with G. C. Shephard (in the papers mentioned above, and in ongoing research), as were aspects of hollow-faced polyhedra which are not discussed here and which will be presented in separate papers.

2. Polygons

Polyhedra, as the term is understood here, are collections of polygons. Hence we lay a foundation for the study of polyhedra by giving a precise definition of "polygon". The definition is more general than the traditional interpretation, but it appears to be the one most appropriate for the present purposes. Moreover, this method of defining polygons will serve as a pattern in the more complicated definition of polyhedra. The central point is that just as a function cannot be considered as given if one knows only its range, so for a polygon it is not sufficient to know only what points lie on its vertices and edges. We require information about the "internal structure" of that set of points. Although other approaches are possible, here we attain this goal by first defining "abstract polygons", and then considering "geometric polygons" as "realizations" of the abstract ones in a Euclidean (or other) setting. In practice, it is convenient to use oriented as well as unoriented polygons. We shall limit ourselves to unoriented polygons, since oriented ones may be easily derived from them, in each case in just two (oppositely directed) ways.

An **abstract polygon** $\mathcal{A}$ is a fixed Eulerian **circuit** $\mathcal{C}$ in a simple Eulerian graph $\mathcal{G}$ (that is, a graph with no loops or multiple edges, and with vertices of even valence). In greater detail: The abstract polygon $\mathcal{A}$ is a system consisting of a set $\mathcal{V}$ of distinct elements, and a cyclically ordered set $\mathcal{C}$ of distinct unordered pairs $\mathcal{E}$ of distinct elements of $\mathcal{V}$, such that

(i) for each pair $\mathcal{E}$ in $\mathcal{C}$ one element of the pair $\mathcal{E}$ is shared with the preceding pair $\mathcal{E}^*$, and the other with the following pair $\mathcal{E}^{**}$; these pairs $\mathcal{E}^*$ and $\mathcal{E}^{**}$ are said to be **adjacent** to the pair $\mathcal{E}$.

(ii) no subsystem of $\mathcal{A}$ which has the same adjacent pairs has property (i).

Note that this implies that card $\mathcal{V} \geq 3$. The definition may be easily modified to apply to infinite polygons $\mathcal{A}$; we do not dwell on this since, except for a brief mention in Section 5, we restrict attention to finite polygons.

The elements of $\mathcal{V}$ are the "vertices" of the polygon, and the pairs represent the "edges" of the polygon. The departure from tradition is evident in the fact that a "vertex" can be visited more than once—but the adjacency relation among the edges is determined once and for all, and each edge is adjacent to precisely two other edges. This is the "structure" to which reference was made earlier.

Abstract polygons can be graphically represented by (undirected) graphs (without loops or multiple edges) provided a labelling is indicated to show which is the cyclic order in which the edges are traversed. If the number of elements of $\mathcal{V}$ is the same as the number of pairs in $\mathcal{C}$, so that each vertex is visited just once, the polygon is said to be **unicursal**.

A **geometric polygon** (or **polygon** for short) $\mathcal{P}$ is the image of an abstract polygon $\mathcal{A}$ under a map ϕ which associates with each element of $\mathcal{V}$ a point (**vertex** of $\mathcal{P}$) in the Euclidean d-space E^d, and with each pair from $\mathcal{C}$ the segment (**edge** of $\mathcal{P}$) having as endpoints the images of the elements of $\mathcal{A}$ that constitute the pair. If $\mathcal{P}$ has n edges we shall call it an **n-gon**.

Clearly, this definition of polygon generalizes naively understood polygons in several respects. First, the polygon may be in a space of arbitrarily large dimension; however, except for brief mentions of skew polygons in Section 5 and 7, each polygon we consider will be a subset of a plane—although when forming polyhedra, the planes of the various polygons will in general be distinct. Second, various kinds of coincidences may occur among the images of the elements of $\mathcal{V}$ and of $\mathcal{C}$, even if $\mathcal{A}$ is unicursal. This includes the possibility that adjacent or other vertices coincide, or that edges overlap or pass through vertices; if any kinds of coincidences are to be excluded in the formulation of a definition or result, this has to be stated explicitly.

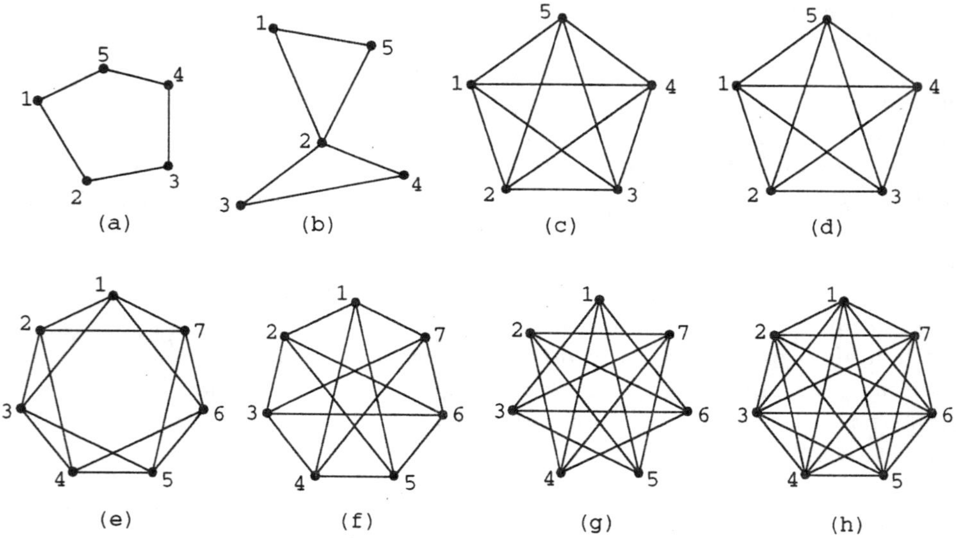

Fig. 1. Examples of polygons. The 5-gon in (a) can be described by the cyclic sequence of vertices 1,2,3,4,5(1); the last label is in parentheses to indicate that it coincides with the first and closes the cycle. Other symbols for the same polygon can be obtained by reversing the listing, or by starting with any other vertex, in either of the two directions. The 6-gon in (b) can have either of the structures 1,2,3,4,2,5(1) or 1,2,4,3,5(1). If the 10-gon in (c) is interpreted as 1,2,4,5,2,3,5,1,3,4(1) and that in (d) as 1,2,5,1,4,5,3,4,2,3(1), they are clearly distinct (non-similar), but the underlying abstract 10-gons are isomorphic. The diagram in (e) leads to two 14-gons constructed analogously: 1,2,4,5,7,1,3,4,6,7,2,3,5,6(1) and 1,2,7,1,6,7,5,6,4,5,3,4,2,3(1) which have non-isomorphic underlying abstract polygons. However, they are isomorphic to the two polygons that can be derived in the same way from (f) and to the two derivable from (g). The 21-gon in (h) can be interpreted in many ways; a very symmetric one is 1,2,4,7,1,3,6,7,2,5,6,1,4,5,7,3,4,6,2,3,5(1). All the diagrams in (c) to (g) can also be interpreted as polygons in several less symmetric ways.

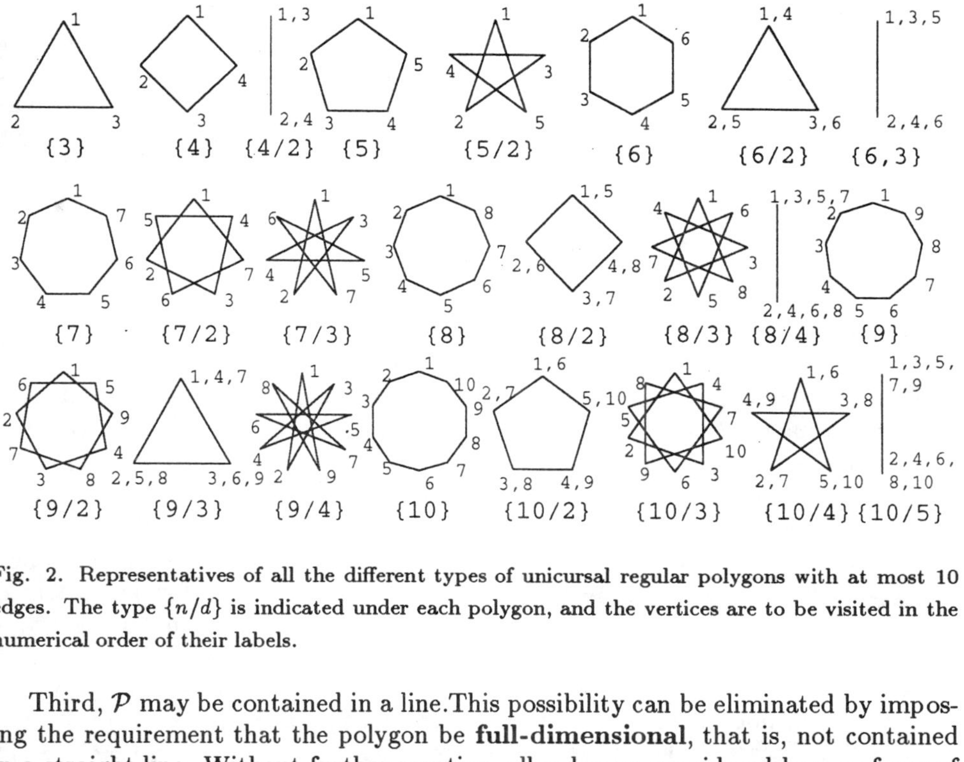

Fig. 2. Representatives of all the different types of unicursal regular polygons with at most 10 edges. The type $\{n/d\}$ is indicated under each polygon, and the vertices are to be visited in the numerical order of their labels.

Third, $\mathcal{P}$ may be contained in a line.This possibility can be eliminated by imposing the requirement that the polygon be **full-dimensional**, that is, not contained in a straight line. Without further mention, all polygons considered here as faces of polyhedra will be assumed full-dimensional.

In Figures 1 and 2 are shown several polygons, meant to illustrate the concepts introduced above. In most situations, there is no need to distinguish polygons that can be mapped onto each other by a similarity; hence we may speak of "the square", etc.

Since an abstract polygon $\mathcal{A}$ is combinatorially just a circuit, the group of incidence-preserving automorphisms of a unicursal n-gon is the **dihedral group** dn, which consists of n stepwise advances along the circuit, together with n maps which reverse the direction. In all other cases, the utomorphism group of an n-gon is a (proper or improper) subgroup of dn.

A **symmetry** of the geometric polygon $\mathcal{P}$ is a pair (σ, π) where σ is an isometry of the plane and π an edge-preserving permutation of the vertices of $\mathcal{P}$, such that π and σ commute on the set of vertices of $\mathcal{P}$. The symmetry group of any (geometric) polygon with n edges is clearly (isomorphic to) a subgroup of dn. A polygon is called **isogonal** [**isotoxal**, or **regular**] if its symmetry group acts transitively on its vertices [edges, flags consisting of a vertex and an edge incident with it]. It is easily seen that:

(i) A polygon is regular if and only if it is isogonal and isotoxal.

(ii) A unicursal polygon is regular if and only if its symmetry group is isomorphic to dn.

For any regular n-gon, all the vertices are obtained from any one vertex by rotations through an angle of $2\pi d/n$ and its multiples, where without loss of generality we can assume that d is an integer such that $1 \leq d \leq n/2$. Different values of d yield different (that is, non-similar) polygons. A unicursal polygon corresponding to n and d shall be denoted by the "Schläfli symbol" $\{n/d\}$; however, $\{n/1\}$ is usually simplified to $\{n\}$. In a suitable system of coordinates the vertices of the polygon $\{n/d\}$ can be chosen as points with coordinates

$$\left\{ \left(\cos\frac{2d\pi j}{n}, \sin\frac{2d\pi j}{n} \right) |j = 0, 1, \ldots, n-1 \right\}.$$

Hence, for each $n \geq 3$, there exist precisely $[n/2]$ distinct unicursal regular n-gons (usually represented by the symbols $\{n/d\}$, $d = 1, 2, ..., [n/2]$) such that no two are similar, and every unicursal regular n-gon is similar to one of them.

Figure 2 shows all unicursal regular n-gons with $n \leq 10$. In this figure and some of the following ones, the numerals j near the vertices are the labels of the vertices. Even for a unicursal regular polygon, if n and d are not relatively prime then several distinct vertices of the underlying abstract polygon are represented by the same point in the geometric polygon. It should be emphasized that each part of Figure 2 represents a single polygon. For example, $\{6/2\}$ is a hexagon whose vertices happen to coincide in pairs; it is not a pair of triangles. This is rather vividly demonstrated in Figure 3, which shows the "metamorphosis" (an isotopy via isogonal polygons) of $\{6\}$ to $\{6/2\}$. Analogous are the metamorphoses for all $n \equiv 2 \pmod 4$; there are $[n/4]$ different isotopies, from $\{n/d\}$ with $1 \leq d \leq [n/4]$, to $\{n/e\}$, where $e = n/2 - d$. In case $n \equiv 0 \pmod 4$ the metamorphoses are slightly different; this is illustrated for $n = 8$ in Figure 4. For more details, including an illustration in case $n = 14$ and analogous illustrations of metamorphoses between regular polygons via isotoxal polygons, see [16].

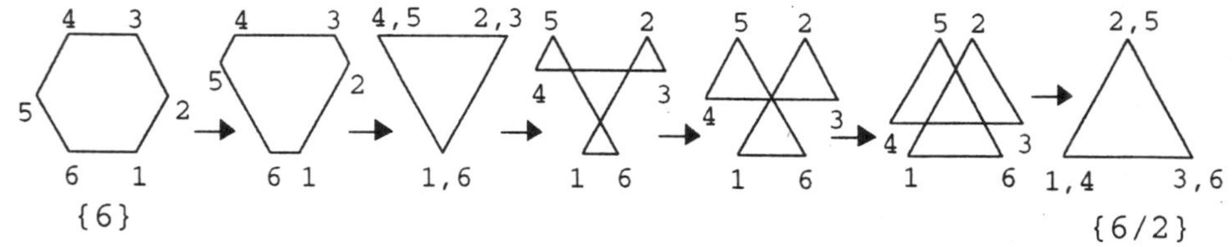

Fig. 3. An isotopy between regular polygons $\{6\}$ and $\{6/2\}$, via isogonal hexagons.

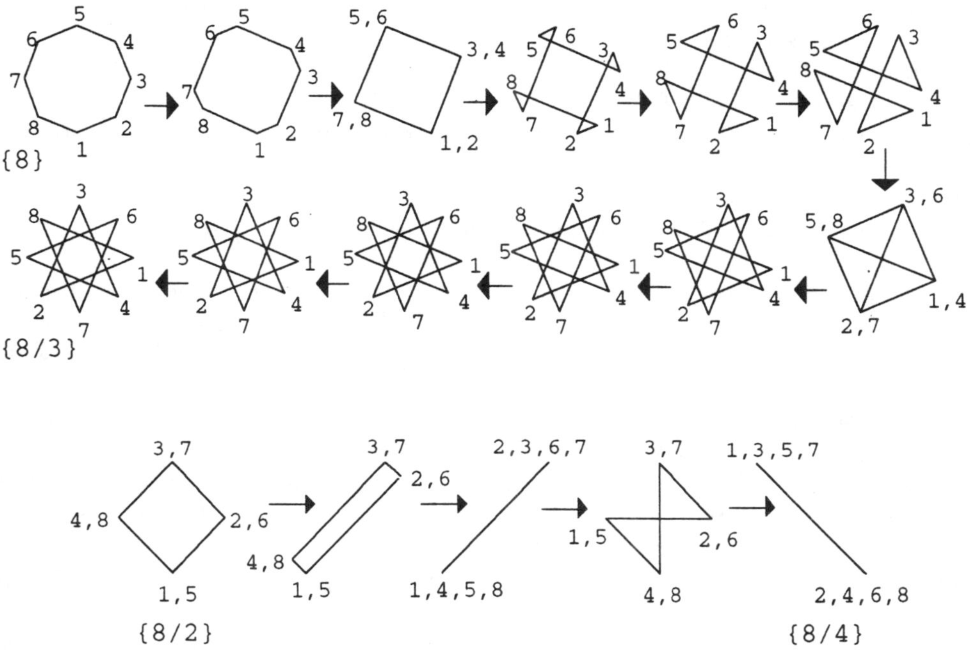

Fig. 4. An isotopy between {8} and {8/3}, and between {8/2} and {8/4}, via isogonal octagons.

3. Polyhedra

Now we proceed to use polygons in order to construct polyhedra. We start with abstract polyhedra; note that our use of this term differs from the meaning it has in other papers in the present volume.

An **edge-sharing family** $\mathcal{P}$ is a triplet $(\mathcal{V}, \mathcal{E}, \mathcal{F})$ consisting of a nonempty set $\mathcal{V}$ of **vertices**, a set $\mathcal{E}$ of **edges**, and a set $\mathcal{F}$ of **faces**, with a relation of incidence satisfying the conditions below:

(i) each E in $\mathcal{E}$ is incident with a pair of distinct vertices in $\mathcal{V}$;

(ii) each F in $\mathcal{F}$ is an abstract unoriented polygon, with vertices in $\mathcal{V}$ and edges in $\mathcal{E}$;

(iii) each edge E in $\mathcal{E}$ is an edge of a positive even number of distinct faces in $\mathcal{F}$; these faces are arranged in pairs, and each member of a pair is said to be **adjacent** to the other member.

If $\mathcal{P} = (\mathcal{V}, \mathcal{E}, \mathcal{F})$ is an edge-sharing family and V is a vertex of $\mathcal{P}$, the vertex figure $C(V)$ of $\mathcal{P}$ at V is a collection of vertices and edges defined as follows: If $F \in \mathcal{F}$ and if $\{V, V_1\}$ and $\{V, V_2\}$ are two adjacent edges of F, then $C(V)$ contains V_1 and V_2 as vertices, and $\{V_1, V_2\}$ as an edge incident with V_1 and V_2. If V_i, V_j and V_k are vertices of $C(V)$, while $\{V_i, V_j\}$ and $\{V_j, V_k\}$ are edges of $C(V)$ arising from faces F^* and F^{**}, then $\{V_i, V_j\}$ and $\{V_j, V_k\}$ are said to be adjacent edges of $C(V)$ if and only if the faces F^* and F^{**} are adjacent faces of $\mathcal{P}$.

Now we are ready for the definition. An **abstract polyhedron** is an edge-sharing family $\mathcal{P} = (\mathcal{V}, \mathcal{E}, \mathcal{F})$ such that:
(i) For each $V \in \mathcal{V}$ the vertex figure $C(V)$ is an abstract polygon;
(ii) No subfamily of $\mathcal{P}$ with the same adjacencies has property (i).

A **geometric polyhedron**, or **polyhedron** for short, is the image of an abstract polyhedron in the d-dimensional Euclidean space E^d, such that vertices are mapped into points, and faces into (geometric) polygons (hence edges into segments, possibly of length 0). Each such geometric polyhedron is said to be a **realization**, or a **representation**, of the abstract polyhedron; in turn, the abstract polyhedron is said to be **underlying** the geometric polyhedron. Note that, in general, there is no requirement that the faces be planar polygons. A geometric polyhedron is called **epipedal** if each of its faces is a planar polygon, and it is called **full-dimensional** if it is not contained in any plane. Traditionally, epipedal polyhedra were the only polyhedra studied; however, polyhedra that are not epipedal can be interesting as well as useful in the study of epipedal polyhedra (see Section 5). Since most of our discussion will concern epipedal polyhedra, unless the opposite is explicitly stated by "polyhedron" we shall mean "full-dimensional epipedal polyhedron".

The polyhedra described by the above restrictions will be called **hollow-faced**. Two hollow-faced polyhedra are said to be **combinatorially equivalent** if they are realizations of the same abstract polyhedron, and **isomorphic** if they are combinatorially equivalent in such a way that the same coincidences of corresponding vertices, as well as the same crossings or coincidences of corresponding edges, take place in both. A polyhedron (abstract or geometric) is called **oriented** if its faces are oriented polygons, such that each edge is oriented in opposite directions in the pair of adjacent polygons that contain it. A polyhedron is **orientable** if its faces can be assigned orientations in such a way that the polyhedron is oriented. Unless the contrary is explicitly mentioned, whenever we present the faces of an orientable polyhedron by lists of vertices, these will be selected in such a way as to make the polyhedron oriented.

Symmetries of polyhedra are defined in complete analogy to those of polygons, as are concepts of **isogonal** or **isotoxal** polyhedra. A polyhedron is **isohedral** [**regular**] if its faces [flags consisting of a vertex, an edge, and a face, all mutually incident] form one orbit under its symmetries.

An abstract polyhedron is **unicursal** if it has unicursal faces and vertex figures. Each unicursal polyhedron $\mathcal{P}$ can be interpreted as a **map**, that is, a cell-complex decomposition of a compact 2-manifold, by considering each of the faces of $\mathcal{P}$ as the boundary of a 2-cell. If $\mathcal{P}$ is orientable so is the map, and the genus of the map is taken as the genus of $\mathcal{P}$. Abstract polyhedra that are not unicursal correspond to cell complexes more general than 2-manifolds.

It is clear that polyhedra as defined here are generalizations of the usual concept of polyhedra. If the faces of a polyhedron are simple polygons, they determine polygonal regions; if these regions have no unsuitable intersections we have an acoptic polyhedron. This is the basis of constructing cardboard models of such polyhedra. In most illustrations of polyhedra (even non-acoptic ones) the same approach is used; however, for polyhedra with selfintersections its usefulness is limited at best, and at times can be badly misleading. Our illustrations will rely on "skeletal" representations, which show only the segments that form the faces of the polyhedra in question.

A few examples will illustrate the concepts introduced above. The polyhedron $\mathcal{P}$ shown in Figure 5a and described in the caption, is isogonal and oriented; three of its faces are selfintersecting quadrangles. The map $M(\mathcal{P})$ associated with $\mathcal{P}$ is toroidal, the genus of $\mathcal{P}$ is 1.

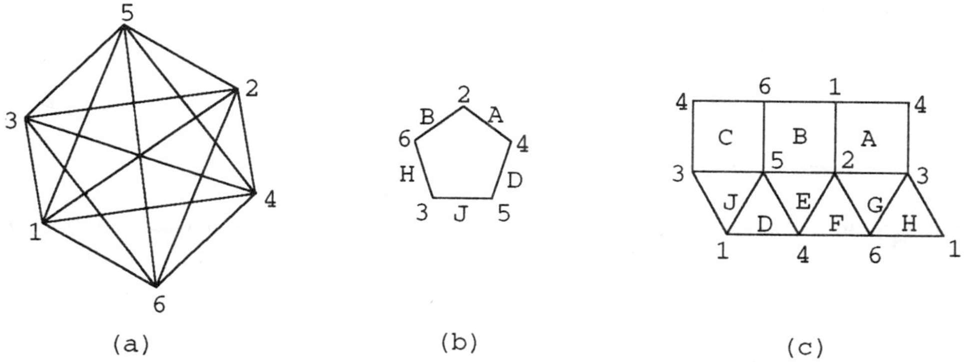

Fig. 5. The vertices of the polyhedron $\mathcal{P}$ shown in (a) are labelled 1,2,3,4,5,6; they coincide with the vertices of a regular octahedron. The nine faces of $\mathcal{P}$ are A = 1,2,3,4(1); B = 2,1,6,5(2); C = 4,3,5,6(4); D = 1,4,5(1); E = 2,5,4(2); F = 2,4,6(2); G = 2,6,3(2); H = 1,3,6(1); J = 1,5,3(1); it follows that $\mathcal{P}$ is isogonal. The symmetry group of $\mathcal{P}$ is isomorphic to $[2^+, 6]$ in the notation of Coxeter-Moser [8]. The map associated with $\mathcal{P}$, shown in (c), is a torus, hence $\mathcal{P}$ has genus 1. The vertex figure of $\mathcal{P}$ at the vertex 1 is the abstract polygon 2,4,5,3,6(2) schematically shown in (b).

Next, consider the polyhedron $\mathcal{P}$ shown in Figure 6, and described in its caption. It is oriented and isogonal, and is combinatorially equivalent (but not isomorphic) to the (uniform, Archimedean) truncated cube (which is usually denoted by (3.8.8)); thus the map of the polyhedron is spherical, and so the genus is 0. We note that while it may require some effort to understand the structure of this polyhedron, this task is much easier if we consider the isotopy of the isogonal family of polyhedra to which it belongs (shown in Figure 10 and discussed in Section 4); the labels in one of the diagrams in Figure 10 correspond to the labels in Figure 6(a).

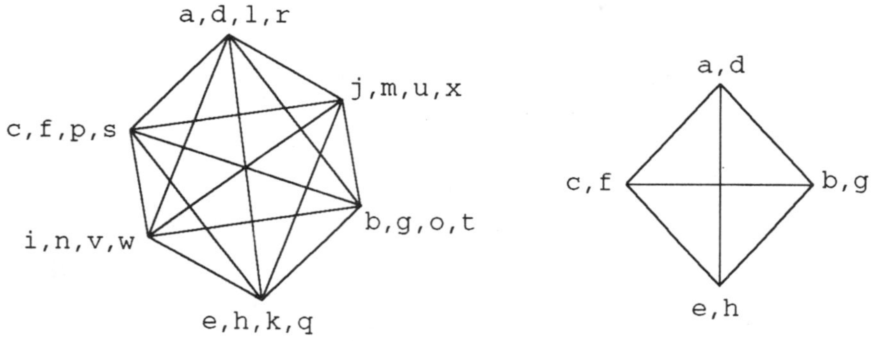

(a) (b)

Fig. 6. Quadruplets of vertices of the polyhedron $\mathcal{P}$ coincide with each vertex of a regular octahedron; their labels are *a,b,c,d,e,f,g,h,i,j,k,l,m,n,o,p,q,r,s,t,u,v,w,x,* as indicated in (a). The faces of $\mathcal{P}$ are *a,b,c,d,e,f,g,h(a); a,h,i,j,k,l,m,n(a); k,o,p,q,r,s,t,l(k); q,x,w,e,d,v,u,r(q); b,n,m,t,s,u,v,c(b); f,w,x,p,o,j,i,g(f); a,n,b(a); c,v,d(c); e,w,f(e); g,i,h(g); j,o,k(j); l,t,m(l); p,x,q(p); r,u,s(r).* The shape of one of the octagons is indicated in (b), and the other five octagons are congruent to it; all eight triangles are equilateral. All vertex figures are triangles. The symmetry group is isomorphic to [3, 4].

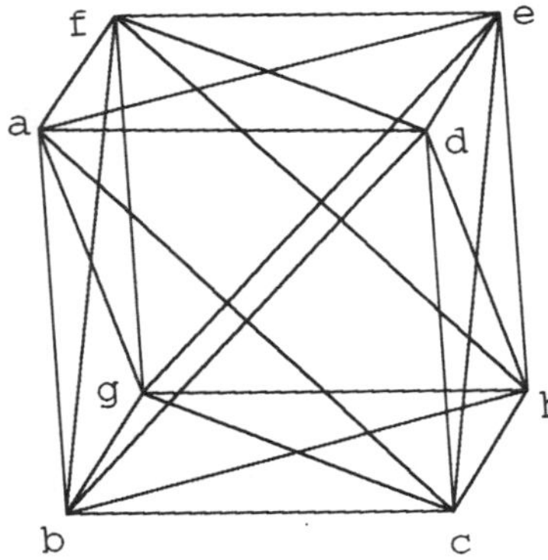

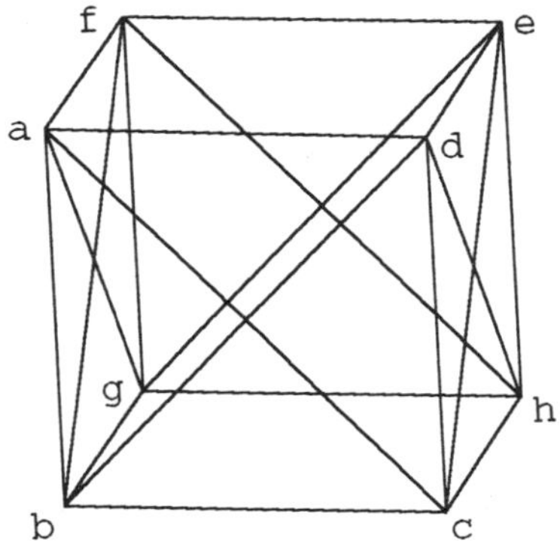

Fig. 7 Fig. 8

Fig. 7. The eight vertices of $\mathcal{P}$ are at the vertices of a cube. The twelve faces are *a,b,d,c(a); a,c,b,d(a); a,b,f,g(a); a,f,b,g(a); a,d,f,e(a); a,e,d,f(a); b,c,g,h(b); b,g,c,h(b); c,d,h,e(c); c,e,d,h(c); e,f,h,g(e); e,g,f,h(e).* All faces are congruent selfintersecting quadrangles. This polyhedron $\mathcal{P}$ is isogonal and isohedral, but it is not orientable; its Euler characteristic is -4, and its symmetry group is isomorphic to [3, 4].

Fig. 8. The vertices of this polyhedron $\mathcal{P}$ are the same as in Figure 7; its faces are *a,c,b,d(a); a,d,h,g(a); a,f,h,c(a); a,g,b,f(a); b,c,e,f(b); b,g,e,d(b); c,h,d,e(c); e,g,h,f(e).* This $\mathcal{P}$ is oriented and isogonal, but not isohedral; its symmetry group is isomorphic to [2, 4]. The associated map is one of the regular maps on the torus.

The isogonal and isohedral polyhedron $\mathcal{P}$ shown in Figure 7 is not orientable; its Euler characteristic is -4. Note that some pairs of faces (such as the first two) have two disjoint edges in common, as well as all four vertices. Neither of these coincidences is forbidden by the definition of polyhedra.

The map corresponding to the polyhedron $\mathcal{P}$ in Figure 8 is a regular map on the torus. In other words, the underlying polyhedron of $\mathcal{P}$ is regular; however, this particular geometric realization is isogonal but not regular. It is not known whether any realization of this map by a full-dimensional, epipedal geometric polyhedron is a regular polyhedron.

A family of non-unicursal polyhedra, which we call corrugated pyramids, is illustrated in Figure 9. A pyramid over any non-unicursal polygon provides another example of a non-unicursal polyhedron. More interesting examples will be encountered in Section 6.

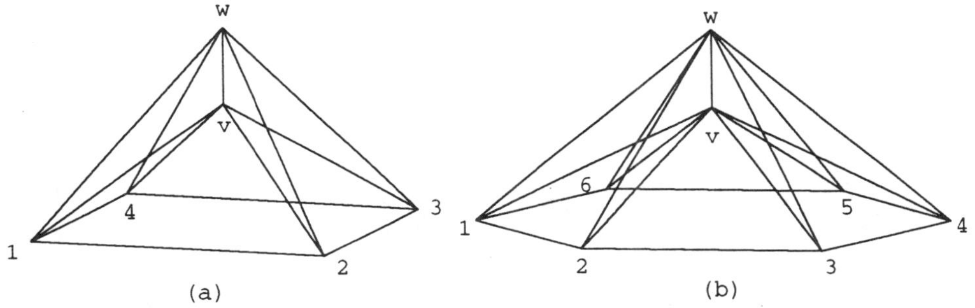

Fig. 9. Examples of corrugated pyramids: 4-sided in (a), 6-sided in (b). In general, a $2m$-sided corrugated pyramid has as basis a $(2m)$-gon with vertices $1, 2, ..., 2m(1)$, and two additional vertices v and w. Besides the basis, its faces are $j, v, w(j)$ for $j = 1, 2, ..., 2m$, as well as $2j$, $2j+1$, $v(2j)$ and $2j-1$, $2j$, $w(2j-1)$ for $j = 1, 2, ..., m$. For each $j = 1, 2, ..., m$ and faces $j, v, w(j)$ and $j+m, v, w(j+m)$ are adjacent along the common edge v, w. The corrugated pyramids are not unicursal since the vertex figures at v and w are not unicursal polygons. The $2m$-sided corrugated pyramid is orientable for odd m, non-orientable for even m. The lists of faces given are not arranged to show orientability.

4. Isogonal polyhedra

We now turn to the consideration of some special hollow-faced polyhedra. As mentioned in Section 2, for each n, all isogonal n-gons fit into a finite number of families within which the members are related by isotopy (via isogonal n-gons). Isogonal polyhedra can be arranged in families of a similar nature. (For convex isogonal polyhedra such isotopies have been described by Robertson [29] and others.) An example of such an isotopy is shown in Figure 10; this extends the isotopy on page 92 of [29]. It should be noted that the last polyhedron in the isotopy is inverse to (that is, negative of) the starting polyhedron. It can be shown that every realization (as isogonal polyhedron) of the abstract polyhedron which underlies those in Figure 10 has the same symmetry group [3,4] as those in Figure 10, and belongs to the same isotopy arc as shown there (or its mirror image).

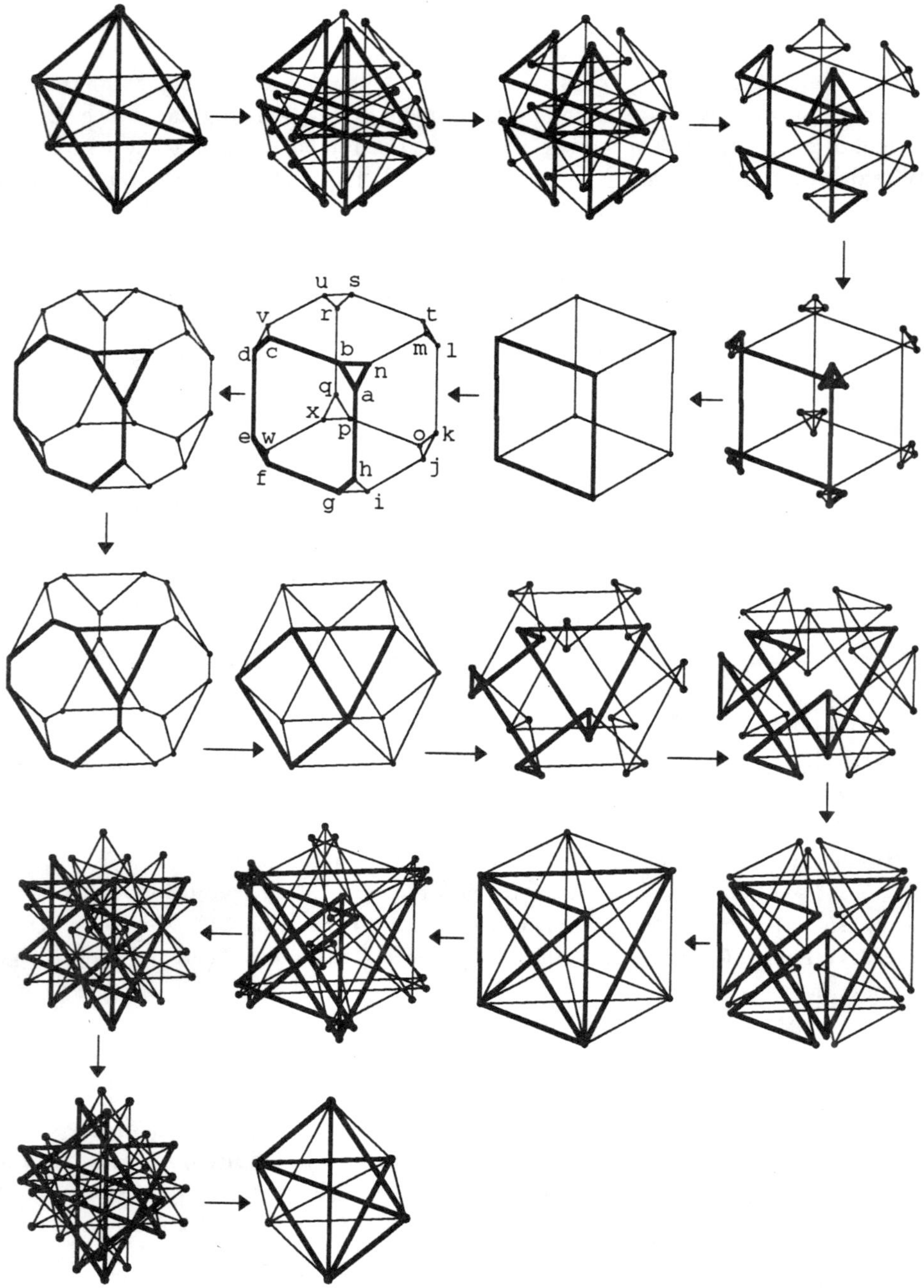

Fig. 10. An illustration of the isotopy connecting all isogonal polyhedra isomorphic with the truncated cube; note that at certain stages pairs, triplets or quadruplets of vertices are represented by the same point. In particular, the polyhedra appearing to be a cube, an octahedron and a cuboctahedron are not isomorphic to these classical polyhedra. Only the polyhedra strictly between the "cube" and the "cuboctahedron" are convex. One representative of each orbit of faces is emphasized.

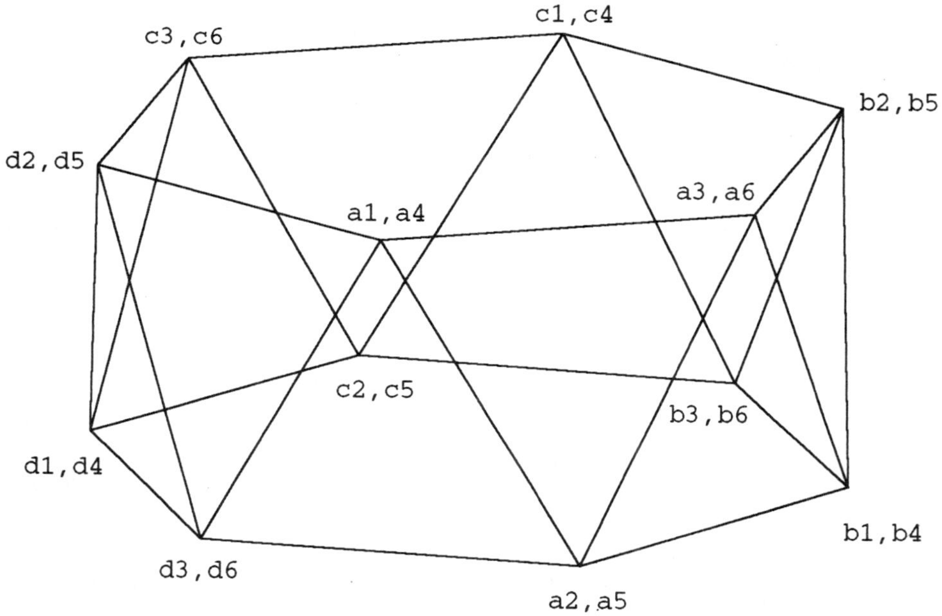

Fig. 11. A uniform polyhedron p of genus 5, all faces of which are regular hexagons: four faces of type $\{6\}$, and twelve faces of type $\{6/2\}$. Details of the construction are indicated in the text. The symmetry group is the direct product of $[2^+, 12]$ and the cyclic groups $c2$.

Similar isotopies exist for other types of isogonal polyhedra, as well as for isohedral polyhedra. Most of them have not been investigated in detail, and the topological character of their realization space is not known. Also, in only a few cases other than the one shown in Figure 10 is it known how many disjoint isotopic families there are of isogonal (or of isohedral) polyhedra which are realizations of the same abstract polyhedron. Since in the case of n-gons (for even n) the number of isotopic families grows linearly with n, prisms over such polygons show that there is no finite upper bound on the number of isotopic families of isogonal polyhedra combinatorially equivalent to each other.

While it is obvious that the vertices of a uniform polyhedron must all be located at sets of points which form one orbit under symmetries, this does not limit the types of isogonal polyhedra to the traditional ones since in the present context several vertices may be represented by the same point. As an example of such a "new" type consider the following polyhedron $\mathcal{P}$ which is not merely isogonal but uniform (since all its faces are regular polygons). The $n = 12m$ vertices of $\mathcal{P}$ are located at the vertices of a uniform antiprism based on a regular $(3m)$-gon $\{3m\}$. Each vertex of the antiprism is the image of two vertices of $\mathcal{P}$. A sketch of $\mathcal{P}$ (for $m = 2$) is shown in Figure 11. The peculiar labelling is meant to make obvious the formation of the faces, the claim of isogonality, and the generalization to all values of m. The "bases" are regular hexagons $\{6\}$ (regular polygons $\{3m\}$ in general); the two upper bases are $a1, a6, b5, c4, c3, d2(a1)$ and $a4, a3, b2, c1, c6, d5(a4)$, while the two lower bases are $a2, d3, d4, c5, b6, b1(a2)$ and $a5, d6, d1, c2, b3, b4(a5)$. All "mantle faces" are regular hexagons $\{6/2\}$, formed by vertices with labels (disregarding the letters) $1, 2, 3, 4, 5, 6(1)$ if they have edges on upper bases, and $6, 5, 4, 3, 2, 1(6)$ if they have edges on lower bases. The polyhedron $\mathcal{P}$ is oriented and of genus 5 (in the general case the genus is $3m - 1$).

Many other types of isogonal (or even uniform) polyhedra exist. There seems to be no detailed information available on the possibilities. It is even not known whether every isogonal type is combinatorially equivalent to a uniform polyhedron—a fact that is well known for convex polyhedra. Prime candidates for isogonal polyhedra not isomorphic to uniform ones are the various isogonal polyhedra of positive genus described in [17].

5. Regular polyhedra

Turning now to the question what are the regular polyhedra that can be constructed with hollow faces, we note first that they all must "look like" the "old" regular polyhedra; however, this does not preclude the existence of infinitely many "new" ones. (See Section 7, as well as [15], for remarks about the "new" regular polyhedra.) We start by recalling that the four well-known Kepler-Poinsot polyhedra are, in fact, polyhedra with hollow faces, and are best understood as such. Next, there is a set of eight new regular polyhedra which are obtainable by a certain construction from the Platonic and Kepler-Poinsot polyhedra (other than the cube). To obtain one of the new polyhedra, we replace each vertex v of the original polyhedron by two vertices, v^+ and v^-. The faces of the new polyhedra are obtained by going around the original faces, but alternating between $+$ and $-$ labelled vertices. Since the number of edges of each face of P is odd, to complete a face of the new polyhedron we need to go twice around the old one. Hence if P was the regular polyhedron $\{n, k\}$, with n odd, the new polyhedron is $\{(2n)/2, k\}$. (We use the natural extension of the familiar Schläfli notation.) The polyhedra $\{6/2, 3\}$ and $\{6/2, 4\}$ are shown in Figure 12 together with the corresponding regular maps. Here and in later captions maps are identified by a symbol of the type W#$e.f$, which means that it is entry f among maps with e edges listed in the catalogue of S. E. Wilson [32]. Although this construction does not work for the cube, there exists a regular polyhedron $\{8/2, 3\}$. It is shown in Figure 13, and its regularity is easy to verify.

One way to get an understanding of the method used in constructing the polyhedra in Figure 12 and the six analogous ones is the following. Start with any one of the classical regular polyhedra (except the cube). Replace every n-gonal face $\{n/d\}$ by a regular skew prismatic $(2n)$-gon (that is, $\{2.n^\alpha/d\}$ in the notation of [14]), identifying the edges of adjacent polygons in the obvious way. It is clear that the polygons do not really fit—but the fit improves as α decreases; in the limit, "squashing" the prismatic polygons to $\alpha = 0$, we get the polyhedra as described above.

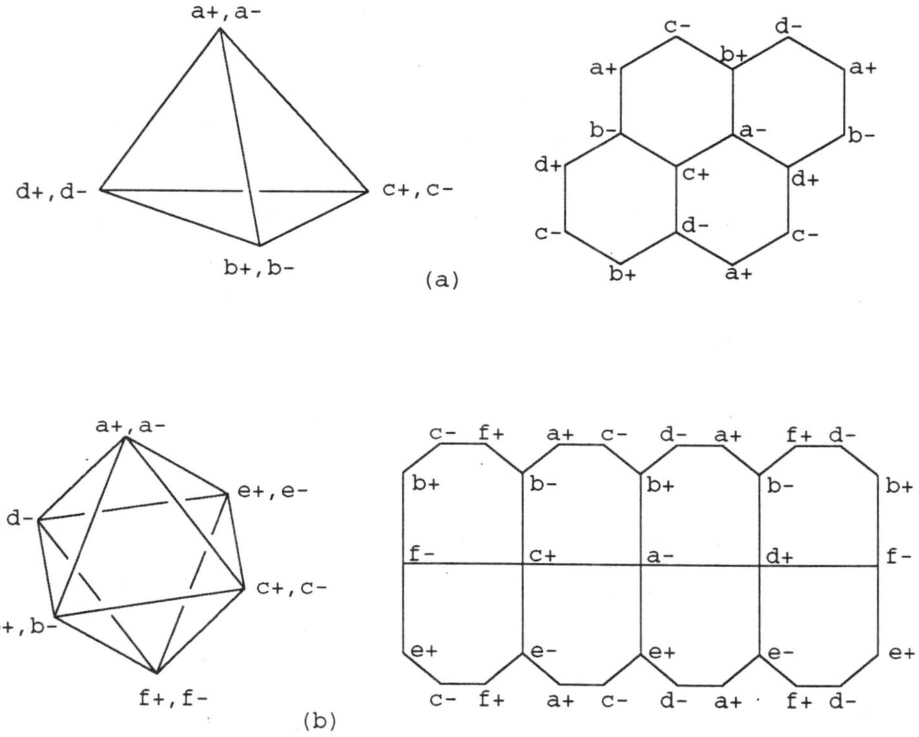

Fig. 12. **Two regular polyhedra with hollow faces, and the corresponding regular maps. The construction is explained in the text. The polyhedron in (a) is {6/2,3} with symmetry group isomorphic to $c2 \otimes [3,3]$ and map W#12.3 of genus 1, that in (b) is {6/2,4} with symmetry group isomorphic to $c2 \otimes [3,4]$ and map W#24.27 of genus 3.**

The reason for bringing up this explanation (and the skew polygons that appear in it) is that a similar construction applied to the regular octahedron leads to infinitely many new regular polyhedra. We start with the regular octahedron $\{3,4\}$, and replace (abstractly) each triangle by a regular helical polygon $\{\infty^{a,b}\}$ (see [14]) with $b = 2p/3$ (that is, by an infinite triangular helix); adjacent helices should wind in opposite directions. This clearly results in a regular polyhedron— admittedly, an abstract one, and one which uses infinite skew helices as faces. Selecting any integer $d \geq 2$ that is not a multiple of 3, identifying in each helical polygon vertices that are at distance $3d$ along the helix from each other, and finally squashing the helices to $\alpha = 0$ we get an epipedal, regular polyhedron $\{(3d)/d, 4\}$. For $d = 2$ we obtain again the polyhedron $\{6/2, 4\}$ shown in Figure 12(b). The smallest new polyhedron $\{12/4, 4\}$ is obtained for $d = 4$; it is shown, together with the corresponding map, in Figure 14.

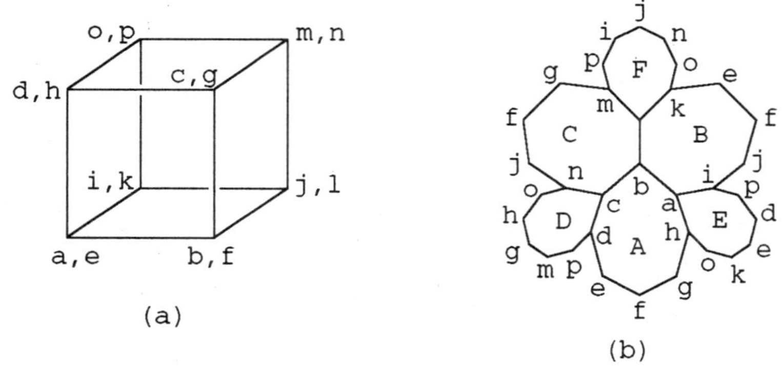

Fig. 13. (a) An orientable regular polyhedron {8/2,3} with 16 vertices and six octagonal facaes: $A = a,b,c,d,e,f,g,h(a)$; $B = a,i,j,f,e,k,l,b(a)$; $C = b,l,m,g,f,j,n,c(b)$; $D = c,n,o,h,g,m,p,d(c)$; $E = a,h,o,k,e,d,p,i(a)$; $F = i,p,m,l,k,o,n,j(i)$. (b) A topologically regular map (W#24.21) of genus 2, which is isomorphic to the regular polyhedron in (a).

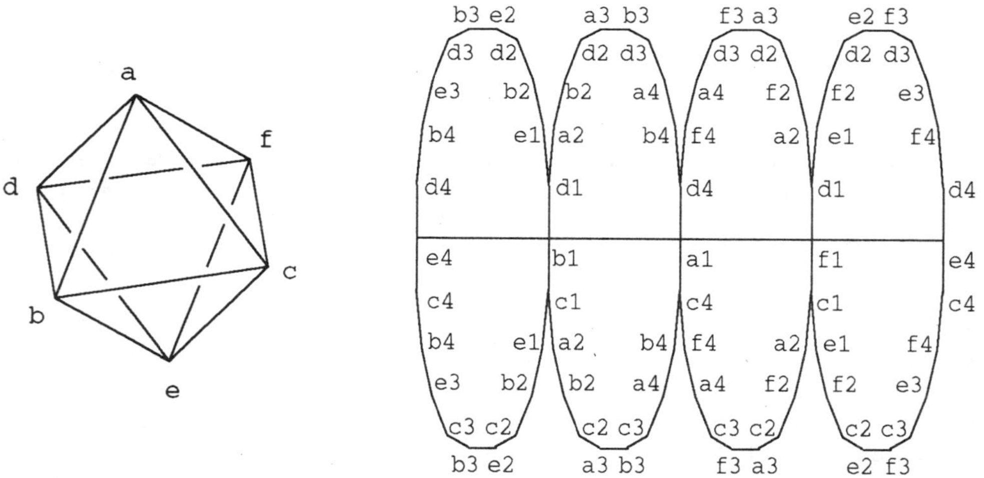

Fig. 14. The regular polyhedron {12/4,4} obtained by the construction explained in the text. Each letter near a vertex in the diagram at left stands for labels of four vertices, distinguished by subscripts 1,2,3,4. The corresponding map is W#48.45, of genus 9.

Although this method applies only to the octahedron, there exist other regular polyhedra in which faces are of the type $\{(kd)/d\}$. As an example, a polyhedron $\{12/3,3\}$ is shown in Figure 15. On the other hand, it seems that not every symbol of the type $\{(ab)/b, c\}$ corresponds to a regular polyhedron. For example, there appears to exist no hollow-faced polyhedron $\{9/3, 3\}$.

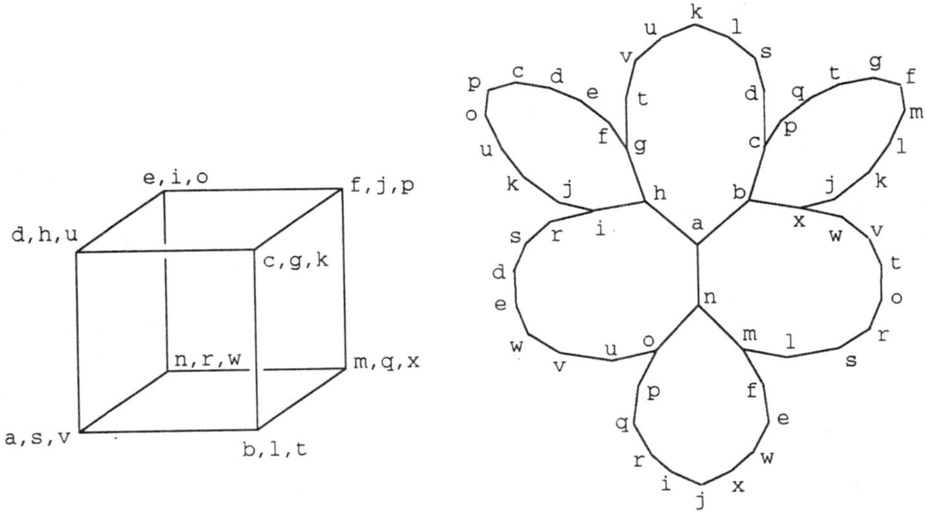

Fig. 15. A regular polyhedron {12/3,3} and the corresponding map W#36.13 of genus 4.

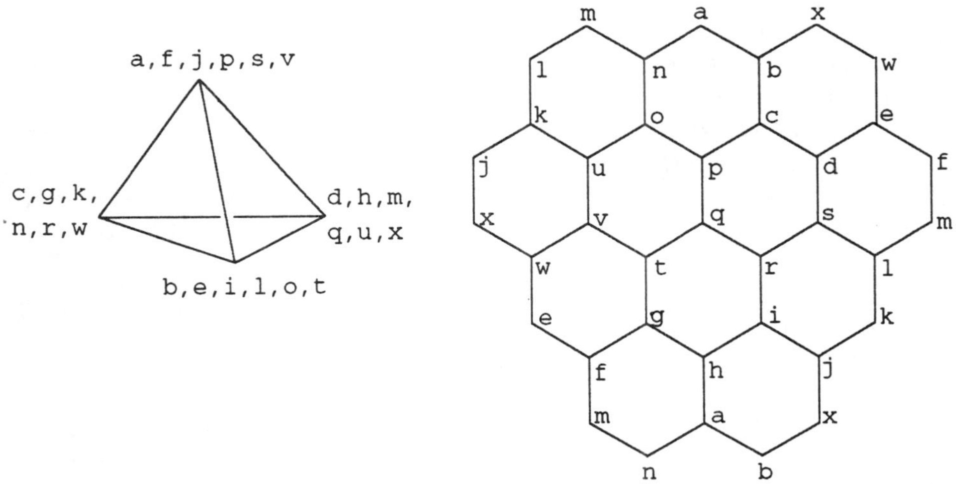

Fig. 16. Another orientable regular polyhedron {6/2,3}. It has 12 faces and 24 vertices, and the corresponding map is W#36.12 of genus 1.

We conclude this brief introduction to regular polyhedra by noting that the Schläfli symbols we use are inadequate to characterize these polyhedra. For example, the polyhedron shown in Figure 16 has symbol $\{6/2, 3\}$, the same as the polyhedron in Figure 12(a)—although the former has 12 faces and the latter only 4. Thus, there is a need for a better notation.

Additional comments concerning regular polyhedra will be made in Section 7.

6. Noble polyhedra

The last topic we shall discuss concerns an interesting family of polyhedra that briefly emerged in the works of Hess [24], [25], [26], and Brückner [4], [5], [6], [7], but seems to have been completely forgotten since the early years of this century: the hollow-faced polyhedra which are both isogonal and isohedral. In order to simplify the syntax, we shall call such polyhedra "noble". Naturally, all regular polyhedra are noble, as are all (necessarily convex) "sphenoids" (that is, tetrahedra with all faces mutually congruent). In fact, among convex polyhedra these are the only possibilities. However, among hollow-faced polyhedra there are many additional noble polyhedra, and their complete enumeration with the definition of polyhedron adopted here is still outstanding. The papers by Hess and Brückner may have choked off the investigation of noble polyhedra by their vagueness, inconsistency and errors. Some examples of these shortcomings will be given below, while the full discussion must await another occasion.

We begin with the polyhedra that appear to have been discovered by Hess, called "stephanoids" by Hess and Brückner. The name comes from the Greek word for "crown", and was chosen by Hess on account of the appearance of these polyhedra. They form two infinite families, related to the prisms and antiprisms, which we shall call **prismatic** (resp. **antiprismatic**) **crown** polyhedra. Combinatorially, each such polyhedron depends on three positive integer parameters. To obtain the prismatic crown polyhedron PC(n,p,q), where $2p - n < 2q < p < n$, we start from an n-sided prism with vertices $a_1, \ldots, a_n, b_1, \ldots, b_n$ (see Figure 17(a)), and for $j = 1, \ldots, n$ we form the selfintersecting quadrangles $a_j, b_{j+q}, a_{j+p}, b_{j+p-q}(a_j)$ and $b_j, a_{j+q}, b_{j+p}, a_{j+p-q}(b_j)$. It is easily checked that these $2n$ faces indeed define a polyhedron, that the polyhedron is orientable of genus 1, and that it is both isogonal and isohedral. The antiprismatic crown polyhedron $AC(n,p,q)$, where $2p - n < q < p < n$ and q is odd, results from the n-sided antiprism with vertices of the mantle labelled consecutively $a_1, \ldots, a_{2n}$ (see Figure 17(b)) by constructing all the quadrangles $a_j, a_{j+q}, a_{j+2p}, a_{j+2p-q}(a_j)$, for $j = 1, \ldots, 2n$. The resulting polyhedron is again orientable of genus 1, and it is isogonal and isohedral. The crown polyhedra depend also on a continuous parameter, determined by the metric proportions of the starting prism or antiprism. While all crown polyhedra are isomorphic to their polars (polarity with respect to a unit sphere with the same center), for suitable choices of the continuous parameter the polar polyhedron is congruent to the original. More interestingly, however, for some sets of the discrete parameters (for even p in the prismatic case, or even $p - q$ in the antiprismatic case), the continuous parameter can be chosen so that the polar polyhedron is identical with the starting one. A polyhedron which is selfpolar in this most restricted sense will be called **autopolar**. Clearly, no convex polyhedron is autopolar.

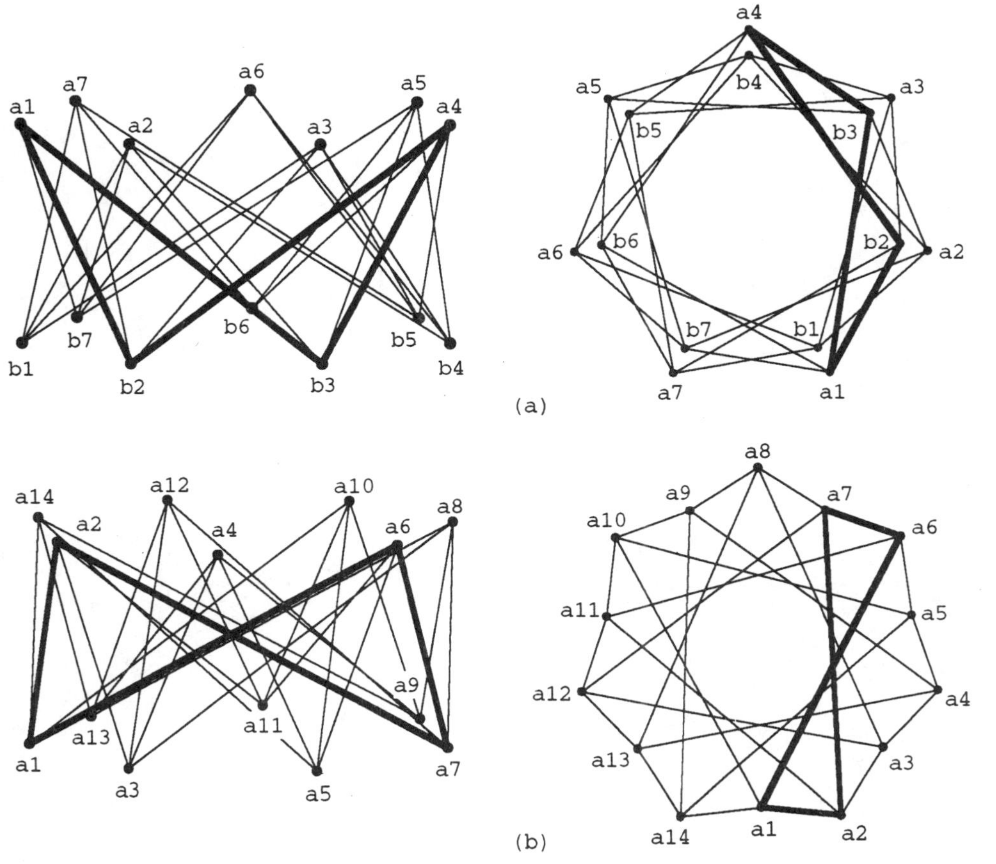

Fig. 17. Crown polyhedra, $PC(7,3,1)$ in (a) and $AP(7,3,2)$ in (b), each in two perspective views; the views at right are from a point on the axis of the polyhedron, not very high above the polyhedron itself.

Other noble polyhedra arise if each of the quadrangles that form a crown polyhedron is decomposed into two triangles by one of its diagonals (either the short one, or the long one, in all faces). We obtain a triangle-faced noble polyhedron which we call a prismatic or antiprismatic **wreath** polyhedron; its vertices are 6-valent, and it is orientable of genus 1. Using a notation analogous to the one for crown polyhedra, the examples shown in Figure 18 are PW(7,3,1) and AW(7,3,2). Since the vertices of wreath polyhedra are 6-valent, infinitely many other noble polyhedra may be obtained by applying to wreath polyhedra with even n an analogue of the construction using helical polygons described in Section 5.

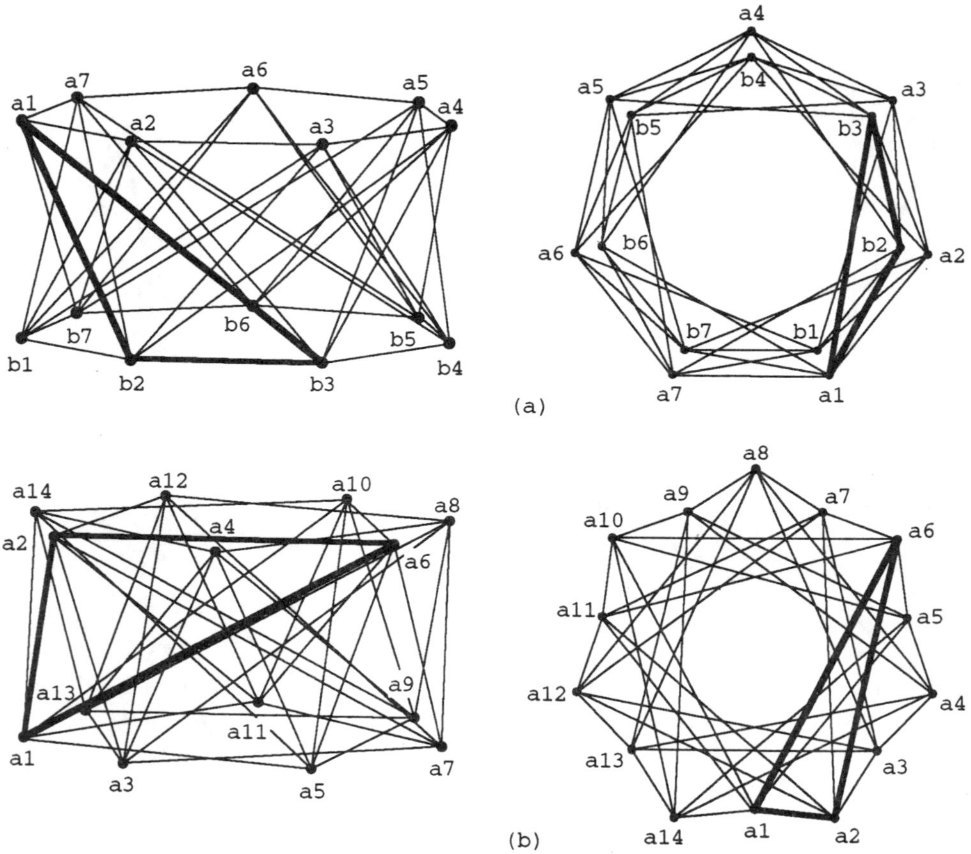

Fig. 18. Two wreath polyhedra, each shown in two perspective views. (A) A prismatic polyhedron $PW(7,3,1)$; (b) an antiprismatic polyhedron $AW(7,3,2)$.

An additional infinite class of noble polyhedra, the family of V-faced polyhedra, is illustrated in Figure 19. The vertices of V-faced polyhedra are situated at the vertices of (skew) antiprismatic and prismatic polygons (see [14]); their faces are quadrangles in which one pair of nonadjacent vertices is represented by the same point. In the notation used in Figure 19, the faces are $a_j, a_{j+1}, b_j, a_{j-1}(a_j)$ and $b_j, b_{j+1}, a_j, b_{j-1}(b_j)$, for $j = 1, 2, \ldots, n$.

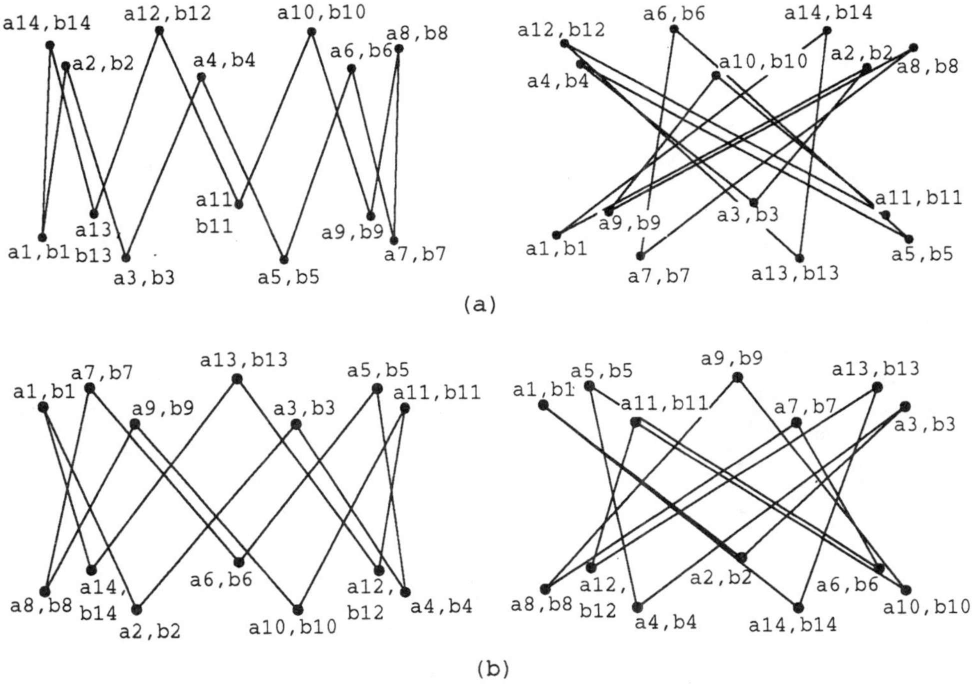

Fig. 19. Antiprismatic and prismatic V-faced polyhedra in which each face is a quadrangle with two nonadjacent vertices at the same point.

We next show a few examples of noble polyhedra related to regular or Archimedean polyhedra, in hope that they will whet the reader's appetite for a more thorough investigation of such polyhedra. Figure 20(a) shows an autopolar noble polyhedron that has 20 hexagonal faces (one of which is indicated by heavy lines) and 20 vertices. Figure 20(b) can be interpreted as illustrating several remarkable noble polyhedra; we denote the first two by $\mathcal{P}_1$ and $\mathcal{P}_2$. We start with 12 points at the vertices of a regular icosahedron as vertices, and as edges we take all the segments determined by the vertices, except the diameters of the icosahedron; see Figure 20(b). The faces of $\mathcal{P}_1$ and $\mathcal{P}_2$ are the 10-gons shown in Figure 1(c) and 1(d), respectively; the edges of one face are heavily drawn in Figure 20(b). It is easily checked that both $\mathcal{P}_1$ and $\mathcal{P}_2$ are autopolar. To obtain two other noble polyhedra, we interpret each point v which is a vertex of the icosahedron in Figure 20(b) as representing two vertices v^+ and v^- of the polyhedron we are constructing. Each of $\mathcal{P}_3$ and $\mathcal{P}_4$ has 24 isogonal decagons as faces. One face of $\mathcal{P}_3$ is $b^+, c^-, e^+, f^-, c^+, d^-, f^+, b^-, d^+, e^-(b^+)$, another is obtained from this by interchanging $+$ and $-$; note that the vertices of either of these faces are at points that in the original icosahedron were adjacent to one vertex. The other 22 faces of $\mathcal{P}_3$ are obtained analogously, from points adjacent to other vertices of the original icosahedron. The polyhedron $\mathcal{P}_4$ is obtained similarly, except that

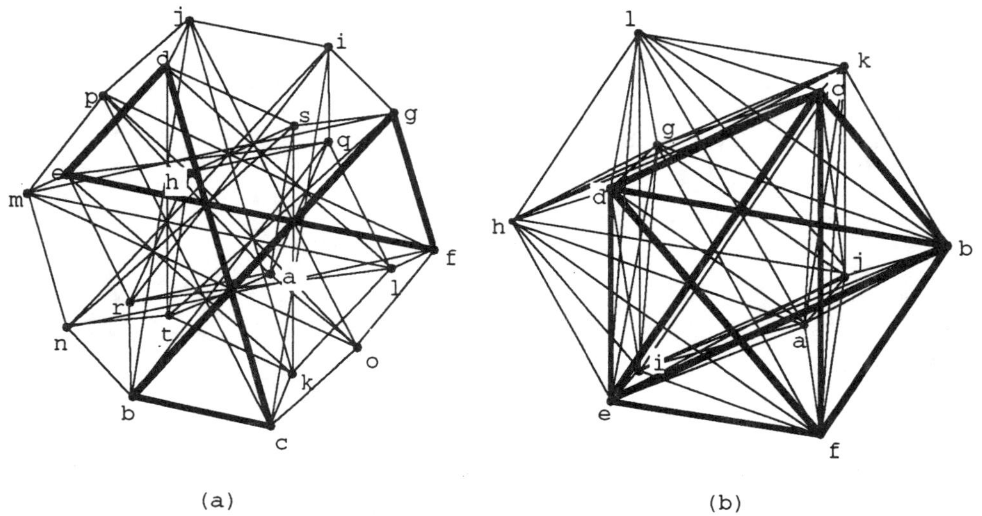

Fig. 20. (a) An autopolar noble polyhedron described by Brückner [6], with 20 vertices and 20 hexagonal faces. (b) A sketch of two noble polyhedra with 24 vertices and 24 decagonal faces. Details are given in the text.

the faces are now all congruent to $b^+, c^-, f^+, b^-, e^+, f^-, d^+, e^-, c^+, d^-(b^+)$. The isogonality and isohedrality of the polyhedra is easily established in all cases.

Our final examples of noble polyhedra are shown in Figure 21. They can most easily be understood as being constructed on a rhombicuboctahedral scaffold (indicated by the dotted lines). The edges of all four are the same, but their organization into faces is different; the faces of the first and third polyhedron are polygons of rotation number 0, those of the second and fourth have rotation number ± 2. Details of the construction are indicated in the caption.

7. Remarks and comments

The desirability and necessity of formulating a consistent framework for a theory of polygons and polyhedra in which selfintersections are permitted becomes obvious when one tries to find existing writings on this topic. Meister [27] started developing such a theory in 1769, but the influence of his writings — often quoted but apparently hardly ever looked at — was negligible. At least in part this is due to a misrepresentation of his definitions by later authors, such as Günther [22] and Brückner [4]; see [15] and [16] for details. In fact, our definition is nothing but Meis-

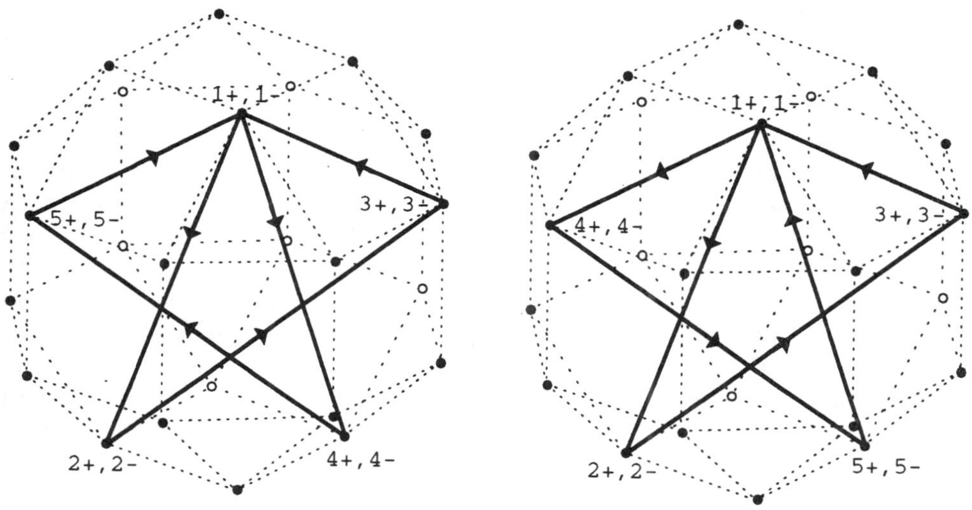

Fig. 21. Schemes for the construction of four noble polyhedra. For the first two, the vertices are those of the rhombicuboctahedron (which is shown by dotted lines). One face of each polyhedron is 1,2,3,1,4,5(1); note the difference in labelling of the two cases. As an aid to visualization, the selected face was shown as an oriented polygon. The other 23 faces are obtained by symmetries of the rhombicuboctahedron. Two additional noble polyhedra result if every vertex of the rhombicuboctahedron stands for two vertices of the noble polyhedron; the two vertices are distinguished by + or - in their labels. One face of each polyhedron is 1+, 2-, 3+, 1-, 4+, 5-(1+); the other 47 faces are obtained by symmetries (including those interchanging + and -).

ter's put in more modern language. Brückner's book, which is supposed to be a survey of previous knowledge concerning nonconvex polyhedra, is so deeply flawed precisely in presenting this topic, that it is useless as a basis for any research on hollow-faced polyhedra. The worst offense is the book's internal inconsistency, even concerning concepts as basic as the definition of a polygon or polyhedron. This lack of precise concepts in [4] and in most other writings on the topic of polyhedra more general than convex polyhedra becomes particularly aggravating, and impedes any resolution of basic questions, when one is considering regular polyhedra or noble ones. A detailed description of the faults has no place here, but we have to mention the fact that on page 2 of [4] a polygon is defined by requiring (among other properties) that each endpoint of each edge is the endpoint of one and only one other edge; however, later on (for example, on page 215) he considers a 10-gon in which pairs of vertices coincide; this is the polygon in our Figure 1(d). It is interesting to note that Brückner never mentions the 10-gon in our Figure 1(c), which should have been considered at the same place, since it leads to a noble polyhedron just as the former does (see Figure 21). Brückner's declared intention that the polygons considered as faces of polyhedra be unicursal could have been met by doubling up the vertices as we did in the construction of the third and fourth noble polyhedra described in

Figure 21. But Brückner remained inconsistent and imprecise concerning allowed polygons, and polyhedra; neither Brückner nor anyone else explained, at any time, precisely what is to be understood by a "polyhedron" in which some or all faces have vertices at which four edges of one face meet, or have multiple vertices at the same point. Until such a definition is adopted, statements about such "polyhedra" are devoid of meaning. After a definition is adopted (be it the one we used, or some other) it is possible to decide whether a bunch of polygons is, or is not, a polyhedron of a specified kind.

One other shortcoming of [4] is the lack of consideration of regular polygons $\{n/d\}$, with n and d not relatively prime. Clearly, one can require from the start that all vertices of a polygon should not only be distinguishable by the role they play in the polygon (that is, by their labels), but be represented by distinct points. With this additional condition, the classical enumeration of regular polyhedra becomes valid, as do various other results. However, it should be realized that this is an arbitrary restriction which should not be imposed for several reasons: (i) a more satisfactory and far richer theory can be developed without the restriction — with the restriction, various polyhedra discussed elsewhere in [4] or [5] are inadmissible; (ii) continuity and consistency are lost in such situations as the isotopies considered in our Section 4, or in considerations of winding numbers and rotation numbers; (iii) some very beautiful results are lost if this restriction is imposed. The validity of (i) is evident even from the modest efforts presented here. Some aspects of (ii) are discussed in [16] and [19], but it is strange that Hess, by failing to adhere consistently to his own definitions, was unable—in an article [23] of well over a hundred pages—to rid himself of the confusion between winding numbers and rotation numbers of polygons. However, (iii) deserves a short explanation.

It is a curious fact that a set of beautiful results concerning polygons has been developed over the last 150 or so years, without attracting any attention on the part of most geometers; the author regrets to have been part of that majority till very recently. The results in question deal with various kinds of "smoothing" operations on polygons; some of them replace a given polygon by one in which vertices are weighted averages of sets of vertices of the starting one, or points associated in other ways with the original polygon. Some of the results can be found in [1], [3], [9], [28] and a variety of other books and articles (a survey of the topic, with extensive references, is in preparation); part of the results are often said to be generalizations of "Napoleon's theorem". But the one aspect of all this which is relevant to the present discussion is that these results could not even be formulated, much less established, without using regular polygons in the generality adopted here. The historical account and explanations given in [15] concerning the almost two centuries of narrow interpretation of "regular polygons" in the theory of regular polyhedra remain valid. However, one should add that throughout this period, a group of mathematicians used the more general definition of polygons, without any reservations and, probably, without realizing that their usage is not the one generally adhered to. Still without a satisfactory explanation is the fact that prior to [15] there was no attempt to use regular polygons such as $\{6/2\}$ or $\{8/2\}$ in the construction of regular or other polyhedra.

At this time, there are several open challenges concerning hollow-faced polyhedra.

The most obvious one is to find a complete characterization of the regular ones among them, and to determine the corresponding groups of symmetries. This, in fact, consists of several different subproblems, depending on the generality considered (see below). Other related directions of desirable investigation concern polyhedra that are isogonal, or noble, or have some other transitivity properties; also, enumeration of all types of hollow polyhedra with a given (small) number of edges should be interesting.

The family of hollow-faced polyhedra reestablishes in part the correspondence between polyhedra and maps on manifolds that is so useful in the study of convex polyhedra. But it should be noted that in order to get a more satisfactory relationship, a class of polyhedra more general than the hollow-faced ones should be considered. In our presentation, two faces may have several vertices (or edges) in common, while two vertices may not have more than one edge connecting them. A generalization admitting such and other incidences forbidden for hollow-faced polyhedra can be developed, but this has not been done so far. One other benefit of the introduction of such generality would be the unlimited possibility of polarity between polyhedra in question; as is easily seen, the family of hollow-faced polyhedra is not closed under duality.

The noble polyhedron in Figure 20(a) and the polyhedron $\mathcal{P}_2$ of Figure 20(b) were briefly mentioned by Hess [25], and described in detail by Brückner [6]. Both failed to notice the noble polyhedron $\mathcal{P}_1$. Neither Hess nor Brückner appear concerned about the fact that $\mathcal{P}_2$ is not a polyhedron according to the definitions they purport to use. Naturally, they fail to discuss polyhedra such as $\mathcal{P}_3$ or $\mathcal{P}_4$, in which distinct vertices are represented by the same point. The first two noble polyhedra in Figure 21 are also due to Hess and Brückner; these polyhedra are remarkable in that they are noble although their faces and vertex figures are rather irregular. Many other possibilities for noble polyhedra exist (several of which were investigated by Hess and Brückner), but no complete enumeration is available. Particularly interesting would be a complete list of autopolar noble polyhedra.

Also deserving consideration are **compounds** of regular or noble polyhedra, that is, families of separate congruent polyhedra with a large degree of symmetry of the union. In the old literature (and in some more recent works) these are called "discontinuous polyhedra"; however, in modern contexts it appears more reasonable to restrict the concept of "polyhedron" to connected families of polygons, as done in this note.

The topic of polyhedra that are not epipedal is almost completely unexplored. Such polyhedra appear very naturally as Petrie-polyhedra of other (including epipedal) polyhedra, and are probably the most general type of object in 3-space one may wish to call "polyhedron". However, there has been no work done on such polyhedra beyond the initial discussion in [15], the completion of the enumeration of regular ones among these polyhedra by Dress [9], [11], and the papers of Farris [12], [13] on "fully transitive" polyhedra. Moreover, even in these papers the generality is restricted by insisting that the vertices of each skew polygon are represented by distinct points. Thus there is great scope for further work in this direction as well.

8. Comments added in proof (13 December 1993).

Two additional facts concerning the autopolar noble polyhedron in Figure 20(a) have come to the author's attention.

(i) The polyhedron was first described, and shown in a pair of stereoscopic diagrams, by Theodor Hugel, in a booklet "Die regulären und halbregulären Polyeder" (published by Gottschick-Witter, Neustadt a. d. H., 1876). In fact, in Hugel's view this polyhedron should be considered to be as regular as the Kepler-Poinsot polyhedra. Naturally, this depends on the definitions adopted, and the ones proposed by Hugel are not considered appropriate today.

(ii) J. M. Wills observed (in "The combinatorially regular polyhedra of index 2", Aequationes Math. 34(1987), 206 - 220; see p. 212) that the map which corresponds to the polyhedron in Figure 20(a) is, in fact, the regular map W#60.57. No doubt, Hugel would have been pleased!

Not much attention was given in the present paper to non-unicursal polygons and polyhedra, hence a few additional words may be appropriate. A description of the regular polygons of this kind is quite simple. We start from a regular unicursal polygon $\{n/d\}$, with edges e_i, where e_i is adjacent to e_{i+1} for $i = 1, \ldots, n-1$, and e_n is adjacent to e_1. We replace each edge e_i by k edges $e_{i,1}, \ldots, e_{i,k}$, and declare as adjacent edges $e_{i,j}$ and $e_{i+1,j}$ for $i = 1, \ldots, n-1$, as well as $e_{n,j}$ as adjacent to $e_{1,j+1}$. This leads to a regular polygon with n vertices and nk edges, that can be denoted by $\{n_k/d\}$. Examples of isogonal (but not regular) non-unicursal polygons are shown in parts (c) to (h) of Figure 1. The complete characterization of such polygons appears to be not too complicated, but has not been carried out so far.

Regarding polyhedra, one can start from a unicursal polyhedron in which sets of vertices are represented by the same point, and identify all the vertices in some or all such sets; whether or not a polyhedron results from this procedure depends on how the various adjacencies are defined. In the case of the regular polyhedra in Figure 12, defining the adjacencies is easy to accomplish in such a way that the resulting polyhedra are again regular. Convenient symbols for these regular polyhedra are $\{3_2/1, 3\}$ and $\{3_2/1, 4\}$. However, only a few such examples are known; no results of a general nature are available at present.

References

[1] Bachmann, F. and Schmidt, E., *n-gons.* Translated from German by C. W. L. Garner. Mathematical Expositions No. 18, Toronto Univ. Press, 1975.

[2] Barnette, D.W., Gritzmann, P. and Höhne, R., *On valences of polyhedra.* J. Combinat. Theory A 58(1991), 279 - 300.

[3] Berlekamp, E.R., Gilbert, E. N. and Sinden, F. W., *A polygon problem.* Amer. Math. Monthly 72(1965), pp. 233-241.

[4] Brückner, M., *Vielecke und Vielflache. Theorie und Geschichte.* Teubner, Leipzig 1900.

[5] Brückner, M., *Über die diskontinuierlichen and nicht-konvexen gleicheckig-gleich-flächigen Polyeder.* Verh. des dritten Internat. Math.-Kongresses Heidelberg 1904. Teubner, Leipzig 1905, pp. 707 - 713.

[6] Brückner, M., *Über die gleicheckig-gleichflächigen, diskontinuirlichen und nichtkonvexen Polyeder.* Nova Acta Leop. 86(1906), No. 1, pp. 1 - 348 + 29 plates.

[7] Brückner, M., *Zur Geschichte der Theorie der gleicheckig-gleichflächigen Polyeder.* Unterrichtsblätter für Mathematik und Naturwissenschaften, 13(1907), 104 - 110, 121 - 127 + plate.

[8] Coxeter, H. S. M. and Moser, W. O. J., *Generators and Relations for Discrete Groups.* 4th ed. Springer, Berlin 1980.

[9] Douglas, J., *Geometry of polygons in the complex plane.* J. Math. Phys. 19(1940), pp. 93-130.

[10] Dress, A. W. M., *A combinatorial theory of Grünbaum's new regular polyhedra, Part I: Grünbaum's new regular polyhedra and their automorphism group.* Aequationes Math. 23(1981), 252 - 265.

[11] Dress, A. W. M., *A combinatorial theory of Grünbaum's new regular polyhedra, Part II: Complete enumeration.* Aequationes Math. 29(1985), 222 - 243.

[12] Farris, S. L., *Completely classifying all vertex-transitive and edge-transitive polyhedra, Part I: necessary class conditions.* Geometriae Dedicata 26(1988), 111–124.

[13] Farris, S. L., *Completely classifying all vertex-transitive and edge-transitive polyhedra, Part II: finite, fully-transitive polyhedra.* J. of Geometry (to appear).

[14] Grünbaum, B., *Regular polyhedra – old and new.* Aequationes Math. 16(1977), 1 - 20.

[15] Grünbaum, B., *Regular polyhedra.* Companion Encyclopaedia of the History and Philosophy of the Mathematical Sciences, I. Grattan-Guinness, ed. Routledge, London 1993 (to appear).

[16] Grünbaum, B., *Metamorphoses of polygons.* In: "The Lighter Side of Mathematics", Proc. Strens Conference, R. K. Guy et al. eds. Math. Assoc. of America (to appear).

[17] Grünbaum, B. and Shephard, G. C., *Polyhedra with transitivity properties.* C. R. Math. Rep. Acad. Sci. Canada, 6(1984), 61 - 66.

[18] Grünbaum, B. and Shephard, G. C., *Duality of polyhedra.* In: "Shaping Space: A Polyhedral Approach", Proc. "Shaping Space" Conference, Smith College, April 1984. M. Senechal and G. Fleck, eds. Birkhäuser, Boston 1988, pp. 205 - 211.

[19] Grünbaum, B. and Shephard, G. C., *Rotation and winding numbers for planar polygons and curves.* Trans. Amer. Math. Soc. 322(1990), 169 - 187.

[20] Grünbaum, B. and Shephard, G. C., *Isohedra with non-convex faces.* J. of Geometry (to appear).

[21] Grünbaum, B. and Shephard, G. C., *A new look at Euler's theorem for polyhedra.* Amer. Math. Monthly (to appear).

[22] Günther, S., *Vermischte Untersuchungen zur Geschichte der mathematischen Wissenschaften.* Teubner, Leipzig 1876.

[23] Hess, E., *Über gleicheckige und gleichkantige Polygone.* Schriften der Gesellschaft zur Beförderung der gesammten Naturwissenschaften zu Marburg, Band 10, Abhandlung 12, pp. 611 - 743, 29 figures. Th. Kay, Cassel 1874.

[24] Hess, E., *Ueber zwei Erweiterungen des Begriffs der regelmässigen Körper.* Sitzungsberichte der Gesellschaft zur Beförderung der gesammten Naturwissenschaften zu Marburg 1875, pp. 1 - 20.

[25] Hess, E., *Ueber die zugleich gleicheckigen und gleichflächigen Polyeder.* Schriften der Gesellschaft zur Beförderung der gesammten Naturwissenschaften zu Marburg, Band 11, Abhandlung 1, pp. 1 - 97, 11 figures. Th. Kay, Cassel 1876.

[26] Hess, E., *Ueber einige merkwürdige nichtkonvexe Polyeder.* Sitzungsberichte der Gesellschaft zur Beförderung der gesammten Naturwissenschaften zu Marburg 1877, pp. 1 - 13.

[27] Meister, A. L. F., *Generalia de genesi figurarum planarum et inde pendentibus earum affectionibus.* Novi Comm. Soc. Reg. Scient. Gotting. 1(1769/70), pp. 144 - 180 + plates.

[28] Neumann, B. H., *Some remarks on polygons.* J. London Math. Soc. 16(1941), pp. 230-245.

[29] Robertson, S. A., *Polytopes and Symmetry.* London Math. Soc. Lecture Note Series No. 90. Cambridge Univ. Press 1984.

[30] Stewart, B. M., *Adventures Among The Toroids.* 2nd ed. Okemos MI, 1980.

[31] Szillasi, L., *Regular toroids.* Structural topology 13(1986), 69 - 80.

[32] Wilson, S. E., *New techniques for the construction of regular maps.* Ph. D. thesis, University of Washington, Seattle 1976.

A HIERARCHICAL CLASSIFICATION OF EUCLIDEAN POLYTOPES WITH REGULARITY PROPERTIES

HORST MARTINI
Fachbereich Mathematik
Technische Universitaet Chemnitz
Chemnitz, Germany D-09009

1. Introduction

During the last decades, many mathematicians investigated classes of polytopes which are determined by natural weakenings of the definitions of regular polytopes. Examples are the sets of regular-faced polytopes, congruent-faced polytopes and isogonal polytopes. The aim of the present article is to show one possibility of bringing these classes into one hierarchical structure, perhaps stimulating further research with respect to gaps in this structure, natural extensions of its parts or new combinations of the definitions contained. In addition, we give a survey regarding important results on each of these polytope classes. (More related material can be found in a recent survey by the author which, however, does not show this hierarchical structure and the related connections between the polytope classes discussed here, see Martini (1994).)

Our main viewpoint is a metrical one. This means that the rapid development of the theory of combinatorially regular polytopes (in the widest sense) is ignored. For this line of research, referring to polyhedral manifolds with regularity properties, regular incidence complexes, abstract regular polytopes, buildings and further aspects, we cite only a few surveys or extensive articles, such as Tits (1981), Bilinski (1988), McMullen (1989), Ivić-Weiss (1989), McMullen and Schulte (1990_1, 1990_2) Bokowski (1991), Schulte and Wills (1991), Schulte and Ivić-Weiss (1991), Brehm and Wills (1993, §4) and a monograph by P. McMullen and E. Schulte on abstract regular polytopes in preparation. In addition, the reader should consult the articles in this book by B. Grünbaum, P. McMullen and E. Schulte.

2. Basic Notions

Our considerations mainly refer to *convex d–polytopes*; that is, to d–dimensional compact, convex subsets of Euclidean d–space ($d \geq 3$) with finitely many extreme points (see Grünbaum (1967) for further background). The k–dimensional intersections ($0 \leq k \leq d-1$) of these d–polytopes with supporting hyperplanes are called their *k–faces*; in particular, for $k \in \{0, 1, d-1\}$ we have *vertices, edges* and *facets*, respectively. For $d = 3$, we sometimes use the term *polyhedron* instead of 3–polytope. We shall also consider *non–convex d–polytopes*; that is, compact sets with interior points and piecewise linear boundary structure, homeomorphic to the d–ball or even having higher genus. Moreover, self–intersections of the boundary can be considered as well. (Obviously, this is not covered by the definitions above. For that reason,

T. Bisztriczky et al. (eds.), POLYTOPES: Abstract, Convex and Computational, 71–96.

we emphasize that there is no general definition of what is to be understood as a d-dimensional polytope if no special assumptions are made. A critical discussion of this problem is presented by B. Grünbaum in this book.) Coxeter (1948, p. 16, p. 68) defines a *vertex-figure* of a convex d-polytope by using edges containing a fixed vertex. The segments joining the midpoints of adjacent edges of this edge system form the corresponding vertex figure. Then a convex d-polytope ($d \geq 3$) is *regular* if all its facets and the convex hulls of all its vertex figures are regular ($d-1$)-polytopes, cf. Coxeter (1948) and L. Fejes Tóth (1965). (For the starting point of this inductive definition in 3-space, the facets have to be regular polygons; that is, plane, convex polygons with equal sides and equal angles.) Many further definitions of regular d-polytopes (all equivalent in the convex case) were discussed or cited by DuVal (1964), Grünbaum (1967, §19. 1), P. McMullen (in his Ph. D. thesis, University of Birmingham, 1968) and others; see also Grünbaum (1991_1). Besides convexity, the definition used here needs three properties each of which can be weakened in some situations. However, they are useful in view of getting more and more extensive polytope classes in our hierarchical structure, by successive weakenings or omissions of them.

Thus, we say that a convex d-polytope P is *isogonal* (or vertex-transitive) if the group of isometric symmetries of P acts transitively on the vertices of P. Further, P is said to be *regular-faced* (*congruent-faced*) if all its facets are regular (congruent). For these three properties we shall use the abbreviations (i), (r) and (c), respectively.

Now we define a convex d-polytope ($d \geq 3$) to be *regular* if it satisfies all three conditions (i,r,c). It is easily checked (for example, by successive reduction of the completely known set of regular-faced polytopes) that this definition of regular polytopes is equivalent to other usual ones.

3. The Hierarachical Classification, Its Refinements and Extensions

The announced hierarchical structure for special classes of convex d-polytopes is given by the following diagram which can be graphically realized by the edge–skeleton of a 3–cube.

It should be noted that (as will be shown in the following sections) the classes (i,r,c), (i,r), (r,c) and (r) are completely known, for arbitrary d. Possibilities for suitable weakenings of the three basic criteria, leading to a finer structure, are given below. Clearly, they can be combined in several ways, too.

(I) Isogonality

a) The consideration of *k–isogonal polytopes* (that is, of convex d–polytopes having $k \geq 1$ transitivity classes of vertices) yields polytope classes containing each set marked with (i) in our diagram. This notion goes back to Krötenheerdt (1969); see also Grünbaum and Shephard (1987, pp. 65–70). As an example, from this viewpoint, one can classify the regular–faced polytopes step by step, and then have the semiregular ones as the most specific subset.

b) One can weaken isogonality to the requirement of congruent vertex figures or even combinatorially equivalent vertex figures. For instance, the first step is adding precisely one polyhedron to the set of semiregular 3–polytopes (see section 5).

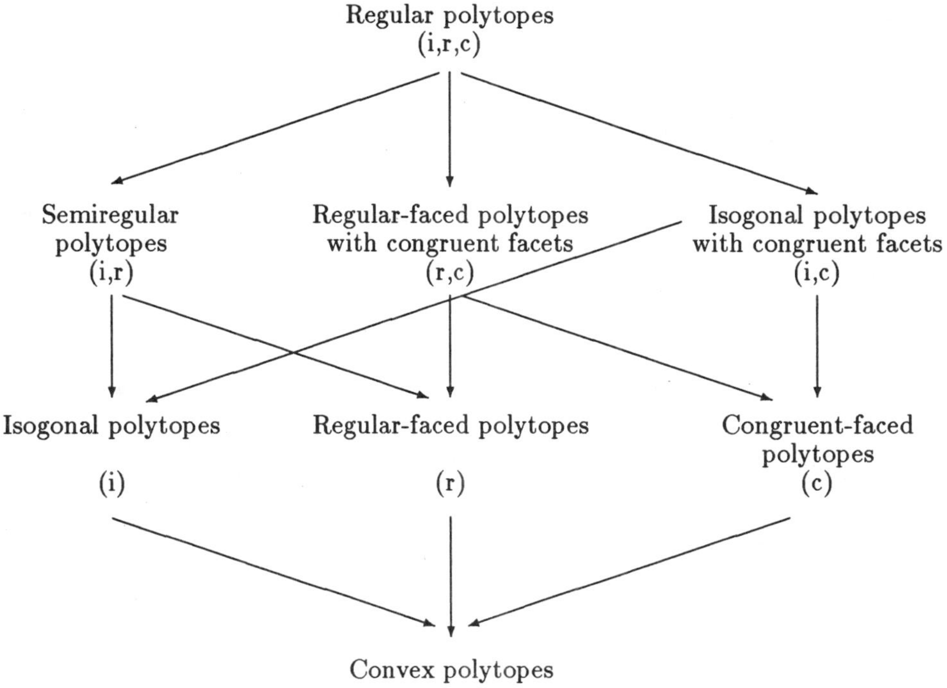

c) A third point of view is given by the requirement that the whole vertex set lies on a sphere.

(II) Regularity of facets

a) Here our first possibility is based on the inductive character of some definitions in this article. Namely, one can extend the property (r) to having semiregular, uniform or (in the spirit of our considerations) more general facets. Among other things, this yields the known class of uniform polytopes, in our diagram "between" (i,r) and (i) (see section 6).

b) This inductive generalization of types of facets is partially connected with the question about polytopes whose k-faces $(2 \leq k \leq d-2)$ are regular (or more general, in our sense).

c) Furthermore, one can generalize the metrical regularity of facets (or k-faces) to affine, projective or even combinatorial regularity. For example, McMullen (1967) proved that any combinatorially regular convex d-polytope is combinatorially isomorphic to a metrically regular d-polytope; cf. also McMullen (1968_2), McMullen's Ph. D. thesis (Univ. of Birmingham, 1968) and Dress and Schulte (1987) for other proofs.

(III) Congruence of facets

a) A natural extension is the consideration of polytopes with $k \geq 1$ congruence classes of facets, leading to polytope classes having each set marked with (c) as a subset. For example, a further classification of regular-faced polytopes, with the class (r,c) as the most special subset, is possible.

b) Analogously, one can investigate polytopes whose k–faces ($1 \leq k \leq d-2$) are congruent. So the construction of d–zonotopes with congruent 2–faces is connected with equiangular line systems; see, for example, Bilinski (1960), Lemmens and Seidel (1973).

c) Finally, the congruence of facets (or k–faces) can be also generalized to affine, projective or combinatorial equivalence. With respect to the latter viewpoint, the cubical polytopes and the simplicial polytopes represent important classes; see Grünbaum (1967, §4. 5, 4. 6, 9. 2, 9. 4). Next, an interesting series of recent results on polytopes with combinatorially equivalent facets should be mentioned here: For $d \geq 2$, a convex d–polytope P is *facet forming* if there exists a convex $(d+1)$–polytope Q whose facets are combinatorially equivalent to P; if there is no such polytope Q then P is called a *non–facet*. Every convex n–gon with $n \in \{3,4,5\}$ is facet forming, and for each $n \geq 6$, it is a non–facet. Perles and Shephard (1967) gave examples for both cases in higher dimensions, see also Barnette (1969).

Barnette (1980) showed interesting relations of this decision problem to shellability questions, and Schulte (1984, 1985) investigated its connections with the theory of tilings. Kalai (1990) sharpened several results of Perles, Shephard (1967) and Schulte (1984) to the following recent state: From the set of convex regular d–polytopes ($d \geq 4$), the 24-, 120-, and 600–cell as well as the cross–polytope are non–facets. Thus, for $d \geq 4$ and in this set, only the cube and the simplex are facet forming. Since four Platonic solids are realizable as facets of regular 4–polytopes, the only undecided case is the icosahedron. With this possible exception, each regular d–polytope which is not a facet of a regular $(d+1)$–polytope is itself a non–facet. (For a detailed description of all these special polytope types, the reader is referred to the sections below.)

Finally, we take a short look at additional criteria (different from (i,r,c)), usable for refinements or extensions of the present hierarchy. First, the property of vertex–transitivity can be replaced or completed by *edge–transitivity* or *facet–transitivity*. For example, the edge–transitive semiregular 3–polytopes are the so–called *quasiregular polyhedra* (that is, besides the Platonic solids, the cuboctahedron and icosidodecahedron) whose position in our diagram is therefore "between"(i,r,c) and (i,r).

The facet–transitivity, as a sharpening of (c), defines the class of *isohedral polytopes* (cf. sections 9 and 10). More generally, one can consider *k-transitivity* for each $k \in \{0,1,...,d-1\}$. Among the various questions concerning this notion, the relations between the transitivities of different dimensions deserve to be investigated.

On the other hand, one can omit convexity and even permit self–intersections of the piecewise linear boundary. For this point of view, we refer to the addenda to each of the following sections, in every case with the headline "Related polytope classes". Here one can mention three other concepts for polyhedra with regularity properties. The first (*skew vertex-figures*) yields the *Petrie–Coxeter polyhedra*; see Coxeter (1968_2), Schulte and Wills (1986), Wills (1987) and Grünbaum (1993_1). The second, neglecting the planarity of the facial polygons, is due to Grünbaum (1977); cf. also Dress (1981, 1985) and Farris (1986, 199?). It yields the set of *Grünbaum-Dress polyhedra*. The third, the concept of *polystromas*, is also due to Grünbaum (1976) and refers to a class of hierarchical sets which contain honeycombs, twisted honeycombs, regular configurations and toroidal maps. For an enumeration of 3-

polystromas arising from 3-honeycombs we refer to Colbourn and Ivić-Weiss (1984).

4. Regular Polytopes

With the properties of being isogonal and having congruent, regular facets, there are precisely five types of convex 3–polytopes: the *regular tetrahedron, octahedron, icosahedron, dodecahedron,* and the *cube.* Since Plato (1952) mentioned them in his "Timaeus", they are usually called the *Platonic solids.* Further historical information on that class of polyhedra was given by Lindemann (1896/97), Brückner (1900), Tropfke (1924), Waterhouse (1972), Critchlow (1979), Emmer (1982), and Toepell (1991).

By weaker criteria (such as 2–faces of equal vertex numbers and vertices of equal valences), one can combinatorially exclude further candidates. On the other hand, there are various construction possibilities. An elegant construction of the Platonic icosahedron was proposed by Luca Paciolo in 1509; see Stillwell (1989, p. 16). Other methods (by using nets, paper folding, descriptive geometry and so on) were described by Locher–Ernst (1955), Unger (1962), Roman (1968), Wenninger (1971), Brauner (1986, §6. 1. 4), Hilton and Pedersen (1988) and many others. Different analytical possibilities were discussed by Schoute (1905, §6), Hill (1940), Monson (1988), Vohla (1989), and Buchholz (1991).

A few results and problems from the last three decades will demonstrate the undiminished topicality of the Platonic solids in geometry: Jeger (1975) and Hippenmeyer (1979) obtained results on the classification of their planar nets. Graphs of the edge–skeletons were investigated by Lehmer (1986), Harborth, Kemnitz, Möller, and Süssenbach (1987). Dorwart (1988) reinvestigated relations to plane configurations and circle geometry. Spatial arrangements of Platonic solids were considered by Mason (1972), Stein (1978), Rossat and Sloane (1988). The latter paper refers to the maximum number of congruent Platonic polyhedra that have a common vertex but disjoint interiors. For tetrahedra and octahedra, this is an open problem with the correspondingly conjectured numbers of 20 and 7. On the other hand, Wagon (1985, pp. 68-72) discussed the group-theoretical background of problems referring to tetrahedral snakes (that is, sequences of regular tetrahedra in 3-space, where two consecutive tetrahedra share exactly one facet, and each tetrahedron is distinct from its predecessor's predecessor). Among other things, one might ask whether two tetrahedra in such sequences can be translates of each other or even can coincide. Results in that direction, also related to famous toys invented by Ernö Rubik in the 1970s, were obtained by Świerczkowski (1958) and Dekker (1959). Mason (1972) as well as Stein (1978) rediscovered some of them. Special possibilities of inscribing a regular polyhedron in a non–trivial manner into another one were described by Bakos (1959) and Bourgin and Mendel (1968). Croft (1980) considered pairs $\{P_1, P_2\}$ of Platonic solids with $P_1 \subseteq P_2$ and maximal volume ratios $V(P_1)/V(P_2)$. The six unsolved cases are given by P_1 = tetrahedron, octahedron or cube with P_2 = icosahedron, P_1 = dodecahedron with P_2 = tetrahedron or octahedron, P_1 = icosahedron with P_2 = dodecahedron; see also Croft, Falconer and Guy (1991, B 3). Croft (1962) also investigated metrical properties of arrow systems in the facets of regular polyhedra. Many extremal properties of Platonic solids in the spirit of isoperimetric and related questions are cited in the surveys of Florian (1979, 1993); see also L. Fejes

Tóth (1965, §II. 4 and II. 5; 1972 §V). Different applications of regular polyhedra in physics, chemistry and biology refer to Dirac's wave equation (Haenzel (1941)), the capacitance of polyhedral surfaces (Brown (1990)), molecular and crystal structures (I. and M. Hargittai (1988), Chieh (1988) and Brandmüller (1990, 1992)), the shapes of several microorganisms (Martin (1978) and Brandmüller (1990, 1992)), and other subjects.

For $d \geq 4$, the *regular polytopes* were discovered by Schläfli (1860, 1901). The analogues to the regular tetrahedra, octahedra and cubes exist for each $d \geq 4$; namely, *regular simplices, cross-polytopes* and *(hyper-)cubes*. In addition, there are precisely three regular polytopes in 4-space: the *regular 600-cell* (with 600 tetrahedral facets), the *regular 120-cell* (with 120 dodecahedral facets) and the *24-cell* (bounded by 24 regular octahedra).

Since the 600-cell and the 120-cell can be regarded as analogues of the Platonic solids with icosahedral symmetry, the 24-cell is the only regular polytope having no analogue in any dimension $d \geq 3$. Although it has a centre of symmetry, it is self-dual. For alternate definitions, necessary conditions and construction methods regarding the higher-dimensional regular polytopes, we refer to Coxeter (1948, §7; 1981, §22), L. Fejes Tóth (1965, §1. 5), Franz (1971), Rosenfeld and Yaglom (1971), Berger (1987, §12) and Böhm and Quaisser (1991, §3. 1). Figures of these polytopes (such as parallel projections, Schlegel diagrams or nets) were for example given by Hilbert and Cohn-Vossen (1932, §32), Coxeter (1948, plates I-VIII; 1981_1, §22), Rosenfeld and Yaglom (1971), Franz (1971), Berger (1987, §12), and Banchoff (1990). An interesting refinement of Schlegel's projection method with a special view on the 24-cell was presented by Banchoff (1988). Historical comments can be found in the literature cited above as well as a paper of Coxeter (1978).

As an example of recent research on metrical properties of these higher-dimensional regular polytopes, we mention the following question: For which dimensions d is it possible to inscribe the regular d-simplex into the regular d-cube such that all vertices of the simplex are vertices of the cube? (For all other cases of such pairings in the set of regular polytopes, we refer to Schoute (1905).) For one historical source of this question, connected with a work of Barrau (1906), the reader should consult Coxeter (1978, §3). Equivalent questions refer to the existence of Hadamard matrices, Jung constants of the form $d/(d+1)$ in l_1^d and coding theory; cf. Hadwiger (1956), Dol'nikov (1987) and Wallis (1988). The necessary condition $d \equiv 3 \pmod 4$ is known; see Kibble (1949) and Medjanik (1973). By using Hadamard matrices, the sufficiency of that criterion was proved for all $d \leq 423$; see Wallis (1988, p. 139), Yamada (1989) and with respect to further constructions and open cases for $d > 423$, Miyamoto (1991) and Seberry and Yamada (1992). The latter paper contains a list of all unsettled cases for $d < 2000$, which are given (in terms of the simplex-cube embedding problem) by $d \in \{427, 651, 667, 715, 763, 851, 891, 955, 1003, 1131, 1243, 1387, 1435, 1675, 1771, 1851, 1911, 1915, 1947, 1963\}$. Related problems were investigated by Schoenberg (1937), Pelling (1977), Jarratt and Seidel (1986), Scott (1987), and Macdonald (1987). The question of embedding the vertex sets of regular d-polytopes into regular r-lattices ($r \geq d$) was completely solved by Patruno (1983).

RELATED POLYTOPE CLASSES

Omitting the convexity property in the definition of regular d-polytopes, one is led to consider regular star polytopes with self–intersections. For $d = 3$, this extension yields precisely four star polyhedra. Two of them (with pentagrams as facets) were found by Kepler (1619), the remaining dual ones by Poinsot (1810). (To be precise, we mention that one of Kepler's star polyhedra was found already earlier, probably by P. Ucello who was in Venice during the years 1425–1430. It is contained in a mosaic in San Marco, see also Emmer (1982) and Toepell (1991, §5.3) for more details. On the other hand, the dual star polyhedron was anticipated in a slightly imprecise form already by W. Jamnitzer, cf. Jamnitzer (1568).) The completeness of this set of *Kepler–Poinsot solids* was shown by Cauchy (1813). The following names go back to Cayley (1859): *small stellated dodecahedron, great dodecahedron, great stellated dodecahodron*, and *great icosahedron*. All of them have icosahedral symmetry. For properties, constructions and figures, the reader is referred to Coxeter (1948, §6), Roman (1968), Wenninger (1971), Berger (1987, §12. 6), and Böhm and Quaisser (1991, §3. 2). The history was discussed, for example, by Coxeter (1948, §6; 1978), Field (1979) and Toepell (1991, §5. 3).

About 1850, the 4–dimensional analogues of the Kepler–Poinsot solids were discovered by L. Schläfli; namely, the *great 120–cell, grand 600–cell, great stellated 120–cell*, and *grand great stellated 120–cell*; see Coxeter (1974, pp. 42–46). Six further regular star polytopes in 4–space were added by Hess (1885). The first systematic study of this polytope set was given by Mulder (1907), and van Oss (1915) verified the completeness. All ten polytopes have the symmetry group of the regular 600–cell with 14400 elements; eight of them are pairwise dual, and the remaining two are self–dual.

For representations of this polytope class, usually called the set of *Schläfli–Hess polytopes*, we refer to Coxeter (1948, §14), McMullen (1968_1), Berger (1987, §12. 6), and Böhm and Quaisser (1991, §3. 2). For example, McMullen (1968_1) gave a theoretical proof of an observation of Hess: Each regular star polytope has the same vertex set and symmetry group as some convex regular polytope. (For $d = 3$, this idea was used already by Cauchy.) More recent results on regular star polytopes were obtained by Coxeter (1989) and McMullen (1991), for example, leading to new insights about the volume computation of spherical simplices.

For $d > 4$, there are no regular star polytopes. This can be verified on the lines of van Oss; cf. Coxeter (1948, pp. 277–278) and McMullen (1968_1).

5. Semiregular Polytopes

In our terminology, a convex d–polytope ($d \geq 3$) is called *semiregular* if it is isogonal and regular–faced. (Note that for $d = 3$, the definitions of *semiregular* and *uniform d*–polytopes coincide. We discuss uniform polytopes in section 6.) In 3–space, there are (in addition to the Platonic solids) exactly 13 convex polyhedra and two infinite series of prisms and antiprisms satisfying that definition. Since Archimedes described the 13 polyhedra mentioned above, they are called *Archimedean solids*. Kepler (1619) observed that the series of prisms and antiprisms have to be added. For further historical studies, we refer to Tropfke (1924, p. 52), Hofmann (1963), Federico (1982), and Malkevitch (1988). (Some authors define the Archimedean solids with the weaker criterion of congruent vertex figures instead of isogonality. This is supported by the argument that Archimedes certainly did not know group–theoretic concepts. On the other hand, this modified definition is satisfied by precisely one additional convex polyhedron – the so-called *pseudo rhombi-cuboctahedron* – which Archimedes did not know. For this polyhedron, the priority is mostly given to J. C. P. Miller who discovered it in 1930, see Aškinuze (1957), Ball and Coxeter (1987, §V). However, in an earlier paper of Sommerville (1905), a Schlegel diagram of that polyhedron is given. Moreover, even Kepler mentioned this polyhedron, as "Archimedeis quattuordecim", in his "Strena seu de nive sexangula", cf. Kepler (1619).)

Complete representations of the class of semiregular 3-polytopes can be found in books or book chapters of Cundy and Rollett (1961), Lyusternik (1963), L. Fejes Tóth (1965), Roman (1968), Aškinuze (1969), Wenninger (1971), and Federico (1982). A collection of largely different approaches can be taken from the references above and papers of Coxeter (1940), Shephard (1954), Pargeter (1959), Robertson and Carter (1970), Walsh (1972), Dragomir and Gheorgiu (1973), Vivarelli (1974), and Angell and Moore (1987). For example, approaches are truncations of Platonic and Archimedean solids, combinatorial determination (using Euler's theorem), intersections of convex cylinders, symmetry groups and various other methods. In view of midpoints or contact points of spheres, special sphere packings reproduce the shapes of Archimedean solids (and their polars); cf. Hilbert and Cohn-Vossen (1932, §7), Steinhaus (1959, p. 207), L. Fejes Tóth (1965, §II. 4; 1972, §III), Coxeter (1975), and Propp (1988). This property yields several applications in natural sciences such as crystallography, virology, molecular chemistry and so on; see Tarnai (1974), Brun (1978), Martin (1978), Brieskorn (1983, §1), King (1987), Senechal (1988), I. and M. Hargittai (1988), Chieh (1988), Brandmüller (1990, 1992), and for further aspects, Schneebeli (1993). In particular, we mention a recently discovered molecular form of the element carbon, known as the "Buckminsterfullerene" after the famous architect. Its 60 carbon atoms represent the vertex set of an Archimedean solid, the truncated icosahedron. It is proving unusually stable and is believed to exist in interstellar space. Further, it was called the "molecule of the year 1991". Interesting relations between geodesic domes (see also section 10), circle coverings of the 2-sphere, polyhedral structures and Buckminsterfullerenes were investigated by Tarnai (1993) and Schneebeli (1993).

An important subset of the Archimedean solids is given by the *quasiregular polyhedra*. They are edge-transitive and can be obtained as convex hulls of the edge midpoints of the regular octahedron (yielding the *cuboctahedron*) and of the regular icosahedron (yielding the *icosidodecahedron*); cf. Coxeter (1948, §2. 3), L. Fejes Tóth (1965, p. 108) and Robertson (1984, §6, §7). Another remarkable subset consists of those Archimedean solids which have octahedral or icosahedral symmetry, but no symmetry plane; namely, the *snub cube* and the *snub dodecahedron*. Their metrical and projective properties give a lot of interesting problems and generalizations; see Huybers and Coxeter (1979), Hohenberg (1980_1, 1981, $1982_{1,2}$, 1983, $1984_{1,2}$), and Wyss (1986). For example, both polyhedra contain segments connecting vertices and having lengths which are only describable by irreducible cubic equations.

The higher-dimensional semiregular polytopes (that is, all convex, isogonal d-polytopes ($d \geq 4$) with regular facets) are completely known. More precisely, the completeness of a list given by Gosset (1900) was verified by Makarov (1988) for $d = 4$ and by G. Blind and R. Blind (1991) for $d \geq 4$. Except for the regular polytopes, there exist three such polytopes in 4–space and precisely one for each $d \in \{5, ..., 8\}$. The sets of facets of the 4–dimensional polytopes are: {5 tetrahedra, 5 octahedra}, {24 icosahedra, 120 tetrahedra}, {120 icosahedra, 600 octahedra}. For $5 \leq d \leq 8$, the facets consist only of $(d-1)$–dimensional cross–polytopes and simplices.

Following a definition of Coxeter (1974, p. 18) for higher–dimensional quasiregular polytopes, Makarov (1987) showed that there are only two such polytopes in 4–space: the convex hulls of the edge midpoints of the regular 4–simplex and of the regular 600–cell, respectively.

RELATED POLYTOPE CLASSES

Omitting the convexity demand, Coxeter, Longuet–Higgins and Miller (1954) enumerated the semiregular polyhedra with self–intersections in 3–space. Aside from the Platonic, Archimedean and Kepler–Poinsot solids as well as the infinite series of prisms and antiprisms, they found 53 non–convex members of this class. Except for some representatives, Wythoff's construction (by reflections at the sides of suitable spherical triangles) was sufficient for their generation. The paper contains figures, even photographs and lists of metrical properties of these polyhedra. The completeness of this enumeration was independently shown by Sopov (1970) and Skilling (1975). Szepesvari (1981) gave a third proof of this completeness, using modified definitions.

Impressive figures of non–convex semiregular polyhedra and algorithms for their construction were given by Brückner (1900), Wenninger (1971), Hudson and Kingston (1988), and from the historical point of view, the work of Hess (1878), Badoureau (1881) and Pitsch (1881) should be mentioned.

Skilling (1976) investigated *compounds* of semiregular polyhedra and, in particular, their symmetry relations to the original semiregular solids from which they are constructed. Further results in that direction were obtained by Smith (1973, 1974) as well as Norman and Smith (1973).

For non–convex semiregular d–polytopes when $d \geq 4$, the reader should, for example, consult the manuscript of Johnson (199?, §4, §7).

Further extensions were considered by Sopov ($1966_{1,2}$, 1968, 1969). He investigated isogonal polyhedra with regular facets and non–negative Euler characteristic. For example, if this Euler characteristic is zero, there exists exactly one representative.

6. Isogonal Polytopes

A convex d–polytope is *isogonal* if its group of isometric symmetries acts transitively on the vertices. For a complete classification of isogonal 3–polytopes (by combinatorial type and symmetry group) the reader is referred to Grünbaum and Shephard (1981_1, 1981_2). Earlier results on that polytope class are due to Hess (1883), Fedorov (1885) and Brückner (1900), the latter book containing further references. More recent approaches were taken by Robertson and Carter (1970) and Robertson (1984, §6). The paper of Robertson and Carter contains a slightly finer classification than that of Grünbaum and Shephard, using a topological space T which parametrizes the similarity classes of isogonal polyhedra of the same symmetry type. The topology of T is designed to express the relation between any two isogonal polyhedra; for example, the fact that T is connected implies that any isogonal polyhedron can be deformed continuously through isogonal polyhedra into any other. Surprisingly, the regular icosahedron itself has to be passed if the given polyhedron has icosahedral symmetry. Exactly this phenomenon yields one "distorted icosahedron" in the list of Robertson and Carter, which is not an extra type in the classifiction of Grünbaum and Shephard. Further, it should be noted that each spherical isogonal polyhedron is combinatorially isomorphic to an isogonal polyhedron with regular facets. Hohenberg ($1970_{1,2}$, 1971, 1980_2) presented an interesting way of generating isogonal 3–polytopes by line systems transitive with respect to special finite groups.

Perhaps the most important subset of the set of isogonal polytopes is determined by a definition which inductively generalizes that of semiregular polytopes. With the 3–dimensional semiregular polytopes as a starting point, an isogonal d–polytope ($d \geq 4$) is *uniform* if its facets are uniform (that is, for $d = 4$, the facets are semiregular). This polytope class is far from completely known. However, many examples were described by Gosset (1900), Andreini (1905) Schoute (1908, 1911, 1913), Alicia Boole Stott (1900, 1910), Elte (1912), Coxeter (1928, 1930, 1932, 1940, 1985, 1988_1), Conway (1967) and Johnson (199?), and for $d = 4$, the list is complete. Moreover, Coxeter (1978) gave a historical account.

We describe an example.

By arranging suitable polytopes around one point in 4–space, Conway (1967) (together with M. Guy) found all the uniform 4–polytopes described here plus an additional one. This *grand antiprism* has 100 of the 120 vertices of the regular 600–cell, 500 edges, 20 pentagonal and 700 triangular 2–faces, 20 pentagonal antiprisms and 300 tetrahedra as facets. Each vertex is surrounded by two pentagonal antiprisms and 12 tetrahedra, the symmetry group is of order 400 and can be generated by two reflections and two half–turns (cf. Coxeter (1985, pp. 588-590) and Johnson (199?, §4. 3) for a more detailed description). In view of this example, one can expect a lot of complicated uniform polytopes which are not constructable by known methods such as the reflection method of Wythoff.

RELATED POLYTOPE CLASSES

As for semiregular 3–polytopes (cf. Coxeter, Longuet–Higgins and Miller (1954), Skilling (1975) and others), one can also ask for higher–dimensional non–convex uniform polytopes with self–intersections. We give an example. Namely, instead of a decagonal double prism as a starting point for the construction of the grand antiprism, one can start with a decagrammatic double prism to generate a 4–dimensional uniform star polytope, the so–called *pentagrammatic double inverted antiprismoid.* Its relations to the grand 600–cell (being a Schläfli–Hess polytope generated from the regular 600–cell) are analogous to those of the grand antiprism to the regular 600–cell, and its facets are 20 pentagrammatic inverted antiprisms and 300 tetrahedra. For more information on such polytopes, the reader should consult Coxeter (1985, 1988_1), Johnson (199?) and the references given at the beginning of the section on uniform polytopes.

At an Oberwolfach meeting in 1978, U. Brehm presented the first example of an isogonal polyhedron of genus 1; see Brehm and Kühnel (1982). Later, Grünbaum and Shephard (1984) gave examples of isogonal polyhedra which are toroids or have even higher genus (cf. the discussion in the section on isohedral polyhedra as a subset of congruent– faced polyhedra).

An interesting discussion of infinite polyhedral sets with isogonality properties is due to Grünbaum (1993_1, 1993_2).

7. Convex Polytopes with Regular and Congruent Facets

Omitting the isogonality, one extends the set of regular d–polytopes to all *convex d–polytopes with regular and congruent facets.* Aside from the Platonic solids and bipyramids over regular triangles and pentagons, there are three convex 3–polytopes satisfying that definition. They have 12, 14, and 16 triangular facets and were derived by Freudenthal and van der Waerden (1947); see also Galafassi (1960) and Trigg and Hopkins (1976).

The class of 3–polytopes discussed here is usually called the class of *deltahedra.* From the historical point of view one should mention a paper of Rausenberger (1915), older than that of Freudenthal and van der Waerden and therefore containing an earlier systematic study of convex deltahedra. (Moreover, already Hermes (1894) described necessary criteria for the existence of polyhedra which are dual to the deltahedra.) Nets for construction and further information can be found in the books of Cundy and Rollett (1961, §3. 11) and Beck, Bleicher and Crowe (1982, §3). For $d \geq 4$, the complete set of convex polytopes described here is known. Aside from the regular d–polytopes, only bipyramids over $(d-1)$–simplices and, for $d = 4$, bipyramids over Platonic icosahedra belong to it (cf. R. Blind (1979_2), G. Blind and R. Blind (1980)).

RELATED POLYTOPE CLASSES

Neglecting the convexity for deltahedra, Cundy (1952) found 17 types of non–convex polyhedra with congruent triangular facets and only two types of vertices with respect to the symmetry group of those polyhedra. Additional results in that direction were obtained by Trigg (1978), Powarzynski and Spiegel (1979). Further, Mason (1972) and Stein (1978) showed that one cannot construct a ring of regular tetrahedra with neighbourly common facets. (It should be remarked that analogous constructions of polyhedral tori are realizable if one uses any other Platonic solid.) Bernal (1960) maintained that molecules in liquids form deltahedral structures.

8. Regular-Faced Polytopes

A convex d–polytope is *regular–faced,* if all its facets are regular $(d-1)$–polytopes. Grünbaum and Johnson (1965) showed that the facets of a regular–faced 3–polytope must be regular n–gons with $n \in \{3,4,5,6,8,10\}$. Later Johnson (1966) gave a list of all regular–faced 3–polytopes, the completeness of which was verified by Zalgaller (1967, 1969). Aside from the Platonic and Archimedean solids and the infinite classes of prisms and antiprisms, there are 92 representatives without isogonality. Upper and lower bounds for the numbers of facets, edges and vertices within that class of 3–polytopes are $\{4;\ 92\}$, $\{6;\ 180\}$, and $\{4;\ 120\}$, respectively. Zalgaller and Berman (1971) showed that all 112 types of regular–faced 3–polytopes can be obtained from combinations of suitable prisms, antiprisms and 28 additional 3–polytopes with only

3–valent vertices. Besides a complete list of names (which is due to N.W. Johnson), Berman (1971) gave photographs of models, nets of all types, and a lot of metrical information (such as volume, sphericity, dihedral angles and so on); see also Zalgaller (1969), Gagnon (1982) and Pugh (1976).

Valette (1969) investigated the subset of those 25 regular–faced 3–polytopes which are neither regular nor semiregular but have their complete vertex set on a 2–sphere. All regular–faced 3–polytopes have a non–trivial symmetry group. By the congruence of all $(d-2)$–faces of a regular–faced d–polytope $(d \geq 4)$, it follows that if the facets of such a polytope are not all congruent then it can have only facets of the following types: regular simplices and cross-polytopes, regular 600–cells $(d = 5)$ and regular icosahedra $(d = 4)$. On that base, R. Blind (1979_1) gave the complete list of regular–faced polytopes in higher dimensions. For each $d \geq 4$, this class consists of all regular and semiregular d–polytopes, d–pyramids over regular cross-polytopes and d–bipyramids over regular simplices. Additional such polytopes can be found only for $d = 4$, namely as

- suitable unions of semiregular polytopes with tetrahedral and octahedral facets and pyramids over their octahedral facets,
- pyramids and bipyramids over regular icosahedra,
- suitable truncations of the regular 600–cell.

The subset of regular–faced 4–polytopes with tetrahedral facets was investigated in detail by G. Blind and R. Blind (1980). The truncations of the 600–cell arise from that regular 4–polytope by the deletion of k non–adjacent vertices (hence $1 \leq k \leq 24$). Thus, as the convex hulls of the remaining $120-k$ vertices, they have k icosahedral and $600-20k$ tetrahedral facets. G. Blind and R. Blind (1989) have shown that all regular–faced d–polytopes $(d \geq 3)$ with trivial symmetry group must belong to this truncation set. Kirrmann (1991) gave a more detailed description of the congruence classes in this set, with complete results for $k \leq 6$ and $21 \leq k$. As an example, for $k = 6$ there exist 45775 congruence types, 41551 of which have trivial symmetry, and the existence of representatives with trivial symmetry group is ensured precisely for each k with $3 \leq k \leq 21$. (It should be noted that the analogous truncation problem for $d = 3$ permits only $1 \leq k \leq 3$ and yields exactly four types of regular–faced truncations of the Platonic icosahedron.)

RELATED POLYTOPE CLASSES

A natural extension of the polytope class described above was considered by Zalgaller (1967, 1969); namely, the class S of convex 3–polytopes whose facets are regular or can be divided into regular polygons by diagonals called *conditional edges.* Aside from prisms and antiprisms, he showed the finiteness of S, and he introduced *simple members* of S by the property that they cannot be cut into two members of S by plane sections only through edges and conditional edges. Zalgaller proved the existence of exactly 28 simple members of S without conditional edges, and Ivanov (1971) completed the set of simple members with five 3–polytopes having conditional edges.

An extension of S was introduced by Prjahin (1973, 1974): all facets or resulting polygons (after the subdivision of these facets by conditional edges) are required only to be equiangular. Except for prisms and antiprisms, this class of convex 3–polytopes is finite.

In a series of papers, Gurin (1983, 1984, 1985, 1986, 1987) investigated convex polyhedra with regular or congruent vertices. In particular, he looked for polars of regular–faced polyhedra.

Stewart (1970) dispersed with the convexity property and considered regular–faced, orientable polyhedral manifolds of non-negative genus. His book is also interesting for architects and designers; for example, the rich material contains also the construction of crystal lattices from Archimedean solids. Note also a related paper of Szilassi (1986).

9. Isogonal Polytopes with Congruent Facets

Clearly, a convex polytope from (i,c) is vertex–transitive and has congruent facets. For $d = 3$, this class of polyhedra can be easily derived as a subset of the convex isogonal polyhedra, completely presented in table 3 and figure 3 of Grünbaum and Shephard (1981_1); see also Robertson and Carter (1970). It turns out that the *isosceles tetrahedron* or *disphenoid* (that is, the non–regular tetrahedron with congruent facets) is particularly interesting because it is the only non–regular convex polyhedron which is isogonal and isohedral. (Note that the isohedrality is a natural sharpening of (c).) Further polyhedra having these two properties are only realizable with self–intersections and were studied already by Hess (1875, 1876, 1877) and Brückner (1900, 1905, 1906, 1907).

The disphenoid has a number of remarkable characterizations; for example, by important points, geodesic lines in its boundary and so on; see Bottema (1971/72), Devide (1984), Post (1970), Strantzen and Lu (1992), Kupitz and Martini (1993) as well as the book of Court (1964, §IV. 6). Moreover, there is no non–regular d–simplex ($d \geq 4$) having only disphenoids as 3–faces. This yields the characterization of regular d–simplices ($d \geq 4$) by 2–faces of equal 2–volume (even in spaces of constant curvature); see Frankl, Maehara (1990) and Martini (1993). In addition, it should be remarked that there are two types of disphenoids, with different symmetries: the tetragonal disphenoid (having isosceles facets) and the rhombic one.

The characterization of all convex d–polytopes ($d \geq 4$) which are isogonal and isohedral seems to be an interesting problem.

RELATED POLYTOPE CLASSES

For star polytopes satisfying (i, c) one should look at the literature referring to the property (i) and reduce the respective set of polytope types; see section 6.

Infinite isogonal polyhedra with equal polygons as faces were considered, for example, by ApSimon (1950); see also Coxeter (1968) and Grünbaum (1993_1).

10. Congruent-Faced Polytopes

In this section, we discuss convex d–polytopes with congruent facets that are also called *congruent–faced* (or *monohedral*) *polytopes*.

Giving a negative answer to a question of H. Steinhaus (cf. Problem 193 of Fast and Świerczkowski (1957)), Grünbaum (1960) proved that a congruent–faced 3–polytope must have an even facet number. The investigation of convex 3–polytopes with triangular congruent facets already yields sufficiently interesting problems.

Davies (1967) investigated such polyhedra with all vertices on a 2–sphere, and Min (1989) studied special polyhedra with one or two types of isosceles triangles as facets, with a view on applications to quasicrystals with 8– or 12–fold symmetry. Croft, Falconer and Guy (1991, B 22) asked for the types of triangles realizable as facets of a monohedral 3–polytope, either with or without reflections allowed. Grünbaum and Shephard (1992) gave a partial answer: by suitable reflections, one can generate monohedral polyhedra of octahedral type from an arbitrary triangle. They also investigated the class of monohedral polyhedra with triangular faces and the same number of facets at each vertex. Further, the great circle patterns on 2–spheres related to special finite groups (cf. the corresponding figures drawn by DuVal (1964) and Wenninger (1979)) and the geodesic domes of Buckminster Fuller (see Edmondson (1987), Tarnai (1993) and again Wenninger (1979)) can give information on interesting monohedral 3–polytopes with triangular facets, also regarding natural generalizations of the Platonic solids (cf. the discussion of Coxeter (1988_2)).

Special zonohedra (that is, vector sums of line segments) occur as congruent–faced 3–polytopes with rhombic facets; see Coxeter (1948, §2), Bilinski (1960) and Linhart (1986). Surprisingly, Bilinski proved that (besides the classical rhombic dodecahedron) there is a second zonohedron with 12 congruent rhombic facets.

An important subset of the set of monohedral polyhedra is the set of *isohedral polyhedra*; that is, those convex 3–polytopes whose symmetry group acts transitively on their facets. A complete classification (by combinatorial isomorphism types and symmetry groups) can be found in a paper of Grünbaum and Shephard (1981_2); for figures, the reader is referred to Wenninger (1979). An application of isohedra is given by games of dice, with respect to polyhedra of homogeneous density being fair in this sense (cf. Diaconis and Keller (1989)). Bigalke (1986) investigated the isohedra with pentagonal facets.

Particularly interesting are the polars of Archimedean solids. These are isohedra whose face angles around one vertex are congruent (that is, they have regular vertices). The polars of the quasiregular polyhedra, called *rhombic dodecahedron* and *rhombic triacontahedron*, play an important role in the theory of (quasi–)crystals. They were known to Kepler (1619), and Catalan (1865) investigated the remaining polars of semiregular 3–polytopes. Therefore the whole class is usually called the set of *Catalan solids* (see also Brückner (1900, p. 160)).

Extensive representations of the Catalan solids were given by Cundy and Rollett (1961), Roman (1968), Aškinuze (1969), Holden (1971), Pearce (1978), and Wenninger (1983).

RELATED POLYTOPE CLASSES

Grünbaum (1991_2) conjectured that his result on the even facet number of a congruent–faced 3–polytope can be analogously extended to non–convex polyhedra (without self–intersections), even if they are of positive genus. Grünbaum and Shephard (199?) also constructed spherical but non-convex monohedral polyhedra, all of whose vertices meet the same number of facets. Their investigations can be considered as starting points for natural extensions of several notions and concepts from convex to non-convex polytopes.

If convexity is eliminated from the definition of isohedral polyhedra, then one can show that there are no isohedra without self–intersections (except for slight weakenings of the conditions; see Wills (1986)) which are dual to corresponding isogonal polyhedra of positive genus. Thus, the "natural duality" of isogonal and isohedral polyhedra (mentioned by several authors in such a general sense) is not correct. Such errors, connected with the confusion of topological and projective duality, were discussed by Grünbaum and Shephard (1988); see Grünbaum (1991_1), Grünbaum and Senechal (1991, p. 380) and the recent paper of Grünbaum and Shephard (199?).

The star polyhedra with self–intersections, obtained by stellation of Catalan solids, and related compounds were studied by Hess (1878), Pitsch (1881), Brückner (1900) and, more recently, Wenninger (1983) and Fleurent (1989).

The author wishes to thank B. Grünbaum and N.W. Johnson for helpful discussions.

References

Hint: The following abbreviations are used:

AMM: Amer. Math. Monthly, **CF:** The Geometric Vein (The Coxeter Festschrift), Eds., C. Davis, B. Grünbaum and F. A. Sherk, Springer–Verlag, New York et al., 1981, **EM:** Elem. Math., **MG:** Math. Gaz., **GD:** Geom. Dedicata, **IJ:** Israel J. Math., **CJ:** Canad. J. Math., **JL:** J. London Math. Soc., **ÖAW:** Sitzungsber. Österr. Akad. Wiss., Math.–Naturwiss. Kl., Abt.II., **HCG:** Handbook of Convex Geometry. Eds. P. M. Gruber and J. M. Wills, North–Holland, 1993.

Andreini, A.: *Sulle reti di poliedri regolari e semiregolari.* Mem. Soc. Ital. Sci. (3) 14 (1905), 75-129.

Angell, I. O., Moore, M.: *Symmetrical intersections of cylinders.* Acta Cryst. Sect. A 43 (1987), no. 2, 244–250.

ApSimon, H.: *Three facially–regular polyhedra.* CJ 2 (1950), 326–330.

Aškinuze, V. G.: *Über die Anzahl halbregulärer Polyeder (Russ.)* Mat. Prosvešč.. 1 (1957), 107–118.

Vielecke und Vielflache. In: Enzykl. Elementarmath. IV, Verlag der Wiss., Berlin 1969.

Badoureau, A.: *Mémoire sur les figures isoscèles.* J. École Polytechn. 49 (1881), 47–172.

Bakos, T: *Octahedra inscribed in a cube.* MG 43 (1959), 17–20.

Ball, W. W. R., Coxeter, H.S.M.: *Mathematical Recreations and Essays*, Dover, New York 1987.

Banchoff, T. F.: *Torus decomposition of regular polytopes in 4–space.* In: Shaping Space. A Polyhedral Approach. Eds. M. Senechal and G. Fleck, Birkhäuser, Boston 1988, 221–230.

Beyond the Third Dimension. Scient. Amer. Library, Freeman, New York 1990.

Barnette, D.: *A simple 4–dimensional nonfacet.* IJ 7 (1969), 16–20.

Nonfacets for shellable spheres. IJ 35 (1980), 286-288.

Barrau, J. A.: *Die zentrische Zerlegung der regulären Polytope.* Nieuw Arch. Wiskd. (2) 7 (1906), 250–270.

Beck, A., Bleicher, M. N., Crowe, D. W.: *Excursions into Mathematics.* Worth Publ., Inc., New York 1982.

Berger, M.: *Geometry, I, II.* Springer, Berlin et al., 1987.

Berman, M.: *Regular-faced convex polyhedra.* J. Franklin Inst. 291 (1971), 329–352.

Bernal, J. D.: *The structure of liquids.* Scientific American, August 1960, 124-130.

Bigalke, H. G.: *Die flächenäquivalenten Pentagondodekaeder.* Did. Math. 3 (1986), 204–221.

Bilinski, S.: *Über Rhombenisoeder.* Glasnik Mat. –Fiz. Astron. Društvo Mat. –Fiz. Hrvatske, Ser. II, 15 (1960), 251–263.

Die windschiefen Archimedischen Polyeder höheren Geschlechts. ÖAW 197 (1988), 315–326.

Blind, G., Blind, R.: *Die konvexen Polytope im R^4, bei denen alle Facetten reguläre Tetraeder sind.* Monatsh. Math. 89 (1980), 87–93.

Über die Symmetriegruppen von regulärseitigen Polytopen. Monatsh. Math. 108 (1989), 103–114.

The semiregular polytopes. Comment. Math. Helv. 66 (1991), 150–154.

Blind, R.: *Konvexe Polytope mit regulären Facetten im $R^n (n \geq 4)$.* In: Contr. to Geom., Eds. J. Tölke and J. M. Wills. Birkhäuser, Basel 1979_1, 248–254.

Konvexe Polytope mit kongruenten regulären $(n-1)$-Seiten im $R^n (n \geq 4)$. Comment. Math. Helv. 54 (1979_2), 304–308.

Böhm, J., Quaisser, E.: *Schönheit und Harmonie geometrischer Formen – Sphäroformen und symmetrische Körper.* Akademie–Verlag, Berlin 1991.

Bokowski, J.: *On the geometric flat embedding of abstract complexes with symmetries.* In: Symmetry of Discrete Math. Structures and Their Symmetry Groups, Eds. K. H. Hofmann, R. Wille. Heldermann–Verlag, Berlin 1991, 1–48.

Bottema, A.: *The centroids of a simplex (Dutch).* Euclides (Groningen) 31 (1971/72), 206–210.

Bourgin, D. G., Mendel, C. W.: *Cubes in cubes.* Rend. Circ. Mat. Palermo (2) 117 (1969), 313–327.

Brandmüller, J.: *Die fünfzählige Symmetrie in der Mathematik, Physik, Chemie, Biologie und darüber hinaus.* Erweiterte Vortragsfassung, Jahrestagung MNU, München 1990.

Fivefold Symmetry in Mathematics, Physics, Chemistry, Biology and Beyond. In: Fivefold Symmetry (Ed. I. Hargittai), World Scient., Singapore et al., 1992.

Brauner, H.: *Lehrbuch der konstruktiven Geometrie.* Fachbuchverlag Leipzig, Springer–Verlag Wien, 1986.

Brehm, U., Kühnel, W.: *Smooth approximation of polyhedral surfaces regarding curvature.* GD 12 (1982), 435–461.

Brehm, U., Wills, J. M.: *Polyhedral manifolds.* In: HCG, 1993.

Brieskorn, E.: *Lineare Algebra und Analytische Geometrie, Bd. I.* Vieweg, Braunschweig–Wiesbaden, 1983.

Brown, C. S.: *Capacity of the regular polyhedra.* Comp. Math. Appl. 20 (1990), 43–56.

Brückner, M.: *Vielecke und Vielflache – Theorie und Geschichte.* Teubner, Leipzig 1900.

Über die diskontinuierlichen und nicht-konvexen gleicheckig-gleichflächigen Polyeder. Verh. des dritten Internat. Math.-Kongresses Heidelberg 1904. Teuber, Leipzig 1905, pp. 707-713.

Über die gleicheckig-gleichflächigen, diskontinuierlichen und nichtkonvexen Polyeder. Nova Acta Leop. 86 (1906), no. 1, pp. 1-384, 24 plates.

Zur Geschichte der Theorie der gleicheckig-gleichflächigen Polyeder. Unterrichtsblätter Math. Naturwiss. 13 (1907), 104-110, 121-127; 1 plate.

Brun, V.: *On some problems in solid geometry inspired by virology.* Nordisk. Mat. Tidskr. 25–26 (1978), 113–119.

Buchholz, I.: *Zur vektorgeometrischen Behandlung der regulären Polyeder.* Der Mathematikunterricht 37 (1991), no. 4, 30–44.

Catalan, F. C.: *Mémoire sur la théorie des polyèdres.* J. École Polytech. 41 (1865), 1–71.

Cauchy, L.: *Recherches sur les polyèdres.* J. École Polytechn. 16 (1813), 68–86. Transl.: Untersuchungen über Vielflache. In: R. Haussner (ed.), Ostwald's Klassiker der exakten Wissenschaften, No. 151, pp. 49–72, Engelmann, Leipzig 1906.

Cayley, A.: *On Poinsot's four new regular solids.* Philos. Mag. (4) 17 (1859), 123–128. Reprinted in: Collected Math. Papers, Vol. IV, Cambridge Univ. Press, Cambridge, 1891. Transl.: Über Poinsot's vier neue regelmäßige Körper. In: R. Haussner (ed.), Ostwald's Klassiker der exakten Wissenschaften, No. 151, pp. 95–97, Engelmann, Leipzig 1906.

Chieh, C.: *Polyhedra and crystal structures.* In: Shaping Space. A Polyhedral Approach. Eds. M. Senechal, G. Fleck. Birkhäuser, Boston–Basel 1988, 93–105.

Colbourn, C. J., Ivić-Weiss, A.: *A census of regular 3-polystromas arising from honeycombs.* Discrete Math. 50 (1984), 29-36.

Conway, J. H.: *Four-dimensional Archimedean polytopes.* In: Proc. Coll. Convexity, Copenhagen 1965, Kobenhavns Univ. Mat. Inst. (1967), 38–39.

Court, N.A.: *Modern Pure Solid Geometry.* Chelsea, New York 1964.

Coxeter, H. S. M.: *The pure Archimedean polytopes in six and seven dimensions.* Proc. Cambr. Philos. Soc. 24 (1928), 1–9.

The polytopes with regular-prismatic vertex figures, I. Trans. Roy. Soc. London A 229 (1930), 329–425.

The polytopes with regular–prismatic vertex figures, II. Proc. London Math. Soc. (2) 34 (1932), 126–189.

Regular and semiregular polytopes, I. Math. Z. 46 (1940), 380–407.

Regular polytopes. London, 1948.

Regular skew polyhedra in three and four dimensions, and their topological analogues. In: H. S. M. Coxeter, Twelve Geom. Essays, South. Illinois Univ. Press, 1968, 75–105.

Regular Complex Polytopes. Cambridge Univ. Press, London – New York, 1974.

Polyhedral numbers. In: For Dirk Struik, Eds. R. S. Cohen et al., Reidel Publ. Co., Dordrecht 1975.

Polytopes in the Netherlands. Nieuw Arch. Wiskd., Ser. III, 26 (1978), 116–141.

Unvergängliche Geometrie. Birkhäuser, Basel 1981.

Regular and semi-regular polytopes, II. Math. Z. 188 (1985), 559–591.

Regular and semi-regular polytopes, III. Math. Z. 200 (1988_1), 3–45.

Regular and semi-regular polyhedra. In: Shaping Space. A Polyhedral Approach. Eds. M. Senechal and G. Fleck. Birkhäuser, Boston–Basel, 1988_2, 67–79.

Star polytopes and the Schläfli function $f(\alpha\beta,\gamma)$. EM 44 (1989), 25–36.

Coxeter, H. S. M., Longuet-Higgins, M. S., Miller, J. C. P.: *Uniform polyhedra.* Philos. Trans. Roy. Soc. London, Ser. A 246 (1954), 401–450.

Critchlow, K.: *Time stands still, new light on Megalithic science.* Fraser, London 1979.

Croft, H. T.: *Arrows on polyhedra.* JL 37 (1962), 287–300.

On maximal regular polyhedra inscribed in a regular polyhedron. Proc. London Math. Soc.(3), 41 (1980), 279–296.

Croft, H. T., Falconer, K. J., Guy, R. K.: *Unsolved Problems in Geometry.* Springer, New York et al., 1991.

Cundy, H. M.: *Deltahedra.* MG 36 (1952), 263–266.

Cundy, H. M., Rollett, A. P.: *Mathematical Models.* Oxford Univ. Press 1961.

Davies, H. L.: *Packings of spherical triangles and tetrahedra.* Proc. Coll. Convexity (Copenhagen 1965), 42–51, Kobenhavns Univ. Mat. Inst., Copenhagen 1967.

Dekker, T. J.: *On reflections in Euclidean spaces generating free products.* Nieuw Arch. Wiskd. (3) 7 (1959), 57-60.

Devide, V.: *Über eine Klasse von Tetraedern.* Rad. Jugosl. Akad. Znan. Umjet. 408 (1984), 45–50.

Diaconis, P., Keller, J. B.: *Fair dice.* AMM 96 (1989), 337–339.

Dol'nikov, V. L.: *On the Jung constant in* l_1^n *(Russ.).* Mat. Zametki 42 (1987), 519–526. Transl.: Math. Notes 42 (1987), 787–791.

Dorwart, H. L.: *Point and circle configurations; a new theorem.* Math. Mag. 61 [1988), 253–259.

Dragomir, V., Gheorghiu, A.: *Genése et construction des polyèdres semi-reguliers.* Bull. Soc. Roy. Sc. Liège 42 (1973), 5–20.

Dress, A. W. M.: *A combinatorial theory of Grünbaum's new regular polyhedra. Part I: Grünbaum's new regular polyhedra and their automorphism group.* Aequat. Math. 23 (1981), 252–265.

Part II: Complete enumeration. Aequat. Math. 29 (1985), 222–243.

Dress, A. W. M., Schulte, E.: *On a theorem of McMullen about combinatorially regular polytopes.* Simon Stevin 61 (1987), 265–273.

DuVal, P.: *Homographies, Quaternions and Rotations.* Oxford, 1964.

Edmondson, A. C.: *A Fuller Explanation.* Birkhäuser, Boston et al., 1987.

Elte, E. L.: *The Semiregular Polytopes of the Hyperspaces.* Hoitsema, Groningen, 1912.

Emmer, M.: *Art and Mathematics: The Platonic solids.* Leonardo 15 (1982), no. 4, 277-282.

Farris, S. L.: *Completely classifying all vertex-transitive and edge-transitive polyhedra, part I: necessary class conditions.* GD 26 (1988), 111-124.

Completely classifying all vertex-transitive and edge-transitive polyhedra, part II: finite, fully-transitive polyhedra. Journal of Geometry, to appear 199?.

Fast, H., Świerczkowski, S. (eds.): *The New Scottish Book.* Wroclaw, 1957.

Federico, P. J.: *Descartes on Polyhedra.* Springer, New York et al., 1982.

Fedorov, E. S.: *Foundations of the Theory of Figures (Russian).* Acad. Sci. Petersburg, 1885. Republished with comments: Akad. Nauk SSSR, Moscow 1953.

Fejes Tóth, L.: *Reguläre Figuren.* Teubner, Leipzig 1965.

Lagerungen in der Ebene, auf der Kugel und im Raum. Springer, Berlin et al., 1972.

Field, J. V.: *Kepler's star polyhedra.* Vistas in Astronomy 23 (1979), 109–141.

Fleurent, G. M.: *Symmetry and polyhedral stellation, Ia und Ib.* Comput. Math. Appl. 117 (1989), 167–193.

Florian, A.: *Neuere Entwicklungen über reguläre Polyeder.* In: Contr. to Geometry. Eds. J. Tölke and J. M. Wills, Birkhäuser, Basel 1979, 238–247.

Extremum problems for convex discs and polyhedra. In: HCG, 1993.

Frankl, P., Maehara, H.: *Simplices with given 2-face areas.* Europ. J. Combin. 11 (1990), 241–247.

Franz, W.: *Dreidimensionale und mehrdimensionale Geometrie: Die regulären Polytope.* Sitzungsber. Wiss. Ges. Goethe-Univ. Frankfurt/Main 9 (1970), no. 3, 67–104, Steiner–Verlag, Wiesbaden 1971.

Freudenthal, H., van der Waerden, B. L.: *Over een bewering van Euclides.* Simon Stevin 25 (1947), 115–121.

Gagnon, S.: *Convex polyhedra with regular faces.* Structural Topology 6 (1982), 83–95.

Galafassi, V. E.: *I poliedri convessi con facce regolari eguali.* Archimede 12 (4) (1960), 169–177.

Gosset, T.: *On the regular and semiregular figures in space of n dimensions.* Mess. Math. 29 (1900), 43–48.

Grünbaum, B.: *On polyhedra in E^3 having all faces congruent.* Bull. Res. Council Israel Sect. F., F 8 (1960), 215–218.

Convex Polytopes. Wiley & Sons, London et al., 1967.

Regularity of graphs, complexes and designs (French summary). In: Problèmes combinatoires et théorie des graphes (Coll. Internat. C.N.R.S., Orsay 1976, C.N.R.S., Paris 1978), 191-197.

Regular polyhedra – old and new. Aequat. Math. 16, (1977), 1–20.

Regular polyhedra. Manuscript 1991_1, to appear in: Encyclopaedia of the History and Philosophy of the Math. Sciences, Routledge, London.

Odd polyhedra. Geombinatorics 1 (1991_2), 4–5.

Infinite uniform polyhedra. Geombinatorics 2 (1993_1), 53–60.

Holey isogonal columns. Geombinatorics 2 (1993_2), 75-78.

Grünbaum, B., Johnson, N. W.: *The faces of a regular-faced polyhedron.* JL 40 (1965), 577–586.

Grünbaum, B., Senechal, M.: *Review on "'Symmetrie–Gruppe–Dualität"' by E. Scholz.* (Birkhäuser, Basel (1989)), Historia Math. 18 (1991), 377–380.

Grünbaum, B., Shephard, G. C.: *Patterns on the 2-sphere.* Mathematika 28 (1981_1), 1–35.

Spherical tilings with transitivity properties. In: CF, 1981_2, 65–98.

Polyhedra with transitivity properties. C. R. Math. Rep. Acad. Sci. Canada 6 (1984), 61–66.

Tilings and Patterns. Freeman, New York, 1987.

Duality of polyhedra. In: Shaping Space. A Polyhedral Approach. Eds. M. Senechal and G. Fleck, Birkhäuser, Boston 1988, 205–211.

Triangle-faced polyhedra. Preliminary version, 13 pp., 15 figures, 1992.

Isohedra with non-convex faces. Journal of Geometry, to appear 199?.

Gurin, A. M.: *Realizations of nets of convex polyhedra with equiangular vertices (Russ.).* Ukr. Geom. Sbornik.I: 26 (1983), 41–48; II: 27 (1984), 22–26; III: 28 (1985), 26–43; IV: 29 (1986), 32–47; V: 30 (1987), 22–36.

Hadwiger, H.: *Ungelöste Probleme.* Nr. 10. EM 11 (1956), 36–37.

Haenzel, G.: *Die Diracsche Wellengleichung und das Ikosaeder.* J. Reine Angew. Math. 183 (1941), no. 4, 232–242.

Harborth, H., Kemnitz, A., Möller, M., Süssenbach, A.: *Ganzzahlige planare Darstellungen der platonischen Körper.* EM 42 (1987), 118–122.

Hargittai, I., Hargittai, M.: *Polyhedral molecular geometries.* In: Shaping Space. A Polyhedral Approach. Eds.: M. Senechal and G. Fleck, Birkhäuser, Boston, 1988, 172–188.

Hermes, O.: *Über Anzahl und Form von Vielflächnern.* Wiss. Beilage zum Jahresbericht des Köllnischen Gymnasiums zu Berlin, Ostern 1894; Berlin, 1894.

Hess, E.: *Über zwei Erweiterungen des Begriffs der regelmässigen Körper.* Sitzungsber. Ges Beförd. Naturwiss. Marburg, 1875, 1-20.

Über die zugleich gleicheckigen und gleichflächigen Polyeder. Schriften Ges. Beförd. Naturwiss. Marburg, Band 11, Abh. 1, pp. 1-97; 11 figures. Th. Kay, Cassel 1876.

Über einige merkwürdige nichtkonvexe Polyeder. Sitzungsber. Ges. Beförd. Naturwiss. Marburg, 1877, 1-13.

Über die vier archimedischen Polyeder höherer Art. Sitzungsber. Ges. Beförd. Naturwiss. Marburg 11 (1878), 261-271.

Einleitung in die Lehre von der Kugelteilung. Teubner, Leipzig 1883.

Die regulären Polytope höherer Art. Sitzungsber. Ges. Beförd. Naturwiss. Marburg 20 (1885), 31-57.

Hilbert, D., Cohn–Vossen, S.: *Anschauliche Geometrie.* Springer, Berlin 1932.

Hill, L. S.: *Notes on the regular icosahedron and the regular dodecahedron.* Scripta Math. 7 (1940), 99–109.

Hilton, P., Pedersen, J.: *Build Your Own Polyhedra.* Addison–Wesley, Menlo Park, CA, 1988.

Hippenmeyer, C.: *Aufgabe 804.* EM 34 (1979), 61–63.

Hofmann, J. E.: *Über Archimedes' halbregelmäßige Körper.* Arch. Math. 14 (1963), 212–216.

Hohenberg, F.: *Einige Geradensysteme der erweiterten Ikosaedergruppe.* ÖAW 178 (1970_1), 285–297.

Einige Figuren der erweiterten Oktaedergruppe. EM 25 (1970_2), 55–60.

Die Geradensysteme der erweiterten Polyedergruppen. ÖAW 179 (1971), 63–92.

Besondere Bilder des abgestumpften Würfels. Ber. Math. -Statist. Sekt. Forschungszentr. Graz 146 (1980_1), 1–14.

Die Geradensysteme der erweiterten Polyedergruppen. In: Tagungsband 2. Koll. Diskrete Geom., Univ. Salzburg, (1980_2), Ed. A. Florian, 101–106.

Projektive Eigenschaften des abgestumpften Würfels. EM 36 (1981), 49–58.

Projektive Eigenschaften eines besonderen Systems von Polyedern der Hexaedergruppe. ÖAW 191 (1982_1), 173–186.

Metrische und projektive Verallgemeinerungen des abgestumpften Würfels des Archimedes. ÖAW 191 (1982_2), 165–172.

Das abgestumpfte Dodekaeder des Archimedes und seine projektiven Eigenschaften. ÖAW 192 (1983), 143–159.
Vier Verallgemeinerungen des abgestumpften Dodekaeders. ÖAW 193 (1984_1), 177–184.
Projektive Eigenschaften zweier besonderer Systeme von Polyedern der Dodekaedergruppe. ÖAW 193 (1984_2), 185–191.
Holden, A.: *Shapes, Space and Symmetry.* Columbia Univ. Press, New York 1971.
Hudson, J. L., Kingston, J. G.: *Stellating polyhedra.* Math. Intell. 10 (1988), 50–61.
Huybers, P., Coxeter, H. S. M.: *A new approach to the chiral Archimedean solids.* C. R. Math. Rep. Acad. Sci. Canada 1 (1979), 259–274.
Ivanov, B. A.: *Polyhedra with faces that are composed by regular polygons (Russ.).* Ukr. Geom. Sb. 10 (1971), 20–34.
Ivić–Weiss, A.: *Incidence polytopes with toroidal cells.* Discrete Comput. Geom. 4 (1989), 55–73.
Jamnitzer, W.: *Perspective Corporum Regularium.* Nürnberg 1568, Neudruck: Frankfurt 1972, Graz 1975; short description by Locher–Ernst in: EM 11 (1956), 97–100.
Jarratt, J. D., Seidel, J. J.: *A note on the equivalence of the problem of embedding a regular N–simplex in the N–dimensional hypercubic lattice to a certain integral matrix completion problem.* Indian J. Math. 27 (1985) no. 1–3, 57–62 (1986).
Jeger, M.: *Über die Anzahl der inkongruenten ebenen Netze des Würfels und des regulären Oktaeders.* EM 30 (1975), 73–96.
Johnson, N. W.: *Convex polyhedra with regular faces.* CJ 18 (1966), 169–200.
Uniform Polytopes. Monograph (manuscript), to appear 199?.
Kalai, G.: *On low–dimensional faces that high–dimensional polytopes must have.* Combinatorica 10 (1990), 271–280.
Kepler, J.: *Harmonices Mundi.* J. Planck, Linz 1619. Reprinted in: Opera Omnia, Vol. 5, Heider & Zimmer, Frankfurt 1864; Gesammelte Werke, Bd. 6, Beck, München, 1940.
Strena seu de nive sexangula (Neujahrsgabe oder vom sechseckigen Schnee). J. Planck, Linz 1619. Gesammelte Werke, Bd. 4, Berlin, 1943.
Kibble, W. F.: *Regular polytopes inscribed in other regular polytopes.* Math. Student 17 (1949), 26–31.
King, R. B.: *Chemical applications of topology and group theory.* XXIV: Chiralization of chemically significant polyhedra. J. Math. Chem. 1 (1987), 415–421.
Kirrmann, G. J.: *Regulärseitige Polytope, die durch Abschneiden von Ecken eines 600–Zells entstehen.* Diss. Univ. Stuttgart, 1991.
Krötenheerdt, O.: *Die homogenen Mosaike n–ter Ordnung in der euklidischen Ebene, Teil I.* Wiss. Z. Martin–Luther–Universität Halle–Wittenberg, Math. – Naturwiss. Reihe 19 (1969), no. 4, 273–290.
Kupitz, Y., Martini, H.: *The Fermat–Torricelli problem and isosceles tetrahedra.* Manuscript, 16 pp., 1993, to appear in: Journal of Geometry.
Lehmer, D. H.: *Coloring the Platonic solids.* AMM 93 (1986), 288–292.

Lemmens, P. W. H., Seidel J. J.: *Equiangular lines.* J. Algebra 24 (1973), 494–512.
Lindemann, F.: *Zur Geschichte der Polyeder und der Zahlzeichen.* Sitzungsber. Bayer. Akad. d. Wiss., Math. -Phys. Kl. 26 (1896/97), H. 4, 625–783.
Linhart, J.: *Extremaleigenschaften der regulären 3-Zonotope.* Studia Sci. Math. Hungar. 21 (1986), 181–188.
Locher-Ernst, L.: *Konstruktionen des Dodekaeders und Ikosaeders.* EM 10 (1955), 73–81.
Lyusternik, L. A.: *Convex Figures and Polyhedra.* Dover Publ., New York 1963.
Macdonald, I. G.: *Regular simplexes with integral vertices.* C. R. Rep. Acad. Sc. Canada 9 (1987), No. 4, 189–193.
Makarov, P. V.: *Four-dimensional quasi-regular polytopes (Russ.).* Izv. Akad. Nauk Mold. SSR, Ser. Fiz. -Tekh. Mat. Nauk No. 2 (1987), 10–14.
On derivation of four-dimensional semiregular polyhedra (Russ.). Mat. Issled. 103 (1988), 139–150.
Malkevitch, J.: *Milestones in the history of polyhedra.* In: Shaping Space. A Polyhedral Approach. Eds. M. Senechal and G. Fleck, Birkhäuser, Boston-Basel, 1988, 80–92.
Martin, S. J.: *The Biochemistry of Viruses.* Cambridge Univ. Press, Cambridge 1978.
Martini, H.: *Regular simplices in spaces of constant curvature.* AMM 100 (1993), 169–171.
Reguläre Polytope und Verallgemeinerungen. Manuscript, 35 pp., to appear in: Geometrie und ihre Anwendungen, Eds. O. Giering and J. Hoschek, Carl Hanser Verlag, München 1994.
Mason, J. H.: *Can regular tetrahedra be glued together face to face to form a ring?* MG 56 (1972), 194–197.
McMullen, P.: *Combinatorially regular polytopes.* Mathematika 14 (1967), 142–150.
Regular star-polytopes, and a theorem of Hess. Proc. London Math. Soc. (3) 18 (1968_1), 577–596.
Affinely and projectively regular polytopes. JL 43 (1968_2), 755–757.
Realizations of regular polytopes. Aequat. Math. 37 (1989), 38–56.
The order of a finite Coxeter group. EM 46 (1991), 121–130.
McMullen, P., Schulte, E.: *Constructions of regular polytopes.* J. Combin. Theory, Ser. A, 53 (1990_1), 1–28.
Regular polytopes from twisted Coxeter groups and unitary reflexion groups. Advances Math. 82 (1990_2), 35–87.
Medjanik, A. I.: *A regular simplex inscribed in the cube (Russ.).* Ukrain. Geom. Sb. 13 (1973), 109–112.
Min, L. Q.: *Geometric parameters of polyhedra and quasicrystals with eight- and twelve-fold rotational symmetry (Chinese).* Math. Practice Theory 1989, no. 3, 36–43.
Miyamoto, M.: *Construction of Hadamard matrices.* J. Combin. Theory, Ser. A, 57 (1991), 86–108.
Monson, B.: *Uniform polyhedra from Diophantine equations.* In: Shaping Space. A Polyhedral Approach. Eds. M. Senechal and G. Fleck, Birkhäuser, Boston 1988, 219–220.
Mulder, P.: *Stervormige Polytopen.* Nieuw Arch. Wiskd. (2) 7 (1907), 283–311.

Norman, A. C., Smith, A.: *Computer drawings of compounds of star-polyhedra.* MG 57 (1973), 303–306.

van Oss, S. L.: *Die regelmäßigen vierdimensionalen Polytope höherer Art.* Verh. Konink. Akad. Wetensch. Amsterdam (eerste sectie) 12 (1915), no. 1, 13pp., 6 fig.

Pargeter, A. R.: *Plaited polyhedra.* Math. Mag. 43 (1959), 88–161.

Patruno, G. N.: *The lattice polytope problem.* EM 38 (1983), 69–71.

Pearce, P.: *Structure in Nature is a Strategy of Design.* MIT Press, Cambridge 1978.

Pelling, M. J.: *Regular simplices with rational vertices.* Bull. London Math. Soc. 9 (1977), 199–200.

Perles, M. A., Shephard, G. C.: *Facets and nonfacets of polytopes.* Acta Math. 119 (1967), 113–145.

Pitsch, J.: *Über halbreguläre Sternpolyeder.* Z. Realschulwesen 6 (1881) 9–24, 64–65, 72–89, 216.

Plato: *Timaeus.* In: The Great Books of Western World. London, Encyclop. Britannica, 1952, Vol. 7, 442–477.

Poinsot, L.: *Mémoire sur les polygones et les polyèdres.* J. École Polytechn. 10 (1810), 16–48. Transl.: Abhandlung über die Vielecke und Vielflache. In: R. Haussner (ed.), Ostwald's Klassiker der exakten Wissenschaften, No. 151, pp. 3–48, Engelmann, Leipzig 1906.

Post, K. A.: *Geodesic lines on a closed convex polyhedron.* Studia Sci. Math. Hungar. 5 (1970), 411–416.

Powarzynski, R., Spiegel, H.: *Über zwei besondere Arten von Dreiecksflächnernetzen.* EM 34 (1979), 140–146.

Prjahin, J. A.: *Convex polyhedra with regular faces (Russ.).* Ukr. Geom. Sb. 14 (1973), 83–88, 115.

Convex polyhedra whose faces are equiangular polygons or combinations of them (Russ.). In: Problems in Global Geom., Zap. Naučn. Sem. Leingr. Otdel. Math. Inst. Steklov, (LOMI) 45 (1974), 111–112, 119–120.

Propp, J. G.: *Kepler's spheres and Rubik's cube.* Math. Mag. 61 (1988), 231–239.

Pugh, A.: *Polyhedra: a visual approach.* Univ. of Calif. Press, Berkeley 1976.

Rausenberger, O.: *Konvexe pseudoreguläre Polyeder.* ZMNU 46 (1915), no. 1, 135–142.

Robertson, S. A.: *Polytopes and Symmetry.* London Math. Soc. Lect. Notes Series No. 90, Cambridge, Univ. Press, 1984.

Robertson, S. A.; Carter, S.: *On the Platonic and Archimedean solids.* J. London Math. Soc. (2) 2 (1970), 125–132.

Roman, T.: *Reguläre und halbreguläre Polyeder.* Deutscher Verlag der Wiss., Berlin 1968.

Rosenfeld, B. A., Yaglom, I. M.: *Mehrdimensionale Räume.* In: Enzykl. Elementarmath., Bd. 5, Deutscher Verlag der Wiss., Berlin 1971.

Rossat, H., Sloane, J. A.: *Four icosahedra can meet at a point.* GD 27 (1988), 219–222.

Schläfli, L.: *On the multiple integral* $\int^n dx\,dy \ldots dz$, *whose limits are* $p_1 = a_1x + b_1x + \ldots + h_1z > 0, p_2 > 0, \ldots, p_n > 0$, *and* $x^2 + y^2 + \ldots + z^2 < 1$ *(Part 3).* Quart. J. Pure Appl. Math. 3 (1860), 97–108.

Theorie der vielfachen Kontinuität. Neue Schweiz. Ges. Naturwiss. 38, I (1901), 1–237 (memorial volume). Reprinted in: Ges. Abh., Bd. I, Birkhäuser, Basel 1950, 167–387.

Schneebeli, H. R.: *Zur Geometrie der Mikrocluster.* EM 48 (1993), 1–16.

Schoenberg,I. J.: *Regular simplices and quadratic forms.* JL. 12 (1937), 48–55.

Schoute, P. H.: *Mehrdimensionale Geometrie, Bd. 2 (Die Polytope).* Leipzig, 1905.

The sections of the net of measure polytopes M_n of space Sp_n with a space Sp_{n-1} normal to a diagonal. Konink. Akad. Wetensch. Amsterdam Proc. Sect. Sci. 10 (1908), 688–698.

Analytical treatment of the polytopes regularly derived from the regular polytopes (section I). Verh. Konink. Akad. Wetensch. Amsterdam (eerste sectie) 11 (1911), no. 3, 82 pp., 1 plate, 3 tables.

Analytical treatment of the polytopes regularly derived from the regular polytopes (section II, III, IV). Verh. Konink. Akad. Wetensch. Amsterdam (eerste sectie) 11 (1913), no. 5, 108 pp., 1 plate, 8 tables.

Schulte, E.: *Nontiles and nonfacets for the Euclidean space, spherical complexes and convex polytopes.* J. Reine Angew. Math. 352 (1984), 161–183.

The existence of nontiles and nonfacets in three dimensions. J. Combin. Theory, Ser. A, 38 (1985), 75–81.

Schulte, E., Ivić–Weiss, A.: *Chiral polytopes.* DIMACS Ser. Discrete Math. & Computer Sci., Vol. 4 (1991), 493–516.

Schulte, E., Wills, J. M.: *On Coxeter's regular skew polyhedra.* Discrete Math. 60 (1986), 253–262.

Combinatorially regular polyhedra in 3–space. In: Symmetry and Discrete Math. Structures and Their Symmetry Groups, Eds. K. H. Hofmann and R. Will, Heldermann–Verlag, Berlin 1991, 49–88.

Scott, P. R.: *Equiangular lattice polygons and semiregular lattice polyhedra.* College Math. J. 18 (1987), 300-306.

Seberry, J.R., Yamada, M.: *Hadamard matrices, sequences, and block designs.* In: Contemporary Design Theory - A Collection of Surveys. Eds. J.H. Dinitz and D.R. Stinson. Wiley, New York 1992.

Senechal, M.: *A visit to the Polyhedral Kingdom.* In: Shaping Space. A Polyhedral Approach. Eds. M. Senechal and G. Fleck, Birkhäuser, Boston 1988, 3–43.

Shephard, G. C.: *A construction of Wythoffian polytopes.* CJ 6 (1954), 128–134.

Skilling, J.: *The complete list of uniform polyhedra.* Philos. Trans. Roy. Soc. London, Ser. A 278 (1975), 111–135.

Uniform compounds of uniform polyhedra. Math. Proc. Cambridge Phil. Soc. 79 (1976), 447–457.

Smith, A.: *Some regular compounds of star–polyhedra.* MG 57 (1973), 39–46.

Uniform compounds and the group A_4. Proc. Cambr. Philos. Soc. 75 (1974), 115–117.

Sommerville, D. M. Y.: *Semi–regular networks of the plane in absolute geometry.* Trans. Roy. Soc. (Edinburgh) 41 (1905), 725–747; plates I–XII.

Sopov, S. P.: *The number of uniform polyhedra with non–negative Euler characteristic (Russ.).* Ukr. Geom. Sb. 3 (1966_1), 123–129.

A theorem on uniform polyhedra (Russ.). Ukr. Geom. Sb. 2 (1966_2), 98–106.

The finiteness of the number of elementary uniform polyhedra with non-zero density (Russ.). Ukr. Geom. Sb. 5–6 (1968), 160–166.

A certain class of homogeneous polyhedra (Russ.). Ukr. Geom. Sb. 7 (1969), 130–140, 186.

A proof of the completeness of the list of elementary homogeneous polyhedra (Russ.). Ukr. Geom. Sb. 8 (1970), 139–156.

Stein, S. K.: *The planes obtainable by gluing regular tetrahedra.* AMM 85 (1978), 477–479.

Steinhaus, H.: *Kaleidoskop der Mathematik.* Deutscher Verlag der Wiss., Berlin 1959.

Stewart, B. M.: *Adventures Among the Toroids.* A Study of Orientable Polyhedra with Regular Faces. (Illustrated, hand-lettered and published by the author.) 2nd ed., B. M. Stewart, 4494 Wausau Road, Okemo, Michigan 48864, 1970, 206 pp.

Stillwell, J.: *Mathematics and Its History.* Springer, New York 1989.

Stott, A. Boole: *On certain series of sections of the regular four-dimensional hypersolids.* Verh. Konink. Akad. Wetensch. Amsterdam (eerste sectie) 7 (1900), no. 3, 21 pp., 5 plates.

Geometrical deduction of semi-regular from regular polytopes and space-fillings. Verh. Konink. Akad. Wetensch. Amsterdam (eerste sectie) 11 (1910), no. 1, 24 pp., 3 plates, 3 tables.

Strantzen, J., Lu, Y.: *Regular simple geodesic loops on a tetrahedron.* GD 42 (1992), 139–153.

Świerczkowski, S.: *On a free group of rotations of the Euclidean space.* Indag. Math. 20 (1958), 376-378.

Szepesvari, I.: *On the number of uniform polyhedra I, II.* Mat. Lapok 29 (1981), 273–328 (Hungarian).

Szilassi, L.: *Regular toroids.* Structural Topology 13 (1986), 69–80.

Tarnai, T.: *Spherical grids of triangular networks.* Acta Tech. Acad. Sci. Hungar. 76 (1974), 307–336.

Geodesic domes and fullerenes. Phil. Trans. Roy. Soc. London A (1993), 343, 145-154.

Tits, J.: *A local approach to buildings.* In: CF, 1981, 519–547.

Toepell, M.: *Platonische Körper in Antike und Neuzeit.* Der Mathematikunterricht 37 (1991), no. 4, 45–79.

Trigg, C.: *An infinite class of deltahedra.* Math. Mag. 51 (1978), 55–57.

Trigg, C. W., Hopkins L. M.: *Problem 929.* Math. Mag. 49 (1976), 97.

Tropfke, J.: *Geschichte der Elementar-Mathematik.* Bd. 7, 2. Ausgabe, de Gruyter, Berlin 1924.

Unger, G.: *Eine stereometrische Dodekaeder-Konstruktion.* EM 17 (1962), 38–40.

Valette, G.: *Les polyèdres inscriptibles à faces régulières.* Acad. Roy. Belg. Bull. Cl. Sci. (5) 55 (1969), 916–928.

Vivarelli, M. D.: *Su una relazione caratteristica dei poliedre semiregolari archimedei e convessi.* 1st. Lombardo Accad. Sci. Lett. Rend. A 108 (1974), 839–848.

Vohla, H.: *Über die Berechnung des regelmäßigen Dodekaeders und des regelmäßigen Ikosaeders ohne Verwendung sphärischer Trigonometrie.* Vortragsskriptum, Jahrestagung Deutsch–Österr. Math. –Ver., Wien 1989.

Wagon, S.: *The Banach-Tarski Paradox.* Cambridge Univ. Press, 1985.
Wallis, W. S.: *Combinatorial Designs.* Marcel Dekker, Inc., New York and Basel, 1988.
Walsh, T. R. S.: *Characterizing the vertex neighbourhoods of semi-regular polyhedra.* GD 1 (1972), 117–123.
Waterhouse, W. C.: *The discovery of the regular solids.* Archive Hist. Exact Sci. 9 (1972), 212–221.
Wenninger, M. J.: *Polyhedron Models.* Cambridge Univ. Press 1971.
Spherical Models. Cambridge Univ. Press 1979.
Dual Models. Cambridge Univ. Press 1983.
Wills, J. M.: *On polyhedra with transitivity properties.* Discrete Comput. Geom. 1 (1986), 195–199.
The combinatorially regular polyhedra of index 2. Aequat. Math. 34 (1987), 206–220.
Wyss, A.: *Die Sonderlinge.* Verlag P. Haupt, Bern 1986.
Yamada, M.: *Some new series of Hadamard matrices.* J. Austr. Math. Soc. 46 (1989), 371–383.
Zalgaller, V. A.: *Convex polyhedra with regular faces (Russ.).* Zap. Naučn. Sem. Leningr. Otdel. Mat. Inst. Steklov. (LOMI) 2 (1967), 220 pp.
Convex polyhedra with regular faces. Consultants Bureau, New York 1969 (transl. from the Russ.).

MODERN DEVELOPMENTS IN REGULAR POLYTOPES

PETER MCMULLEN
Department of Mathematics, University College London,
Gower Street, London WC1E 6BT, England.

1. Historical Introduction

While the aim of these lectures is to give an overview of modern developments in the theory of abstract regular polytopes, it seeems helpful first to place the subject in its historical context. We shall do that briefly here.

1.1. The Classical Period

1.1.1. *Early Occurrences*

Regular planar and solid figures occur from very early times—indeed, it would be curious if at least the square and cube did not. Cubes turn up as dice, as do (regular) dodecahedra in Etruscan remains. However, as yet we are far from any mathematical treatment.

Despite various recent claims to the contrary, it seems clear that the Greeks were the first to conceive of mathematics as we now understand it. (The mere listing of, for example, Pythagorean triples does not constitute mathematics; a proof of the Pythagoras theorem obviously does.) The school of Pythagoras (6th century bce) is reported to have known about the five regular solids, and about the regular pentagram (apparently using it as a symbol of good luck).

The regular solids in ordinary space are named after Plato (Aristocles son of Ariston, 427–347bce); as often happens, this is a misattribution (by Heron); their first mathematical treatment was by Theaetetus (d.369bce). In his *Timaeus*, Plato does discuss the regular solids, but while he constantly professes enthusiasm for and appreciation of mathematics, it is apparent there that his understanding of it fell short of his aspirations. At one point, to illustrate his theme he mentions stuffed leather balls made out of twelve pentagons. More interestingly, he describes the three triangular solids (tetrahedron, octahedron and icosahedron) as built up from equilateral triangles themselves divided into six right-angled triangles (halves of equilateral triangles), thereby anticipating the kaleidoscope of their reflexion planes. Unfortunately, Plato does not spot the general pattern, since the square faces of the cube are only divided into four right-angled triangles (and the dodecahedron is not seen in this way at all).

Plato's pupil Aristotle (384–322bce) deserves only the briefest mention, for his mistaken assertion that the regular tetrahedron tiles ordinary space. Unfortunately, that opinion was not challenged until comparatively recent times, although its falsity could have been established at the time it was made.

T. Bisztriczky et al. (eds.), POLYTOPES: Abstract, Convex and Computational, 97–124.

1.1.2. *Euclid*

Euclid's *Elements* ($\Sigma\tau o\iota\chi\epsilon\iota\alpha$) is undoubtedly the earliest true mathematics book, in the sense that it fully recognizes the paradigm of axiom-definition-theorem-proof. Until early this century, parts of it, notably the first six Books ($\Sigma\chi o\lambda\iota\alpha$), provided, essentially unchanged, a fine introduction to basic geoemtry. It is unclear to what extent Euclid discovered his material or merely compiled it; our ignorance of Euclid himself extends to our being uncertain of more than that he probably lived and worked in Alexandria at the time of Ptolemy I Soter (reigned 323–283bce).

The regular solids themselves are discussed in Book XIII. All five are described, and, using a proof which is much like one that would still be used today (namely, the fact that the angle sum of the polygons at a vertex is less than 2π), it is shown that there are only five.

Further results about the regular solids occur in Books XIV and XV (which were not by Euclid), in particular some anticipation of duality. The details, and modern explanations of the results, are described in [5].

1.2. The Mediaeval Period

1.2.1. *The Early Middle Ages*

The Christian Roman Empire got off to a poor start in its treatment of learning, when the last of the Alexandrine philosophers, the talented and beautiful Hypatia (375–415ce), was carved apart with oyster shells by a Christian mob, instigated by Bishop Cyril (later, of course, canonized). It is still a lively subject of historical debate as to which group of religious fanatics burnt the great library of Alexandria; whether Christians or the Muslims under Omar were responsible (if that is the appropriate word), the loss to scholarship must have been considerable.

The attitude of the Byzantine (Eastern Roman) Empire to mathematics (and the other sciences) was distinctly ambiguous, alternating between encouragement and suppression. Justinian I (reigned 527–565) initially seemed supportive, but soon closed the Academy at Athens in 529 (although there was little loss to mathematics in the act). It is due to a few people in the 9th (particularly Leo "the mathematician") and 10th centuries, that we have the Greek manuscripts of Euclid which survive. However, they preserved rather than added to the corpus of knowledge.

The mathematical torch was also carried on by the Arabs, but they too seem to have added little to geometry. (The contributions of the Islamic world to algebra are a quite different matter.) They had a good empirical knowledge of symmetry groups; in the Alhambra there are planar patterns which exemplify all 17 possible symmetry groups.

1.2.2. *The Later Middle Ages*

From about the 12th century, mathematical knowledge began seeping back into western Europe. Thomas Bradwardine "the Profound Doctor" (c.1290–1349), Archbishop of Canterbury for just forty days after his consecration (he died of plague), systematically investigated star-polygons, obtaining $\{\frac{n}{d}\}$ by stellating $\{\frac{n}{d-1}\}$.

Although Kepler and Poinsot (see below) are credited with discovering the regular star-polyhedra in three dimensions, the polyhedron $\{\frac{5}{2},5\}$ was depicted by Paolo Uccello in 1420, while $\{5,\frac{5}{2}\}$ occurs in an engraving of 1568 by Wentzel Jamnitzer.

1.3. The Modern Period

1.3.1. *Before Schläfli*

Johannes Kepler (1571–1630) began the modern investigation of regular polytopes by his discovery of the two star-polyhedra $\{\frac{5}{2},5\}$ (strictly, a rediscovery) and $\{\frac{5}{2},3\}$. He also investigated various regular star-polygons, particularly the heptagons. However, the lingering effect of mediaeval (or perhaps even classical) thought on him shows up in his attempt to identify the relative sizes of the orbits of the planets to the ratios of in- and circumradii of the regular polyhedra.

In 1809, Louis Poinsot rediscovered the first two regular star-polyhedra, and found their duals $\{5,\frac{5}{2}\}$ (again, really a rediscovery) and $\{3,\frac{5}{2}\}$. Very soon afterwards, Cauchy proved that there were no more such regular star-polyhedra.

1.3.2. *Schläfli*

At a time when very few mathematicians had any concept of working in higher dimensional spaces, Ludwig Schläfli (1814–1895) discovered regular polytopes and honeycombs in four and more dimensions around 1850 (see [26]). In fact, in effect he found all the regular polytopes whose symmetry groups are generated by reflexions in hyperplanes in euclidean spaces, or, rather, he found all their groups. But against all his evidence he refused to recognize the dual pair $\{\frac{5}{2},5\}$ and $\{5,\frac{5}{2}\}$ as "genuine" polyhedra (because they have non-zero genus), and so would not accept either the regular 4-polytopes which have these as facets or vertex- figures, even though calculating the spherical volumes of corresponding tetrahedra on the 3-sphere was a central part of his treatment.

From 1880 onwards, the regular polytopes in higher dimensions were rediscovered many times, beginning with Stringham [32]. We refer to [5, p.144] for the relevant details. Edmund Hess [14] found the remaining regular star-polytopes, and S.L. van Oss [25] proved that the enumeration was complete (for an argument avoiding consideration of each separate case, see [16]).

1.3.3. *Coxeter*

The subject of regular polytopes had gone into somewhat of a decline when it was taken up by H.S.M. (Donald) Coxeter (b.1907). His investigations and consolidation of the theory culminated in his famous book *Regular Polytopes* [5], first published in 1948. His contributions are too numerous to list here, but we should at least mention Coxeter diagrams, and the complete classification of the discrete euclidean reflexion groups among all Coxeter groups.

But Coxeter also pointed towards later developments of the theory. In particular, when J.F. Petrie (1907–1972) (the inventor of the skew polygon which bears his name) found the two regular skew apeirohedra $\{4,6\,|\,4\}$ and $\{6,4\,|\,4\}$, he immediately found the third $\{6,6\,|\,3\}$, and set the whole theory in a general context (see [3]). He also looked at regular maps and their automorphism groups, regarding the star-polyhedra as particular examples; he first observed that the Petrie polygons of a regular map themselves (usually) form another regular map (see [6]).

In 1975, Grünbaum (see [12]) gave the theory a further impetus. He generalized the regular skew polyhedra, by allowing skew polygons as faces as well as vertex-figures. He found eight more individual examples and twelve infinite families (with non-congruent realizations of isomorphic apeirohedra), and Dress [8, 9] completed the classification by finding the final case.

2. Abstract Regular Polytopes

In this section, we give a brief outline of the theory of abstract regular polytopes. Naturally, the definitions adopted are designed to generalize those for the familiar regular polytopes of [5]. This more recent theory received its initial impetus from [12], as explained already in §1.3. The general concept of regular incidence complex was developed in [7, 27, 28]; regular incidence polytopes (as abstract regular polytopes were earlier called) are special cases of these, and were further investigated in papers such as [21, 22] (see also [24, Chapter 2]). Some of the basic notions towards an abstract approach to the theory also occur in [10, 15, 16].

2.1. Abstract Polytopes

2.1.1. *Basic Definitions*

An *abstract n-polytope* is a *poset* (partially ordered set) $\mathcal{P}$, with elements called *faces*, which satisfies the following properties (P1)–(P4).

(P1) $\mathcal{P}$ has a unique minimal face F_{-1} and unique maximal face F_n.

The faces F_{-1} and F_n are the *improper* faces of $\mathcal{P}$; the other faces are *proper.* A *chain* of a poset is a totally ordered subset; a maximal chain is called a *flag.*

(P2) The flags of $\mathcal{P}$ all contain exactly $n+2$ faces (including F_{-1} and F_n).

As a consequence of (P1) and (P2), we have

LEMMA **2.1** *An abstract n-polytope has a strictly monotone rank function* rank$(\cdot)$ *with range* $\{-1, 0, \ldots, n\}$.

We shall call a face F of rank j (that is, with rank$F = j$) a *j-face*; we denote by $\mathcal{P}_j$ the family of all j-faces of $\mathcal{P}$. Further, 0-, 1- and $(n-1)$-faces of an n-polytope are called *vertices, edges* and *facets*, respectively. Unless we need to emphasize the fact, we shall often drop the qualification "abstract" in what follows.

We call two faces F and G of $\mathcal{P}$ *incident* if $F \leq G$ or $F \geq G$. We say that $\mathcal{P}$ is *connected* if, given any two of its proper faces F and G, there is a sequence $F = F_0, F_1, \ldots, F_k = G$ of proper faces of $\mathcal{P}$ (for some k) such that F_{i-1} and F_i are incident for each $i = 1, \ldots, k$. Further, $\mathcal{P}$ is *strongly* connected if, for each pair of faces $F < G$ with rank$G -$ rank$F \geq 3$, the *section*

$$G/F := \{J \in \mathcal{P} \mid F \leq J \leq G\}$$

is connected. Note that we can usually safely identify a face F with the section F/F_{-1}. As a special case, if F is a face, then F_n/F is called the *coface* of $\mathcal{P}$ at F, or the *vertex-figure* if F is a vertex. Then

(P3) $\mathcal{P}$ is strongly connected.

Finally, we have the condition which makes abstract polytopes analogues of convex polytopes.

(P4) If $F < G$ with rank$F + 1 = j =$ rank$G - 1$, then there are exactly two j-faces J such that $F < J < G$.

Two flags Φ and Ψ of $\mathcal{P}$ are called *adjacent* if they differ by exactly one face. Necessarily, this is a j-face for some $j = 0, \ldots, n-1$; we then say that Φ and Ψ are *j-adjacent*, and write $\Psi = \Phi^j$. Note that Φ^j is unique by (P4), and hence $\Psi^j = \Phi$.

We call $\mathcal{P}$ *flag-connected* if, given any two flags Φ and Ψ of $\mathcal{P}$, there is a sequence $\Phi = \Phi_0, \Phi_1, \ldots, \Phi_k = \Psi$ of flags, such that Φ_{i-1} and Φ_i are adjacent for $i = 1, \ldots, k$. Further, $\mathcal{P}$ is *strongly* flag-connected if, in addition, each Φ_i can be chosen so that $\Phi \cap \Psi \subseteq \Phi_i$. Then we have

THEOREM **2.2** *An abstract polytope is strongly flag-connected. Moreover, strong flag-connectedness implies strong connectedness.*

The argument is straightforward, and we shall only sketch it. Let $\Omega := \Phi \cap \Psi$. If $\Omega \neq \{F_{-1}, F_n\}$, say $\Omega = \{F_{-1}, G_1, \ldots, G_r, F_n\}$ with $r \geq 1$ and $G_0 < \cdots < G_r$, we can use induction on n to move the section of Φ in G_{s+1}/G_s (for $s = 0, \ldots, r$; we take $G_0 := F_{-1}$ and $G_{r+1} := F_n$) through adjacent flags to the corresponding section of Ψ. These sectional movements are then fitted together to give the corresponding sequence of adjacent flags from Φ to Ψ. If $\Omega = \emptyset$, connect a proper face F of Φ to a proper face G of Ψ by a sequence $F = F_0, F_1, \ldots, F_k = G$ of proper faces, with F_{j-1} and F_j incident for $j = 1, \ldots, k$. If Φ_j is a flag containing $\{F_{j-1}, F_j\}$, then with $\Phi_0 := \Phi$ and $\Phi_{k+1} := \Psi$, we can apply the above argument in turn to Φ_j and Φ_{j+1} for $j = 0, \ldots, k$. The reverse argument is even easier. □

The *point* and *segment* are the (essentially) unique polytopes of rank 0 and 1. The 2-polytopes are the (finite) polygons, and the (infinite) apeirogon.

2.1.2. *Homomorphisms*

A *homomorphism* from one poset $\mathcal{P}$ to another $\mathcal{Q}$ is a mapping $\varphi : \mathcal{P} \to \mathcal{Q}$ which preserves the partial order; that is, if $F, F' \in \mathcal{P}$, then $F \leq F'$ implies $F\varphi \leq F'\varphi$. As usual, an *isomorphism* is a homomorphism for which there exists an inverse homomorphism. We write $\mathcal{P} \cong \mathcal{Q}$ if $\mathcal{P}$ and $\mathcal{Q}$ are isomorphic.

In general, the homomorphic image of a polytope will not be a polytope. By a *pre-polytope* (of rank n) we shall mean a poset which satisfies properties (P1),(P2) and (P4) of an n-polytope. That is, we drop the condition (P3) of strong connectedness; in fact, we need not even assume that a pre-polytope is connected. The remaining language of polytopes carries over naturally.

Let $\mathcal{P}$ and $\mathcal{Q}$ be pre-polytopes of the same rank. A homomorphism $\varphi : \mathcal{P} \to \mathcal{Q}$ is called a *rap-map* if it is rank- and adjacency-preserving, the latter meaning that φ takes adjacent flags of $\mathcal{P}$ onto (distinct) adjacent flags of $\mathcal{Q}$. Thus, if $\mathcal{Q}$ is flag-connected, then a rap-map $\varphi : \mathcal{P} \to \mathcal{Q}$ is surjective. Such a surjective rap-map is called a *covering*; we then also say that $\mathcal{P}$ *covers* $\mathcal{Q}$. Then we have

THEOREM **2.3** *Let $\mathcal{P}$ be a polytope, $\mathcal{Q}$ a pre-polytope, and $\varphi : \mathcal{P} \to \mathcal{Q}$ a rap-map. Then $\mathcal{Q}$ is a polytope (and φ a covering) if and only if the image of each section of $\mathcal{P}$ is a section of $\mathcal{Q}$.*

For the necessity of the condition, let $G < G'$ be faces of $\mathcal{Q}$. Write G_{-1} and G_n for the minimal and maximal faces of $\mathcal{Q}$. Then $G_j = F_j\varphi$ for $j = -1, n$. Since $G' \in \mathcal{Q} = G_n/G_{-1}$, we have $G' = F'\varphi$ for some $F' \in \mathcal{P}$. Since $G \in G'/G_{-1}$, we have

$G = F\varphi$ for some $F \in F'/F_{-1}$. Finally, $G'/G = (F'/F)\varphi$. Since F'/F is connected, so is G'/G. Because G and G' are arbitrary, this shows that $\mathcal{Q}$ is strongly connected, and so a polytope.

Conversely, suppose that the condition does not hold. Then there is some minimal section F'/F of $\mathcal{P}$, such that $(F'/F)\varphi$ is not a section of $\mathcal{Q}$. Writing $G := F\varphi$, and so on, we then have $(F'/F)\varphi \subset G'/G$, so that there is some face G'' of $\mathcal{Q}$, such that $G < G'' < G'$, but $G'' \notin (F'/F)\varphi$. Then no flag of G'/G which contains G'' can meet $(F'/F)\varphi$ except at G or G', otherwise F'/F would not be minimal, so that G'/G is not flag-connected, and hence $\mathcal{Q}$ is not a polytope. □

2.1.3. *Duality*

Let $\mathcal{P}$ be an abstract polytope. If we define a new partial ordering on the same underlying face-set $\mathcal{P}$ by

$$F \leq^* G \Longleftrightarrow F \geq G,$$

we obtain a new poset, which we denote by $\mathcal{P}^*$, and call the *dual* of $\mathcal{P}$. Then we have the obvious

THEOREM **2.4** *The dual $\mathcal{P}^*$ of a n-polytope $\mathcal{P}$ is an n-polytope, and $\mathcal{P}^{**}(:= (\mathcal{P}^*)^*)$ is isomorphic to $\mathcal{P}$. To each j-face of $\mathcal{P}$ (with $j = -1, 0, \ldots, n$) corresponds a unique $(n-j-1)$-face of $\mathcal{P}^*$.*

If $\mathcal{P}^*$ is isomorphic to $\mathcal{P}$, then we call $\mathcal{P}$ *self-dual*. A *polarity* is an involutory duality of a self-dual polytope; a polytope may be self-dual without possessing a polarity. We shall return to these concepts in §2.2.1.

2.2. AUTOMORPHISM GROUPS

2.2.1. *Regularity*

The *automorphism group* $\Gamma(\mathcal{P})$ of an abstract n-polytope $\mathcal{P}$ consists of the order-preserving permutations of $\mathcal{P}$. We begin with a useful observation.

THEOREM **2.5** *Let $\mathcal{P}$ be a polytope, and Φ a flag of $\mathcal{P}$. Then an element of $\Gamma(\mathcal{P})$ is completely determined by its effect on Φ.*

The reason is simple. If $\gamma \in \Gamma := \Gamma(\mathcal{P})$, then for each flag Ψ and each j we have $\Psi^j\gamma = (\Psi\gamma)^j$, since both sides of the equation are the unique flag of $\mathcal{P}$ which is j-adjacent to $\Psi\gamma$. The result now follows from the flag-connectedness of $\mathcal{P}$. □

Note that we have not actually used the strong flag-connectedness of $\mathcal{P}$. If we replace condition (P3) above by the weaker condition of flag-connectedness (compare Theorem 2.2), then we obtain the *polystromas* of Grünbaum [13].

Since $\Gamma(\mathcal{P})$ thus splits the flags of $\mathcal{P}$ into orbits of the same size, as a direct consequence we have

COROLLARY **2.6** *The order of $\Gamma(\mathcal{P})$ is a divisor of the number of flags of $\mathcal{P}$.*

We say that the polytope $\mathcal{P}$ is *regular* if $\Gamma(\mathcal{P})$ is transitive on the flags of $\mathcal{P}$. From Theorem 2.5 and Corollary 2.2, we see that, in this case, $\Gamma(\mathcal{P})$ is simply transitive on the flags of $\mathcal{P}$, and its order is the number of these flags. Let $\Phi := \{F_{-1}, F_0, \ldots, F_n\}$ be a fixed or *base* flag of $\mathcal{P}$; often in what follows we do not mention F_{-1} and F_n, because they belong to all flags.

THEOREM **2.7** *Let $\mathcal{P}$ be a regular n-polytope, and for each $j = 0, \ldots, n-1$, let $\rho_j \in \Gamma(\mathcal{P})$ be the automorphism such that $\Phi\rho_j = \Phi^j$. If $\Omega \subseteq \Phi$, and $\mathcal{F}_\Omega(\mathcal{P})$ is the set of flags containing Ω, then*

$$\Gamma_\Omega(\mathcal{P}) := \langle \rho_j \mid F_j \notin \Omega \rangle$$

is simply transitive on $\mathcal{F}_\Omega(\mathcal{P})$.

As in Theorem 2.5, we see that, for each $j = 0, \ldots, n-1$ and each $\gamma \in \Gamma(\mathcal{P})$, we have $(\Phi\gamma)^j = \Phi^j\gamma = \Phi(\rho_j\gamma)$. So, let $\Psi \in \mathcal{F}_\Omega(\mathcal{P})$; there is then a unique $\gamma \in \Gamma(\mathcal{P})$ such that $\Psi = \Phi\gamma$. By the strong flag-connectedness of $\mathcal{P}$, we can find a sequence $\Phi = \Phi_0, \Phi_1, \ldots, \Phi_k = \Psi$ of flags in $\mathcal{F}_\Omega(\mathcal{P})$, such that Φ_{i-1} and Φ_i are adjacent for $i = 1, \ldots, k$, say $\Phi_i = \Phi_{i-1}^{j(i)}$ with $F_{j(i)} \notin \Omega$ for each i. That is,

$$\Psi = \Phi^{j(1)\ldots j(k)},$$

or

$$\Psi = \Phi(\rho_{j(k)} \cdots \rho_{j(1)}),$$

using the above remark. Then $\gamma = \rho_{j(k)} \cdots \rho_{j(1)} \in \Gamma_\Omega(\mathcal{P})$ by Theorem 2.5, which yields the result. □

The ρ_j are called the *distinguished* generators of $\Gamma(\mathcal{P})$, relative to the base flag Φ. We write $p_{ij} = p_{ji}$ for the period of $\rho_i\rho_j$ for each $i, j = 0, \ldots, n-1$, that is, the smallest p such that

$$(\rho_i\rho_j)^{p_{ij}} = \varepsilon,$$

where ε denotes the identity. We allow $p_{ij} = \infty$ when the product has infinite period. Then we have

LEMMA **2.8** *For the distinguished generators of the regular n-polytope $\mathcal{P}$, there hold $p_{jj} = 1$ for $j = 0, \ldots, n-1$, and $p_{ij} = 2$ if $0 < i+1 < j < n$.*

The results are equivalent (in the two cases) to $\Phi^{jj} = \Phi$ and $\Phi^{ij} = \Phi^{ji}$, where we write $\Phi^{ij} := (\Phi^i)^j$. □

If $\mathcal{P}$ is a regular n-polytope, then for $j = 1, \ldots, n-1$ we define $p_j := p_{j-1,j}$; the p_j's then determine the *Schläfli type* $\{p_1, \ldots, p_{n-1}\}$ of $\mathcal{P}$. The symbol $\{p_1, \ldots, p_{n-1}\}$ (with no qualification) refers to the *universal* regular polytope whose group is solely determined by the relations of Lemma 2.8 and the numbers p_j.

We say that a group $\Gamma := \langle \rho_0, \ldots, \rho_{n-1} \rangle$ has the *intersection property* with respect to its distinguished generators $\rho_0, \ldots, \rho_{n-1}$ if

$$\langle \rho_i \mid i \in I \rangle \cap \langle \rho_i \mid i \in J \rangle = \langle \rho_i \mid i \in I \cap J \rangle,$$

for all $I, J \subseteq \{0, \ldots, n-1\}$. (We are thinking about the automorphism groups of regular polytopes here, but the definition does not demand that the ρ_j be involutions.) A group with involutory generators satisfying the intersection property is called a *C-group*; here, "C" stands for "Coxeter". If, in addition, $p_{ij} = 2$ for $i \leq j-2$ (as in Lemma 2.8 above), we call Γ a *string* C-group. More generally, a (*string*) *C-group* is such a group relative to some distinguished generators. Since we usually discuss no other kind of C-group here, for brevity we shall sometimes drop the qualifier "string".

By the results above, the automorphism group of a regular polytope is a string C-group. A central fact in the theory is the following (see [21, 24, 27]).

THEOREM 2.9 *A string C-group* Γ *is the automorphism group of a regular polytope* $\mathcal{P} := \mathcal{P}(\Gamma)$, *so that* $\Gamma \cong \Gamma(\mathcal{P})$.

If Γ is a string C-group, say $\Gamma = \langle \rho_0, \dots, \rho_{n-1} \rangle$ with the ρ_j as distinguished generators, then for each $j = -1, 0, \dots, n$ we define

$$\Gamma_j := \langle \rho_i \mid i \neq j \rangle.$$

Thus $\Gamma_{-1} = \Gamma_n = \Gamma$. For each j, the j-faces of $\mathcal{P}$ are the right cosets $\Gamma_j \sigma$ of Γ_j (with $\sigma \in \Gamma$), and for $j \leq k$ and $\sigma, \tau \in \Gamma$, we set $\Gamma_j \sigma \leq \Gamma_k \tau$ if $\Gamma_j \sigma \cap \Gamma_k \tau \neq \emptyset$.

Obviously, the unique (-1)- and n-faces are incident with all others, so we henceforth confine our attention to proper faces. For $j \leq k$, the condition $\Gamma_j \sigma_j \leq \Gamma_k \sigma_k$ is equivalent to

$$\begin{aligned} \sigma_j \sigma_k^{-1} &\in \Gamma_j \Gamma_k \\ &= \langle \rho_0, \dots, \rho_{j-1} \rangle \langle \rho_{j+1}, \dots, \rho_{n-1} \rangle \cdot \langle \rho_0, \dots, \rho_{k-1} \rangle \langle \rho_{k+1}, \dots, \rho_{n-1} \rangle \\ &= \langle \rho_{j+1}, \dots, \rho_{n-1} \rangle \langle \rho_0, \dots, \rho_{k-1} \rangle, \end{aligned}$$

as may easily be seen (recall that non-adjacent ρ_i commute). The transitivity of $\leq$ is an immediate consequence: if $i \leq j \leq k$ and $\Gamma_i \sigma_i \leq \Gamma_j \sigma_j \leq \Gamma_k \sigma_k$, then

$$\begin{aligned} \sigma_i \sigma_k^{-1} &= \sigma_i \sigma_j^{-1} \cdot \sigma_j \sigma_k^{-1} \\ &\in \langle \rho_{i+1}, \dots, \rho_{n-1} \rangle \langle \rho_0, \dots, \rho_{j-1} \rangle \cdot \langle \rho_{j+1}, \dots, \rho_{n-1} \rangle \langle \rho_0, \dots, \rho_{k-1} \rangle \\ &= \langle \rho_{i+1}, \dots, \rho_{n-1} \rangle \langle \rho_0, \dots, \rho_{k-1} \rangle, \end{aligned}$$

as required. The remaining properties of an abstract regular polytope are easily established. □

Let $\mathcal{P}, \mathcal{Q}$ be regular polytopes, with groups $\Gamma(\mathcal{P}) = \langle \rho_0, \dots, \rho_{n-1} \rangle$ and $\Gamma(\mathcal{Q}) = \langle \sigma_0, \dots, \sigma_{m-1} \rangle$. If the mapping $\rho_j \mapsto \sigma_j$ for $j = 0, \dots, n-1$, with $\sigma_j := \varepsilon$ for $m \leq j < n$, induces a homomorphism of $\Gamma(\mathcal{P})$ onto $\Gamma(\mathcal{Q})$, we say that $\mathcal{Q}$ is a *quotient* of $\mathcal{P}$. Strictly speaking, we should talk about *ordinary* quotients, but here no more general kind will be needed. We may observe that such a quotient is alternatively defined by means a of suitable normal subgroup N of $\Gamma(\mathcal{P})$. General conditions on such a subgroup N which guarantees that the quotient, which is then written $\mathcal{P}/N$, be a polytope (that is, that $\Gamma(\mathcal{P})/N$ be a C-group) are discussed in [24, Chapter 2]; see also Theorem 2.12 below. If $m = n$, then a quotient is a cover in the earlier sense; when $m < n$, we also say that $\mathcal{P}$ *collapses* onto $\mathcal{Q}$.

We end the section with some brief remarks on duality. Let $\mathcal{P}$ be a regular polytope, with $\Gamma(\mathcal{P}) = \langle \rho_0, \dots, \rho_{n-1} \rangle$. Then its dual $\mathcal{P}^*$ is also regular, with $\Gamma(\mathcal{P}^*) = \langle \rho_{n-1}, \dots, \rho_0 \rangle$; that is, the groups are the same, but the distinguished generators are listed in the reverse order. If, in addition, $\mathcal{P}$ is self-dual, then it has a polarity. Indeed, this is given by the outer automorphism ω of $\Gamma(\mathcal{P})$, induced by $\rho_j \omega = \rho_{n-j-1}$ for $j = 0, \dots, n-1$.

2.2.2. *Polytopality*

It often occurs that we construct an abstract or geometric object that looks very much like a regular polytope, for example, by identifying faces of a known regular polytope geometrically, or by means of a normal subgroup N of its automorphism group $\langle \sigma_0, \ldots, \sigma_{n-1} \rangle$, say. We then ask whether the new object is actually a polytope, that is, we wish to verify its *polytopality*. Of course, this is just another way of asking whether the quotient group by N satisfies the intersection property with respect to the natural distinguished generators $\rho_j := N\sigma_j$ for $j = 0, \ldots, n-1$ (we suppose that the other properties of a C-group have already been verified), but it is often more natural to pose the problem in such a geometric way. We now discuss this question.

We call a group $\Gamma = \langle \rho_0, \ldots, \rho_{n-1} \rangle$ a *pre-C-group* (as usual, with respect to the ρ_j as distinguished generators) if the ρ_j are involutions satisfying $(\rho_j \rho_k)^2 = \varepsilon$ whenever $j < k - 1$. That is, we drop the intersection property from the definition of a string C-group. We then have

THEOREM **2.10** *Let* $\Gamma = \langle \rho_0, \ldots, \rho_{n-1} \rangle$ *be a pre-C-group. Then* Γ *is a C-group if and only if each of* $\Gamma_0 := \langle \rho_1, \ldots, \rho_{n-1} \rangle$ *and* $\Gamma_{n-1} := \langle \rho_0, \ldots, \rho_{n-2} \rangle$ *is a C-group, and*

$$\Gamma_0 \cap \Gamma_{n-1} = \langle \rho_1, \ldots, \rho_{n-2} \rangle.$$

The necessity of the conditions is clear, so we just prove their sufficiency. We first note that for $k = 0, \ldots, n-2$,

$$\begin{aligned} \Gamma_0 \cap \langle \rho_0, \ldots, \rho_k \rangle &= (\Gamma_0 \cap \Gamma_{n-1}) \cap \langle \rho_0, \ldots, \rho_k \rangle \\ &= \langle \rho_1, \ldots, \rho_{n-2} \rangle \cap \langle \rho_0, \ldots, \rho_k \rangle \\ &= \langle \rho_1, \ldots, \rho_k \rangle, \end{aligned}$$

using the intersection property for Γ_{n-1}.

We now show the result under the assumption that Γ_0 alone is a C-group, and that the property just shown holds (this alternative will be used below). For each j, define $\Gamma_j := \langle \rho_i \mid i \neq j \rangle$ as usual, let $\Delta := \Gamma_0$ and $\Delta_j := \langle \rho_i \mid i \neq 0, j \rangle$. Then for $j = 1, \ldots, n-1$,

$$\begin{aligned} \Delta \cap \Gamma_j &= \Delta \cap \langle \rho_0, \ldots, \rho_{j-1} \rangle \langle \rho_{j+1}, \ldots, \rho_{n-1} \rangle \\ &= (\Delta \cap \langle \rho_0, \ldots, \rho_{j-1} \rangle) \langle \rho_{j+1}, \ldots, \rho_{n-1} \rangle \\ &= \langle \rho_1, \ldots, \rho_{j-1} \rangle \langle \rho_{j+1}, \ldots, \rho_{n-1} \rangle = \Delta_j. \end{aligned}$$

(Note that $\langle \rho_{j+1}, \ldots, \rho_{n-1} \rangle \subseteq \Delta$.)

To verify the intersection property, it is enough to prove

$$\langle \rho_j \mid j \notin J \rangle = \bigcap_{j \in J} \Gamma_j$$

for each $J \subseteq \{0, \ldots, n-1\}$. Write $k := \min J$. If $k = 0$, then

$$\begin{aligned} \bigcap_{j \in J} \Gamma_j &= \bigcap_{j \in J \setminus \{0\}} (\Delta \cap \Gamma_j) \\ &= \bigcap_{j \in J \setminus \{0\}} \Delta_j \\ &= \langle \rho_j \mid j \notin J \rangle, \end{aligned}$$

since $\Delta = \Gamma_0$ is a C-group. If $k > 0$, then

$$\begin{aligned}\bigcap_{j\in J} \Gamma_j &= \langle \rho_0, \ldots, \rho_{k-1}\rangle\langle \rho_{k+1}, \ldots, \rho_{n-1}\rangle \cap \bigcap_{j\in J\setminus\{k\}} \Gamma_j \\ &= \langle \rho_0, \ldots, \rho_{k-1}\rangle(\Gamma_0 \cap \langle \rho_{k+1}, \ldots, \rho_{n-1}\rangle \cap \bigcap_{j\in J\setminus\{k\}} \Gamma_j) \\ &= \langle \rho_0, \ldots, \rho_{k-1}\rangle(\Gamma_0 \cap \bigcap_{j\in J} \Gamma_j) \\ &= \langle \rho_0, \ldots, \rho_{k-1}\rangle\langle \rho_j \mid j \notin J \cup \{0\}\rangle \\ &= \langle \rho_j \mid j \notin J\rangle,\end{aligned}$$

as claimed. This completes the proof. □

Another useful criterion for polytopality arises from quotients; indeed, it is known as the *quotient lemma.*

THEOREM **2.11** *Let* $\Gamma = \langle \rho_0, \ldots, \rho_{n-1}\rangle$ *be a pre-C-group, let* $\Delta = \langle \sigma_0, \ldots, \sigma_{n-1}\rangle$ *be a string C-group, and let* $\varphi : \Gamma \to \Delta$ *be a homomorphism, such that* $\rho_j\varphi = \sigma_j$ *for* $j = 0, \ldots, n-1$. *If* φ *is one-to-one on either* Γ_0 *or* Γ_{n-1}, *then* Γ *is also a C-group.*

The notation of the theorem is that employed above. Assume that φ is one-to-one on Γ_0, say. By Theorem 2.10 (or rather its proof), it is enough to check that

$$\Gamma_0 \cap \langle \rho_0, \ldots, \rho_j\rangle = \langle \rho_1, \ldots, \rho_j\rangle$$

for each $j = 0, \ldots, n-2$. Let $\gamma \in \Gamma_0 \cap \langle \rho_0, \ldots, \rho_j\rangle$. Then

$$\gamma\varphi \in \langle \sigma_1, \ldots, \sigma_{n-1}\rangle \cap \langle \sigma_0, \ldots, \sigma_j\rangle = \langle \sigma_1, \ldots, \sigma_j\rangle,$$

because Δ is a C-group. Thus $\gamma\varphi$ has a preimage in $\langle \rho_1, \ldots, \rho_j\rangle$. But φ is one-to-one on Γ_0, so that γ is the only preimage of $\gamma\varphi$ in Γ_0; thus $\gamma \in \langle \rho_1, \ldots, \rho_j\rangle$, as required. This proves one inclusion, and the other is trivial. Thus Γ is a string C-group. □

Of course, under the conditions of Theorem 2.11, φ will induce a covering from $\mathcal{P}(\Gamma)$ to $\mathcal{P}(\Delta)$.

Finally, we have a result in the other direction.

THEOREM **2.12** *Let* $\Gamma = \langle \rho_0, \ldots, \rho_{n-1}\rangle$ *be a string C-group, and let* φ *be a homomorphism on* Γ *such that* $\ker\varphi \cap \Gamma_0\Gamma_{n-1} = \{\varepsilon\}$. *Then* $\Delta := \Gamma\varphi$ *is a string C-group with distinguished generators* $\sigma_j := \rho_j\varphi$ *for* $j = 0, \ldots, n-1$.

The notation is again that used previously. First note that $\ker\varphi \cap \Gamma_0 = \{\varepsilon\} = \ker\varphi \cap \Gamma_{n-1}$, so that φ is one-to-one on both Γ_0 and Γ_{n-1}. Hence both Δ_0 and Δ_{n-1} are string C-groups (with the obvious notation). Using Theorem 2.10, it suffices to check that $\Delta_0 \cap \Delta_{n-1} = \langle \sigma_1, \ldots, \sigma_{n-2}\rangle$. Let $\beta \in \Gamma_0$ and $\gamma \in \Gamma_{n-1}$ satisfy $\beta\varphi = \gamma\varphi$. Then $\beta\gamma^{-1} \in \ker\varphi \cap \Gamma_0\Gamma_{n-1} = \{\varepsilon\}$, so that $\beta = \gamma \in \Gamma_0 \cap \Gamma_{n-1} = \langle \rho_1, \ldots, \rho_{n-2}\rangle$, and the proof follows at once. □

2.2.3. *Chirality*

There is an important class of nearly regular polytopes. If $\mathcal{P}$ is an n-polytope such that $\Gamma(\mathcal{P})$ is transitive on exactly half the flags of $\mathcal{P}$, with adjacent flags lying in different orbits under $\Gamma(\mathcal{P})$, then we say that $\mathcal{P}$ is *chiral.*

We shall not prove any results in this section, but instead list a few of the major ones, referring the reader to [30] for further details. As before, we take $\Phi = \{F_{-1}, F_0, \ldots, F_n\}$ to be a fixed flag of $\mathcal{P}$.

THEOREM **2.13** *The automorphism group $\Gamma(\mathcal{P})$ of a chiral n-polytope $\mathcal{P}$ has $n-1$ generators σ_j ($j = 1, \ldots, n-1$), where σ_j cyclically permutes the j-faces of the section F_{j+1}/F_{j-2}. The σ_j satisfy (among others) the relations*
a) $(\sigma_j\sigma_{j+1}\cdots\sigma_k)^2 = \varepsilon$ *if* $j < k$,
b) $\sigma_j \rightleftharpoons \sigma_k$ *if* $j \leq k-3$,
where $\rightleftharpoons$ *means "commutes with".*

It is helpful here to think of $\Gamma(\mathcal{P})$ as, in a sense, corresponding to the "rotation group" of a regular polytope, with $\sigma_j = \rho_{j-1}\rho_j$ in our usual notation. If we write p_j for the period of σ_j (allowing $p_j = \infty$), then $\{p_1, \ldots, p_{n-1}\}$ is again the *Schläfli type* of $\mathcal{P}$.

The distinguished generators of $\Gamma(\mathcal{P})$ corresponding to the adjacent flag Φ^0 of $\mathcal{P}$ are

$$\sigma_1^{-1}, \sigma_1^2\sigma_2, \sigma_3, \ldots, \sigma_{n-1}.$$

Thus, if we know that $\mathcal{P}$ is at least chiral, then it is actually regular if and only if we can adjoin an involution ρ_0 such that

$$\begin{aligned}
\rho_0\sigma_1\rho_0 &= \sigma_1^{-1}, \\
\rho_0\sigma_2\rho_0 &= \sigma_1^2\sigma_2, \\
\rho_0\sigma_j\rho_0 &= \sigma_j, \text{ if } j \geq 3.
\end{aligned}$$

There is a criterion corresponding to the intersection property for the groups of regular polytopes, but it is rather more complicated to state, and so we shall omit it. However, one interesting property which should not go unmentioned is

THEOREM **2.14** *Let $\mathcal{P}$ be a chiral n-polytope. Then its $(n-2)$-faces, and its cofaces to edges, are regular.*

Examples of chiral 3-polytopes are the toroids $\{4,4\}_{(s,t)}$, $\{3,6\}_{(s,t)}$ and $\{6,3\}_{(s,t)}$ with $st(s-t) \neq 0$. Examples of higher rank can be found in [2, 4, 31].

2.3. CONSTRUCTIONS

2.3.1. *General Remarks*

The central problem in the theory of regular polytopes is obviously that of classifying all the regular polytopes of a given rank. This is hopelessly difficult, so we might ask instead for all the finite regular polytopes; however, if anything this is more difficult to solve, since determination of finiteness is often very complicated.

It is more fruitful to ask questions which are less all-embracing. In the following sections, we shall address such specific problems.

2.3.2. *The Amalgamation Problem*

One possible way of constructing new regular polytopes from ones that are already known is the following. Suppose that we are given two regular n-polytopes $\mathcal{P}$ and $\mathcal{Q}$, such that the vertex-figure of $\mathcal{P}$ is isomorphic to the facet of $\mathcal{Q}$. Obviously this condition is necessary for there to exist a regular $(n+1)$-polytope, whose facets are isomorphic to $\mathcal{P}$ and whose vertex-figures are isomorphic to $\mathcal{Q}$; the class of all such polytopes will be denoted $\langle \mathcal{P}, \mathcal{Q} \rangle$. However, it is by no means sufficient; we cannot carry out such a construction with $\mathcal{P} = \{4,4\}_{(2,0)}$ and $\mathcal{Q} = \{4,4,\}_{(3,0)}$, for example. (Such examples will be discussed by Schulte [29].)

A useful positive result is

THEOREM **2.15** *Let $\mathcal{P}$ and $\mathcal{Q}$ be regular n-polytopes, such that the vertex-figure of $\mathcal{P}$ is isomorphic to the facet of $\mathcal{Q}$. If the class $\langle \mathcal{P}, \mathcal{Q} \rangle$ is non-empty, then there is a universal member $\{\mathcal{P}, \mathcal{Q}\}$ of the class, which covers all other members.*

The proof is straightforward. We are given that there exists a string C-group $\Delta = \langle \sigma_0, \sigma_1, \ldots, \sigma_n \rangle$, say, such that (in the usual notation) $\Delta_n \cong \Gamma(\mathcal{P})$ and $\Delta_0 \cong \Gamma(\mathcal{Q})$. Let $\Gamma = \langle \rho_0, \rho_1, \ldots, \rho_n \rangle$ be the abstract group, with $\Gamma_n \cong \Gamma(\mathcal{P})$ and $\Gamma_0 \cong \Gamma(\mathcal{Q})$, but satisfying no other relations apart from $(\rho_0 \rho_n)^2 = \varepsilon$. In other words, we impose on the generators of Γ exactly those relations which hold between $\sigma_0, \ldots, \sigma_{n-1}$ or between $\sigma_1, \ldots, \sigma_n$, and which force it to be a pre-C-group, but not any which involve all of $\rho_0, \ldots, \rho_n$. (At this point, it is worth remarking that the relations imposed on the distinguished generators of the group of a regular polytope which are required to present it can always be reduced to ones involving each generator of consecutive blocks of them.)

We now appeal to the quotient lemma (Theorem 2.11). The natural homomorphism given by $\rho_j \mapsto \sigma_j$ for $j = 0, \ldots, n$ is one-to-one on both Γ_0 and Γ_n. Since Δ is a C-group, we conclude that Γ is a C-group also.

Finally, the universal property of Γ is clear from the construction. □

This result enables us to ask much more finely focussed questions. First, given such polytopes $\mathcal{P}$ and $\mathcal{Q}$, when is $\langle \mathcal{P}, \mathcal{Q} \rangle$ non-empty? Sceond, if it is, and $\mathcal{P}$ and $\mathcal{Q}$ are both finite, when is the universal $\{\mathcal{P}, \mathcal{Q}\}$ also finite? Such questions (for special classes of polytopes $\mathcal{P}$ and $\mathcal{Q}$) lie at the heart of the discussion of [29].

2.3.3. *Mixing and Twisting*

We shall only outline material in this section, since we do not wish to trespass into the province of [29].

There are, broadly speaking, two main tools for constructing regular polytopes, or, equivalently, string C-groups. These are mixing and twisting. In both cases, our starting point is usually a C-group, whose properties are well understood.

For *mixing*, we construct a new group whose generators are certain products of generators of the old one. As a single illustration, if $\Gamma = \langle \rho_0, \ldots, \rho_3 \rangle$ is the group of the regular honeycomb $\{4,3,4\}$ of $\mathbb{E}^3$ by cubes, then the mixing operation

$$(\rho_0, \ldots, \rho_3) \mapsto (\rho_0, \rho_1\rho_3, \rho_2) =: (\sigma_0, \sigma_1, \sigma_2)$$

gives the group $\Delta = \langle \sigma_0, \sigma_1, \sigma_2 \rangle$ of Petrie's regular skew apeirohedron $\{4, 6 \mid 4\}$.

For *twisting*, we adjoin to the group certain of its outer automorphisms. Again, we give an example. Let $\Gamma = \langle \rho_0, \rho_1, \rho_2 \rangle$ be the group of the regular tetrahedron $\{3, 3\}$. There is an involutory automorphism σ_2 (say) of Γ, which interchanges ρ_j and ρ_{2-j} for each j. We adjoin σ_2 to Γ, and express this as

$$\sigma_2 \rho_j \sigma_2 = \rho_{2-j},$$

for $j = 0, 1, 2$. Defining $\sigma_0 := \rho_1$ and $\sigma_1 := \rho_0$, the new group $\Delta = \langle \sigma_0, \sigma_1, \sigma_2 \rangle$ is the automorphism group of the regular octahedron $\{3, 4\}$.

A similar involutory automorphism can be adjoined to the group of the (self-dual) honeycomb $\{4, 3, 3, 4\}$ to yield the group of $\{3, 4, 3, 3\}$.

2.3.4. *Polyhedra*

For regular *polyhedra* (3-polytopes), certain of the constructions of §2.3.3 are of considerable efficacy. We list some of them here. Throughout, $\mathcal{P}$ will be a polyhedron of Schläfli type $\{p, q\}$, with group $\Gamma(\mathcal{P}) = \langle \rho_0, \rho_1, \rho_2 \rangle$, and we shall obtain a new polyhedron $\mathcal{Q}$ with group $\Gamma(\mathcal{Q}) = \langle \sigma_0, \sigma_1, \sigma_2 \rangle$.

First, let $2 \leq k \leq \lfloor \frac{1}{2} q \rfloor$. Then the *facetting operation* φ_k is given by the mixing operation

$$(\rho_0, \rho_1, \rho_2) \mapsto (\rho_0, (\rho_1 \rho_2)^{k-1} \rho_1, \rho_2) =: (\sigma_0, \sigma_1, \sigma_2).$$

If $(q, k) = 1$ (greatest common divisor), then $\mathcal{Q} := \mathcal{P}^{\varphi_k}$ (if a polyhedron) will have the same vertices and edges as $\mathcal{P}$. Its faces are formed by the edge-paths which leave a vertex after k steps around the vertex-figure (rather than at the first exit), in a fixed (local) sense. As an example, the regular star-polyhedron $\{5, \frac{5}{2}\}$ is $\{3, 5\}^{\varphi_2}$. If $(q, k) > 1$, then the vertices of $\mathcal{Q}$ may form a proper subset of those of $\mathcal{P}$.

Next, we have the *Petrie operation* π, which is given by

$$(\rho_0, \rho_1, \rho_2) \mapsto (\rho_0 \rho_2, \rho_1, \rho_2) =: (\sigma_0, \sigma_1, \sigma_2).$$

Again, the vertices and edges of $\mathcal{Q} := \mathcal{P}^{\pi}$, which is called the *Petrie dual* or *Petrial* of $\mathcal{P}$, are those of $\mathcal{P}$. This time, the faces are formed by the *Petrie polygons* of $\mathcal{P}$, of which two successive edges, but no three, belong to a face of $\mathcal{P}$.

In this context, it is useful to denote the duality operation by $\mathcal{P} \mapsto \mathcal{P}^{\delta}$, rather than by $\mathcal{P} \mapsto \mathcal{P}^*$. Then we have the following. First, π commutes with each φ_k. Second, $\pi\delta$ has period 3.

Finally, if $\mathcal{P}$ is self-dual, of type $\{p, p\}$ say, then we can apply an outer automorphism to $\Gamma(\mathcal{P})$, as in §2.3.3, to obtain a polyhedron of type $\{p, 4\}$. In certain cases, the reverse operation is available. It is usually applied when $\mathcal{P}$ is of type $\{4, q\}$, by means of the operation

$$(\rho_0, \rho_1, \rho_2) \mapsto (\rho_1, \rho_2, \rho_0 \rho_1 \rho_0) =: (\sigma_0, \sigma_1, \sigma_2)$$

which is called *halving*, to yield the polyhedron $\mathcal{Q} := \mathcal{P}^{\eta}$.

In [21], these operations are employed to derive all the Grünbaum-Dress apeirohedra of [12, 8, 9], which do not belong to an infinite family, from $\{4, 6 \,|\, 4\}$.

3. Realizations

A good part of the appeal of regular polytopes to mathematicians as well as others lies in their symmetry properties as figures in euclidean spaces. This concrete aspect of the subject is connected to the abstract theory by the notion of a *realization* of a regular polytope. We shall outline the theory of such realizations in this section.

3.1. The General Theory

3.1.1. *Realizations*

A *realization* of an (abstract) regular polytope $\mathcal{P}$ consists of a set V of points in some euclidean space E, together with a surjection β from the vertex-set $\mathcal{P}_0$ of $\mathcal{P}$ onto V, such that each automorphism of $\mathcal{P}$ induces a isometric permutation of V. We call V the *vertex-set* of the realization β. In this context, observe that an isometry of V always extends to an isometry of E. We emphasize that our realizations are always finite dimensional.

The remaining structure of the realization is determined by β as follows. Writing 2^X for the family of subsets of the set X, we have

THEOREM **3.1** *Let β be a realization of a regular n-polytope $\mathcal{P}$. With $\beta_0 := \beta$, $V_0 := V$, then for $j = 1, \ldots, n$, β recursively induces a surjection $\beta_j : \mathcal{P}_j \to V_j$, with $V_j \subseteq 2^{V_{j-1}}$, given by*

$$F\beta_j := \{G\beta_{j-1} \mid G \in \mathcal{P}_{j-1} \text{ and } G \leq F\}$$

for each $F \in \mathcal{P}_j$; also, β_{-1} is given by $F_{-1}\beta_{-1} := \emptyset$.

Even though each β_j is determined by $\beta = \beta_0$, it is often helpful to think of the realization as given by all the β_j. We call β *faithful* if each β_j is a bijection; otherwise, β is *degenerate*. If β_0 is bijective, then we say that β is *vertex-faithful*. If β is degenerate, it is often convenient to regard the sets V_j as multiply counted (ordered) sets. An extreme example of a degenerate realization is the *trivial* one, where V consists of a single point.

As in §2, there is little possibility of confusion if we identify a section F/F_{-1} of a polytope $\mathcal{P}$ with the face F. If we regard the sets V_j of Theorem 3.1 as multiply counted, we may make similar identifications in a realization of $\mathcal{P}$. In particular, if $\operatorname{rank}\mathcal{P} = n$, then $V_n = F_n\beta_n$ encapsulates (in this sense) all the structure of $\mathcal{P}$. We shall therefore use a notation such as $P := V_n$ to denote the realization of $\mathcal{P}$, and refer to P informally as a (*euclidean*) *regular polytope*. Furthermore, we shall often not distinguish between a face $F \in \mathcal{P}_j$ and the corresponding $F\beta_j$. If P is a regular polytope (thus understood), we shall then write $V(P) := V$ for its vertex-set.

There is usually no loss of generality in assuming that $E = \operatorname{aff} V$ is the affine hull of the vertex-set $V = V(P)$ of the regular polytope P (or of the realization β). We then call $\dim V := \dim \operatorname{aff} V$ the *dimension* of P, and denote it by $\dim P$.

The isometries of the vertex-set V induced by the automorphisms of $\mathcal{P}$ are called the *symmetries* of the polytope P, and form its *symmetry group* $G(P)$. Clearly, $G(P)$ is a euclidean representation of the automorphism group $\Gamma(\mathcal{P})$; it will be faithful if the realization is faithful, but perhaps also in other cases. We shall discuss $G(P)$ in more detail in §3.1.2 and §3.1.3.

For finite polytopes, vertex-faithful realizations always exist (just map the vertices to those of a regular simplex of the appropriate dimension; see 3.2.1 below), but for apeirotopes even this question remains open. Whether a vertex-faithful realization is faithful is a purely combinatorial condition on the polytope. However, we do have

THEOREM **3.2** *A (faithful) realization of a regular polytope $\mathcal{P}$ induces a (faithful) realization of each section of $\mathcal{P}$.*

We also have the following straightforward result.

THEOREM **3.3** *If $\mathcal{P}$ and $\mathcal{Q}$ are regular polytopes with $\mathcal{Q}$ a quotient of $\mathcal{P}$, then any realization of $\mathcal{Q}$ induces one of $\mathcal{P}$.*

3.1.2. *Symmetries*

Here we describe some properties of the symmetry group $G(P)$ of a realization P of an abstract regular polytope $\mathcal{P}$.

Corresponding to each generating involution ρ_j of the automorphism group $\Gamma(\mathcal{P})$ of $\mathcal{P}$ is an isometry R_j of the euclidean space E, which is either also an involution, that is, a *reflexion*, or the identity, which we denote by the same symbol E as for the space. We identify a reflexion R with its *mirror* of fixed points, namely $\{x \in E \mid xR = x\}$, which explains the reasons for writing the identity as E. First we have

THEOREM **3.4** *If $(R_{j-1}R_j)^2 = E$ for some $j = 1, \ldots, n-1$, then $\mathcal{P}$ collapses onto its face F_j, and the realization of $\mathcal{P}$ is induced by one of F_j.*

This follows easily from the expression of

$$G(P) = \langle R_0, \ldots, R_{j-1}\rangle\langle R_j, \ldots, R_{n-1}\rangle;$$

as a product of two commuting subgroups. Note that $\langle R_j, \ldots, R_{n-1}\rangle$ leaves the vertex of P in the base flag fixed. □

Observe that this will happen if $p_{j-1,j} = 2$, in the notation of §2, or if $R_j = E$.

As we remarked in §3.1.1, to each realization of $\mathcal{P}$ corresponds an euclidean representation of $\Gamma(\mathcal{P})$. In the next section, we discuss the converse of this.

3.1.3. *Wythoff's Construction*

Suppose now that $G := \langle R_0, \ldots, R_{n-1}\rangle$ is an euclidean representation (that is, by isometries) of the string C-group Γ, so that (in particular), for $j = 0, \ldots, n-1$, R_j is a reflexion in the euclidean space E, or possibly $R_j = E$. We now describe *Wythoff's construction*, which yields realizations of $\mathcal{P}(\Gamma)$.

The *Wythoff space* $W = W_G$ of G is defined to be

$$W := R_1 \cap \cdots \cap R_{n-1}.$$

Pick a point $p \in W$, and define $V := pG$; this is the vertex-set of the realization. Setting $F_0 := p$, we define recursively

$$F_j := F_{j-1}\langle R_0, \ldots, R_{j-1}\rangle$$

for $j = 1, \ldots, n-1$; this gives the base flag $\Phi := \{F_{-1}, F_0, \ldots, F_n\}$ (with $F_{-1} := \emptyset$) of a realization P of $\mathcal{P}(\Gamma)$, whose remaining flags are the Φg with $g \in G$.

THEOREM **3.5** *Every euclidean representation of the group $\Gamma(\mathcal{P})$ of a regular polytope $\mathcal{P}$ with a non-empty Wythoff space gives rise to a realization of $\mathcal{P}$.*

However, whether or not the realization is non-trivial (let alone faithful) is not so easy to determine. We mention here a few possibilities.

First, we may suppose that $W \neq \emptyset$, otherwise we obtain no realization at all from the representation.

Next, if $R_0 \cap W \neq \emptyset$, then the realization is necessarily bounded; in fact, it is trivial if W then consists of a single point. If W is not a point-set, then an infinite realization is necessarily non-discrete. Conversely, for the realization to be bounded, all the mirrors R_j must contain a common point.

Third, suppose that $R_0 \cap W = \emptyset$. If the subgroup $G_0 := \langle R_1, \ldots, R_{n-1} \rangle$ is non-trivial, then the realization is unbounded, and hence infinite. The initial vertex p is moved by R_0 to a point whose images under G_0 lie on a sphere centred at p, and these images are themselves centres of such spheres.

3.1.4. *Operations on Realizations*

There are two basic operations on realizations of regular polytopes.

The first construction corresponds to the sum of two representations of a group.

LEMMA **3.6** *Let Γ be a C-group, let L, M be orthogonal complementary subspaces of an euclidean space E, and let $S_0, \ldots, S_{n-1} \subseteq L$ and $T_0, \ldots, T_{n-1} \subseteq M$ be the generating reflexions of representations of Γ in L and M. If $R_j := S_j \times T_j \subseteq L \times M = E$ for $j = 0, \ldots, n-1$, then $\langle R_0, \ldots, R_{n-1} \rangle$ is a representation of Γ.*

We recall here the convention of identifying a reflexion with its mirror.

If we apply Lemma 3.6 to realizations P in L and Q in M of a regular polytope $\mathcal{P}$, there is a natural choice of an associated realization of $\mathcal{P}$ in E. If we write $G(P) := \langle S_0, \ldots, S_{n-1} \rangle$, $G(Q) := \langle T_0, \ldots, S_{n-1} \rangle$ and $G := \langle R_0, \ldots, R_{n-1} \rangle$, then the new Wythoff space is

$$W_G = W_{G(P)} \times W_{G(Q)}.$$

Indeed, if $p \in V(P) \cap W_{G(P)}$ and $q \in V(Q) \cap W_{G(Q)}$ are the initial vertices of P and Q, then $(p, q) \in W_G$ is a suitable initial vertex of the new regular polytope. We denote this polytope by $P\#Q$, and call it the *blend* of P and Q.

By construction, we have $V(P\#Q) \subseteq V(P) \times V(Q)$, and if $\{F_0, \ldots, F_{n-1}\}$ and $\{G_0, \ldots, G_{n-1}\}$ are the base flags of P and Q, then the corresponding base flag of $P\#Q$ is $\{F_0\#G_0, \ldots, F_{n-1}\#G_{n-1}\}$.

If a realization cannot be expressed as a blend in a non-trivial way (we shall discuss this further in §3.1.5), we call it *pure*. We observe that a realization P will be pure precisely when $G(P)$ acts irreducibly on $\operatorname{aff} V(P)$.

If G is a representation of the C-group Γ, λ, μ are constants, and $p, q \in W_G$, then $\lambda p + \mu q \in (\lambda + \mu) W_G$, which is the Wythoff space of the representation of Γ with mirrors $(\lambda + \mu) R_j$ for $j = 0, \ldots, n-1$. In this context, the correct convention is to define $0R_j$ to be the translate of R_j which contains the origin o. If p, q are the initial vertices of the realizations P, Q of $\mathcal{P}(\Gamma)$, then $\lambda p + \mu q$ is the initial vertex of a realization which we denote by $\lambda P + \mu Q$, and call a *linear combination* of P and Q. If $\mu = 0$, we obtain the *scalar multiple* λP of P.

We note that the blend is a special case of a sum, because we can write the direct product as $S_j \times T_j = S_j + T_j$, regarded as subspaces of E. However, as in [17], we prefer to keep the notions of blend and sum separate. Observe, though, that $(\lambda P)\#(\mu P)$ is congruent to νP, where $\nu^2 = \lambda^2 + \mu^2$.

3.1.5. *The Realization Cone*

We now show that the space of realizations forms, in a natural way, a convex cone.

In discussing possible realizations of an abstract regular polytope $\mathcal{P}$, we should not wish to distinguish between *congruent* realizations, where the natural bijection between their vertex-sets, regarded if necessary as ordered sets, is induced by an isometry between the ambient spaces. It is a convenient shorthand to use the same symbol $\mathcal{P}$ to denote the family of all classes under congruence of realizations of $\mathcal{P}$.

A pair of distinct vertices of $\mathcal{P}$ is called a *diagonal* of $\mathcal{P}$. Two diagonals are said to be *equivalent* if they are equivalent under $\Gamma(\mathcal{P})$. The diagonals of $\mathcal{P}$ thus fall into *diagonal classes*, consisting of equivalent diagonals. We assume these countably many diagonal classes ordered in some way; conventionally, we think of the first diagonal class as that of an edge of $\mathcal{P}$.

Let P be a (possibly degenerate) realization of $\mathcal{P}$, so that $P \in \mathcal{P}$ in our new notation. If the j-th diagonal class of $\mathcal{P}$ is represented by the vertex pair $\{p, q\}$ of P, write $\delta_j = \delta_j(P) := \|p - q\|^2$, the square of the length of the diagonal, and define the *diagonal vector* of P to be the vector $\Delta(P) := (\delta_1, \delta_2, \ldots)$.

We first have an obvious remark.

LEMMA **3.7** *$\Delta(P)$ determines the realization P up to congruence.*

The Cayley-Menger criterion for embeddability of a set of points with given mutual distances in some euclidean space is relevant here.

If we identify $P \in \mathcal{P}$ with its diagonal vector $\Delta(P)$, then we easily see that

$$\begin{aligned} \Delta(P\#Q) &= \Delta(P) + \Delta(Q), \\ \Delta(\lambda P) &= \lambda^2 \Delta(P), \end{aligned}$$

for $P, Q \in \mathcal{P}$ and $\lambda \in \mathbb{R}$. We therefore have

THEOREM **3.8** *The family $\mathcal{P}$ has the structure of a convex cone.*

Moreover,

THEOREM **3.9** *The vertex-faithful realizations in $\mathcal{P}$ form a convex subset of $\mathcal{P}$.*

The pure realizations are obviously of some interest; they are easily characterized.

THEOREM **3.10** *The pure realizations of $\mathcal{P}$ correspond to points on extreme rays of the realization cone $\mathcal{P}$.*

3.1.6. *Canonical Expressions*

Let G be an irreducible representation of the automorphism group $\Gamma(\mathcal{P})$ of a regular polytope $\mathcal{P}$, acting on the d-dimensional space E. Even up to similarity, there may be many different (pure) realizations associated with G.

Recall from §3.1.3 that a realization of $\mathcal{P}$ associated with G arises by applying Wythoff's construction to a point p of the Wythoff space W of G. There is no loss of generality in supposing that $o \in W$ (if $W = \emptyset$, then the realization space corresponding to G is also empty). We define

$$w := \begin{cases} \dim W & \text{if } R_0 \cap W \neq \emptyset, \\ \dim W + 1 & \text{if } R_0 \cap W = \emptyset. \end{cases}$$

That is, w is the number of degrees of freedom in the choice of the initial vertex p in realizations with group conjugate to G under similarity. Note that, since G is irreducible, R_0 can meet W in no more than a single point. Indeed, even more is true: if $R_0' := R_0 - R_0$ is the linear subspace of E parallel to R_0, then we must have $\dim(R_0' \cap W) = 0$ as well. The realization will be pure, unless it is trivial. We define $\mathcal{P}_G$ to be the subfamily of $\mathcal{P}$ consisting of all possible blends of these realizations with symmetry group G. Clearly

LEMMA **3.11** *$\mathcal{P}_G$ is a convex subcone of $\mathcal{P}$.*

The expression of a polytope in $\mathcal{P}_G$ as a blend of pure realizations need not be unique (even excepting changes in the order of the components).

LEMMA **3.12** *Let $P, Q \in \mathcal{P}$ have the same symmetry group G (acting on E), and let $\lambda^2 + \mu^2 = 1$. Then*

$$P \# Q = (\lambda P + \mu Q) \# (\mu P - \lambda Q).$$

In this context, G need not act irreducibly on E. Since G acts on $E \times E$ here by

$$(x, y)g = (xg, yg),$$

with $x, y \in E$ and $g \in G$, the subspace

$$L := \{(\mu x, -\lambda x) \mid x \in E\}$$

of $E \times E$ is invariant under G. The orthogonal projection on L is (essentially) $(x, y) \mapsto \lambda x + \mu y$, so that $\lambda P + \mu Q$ is a component of $P \# Q$. The orthogonal complement of L is

$$M := \{(\lambda x, \mu x) \mid x \in E\},$$

and the complementary component of $P \# Q$ is $\mu P - \lambda Q$. The statement of the lemma follows. □

Lemma 3.12 is our main tool for investigating $\mathcal{P}_G$. We next need an analogue of a well-known result on matrices. We call the $k \times w$ matrix $\Lambda = (\lambda_{ij})$ *canonical* if

(C1) each zero row of Λ lies beneath each non-zero row;

(C2) the leading (non-zero) entry $\lambda_{i,j(i)}$ in a non-zero row i (and in column $j(i)$) is positive;

(C3) if Λ has s non-zero rows, then $1 \leq j(1) < j(2) < \cdots < j(s) \leq w$.

We then have

LEMMA **3.13** *Let A be a real $k \times w$ matrix. Then there are an orthogonal $k \times k$ matrix U, and a unique canonical $k \times w$ matrix Λ, such that $\Lambda = UA$.*

As a direct consequence, we have

THEOREM **3.14** *A polytope P in the subcone $\mathcal{P}_G$ is a blend of at most w pure polytopes. Moreover, P has dimension at most wd, and wd is possible.*

To show this, we pick any basis $\{p_1, \ldots, p_w\}$ of W (linear or affine, as appropriate), and let $\{P_1, \ldots, P_w\} \subseteq \mathcal{P}_G$ be the corresponding (pure) realizations of $\mathcal{P}$. The general pure polytope in $\mathcal{P}_G$ is of the form $\alpha_1 P_1 + \cdots + \alpha_w P_w$, arising from $\alpha_1 p_1 + \cdots + \alpha_w p_w \in (\alpha_1 + \cdots + \alpha_w)W$, and so the general polytope in $\mathcal{P}_G$ is of the form

$$P = \#_{i=1}^{k} (\sum_{j=1}^{w} \alpha_{ij} P_j).$$

Consider the $k \times w$ matrix $A := (\alpha_{ij})$. Because of Lemma 3.12, and the facts that $-P = P$ (up to congruence) and that $\#$ is commutative (as well as associative), we see that P is unchanged by multiplying A on the left by orthogonal matrices. If Λ is the unique canonical matrix corresponding to A (as guaranteed by Lemma 3.13), we thus have

$$P = \#_{i=1}^{s} (\sum_{j=1}^{s} \lambda_{ij} P_j),$$

which is a blend of $s \leq w$ pure polytopes in $\mathcal{P}_G$.

Finally, note that, in this case, P has dimension $sd \leq wd$. Further, $P_1 \# \cdots \# P_w$ clearly has dimension wd, so that the bound is attained. □

This expression for P is called *canonical* in terms of the ordered set $(P_1, \ldots, P_w)$ of pure realizations. We can deduce more from the notions above.

THEOREM **3.15** *The cone $\mathcal{P}_G$ has dimension $\frac{1}{2}w(w+1)$.*

With the notation of Theorem 3.14, the diagonal vector of $\alpha_1 P_1 + \cdots + \alpha_w P_w$ is a quadratic polynomial in $\alpha_1, \ldots, \alpha_w$, and so a linear combination (with coefficients the α_i^2 and $\alpha_i \alpha_j$) of at most $\frac{1}{2}w(w+1)$ vectors, which depend only on $P_1, \ldots, P_w$. The same is therefore true of any blend of such polytopes, and hence $\dim \mathcal{P}_G \leq \frac{1}{2}w(w+1)$.

On the other hand, the diagonal map $P \mapsto \Delta(P)$ is one-to-one on the canonical expressions for the $P \in \mathcal{P}_G$. In particular, a neighbourhood of the identity matrix in the set of real upper triangular $w \times w$ matrices is mapped into the cone $\mathcal{P}_G$ by a one-to-one quadratic polynomial, from which it follows that $\dim \mathcal{P}_G \geq \frac{1}{2}w(w+1)$. The theorem follows at once. □

A representative of each similarity class in $\mathcal{P}_G$ corresponds to a point of the unit sphere in the space coordinatized by the canonical expressions. The diagonal map $P \mapsto \Delta(P)$ identifies antipodal points of this sphere, but no others (because of the uniqueness of the canonical expression). There then follows

THEOREM **3.16** *If $w \geq 2$, then the cone $\mathcal{P}_G$ has base the convex hull of a $(w-1)$-dimensional real projective space embedded in an ellipsoid of dimension $\frac{1}{2}w(w+1)-2$.*

We should note that, when $w \geq 2$, a polytope $P \in \mathcal{P}_G$ will be a blend of strictly fewer than $\dim \mathcal{P}_G$ pure polytopes, so that the bound of Theorem 3.10 is not usually sharp.

3.2. Finite Regular Polytopes

3.2.1. *Numerical Relationships*

When the abstract regular polytope $\mathcal{P}$ has finitely many vertices (in particular, if it is finite), then we can describe the realization cone more exactly. We shall write v for the number of vertices of $\mathcal{P}$, and r for the number of its diagonal classes. If G is an irreducible orthogonal representation of $\Gamma := \Gamma(\mathcal{P})$, then we write d_G and w_G for the degree and dimension of the Wythoff space. If $w_G = 0$, then G gives rise only to the trivial realization of $\mathcal{P}$, and so we shall discount these cases in what follows.

We first remark

THEOREM **3.17** *$\mathcal{P}$ is the direct sum of the cones $\mathcal{P}_G$ for which $w_G > 0$. Moreover, $\mathcal{P}$ is closed and* $\dim \mathcal{P} = r$.

First, $\mathcal{P}$ is the direct sum of the cones $\mathcal{P}_G$, since any representation of Γ is expressible (essentially uniquely) as a sum of irreducible representations. Next, $\dim \mathcal{P} \leq r$, since the diagonal vectors lie in $\mathbb{E}^r$; further, a limit of such diagonal vectors is clearly another, so that $\mathcal{P}$ is closed. We now consider the *simplex realization* T of $\mathcal{P}$, whose vertex-set is that of a regular $(v-1)$-simplex with unit edge-lengths; thus, $\Delta(T) = (1, \ldots, 1)$. If $\Delta = (\delta_1, \ldots, \delta_r)$ is near enough $\Delta(T)$, then there will be a $(v-1)$-simplex whose squared edge-lengths are the appropriate δ_j, so that Δ is also the diagonal vector of a realization of $\mathcal{P}$. Hence $\dim \mathcal{P} \geq r$ also. □

Let $\overline{W}$ be the Wythoff space of the simplex realization T, and let $\overline{w}$ be its dimension. We then have

THEOREM **3.18** *The following relationships hold for $\mathcal{P}$, where in each case the sum is over the distinct irreducible orthogonal representations of Γ.*

a) $\sum_G w_G d_G = v - 1$,
b) $\sum_G \frac{1}{2} w_G(w_G + 1) = r$,
c) $\sum_G w_G^2 = \overline{w}$,

These results are direct consequences of Theorem 3.17, and the observation from Theorem 3.14 that the maximal dimension of a polytope in $\mathcal{P}_G$ is $w_G d_G$. □

We can calculate the numbers v, r and $\overline{w}$ in a different way. A vertex of $\mathcal{P}$ can be identified with a right coset $\Gamma_0\sigma$ of the subgroup $\Gamma_0 := \langle \rho_1, \ldots, \rho_{n-1} \rangle$ of Γ. We label the vertices of T by these cosets, using the obvious bijection.

A typical diagonal of $\mathcal{P}$ is (using transitivity of Γ) a pair $\{\Gamma_0, \Gamma_0\sigma\}$ with $\sigma \notin \Gamma_0$; two such diagonals $\{\Gamma_0, \Gamma_0\sigma_1\}$ and $\{\Gamma_0, \Gamma_0\sigma_2\}$ are equivalent just when $\{\Gamma_0, \Gamma_0\sigma_1\}\tau = \{\Gamma_0, \Gamma_0\sigma_2\}$ for some $\tau \in \Gamma$. Considering the two possibilities $\Gamma_0\tau = \Gamma_0$ or $\Gamma_0\tau = \Gamma_0\sigma_2$ and eliminating τ, we see that the diagonals are equivalent if and only if

$$\sigma_2 \in \Gamma_0\sigma_1\Gamma_0 \cup \Gamma_0\sigma_1^{-1}\Gamma_0.$$

Last, the Wythoff space $\overline{W}$ of T is the set of points invariant under Γ_0, and so is the subspace spanned by the centroids of the orbits $\Gamma_0\sigma\Gamma_0$ of the vertices of T, which are affinely independent. Thus we have

THEOREM **3.19** *Let $\mathcal{P}$ be an abstract regular polytope with finitely many vertices. With the notation above*

a) $v - 1 = \operatorname{card}\{\Gamma_0 \mid \sigma \in \Gamma \setminus \Gamma_0\}$;

b) $r = \operatorname{card}\{\Gamma_0\sigma\Gamma_0 \cup \Gamma_0\sigma^{-1}\Gamma_0 \mid \sigma \in \Gamma \setminus \Gamma_0\}$;

c) $\overline{w} = \operatorname{card}\{\Gamma_0\sigma\Gamma_0 \mid \sigma \in \Gamma \setminus \Gamma_0\}$.

3.2.2. *Examples*

We briefly give some examples to illustrate §3.2.1; any notation is that used there.

A regular p-gon has $\lfloor \frac{1}{2}p \rfloor$ diagonal classes, and for each $k = 1, \ldots, \lfloor \frac{1}{2}p \rfloor$ we have a plane regular polygon $\{\frac{p}{k}\}$, which gives a degenerate realization if $(p, k) > 1$, reducing to a line-segment if $p = 2k$. An arbitrary regular p-gon is a blend of these pure realizations.

The regular n-simplex has a unique realization up to similarity, since all its diagonals are edges.

If the regular polytope $\mathcal{P}$ has a central involutory automorphism, this pairs up the vertices, so that $\mathcal{P}$ has centrally symmetric realizations, and in particular the *cross-polytope realization* X of dimension $\frac{1}{2}v$. There are also degenerate realizations obtained by identifying these opposite vertices, which yields the polytope denoted by $\mathcal{P}/2$. Pure realizations of $\mathcal{P}$ will be components of X or of $X/2$ (which is a $(\frac{1}{2}v - 1)$-simplex). The regular cross-polytope itself, with two pure realizations, is such an example.

The regular n-cube C^n has n diagonal classes, and n pure realizations (up to similarity), of dimensions $\binom{n}{k}$ $(k = 1, \ldots, n)$. The first is C^n itself, while the last is its collapse onto the line-segment. Further, those for even k are realizations of $C^n/2$.

The regular icosahedron $\{3, 5\}$ has three diagonal classes. Apart from $\{3, 5\}$ itself, another faithful realization is the great icosahedron $\{3, \frac{5}{2}\}$. The final pure realization is that of $\{3, 5\}/2 = \{3, 5\}_5$, all of whose diagonals are edges; thus it has the vertices of the 5-simplex.

The regular dodecahedron has five pure realizations: three faithful of dimensions 3, 3 and 4, and two of $\{5, 3\}/2 = \{5, 3\}_5$ of dimensions 4 and 5.

For all the above, we have $w_G = 1$. For an example with $w_G \geq 2$ for at least one realization, we turn to the Klein polyhedron $\{3, 7\}_8$. This has four diagonal classes. There is a degenerate 7-dimensional realization, in which the 24 vertices coincide in threes at the vertices of a regular 7-simplex. There are also (faithful) pure 8-dimensional realizations, with $w_G = 2$. Theorem 3.14 then looks like

a) $7 + 2.8 = 24 - 1$,

b) $1 + 3 = 4$,

c) $1 + 4 = 5$.

3.3. Apeirotopes

3.3.1. *The Apeirogon*

The example of the *apeirogon*, the infinite (regular) polytope $\{\infty\}$ of rank 2, illustrates some of the salient features of the realizations of apeirotopes, that is, infinite regular polytopes.

Suppose that P is an infinite realization of $\{\infty\}$ (the finite realizations are those of its coverings of polygons or the segment). Its group G is generated by two reflexions R_0 and R_1; let $T := R_0R_1$. Then T acts as a (combinatorial) translation, moving one vertex of P to the next, and the diagonal vector $\Delta(P)$ of P is clearly determined by T and the initial vertex of P.

Now, by a suitable choice of coordinate system, we can express T, in a unique way, as a commuting product of a translation (by a vector t, say, which may be o) and an orthogonal transformation Φ acting on a hyperplane orthogonal to t (if $t \neq o$). Further, Φ can be decomposed, again uniquely, into a product of rotations, by distinct angles $\vartheta_1, \ldots, \vartheta_k$ say, and possibly a reflexion, corresponding to an eigenvalue -1; if $t \neq o$, these may all be absent. Note that eigenvalues of Φ are not repeated, because of our assumption that the ambient space E is the affine hull of P.

This analysis demonstrates that the apeirogon P is a blend of pure polygons and apeirogons, with at least one of the latter. We have a polygon corresponding to each rational angle ϑ_j (by *rational* we mean, of course, a rational multiple of π). To a non-zero translation vector t, and each irrational angle ϑ_j, corresponds an apeirogon, in the latter case bounded and non-discrete.

It is now clear that the realization cone of $\{\infty\}$ has uncountable infinite (algebraic) dimension, and cannot be closed. To see this, let $\vartheta_1, \vartheta_2, \ldots$ be distinct angles, and let P_j be the apeirogon with vertices on the unit circle in $\mathbb{E}^2$ corresponding to the rotation though ϑ_j. Then the limit of the diagonal vector of

$$P_1 \# \frac{1}{2} P_2 \# \cdots \# \frac{1}{n} P_n$$

certainly exists, but cannot be that of a realization. The reason is simple; the Cayley-Menger criterion for embedding a set of points in some euclidean space shows that the supposed limit apeirogon would have to have an infinite affinely independent set of vertices. Notice also that if we blend this example with the ordinary linear apeirogon, we have a sequence of discrete realizations whose diagonal vector converges pointwise, but with limit which is not a diagonal vector of a realization.

3.3.2. *Realizations with Bounded Components*

In §3.3.1 we saw that a genuinely infinite apeirotope may have faithful bounded realizations. We thus first consider when a realization has a bounded component with respect to blending.

Theorem 3.20 *Let P be an unbounded realization of a regular apeirotope $\mathcal{P}$ with vertex-set V. Then* $\operatorname{conv} V \neq \operatorname{aff} V$ *if and only if P is a non-trivial blend, one of whose components is bounded.*

The sufficiency of the condition is obvious. For the necessity, write $K := \mathrm{conv}V$. As usual, we assume that E is the ambient space of P, so that K is full-dimensional. Since all points of V are equivalent under $G := G(P)$, it follows that $V \subseteq \mathrm{bd}K$, the boundary of K. (If one vertex in V lies in the interior of K, then by the equivalence they all do, which contradicts $K = \mathrm{conv}V$).

Now consider the recession cone $\mathrm{rec}K$ of K, which consists of the vectors $a \in E$ such that $x+a \in K$ for every $x \in \mathrm{int}K$. As a set of directions, $\mathrm{rec}K$ is invariant under G. If $\mathrm{rec}K$ is not a linear subspace, we let g be the centroid of the intersection of $\mathrm{rec}K$ with the unit ball of E (in its linear hull), so that g gives an invariant direction under G. Then K will have a support hyperplane H normal to g, which is also invariant under G. The points of V must then lie at the same distance from H (and so lie in H), which leads at once to a contradiction.

It follows that $L := \mathrm{rec}K$ is a subspace. Considering the projection on the orthogonal complement of L shows that the points of V are equidistant from a translate of L, and the condition follows at once, the projection giving the bounded component of the realization. □

3.3.3. *Discrete Realizations*

The case of the apeirogon shows that we cannot avoid considering non-discrete realizations, because these may occur as components under blending of discrete realizations, but there is naturally greater interest in discrete realizations. We shall investigate these first. For many examples of discrete regular apeirotopes in euclidean spaces, see [5, 8, 9, 20, 21, 22, 23].

Let P be an infinite discrete realization of a regular apeirotope $\mathcal{P}$. There are two cases, according as the ambient space E is or is not the convex hull of P, or, rather, of its vertex-set V.

Suppose first that $\mathrm{conv}V \neq E$. As we saw in §3.3.2, we can translate so that the points of V are equidistant from some linear subspace L of E. The orthogonal projection of V on L yields another realization of $\mathcal{P}$.

LEMMA **3.21** *The projection of V on L is a discrete set.*

The points of V are at a fixed distance $\rho > 0$ from L. Since the sphere is compact, any cluster point of the projection lifts to a cluster point of V itself. But V is discrete; hence no such cluster point can exist. □

Note that the projection along L need not be discrete (it is certainly bounded), as the examples of realizations of the apeirogon demonstrate; compare §3.3.1.

We may now suppose that $\mathrm{conv}V = E$. The symmetry group G of P is transitive on V, and so, by a famous theorem of Bieberbach (see [1, 11]), G contains a subgroup T of translations of full rank $\dim E$. This subgroup T is discrete, and so can be identified with a lattice in E. Note that G will then act as a group of automorphisms (that is, symmetries) of T.

3.3.4. *Translation-Free Realizations*

We now come to a curiosity about discrete realizations. We call a realization P of a regular apeirotope $\mathcal{P}$ *translation-free* if its symmetry group G contains no (nonzero) translations. Of course, a bounded realization must be translation-free, but we also observed in §3.3.1 that we have unbounded translation-free realizations of the regular apeirogon. We shall prove the following somewhat surprising result.

THEOREM **3.22** *If a regular apeirotope $\mathcal{P}$ has a discrete realization, then it has a translation-free realization.*

In view of §3.3.3, we may suppose that we have a discrete realization P of $\mathcal{P}$, whose vertex-set V satisfies $\operatorname{conv} V = \operatorname{aff} V = E$, the ambient space of the realization. So, the symmetry group G of P contains a lattice Λ of translations, of rank $\dim E$. If we take the initial vertex of P to be the origin o, then $\Lambda \subseteq V$, in the obvious sense. In fact, we see that the vertex-set V decomposes into finitely many translates of Λ.

Now G acts on Λ by conjugation; the effect is that of applying the rotation part of an element of G to the translation vectors in Λ. Let Λ^* denote the reciprocal lattice of Λ; that is,

$$\Lambda^* := \{x^* \in E \mid \langle x, x^* \rangle \in \mathbb{Z} \text{ for all } x \in \Lambda\},$$

where $\langle .,. \rangle$ is the usual inner product on E. Then G similarly acts on Λ^*.

We now lift Λ^* in a symmetry-preserving way. Specifically, suppose that there are $2n$ adjacent points $\pm p_1, \ldots, \pm p_n$ of Λ^* to the origin o, corresponding to the $2n$ facets of the Voronoĭ region of Λ^*. Then these are the images of the basis points $\pm e_1, \ldots, \pm e_n$ of the integer lattice $\mathbb{Z}^n$ under the linear map $\Phi : \mathbb{E}^n \to E$, with $e_i\Phi = p_i$ for $i = 1, \ldots, n$. The kernel of Φ is given by the linear relations between the p_i, so that

$$\ker \Phi = \{(\alpha_1, \ldots, \alpha_n) \in \mathbb{E}^n \mid \sum_{i=1}^{n} \alpha_i p_i = o\}.$$

Now $\ker \Phi$ is a rational subspace of $\mathbb{E}^n$, and so spanned by points of $\mathbb{Z}^n$, since the p_i clearly generate the lattice Λ^*. We can now identify E with the orthogonal complement of $\ker \Phi$, and under this identification, $\Lambda = E \cap \mathbb{Z}^n$.

Now let β be an irrational multiple of π, let $\psi : \mathbb{R} \to \mathbb{E}^3$ be given by $\xi\psi := (\xi, \cos(\beta\xi), \sin(\beta\xi))$, and define $\Psi : \mathbb{E}^n \to \mathbb{E}^{3n}$ by $(\xi_1, \ldots, \xi_n)\Psi := (\xi_1\psi, \ldots, \xi_n\psi)$, with the obvious meaning. Then Ψ respects the action of the group G; that is, there is an isomorphic group G' acting on E^{3n}, such that $(xg)\Psi = (x\Psi)g'$ for $x \in E$ and corresponding $g \in G$ and $g' \in G'$. The crucial feature of Ψ, though, is that the lattice translations in $\mathbb{Z}^n$ are destroyed, and so the group G' acting on the new realization $V\Psi$ contains no translations. This proves the assertion of the theorem. □

In fact, the proof above yields a little more.

COROLLARY **3.23** *If a regular apeirotope has a discrete infinite realization, then it has uncountably many translation-free discrete realizations.*

3.3.5. *Non-Discrete Realizations*

The situation for unbounded non-discrete realizations of an apeirotope is quite different from that of §3.3.3. We present a variety of examples to illustrate the kinds of thing that can happen; other examples can be found in [18]. We shall omit details of the proofs.

As we remarked earlier, we cannot avoid non-discrete realizations. For instance, in the plane $\mathbb{E}^2$, the tessellation $\{8, \frac{8}{3}\}$ is a non-discrete realization of the hyperbolic honeycomb $\{8, 8\}$, as is its isomorph $\{\frac{8}{3}, 8\}$. If we blend these two realizations together, with an appropriate choice of scaling of each, the resulting set of vertices of the blended realization in $\mathbb{E}^4$ is $\mathbb{Z}^4$. Indeed, every non-discrete regular tessellation $\{p, q\}$ in $\mathbb{E}^2$ (with at least one of p, q fractional) lifts to a discrete realization in this way.

However, let β be an irrational multiple of π, and consider the reflexions in $\mathbb{E}^2$ defined by

$$\begin{aligned} xR_0 &:= (1-\xi_1, \xi_2), \\ xR_1 &:= (\xi_1 \cos\beta + \xi_2 \sin\beta, \xi_1 \sin\beta - \xi_2 \cos\beta), \\ xR_2 &:= (\xi_1, -\xi_2), \end{aligned}$$

where $x = (\xi_1, \xi_2)$. Then $G := \langle R_0, R_1, R_2 \rangle$ is the symmetry group of a regular tessellation P of type $\{\infty, \infty\}$. It is not the universal $\{\infty, \infty\}$, because G contains translations (the product of any two conjugates of R_0R_2); indeed, the translations in G form an abelian group of infinite rank, because the rotation of one translation by a multiple of β is another. Of course, P must be non-discrete, but in contrast to the previous examples, P cannot be lifted to a discrete realization, since its vertex-figure is infinite, and in any realization, the points of the vertex-figure at a vertex v are equidistant from v.

Similar examples can be constructed in all dimensions greater than 2. If we fit copies of a centrally symmetric convex regular polytope Q with an irrational dihedral angle facet-to-facet, we obtain a regular apeirotope P whose vertex-figure is infinite. Since the reflexion in the centre of each copy of Q is a symmetry, we see that the symmetry group of P contains a translation group of infinite rank. In $\mathbb{E}^n$ with $n \neq 4$, the cross-polytope provides an example; in $\mathbb{E}^4$, we may use $\{3, 3, 5\}$.

We now come to an example of a quite different nature in $\mathbb{E}^3$. Let $\alpha > 0$ be a number to be chosen later. We define the following three half-turns:

R_0: about the line through $(\frac{2\alpha^2+1}{\alpha}, 0, 0)$ in direction $(0, 0, 1)$;
R_1: about the line through o in direction $(\alpha, \sqrt{\alpha^2+1}, 1)$;
R_2: about the line through o in direction $(0, 1, 0)$.

Then R_0R_1 is a rotatory-translation, with twist ϑ given by

$$2\cos(\tfrac{1}{2}\vartheta) = \frac{1}{\sqrt{2(\alpha^2+1)}}.$$

Clearly ϑ will be an irrational multiple of π for all but countably many α. Next, $(R_1R_2)^4 = E$, the identity. Finally, R_0 obviously commutes with R_2.

Thus $G := \langle R_0, R_1, R_2 \rangle$ is a homomorphic image of the Coxeter group $[\infty, 4]$. However, if R_0' is the half-turn about the line through o parallel to the axis of R_0, then the group $\langle R_0', R_1, R_2 \rangle$ can be made to be isomorphic to $[\infty, 4]$. To see this, consider the reflexion group obtained by replacing R_0', R_1, R_2 by the planes

S_0, S_1, S_2 orthogonal to them; applying Wythoff's construction to $S_0 \cap S_1$ yields a spherical tessellation by squares $\{4\}$. Then for only countably many values of α can fortuitous identifications of vertices of this tessellation occur; otherwise, it is isomorphic to $\{4, \infty\}$. Because the related group is a quotient of G, we see that $G \cong [\infty, 4]$.

If we apply Wythoff's construction with o as initial vertex, it is clear that the vertex-set V of the resulting regular apeirotope $P := \{\infty, 4\}$ satisfies $\text{conv}V = \mathbb{E}^3$. However, G can contain no translations, since $[\infty, 4]$ contains no commuting elements of infinite order (alternatively, a translation would have to correspond to a non-trivial relation in the quotient group $\langle R_0', R_1, R_2\rangle$). We conclude that there is no analogue of Bieberbach's theorem for the non-discrete case.

Between these two extremes lie the cases where the apeirotope has a translation group of finite positive rank. We now discuss this.

LEMMA **3.24** *Let P be a realization of a regular apeirotope $\mathcal{P}$ whose symmetry group G contains a translation group Λ of finite positive rank. Then $\mathcal{P}$ has a realization whose group contains a translation subgroup isomorphic to Λ whose vectors span the ambient space.*

Indeed, the linear subspace L spanned by the translation vectors in Λ is invariant under G, and so orthogonal projection of P on L will yield the required realization.

Henceforth, we shall work with realizations having the characteristic property of Lemma 3.24. Our last main result is

THEOREM **3.25** *Let $\mathcal{P}$ be a regular apeirotope, which has a realization P whose translation group is of finite positive rank, with its vectors spanning the ambient space. Then $\mathcal{P}$ has a discrete infinite realization.*

In a sense, we lift P to a discrete realization. Each element of the symmetry group G of P can be written in the form $x \to x\Phi + t$, where Φ is an orthogonal mapping, and t is a translation vector; these associated orthogonal mappings Φ form a group $\overline{G}$, which is a quotient of G, called the *point-group* of G. Since $\overline{G}$ acts on the translation subgroup Λ of G, which is of finite rank m say, with vectors spanning the ambient space E, we see that $\overline{G}$ must be finite. We now lift a finite $\mathbb{Z}$-spanning subset $\{p_1, \ldots, p_n\}$ of Λ which is invariant under $\overline{G}$; the image of the integer lattice $\mathbb{Z}^n$ (with usual basis $\{e_1, \ldots, e_n\}$) under the orthogonal projection in $\mathbb{E}^n$ with kernel

$$\{\sum_{i=1}^{n} \alpha_i e_i \mid \alpha_i \text{ rational, and } \sum_{i=1}^{n} \alpha_i p_i = o\}$$

is a discrete lattice $\widehat{\Lambda}$, on which acts a group isomorphic to $\overline{G}$ (compare Theorem 3.22; our projection is designed to preserve these symmetries).

Lifting the whole of G is now comparatively straightforward; we omit the details, referring the reader to [19] instead. □

As a consequence of the above results, we have a generalization of Theorem 3.22:

THEOREM **3.26** *If a regular apeirotope $\mathcal{P}$ has an infinite realization, such that the translation subgroup of its symmetry group has finite rank, then $\mathcal{P}$ has an infinite translation-free realization.*

References

1. Bieberbach, L., Über die Bewegungsgruppen der euklidischen Räume: erste Abhandlungen. *Math. Ann.* **70** (1910), 297–336.
2. Colbourn, C.L. and Weiss, A.I., A census of regular 3-polystroma arising from honeycombs. *Discrete Math.* **50** (1984), 29–36.
3. Coxeter, H.S.M., Regular skew polyhedra in 3 and 4 dimensions and their topological analogues. *Proc. London Math. Soc.* (2) **43** (1937), 33–62. (Reprinted with amendments in *Twelve Geometric Essays*, Southern Illinois Univ. Press (Carbondale, 1968), 76–105.)
4. Coxeter, H.S.M., *Twisted honeycombs.* Regional Conf. Ser. in Math., Amer. Math. Soc. (Providence, 1970).
5. Coxeter, H.S.M., *Regular Polytopes* (3rd edition). Dover (New York, 1973).
6. Coxeter, H.S.M. and Moser, W.O.J., *Generators and Relations for Discrete Groups* (4th edition). Springer (Berlin, 1980).
7. Danzer, L. and Schulte, E., Reguläre Inzidenzkomplexe, I. *Geom. Ded.* **13** (1982), 295–308.
8. Dress, A.W.M., A combinatorial theory of Grünbaum's new regular polyhedra, Part I: Grünbaum's new regular polyhedra and their automorphism group. *Aequationes Math.* **23** (1981), 252–265.
9. Dress, A.W.M., A combinatorial theory of Grünbaum's new regular polyhedra, Part II: Complete enumeration. *Aequationes Math.* **29** (1985), 222–243.
10. Du Val, P., *Homographies, Quaternions and Rotations*, Oxford University Press (1964).
11. Engel, P., *Geometric Crystallography.* Reidel (Dordrecht etc., 1986).
12. Grünbaum, B., Regular polyhedra—old and new. *Aequationes Math.* **16** (1977), 1–20.
13. Grünbaum, B., Regularity of graphs, complexes and designs. *Colloques CNRS, Problèmes Combinatoires et Theorie des Graphes*, CNRS (Orsay, 1976), 191–198.
14. Hess, E., Über die regulären Polytope höherer Art. *Sitz. Ges. Bef. ges. Naturwiss. Marburg* (1885), 31–57.
15. McMullen, P., Combinatorially regular polytopes. *Mathematika* **14** (1967), 142–150.
16. McMullen, P., Regular star-polytopes, and a theorem of Hess. *Proc. London Math. Soc.* (3) **18** (1968), 577–596.
17. McMullen, P., Realizations of regular polytopes. *Aequationes Math.* **37** (1989), 38–56.
18. McMullen, P., Nondiscrete regular honeycombs. Chapter 10 in *Quasicrystals, Networks, and Molecules of Fivefold Symmetry*, ed. I. Hargittai (VCH Publishers, New York etc., 1990), 159–179.
19. McMullen, P., Realizations of regular apeirotopes. *Aequationes Math.* (in press).
20. McMullen, P. and Schulte, E., Regular polytopes from twisted Coxeter groups. *Math. Z.* **201** (1989), 209–226.
21. McMullen, P. and Schulte, E., Constructions for regular polytopes. *J. Combinat. Theory A* **53** (1990), 1–28.

22. McMullen, P. and Schulte, E., Regular polytopes from twisted Coxeter groups and unitary reflexion groups. *Advances Math.* **82** (1990), 35–87.
23. McMullen, P. and Schulte, E., Hermitian forms and locally toroidal regular polytopes. *Advances Math.* **82** (1990), 88–125.
24. McMullen, P. and Schulte, E., *Abstract Regular Polytopes.* (In preparation.)
25. van Oss, S.L., Die regelmässige fierdimensionalen Polytope höherer Art. *Verh. Konink. Akad. Weten. Amsterdam* (eerste sectie) **12.1** (1915).
26. Schläfli, L., Theorie der vielfachen Kontinuität. *Denkschriften der Schweizerlichen Naturforschenden Gesellschaft* **38** (1901), 1–237.
27. Schulte, E., Reguläre Inzidenzkomplexe, II. *Geom. Ded.* **14** (1983), 33–56.
28. Schulte, E., Reguläre Inzidenzkomplexe, III. *Geom. Ded.* **14** (1983), 57–79.
29. Schulte, E., Classification of locally toroidal regular polytopes. *This volume,*
30. Schulte, E. and Weiss, A.I., Chiral polytopes. In *Applied Geometry and Discrete Mathematics,* DIMACS Series in Discrete Mathematics and Theoretical Computer Science **4** (1991), 493–516.
31. Schulte, E. and Weiss, A.I., Chirality and projective linear groups. *Discrete Math.* (to appear).
32. Stringham, W.I., Regular figures in n-dimensional space. *Amer. J. Math.* **3** (1880), 1–14.

CLASSIFICATION OF LOCALLY TOROIDAL REGULAR POLYTOPES

EGON SCHULTE * †
Department of Mathematics
Northeastern University
Boston, MA 02115, USA

Abstract. A central problem in classical geometry is the classification of all regular polytopes and tessellations in spherical, euclidean or hyperbolic space. When asked within the theory of abstract regular polytopes, the classification problem must necessarily take a different form, because a priori an abstract polytope is not embedded into the geometry of an ambient space. The appropriate substitute now calls for the classification of abstract regular polytopes by their local or global topological type. The classical theory of regular polytopes is concerned with, and solves, the spherical case. In recent years, much work has been done on the toroidal case and a complete classification is now within reach.

Key words: Abstract regular polytopes, Toroidal polytopes, Reflection groups, Coxeter groups.

1. Introduction

Symmetry of geometric figures is a fascinating phenomenon which makes a powerful appearance in the classical theory of regular polytopes (Coxeter [9]). These figures have an outstanding history of study unmatched by almost any other geometric object. For a more detailed discussion on this history the reader is referred to the article by Peter McMullen in this volume.

In the past 15 years this area of classical geometry has been extended in several directions which are all centered around an abstract combinatorial polytope theory and a combinatorial notion of regularity. Abstract regular polytopes generalize the classical notion of a regular polytope and tessellation to more complicated combinatorial structures with a distinctive geometric and topological flavour. The notion of an abstract polytope was introduced in Grünbaum [18] and Danzer & Schulte [12], with a more systematic approach starting in [36]. For related concepts which occurred earlier or at about the same time, we also refer to Buekenhout [3], Dress [13], McMullen [21], Tits [46] and Vince [47]. See again the article by Peter McMullen for more details on the history of this subject.

A central problem in the classical theory is the complete classification of all regular polytopes and tessellations in spherical, euclidean or hyperbolic space. The solution to this problem is well-known and is closely related to the classification of Coxeter groups of spherical, euclidean or hyperbolic type; see Coxeter & Moser [10], Humphreys [19], or the article by Arjeh Cohen in this volume. When asked within the theory of abstract polytopes, the classification problem must necessarily take

* Supported by NSF grant DMS-9202071
† Email address: schulte@northeastern.edu

T. Bisztriczky et al. (eds.), POLYTOPES: Abstract, Convex and Computational, 125–154.

a different form, because a priori an abstract polytope is not embedded into the geometry of an ambient space. The appropriate substitute now calls for the classification of abstract regular polytopes by their local or global topological type. In the first place this requires to associate with abstract polytopes a natural topology, a problem which is very subtle and which in general cannot be uniquely solved. On the other hand, many polytopes admit a natural topology and so are subject to the classification with respect to this topology.

The classical theory of regular polytopes is concerned with, and solves, the spherical case. Convex polytopes and tessellations are locally spherical in the sense that their local building blocks (facets or tiles, and vertex-figures) are topologically spheres; and convex polytopes are also globally spherical ([17]). Using terminology introduced further below we can restate this by saying that the only universal abstract regular polytopes which are locally spherical are the classical regular tessellations in spherical, euclidean or hyperbolic space; among those, only the spherical regular tessellations (convex regular polytopes) are finite and globally spherical ([8, 9, 21, 14]).

A major concern is now to extend this classification to polytopes with more sophisticated topologies like that of arbitrary spherical, euclidean or hyperbolic space-forms (Wolf [52]). In this generality the classification problem is wide open, yet significant progress has been made in the case where the euclidean space-form is a torus (McMullen & Schulte [25, 27, 28, 29, 30]). In this paper we shall mainly discuss this toroidal case. For a clarification of what classification means in this context we refer to Section 3.

In the 70's, Grünbaum [18] triggered the theory of abstract polytopes by posing the challenging problem, as yet unsolved, of completely classifying the locally toroidal regular polytopes in each rank $n \geq 4$. As a first step this requires, for each rank n, the classification of the globally toroidal regular polytopes, the toroids; see Section 4. For rank 3 the toroids are the well-known regular (reflexible) maps on the 2-torus (Coxeter & Moser [10]). An abstract polytope of rank n is now called locally toroidal if its facets and vertex-figures are (globally) spherical or toroidal, with at least one kind toroidal. Locally toroidal regular polytopes can exist in ranks 4, 5 and 6 alone, because in higher ranks there are no suitable hyperbolic honeycombs to derive them from.

The situation is currently best understood in ranks 4 and 5. In rank 4, the classification involves analysis of the Schläfli types $\{4,4,r\}$ with $r = 3,4$, $\{6,3,p\}$ with $p = 3,4,5,6$, and $\{3,6,3\}$, and their duals. A complete classification is known for all types except $\{4,4,4\}$ and $\{3,6,3\}$ ([25, 27, 28]). For $\{4,4,4\}$ the classification is almost complete, and for $\{3,6,3\}$ partial results were obtained. The picture is particularly satisfactory for the types $\{6,3,p\}$ and the known cases of $\{3,6,3\}$ ([25]). Here the structure of the polytopes is governed by a complex hermitian form. In particular, the polytope is finite if and only if the corresponding form is positive definite. This generalizes the well-known classical situation where the structure of a regular convex polytope or regular tessellation is determined by a real quadratic form which determines the geometry of the ambient space; this correspondence sets up a beautiful link between geometry and algebra ([9]).

In rank 5, only one Schläfli type occurs, $\{3,4,3,4\}$ (and its dual). The locally toroidal polytopes of this type are completely classified ([29, 30]).

In rank 6, the types are $\{3,3,3,4,3\}$, $\{3,3,4,3,3\}$ and $\{3,4,3,3,3\}$, and their duals. There is a list of known finite polytopes which is conjectured to be complete ([29, 30]). In general, these polytopes are huge and wild, and so their classification is difficult. Honoring our old friends who used to inhabit our planet millions of years ago, we may wish to call them dinotopes. However, in contrast to those, they are still giving us a hard time.

On the group level, the classification of toroidal and locally toroidal polytopes amounts to the classification of certain groups which are defined in terms of generators and relations. These groups are quotients of euclidean or hyperbolic Coxeter groups and are obtained from those by either one or two extra relations. These extra relations force the toroidal structure upon the whole polytope or its facets or vertex-figures.

In contrast to regular polytopes, relatively little is known about chiral polytopes (Schulte & Weiss [39, 40], Monson & Weiss [33], Nostrand [34]). While regular polytopes have maximal symmetry with respect to (combinatorial) reflection, chiral polytopes are abstract polytopes with maximal rotational symmetry. Chirality of polytopes is a fascinating phenomenon which does not occur in the classical theory. In rank 3, the chiral polytopes are the irreflexible maps on surfaces ([10]); there are infinitely many such maps of genus 1 but for higher genus the occurrence is rather sporadic. For ranks $n \geq 4$ there are no chiral toroids, so the classification of locally toroidal chiral polytopes makes sense in rank 4 alone. However, here our knowledge is rather incomplete and is complicated by the fact that chiral polytopes occur in two enantiomorphic (mirror image) forms. The methods used for the construction of chiral polytopes of rank 4 all employ representations of hyperbolic rotation groups as projective linear groups over finite rings.

2. Basic Notions

In this section we give a brief introduction to the theory of abstract regular and chiral polytopes. For more details the reader is referred to [31, 39] or the article by Peter McMullen in this volume.

An *(abstract) polytope of rank* n, or simply an n*-polytope*, is a partially ordered set $\mathcal{P}$ with a strictly monotone rank function with range $\{-1, 0, \ldots, n\}$. The elements of rank i are called the i*-faces* of $\mathcal{P}$, or *vertices*, *edges* and *facets* of $\mathcal{P}$ if $i = 0, 1$ or $n-1$, respectively. The *flags* (maximal totally ordered subsets) of $\mathcal{P}$ all contain exactly $n+2$ faces, including the unique minimal face F_{-1} and unique maximal face F_n of $\mathcal{P}$. Further, $\mathcal{P}$ is *strongly flag-connected*, meaning that any two flags Φ and Ψ of $\mathcal{P}$ can be joined by a sequence of flags $\Phi = \Phi_0, \Phi_1, \ldots, \Phi_k = \Psi$, which are such that Φ_{i-1} and Φ_i are *adjacent* (differ by just one face), and such that $\Phi \cap \Psi \subseteq \Phi_i$ for each i. Finally, if F and G are an $(i-1)$-face and an $(i+1)$-face with $F < G$, then there are exactly *two* i-faces H such that $F < H < G$.

If F and G are faces with $F \leq G$, we call $G/F := \{H \mid F \leq H \leq G\}$ a *section* of $\mathcal{P}$. We can usually safely identify a face F with the section F/F_{-1}. For a face F, the section F_n/F is called the *coface of* $\mathcal{P}$ *at* F, or the *vertex-figure at* F if F is a vertex.

An abstract n-polytope $\mathcal{P}$ is *regular* if its (*combinatorial automorphism*) *group* $A(\mathcal{P})$ is transitive on its flags. Let $\Phi := \{F_{-1}, F_0, \ldots, F_n\}$ be a fixed or *base* flag of $\mathcal{P}$. The group $A(\mathcal{P})$ of a regular n-polytope $\mathcal{P}$ is generated by *distinguished generators* $\rho_0, \ldots, \rho_{n-1}$ (*with respect to* Φ), where ρ_i is the unique automorphism which keeps all but the i-face of Φ fixed. These generators satisfy relations

$$(\rho_i\rho_j)^{p_{ij}} = \varepsilon \quad (i,j = 0,\ldots,n-1) \tag{1}$$

with

$$p_{ii} = 1\,,\; p_{ij} = p_{ji} \geq 2\,(i \neq j)\,, \tag{2}$$

and

$$p_{ij} = 2 \text{ if } |i-j| \geq 2\,. \tag{3}$$

Here the numbers $p_{i+1} := p_{i,i+1}$ determine the (*Schläfli*) *type* $\{p_1, \ldots, p_{n-1}\}$ of $\mathcal{P}$. Further, $A(\mathcal{P})$ has the *intersection property* (*with respect to the distinguished generators*), namely

$$\langle \rho_i | i \in I \rangle \cap \langle \rho_i | i \in J \rangle = \langle \rho_i | i \in I \cap J \rangle \text{ for all } I, J \subset \{0,\ldots,n-1\}\,. \tag{4}$$

By a *C-group* we mean a group which is generated by involutions such that (1), (2) and (4) hold. If in addition (3) holds, then the group is called a *string C-group*. The group of a regular polytope is a string C-group. Conversely, given a string C-group there is an associated regular polytope of which it is the automorphism group ([36]). Note that Coxeter groups are examples of C-groups ([19]).

Each string C-group is a quotient of the Coxeter group $[p_1, \ldots, p_{n-1}]$ with the string diagram

(5)

with the integers p_j defined as above. For any $p_1, \ldots, p_{n-1} \geq 2$, $[p_1, \ldots, p_{n-1}]$ is the group of the *universal* regular polytope $\{p_1, \ldots, p_{n-1}\}$ ([36, 46]). This polytope is denoted by $\{p_1, \ldots, p_{n-1}\}$ and covers any other regular polytope of type $\{p_1, \ldots, p_{n-1}\}$.

For a regular polytope $\mathcal{P}$ the rotations

$$\sigma_j := \rho_j\rho_{j-1} \quad (j = 1,\ldots,n-1)$$

generate the *rotation subgroup* $A^+(\mathcal{P})$ of $A(\mathcal{P})$, which is of index at most 2. These rotations σ_j fix all faces in $\Phi \setminus \{F_{j-1}, F_j\}$ and cyclically permute consecutive j-faces of $\mathcal{P}$ in the section F_{j+1}/F_{j-2} of $\mathcal{P}$ of rank 2. A regular polytope $\mathcal{P}$ is called *directly regular* if $A^+(\mathcal{P})$ has index 2 in $A(\mathcal{P})$. For a regular polytope $\mathcal{P}$, direct regularity is equivalent to orientability of its *order complex* $\Delta(\mathcal{P})$, the simplicial complex whose simplices are given by the totally ordered subsets of $\mathcal{P}$ not containing F_{-1} and F_n ([45]). Note that for $\mathcal{P} = \{p_1, \ldots, p_{n-1}\}$ the rotation subgroup $A^+(\mathcal{P})$ is the even subgroup $[p_1, \ldots, p_{n-1}]^+$ of $[p_1, \ldots, p_{n-1}]$ ([10]).

Now let $\mathcal{P}$ be a polytope of rank $n \geq 3$. Then $\mathcal{P}$ is said to be *chiral* if $\mathcal{P}$ is not regular, but if for some base flag $\Phi = \{F_{-1}, F_0, \ldots, F_n\}$ of $\mathcal{P}$ there still exist automorphisms $\sigma_1, \ldots, \sigma_{n-1}$ of $\mathcal{P}$ such that σ_j fixes all faces in $\Phi \setminus \{F_{j-1}, F_j\}$ and cyclically permutes consecutive j-faces of $\mathcal{P}$ in the section F_{j+1}/F_{j-2} of rank 2. These automorphisms $\sigma_1, \ldots, \sigma_{n-1}$ (when suitably oriented) are called the *distinguished generators* of $A(\mathcal{P})$. Then a polytope $\mathcal{P}$ is chiral if and only if its group $A(\mathcal{P})$ has precisely two orbits on the flags with adjacent flags belonging to different orbits.

Each chiral polytope occurs in two *enantiomorphic forms*, in a sense in a *right* and a *left* version. In terms of groups and generators, these can be represented by two distinct systems of generators for $A(\mathcal{P})$, $\{\sigma_1, \ldots, \sigma_{n-1}\}$ and $\{{\sigma_1}^{-1}, \sigma_2{\sigma_1}^2, \sigma_3, \ldots, \sigma_{n-1}\}$, belonging to Φ and its adjacent flag with another vertex, respectively. Note that for a directly regular polytope $\mathcal{P}$ the corresponding systems are equivalent under conjugation in $A(\mathcal{P})$ by the "reflection" ρ_0; that is, there is no distinction between a left and right version of $\mathcal{P}$ or, equivalently, the two enantiomorphic forms are the same. An *oriented chiral* or *oriented directly regular* polytope is a chiral or directly regular polytope together with a distinguished enantiomorphic form; in the chiral case there are two "orientations", in the directly regular case only one. We shall often drop the qualification "oriented" when confusion is not possible.

3. The Classification Problem

A main thrust in regular polytopes is the amalgamation of polytopes of lower rank. Let $\mathcal{P}_1$ and $\mathcal{P}_2$ be two polytopes of rank n such that the vertex-figures of $\mathcal{P}_1$ are isomorphic to the facets of $\mathcal{P}_2$.

If $\mathcal{P}_1$ and $\mathcal{P}_2$ are regular, we denote by $\langle \mathcal{P}_1, \mathcal{P}_2 \rangle$ the *class* of all regular polytopes $\mathcal{P}$ of rank $n+1$ whose facets are isomorphic to $\mathcal{P}_1$ and whose vertex-figures are isomorphic to $\mathcal{P}_2$. Each non-empty class $\langle \mathcal{P}_1, \mathcal{P}_2 \rangle$ contains a member, denoted by $\{\mathcal{P}_1, \mathcal{P}_2\}$, which is *universal* in the sense that it covers any other polytope in the class $\langle \mathcal{P}_1, \mathcal{P}_2 \rangle$ ([38]). By $[\mathcal{P}_1, \mathcal{P}_2]$ we denote the group of $\{\mathcal{P}_1, \mathcal{P}_2\}$. If $\mathcal{P}_1$ and $\mathcal{P}_2$ are directly regular, then so is $\{\mathcal{P}_1, \mathcal{P}_2\}$. Note that there are examples where $\langle \mathcal{P}_1, \mathcal{P}_2 \rangle$ is empty.

These universal polytopes are our main object of study. The following simple example illustrates some natural questions about these polytopes. Assume that we wish to construct a triangulated surface in which every vertex of the triangulation is contained in 5 triangles; that is, the vertex-figures are pentagons $\{5\}$. This can be done in only two ways both leading to finite triangulations. If the triangulation is "freely" generated , then the resulting surface is the 2-sphere and the triangulation is isomorphic to the icosahedron $\{3,5\}$. However, if additional identifications are allowed to be made, we can also construct the hemi-icosahedron $\{3,5\}/2$, the triangulation of the real projective plane obtained from $\{3,5\}$ by identifying antipodal points. In the above notation, $\{3,5\}$ and $\{3,5\}/2$ are members of $\langle\{3\},\{5\}\rangle$, and $\{3,5\} = \{\{3\},\{5\}\}$, the universal 3-polytope with triangular facets and pentagonal vertex-figures. The important point to make here is that this universal polytope is finite.

The picture changes completely if we require exactly 6 triangles around a vertex. Now there are many ways to generate triangulations including the maps $\{3,6\}_{(b,c)}$ on the torus (described in Section 4) and the (freely generated) triangular tessellation

$\{3,6\}$ in the euclidean plane. All these are members of $\langle\{3\},\{6\}\rangle$, and $\{3,6\} = \{\{3\},\{6\}\}$ which is now infinite.

These examples address the following problems about general universal polytopes $\{\mathcal{P}_1,\mathcal{P}_2\}$ for given regular n-polytopes $\mathcal{P}_1$ and $\mathcal{P}_2$.

- When is $\langle\mathcal{P}_1,\mathcal{P}_2\rangle \neq \emptyset$? Or, equivalently, when does $\{\mathcal{P}_1,\mathcal{P}_2\}$ exist?
- When is $\{\mathcal{P}_1,\mathcal{P}_2\}$ finite? (That is, when does it behave like a convex polytope, when like an infinite tessellation?)
- Identify the group $[\mathcal{P}_1,\mathcal{P}_2]$ of $\{\mathcal{P}_1,\mathcal{P}_2\}$. (That is, construct $\{\mathcal{P}_1,\mathcal{P}_2\}$ explicitly.)

In this paper, when we use the term "classification" of polytopes, then in the given context we mean the classification of all the *finite universal* polytopes.

Given $\mathcal{P}_1$ and $\mathcal{P}_2$ the search for the universal polytope in $\langle\mathcal{P}_1,\mathcal{P}_2\rangle$ involves analysis of the group A generated by involutions $\rho_0,\ldots,\rho_n$ subject to the relations dictated by $A(\mathcal{P}_1)$ (for $\rho_0,\ldots,\rho_{n-1}$) and $A(\mathcal{P}_2)$ (for $\rho_1,\ldots,\rho_n$) together with $(\rho_0\rho_n)^2 = \varepsilon$ ([38]). This group is a quotient of the free amalgamated product of $A(\mathcal{P}_1)$ and $A(\mathcal{P}_2)$ with amalgamation along their joint subgroup which is the group of the vertex-figure of $\mathcal{P}_1$ (and the facet of $\mathcal{P}_2$), the quotient being defined by the additional relation $(\rho_0\rho_n)^2 = \varepsilon$. Now, the universal polytope $\{\mathcal{P}_1,\mathcal{P}_2\}$ exists if and only if this group A has the intersection property (4) and its subgroups $\langle\rho_0,\ldots,\rho_{n-1}\rangle$ and $\langle\rho_1,\ldots,\rho_n\rangle$ are isomorphic to $A(\mathcal{P}_1)$ and $A(\mathcal{P}_2)$, respectively. It is usually difficult to verify these conditions.

It is easy to see that in rank 3 the universal polytopes $\{\mathcal{P}_1,\mathcal{P}_2\}$ are precisely the regular tessellations $\{p,q\}$ on the 2-sphere, in the euclidean plane or in the hyperbolic plane. However, in higher ranks the structure of abstract regular polytopes is far less obvious and is complicated by the lack of easily accessible non-classical examples. To give an example in rank 4, let $\mathcal{P}_1$ be the torus map $\{6,3\}_{(s,s)}$ and $\mathcal{P}_2$ the tetrahedron $\{3,3\}$. Then $\{\mathcal{P}_1,\mathcal{P}_2\} = \{\{6,3\}_{(s,s)},\{3,3\}\}$ is a 4-polytope with toroidal facets and spherical vertex-figures. Its group $[\{6,3\}_{(s,s)},\{3,3\}]$ has the presentation

$$\rho_0^2 = \rho_1^2 = \rho_2^2 = \rho_3^2 = \varepsilon\ ,$$

$$(\rho_0\rho_1)^6 = (\rho_1\rho_2)^3 = (\rho_2\rho_3)^3 = (\rho_0\rho_2)^2 = (\rho_0\rho_3)^2 = (\rho_1\rho_3)^2 = \varepsilon\ ,$$

$$\left(\rho_0(\rho_1\rho_2)^2\right)^{2s} = \varepsilon\ .$$

The relations in the first two rows are the standard relations for the Coxeter group $[6,3,3]$, and the one extra relation in the third row corresponds to (7) below and causes the collapse of $\{6,3\}$ to the torus map $\{6,3\}_{(s,s)}$. In Section 6.2 we shall use hermitian forms to study these groups and the related universal polytopes $\{\{6,3\}_{(s,s)},\{3,3\}\}$.

For chiral polytopes the definition of classes is more subtle and involves taking care of the two enantiomorphic forms in which a polytope can occur. More precisely, if $\mathcal{P}_1$ and $\mathcal{P}_2$ are oriented chiral or directly regular n-polytopes, then $\langle\mathcal{P}_1,\mathcal{P}_2\rangle^{ch}$ denotes the *class* of all oriented chiral $(n+1)$-polytopes $\mathcal{P}$ with (oriented) facets isomorphic to $\mathcal{P}_1$ and (oriented) vertex-figures isomorphic to $\mathcal{P}_2$. Again, if $\mathcal{P}_1$ or $\mathcal{P}_2$ is chiral and the class $\langle\mathcal{P}_1,\mathcal{P}_2\rangle^{ch}$ is non-empty, then it also contains a universal member denoted by $\{\mathcal{P}_1,\mathcal{P}_2\}^{ch}$. Note that if the orientations of both $\mathcal{P}_1$ and $\mathcal{P}_2$ get changed, then the orientations of all members in the class get changed; and hence that of $\{\mathcal{P}_1, \mathcal{P}_2\}^{ch}$. However, the classes seem to be unrelated if the orientations of only one polytope is changed ([40]).

An abstract n-polytope $\mathcal{P}$ is called *spherical* if it is isomorphic to the partially ordered set of faces of a spherical complex on the euclidean $(n-1)$-sphere $\mathbb{S}^{n-1}$ ([17]). If a spherical polytope $\mathcal{P}$ has a Schläfli symbol $\{p_1, \ldots, p_{n-1}\}$ (with $p_1, \ldots, p_{n-1} \geq 3$), then it is regular and is isomorphic to the face-lattice of a regular convex polytope ([21, 14]). In particular this rules out the existence of chiral spherical polytopes.

A *toroidal* polytope or, more briefly, a *toroid*, of rank $n+1$ is an abstract $(n+1)$-polytope which is the quotient of a periodic tessellation $\mathcal{T}$ of euclidean n-space $\mathbb{E}^n$ by a subgroup Λ of its translational symmetries generated by n independent translations; the resulting toroid is written $\mathcal{T}/\Lambda$ ([29]). For a regular (resp. chiral) toroid $\mathcal{P}$ we may also assume $\mathcal{T}$ to be regular and then Λ must be normal in the symmetry group $A(\mathcal{T})$ of $\mathcal{T}$ (resp. the rotation subgroup $A^+(\mathcal{T})$ of $A(\mathcal{T})$). We shall also refer to Λ as the *identification lattice* for $\mathcal{P}$.

For a classification of the regular and chiral toroids see Section 4. It would be interesting to extend this classification to polytopes on arbitrary spherical, euclidean or hyperbolic space-forms ([52]). In rank 3 this (essentially) amounts to the classification of regular and chiral maps on surfaces; in the orientable case such a classification is known up to genus 6 ([10, 44, 16]). For higher ranks see also Section 12.

Let $\mathcal{P}$ be an abstract polytope. We call $\mathcal{P}$ *locally spherical* if both its facets and vertex-figures are spherical. We say that $\mathcal{P}$ is *locally toroidal* if its facets and vertex-figures are spherical or toroidal, with at least one kind toroidal. Our use of the term "locally of some type" always refers to the sections of rank $n-1$ of the polytope. More general terminology may only require the minimal sections which are not spherical to be of the required topological type; see for instance ([23]).

In general it is a very subtle problem to define the global topology of an abstract polytope $\mathcal{P}$. Clearly, since $\mathcal{P}$ is a partially ordered set, its order complex $\Delta(\mathcal{P})$ is a simplicial complex which provides a topological space $|\Delta(\mathcal{P})|$ on which the full automorphism group $A(\mathcal{P})$ acts as a group of homeomorphisms ([45]). However, unless all facets and vertex-figures of $\mathcal{P}$ are spherical, this space distorts some of the topological features of $\mathcal{P}$ which we may wish to preserve.

For example, if $\mathcal{P}$ is a locally toroidal 4-polytope in $\langle\{6,3\}_{(s,s)}, \{3,3\}\rangle$, then in $|\Delta(\mathcal{P})|$ each facet is realized as a cone over the 2-torus but not as a solid torus as may be desirable. However, we can overcome this problem at the price of ambiguity. In fact, given $\mathcal{P}$ we can construct a closed real 3-manifold $\mathcal{M}$ and a decomposition $\mathcal{P}'$ of $\mathcal{M}$ into solid tori, each equipped with a map $\{6,3\}_{(s,s)}$ on its boundary, such that $\mathcal{P}'$ is isomorphic to $\mathcal{P}$. In a sense, $\mathcal{P}'$ is a combinatorially regular generalized Heegaard splitting of $\mathcal{M}$ of genus 1 ([41]). But as for ordinary Heegaard splittings of genus 1 (which involve only two tori), there are many different ways in which the solid tori can be glued together to give a manifold. (In fact, for two tori the resulting manifolds are known to be $\mathbb{S}^3$, $\mathbb{S}^2 \times \mathbb{S}^1$ and the lens spaces.)

These examples illustrate some of the difficulties in defining the global topological type of an abstract polytope. For a more systematic approach to these problems as well as some classification results on the possible manifolds $\mathcal{M}$, the reader is referred to Brehm, Kühnel & Schulte [2]. Examples with $\mathcal{M} = \mathbb{S}^3$ were also discovered in Grünbaum [18], Coxeter & Shephard [11].

It is worth noting that our definition of spherical or toroidal polytopes avoids any of the problems just mentioned. In fact, by definition a spherical or toroidal polytope $\mathcal{P}$ has spherical facets and vertex-figures, and so the sphere or torus is its natural topological space (which also coincides with $|\Delta(\mathcal{P})|$). A more general notion of spherical or toroidal polytope $\mathcal{P}$ may also allow the sphere or torus to be decomposed into handlebodies which are then the facets of $\mathcal{P}$. This wider use of terminology adds on considerable complications and it may well be that the corresponding classification is then completely intractable. However, in this paper we shall not pursue these lines.

4. The Toroids

The regular and chiral toroids of rank 3 are well-known, and have been much discussed in the literature ([10]). They are the reflexible and irreflexible maps on the 2-torus and are of types $\{3,6\}$, $\{6,3\}$ and $\{4,4\}$. We begin with the first type.

Consider the euclidean plane tessellation $\mathcal{T} = \{3,6\}$. Its translation group is generated by translations τ_1, τ_2 along unit vectors x_1, x_2 inclined at $\pi/3$. If $A(\mathcal{T}) = \langle \rho_0, \rho_1, \rho_2 \rangle$ and $A^+(\mathcal{T}) = \langle \sigma_1, \sigma_2 \rangle$, we can take

$$\tau_1 = (\rho_2\rho_1\rho_0)^2 = \sigma_2^2\sigma_1^{-1}, \quad \tau_2 = (\rho_0\rho_2\rho_1)^2 = \sigma_2\sigma_1^{-1}\sigma_2\,. \tag{6}$$

For each pair $s = (b,c)$ of non-negative integers the fundamental region of the subgroup $\Lambda_s = \Lambda_{(b,c)} := \langle \tau_1^b\tau_2^c, \tau_1^{-b}\tau_2^{b+c} \rangle$ is a parallelogram with vertices (b,c), $(0,0)$, $(-c, b+c)$, $(b-c, b+2c)$ (with coordinates relative to x_1, x_2). We define $\{3,6\}_s = \{3,6\}_{(b,c)} := \mathcal{T}/\Lambda_s$, the quotient of $\mathcal{T}$ by Λ_s. If $(b,c) \neq (1,0), (0,1)$, this is a toroid of rank 3, which is regular if $bc(b-c) = 0$ and chiral otherwise; in the excluded cases the map on the torus is not a polytope in our sense. We give the details of the toroids in Table 1. The most important things we need subsequently are the numbers v of their vertices and f of their facets, and the orders g of their groups. In the regular case we usually write $s = (b,c)$ in the form $s = (s^k, 0^{2-k})$ with $s \geq 1$ and $k = 1$ or 2. In the chiral case the maps $\{3,6\}_{(b,c)}$ and $\{3,6\}_{(c,b)}$ are enantiomorphic.

We shall also write $[3,6]_s$ for the group of $\{3,6\}_s$, and $[3,6]_s^+$ for its rotation subgroup. Then we have

Theorem 1 (*a*) *For each* $s = (s^k, 0^{2-k})$ *with* $s \geq 2, k = 1$ *or* $s \geq 1, k = 2$, *the group* $[3,6]_s$ *of the regular toroid* $\{3,6\}_s$ *is the Coxeter group* $[3,6] = \langle \rho_0, \rho_1, \rho_2 \rangle$, *factored out by the relation*

$$\begin{cases} (\rho_0\rho_1\rho_2)^{2s} = \varepsilon & \text{if } k = 1, \\ (\rho_0(\rho_1\rho_2)^2)^{2s} = \varepsilon & \text{if } k = 2. \end{cases} \tag{7}$$

(*b*) *For each* $s = (b,c)$ *with* $b, c \geq 0$ *and* $(b,c) \neq (0,0), (1,0), (0,1)$, *the rotation subgroup* $[3,6]_{(b,c)}^+$ *of the (regular or chiral) toroid* $\{3,6\}_{(b,c)}$ *is the even subgroup* $[3,6]^+ = \langle \sigma_1, \sigma_2 \rangle$ *(defined by* $\sigma_1^3 = \sigma_2^6 = (\sigma_1\sigma_2)^2 = \varepsilon$*) of the Coxeter group* $[3,6]$,

s	v	f	g
$(s,0)$	s^2	$2s^2$	$12s^2$
(s,s)	$3s^2$	$6s^2$	$36s^2$

Table 1. The regular toroids $\{3,6\}_s$

factored out by the relations

$$(\sigma_2^2\sigma_1^{-1})^b(\sigma_2\sigma_1^{-1}\sigma_2)^c = \varepsilon. \tag{8}$$

For Theorem 1 and similar situations below, note that the extra relations correspond to the defining translation, here in the direction of $s = (b,c)$; this translation and its conjugates in the group span the identification lattice.

The toroid $\{6,3\}_{(b,c)}$ is the dual of $\{3,6\}_{(b,c)}$. The corresponding presentations for $[6,3]_{(b,c)}$ and $[6,3]^+_{(b,c)}$ can be obtained by dualizing the above relations; that is, by replacing ρ_0,ρ_1,ρ_2 by ρ_2,ρ_1,ρ_0, and σ_1,σ_2 by $\sigma_2^{-1},\sigma_1^{-1}$.

The toroids of type $\{4,4\}$ are constructed in a similar way from the euclidean plane tessellation $\mathcal{T} = \{4,4\}$ with vertex-set $\mathbb{Z}^2$, the set of points with integer cartesian coordinates. Now the translation group is generated by the unit translations τ_1,τ_2 along the cartesian axes. If $A(\mathcal{T}) = \langle\rho_0,\rho_1,\rho_2\rangle$ and $A^+(\mathcal{T}) = \langle\sigma_1,\sigma_2\rangle$, we can take

$$\tau_1 = \rho_0\rho_1\rho_2\rho_1 = \sigma_1^{-1}\sigma_2,\ \ \tau_2 = \rho_2\rho_1\rho_0\rho_1 = \sigma_2\sigma_1^{-1}\ .$$

For each pair $s = (b,c)$ of non-negative integers we set $\{4,4\}_s = \{4,4\}_{(b,c)} := \mathcal{T}/\Lambda_s$, with $\Lambda_s = \Lambda_{(b,c)} := \langle\tau_1^b\tau_2^c, \tau_1^{-c}\tau_2^b\rangle$ whose fundamental region is the square with vertices (b,c), $(0,0)$, $(-c,b)$ and $(b-c,b+c)$. If $(b,c) \neq (0,0),(1,0),(0,1),(1,1)$ this is a toroid of rank 3, which is regular if $bc(b-c) = 0$ and chiral otherwise. In the chiral case, $\{4,4\}_{(b,c)}$ and $\{4,4\}_{(c,b)}$ are enantiomorphic.

The regular toroids $\{4,4\}_s$ with $s = (s^k, 0^{2-k})$ with $s \geq 2$, $k = 1,2$ are the first instance of a series of toroids $\{4,3^{n-2},4\}_s$ of rank $n+1$. These will be discussed further below.

Theorem 2 *For each $s = (b,c)$ with $b,c \geq 0$ and $(b,c) \neq (0,0)$, $(1,0)$, $(0,1)$, $(1,1)$, the rotation subgroup $[4,4]^+_{(b,c)}$ of the (regular or chiral) toroid $\{4,4\}_{(b,c)}$ is the even subgroup $[4,4]^+ = \langle\sigma_1,\sigma_2\rangle$ (defined by $\sigma_1^4 = \sigma_2^4 = (\sigma_1\sigma_2)^2 = \varepsilon$) of the Coxeter group $[4,4]$, factored out by the relations*

$$(\sigma_1^{-1}\sigma_2)^b(\sigma_2\sigma_1^{-1})^c = \varepsilon. \tag{9}$$

For a detailed discussion of the toroids of higher rank we refer to [29]. Here we recall some important facts. We begin with the following observation.

Theorem 3 *There are no chiral toroids of rank greater than 3.*

For notational reasons, in the remainder of this section we prefer to denote the rank by $n+1$. To construct a regular toroid of rank $n+1 \geq 4$, we must begin

s	v	f	g
$(s,0^{n-1})$	s^n	s^n	$(2s)^n \cdot n!$
$(s^2,0^{n-2})$	$2s^n$	$2s^n$	$2^{n+1}s^n \cdot n!$
(s^n)	$2^{n-1}s^n$	$2^{n-1}s^n$	$2^{2n-1}s^n \cdot n!$

Table 2. The regular toroids $\{4,3^{n-2},4\}_s$

with a regular honeycomb of $\mathbb{E}^n$. Except for $n = 4$, the only such honeycomb is the tessellation $\{4,3^{n-2},4\}$ of $\mathbb{E}^n$ by cubes; here and below, r^k will be used to denote a string of k consecutive r's. In $\mathbb{E}^4$, there are two other regular honeycombs $\{3,3,4,3\}$ and $\{3,4,3,3\}$, which are duals.

We first consider the cubic tessellation $\{4,3^{n-2},4\}$ (with $n \geq 2$). Its vertex set may be taken to be $\mathbb{Z}^n$; this set can also be regarded as its translation group. Because we wish the resulting toroid to be regular, if the translation by $s \in \mathbb{Z}^n$ occurs in the identification lattice Λ, then so must all its conjugates under the group $[4,3^{n-2},4]$ of the honeycomb, or, what amounts to the same thing, under the group $[3^{n-2},4]$ of its vertex-figure, which consists of all permutations of the coordinates of vectors with all changes of signs. We shall write Λ_s for the translation group generated by $s := (s^k,0^{n-k})$ and its images under permutation and changes of sign of coordinates, where $s \geq 1$ is an integer and $1 \leq k \leq n$. We shall see that the only allowed values of k are $k = 1, 2$ or n. The regular polytope which results by this factorization is denoted by $\{4,3^{n-2},4\}_s := \{4,3^{n-2},4\}/\Lambda_s$. In order that the corresponding group, which we write as $[4,3^{n-2},4]_s$, satisfy the intersection property, we must actually have $s \geq 2$, but otherwise there are no further restrictions; see Table 2.

Theorem 4 *For each $n \geq 2$, and $s = (s^k,0^{n-k})$ with $s \geq 2$ and $k = 1,2$ or n, there is a (self-dual) regular toroid $\{4,3^{n-2},4\}_s$ of rank $n+1$. Its group $[4,3^{n-2},4]_s$ is the Coxeter group $[4,3^{n-2},4] = \langle \rho_0,\ldots,\rho_n \rangle$, factored out by the single extra relation*

$$(\rho_0\rho_1\cdots\rho_n\rho_{n-1}\cdots\rho_k)^{sk} = \varepsilon \,. \tag{10}$$

As we said above, the only other toroids are dual pairs derived from $\{3,3,4,3\}$ and $\{3,4,3,3\}$. We just consider the former. We may take the vertex set to be $\mathbb{Z}^4 \cup (\mathbb{Z}^4 + (\frac{1}{2},\frac{1}{2},\frac{1}{2},\frac{1}{2}))$, the set of points of $\mathbb{E}^4$ whose cartesian coordinates are all integers or all halves of odd integers. These points also correspond to the integer quaternions; in this context, the symmetry group $[3,3,4,3]$ consists of the mappings $x \mapsto q_1xq_2 + h$ and $x \mapsto q_1\overline{x}q_2 + h$, where q_1, q_2 are unit integer quaternions, h is an integer quaternion, and $\overline{x}$ is the (quaternion) conjugate of x ([15]). Much the same analysis as above applies, and, initially bearing only the vertices of $\{3,3,4,3\}$ in $\mathbb{Z}^4$ in mind, we conclude that the identification is by a vector $(s^k,0^{4-k})$ (and its images under permutation and changes of sign of cordinates) for some integer $s \geq 2$ and some $k = 1,2$ or 4. However, taking the full group of symmetries of $\{3,3,4,3\}$ into account, we observe that (s^4) is equivalent to $(2s,0^3)$, and so the last case has already been counted. Using the same notation as for the cubic toroids, and denoting the dual by the same suffix, we thus obtain

s	v	f	g
$(s,0,0,0)$	s^4	$3s^4$	$1152s^4$
$(s,s,0,0)$	$4s^4$	$12s^4$	$4608s^4$

Table 3. The regular toroids $\{3,3,4,3\}_s$

Theorem 5 *For each $s = (s^k, 0^{4-k})$ with $s \geq 2$ and $k = 1$ or 2, there is a regular toroid $\{3,3,4,3\}_s$ (and its dual $\{3,4,3,3\}_s$) of rank 5. The group $[3,3,4,3]_s$ is the Coxeter group $[3,3,4,3] = \langle \rho_0, \ldots, \rho_4 \rangle$, factored out by the extra relation*

$$\begin{cases} (\rho_0\sigma\tau\sigma)^s = \varepsilon & \text{if } k = 1, \\ (\rho_0\sigma\tau)^{2s} = \varepsilon & \text{if } k = 2, \end{cases} \tag{11}$$

where $\sigma := \rho_1\rho_2\rho_3\rho_2\rho_1$, $\tau := \rho_4\rho_3\rho_2\rho_3\rho_4$.

We list the details of these polytopes in Table 3. However, since the number of vertices of $\{3,3,4,3\}_s$ is the same as the number of facets of its dual $\{3,4,3,3\}_s$, and vice versa, we need only consider the former.

There are various quotient and subgroup relations between the groups of these toroids ([29]). The quotient relations arise from corresponding subgroup relations between the translation groups Λ_s. For $[4,3^{n-2},4]_s$ we have

$$\Lambda_{(2s,0^{n-1})} \leq \begin{Bmatrix} \Lambda_{(s^n)} \\ \Lambda_{(s^2,0^{n-2})} \end{Bmatrix} \leq \Lambda_{(s,0^{n-1})} \, ,$$

for all $s \geq 2$. If n is even, there is also the relation

$$\Lambda_{(s^n)} \leq \Lambda_{(s^2,0^{n-2})} \, .$$

Moreover, if p is an odd prime, we obviously have $\Lambda_{ps} \leq \Lambda_s$ for every s. It may be seen that every other subgroup relationship is a consequence of these. We deduce

Theorem 6 *Let $n \geq 3$. For each $s \geq 2$, there are coverings*

$$\{4,3^{n-2},4\}_{(2s,0^{n-1})} \searrow \begin{Bmatrix} \{4,3^{n-2},4\}_{(s^n)} \\ \{4,3^{n-2},4\}_{(s^2,0^{n-2})} \end{Bmatrix} \searrow \{4,3^{n-2},4\}_{(s,0^{n-1})} \, .$$

In addition, if n is even, there is a covering

$$\{4,3^{n-2},4\}_{(s^n)} \searrow \{4,3^{n-2},4\}_{(s^2,0^{n-2})} \, .$$

Lastly, for each $s = (s^k, 0^{n-k})$ (with $s \geq 2$ and $k = 1, 2$ or n) and every odd prime p, there is a covering

$$\{4,3^{n-2},4\}_{ps} \searrow \{4,3^{n-2},4\}_s \, .$$

Exactly similar considerations apply to the polytopes of type $\{3,3,4,3\}$, and we obtain

Theorem 7 *Let $s \geq 2$. Then there are coverings*

$$\{3,3,4,3\}_{(2s,0,0,0)} \searrow \{3,3,4,3\}_{(s,s,0,0)} \searrow \{3,3,4,3\}_{(s,0,0,0)}\,.$$

Further, if p is an odd prime, there is a covering

$$\{3,3,4,3\}_{p\mathbf{s}} \searrow \{3,3,4,3\}_{\mathbf{s}}\,,$$

with $\mathbf{s} = (s,0,0,0)$ *or* $(s,s,0,0)$.

5. Hyperbolic Honeycombs

In preparation for our investigation of the locally toroidal regular polytopes, we now recall some facts about regular honeycombs in hyperbolic space $\mathbb{H}^n$ of dimension $n \geq 3$ ([8]). Since the facets and vertex-figures of a locally toroidal polytope are spherical or quotients of euclidean tessellations, the polytope itself must necessarily be a quotient of a hyperbolic honeycomb with spherical or euclidean facets or vertex-figures.

In $\mathbb{H}^3$, there are 15 regular honeycombs. The honeycombs $\{3,4,4\}$, $\{3,3,6\}$, $\{4,3,6\}$ and $\{5,3,6\}$ have spherical facets and have all their vertices on the absolute. Their duals, $\{4,4,3\}$, $\{6,3,3\}$, $\{6,3,4\}$ and $\{6,3,5\}$ have spherical vertex-figures and all their facets are inscribed in horospheres instead of finite spheres. The self-dual honeycombs $\{4,4,4\}$, $\{6,3,6\}$ and $\{3,6,3\}$ have both their vertices at infinity and their facets inscribed in horospheres. All these eleven types occur as Schläfli symbols of locally toroidal regular polytopes of rank 4. The remaining four honeycombs $\{3,5,3\}$, $\{4,3,5\}$, $\{5,3,4\}$ and $\{5,3,5\}$ are locally spherical and are (locally finite) tessellations in $\mathbb{H}^3$.

In $\mathbb{H}^4$, there are 7 regular honeycombs. Of those, only $\{3,4,3,4\}$ and its dual $\{4,3,4,3\}$ are not locally spherical and can occur as the type of some locally toroidal regular polytope of rank 5. The first has 24-cells as facets and its vertices are all on the absolute, and the second has 24-cells as vertex-figures and its facets are cubic tessellations inscribed into horospheres.

In $\mathbb{H}^5$, there are 5 regular honeycombs, all of which are not locally spherical and have euclidean tessellations as facets inscribed into horospheres or all their vertices at infinity. These are $\{3,3,3,4,3\}$, $\{4,3,3,4,3\}$, $\{3,3,4,3,3\}$, $\{3,4,3,3,4\}$ and $\{3,4,3,3,3\}$. Only the first has spherical facets (which are crosspolytopes), and only the last, the dual of the first, has spherical vertex-figures (which are cubes). These are the only types for locally toroidal regular polytopes of rank 6.

In $\mathbb{H}^n$ with $n \geq 6$, there are no regular honeycombs. As a consequence, locally toroidal regular polytopes can exist in ranks 4, 5 and 6 alone.

6. Polytopes of Rank 4

In constructing regular polytopes from groups, the following *twisting technique* has proved to be extremely useful ([24, 30]).

Let W be a group generated by k involutions $\sigma_1, \ldots, \sigma_k$; usually W is a C-group, for example, a Coxeter group or unitary reflection group. A *twisting operation* shall only be defined for those groups W which admit certain automorphisms τ permuting the generators σ_i. If these automorphisms τ are themselves involutions, we can augment W by their addition and in suitable cases obtain a new group A with certain distinguished generators $\rho_0, \ldots, \rho_{n-1}$. Writing B for the group of automorphisms of W generated by these τ, we have $A = W \ltimes B$, the semi-direct product of W by B. We shall write such a twisting operation as

$$(\sigma_1, \ldots, \sigma_k; \tau\text{'s}) \mapsto (\rho_0, \ldots, \rho_{n-1}) .$$

In applications, B may be of order 2 generated by just one involutory automorphism τ (as in (12)), or, in the other extreme case, may itself be the group of any regular polytope of higher rank suitably acting on W. In many examples the group W can actually be represented by a diagram and the automorphism τ can be realized by symmetries of this diagram.

We now discuss the locally toroidal regular polytopes of rank 4 and begin with those of Schläfli type $\{4,4,r\}$ (or $\{r,4,4\}$) with $r = 3$ or 4.

6.1. Types $\{4,4,r\}$

The universal regular 4-polytopes

$${}_1\mathcal{T}^4_{\mathbf{s}} := \{\{4,4\}_{\mathbf{s}}, \{4,3\}\} ,$$

$\mathbf{s} = (s^k, 0^{2-k})$ with $s \geq 2$ and $k = 1$ or 2, can be constructed directly from twisting operations on Coxeter groups. Since these are simple, we shall include them here. If $k = 1$, we can simply take the group $W = \langle \sigma_0, \ldots, \sigma_4 \rangle$ with diagram

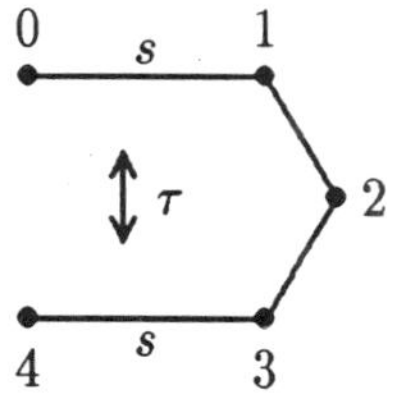

(12)

and apply the operation

$$(\sigma_0, \ldots, \sigma_4; \tau) \mapsto (\sigma_0, \tau, \sigma_3, \sigma_2) =: (\rho_0, \ldots, \rho_3) . \tag{13}$$

It is straightforward to verify the defining relations for the corresponding group, which in turn implies the universality of the polytope.

If $k = 2$, we can work instead with $W = \langle \sigma_0, \ldots, \sigma_5 \rangle$ with diagram

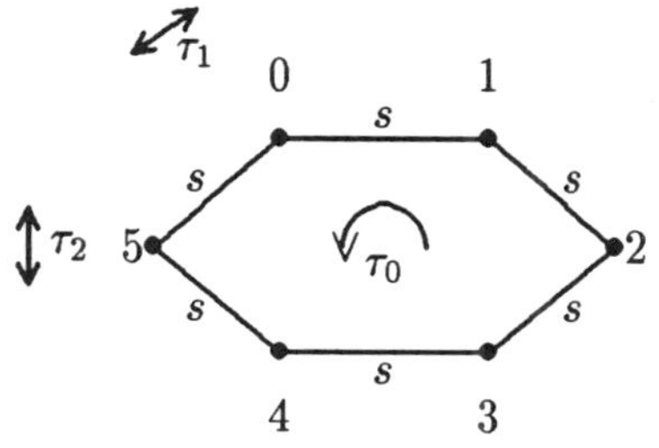

(14)

and use the operation

$$(\sigma_0, \ldots, \sigma_5; \tau_0, \tau_1, \tau_2) \mapsto (\tau_0, \sigma_2, \tau_1, \tau_2) =: (\rho_0, \ldots, \rho_3) .$$

This proves

Theorem 8 *The regular 4-polytope* ${}_1\mathcal{T}_s^4 = \{\{4,4\}_s, \{4,3\}\}$ *exists for all* $s = (s^k, 0^{2-k})$ *with* $s \geq 2$ *and* $k = 1, 2$. *The only finite instances occur for* $s = (2,0), (3,0)$ *and* $(2,2)$, *with groups* $D_4 \times S_4$ *of order* 192, $S_6 \times C_2$ *of order* 1440, *and* $C_2 \wr D_6$ *(wreath product) of order* 768, *respectively.*

The classification of the universal regular 4-polytopes

$${}_2\mathcal{T}_{s,t}^4 := \{\{4,4\}_s, \{4,4\}_t\} ,$$

$s = (s^k, 0^{2-k}), t = (t^l, 0^{2-l})$ with $s, t \geq 2$ and $k, l = 1$ or 2, is more difficult and requires more sophisticated twisting operations ([27]). The classification is complete except when $k = l = 1$ and s, t are odd and distinct.

Theorem 9 *The regular 4-polytope* ${}_2\mathcal{T}_{s,(t,t)}^4 = \{\{4,4\}_s, \{4,4\}_{(t,t)}\}$ *exists for all* $s = (s^k, 0^{2-k})$ *with* $s \geq 2$ *and* $k = 1, 2$ *and all* $t \geq 2$. *The only finite instances occur for:* $s = (2,0), t \geq 2$; $s = (3,0), t = 2$; *and* $s = (2,2), t = 2$ *or* 3. *The corresponding groups are:* $(D_t \times D_t \times C_2 \times C_2) \ltimes (C_2 \times C_2)$ *of order* $64t^2$; $(S_4 \times S_4) \ltimes (C_2 \times C_2)$ *of order* 2304; $C_2^4 \ltimes [4,4]_{(2,2)}$ *of order* 1024; *and* $C_2^6 \ltimes [4,4]_{(3,3)}$ *of order* 9216, *respectively.*

Theorem 10 *Let* $2 \leq s \leq t$, *and let* s, t *not be both odd and distinct. Then the regular 4-polytope* ${}_2\mathcal{T}_{(s,0),(t,0)}^4 = \{\{4,4\}_{(s,0)}, \{4,4\}_{(t,0)}\}$ *exists except when* $s = 2$ *and* t *is odd. The only finite instances occur for* $s = 2$ *and* $t = 2m$ *even, and* $(s,t) = (3,3)$ *or* $(3,4)$. *The corresponding groups are* $(D_m \times D_m) \ltimes [4,4]_{(2,0)}$ *of order* $128m^2$ *(with* $D_1 = C_2$ *if* $m = 1$*),* $S_6 \times C_2$ *of order* 1440, *and* $C_2 \wr [4,4]_{(3,0)}$ *of order* 36864, *respectively.*

In the exceptional case when s, t are odd and distinct, the cut method of Section 7 below supports the following

Conjecture 1 *Let* $s, t \geq 3$ *be odd and distinct. Then the regular 4-polytope* ${}_2\mathcal{T}_{(s,0),(t,0)}^4 = \{\{4,4\}_{(s,0)}, \{4,4\}_{(t,0)}\}$ *exists and is infinite if* $(s,t) \neq (3,5), (5,3)$.

Note that an application of the Coxeter-Todd coset enumeration algorithm suggests that the polytope is also infinite in the two cases excluded in the conjecture (even though the corresponding cuts are finite).

6.2. Types $\{6,3,p\}$

In this section we classify the universal regular 4-polytopes

$${}_p\mathcal{T}_s^4 = \{\{6,3\}_s, \{3,p\}\}$$

with $p = 3,4,5$ and $s = (s^k, 0^{2-k})$, with $s \geq 2$ if $k = 1$ and $s \geq 1$ if $k = 2$, as well as

$$_6\mathcal{T}^4_{s,t} = \{\{6,3\}_s, \{3,6\}_t\}$$

with $s = (s^k, 0^{2-k}), t = (t^l, 0^{2-l})$, with $s,t \geq 2$ if $k,l = 1$ and $s,t \geq 1$ if $k,l = 2$. Note that the left suffix (3, 4, 5 or 6) in our notation is the same as the last entry in the Schläfli symbol. We write $_pA^4_s$ and $_6A^4_{s,t}$ for the quotient $\langle \rho_0, \rho_1, \rho_2, \rho_3 \rangle$ of $[6,3,p]$ or $[6,3,6]$ which is defined by the extra relations for $[6,3]_s$ and $[3,6]_t$; see Theorem 1. Then, if $_p\mathcal{T}^4_s$ and $_6\mathcal{T}^4_{s,t}$ exist, then $_pA^4_s = [\{6,3\}_s, \{3,p\}]$ and $_6A^4_{s,t} = [\{6,3\}_s, \{3,6\}_t]$, with the notation as in Section 3.

In classifying these polytopes $\mathcal{P}$ the following strategy proved to be successful.

1. Find a "suitable" normal subgroup W of A $(= {_pA^4_s}, {_6A^4_{s,t}})$ of finite index.
2. Construct a "locally unitary" representation of W over the complex numbers $\mathbb{C}$, $f : W \mapsto GL_m(\mathbb{C})$ (say) with m determined by the vertex-figure $\{3,p\}$ or $\{3,6\}_t$. This representation f will support a hermitian form $\langle\ ,\ \rangle$ on $\mathbb{C}^m$.
3. Use $\langle\ ,\ \rangle$ to determine the structure of $\mathcal{P}$ and A.

The construction of W and f is based on the following observation on unitary reflection groups (Shephard & Todd [43], Coxeter [7], Cohen [4]).

Consider the group $[1\ 1\ 1]^s$ $(s \geq 2)$ which is generated by involutions $\sigma_1, \sigma_2, \sigma_3$ and abstractly defined by the presentation

$$\sigma_1^2 = \sigma_2^2 = \sigma_3^2 = (\sigma_1\sigma_2)^3 = (\sigma_2\sigma_3)^3 = (\sigma_1\sigma_3)^3 = (\sigma_1\sigma_2\sigma_3\sigma_2)^s = \varepsilon\,. \qquad (15)$$

This group can be represented by a triangular diagram

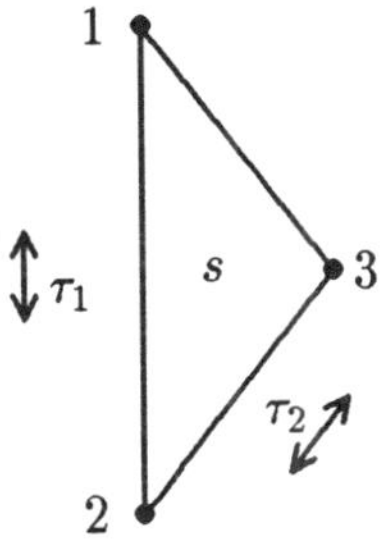

(16)

(the underlying Coxeter diagram), with a mark s inside the triangle to indicate the rightmost extra relation. Now, using the two group automorphisms τ_1 and τ_2 which act as indicated by the diagram symmetries, we can extend $[1\ 1\ 1]^s$ in two ways, both simple examples of twisting operations. First, the operation

$$(\sigma_1, \sigma_2, \sigma_3; \tau_1) \mapsto (\tau_1, \sigma_2, \sigma_3) =: (\rho_0, \rho_1, \rho_2)$$

recognizes the group $[6,3]_{(s,0)}$ as $[1\ 1\ 1]^s \ltimes C_2$. And second,

$$(\sigma_1, \sigma_2, \sigma_3; \tau_1, \tau_2) \mapsto (\sigma_1, \tau_1, \tau_2) =: (\rho_0, \rho_1, \rho_2)$$

gives $[6,3]_{(s,s)} \cong [1\ 1\ 1]^s \ltimes S_3$.

Geometrically the generators σ_i can be described as follows ([7, 25]). Let e_1, e_2, e_3 be the canonical base of complex 3-space $\mathbb{C}^3$. Define the linear mapping $S_i : \mathbb{C}^3 \mapsto \mathbb{C}^3$ by

$$xS_i = x - 2\langle x, e_i\rangle e_i \quad (x \in \mathbb{C}^3), \qquad (17)$$

where $\langle\ ,\ \rangle$ is a hermitian form on $\mathbb{C}^3$ defined by

$$\langle x, y\rangle = \sum_{i=1}^{3} x_i\overline{y_i} \ - \ \tfrac{1}{2} \sum_{i,j=1, i\neq j}^{3} c_{ij}x_i\overline{y_j}\,. \tag{18}$$

There are several choices for the coefficients c_{ij} each of which gives a positive definite form $\langle\ ,\ \rangle$ (defining a unitary geometry) such that $\sigma_i \mapsto S_i$ $(i = 1,2,3)$ defines a unitary representation of $[1\ 1\ 1]^s$. Write $\gamma_{123} := c_{12}c_{23}c_{31}$ and $c_s := e^{2\pi i/s}$. Then one such choice requires that both each c_{ij} and γ_{123} are equal to c_s or $\overline{c_s}$. Note that this is a symmetrical version of the choice in [7].

For each of the groups ${}_pA^4_s$ and ${}_6A^4_{s,t}$ it is now possible to identify the group W and representation f. In each case the choice depends on the parameters s, p and t. We shall illustrate the method by an example rather than discussing the construction in full generality.

Consider the group ${}_3A^4_{(s,s)}$ of ${}_3\mathcal{T}^4_{(s,s)} = \{\{6,3\}_{(s,s)}, \{3,3\}\}$. Then the vertex-figure is a tetrahedron $\{3,3\}$ and has 4 vertices. Take the group $W = W_{(s,s)}$ with 4 generators $\sigma_1, \ldots, \sigma_4$ abstractly defined by

$$\sigma_i^2 = (\sigma_i\sigma_j)^3 = (\sigma_i\sigma_j\sigma_k\sigma_j)^s = \varepsilon \quad (1 \leq i,j,k \leq 4;\ \text{distinct}). \tag{19}$$

This group can be represented by the tetrahedral diagram

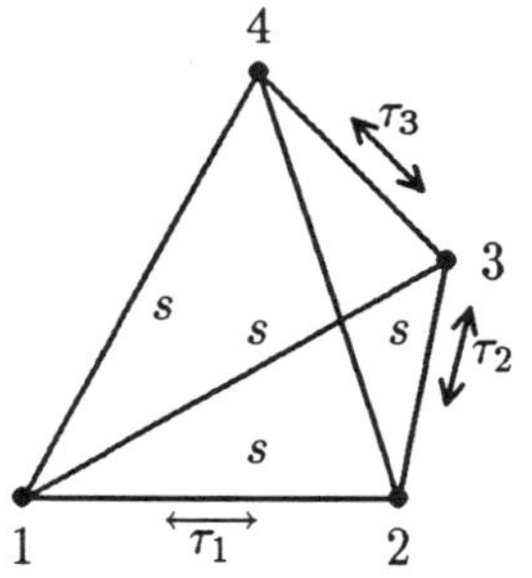

(20)

in which each triangular 2-face is marked by s. (The number of generators is 4 because the vertex-figure has 4 vertices, not because the rank of the locally toroidal polytope is 4. If the vertex-figure is an icosahedron $\{3,5\}$, then there are 12 generators and the hermitian form has 12 variables.) Now W admits three group automorphisms τ_1, τ_2, τ_3 each represented by a transposition. Adjoining these to W and using the twisting operation

$$(\sigma_1, \ldots, \sigma_4; \tau_1, \tau_2, \tau_3) \mapsto (\sigma_1, \tau_1, \tau_2, \tau_3) =: (\rho_0, \ldots, \rho_3) \tag{21}$$

we can now recognize ${}_3A^4_{(s,s)}$ as $W_{(s,s)} \ltimes S_4$; in fact, the defining relations for the two groups correspond to each other.

Next we construct a complex representation $f : W_{(s,s)} \mapsto GL_4(\mathbb{C})$ which supports a hermitian form $\langle\ ,\ \rangle$ on $\mathbb{C}^4$. We define S_i as in (17) (with $\mathbb{C}^3$ replaced $\mathbb{C}^4$) and $\langle x, y\rangle$ as in (18) (with 3 replaced by 4). Writing $\gamma_{ijk} := c_{ij}c_{jk}c_{ki}$ $(i,j,k$ distinct) we impose the condition that each c_{ij} and each γ_{ijk} is equal to c_s or $\overline{c_s}$. (In the case of an arbitrary vertex-figure modifications to this rule are needed for index sets $\{i,j\}$

or $\{i,j,k\}$ which are non-edges or non-faces of the vertex-figure.) For instance, we can take

$$c_{12} = c_{34} = c_{31} = c_s\,,\; c_{23} = c_{24} = c_{41} = \overline{c_s}\,.$$

The condition on the c_{ij}'s and γ_{ijk}'s implies that any restriction of $\langle\ ,\ \rangle$ to 3 variables is a positive definite form; that is, h is *locally unitary*. In particular,

$$\langle\sigma_i,\sigma_j,\sigma_k\rangle \cong \langle S_i,S_j,S_k\rangle \cong [1\,1\,1]^s$$

for all $\{i,j,k\}$, and $\langle S_1,\ldots,S_4\rangle$ acts irreducibly on $\mathbb{C}^4$.

We can now decide which groups $W_{(s,s)}$ are finite. If h is a non-degenerate and indefinite form, then $\langle S_1,\ldots,S_4\rangle$ acts irreducibly on $\mathbb{C}^4$ and is infinite; it follows that $W_{(s,s)}$ must also be infinite. If h is positive definite, then $\langle S_1,\ldots,S_4\rangle$ is a finite unitary reflection group and the representation f is faithful. (In the general case it is not known if f is always faithful.) It follows that $W_{(s,s)}$ is finite if and only if h is positive definite. The same is now also true for ${}_3A^4_{(s,s)}$ and its polytope ${}_3\mathcal{T}^4_{(s,s)}$. But

$$det(h) = (-9 - 16\cos(2\pi/s) - 2\cos(4\pi/s))/16\,,$$

so that positive definiteness occurs exactly for $s = 2$; in particular, $W_{(2,2)} \cong S_5$ and ${}_3A^4_{(2,2)} \cong S_5 \times S_4$, so we actually have real groups here.

In a similar fashion we can classify (almost) all universal locally toroidal regular 4-polytopes of types $\{6,3,p\}$ with $p = 3,4,5,6$. We now summarize the results. For notation for unitary reflection groups we refer to [7, 25].

Theorem 11 *The regular 4-polytope ${}_3\mathcal{T}^4_s = \{\{6,3\}_s,\{3,3\}\}$ exists for all $s = (s^k,0^{2-k})$ with $s \geq 2$ and $k = 1,2$ (but not for $s = 1, k = 2$). The only finite instances occur for $s = (2,0)$, $(3,0)$, $(4,0)$ and $(2,2)$. In the first case, its group is $S_5 \times C_2$ of order 240, and in the last case it is $S_5 \times S_4$ of order 2880. In the second and third case (where $s = 3,4$), the group is $[1\,1\,2]^s \ltimes C_2$, of order 1296 or 25360 respectively; here $[1\,1\,2]^s$ is the finite unitary reflection group in $\mathbb{C}^4$ whose diagram is obtained by attaching at vertex 3 of (16) a tail consisting of one unmarked branch.*

Note that Theorem 11 confirms a conjecture of Grünbaum [18] on the finiteness of these polytopes (which he denoted by $\mathcal{H}_{(s,0)}$ and $\mathcal{H}_{(s,s)}$, respectively). See also Altshuler [1] for the construction of 3-dimensional simplicial complexes whose vertex links are preassigned torus maps; the duals of the corresponding face lattices are abstract 4-polytopes with toroidal facets and simplicial vertex-figures; however, in general, these are neither regular or chiral.

Theorem 12 *The regular 4-polytope ${}_4\mathcal{T}^4_s = \{\{6,3\}_s,\{3,4\}\}$ exists for all $s = (s^k,0^{2-k})$, with $s \geq 2$ if $k = 1$ and $s \geq 1$ if $k = 2$. The only finite instances occur for $s = (1,1)$ and $(2,0)$, with groups $S_3 \ltimes [3,4]$ of order 288 and $[3,3,4] \ltimes C_2$ of order 768, respectively.*

Theorem 13 *The regular 4-polytope ${}_5\mathcal{T}^4_s = \{\{6,3\}_s,\{3,5\}\}$ exists for all $s = (s^k,0^{2-k})$ with $s \geq 2$ and $k = 1,2$ (but not for $s = 1, k = 2$). The only finite instance occurs for $s = (2,0)$, in which case the group is $[3,3,5] \ltimes C_2$ of order 28800.*

Theorem 14 (*a*) *The regular 4-polytope* ${}_6\mathcal{T}^4_{(s,s),t} = \{\{6,3\}_{(s,s)}, \{3,6\}_t\}$ *exists for all* $s \geq 1$ *and all* $t = (t^l, 0^{2-l})$, *except when* $s = l = 1$ *and* $3 \nmid t$. *The only finite instances occur for* $s = 1$, *or* $s = 2$ *and* $t = (2,0)$. *In the first case the group is* $S_3 \ltimes [3,6]_t$ *which is of order* $72t^2$ *if* $l = 1$ *or* $216t^2$ *if* $l = 2$, *and in the second case the group is* $S_5 \times S_4 \times C_2$ *of order* 5760.
(*b*) *The regular 4-polytope* ${}_6\mathcal{T}^4_{(s,0),(t,0)} = \{\{6,3\}_{(s,0)}, \{3,6\}_{(t,0)}\}$ *exists for all* s,t *with* $s,t \geq 2$. *The only finite instances occur for* $t = 2 \leq s \leq 4$ (*or* $s = 2 \leq t \leq 4$), *in which case the group is* $[1\,1\,2]^s \ltimes (C_2 \times C_2)$, *of order* 480, 2592 *and* 30720 *respectively, with* $[1\,1\,2]^s$ *as in Theorem 11.*

Many polytopes in the above theorems admit geometric realizations in euclidean spaces, and for several finite examples explicit coordinates of the vertices of these realizations are known ([24, 25]). For a general discussion on realizations we refer to [22] or the article by Peter McMullen in this volume.

6.3. Type $\{3,6,3\}$

Relatively little is known about the universal regular 4-polytopes

$${}_7\mathcal{T}^4_{s,t} = \{\{3,6\}_s, \{6,3\}_t\}$$

with $s = (s^k, 0^{2-k}), t = (t^l, 0^{2-l})$, with $s,t \geq 2$ if $k,l = 1$ and $s,t \geq 1$ if $k,l = 2$. Except for some specific parameter values like those mentioned in [5, 49], the only results known are those obtained by the method in the previous section and some variants of this ([25, 28]). However, these methods are not strong enough to settle the general case for the type $\{3,6,3\}$. In particular, one can prove

Theorem 15 *The regular 4-polytopes* ${}_7\mathcal{T}^4_{(s,s),(s,0)} = \{\{3,6\}_{(s,s)}, \{6,3\}_{(s,0)}\}$ *and* ${}_7\mathcal{T}^4_{(s,s),(3s,0)} = \{\{3,6\}_{(s,s)}, \{6,3\}_{(3s,0)}\}$ *exist for all* $s \geq 2$, *the latter* (*but not the former*) *also for* $s = 1$. *Among these, the only finite instances are* ${}_7\mathcal{T}^4_{(1,1),(3,0)}$ *with group* $[1\,1\,1]^3 \ltimes S_3$ *of order* 324 *and* ${}_7\mathcal{T}^4_{(2,2),(2,0)}$ *with group* $S_5 \times S_3$ *of order* 720.

There are various quotient and subgroup relations between the locally toroidal regular 4-polytopes of types $\{6,3,3\}$, $\{6,3,4\}$, $\{6,3,6\}$ and $\{3,6,3\}$; see [28] for a detailed discussion. These are based on relations between the symmetry groups of the corresponding hyperbolic honeycombs. For example, the polytopes in the next theorem are related to ${}_3\mathcal{T}^4_{(s,0)} = \{\{6,3\}_{(s,0)}, \{3,3\}\}$.

Theorem 16 *The regular 4-polytope* ${}_7\mathcal{T}^4_{(s,0),(s,0)} = \{\{3,6\}_{(s,0)}, \{6,3\}_{(s,0)}\}$ *exists at least for all* s *with* $3 \nmid s$. *It is infinite when* $s \geq 5$ *and* $3 \nmid s$ (*and most likely also when* $s = 4$). *If* $s = 2$ *it is finite and its group is* $S_5 \times C_2$ *of order* 240.

7. The Cut Method

Before we proceed with the discussion in ranks 5 and 6 we illustrate a powerful geometric method, the *cut method*, which sheds some light on why certain parameter vectors s, t give finite polytopes while others give infinite ones. This method reduces problems on polytopes of a given rank to similar problems on polytopes in lower ranks. At present the method is only developed to the extent that it explains many of our results rather than that it proves them. It is on open problem whether or not the cut method can actually be turned into a proof technique. What is needed is a *cut theorem* which gives a criterion for the universality of the cut of a universal polytope.

We shall deliberately be vague with our definition of a cut. By a *cut* of a regular polytope $\mathcal{P}$ we mean a regular polytope $\mathcal{K}$ of lower rank with $A(\mathcal{K}) \leq A(\mathcal{P})$, whose vertices are vertices of $\mathcal{P}$, and are just those invariant under some distinguished subgroup of $A(\mathcal{P})$.

To give a simple example, let $\mathcal{P} := \{4, 3^{n-2}, 4\}_s$ with $s := (s^k, 0^{n-k})$ and $s \geq 2$, $k = 1, 2$; this is a cubic toroid of rank $n+1$. Let $A(\mathcal{P}) = [4, 3^{n-2}, 4]_s = \langle \rho_0, \dots, \rho_n \rangle$ and write $\tau := \rho_2\rho_3 \dots \rho_n\rho_{n-1} \dots \rho_2$. Then (10) becomes

$$\begin{cases} (\rho_0\rho_1\tau\rho_1)^s = \varepsilon & \text{if } k = 1 \\ (\rho_0\rho_1\tau)^{2s} = \varepsilon & \text{if } k = 2. \end{cases} \tag{22}$$

Now, the subgroup $\langle \rho_0, \rho_1, \tau \rangle$ of $A(\mathcal{P})$ is the group of the cut of $\mathcal{P}$ invariant under $\rho_3, \dots, \rho_n$. More specifically, if $\widetilde{s} := (s^k, 0^{2-k})$ for each $s := (s^k, 0^{n-k})$ with $k =$ 1 or 2, this cut is the toroid $\{4, 4\}_{\widetilde{s}}$ of rank 3.

Such a cut is called *universal*, because it is the universal polytope in its "class". The question whether or not a cut is universal can be very subtle. Usually it is necessary, but generally not sufficient, that the relations on the corresponding cut of the universal polytope (in the above case, $\{4, 4\}$) which determine it are just those which determine the whole polytope as a quotient of its universal polytope (which here is $\{4, 3^{n-2}, 4\}$). That the cut in the above example is indeed universal can easily be seen geometrically. However, in general, universality of cuts is much less obvious.

As another example, consider the 4-polytope $\mathcal{P} = \{\{3, 4\}, \{4, 4\}_{(s,0)}\}$, the dual of ${}_1\mathcal{T}^4_{(s,0)}$ from Section 6.1. Write $A(\mathcal{P}) = \langle \rho_0, \dots, \rho_3 \rangle$ and consider the "(mixing) operation"

$$(\rho_0, \dots, \rho_3) \mapsto (\rho_0, \rho_1, \rho_2\rho_3\rho_2) =: (\varphi_0, \varphi_1, \varphi_2)$$

(which is not a twisting operation). Then $\langle \varphi_0, \varphi_1, \varphi_2 \rangle$ is the group of a regular map $\mathcal{M}$ of type $\{3, s\}$, which is the cut of $\mathcal{P}$ on the "reflection wall" of ρ_3 (since ρ_3 commutes with $\varphi_0, \varphi_1, \varphi_2$). If we span topological discs into the 2-faces of $\mathcal{M}$, we can think of $\mathcal{M}$ as a surface which (in a sense) cuts right through $\mathcal{P}$, but usually only locally. Now, dualizing the construction for $\mathcal{P}^*$ in Section 6.1 we find that

$$\langle \varphi_0, \varphi_1, \varphi_2 \rangle = \langle \sigma_2, \sigma_3, \tau\sigma_0\tau \rangle = \langle \sigma_2, \sigma_3, \sigma_4 \rangle \cong [3, s] \, ,$$

so that $\mathcal{M}$ is indeed the (universal) tessellation $\{3, s\}$ ([27]). But clearly, if the cut is infinite, then so must be the whole polytope. This explains why $\mathcal{P}$ is infinite if $s \geq 6$. But $\mathcal{P}$ is finite only if $s = 2$ or 3, and so the finiteness of $\mathcal{M}$ for $s = 4, 5$ is not sufficient to imply finiteness for $\mathcal{P}$, in agreement with our claim that $\mathcal{M}$ cuts only locally through $\mathcal{P}$.

s	v	f	g
(2,0,0)	24	8	9216
(2,2,0)	48	32	36864
(2,2,2)	1536	2048	2359296

Table 4. The finite polytopes $\mathcal{T}_s^5 = \{\{3,4,3\},\{4,3,4\}_s\}$

It would be very helpful to be able to preserve some of the arguments in this example and prove universality of the cut $\mathcal{M}$ without using an explicit construction for $\mathcal{P}$. In fact, this would immediately imply non-finiteness results for $\mathcal{P}$, which would be especially useful in higher ranks.

8. Polytopes of Rank 5

The only candidates for regular 5-polytopes whose facets and vertex-figures are spherical or toroidal (with at least one of the latter kind) are those of type $\{3,4,3,4\}$ and their duals. Confining our attention to the first of each dual pair, we shall write

$$\mathcal{T}_s^5 := \{\{3,4,3\},\{4,3,4\}_s\},$$

with the convention that $s = (s^k, 0^{3-k})$ with $s \geq 2$ and $k = 1,2$ or 3. Then we have ([29, 30])

Theorem 17 *The universal regular 5-polytope $\mathcal{T}_s^5 = \{\{3,4,3\},\{4,3,4\}_s\}$ exists for all $s = (s^k, 0^{3-k})$ with $s \geq 2$ and $k = 1,2,3$. It is finite when $s = 2$, and infinite when $s \geq 3$. If $s = 2$ and $k = 1,2,3$, the corresponding groups are $C_2^3 \ltimes [3,4,3]$ of order 9216, $C_2^5 \ltimes [3,4,3]$ of order 36864, and $(C_2^6 \ltimes C_2^5) \ltimes [3,4,3]$ of order 2359296 respectively.*

For $k = 1$ the polytopes can be constructed by a direct twisting argument, as indicated in

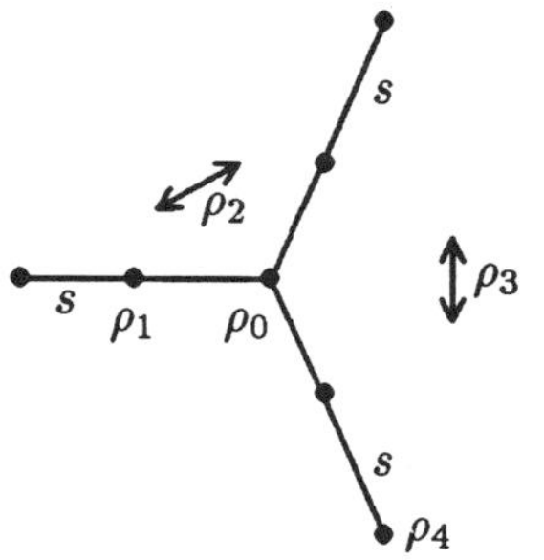

(23)

If W_s is the underlying Coxeter group, then $A(\mathcal{T}^5_{(s,0,0)}) \cong W_s \ltimes S_3$. Hence, finiteness occurs exactly for $s = 2$. We list all the finite polytopes in Table 4.

Let us also note that, when $k = 1$, we have a cut $\{\{3,4\},\{4,4\}_{\widetilde{s}}\}$ of $\mathcal{T}_s^5$, where $\widetilde{s} := (s^k, 0^{2-k})$ with $s \geq 2$, induced by a corresponding cut of $\{3,4,3,4\}$. In fact, this cut is defined by the operation

$$(\rho_0, \ldots, \rho_4) \mapsto (\rho_0, \rho_1, \rho_2, \rho_3\rho_4\rho_3) =: (\varphi_0, \ldots, \varphi_3)$$

which employs only the generators of the left and upper right branches of the diagram. But from Section 7 we know that this operation gives $\{\{3,4\},\{4,4\}_{(s,0)}\}$, so that this cut is universal. We conjecture that the corresponding cut is also universal if $k = 2$, because again the relations on the latter cut which yield $\{\{3,4\},\{4,4\}_{\tilde{s}}\}$ are just those which need to be imposed on $\{3,4,3,4\}$ to yield $\mathcal{T}_s^5$. Note that the analogous cut in case $k = 3$ is not universal.

9. Polytopes of Rank 6

For locally toroidal regular polytopes of rank 6, there is as yet no case completely settled. However, for each of the three possible types $\{3,3,3,4,3\}$, $\{3,3,4,3,3\}$ and $\{3,4,3,3,4\}$ (and their duals) there is a list of the known finite universal regular 6-polytopes which is conjectured to be complete.

9.1. Type $\{3,3,3,4,3\}$

We begin with the type $\{3,3,3,4,3\}$, and explain why we conjecture that there are precisely three finite examples of universal regular 6-polytopes

$${}_1\mathcal{T}_s^6 := \{\{3,3,3,4\},\{3,3,4,3\}_s\},$$

with $s := (s^k, 0^{4-k})$, $s \geq 2$ and $k = 1,2$ ([29]).

Write $A({}_1\mathcal{T}_s^6) = \langle \rho_0, \ldots, \rho_5 \rangle$, $\sigma := \rho_2\rho_3\rho_4\rho_3\rho_2$ and $\tau := \rho_5\rho_4\rho_3\rho_4\rho_5$. Then the operation

$$(\rho_0, \ldots, \rho_5) \mapsto (\rho_0, \rho_1, \sigma, \tau) =: (\varphi_0, \varphi_1, \varphi_2, \varphi_3)$$

defines a cut of ${}_1\mathcal{T}_s^6$ with group $\langle \varphi_0, \ldots, \varphi_3 \rangle$ which belongs to $\langle \{3,4\},\{4,4\}_{\tilde{s}} \rangle$, with $\tilde{s} := (s^k, 0^{2-k})$ when $s := (s^k, 0^{4-k})$. Note that the corresponding cut of the honeycomb $\{3,3,3,4,3\}$ is just $\{3,4,4\}$ itself. Now, by Theorem 5 the defining relations in terms of $\varphi_0, \ldots, \varphi_3$, added to those of $[3,4,4]$, which yield $({}_1\mathcal{T}_{\tilde{s}}^4)^* = \{\{3,4\},\{4,4\}_{\tilde{s}}\}$ are just those which give the quotient ${}_1\mathcal{T}_s^6$ of $\{3,3,3,4,3\}$ in which we are interested. This leads us to conjecture that the cut is indeed universal. But from Section 6.1 we know that the only finite cases of regular 4-polytopes $({}_1\mathcal{T}_{\tilde{s}}^4)^*$ are those with $\tilde{s} = (2,0), (2,2)$ or $(3,0)$. Then we have

Conjecture 2 *The regular 6-polytope ${}_1\mathcal{T}_s^6 = \{\{3,3,3,4\},\{3,3,4,3\}_s\}$ exists for all $s = (s^k, 0^{4-k})$ with $s \geq 2$ and $k = 1,2$. The only finite polytopes ${}_1\mathcal{T}_s^6$ are those for $s = (2,0,0,0), (2,2,0,0)$ and $(3,0,0,0)$.*

We remark that the existence of ${}_1\mathcal{T}_s^6$ is actually known if s is odd or $s = 2m$ with m odd ([29]). In Table 5 we list the details of the known finite polytopes.

s	v	f	g
$(2,0,0,0)$	20	960	368640
$(2,2,0,0)$	160	30720	11796480
$(3,0,0,0)$	780	189540	72783360

Table 5. The known finite polytopes ${}_1\mathcal{T}_s^6 = \{\{3,3,3,4\},\{3,3,4,3\}_s\}$

9.2. Type $\{3,3,4,3,3\}$

The situation for the remaining two types of locally toroidal regular polytopes of rank 6 is somewhat similar. We can appeal in each case to known results about which of the regular 4-polytopes of type $\{4,4,4\}$ exists and is finite, but since these do not, at present, cover all possibilities, our knowledge of the polytopes of rank 6 is correspondingly incomplete ([29]).

Consider the universal regular 6-polytope

$$ {}_2\mathcal{T}_{s,t}^6 := \{\{3,3,4,3\}_s, \{3,4,3,3\}_t\}, $$

with $s = (s^k, 0^{4-k}), t = (t^l, 0^{4-l})$ with $s,t \geq 2$ and $k,l = 1,2$. Now, if $A({}_2\mathcal{T}_{s,t}^6) = \langle \rho_0, \ldots, \rho_5 \rangle$, $\sigma := \rho_1\rho_2\rho_3\rho_2\rho_1$ and $\tau := \rho_4\rho_3\rho_2\rho_3\rho_4$, then the operation

$$ (\rho_0, \ldots, \rho_5) \mapsto (\rho_0, \sigma, \tau, \rho_5) =: (\varphi_0, \ldots, \varphi_3) $$

yields a cut of ${}_2\mathcal{T}_{s,t}^6$ (with group $\langle \varphi_0, \ldots, \varphi_3 \rangle$) in the class $\langle \{4,4\}_{\tilde{s}}, \{4,4\}_{\tilde{t}} \rangle$ (as usual, $\tilde{s} = (s^k, 0^{2-k})$ when $s = (s^k, 0^{4-k})$, and so on). Evidence indicates that this cut is indeed ${}_2\mathcal{T}_{\tilde{s},\tilde{t}}^4 = \{\{4,4\}_{\tilde{s}}, \{4,4\}_{\tilde{t}}\}$ (that is, the cut is universal), but so far we have not been able to prove this. Now, most cases of the regular polytopes of type $\{4,4,4\}$ are completely settled, and as a consequence, we have the following conjecture and subsequent theorem. The known finite polytopes are listed in Table 6.

Conjecture 3 *The regular 6-polytope* ${}_2\mathcal{T}_{s,t}^6 = \{\{3,3,4,3\}_s, \{3,4,3,3\}_t\}$ *exists for each* $s = (s^k, 0^{4-k}), t = (t^l, 0^{4-l})$ *with* $s,t \geq 2$ *and* $k,l = 1,2$, *except when* $s = (2,0,0,0)$ *and t is odd, or* $t = (2,0,0,0)$ *and s is odd.*

Theorem 18 *Under the assumption that the cut above is universal, then if the polytope* ${}_2\mathcal{T}_{s,t}^6$ *exists, it is infinite in at least the following cases:*
(a) $s = (s,0,0,0)$, $t = (t,t,0,0)$ *and* $\frac{1}{s} + \frac{1}{2t} \leq \frac{1}{2}$.
(b) $s = (s,s,0,0)$, $t = (t,t,0,0)$ *and* $\frac{1}{s} + \frac{1}{2t} \leq \frac{1}{2}$ *or* $\frac{1}{2s} + \frac{1}{t} \leq \frac{1}{2}$.
(c) $s = (s,0,0,0)$, $t = (t,0,0,0)$, *with s or t even, or $s = t$ odd, and* $\frac{1}{s} + \frac{1}{t} \leq \frac{1}{2}$.

We shall denote by ${}_1A_s^6$ and ${}_2A_{s,t}^6$ the groups abstractly defined by the presentation belonging to the polytopes ${}_1\mathcal{T}_s^6$ and ${}_2\mathcal{T}_{s,t}^6$; these are the quotients of $[3,3,3,4,3]$ and $[3,3,4,3,3]$ defined by the extra relations for the facets or vertex-figures. If the two polytopes exist, then ${}_1A_s^6$ and ${}_2A_{s,t}^6$ are their groups, respectively.

It is known that $[3,3,4,3,3]$ is a subgroup of index 5 in $[3,3,3,4,3]$ ([29, 30]). The corresponding relationship between the groups of the locally toroidal polytopes is

s	t	v	f	g
$(2,0,0,0)$	$(t,0,0,0)$ (t even)	32	$2t^4$	$36864t^4$
$(2,0,0,0)$	$(t,t,0,0)$ (t even)	32	$8t^4$	$147476t^4$
$(2,2,0,0)$	$(2,2,0,0)$	2048	2048	150994944
$(3,0,0,0)$	$(3,0,0,0)$	2340	2340	218350080

Table 6. The known finite polytopes ${}_2\mathcal{T}^6_{s,t} = \{\{3,3,4,3\}_s, \{3,4,3,3\}_t\}$

Theorem 19 *Under the assumption that Conjectures 2 and 3 hold, for each* $s \geq 2$, ${}_2A^6_{(s,s,0,0),(s,0,0,0)}$ *is a subgroup of index* 5 *in* ${}_1A^6_{(s,s,0,0)}$, *while* ${}_2A^6_{(2s,0,0,0),(s,s,0,0)}$ *is a subgroup of index* 5 *in* ${}_1A^6_{(2s,0,0,0)}$.

9.3. Type $\{3,4,3,3,4\}$

In the case of the universal regular 6-polytope

$$ {}_3\mathcal{T}^6_{s,t} := \{\{3,4,3,3\}_s, \{4,3,3,4\}_t\} $$

with $s = (s^k, 0^{4-k}), t = (t^l, 0^{4-l})$, with $s,t \geq 2$, $k = 1,2$ and $l = 1,2,4$, we can again use a cut of type $\{4,4,4\}$ but now, strictly speaking, of the dual $({}_3\mathcal{T}^6_{s,t})^* = \{\{4,3,3,4\}_t, \{3,3,4,3\}_s\}$. If $A(({}_3\mathcal{T}^6_{s,t})^*) = \langle \rho_0, \ldots, \rho_5 \rangle$, this cut is induced by the operation

$$ (\rho_0, \ldots, \rho_5) \mapsto (\rho_0, \rho_1, \rho_2\rho_3\rho_4\rho_3\rho_2, \rho_5\rho_4\rho_3\rho_4\rho_5) $$

and, if $l = 1$ or 2, belongs to the class $\langle \{4,4\}_{\tilde{t}}, \{4,4\}_{\tilde{s}} \rangle$, where the notation for suffixes is that introduced earlier. In fact, we conjecture again that the cut is universal, so that it is isomorphic to ${}_2\mathcal{T}^4_{\tilde{t},\tilde{s}} = \{\{4,4\}_{\tilde{t}}, \{4,4\}_{\tilde{s}}\}$; on the other hand, if $l = 4$, this cut is not universal. Now, employing the results on polytopes of type $\{4,4,4\}$ this supports the following conjecture.

Conjecture 4 *The regular* 6*-polytope* ${}_3\mathcal{T}^6_{s,t} := \{\{3,4,3,3\}_s, \{4,3,3,4\}_t\}$ *exists for all* $s = (s^k, 0^{4-k}), t = (t^l, 0^{4-l})$ *with* $s,t \geq 2$, $k = 1,2$ *and* $l = 1,2,4$, *except when*
(*a*) $s = (2,0,0,0)$ *and* $t = (t,0,0,0)$ *with* t *odd,*
(*b*) $t = (2,0,0,0)$ *and* $s = (s,0,0,0)$ *with* s *odd,*
(*c*) $t = (2,2,0,0)$ *and* s *odd.*

Note that only the first two excluded cases of this conjecture correspond to cuts which do not give 4-polytopes. The third case of collapse can be proved directly ([29, 30]).

Conjecture 4 was confirmed in [30] for all s,t with $l = 1$ and $t = 2m$ even using a rather sophisticated twisting argument. To explain this, let $\mathcal{K} := \{3,3,4,3\}_s$, a toroid of rank 5 whose facets are 4-crosspolytopes $\{3,3,4\}$. Consider a Coxeter diagram $\mathcal{D}_{s,m}$ whose nodes are the vertices of $\mathcal{K}$ and whose branches are of two types: one connects antipodal vertices of facets of $\mathcal{K}$ and its branches are marked $m(\geq 2)$, and the other connects pairs of vertices which are not vertices of a common facet of $\mathcal{K}$ and its branches are marked ∞. Then the group of the universal 6-polytope

s	t	v	f	g
$(s,0,0,0)$ (s even)	$(2,0,0,0)$	$3s^4$	16	$18432s^4$
$(s,s,0,0)$	$(2,0,0,0)$	$12s^4$	16	$73728s^4$
$(s,0,0,0)$ (s even)	$(2,2,0,0)$	$6s^4$	64	$73728s^4$
$(s,s,0,0)$ (s even)	$(2,2,0,0)$	$24s^4$	64	$294912s^4$
$(2,0,0,0)$	$(2,2,2,2)$	384	1024	18874368
$(2,0,0,0)$	$(4,0,0,0)$	12288	65536	1207959552
$(3,0,0,0)$	$(3,0,0,0)$	2340	780	72783360

Table 7. The known finite polytopes ${}_3\mathcal{T}^6_{s,t} = \{\{3,4,3,3\}_s, \{4,3,3,4\}_t\}$

$({}_3\mathcal{T}^6_{s,(2m,0^3)})^*$ can be constructed by a twisting operation on the corresponding Coxeter group $W_{s,m}$ with diagram $\mathcal{D}_{s,m}$, using the fact that $A(\mathcal{K})$ acts on $\mathcal{D}_{s,m}$ as a group of diagram automorphisms; in particular, $W_{s,m} \ltimes A(\mathcal{K})$ is the group of the resulting polytope. The case $m = 1$ requires a variant of this construction.

Theorem 20 *The 6-polytope* ${}_3\mathcal{T}^6_{s,(2m,0^3)} = \{\{3,4,3,3\}_s, \{4,3,3,4\}_{(2m,0^3)}\}$ *exists for each* $s = (s^k, 0^{4-k})$ *with* $s \geq 2$ *and* $k = 1,2$ *and each* $m \geq 1$. *The only finite instances occur for* $m = 1$, *and* $s = (2,0,0,0), m = 2$. *In the first case the group is* $C_2^4 \ltimes [3,3,4,3]_s$, *of order* $18432s^4$ *or* $73728s^4$ *if* $k = 1$ *or* 2, *respectively; in the second case it is* $C_2^{16} \ltimes [3,3,4,3]_{(2,0,0,0)}$, *of order* 1207959552.

Table 7 lists all the known finite polytopes ${}_3\mathcal{T}^6_{s,t}$, and it is conjectured that this list is complete. The table entries were checked (or obtained) by an application of the Coxeter-Todd coset enumeration algorithm. For all polytopes, except the one in the last row, the structure of the group is explicitly known ([30]). The first two rows and the next to last row are covered by Theorem 20. The group in the third and fourth row is $C_2^6 \ltimes [3,3,4,3]_s$ with $s = (s,0,0,0)$ or $(s,s,0,0)$, respectively; and in the fifth row it is $C_2^{10} \ltimes [3,3,4,3]_{(2,0,0,0)}$. See [30] for more general results on polytopes with small facets or vertex-figures like those discussed here.

Writing ${}_3A^6_{s,t}$ for the group abstractly defined by the presentation belonging to the polytope ${}_3\mathcal{T}^6_{s,t}$, and using the fact that $[3,4,3,3,4]$ is a subgroup of index 10 in $[3,3,3,4,3]$, we now have

Theorem 21 *Under the assumption that Conjectures 2 and 4 hold, for each* $s \geq 2$, ${}_3A^6_{(s,0,0,0),(s,s,0,0)}$ *is a subgroup of index* 10 *in* ${}_1A^6_{(s,s,0,0)}$, *while* ${}_3A^6_{(s,s,0,0),(2s,0,0,0)}$ *is a subgroup of index* 10 *in* ${}_1A^6_{(2s,0,0,0)}$.

Concerning the polytope in the last row of Table 7, it is interesting to note that the two groups ${}_1A^6_{(3,0,0,0)}$ and ${}_3A^6_{(3,0,0,0),(3,0,0,0)}$ are isomorphic and that the corresponding polytopes ${}_1\mathcal{T}^6_{(3,0,0,0)}$ and ${}_3\mathcal{T}^6_{(3,0,0,0),(3,0,0,0)}$ have the same number of vertices. Further, ${}_1A^6_{(3,0,0,0)}$ is a quotient of ${}_2A^6_{(3,0,0,0),(3,0,0,0)}$ by a normal subgroup of order 3, so that the number of vertices of ${}_1\mathcal{T}^6_{(3,0,0,0)}$ is only one third of that of ${}_2\mathcal{T}^6_{(3,0,0,0),(3,0,0,0)}$.

10. Finite Quotients

It is well known that the euclidean and hyperbolic regular tessellations $\{p, q, r\}$ are the universal coverings for infinitely many finite regular tessellations of the same type on closed compact 3-manifolds. This generalizes also to many classes of abstract regular polytopes. The previous sections were aiming at the classification of all the finite polytopes among the universal locally toroidal regular polytopes $\{\mathcal{P}_1, \mathcal{P}_2\}$. The question remains whether the infinite polytopes of this kind cover in fact infinitely many finite regular polytopes in the same class, namely $\langle\mathcal{P}_1, \mathcal{P}_2\rangle$. For most classes $\langle\mathcal{P}_1, \mathcal{P}_2\rangle$ this can indeed be proved by generalizing a technique used in Vince [47]; see [26]. This seems to indicate that in general it is very hard to fully describe these classes.

Let U be any group. Then U is called *residually finite* if for each finite subset $\{\varphi_1, \ldots, \varphi_m\}$ of $U \setminus \{\varepsilon\}$ there exists a homomorphism f of U onto a finite group such that $\varphi_j f \neq \varepsilon$ for $j = 1, \ldots, m$. By a central result in the theory of linear groups, due to Malcev [20], every finitely generated linear group is residually finite; see also [48]. In particular, every (finitely generated) Coxeter group is residually finite.

Theorem 22 *Let $\mathcal{P}_1$ and $\mathcal{P}_2$ be finite regular n-polytopes with $\langle\mathcal{P}_1, \mathcal{P}_2\rangle \neq \emptyset$. Let $\mathcal{P}$ be an infinite regular $(n+1)$-polytope in $\langle\mathcal{P}_1, \mathcal{P}_2\rangle$ whose group $A(\mathcal{P})$ is residually finite. Then $\langle\mathcal{P}_1, \mathcal{P}_2\rangle$ contains infinitely many regular $(n+1)$-polytopes which are finite and are covered by $\mathcal{P}$.*

Clearly the groups of regular polytopes are finitely generated. But then, by Malcev's result, Theorem 22 applies if $\mathcal{P}$ is an infinite member in $\langle\mathcal{P}_1, \mathcal{P}_2\rangle$ whose group is isomorphic to a linear group. This is the form in which Theorem 22 can be applied to many classes of polytopes. We give two examples in rank 4, the first being the well-known classical case. There are similar results for other types and for higher ranks.

Theorem 23 *Let $\{p, q, r\} = \{4,3,4\}, \{3,5,3\}, \{5,3,5\}, \{4,3,5\}$ or $\{5,3,4\}$. Then $\langle\{p, q\}, \{q, r\}\rangle$ contains infinitely many regular polytopes which are finite.*

Theorem 24 *Let $\mathcal{P}_1$ and $\mathcal{P}_2$ be regular toroidal maps for which the universal $\{\mathcal{P}_1, \mathcal{P}_2\}$ exists, is infinite, and is of type $\{4,4,3\}$ or $\{6,3,p\}$ with $p = 3,4,5$ or 6. Then $\langle\mathcal{P}_1, \mathcal{P}_2\rangle$ contains infinitely many regular polytopes which are finite.*

11. Chiral Polytopes

Relatively little is known about locally toroidal chiral polytopes. By Theorem 3, there is no chiral toroid of rank ≥ 4, and so the classification of such polytopes makes sense in rank 4 alone. The corresponding Schläfli types $\{p, q, r\}$ are the same as for regular 4-polytopes but now the parameter vectors s and t for the facets and vertex-figures are less restricted; see Section 4. Another complication is added through enantiomorphism.

We shall introduce notation similar to that of universal regular polytopes, except that now we drop the superfix "4" for the rank and replace it by "*ch*". For example, we write

$$ {}_1\mathcal{T}_s^{ch} := \{\{4,4\}_s, \{4,3\}\}^{ch} , $$

$s = (b, c)$ with $bc(b - c) \neq 0$, for the oriented universal chiral 4-polytope whose oriented facets are maps $\{4,4\}_s$ and whose vertex-figures are (oriented) cubes $\{4,3\}$. (Recall that for directly regular polytopes the two orientations can canonically be identified.) Note that interchanging b and c in s changes ${}_1\mathcal{T}_s^{ch}$ to the other enantiomorphic form, $\overline{{}_1\mathcal{T}_s^{ch}}$ (say), of the same underlying polytope. In short, if $\overline{s} := (c, b)$ when $s = (b, c)$, then ${}_1\mathcal{T}_{\overline{s}}^{ch} = \overline{{}_1\mathcal{T}_s^{ch}}$. The situation is similar for the chiral polytopes

$$ {}_p\mathcal{T}_s^{ch} := \{\{6,3\}_s, \{3,p\}\}^{ch} $$

with $p = 3,4,5$ and $s = (b,c)$, $bc(b-c) \neq 0$; that is, ${}_p\mathcal{T}_{\overline{s}}^{ch} = \overline{{}_p\mathcal{T}_s^{ch}}$ ([39, 40]).

For the three remaining chiral 4-polytopes

$$ {}_2\mathcal{T}_{s,t}^{ch} := \{\{4,4\}_s, \{4,4\}_t\}^{ch}, \quad {}_6\mathcal{T}_{s,t}^{ch} := \{\{6,3\}_s, \{3,6\}_t\}^{ch}, $$

$$ {}_7\mathcal{T}_{s,t}^{ch} := \{\{3,6\}_s, \{6,3\}_t\}^{ch} , $$

with $s = (b, c), t = (d, e)$ and $bc(b - c) \neq 0$ or $de(d - e) \neq 0$, the situation is more complicated, because now both the facets and vertex-figures can be chiral. In general, interchanging the components in only one parameter vector, s (say), does not simply change ${}_i\mathcal{T}_{s,t}^{ch}$ to the other enantiomorphic form $\overline{{}_i\mathcal{T}_{s,t}^{ch}}$. In fact, in general it seems that the two polytopes ${}_i\mathcal{T}_{s,t}^{ch}$ and ${}_i\mathcal{T}_{\overline{s},t}^{ch}$ are unrelated. However, if the components in both parameter vectors are interchanged, then only the enantiomorphic form is changed; that is, ${}_i\mathcal{T}_{\overline{s},\overline{t}}^{ch} = \overline{{}_i\mathcal{T}_{s,t}^{ch}}$ for $i = 2,6,7$.

Problem 1 *Classify all the universal chiral 4-polytopes ${}_i\mathcal{T}_s^{ch}$ for each $i = 1,3,4,5$ and ${}_i\mathcal{T}_{s,t}^{ch}$ for each $i = 2,6,7$.*

Here the term "classification" is used in the same sense as in Section 3. Except for the existence part of the problem and solutions for a few sporadic cases ([5]), no general classification results are known. As usual the existence of the universal polytope can be deduced from the fact that the corresponding class of polytopes is non-empty.

For all seven types $\{p,q,r\}$ of locally toroidal 4-polytopes, chiral polytopes have been constructed from representations of the hyperbolic rotation groups $[p,q,r]^+$ as projective linear groups over finite rings (Weiss [51], Schulte & Weiss [40], Nostrand & Schulte [35]). See also Nostrand [34] for similar such examples of (locally spherical) types $\{3,5,3\}$ and $\{5,3,4\}$. We shall not give the details of these constructions but instead illustrate a typical result for the type $\{3,6,3\}$, which relates to prime decomposition in the ring $\mathbb{Z}[\omega]$ of Eisenstein integers. Let $\mathbb{Z}_m := \mathbb{Z}/m\mathbb{Z}$, the ring of integers modulo m.

Theorem 25 *Let m be a positive integer, let $m = p_1^{e} \cdot \ldots \cdot p_k^{e_k}$ be its prime decomposition, and let $p_j \equiv 1 \,(mod\ 3)$ for each $j = 1,\ldots,k$. Let b,c be positive integers such that $m = b^2 + bc + c^2$, $(b,c) = 1$. Then there exists a self-dual chiral 4-polytope in $\langle\{3,6\}_{(b,c)}, \{6,3\}_{(b,c)}\rangle^{ch}$ whose group is*
(a) $PSL_2(\mathbb{Z}_m)$ if $p_j \equiv 1 \,(mod\ 12)$ for all j;
(b) $PSL_2(\mathbb{Z}_m) \ltimes C_2$ if $p_j \equiv 7 \,(mod\ 12)$ for at least one j.

12. Other Local Topological Types

As remarked in the introduction, there is as yet no comprehensive study of abstract regular or chiral polytopes which are locally of some topological type that is not spherical or toroidal. However, many interesting examples are known, and here we shall mention a few to point out some possible direction of further research.

In rank 4, call a regular (or chiral) polytope $\mathcal{P}$ *locally of genus* g if both its facets and vertex-figures are maps on orientable surfaces of genus at most g, with at least one kind of genus exactly g. For $g = 1$ this gives the locally toroidal polytopes. Variants of this terminology could also include maps on non-orientable surfaces.

Problem 2 *For small* $g \geq 2$*, classify all the regular* 4*-polytopes which are locally of genus* g*.*

Problem 3 *For small* $g \geq 2$*, classify all the regular* 4*-polytopes whose facets are of genus* g *and whose vertex-figures are spherical.*

Clearly, a solution for Problem 2 includes one for Problem 3. Note that the polytopes in Problem 3 are of interest also because they can be realized as combinatorially regular decompositions of certain topological 3-manifolds into handlebodies of genus g ([2]).

An interesting special case is $g = 3$. In [32, 33], Monson and Weiss construct a family of regular or chiral 4-polytopes of type $\{p, 3, 3\}$ with $p \geq 3$ which are related to the Picard group, the projective linear group $PSL(\mathbb{Z}[i])$ over the Gaussian integers $\mathbb{Z}[i]$. For $p = 7$ this gives the universal regular polytope $\{\{7,3\}_8, \{3,3\}\}$ with group $PSL_2(7^2) \ltimes C_2$ of order $7^2(7^4 - 1)$, whose facet is Klein's map $\{3,7\}_8$ of genus 3 ([10, 44]). For many other interesting examples with facets of small genus see also [24, 25].

For higher ranks, the locally projective case has received some attention. For instance, McMullen [23] constructs the regular polytopes $\mathcal{P}$ of rank $n \geq 5$ and type $\{3, 4, 3^{n-3}\}$ which are universal with respect to having their 4-faces isomorphic to the projective polytope $\{3,4,3\}_6$ (which is constructed from the 24-cell $\{3,4,3\}$ by identifying antipodal points); that is, $A(\mathcal{P})$ is the Coxeter group $[3, 4, 3^{n-3}] = \langle \rho_0, \ldots, \rho_{n-1} \rangle$, factored out by the single extra relation

$$(\rho_0\rho_1\rho_2\rho_3)^6 = 1 \ .$$

In particular, $A(\mathcal{P}) \cong S_4 \wr S_{n-1}$ of order $24^{n-1}(n-1)!$. These polytopes are locally projective in the more general sense of the term that each minimal non-spherical section is a projective polytopes (which here is $\{3,4,3\}_6$). Similarly there are locally projective regular polytopes of types $\{3^k, 4, 3, 3, 4, 3^l\}$ all of whose non-spherical 4-sections of types $\{3,4,3\}$, $\{4,3,3\}$ and $\{3,3,4\}$ are isomorphic to $\{3,4,3\}_6$, $\{4,3,3\}_4$ and $\{3,3,4\}_4$, respectively.

We also mention two examples of mixed toroidal-projective type, the polytopes $\{\{6,3\}_{(s,s)}, \{3,5\}_5\}$ of rank 4 with hemi-icosahedral vertex-figures and with $s \geq 2$, and the polytopes $\{\{4,3,4\}_{(2^k,0^{3\ k})}, \{3,4,3\}_6\}$ of rank 5 with $k = 1,2,3$. Using a suitable hermitian form as in Section 6.2 we find that the first polytope is infinite if $s \geq 3$; an application of the Coxeter-Todd coset enumeration algorithm suggests that it is also infinite if $s = 2$ (but there is no proof yet). The group of the second polytope is $C_2^{m_k} \ltimes [3,4,3]_6$, of order $576 \cdot 2^{m_k}$, with $m_k = 3, 5$ or 8 if $k = 1, 2$ or 3, respectively ([30]).

References

1. Altshuler, A.: *3-pseudomanifolds with preassigned links*, Transactions AMS **241** (1978), 213-237.
2. Brehm, U., Kühnel, W. and Schulte, E.: *Manifold structures on abstract regular polytopes*, preprint.
3. Buekenhout, F.:*Diagrams for geometries and groups*, J.Combinatorial Theory A **27** (1979), 121-151.
4. Cohen, A.M.: *Finite complex reflection groups*, Ann. Sci. École Norm. Sup. (4) **9** (1976), 379-436.
5. Colbourn, C.J. and Weiss, A.I.: *A census of regular* 3*-polystroma arising from honeycombs*, Discrete Mathematics **50** (1984), 29-36.
6. Coxeter, H.S.M.: *Regular skew polyhedra in* 3 *and* 4 *dimensions and their topological analogues*, Proc. London Math. Soc. (2) **43** (1937), 33-62 (reprinted with corrections in: *Twelve Geometric Essays*, Southern Illinois University Press, Carbondale (1968), 75-105).
7. Coxeter, H.S.M.: *Groups generated by unitary reflections of period two*, Canadian J. Math. **9** (1957), 243-272.
8. Coxeter, H.S.M.: *Regular honeycombs in hyperbolic space*, in: *Twelve Geometric Essays*, Southern Illinois University Press, Carbondale (1968), 199-214.
9. Coxeter, H.S.M.: *Regular Polytopes*, 3rd edition, Dover, New York, 1973.
10. Coxeter, H.S.M. and Moser, W.O.J.: *Generators and Relations for Discrete Groups*, 4th edition, Springer, Berlin (1980).
11. Coxeter, H.S.M. and Shephard, G.C.: *Regular* 3*-complexes with toroidal cells*, J.Combinatorial Theory B **22** (1977), 131-138.
12. Danzer, L. and Schulte, E.: *Reguläre Inzidenzkomplexe I*, Geometriae Dedicata **13** (1982), 295-308.
13. Dress, A.W.M.: *Regular polytopes and equivariant tessellations from a combinatorial point of view*, in: Algebraic Topology (Göttingen 1984), Lecture Notes in Mathematics 1172, Springer (1985), 56-72.
14. Dress, A.W.M. and Schulte, E.: *On a theorem of McMullen about combinatorially regular polytopes*, Simon Stevin **61** (1987), 265-273.
15. Du Val, P.: *Homographies, Quaternions and Rotations*, Oxford (1964).
16. Garbe, D.: *Über die regulären Zerlegungen geschlossener orientierbarer Flächen*, J. Reine Angewandte Mathematik **237** (1969), 39-55.
17. Grünbaum, B.: *Convex Polytopes*, Interscience Publishers, John Wiley & Sons, London (1967).
18. Grünbaum, B.: *Regularity of graphs, complexes and designs*, Problèmes Combinatoire et Théorie des Graphes, Coll. Int. CNRS No.260, Orsay, 1977, 191-197.
19. Humphreys, J.E.: *Reflection Groups and Coxeter Groups*, Cambridge University Press, Cambridge (1990).

20. Malcev, A.I.: *On faithful representations of infinite groups of matrices*, Mat. Sb. **8** (1940), 405-422 (Russian); English translation in Amer. Math. Soc. Transl. Ser. (2) **45** (1965), 1-18.
21. McMullen, P.: *Combinatorially regular polytopes*, Mathematika **14** (1967), 142-150.
22. P.McMullen: *Realizations of regular polytopes*, Aequationes Math. **37** (1989), 38-56.
23. McMullen, P.: *Locally projective regular polytopes*, J. Combinatorial Theory A, to appear.
24. McMullen, P. and Schulte, E.: *Regular polytopes from twisted Coxeter groups and unitary reflexion groups*, Advances Math. **82** (1990), 35-87.
25. McMullen, P. and Schulte, E.: *Hermitian forms and locally toroidal regular polytopes*, Advances Math. **82** (1990), 88-125.
26. McMullen, P. and Schulte, E.: *Finite quotients of infinite universal polytopes*, in: Discrete and Computational Geometry, edit. by J.Goodman, R.Pollack and W.Steiger, DIMACS Series Vol. 6, AMS-ACM (1991), 231-236.
27. McMullen, P. and Schulte, E.: *Regular polytopes of type* $\{4,4,3\}$ *and* $\{4,4,4\}$, Combinatorica **12** (2) (1992), 203-220.
28. McMullen, P. and Schulte, E.: *Locally toroidal regular polytopes of rank* 4, Commentarii Mathematici Helvetici **67** (1992), 77-118.
29. McMullen, P. and Schulte, E.: *Higher toroidal regular polytopes*, Advances Math., to appear.
30. McMullen, P. and Schulte, E.: *Twisted groups and locally toroidal regular polytopes*, preprint.
31. McMullen, P. and Schulte, E.: *Abstract Regular Polytopes*, monograph, in preparation.
32. Monson, B. and Weiss, A.I.: *Regular* 4*-polytopes related to general orthogonal groups*, Mathematika **37** (1990), 106-118.
33. Monson, B. and Weiss, A.I.: *Polytopes related to the Picard group*, Linear Algebra and Its Applications, to appear.
34. Nostrand, B.: *Chiral honeycombs*, Ph.D.Thesis, Northeastern University, Boston, 1993.
35. Nostrand, B. and Schulte, E.: *Chiral polytopes from hyperbolic honeycombs*, preprint.
36. Schulte, E.: *Reguläre Inzidenzkomplexe II, III*, Geometriae Dedicata **14** (1983), 33-56, 57-79.
37. Schulte, E.: *Regular incidence-polytopes with Euclidean or toroidal faces and vertex-figures*, J.Combinatorial Theory A **40** (1985), 305-330.
38. Schulte, E.:*Amalgamations of regular incidence-polytopes*, Proc. London Math. Soc. (3) **56** (1988), 303-328.
39. Schulte, E. and Weiss, A.I.: *Chiral polytopes*, in: Applied Geometry and Discrete Mathematics - The Victor Klee Festschrift, edit. by P.Gritzmann and B.Sturmfels, DIMACS Series Vol.4, AMS-ACM (1991), 493-516.
40. Schulte, E. and Weiss, A.I.: *Chirality and projective linear groups*, Discrete Mathematics, to appear.

41. Seifert, H. and Threlfall, E.: *Lehrbuch der Topologie*, Teubner 1934 (English translation: *A Textbook of Topology*, Academic Press 1980).
42. Shephard, G.C.: *Unitary groups generated by reflections*, Canad. J. Math. **5** (1953), 364-383.
43. Shephard, G.C. and Todd, J.A.: *Finite unitary reflection groups*, Canad. J. Math. **6** (1954), 274-304.
44. Sherk, F.A.: *The regular maps on a surface of genus 3*, Canadian J. Math. **11** (1959), 452-480.
45. Stanley, R.P.: *Enumerative Combinatorics I*, Wadsworth & Brooks/Cole, Belmont, California, 1986.
46. Tits, J.: *Buildings of Spherical Type and Finite BN-Pairs*, Lecture Notes in Mathematics 386, Springer (1974).
47. Vince, A.: *Regular combinatorial maps*, J.Combinatorial Theory B **35** (1983), 256-277.
48. Wehrfritz, B.A.F.: *Infinite Linear Groups*, Springer-Verlag, New-York, 1973.
49. Weiss, A.I.: *An infinite graph of girth* 12, Transactions AMS **283** (1984), 575-588.
50. Weiss, A.I.: *Incidence-polytopes of type* $\{6,3,3\}$, Geometriae Dedicata **20** (1986), 147-155.
51. Weiss, A.I.: *Incidence-polytopes with toroidal cells*, Discrete & Computational Geometry **4** (1989), 55-73.
52. Wolf, J.: *Spaces of Constant Curvature*, Publish or Perish, Boston, 1974.

FACE NUMBERS AND SUBDIVISIONS OF CONVEX POLYTOPES

MARGARET M. BAYER
Department of Mathematics
University of Kansas
Lawrence, Kansas 66045-2142
USA

Abstract. The first part of the paper surveys results on f-vectors, flag vectors and h-vectors of convex polytopes. These are combinatorial parameters that have been characterized for simplicial polytopes. Many of the results known in the general case depend on the connection between convex polytopes and toric varieties. The second half of the paper looks at polyhedral subdivisions of convex polytopes. The effect of subdivision on the h-vector is studied. The paper discusses the secondary polytope, which encodes the regular subdivisions of a polytope. Fiber zonotopes and the corresponding hyperplane arrangements, called discriminantal arrangements, are studied.

Key words: convex polytope, f-vector, h-vector, subdivision, triangulation, secondary polytope, fiber polytope, hyperplane arrangement, discriminantal arrangement

The combinatorics of convex polytopes is an active area of research. The most important reference in the subject, in spite of its age, is still Grünbaum's book [19]. Highlighting the accomplishments since then are two recent survey articles, by Klee and Kleinschmidt [26], and by Bayer and Lee [8]. In addition Ziegler has a preliminary version of lecture notes on polytopes [43]. All these offer much more extensive bibliographies than is contained here. This paper focuses on three topics in the combinatorial study of polytopes: numbers of faces, subdivisions, and relation to hyperplane arrangements.

1. Numbers of Faces

1.1. f-VECTOR HISTORY

A polytope is the convex hull of a finite point set in $\mathbf{R}^d$. A d-dimensional polytope has faces of dimension 0 (vertices), 1 (edges), and so on, up to $d-1$ (facets). Write f_i for the number of i-dimensional faces. The f-vector of a polytope is the sequence $(f_0, f_1, \ldots, f_{d-1})$. This has been the subject of much study in this century and before. The characterization of f-vectors of d-polytopes for all d is a major open problem.

At the turn of the century Steinitz [41] gave the complete characterization of f-vectors of polytopes of dimension three.

Theorem 1 (Steinitz) *An integer vector* (f_0, f_1, f_2) *is the f-vector of a three-dimensional polytope if and only if*

1. $f_0 - f_1 + f_2 = 2$
2. $f_0 \leq 2f_2 - 4$
3. $f_2 \leq 2f_0 - 4$.

T. Bisztriczky et al. (eds.), POLYTOPES: Abstract, Convex and Computational, 155–171.

No characterization is known for polytopes of any dimension greater than three. For 3-polytopes more detailed combinatorial information has been studied. If P is a 3-polytope, let $p_n(P)$ be the number of faces of P which are n-gons, and call the vector $(p_3, p_4, p_5, \ldots)$ the p-vector. Much is known about p-vectors of 3-polytopes, though it falls short of a complete characterization. The basic theorem is due to Eberhard in 1891 [15].

Theorem 2 (Eberhard) *There exists a 3-polytope with p_n n-gons ($n \geq 3$, $n \neq 6$) and some number of 6-gons if and only if the integer $\sum_{n\geq 3}(6-n)p_n$ is even and is at least 12.*

In the 1970s and 1980s Jendrol' and others studied the possible values of p_6 (see [21]).

A polytope is called *simplicial* if all its faces are simplices. For simplicial polytopes Sommerville [34] first observed a useful invertible linear transformation on the f-vector. Extend the definition of f-vector by writing $f_{-1} = 1$. Then the *h-vector* of a simplicial d-polytope is $(h_0, h_1, \ldots, h_d)$, where

$$h_j = \sum_{i=0}^{j}(-1)^{j-i}\binom{d-i}{d-j} f_{i-1}.$$

In 1971 McMullen [29] conjectured a characterization of h-vectors of simplicial d-polytopes (for all d). The necessity of the "McMullen conditions" was proved in 1980 by Stanley [35], and the sufficiency the same year by Billera and Lee [10]. In 1992 McMullen [30] gave a new proof of necessity, avoiding the algebraic geometry used by Stanley. (Sommerville [34] proved (1) of the theorem below; these equations are known as the Dehn-Sommerville equations.)

Theorem 3 Stanley; Billera and Lee; McMullen
An integer vector $(h_0, h_1, \ldots, h_d)$ is the h-vector of a simplicial d-polytope if and only if

1. $h_i = h_{d-i}$ *for all* i
2. $h_0 = 1$ *and* $h_i \leq h_{i+1}$ *for all* i, $0 \leq i \leq d/2 - 1$
3. $h_{i+1} - h_i \leq (h_i - h_{i-1})^{\langle i\rangle}$ *for all* i, $0 \leq i \leq d/2 - 1$.

The superscript $\langle i\rangle$ represents the pseudopower operation, defined via binomial coefficients as follows. For any positive integers n and i, n has a unique representation

$$n = \binom{n_i}{i} + \binom{n_{i-1}}{i-1} + \cdots + \binom{n_j}{j},$$

where $n_i > n_{i-1} > \cdots > n_j \geq j \geq 1$. The ith pseudopower of n is then

$$n^{\langle i\rangle} = \binom{n_i+1}{i+1} + \binom{n_{i-1}+1}{i} + \cdots + \binom{n_j+1}{j+1}.$$

We also define $0^{\langle i\rangle} = 0$ for any positive integer i.

The h-vector is important not only because it gives a convenient way to write down the conditions, but because it is the counting vector for several different sequences of objects associated with the polytope. The discoveries of these fueled much research in polytopes. Some of them have analogues in the nonsimplicial case, and we turn to that now.

1.2. Flag Vectors

It quickly becomes clear in looking at the general polytope case that the f-vector is not the right object of study. It carries too little of the combinatorial information. We introduce instead the flag vector of a polytope, defined as follows. Consider a sequence of distinct proper faces ordered by inclusion, $F_0 \subset F_1 \subset F_2 \subset \cdots \subset F_k$. This is called an S-flag, where S is the set of dimensions of the faces F_i. The number of S-flags of a polytope is written f_S, and the vector of all such flag numbers is called the *flag vector* of the polytope (here S ranges over all subsets of $\{0, 1, \ldots, d-1\}$). For example, a square based pyramid has flag numbers $f_\emptyset = 1$, $f_0 = f_2 = 5$, $f_1 = 8$, $f_{0,1} = f_{0,2} = f_{1,2} = 16$, and $f_{0,1,2} = 32$.

The big enumeration problem for polytopes is then: characterize the flag vectors of d-dimensional polytopes for all d. The problem is trivially solved for dimension at most 3 because the flag vector then depends only on the f-vector. Also, the flag vector of a simplicial polytope depends only on its f-vector, so there is nothing interesting to say in this special case.

The first general result for flag vectors of polytopes is known as the generalized Dehn-Sommerville equations [5].

Theorem 4 (Bayer and Billera) *The affine dimension of the flag vectors of d-polytopes is $e_d - 1$, where (e_d) is the Fibonacci sequence, $e_d = e_{d-1} + e_{d-2}$, $e_0 = e_1 = 1$. The affine hull of the flag vectors is determined by the equations*

$$\sum_{j=i+1}^{k-1} (-1)^{j-i-1} f_{S\cup\{j\}}(P) = \left(1 - (-1)^{k-i-1}\right) f_S(P),$$

where $i \leq k-2$, $i, k \in S \cup \{-1, d\}$, and S contains no integer between i and k.

The proof that these equations hold for flag vectors of polytopes is the same as Sommerville's proof of the Dehn-Sommerville equations for simplicial polytopes. That they determine the affine hull of flag vectors was proved by exhibiting a basis of polytopes. This was first done in a complicated calculation in [5], and was later done more elegantly by Kalai [24]. In that work Kalai also found the affine hull of flag vectors of k-simplicial d-polytopes (polytopes all of whose k-faces are simplices).

Stanley's proof of the McMullen conditions for f-vectors of simplicial polytopes depends on a connection between rational polytopes and toric varieties. A polytope is rational if all the coordinates of its vertices are rational numbers. Every simplicial polytope is combinatorially equivalent to a rational polytope, but this is not true for nonsimplicial polytopes (see [19, page 94]). For a rational polytope, the affine dependencies among the vertices can be generated by a finite set of affine dependencies with rational coefficients. These are used to define a toric variety. (For an exposition of toric varieties and polytopes see [32].) The h-vector of a simplicial polytope is the sequence of Betti numbers of this variety. For rational, nonsimplicial polytopes, however, the usual Betti numbers no longer depend only on the combinatorial structure of the polytope. The appropriate definition of the h-vector comes from the middle perversity intersection homology. When the polytope is simplicial this definition agrees with the previous definition of h-vector.

A way of calculating the h-vector for general rational polytopes was found independently by several algebraic geometers: Bernstein, Khovanskii and MacPherson. Stanley [36] introduced the formula to combinatorialists; he gives in these proceedings [40] a definition that applies more generally to Eulerian posets. Here is that definition of the *h-vector* of a polytope as a recursion on the face lattice.

For a d-polytope P, from the h-vector $(h_0, h_1, \ldots, h_d) \in \mathbf{N}^{d+1}$ is defined the g-vector $(g_0, g_1, \ldots, g_{\lfloor d/2 \rfloor}) \in \mathbf{N}^{\lfloor d/2 \rfloor + 1}$ by $g_0 = h_0$ and $g_i = h_i - h_{i-1}$ for $1 \leq i \leq d/2$. The generating functions $h(P,t) = \sum_{i=0}^{d} h_i t^{d-i}$ and $g(P,t) = \sum_{i=0}^{\lfloor d/2 \rfloor} g_i t^i$ are defined recursively by

1. $g(\emptyset, t) = h(\emptyset, t) = 1$, and
2. $h(P,t) = \displaystyle\sum_{\substack{G \text{ face of } P \\ G \neq P}} g(G,t)(t-1)^{d-1-\dim G}$.

It is easy to check from this recursion that for a fixed dimension d and any i, h_i is a linear function of the flag vector. Fine [17] has a combinatorial interpretation of the coefficients of the flag numbers in this linear function.

If $S \subseteq \{0, 1, \ldots, d-1\}$, the *S-template* consists of d spaces with a vertical bar after the $(j+1)$st space for each $j \in S$. An *admissible pattern* of weight r for the S-template is a placement of an x in each of r spaces so that to the left of each vertical bar (all the way to the left) there are more blank spaces than xs. The unsigned coefficient of f_S in h_i (for d-polytopes) is the number of admissible patterns of weight $d-i$ for the S-template. The sign of the coefficient of f_S in h_i is $(-1)^{d-i+|S|+1}$.

As an example we compute the formula for the h-vector of a 3-polytope. Here are all the admissible patterns, with signs indicated.

```
(+) x x x
(-) x x _
(-) x _ x
(-) _ x x
(+) _ | x x

(+) x _ _
(+) _ x _
(+) _ _ x
(-) _ | x _
(-) _ | _ x
(-) _ _ | x
(-) x _ _ |
(-) _ x _ |
(-) _ _ x |
(+) _ | _ | x
(+) _ | x _ |
(+) _ | _ x |
(+) _ _ | x |
(-) _ | _ | x |

(-) _ _ _
(+) _ | _ _
(+) _ _ | _
(+) _ _ _ |
(-) _ | _ | _
(-) _ | _ _ |
(-) _ _ | _ |
(+) _ | _ | _ |
```

From these we get the following formulas for the h-vector.

$$\begin{aligned} h_0 &= f_\emptyset \\ h_1 &= f_0 - 3f_\emptyset \\ h_2 &= -f_{012} + f_{01} + 2f_{02} + f_{12} - 2f_0 - f_1 - 3f_2 + 3f_\emptyset \\ h_3 &= f_{012} - f_{01} - f_{02} - f_{12} + f_0 + f_1 + f_2 - f_\emptyset \end{aligned}$$

In summary, associated with any rational d-polytope P is an h-vector $h(P) = (h_0, h_1, \ldots, h_d)$, which depends linearly on the flag vector of P. As in the case of simplicial polytopes, this h-vector is positive, symmetric and unimodal. We don't know if the nonlinear inequalities satisfied by simplicial h-vectors hold in general. Note that the flag vector cannot be computed just from the h-vector; the flag vector contains a Fibonacci number of pieces of information (exponential in dimension d), while the h-vector contains only $d/2$ pieces of information. The unimodality of the h-vector or, equivalently, the nonnegativity of the g-vector gives $d/2$ linear inequalities satisfied by flag vectors of rational d-polytopes. It is believed that these are sharp and that they are satisfied by flag vectors of irrational polytopes. It should be noted that the first two inequalities are known for irrational polytopes as well: one is trivial ($f_0 \geq d+1$), and the other ($f_{02} - 3f_2 + f_1 - df_0 + \binom{d+1}{2} \geq 0$) was proved by Kalai [23] using rigidity.

Kalai [24] made a wonderful observation that enabled him to extend the h-vector to a vector linearly equivalent to the flag vector, and to get many more linear inequalities on flag vectors. This is that the convolution of two linear forms on flag vectors is also a linear form on flag vectors. First we define the convolution of two flag numbers. For $S \subseteq \{0, 1, \ldots, d-1\}$, $T \subseteq \{0, 1, \ldots, e-1\}$, and P a $(d+e+1)$-polytope, let

$$f_S * f_T(P) = \sum_{F \text{ } d\text{-face of } P} f_S(F) f_T(P/F) = f_{S \cup \{d\} \cup (T+d+1)}(P).$$

(Here P/F is the quotient polytope of F, an e-polytope whose face lattice is the interval $[F, P]$ in the face lattice of P.) Repeat to get the convolution of any finite sequence of flag numbers, and extend linearly to define the convolution of a sequence of linear forms in flag numbers. In the following we write g_i^d for the linear form in flag numbers that calculates g_i ($= h_i - h_{i-1}$) for d-polytopes.

Theorem 5 (Kalai) *The following linear forms on flag numbers form a vector of length e_d linearly equivalent to the flag vector:*

$$g_{\ell_1}^{s_1} * g_{\ell_2}^{s_2} * \cdots * g_{\ell_k}^{s_k},$$

where $d = k - 1 + \sum s_i$, $0 \leq \ell_k \leq s_k/2$ and for $1 \leq i < k$, $1 \leq \ell_i \leq s_i/2$.

The convolution of a sequence of linear forms that are nonnegative for all polytopes is itself a linear form that is nonnegative for polytopes of the appropriate dimension. Thus convolutions of g_is are nonnegative for rational polytopes, and hence give linear inequalities on flag numbers that hold for all rational polytopes.

Here is an example. Consider the convolution $g_1^2 * g_1^2$ for 5-polytopes; here $g_1^2(Q) = f_0(Q) - 3 \geq 0$ for a 2-polytope Q. Then for any 5-polytope P

$$g_1^2 * g_1^2(P) = \sum_{F \text{ 2-face of } P} (f_0(F) - 3)(f_0(P/F) - 3) \geq 0.$$

This is equivalent to the linear inequality

$$f_{023}(P) - 3f_{23}(P) - 3f_{02}(P) + 9f_2(P) \geq 0,$$

which holds for all 5-polytopes P. (Note: the inequality $f_0 - 3 \geq 0$ holds for all 2-polytopes, so the resulting inequality holds for irrational as well as rational 5-polytopes.) Each linear form in the flag numbers has a dual linear form, obtained by replacing f_S by $f_{\tilde{S}}$, where $\tilde{S} = \{d - 1 - s : s \in S\}$. By polytope duality the dual of a linear form that is nonnegative for all (rational) d-polytopes is also nonnegative for all (rational) d-polytopes. We thus can use the duals of the g_i^d in convolutions to generate nonnegative linear forms in flag numbers. The following appears in [24].

Conjecture 6 (Kalai) *The nonnegativity of convolutions of the g_i and their duals imply all linear inequalities on the flag numbers of polytopes.*

Meisinger [31] showed that the conjecture is false, because starting in dimension six these inequalities do not imply all the following trivial inequalities: $f_i \geq \binom{d+1}{i+1}$ (that is, the fact that every d-polytope has at least as many i-faces as the d-simplex).

A particular choice of convolutions of g_i^d form a Fibonacci-length vector linearly equivalent to the flag vector. Another such vector, called the *cd*-index, was introduced by Fine (but was first published in [7]). Stanley [39] proved that the *cd*-index is nonnegative for shellable regular CW-spheres (thus proving an extended version of a conjecture of Fine). The resulting inequalities on flag numbers are in particular true for polytopes, but we can expect them to be weaker than the g-vector convolution inequalities, which are special to (rational) polytopes. See the paper by Stanley in these proceedings [40] for more information about the *cd*-index.

Finally let me observe that the intersection homology picture tells us only linear inequalities. We know by convexity considerations that the flag vectors (in dimension four and higher) cannot be characterized only by linear equations and linear inequalities. But we have no conjecture for a set of nonlinear inequalities that are tight and hold for all d-polytopes.

We end this section with a brief look at two special classes of polytopes.

1.3. Cubical Polytopes

A polytope is called *cubical* if and only if all its proper faces are combinatorially equivalent to cubes. As in the simplicial case, the flag vectors of cubical polytopes depend linearly on the f-vectors. Also, the f-vectors of cubical polytopes satisfy $\lceil d/2 \rceil$ linear equations, analogous to the Dehn-Sommerville equations [19]. Adin [2] has defined a special h-vector for cubical polytopes (different from the "toric variety" h-vector), which is symmetric, nonnegative, and computable by shelling the polytope. The symmetry of Adin's h-vector gives the linear equations on the f-vector mentioned above.

A seemingly basic fact was proved only recently by Blind and Blind [13].

Theorem 7 (Blind and Blind) *If P is a cubical d-polytope, then for all i, $f_i(P) \geq f_i(d\text{-cube})$. Equality in a single i implies that P is a d-cube.*

Several people had conjectured that a cubical polytope always has at least as many vertices as facets. This has been disproved recently by Jockusch, who constructs a sequence of cubical polytopes of increasing dimension for which the ratio of number of facets to number of vertices increases without bound [22]. Jockusch also conjectures a lower bound on f_i for cubical polytopes in terms of the number of vertices and the dimension.

1.4. Centrally Symmetric Polytopes

In 1982 Bárány and Lovász conjectured lower bounds on the number of faces of simplicial, centrally symmetric polytopes [3]. This conjecture was later strengthened by Björner. The stronger version was proved by Stanley by studying the effect of symmetries on h-vectors [37].

Theorem 8 (Stanley) *If P is a simplicial, centrally symmetric d-polytope, then for all i, $1 \leq i \leq d/2$,*

$$h_i - h_{i-1} \geq \binom{d}{i} - \binom{d}{i-1}.$$

Adin has given analogous results for rational simplicial polytopes with other types of symmetries [1].

We mention finally a conjecture of Kalai [25] on centrally symmetric polytopes, nonsimplicial as well as simplicial.

Conjecture 9 (Kalai) *For any centrally symmetric polytope*

$$f_0 + f_1 + \cdots + f_{d-1} \geq 3^d - 1$$

2. Subdivision

We now turn to polyhedral subdivisions of polytopes. Initially we will be interested in the effect of subdivision on flag numbers. It is easy to see that as we subdivide the faces of a polytope the flag numbers increase. Also the boundary of an arbitrary polytope can be subdivided to get (combinatorially) a simplicial polytope. This can be done, for example, by barycentric subdivision, but it can also be done without adding any new vertices. The f-vectors and flag vectors of simplicial polytopes are characterized. How can we use a simplicial subdivision to give conditions on flag vectors?

Define a *polyhedral subdivision* of a d-polytope P to be a polyhedral complex whose vertex set is that of P and whose underlying space is P. Note that here we are considering a subdivision of the solid polytope, not of its boundary, but the subdivision of P induces a subdivision of its boundary. Also we assume that no new vertices occur in the subdivision. The subdivision is a *triangulation* if all its faces are simplices. Earlier we defined the h-vector of a d-polytope. More properly we should have called this the h-vector of the boundary complex ∂P of P, that is, the polyhedral complex consisting of the faces of the boundary of P. The computation of an h-vector from a flag vector can be performed for any polyhedral complex. When this is done for the polyhedral complex consisting of the boundary complex of P plus the one d-face P itself, we find that $h_0(P) = h_0(\partial P) = 1$, $h_i(P) = h_i(\partial P) - h_{i-1}(\partial P)$ for $1 \leq i \leq d/2$, and $h_i(P) = 0$ for $i \geq d/2$. (This is the g-vector of the boundary of P.) Stanley showed the following [38].

Theorem 10 (Stanley) *If P is a rational convex polytope and Δ is a (rational) polyhedral subdivision of P, then for all i, $h_i(\Delta) \geq h_i(P)$ and $h_i(\partial\Delta) \geq h_i(\partial P)$.*

This gives a limited amount of inequality information. For example, if we subdivide a rational polytope adding no new vertices and then apply the upper bound theorem, we get inequalities on the h-vector, and thus the flag vector, of the original polytope.

2.1. Shallow Subdivisions

To help understand how subdivision changes the h-vector of a polytope, we study those subdivisions that leave the h-vector unchanged. For ease of presentation we will restrict the discussion to triangulations. (The situation for other subdivisions is understood as well but it is harder to state. See [4] for details.) Let Δ be a triangulation of a polytope P. For a face σ of Δ, define the *carrier* $C(\sigma)$ of σ to be the smallest face of P containing σ. A triangulation Δ of P is *shallow* if and only if for all faces σ of Δ, $\dim C(\sigma) \leq 2 \dim \sigma$. Note that the condition for shallowness applied to vertices of Δ says that every vertex of Δ must be a vertex of P, which we have assumed anyway. A triangulation of a 3-polytope is shallow if and only if every edge of the triangulation is contained in a 2-face, that is, if and only if there are no interior edges.

Here are a few examples. The bipyramid over a triangle has two triangulations, one into two simplices, the other into three. The first is shallow; the second is not. The regular octahedron has only one combinatorial type of triangulation, into four simplices around an edge connecting opposite vertices. It is not shallow. The triangular prism has one combinatorial type of triangulation, into three simplices. It is shallow.

Shallow triangulations are important because of the following theorem [4].

Theorem 11 *1. If Δ is a shallow subdivision of a polytope P, then $h(\Delta) = h(P)$ and $h(\partial\Delta) = h(\partial P)$.*

2. If Δ is a subdivision of a rational *polytope P and $h(\Delta) = h(P)$, then Δ is shallow.*

Thus if P has a shallow triangulation whose boundary is polytopal, then the h-vector of the boundary of P satisfies the McMullen conditions (nonlinear inequalities as well as the linear conditions). It would be interesting, then, to have a nice characterization of polytopes that have shallow subdivisions.

A polytope is *weakly neighborly* if all its triangulations are shallow. (The triangular prism is an example of a weakly neighborly polytope.) There is a nice combinatorial characterization of weakly neighborly polytopes [4]; it explains the choice of the term.

Theorem 12 *A polytope P is weakly neighborly if and only if every set of $k+1$ vertices is contained in a face of dimension at most $2k$ for all k.*

The three-dimensional examples of weakly neighborly polytopes are not very interesting: they are just the triangular prism and all 3-dimensional pyramids. Among simplicial d-polytopes, having a shallow triangulation is equivalent to being $\lfloor d/2 \rfloor$-stacked. The only simplicial, weakly neighborly polytopes are simplices and even-dimensional neighborly polytopes. A large class of weakly neighborly polytopes is the set of Lawrence polytopes. These are the polytopes having symmetric Gale diagrams. Perhaps weakly neighborly polytopes can be characterized as having nearly symmetric Gale diagrams, but a precise statement is known only for polytopes with few vertices [4]. One other example of a weakly neighborly polytope is the Cartesian product of two simplices (of any dimension).

Stanley introduced local h-vectors, which serve as measures of nonshallowness at faces of a polytope [38].

The *local h-vector* of a subdivision Δ of a d-polytope P is the vector of coefficients of the degree $d+1$ polynomial $\ell_P(\Delta, x)$ that satisfies the recursion

1. $\ell_\emptyset(\emptyset, x) = 1$
2. $h(\Delta, x) = \sum_{F \text{ face of } P} \ell_F(\Delta|_F, x) h(P/F, x).$

(Here $\Delta|_F$ is the subdivision of F obtained by restricting Δ; $h(P/F, x)$ is the polynomial whose coefficients form the h-vector of the solid polytope with face lattice equal to the interval $[F, P]$ in the face lattice of P.)

The local h-vector of a subdivision Δ depends on the numbers of flags of Δ and on the dimensions of the carriers of the faces; an explicit formula for the local h-vector reflecting this fact is still lacking.

Here is the most important theorem about local h-vectors, from [38].

Theorem 13 (Stanley) *Let Δ be a subdivision of a d-polytope P, and let $\ell_P(\Delta, x) = \ell_0 + \ell_1 x + \cdots + \ell_{d+1} x^{d+1}$.*

1. *If P is rational, then for all i, $\ell_i = \ell_{d+1-i}$ and $\ell_i \geq 0$.*
2. *If P is rational and Δ is regular, then for $0 \leq i < d/2$, $\ell_i \leq \ell_{i+1}$.*

The first of these is what is used to prove that the h-vector increases under subdivision. We will discuss the regularity condition on subdivisions in the next section.

2.2. Secondary Polytopes

A polyhedral subdivision of a polytope can be obtained in the following way. Given a d-polytope in $\mathbf{R}^d$, embed $\mathbf{R}^d$ in $\mathbf{R}^{d+1}$, and "lift" the polytope by assigning heights $h(x)$ to each of the vertices x. Thus we get a set of points $\{y \in \mathbf{R}^{d+1} : x = (y_1, y_2, \ldots, y_d)$ is a vertex of P and $y_{d+1} = h(x)\}$; call its convex hull Q. Let Q_{top} be the subcomplex of ∂Q that is visible from way out on the $(d+1)$-coordinate axis. Projecting Q_{top} onto $\mathbf{R}^d$ gives a subdivision of P. A subdivision obtained in this way is called *regular*.

Consider the partially ordered set (poset) of regular subdivisions ordered by refinement. The following remarkable result is from [18].

Theorem 14 (Gel'fand, Kapranov and Zelevinsky) *Given any polytope P there exists a polytope $\Sigma(P)$ whose face lattice is isomorphic to the poset of regular subdivisions of P.*

The polytope $\Sigma(P)$ is called the *secondary polytope* of P. Note that its vertices correspond to regular triangulations of P. The case where P is 2-dimensional was proved independently by Carl Lee [27] and Mark Haiman [20]. They called the secondary polytope of a polygon an *associahedron*. For example, the pentagon has five subdivisions into a quadrilateral and a triangle, and five triangulations. The refinement poset is easily seen to be the face lattice of a pentagon. Thus the secondary polytope (associahedron) of a pentagon is a pentagon. The secondary polytope of a hexagon is a simplicial 3-polytope with 14 vertices, 36 edges and 24 2-faces. See [9] for several equivalent descriptions of secondary polytopes.

The secondary polytope can be constructed from a projection of a simplex onto the polytope P. This construction was generalized by Billera and Sturmfels in their definition of *fiber polytopes* [11]. Given polytopes $P \subseteq \mathbf{R}^n$, $Q \subseteq \mathbf{R}^d$, and a projection $\pi : P \to Q$, consider all measurable functions $\gamma : Q \to \mathbf{R}^n$ such that for all x, $\pi \circ \gamma(x) = x$. Define $\Sigma(P,Q) = \frac{1}{\mathrm{vol}\, Q}\{\int_Q \gamma(x)\, dx\} \subseteq \mathbf{R}^n$. Thus $\Sigma(P,Q)$ is an average of the fibers $\pi^{-1}(x)$, as x ranges over the points of Q; it is called the fiber polytope. Billera and Sturmfels proved that $\Sigma(P,Q)$ is a convex polytope of dimension $\dim P - \dim Q$, and that its face lattice is isomorphic to a poset of certain kinds of subdivisions of Q induced from P. We will look at the special case where P is a cube and Q is a zonotope.

2.3. Fiber Zonotopes and Discriminantal Arrangements

This section contains joint work with Brandt [6]. A *zonotope* is the Minkowski sum of a set of intervals $[-a_i, a_i]$. A zonotope has a natural dual object, the central hyperplane arrangement whose hyperplanes have normals a_i. The face lattice of the hyperplane arrangement is dual to the face lattice of the zonotope. We shall study the fiber polytope of a zonotope and see what it says about the hyperplane arrangement.

Fix a set of n nonzero vectors a_i in $\mathbf{R}^d$, no two parallel. Let Z be the zonotope generated by the corresponding intervals $[-a_i, a_i]$. Billera and Sturmfels give an explicit description of the fiber zonotope [11].

Theorem 15 (Billera and Sturmfels) *The fiber polytope $\Sigma(C_n, Z)$ is an $(n-d)$-dimensional zonotope,*

$$\Sigma(C_n, Z) = \frac{1}{vol\, Z} \sum_J [-E_J, E_J],$$

where the sum is over all $(d+1)$-subsets of $\{1, 2, \ldots, n\}$, and for $J = \{j_1, j_2, \ldots, j_{d+1}\}$

$$E_J = \sum_{i=1}^{d+1} (-1)^i \det(a_{j_1}, \ldots, a_{j_{i-1}}, a_{j_{i+1}}, \ldots, a_{j_{d+1}}) \cdot e_{j_i}.$$

(Here e_ℓ is the ℓ'th standard unit vector.) The face lattice of $\Sigma(C_n, Z)$ is isomorphic to the poset of regular zonotopal subdivisions of Z.

Note that the vectors E_J do not have to be nonzero and nonparallel, so the fiber zonotope may have fewer than $\binom{n}{d+1}$ zones.

It remains to define regular zonotopal subdivisions. A *zonotopal subdivision* of the zonotope Z is a polyhedral subdivision of Z each of whose edges is a translation of one of the intervals $[-a_i, a_i]$. Zonotopal subdivisions of Z are ordered by refinement. The most refined subdivide Z into (combinatorial) cubes. Some zonotopal subdivisions of Z can be obtained in the following way. "Lift" the zones of Z by assigning each of them a height h_i. Thus we get a set of vectors $(a_i, h_i) \in \mathbf{R}^{d+1}$, and we take, in addition, the standard unit vector e_{d+1}. Let Z' be the zonotope defined by these vectors, and let Z'_{top} be the subcomplex of $\partial Z'$ that is visible from way out on the $(d+1)$-coordinate axis. Projecting Z_{top} onto $\mathbf{R}^d$ gives a subdivision of Z. A zonotopal subdivision obtained in this way is called *regular*.

Example 1. Take four vectors in $\mathbf{R}^2$. The zonotope is an octagon. It has eight "cubical" subdivisions, having six quadrilaterals each. It has eight subdivisions into one hexagon and three quadrilaterals. All zonotopal subdivisions are regular. Thus the fiber zonotope is also an octagon. A portion of the face lattice of the fiber zonotope is shown in Figure 1.

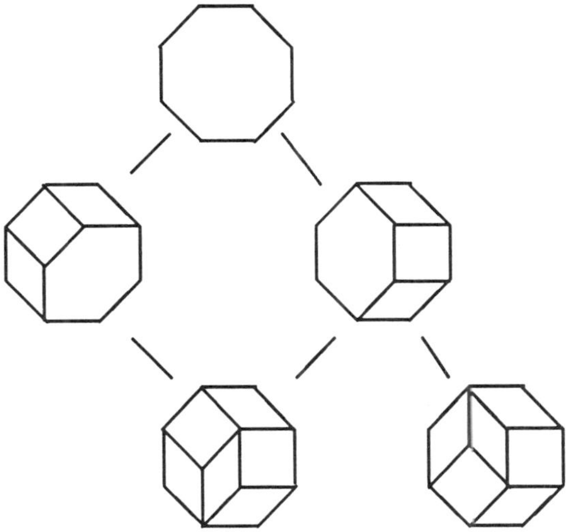

Fig. 1. Portion of face lattice of fiber zonotope

Example 2. Consider the five vectors in $\mathbf{R}^3$: $a_1 = (1,0,0)$, $a_2 = (0,1,0)$, $a_3 = (1,1,0)$, $a_4 = (0,0,1)$ and $a_5 = (1,0,1)$. The zonotope $Z = Z(\mathcal{A})$ is a 3-polytope with four hexagonal faces and eight quadrilateral faces. The drawing in Figure 2 shows the combinatorial structure of one side of this polytope, with half of the 2-faces showing. (Since zonotopes are centrally symmetric, this shows enough to determine the whole polytope.) The fiber zonotope $\Sigma(C_5, Z)$ is a hexagon. The original zonotope has twelve regular zonotopal subdivisions. Four of them each cut Z into two hexagonal prisms and two cubes. Two of them each cut Z into a twelve-sided zonotope and four cubes. Refining these are six cubical subdivisions of Z.

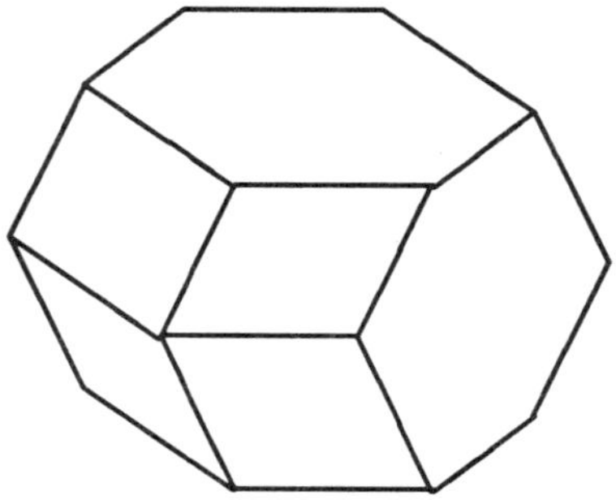

Fig. 2. 3-dimensional zonotope

Write $\mathcal{A}$ for the central hyperplane arrangement, $\mathcal{A} = \{H_1^0, H_2^0, \ldots, H_n^0\}$, where $H_i^0 = \{x \in \mathbf{R}^d : a_i \cdot x = 0\}$. There are two lattices associated with $\mathcal{A}$. The *lattice of intersections* $L(\mathcal{A})$ consists of all distinct intersections $\bigcap_{i \in I} H_i^0$, ordered by reverse inclusion, with a least element $\hat{0}$ adjoined.

To describe the *face lattice* of $\mathcal{A}$, consider the complement of $\mathcal{A}$, that is, $\mathbf{R}^d \setminus \bigcup_{i=1}^{n} H_i^0$. This is a set of open cones. The face lattice $\mathcal{F}(\mathcal{A})$ consists of all faces of the closures of these cones, ordered by inclusion, also with $\hat{0}$ adjoined. The face lattice of $\mathcal{A}$ is dual to the face lattice of the zonotope Z.

For example, for the standard unit vectors $e_1, e_2, \ldots, e_n$ in $\mathbf{R}^n$, Z is the n-cube, $L(\mathcal{A})$ is the Boolean lattice, and $\mathcal{F}(\mathcal{A})$ is the face lattice of the crosspolytope.

How does the fiber zonotope relate to the hyperplane arrangement? Define a new central hyperplane arrangement $B(\mathcal{A})$ by taking as normal vectors a maximal set of nonzero, pairwise nonparallel vectors among the E_J (the vectors of the fiber zonotope). Thus $B(\mathcal{A})$ is the hyperplane arrangement dual to the fiber zonotope. It is an arrangement of hyperplanes in $\mathbf{R}^n$ but it is of essential dimension $n - d$ (that is, all the hyperplanes intersect on a d-dimensional subspace). In the case where the original vectors $a_1, a_2, \ldots, a_n$ are in general position, this is the *discriminantal arrangement* introduced by Manin and Schechtman; we will extend that term to the case of arbitrary a_i. We study the complement and the intersection and face lattices of the discriminantal arrangement.

Consider all *affine* hyperplane arrangements that have normal vectors a_i. These can be parametrized by points of $\mathbf{R}^n$. An easy way to do this is as follows: for $b \in \mathbf{R}^n$, let $\mathcal{A}_b$ be the affine arrangement with hyperplanes $H_i = \{x \in \mathbf{R}^n : a_i \cdot x = b_i\}$. Call $\mathcal{A}_b$ a *parallel translation* of $\mathcal{A}$. It is in *general position* if and only if for all $I \subseteq \{1, 2, \ldots, n\}$, $\bigcap_{i \in I} H_i = \emptyset$ if $|I| > d$, and $\dim \bigcap_{i \in I} H_i = d - |I|$ if $I \leq d$. The central arrangement $\mathcal{A}$ is *generic* if and only if some parallel translation $\mathcal{A}_b$ is in general position. This is the case treated in [28].

Theorem 16 (Manin and Schechtman) *If $\mathcal{A}$ is generic, then the complement of the discriminantal arrangement $B(\mathcal{A})$ is the set of b for which $\mathcal{A}_b$ is in general position.*

To see what the complement is for an arbitrary arrangement $\mathcal{A}$, we need a definition. The affine arrangement $\mathcal{A}_b$ is in *relatively general position* if and only if for all $I \subseteq \{1, 2, \ldots, n\}$, $\bigcap_{i \in I} H_i = \emptyset$ if $|I| > d$, and $\dim \bigcap_{i \in I} H_i \leq d - |I|$ if $I \leq d$. (The idea here is that the high dimensional intersections, but not the parallelisms, of a nongeneral position arrangement can be eliminated by parallel translation of the hyperplanes.)

Theorem 17 *The complement of the discriminantal arrangement of $\mathcal{A}$ is the set of b for which $\mathcal{A}_b$ is in relatively general position.*

Now a hyperplane of the discriminantal arrangement has a vector

$$E_J = \sum_{i=1}^{d+1} (-1)^i \det(a_{j_1}, \ldots, a_{j_{i-1}}, a_{j_{i+1}}, \ldots, a_{j_{d+1}}) \cdot e_{j_i}$$

as normal. Such a vector represents a minimal dependency among the a_i. Corresponding to a minimal dependency among the a_i is a minimal possible violation of relatively general position in the parallel translations of $\mathcal{A}$. Suppose $\{a_i : i \in S\}$ is a minimal dependent set. Define H_S to be the set of points $b \in \mathbf{R}^n$ such that in the affine arrangement $\mathcal{A}_b$, $\dim \bigcap_{i \in S} H_i = d - |S| + 1$.

Theorem 18 *The hyperplanes of the discriminantal arrangement based on $\mathcal{A}$ are the H_S, for all sets S that index minimal dependent subsets of $\{a_i : 1 \leq i \leq n\}$.*

This generalizes the observation of [28] that the hyperplanes of the discriminantal arrangement of a generic arrangement correspond to the $(d+1)$-sets of $\{1, 2, \ldots, n\}$.

We have described the hyperplanes and the complement of the discriminantal arrangement. We now turn to the whole face structure. Every parallel translation $\mathcal{A}_b$ of $\mathcal{A}$ has its own face poset, and we can label each open face with a sign vector in $\{-, 0, +\}$, which indicates its position with respect to each of the n hyperplanes of $\mathcal{A}_b$. This gives a description of the face lattice of the discriminantal arrangement.

Theorem 19 *Two points b and b' are in the same (open) face of the discriminantal arrangement $B(\mathcal{A})$ if and only if the two affine arrangements $\mathcal{A}_b$ and $\mathcal{A}'_b$ have the same labeled face poset.*

Note that each face of the discriminantal arrangement corresponds to a face of the fiber zonotope, which in turn represents a regular zonotopal subdivision of the original zonotope Z. There is a natural way of assigning sign vectors to the faces of a regular zonotopal subdivision, and under the duality between the face lattice of the fiber zonotope and that of the discriminantal arrangement, corresponding faces are labeled with the same sign vectors. The reason for this is that both situations are different interpretations of the same oriented matroid. The base zonotope Z and hyperplane arrangement $\mathcal{A}$ represent a single oriented matroid. Regular zonotopal subdivisions of Z come from projections of a higher dimensional zonotope. Affine parallel translations of $\mathcal{A}$ come from sections of a higher dimensional central hyperplane arrangement. The higher dimensional zonotope and central hyperplane arrangement are duals; that is, they are represented by the same oriented matroid, called a *lifting* of the original oriented matroid. (See [12].)

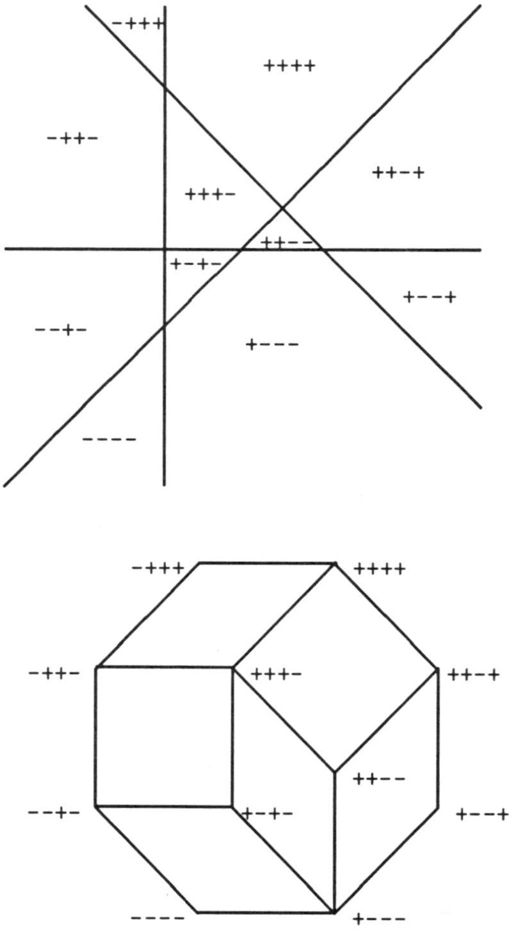

Fig. 3. Labeled faces of an affine arrangement and labeled vertices of the zonotopal subdivision

Example 1 continued. For the arrangement $\mathcal{A}$ of four lines in $\mathbf{R}^2$ choose normals

$a_1 = (1,0)$, $a_2 = (0,1)$, $a_3 = (1,-1)$, and $a_4 = (1,1)$. Let $b = (0,0,2,4)$. In Figure 3 we show $\mathcal{A}_b$ with maximal faces labeled, and the corresponding zonotopal subdivision with vertices labeled.

Manin and Schechtman [28] described the intersection lattice of discriminantal arrangements coming from arrangements of $d+3$ hyperplanes in $\mathbf{R}^d$ in the "most generic" case, that is, for an open Zariski dense subset of all d-arrangements with $d+3$ hyperplanes. Falk [16] showed that not all generic arrangements have an intersection lattice of this type. We conjecture a condition on the a_i guaranteeing "most generic" status, and a description of the intersection lattice $L(B(\mathcal{A}))$ of the discriminantal arrangement for arbitrary n.

For $n \geq d+1 \geq 2$ let $P(n,d)$ be the following poset. The elements are sets $\{S_1, S_2, \ldots, S_m\}$ of subsets of $\{1,2,\ldots,n\}$ satisfying

1. for each i, $|S_i| \geq d+1$
2. for each i,j, $i \neq j$, $|S_i \cap S_j| < d$
3. for each $I \subseteq \{1,2,\ldots,m\}$, $|\bigcup_{i\in I} S_i| > d + \sum_{i\in I}(|S_i| - d)$.

The ordering is given by $\{S_1, S_2, \ldots, S_m\} \preceq \{T_1, T_2, \ldots, T_p\}$ if and only if for each i there exists j such that $S_i \subseteq T_j$. This is a ranked poset, with $\text{rank}\{S_1, S_2, \ldots, S_m\} = \sum_{i=1}^{m}(|S_i| - d)$.

Conjecture 20 *1. If $\mathcal{A}$ is a generic arrangement of n hyperplanes in $\mathbf{R}^d$ with algebraically independent normal vectors, then $L(B(\mathcal{A}))$ is isomorphic to $P(n,d)$.*

2. Among arrangements $\mathcal{A}$ of n hyperplanes in $\mathbf{R}^d$ the number of rank i elements of $L(B(\mathcal{A}))$ is maximized for all i by the arrangements described in (1).

We mention briefly the issue of freeness of discriminantal arrangements. Terao [42] defined a central arrangement to be free if its module of derivations is free (the definition has its source in singularity theory). There has been much interest in determining which arrangements are free. Manin and Schechtman [28] computed the Möbius function for the intersection lattice of the discriminantal arrangement coming from "most generic" arrangements of $d+3$ hyperplanes in $\mathbf{R}^d$. From this Orlik and Terao [33] observed that these discriminantal arrangements are not free. No other general results on freeness of discriminantal arrangements are known. However, known examples of nonfree discriminantal arragements have exhibited a high degree of formality [14], which is a necessary condition for freeness.

2.4. Conclusion

This paper has touched on three interesting topics in the combinatorial study of convex polytopes: numbers of faces, subdivision, and relation to hyperplane arrangements. Results range from the very old to the very new, and we hope the reader is left with an interest in the many open questions in the subject.

References

1. Adin, Ron. On face numbers of rational simplicial polytopes with symmetry. *Adv. in Math.*, to appear.

2. Adin, Ron. A new cubical h–vector. Preprint.
3. Bárány, I. and Lovász, L. Borsuk's theorem and the number of facets of centrally symmetric polytopes. *Acta Math. Acad. Sci. Hungar.*, 40(3–4):323–329, 1982.
4. Bayer, Margaret M. Equidecomposable and weakly neighborly polytopes. *Israel J. Math.*, 81:301–320, 1993.
5. Bayer, Margaret M. and Billera, Louis J. Generalized Dehn-Sommerville relations for polytopes, spheres and Eulerian partially ordered sets. *Invent. Math.*, 79:143–157, 1985.
6. Bayer, Margaret M. and Brandt, Keith A. Discriminantal arrangements, fiber polytopes and formality. In preparation.
7. Bayer, Margaret M. and Klapper, Andrew. A new index for polytopes. *Discrete Comput. Geom.*, 6:33–47, 1991.
8. Bayer, Margaret M. and Lee, Carl W. Combinatorial aspects of convex polytopes. In P. M. Gruber and J. M. Wills, editors, *Handbook of Convex Geometry*, volume A, pages 485–534. North Holland, Amsterdam, 1993.
9. Billera, Louis J., Filliman, Paul and Sturmfels, Bernd. Constructions and complexity of secondary polytopes. *Adv. in Math.*, 83:155–179, 1990.
10. Billera, Louis J. and Lee, Carl W. Sufficiency of McMullen's conditions for f-vectors of simplicial polytopes. *Bull. Amer. Math. Soc.*, 2:181–185, 1980.
11. Billera, Louis J. and Sturmfels, Bernd. Fiber polytopes. *Ann. of Math.*, 135:527–549, 1992.
12. Björner, Anders, Las Vergnas, Michel, Sturmfels, Bernd, White, Neil and Ziegler, Günter M. *Oriented Matroids*, volume 46 of *Encyclopedia of Mathematics and Its Applications.* Cambridge University Press, Cambridge, 1993.
13. Blind, Gerd and Blind, Roswitha. Convex polytopes without triangular faces. *Israel J. Math.*, 71:129–134, 1990.
14. Brandt, Keith A. and Terao, Hiroaki. Free arrangements and relation spaces. *Discrete Comput. Geom.*, to appear.
15. Eberhard, V. *Zur Morphologie der Polyeder.* Leipzig, 1891.
16. Falk, Michael. A note on discriminantal arrangements. *Proc. Amer. Math. Soc.*, to appear.
17. Fine, Jonathan. Intersection homology Betti numbers of algebraic varieties and convex polytopes. Lecture given at conference on Combinatorial Convexity and Algebraic Geometry, Oberwolfach, March–April 1993.
18. Gel'fand, I. M., Kapranov, M. and Zelevinsky, A.V. Newton polyhedra of principal $\mathcal{A}$-determinants. *Soviet Math. Doklady*, 308:20–23, 1989.
19. Grünbaum, Branko. *Convex polytopes*, volume 16 of *Pure and Applied Mathematics.* Interscience Publishers John Wiley & Sons, Inc., New York, 1967. With the cooperation of Victor Klee, M. A. Perles and G. C. Shephard.
20. Haiman, Mark. Constructing the associahedron. Unpublished manuscript.
21. Jendrol', Stanislav. On the face-vectors of trivalent convex polyhedra. *Math. Slovaca*, 33(2):165–180, 1983.
22. Jockusch, William. The lower and upper bound problems for cubical polytopes. *Discrete Comput. Geom.*, 9:159–163, 1993.
23. Kalai, Gil. Rigidity and the lower bound theorem. I. *Invent. Math.*, 88(1):125–151, 1987.

24. Kalai, Gil. A new basis of polytopes. *J. Combin. Theory Ser. A*, 49(2):191–209, 1988.
25. Kalai, Gil. The number of faces of centrally-symmetric polytopes. *Graphs and Combinatorics*, 5:389–391, 1989.
26. Klee, Victor and Kleinschmidt, Peter. Convex polytopes and related complexes. In M. Grötschel, L. Lovász, and R. Graham, editors, *Handbook of Combinatorics*. North Holland, Amsterdam, to appear.
27. Lee, Carl W. The associahedron and triangulations of the n-gon. *Europ. J. Combinatorics*, 10:551–560, 1989.
28. Manin, Yu. I. and Schechtman, V.V. Arrangements of hyperplanes, higher braid groups and higher Bruhat orders. In J. Coates et al., editors, *Algebraic Number Theory—in honor of K. Iwasawa*, volume 17 of *Adv. Studies in Pure Math.*, pages 289–308. Academic Press, 1989.
29. McMullen, Peter. The numbers of faces of simplicial polytopes. *Israel J. Math.*, 9:559–570, 1971.
30. McMullen, Peter. On simple polytopes. *Invent. Math.* 113: 419–444, 1993.
31. Meisinger, Günter. *Automatic Theorem Proving in Polytope Theory.* PhD thesis, University of Passau, 1993.
32. Oda, Tadao. *Convex Bodies and Algebraic Geometry. An Introduction to the Theory of Toric Varieties.* Springer-Verlag, Berlin, 1988.
33. Orlik, Peter and Terao, Hiroaki. *Arrangements of Hyperplanes.* Springer-Verlag, Berlin, 1992.
34. Sommerville, D. M. Y. The relations connecting the angle-sums and volume of a polytope in space of n dimensions. *Proc. Roy. Soc. London Ser. A*, 115:103–119, 1927.
35. Stanley, Richard P. The number of faces of a simplicial convex polytope. *Adv. in Math.*, 35:236–238, 1980.
36. Stanley, Richard P. Generalized H-vectors, intersection cohomology of toric varieties, and related results. In *Commutative algebra and combinatorics (Kyoto, 1985)*, volume 11 of *Adv. Stud. Pure Math.*, pages 187–213, Amsterdam-New York, 1987. North-Holland.
37. Stanley, Richard P. On the number of faces of centrally-symmetric simplicial polytopes. *Graphs Combin.*, 3(1):55–66, 1987.
38. Stanley, Richard P. Subdivisions and local h-vectors. *J. Amer. Math. Soc.*, 5:805–851, 1992.
39. Stanley, Richard P. Flag f-vectors and the cd-index. *Math. Z.*, to appear.
40. Stanley, Richard P. A survey of Eulerian posets. In Tibor Bisztriczky et al., editors, *Polytopes: Abstract, Convex and Computational*, to appear.
41. Steinitz, Ernst. Über die Eulersche Polyederrelationen. *Arch. Math. Phys. (3)*, 11:86–88, 1906.
42. Terao, Hiroaki. Arrangements of hyperplanes and their freeness I, II. *J. Fac. Sci. Univ. Tokyo*, 27:293–320, 1980.
43. Ziegler, Günter M. Lectures on Polytopes. Preliminary Version. Technical Report TR 93-6, Konrad-Zuse-Zentrum für Informationstechnik Berlin, 1993.

APPROXIMATION BY CONVEX POLYTOPES

P. M. GRUBER
Abteilung für Analysis, TU Wien
Wiedner Hauptstraße 8 – 10/1142
A-1040 VIENNA, Austria

Abstract. After some general introductory remarks on approximation in convex geometry we present in the main part of this article asymptotic results on best approximation of convex bodies as the number of vertices, resp. facets of the approximating polytopes tends to infinity. Tools are from (affine) differential geometry. Since the transparent geometric situation in the plane admits much more precise results, this case is treated separately. In higher dimensions the relations to Dirichlet–Voronoi and Delone tilings and to the ball covering problem are indicated. Then algorithmic and, in particular, step-by-step approximation results are discussed. Here tools from number theory are applied. In the material presented emphasis is on the ideas underlying the proofs. Supplementing earlier surveys, the last chapter contains a summary of recent results on approximation of convex bodies.

1. Introduction

1.1. APPROXIMATION PROBLEMS IN CONVEX GEOMETRY

The problem to approximate a convex body in a given class by bodies from a different class, possibly a subclass of the given one, is frequently encountered. Besides the intrinsic interest in such problems it is the application of approximation results and arguments which have induced research in this area.

Examples of types of approximation problems in convexity are the following:

(i) qualitative and quantitative approximation of convex bodies by (classes of) polytopes or convex bodies including analytic convex bodies, simplices, parallelotopes and ellipsoids;

(ii) approximation problem in the context of stability results for geometric inequalities where the bodies for which equality is not attained are approximated by bodies for which equality holds;

(iii) problems concerning the approximate shape of lower-dimensional sections or projections of high-dimensional convex bodies; these investigations belong to the local theory of Banach spaces;

(iv) given a convex body, how fast can we approximate a Euclidean ball by a suitable sequence of geometric transformations of it, e.g. by Steiner or Blaschke symmetrizations;

(v) approximation problems in the sense of approximation theory in finite-dimensional normed spaces (e.g. nearest point mappings) and in the sense of optimization;

(vi) approximation of convex functions;

(vii) estimates of the complexity of algorithms dealing, for example, with the determination of volumes, the enumeration of certain types of convex polytopes or

T. Bisztriczky et al. (eds.), POLYTOPES: Abstract, Convex and Computational, 173–203.

shortest vectors in lattices.

Many approximation results and their applications are difficult to locate since they are hidden in proofs, in some cases in areas remote from convexity.

1.2. APPROXIMATION BY POLYTOPES

Among the different aspects of approximating convex bodies by convex polytopes are the following:

(i) best approximation; general properties and upper and lower bounds;

(ii) asymptotic best approximation as the number of vertices or facets of the approximating polytopes tends to infinity;

(iii) effective algorithms for approximation, in particular step-by-step methods;

(iv) approximation of typical convex bodies (in the sense of Baire categories);

(v) approximation by special (classes of) polytopes such as simplices, parallelotopes, zonotopes, lattice polytopes, or polytopes with special combinatorial properties;

(vi) approximation of polytopes by polytopes with a small number of vertices or facets; such problems are of interest in optimization;

(vii) the large area of random approximation.

Surveys on approximation by polytopes are due to Fejes Tóth [1953] and Gruber [1983a,1993a]. Random approximation is dealt with by Buchta [1985], Schneider [1988] and Affentranger [1992]. For approximation by lattice polytopes see Gruber and Lekkerkerker [1987]. Other selected aspects are contained in various articles in the Handbook of Convex Geometry.

In the following we will concentrate on more recent developments related to (ii) and (iii), explaining the ideas underlying the proofs. This will complement the survey of Gruber [1993a]. In addition, we give references to recent work on approximation.

1.3. NOTATIONS

A *convex body* in Euclidean d-space $\mathbb{E}^d$ is a compact convex subset of $\mathbb{E}^d$ with non-empty interior. Let $\mathcal{C} = \mathcal{C}(\mathbb{E}^d)$ be the space of all convex bodies in $\mathbb{E}^d$. For $n = d+1, \ldots$, denote by $\mathcal{P}_n$ and $\mathcal{P}_{(n)}$ the spaces of all convex polytopes in $\mathcal{C}$ having at most n vertices, resp. facets. Given $C \in \mathcal{C}$, let $\mathcal{P}_n^i = \mathcal{P}_n^i(C)$ and $\mathcal{P}_n^c = \mathcal{P}_n^c(C)$ be the sets of all polytopes in $\mathcal{P}_n$ which are inscribed, resp. circumscribed to C. $\mathcal{P}_{(n)}^i$ and $\mathcal{P}_{(n)}^c$ are defined similarly.

Among the various notions of distance on $\mathcal{C}$ that have been considered in the context of approximation are the *Nikodym* or *symmetric difference metric* δ^S and the (*Pompeiu-*) *Hausdorff* (*-Blaschke*) metric δ^H: let $C, D \in \mathcal{C}$. Then $\delta^S(C,D) = V(C \triangle D)$ where $V(\cdot)$ denotes the volume in $\mathbb{E}^d$ and $\delta^H(C,D)$ is the maximum distance which a point of one of the bodies C, D can have from the other body.

δ^S and δ^H induce the same topology on $\mathcal{C}$. We assume that $\mathcal{C}$ is endowed with this topology. Then $\mathcal{C}$ is locally compact by a version of Blaschke's selection theorem.

For all notions of distance δ on $\mathcal{C}$ considered in this article the following holds: if $C \in \mathcal{C}$ and $\mathcal{Q}_n$ is any of the classes $\mathcal{P}_n, \mathcal{P}_n^i, \ldots, \mathcal{P}_{(n)}^c$, $n = d+1, \ldots$, then there are polytopes $P_n \in \mathcal{Q}_n$, not necessarily unique, such that

$$\delta(C, P_n) = \delta(C, \mathcal{Q}_n) = \inf\{\delta(C, P) : P \in \mathcal{Q}_n\}.$$

P_n is called a *best approximating polytope* of C in $\mathcal{Q}_n$ with respect to δ. If δ is defined only for particular pairs $C, D \in \mathcal{C}$, this holds with suitable modifications.

Let bd stand for boundary, const means a constant which may depend on a given convex body or, simply, on d. If $C \in \mathcal{C}$ is of class $\mathcal{C}^2$, denote its Gaussian curvature by κ_C. For $x \in \mathrm{bd}C$ let $H_C(x)$ be the supporting hyperplane of C at x and $H_C^+(x)$ is the corresponding supporting halfspace.

2. Asymptotic Best Approximation in the Plane

2.1. THE SYMMETRIC DIFFERENCE METRIC

2.1.1. *Earlier Results*

Generalizing a result of Blaschke [1917] on inscribed triangles, Sas [1939] proved that for any $C \in \mathcal{C}(\mathbb{E}^2)$ and $n = 3, 4, \ldots,$

$$(1) \qquad \delta^S(C, \mathcal{P}_n^i) \le A(C)\left(1 - \frac{n}{2\pi}\sin\frac{2\pi}{n}\right) \quad \left(\sim \frac{2\pi^2 A(C)}{3n^2} \text{ as } n \to \infty\right)$$

where $A(\cdot)$ denotes area. Equality is attained precisely when C is an ellipse. Further, there exists a constant depending on C, such that

$$(2) \qquad \delta^S(C, \mathcal{P}_n), \delta^S(C, \mathcal{P}_n^c) \le \frac{\text{const}}{n^2} \text{ for } n = 3, 4, \ldots$$

For $\delta^S(C, \mathcal{P}_n), \delta^S(C, \mathcal{P}_n^c)$ no estimates are known which are similarly precise as the inequality (1) nor do we know the best constants or the extremal bodies in (2). See Fejes Tóth [1953].

Let $C \in \mathcal{C} \cap \mathcal{C}^2$ where $\kappa_C > 0$ and assume that the parameter is the ordinary arclength. The *affine length* $L_a = L_a(C)$ of bd$C$ is defined by

$$L_a = \int_0^L \kappa_C(u)^{1/3} du$$

where L is the ordinary length of bd$C$. For a simple geometric interpretation of L_a see Fejes Tóth [1953], pp. 40 – 42.

Fejes Tóth [1953] conjectured that

$$(3) \qquad \delta^S(C, \mathcal{P}_n^i) \sim \frac{L_a^3}{12n^2}, \quad \delta^S(C, \mathcal{P}_n^c) \sim \frac{L_a^3}{24n^2}, \quad \delta^S(C, \mathcal{P}_n) \sim \frac{L_a^3}{32n^2} \text{ as } n \to \infty.$$

The first two of these asymptotic formulae were confirmed by McClure and Vitale [1975], an alternative proof of the first one is due to Ludwig [1993].

The affine isoperimetric inequality in the plane says that for $C \in \mathcal{C} \cap \mathcal{C}^2$ where $\kappa_C > 0$,

$$L_a(C)^3 \le 8\pi^2 A(C).$$

Here equality holds precisely for ellipses. See Blaschke [1923] and Petty [1985]. Thus, among all convex bodies $C \in \mathcal{C} \cap \mathcal{C}^2$ with $\kappa_C > 0$ and of given area, it is precisely the ellipses that are asymptotically worst approximated with respect to δ^S. This is in agreement with several other approximation results (in part in higher dimensions), including results on random approximation.

2.1.2. *General Remarks*
While in many cases different asymptotic approximation problems require different ideas for their proofs, it is a common feature, that help is provided by choice of appropriate notions of length and/or area on the boundary of the convex body.

In the case of the very precise planar results the analytic estimates form an important part of the proofs, while in higher dimensions the essence of the proofs is more geometrical.

2.1.3. *Affine Arclength and Curvature*
If $C \in \mathcal{C}$ and $P \in \mathcal{P}_n^i$, then for the calculation of $\delta^S(C, P)$ one has to determine the area of each of the n moon-shaped pieces of $C \backslash P$. Since these areas are — trivially — invariant with respect to area-preserving affinities of $\mathbb{E}^2$, it is plausible to look for tools in (equi-)affine differential geometry.

Let $C \in \mathcal{C} \cap \mathcal{C}^4$ where $\kappa_C > 0$ and let $x : [0, L] \to \mathbb{E}^2$ be a parametrization of bd$C$ using ordinary arclength $t \in [0, L]$. Define $s : [0, L] \to [0, L_a]$ by

$$s(t) = \int_0^t \kappa_C(u)^{1/3} du$$

and let $t : [0, L_a] \to [0, L]$ be the inverse function of s. Then $x(t(s)) : s \in [0, L_a]$ is called an *affine arclength parametrization* of bd$C$. For $x(t(s))$ write $x(s)$ and let " $'$ " denote differentiation with respect to s.

For $0 \leq r < s \leq L_a$ let $A(r, s)$ be the area of the moon-shaped piece cut off from C by the line segment $x(r), x(s)$. Then

$$A(r, s) \sim \frac{1}{12}(s - r)^3 \text{ for small } s - r; \tag{4}$$

see Blaschke [1923].

Simple calculations show that

$$\det(x', x'') = 1, \det(x', x''') = 0, \tag{5}$$

but

$$k_a = \det(x'', x''')$$

is non-trivial. k_a is called the *affine curvature* of bd$C$; see Blaschke [1923].

2.1.4. *Asymptotic Expansion of* $\delta^S(C, \mathcal{P}_n^i)$

Theorem 1. *(Ludwig [1993]). Let* $C \in \mathcal{C} \cap C^4$ *where* $\kappa_C > 0$. *Then*

$$\delta^S(C, \mathcal{P}_n^i) \sim \frac{L_a^3}{12n^2} - \frac{L_a^4}{240} \int_0^{L_a} k_a(s) ds \frac{1}{n^4} \text{ as } n \to \infty,$$

$$\delta^S(C, \mathcal{P}_n^c) \sim \frac{L_a^3}{24n^2} + \frac{L_a^4}{240} \int_0^{L_a} k_a(s) ds \frac{1}{n^4} \text{ as } n \to \infty.$$

Outline of the proof for $\delta^S(C, \mathcal{P}_n^i)$

Lemma 1. *For* $0 \leq r < s \leq L_a$,

(6) $\quad A(r,s) = \frac{1}{12}(s-r)^3 - \frac{k_a(r)}{240}(s-r)^5 + o((s-r)^5)$ *uniformly as* $s - r \to 0$.

The proof makes use of the elementary formula

$$A(r,s) = \frac{1}{2} \int_r^s \det(x(v) - x(r), x'(v)) dv.$$

$x(v) - x(r)$ is expanded by Taylor's formula and the identities (5) and the definition of k_a are used for simplifications.

For $n = 3, 4, \ldots$, let $P_n \in \mathcal{P}_n^i$ be best approximating.

Lemma 2. *Let* $0 \leq s_{n1} < s_{n2} < \ldots < s_{nn} < L_a$ *be the values of the affine arclength parameter corresponding to the vertices of* P_n. *Then*

$$a_{ni} = s_{ni} - s_{n,i-1} = \frac{L_a}{n} + o(\frac{1}{n}) \text{ uniformly in } i \text{ as } n \to \infty.$$

To see this first note the following. Since P_n is a best approximating polygon of C, for any three consecutive vertices of P_n, say $x(s_{n,i-1}), x(s_{ni}), x(s_{n,i+1})$, the supporting line of C at $x(s_{ni})$ is parallel to the line segment $x(s_{n,i-1}), x(s_{n,i+1})$; otherwise one may increase the area of P_n by moving $x(s_{ni})$ suitably. Thus

$$\det(x(s_{n,i+1}) - x(s_{n,i-1}), x'(s_{ni})) = 0.$$

Express $x(s_{n,i+1}) - x(s_{n,i-1})$ by means of a Taylor expansion of x at s_{ni}. Insert this into the determinant and use (5). This shows that

$$a_{ni}^2 - a_{n,i+1}^2 \text{ is small,}$$

which in turn yields

(7) $$n^2 a_{ni}^2 = n \sum_{j=1}^n a_{nj}^2 + n b_{ni} \text{ where } b_{ni} \text{ is small.}$$

We estimate the two expressions on the right hand side of (7).

Since $a_{n1} + \ldots + a_{nn} = L_a$, a simple mean inequality shows that

$$L_a^2 \leq n \sum_{j=1}^{n} a_{nj}^2 \leq (n^2 \sum_{j=1}^{n} a_{nj}^3)^{2/3}. \tag{8}$$

This, Lemma 1 and the asymptotic formula (3) for $\delta^S(C, \mathcal{P}_n^i) = \delta^S(C, P_n)$ together imply the following:

$$L_a^3 \leq n^2 \sum_{j=1}^{n} a_{nj}^3 \sim 12n^2\delta^S(C, P_n) \sim L_a^3 \text{ as } n \to \infty. \tag{9}$$

Thus we obtain from (8) that

$$n \sum_{j=1}^{n} a_{nj}^2 \to L_a^2 \text{ as } n \to \infty. \tag{10}$$

A careful estimate of the expression for b_{ni} shows that

$$n|b_{ni}| = o(\max_{j=1,\ldots,n} a_{nj}) n^2 \sum_{j=1}^{n} a_{nj}^3 \text{ as } n \to \infty \text{ uniformly in } i. \tag{11}$$

Since $\delta^S(C, P_n) \to 0$,

$$\max_{j=1,\ldots,n} a_{nj} \to 0 \text{ as } n \to \infty.$$

Hence it follows from (9) and (11) that

$$n \max_{i=1,\ldots,n} |b_{ni}| \to 0 \text{ as } n \to \infty. \tag{12}$$

(7), (10) and (12) give

$$n^2 \max_{i=1,\ldots,n} a_{ni}^2, \quad n^2 \min_{i=1,\ldots n} a_{ni}^2 \to L_a^2,$$

concluding the proof of Lemma 2.

Having available Lemmas 1 and 2, the proof of the asymptotic formula for $\delta^S(C, \mathcal{P}_n^i)$ is simple:

(i) By Lemma 1, inequalities (8) and Lemma 2, the following hold:

$$\begin{aligned}
\delta^S(C, \mathcal{P}_n^i) = \delta^S(C, P_n) &= \sum_{i=1}^{n} \left(\frac{1}{12} a_{ni}^3 - \frac{k_a(s_{n,i-1})}{240} a_{ni}^5 + o(a_{ni}^5)\right) \\
&\geq \frac{L_a^3}{12n^2} - \frac{1}{240} \sum_{i=1}^{n} \left(k_a(s_{n,i-1}) a_{ni} \frac{L_a^4}{n^4} + o\left(\frac{1}{n^5}\right)\right) \\
&= \frac{L_a^3}{12n^2} - \frac{L_a^4}{240} \int_0^{L_a} k_a(s)\, ds \frac{1}{n^4} + o\left(\frac{1}{n^4}\right) \text{ as } n \to \infty.
\end{aligned}$$

(ii) If the polytopes $Q_n \in \mathcal{P}_n^i$ are such that their vertices are $x(0), x(\frac{1}{n}L_a), \ldots, x(\frac{n-1}{n}L_a)$, a similar calculation shows that

$$\delta^S(C, \mathcal{P}_n^i) \leq \delta^S(C, Q_n) \leq \frac{L_a^3}{12n^2} - \frac{L_a^4}{240} \int_0^{L_a} k_a(s) ds \frac{1}{n^4} + o(\frac{1}{n^4}) \text{ as } n \to \infty.$$

2.1.5. *Related Open Problems*

(i) The problem to find a corresponding asymptotic formula for $\delta^S(C, \mathcal{P}_n)$ seems to be difficult. This is reflected in the fact that the simpler formula (3) for $\delta^S(C, \mathcal{P}_n)$ of Fejes Tóth has not been proved so far.

(ii) If C is of class $\mathcal{C}^\infty$ or analytic, it is plausible that $\delta^S(C, \mathcal{P}_n^i), \delta^S(C, \mathcal{P}_n^c)$ and $\delta^S(C, \mathcal{P}_n)$ can be expanded in asymptotic series.

(iii) Do there exist asymptotic formulae where the polygons are replaced by smooth convex bodies the boundaries of which consist of arcs of conics or of parabolae of higher degree? For a first result of this type see the following subsection.

2.1.6. *Asymptotic Best Approximation by Quadratic Splines*

For $n = 3, 4, \ldots$, let $\mathcal{Q}_n$ be the class of convex bodies in $\mathbb{E}^2$ the boundaries of which consist of n parabolic arcs which are smoothly connected at their endpoints. For $C \in \mathcal{C}$ let $\mathcal{Q}_n^c = \mathcal{Q}_n^c(C)$ be the set of all $Q \in \mathcal{Q}_n$ (if any) which contain C and such that the endpoints of the parabolic arcs are on bd$C$. If $k_a > 0$, then $\mathcal{Q}_n^c \neq \emptyset$ and the infimum $\delta^S(C, \mathcal{Q}_n^c)$ is attained.

Without proof we cite the following result.

Theorem 2. *(Ludwig [1993]). Let $C \in \mathcal{C} \cap C^4$ where $\kappa_C, k_a > 0$. Then*

(13) $$\delta^S(C, \mathcal{Q}_n^c) \sim \frac{1}{240}(\int_0^{L_a} k_a(s)^{1/5} ds)^5 \frac{1}{n^4} \text{ as } n \to \infty.$$

By Hölder's inequality, the affine isoperimetric inequality (see 2.1.1) and an extremum property of k_a, the integral in (13) is maximal for all $C \in \mathcal{C} \cap C^4$ where $\kappa_C, k_a > 0$ and which are of given area precisely for ellipses.

There are no closed convex curves in $\mathbb{E}^2$ for which the affine curvature is negative throughout. The problem of asymptotic best approximation of a convex body in $\mathbb{E}^2$ whose affine curvature assumes both positive and negative values by bodies in $\mathcal{Q}_n$ is open.

2.2. THE HAUSDORFF METRIC

2.2.1. *Earlier Results*

It is not difficult to show that for any $C \in \mathcal{C}(\mathbb{E}^2)$ there is a constant such that

$$\delta^H(C, \mathcal{P}_n^i), \delta^H(C, \mathcal{P}_n^c), \delta^H(C, \mathcal{P}_n) \leq \frac{\text{const}}{n^2} \text{ for } n = 3, 4, \ldots$$

It is not known to the author by whom these results were first found. While more precise estimates are due to Popov [1968,1970], best possible results are not yet available.

Let $C \in \mathcal{C} \cap \mathcal{C}^2$ where $\kappa_C > 0$ and assume that the parametrization of bd$C$ is by ordinary arclength. Define

$$L_{\mathrm{II}} = L_{\mathrm{II}}(C) = \int_0^L \kappa_C(u)^{1/2} du.$$

This notion of arclength is related to the deviation of bd$C$ from its tangent; see 2.2.2 and 3.3.1.

Fejes Tóth [1953] stated that

$$\delta^H(C, \mathcal{P}_n^i) \sim \frac{L_{\mathrm{II}}^2}{8n^2} \text{ as } n \to \infty. \tag{14}$$

This is a special case of a result of Fejes Tóth [1948] for curves of class $\mathcal{C}^2$ in $\mathbb{E}^d$ and inscribed polygons. For a different proof see McClure and Vitale [1975]. Since for $C \in \mathcal{C} \cap \mathcal{C}^1$,

$$\delta^H(C, \mathcal{P}_n^i) \sim \delta^H(C, \mathcal{P}_n^c) \sim 2\delta^H(C, \mathcal{P}_n) \text{ as } n \to \infty,$$

further formulae result. Among all convex bodies $C \in \mathcal{C} \cap \mathcal{C}^2$ where $\kappa_C > 0$ and of given perimeter it is only the circular discs for which L_{II} is maximal. This follows from Hölder's inequality.

2.2.2. *The Appropriate Parameter*

Let $C \in \mathcal{C}$ and $P \in \mathcal{P}_n^i$. Then in order to calculate $\delta^H(C, P)$ we have to determine for each edge of P the maximum distance of a point of the corresponding arc of bd$C$ from it. Elementary differential geometric considerations lead to the following choice of parameter well suited for this purpose. (In higher dimensions we proceed in an equivalent, but slightly different way, cf. 3.3.1: We look for a simple way to calculate the distance of a point $v \in$ bd$C$ from the supporting hyperplane of C at a nearby point $x \in$ bd$C$. For this an appropriate tool is the second fundamental form and the corresponding Riemannian metric.)

Let $C \in \mathcal{C} \cap \mathcal{C}^4$ where $\kappa_C > 0$ and consider the ordinary arclength parametrization x of bd$C$. Define $s : [0, L] \to [0, L_{\mathrm{II}}]$ by

$$s(t) = \int_0^t \kappa_C(u)^{1/2} du$$

and let $t : [0, L_{\mathrm{II}}] \to [0, L]$ be the inverse function of s. For $x(t(s))$ we simply write $x(s)$ and " $'$ " denotes differentiation with respect to s.

For $0 \leq r < s \leq L_{\mathrm{II}}$ let $h(r, s)$ be the maximum distance of a point on the arc on bd$C$ with endpoints $x(r), x(s)$ from the line segment $x(r), x(s)$. Then

$$h(r, s) \sim \frac{1}{8}(s - r)^2 \text{ for small } s - r.$$

The following functions $k, l : [0, L_{\mathbb{I}}] \to \mathbb{R}$ will be needed in Theorem 3:

$$k = \kappa_C + \frac{\kappa_C''}{\kappa_C} - \frac{5(\kappa_C')^2}{6\kappa_C^2}, \quad l = \kappa_C + \frac{\kappa_C''}{5\kappa_C} - \frac{7(\kappa_C')^2}{30\kappa_C^2}.$$

2.2.3. *Asymptotic Expansion of* $\delta^H(C, \mathcal{P}_n^i)$

Theorem 3. *(Ludwig [1993]). Let* $C \in \mathcal{C} \cap C^4$ *where* $\kappa_C > 0$. *Then*

$$\delta^H(C, \mathcal{P}_n^i) \sim \frac{L_{\mathbb{I}}^2}{8n^2} - \frac{L_{\mathbb{I}}^3}{384} \int_0^{L_{\mathbb{I}}} k(s) ds \frac{1}{n^4} \quad as\ n \to \infty,$$

$$\delta^H(C, \mathcal{P}_n^c) \sim \frac{L_{\mathbb{I}}^2}{8n^2} + \frac{5L_{\mathbb{I}}^3}{384} \int_0^{L_{\mathbb{I}}} l(s) ds \frac{1}{n^4} \quad as\ n \to \infty.$$

Outline of the proof for $\delta^H(C, \mathcal{P}_n^i)$. The proof of Theorem 3 can be arranged along similar lines as that of Theorem 1. We indicate some steps.

Lemma 3. *For* $0 \le r < s \le L_{\mathbb{I}}$,

$$h(r, s) = \frac{1}{8}(s - r)^2 - \frac{k(r)}{384}(s - r)^4 + o((s - r)^4) \quad uniformly\ as\ s - r \to 0.$$

Next choose best approximating polygons $P_n \in \mathcal{P}_n^i$ of C.

Lemma 4. *Let* $0 \le s_{n1} < s_{n2} < \ldots < s_{nn} < L_{\mathbb{I}}$ *be the values of the parameter which correspond to the vertices of* P_n. *Then*

$$a_{ni} = s_{ni} - s_{n,i-1} = \frac{L_{\mathbb{I}}}{n} + o(\frac{1}{n}) \quad uniformly\ in\ i\ as\ n \to \infty.$$

In the proof of Lemma 4 one makes use of the following general property of the best approximating polygons P_n. If n is sufficiently large, then

$$h(s_{n1}, s_{n2}) = \ldots = h(s_{n,n-1}, s_{nn}) = h(s_{nn}, s_{n1}) = \delta^H(C, \mathcal{P}_n^i).$$

Theorem 3 follows from Lemmas 3 and 4 in a similar way as Theorem 1 was obtained from Lemmas 1 and 2.

2.2.4. *Related Open Problems*

Obviously also in the present case several open problems suggest themselves. Compare the problems in 2.1.5 and the result in 2.1.6.

2.3. OTHER METRICS

It is highly plausible that there are results of the type exhibited in Subsections 2.1.4, 2.1.6 and 2.2.3 also for other metrics. For a list of such metrics and earlier asymptotic formulae see the surveys of Gruber [1983a,1993a].

3. Asymptotic Best Approximation in General Dimensions

In the following we explain the ideas of the proofs of the asymptotic formulae for

$$\delta^S(C,\mathcal{P}^c_{(n)}), \delta^{MW}(C,\mathcal{P}^i_n), \delta^H(C,\mathcal{P}^i_n).$$

Other cases will be mentioned.

In order to make the presentation more intuitive we speak of squares and circular discs when we actually mean cubes or Euclidean balls in $d-1$ dimensions. More generally, a circular disc may also mean a ball in a Riemannian space of dimension $d-1$.

While in some cases the ideas underlying the proofs are rather intuitive the reader is warned that the actual proofs in general are complicated.

3.1. THE SYMMETRIC DIFFERENCE METRIC

3.1.1. *Earlier Results*

A theorem of Macbeath [1951] which extends results of Blaschke and Sas for the planar case (see 2.1.1) says the following. Let $C \in \mathcal{C}(\mathbb{E}^d)$ and let B be a Euclidean ball such that $V(B) = V(C)$. Then

$$\delta^S(C,\mathcal{P}^i_n(C)) \leq \delta^S(C,\mathcal{P}^i_n(B)) \text{ for } n = d+1,\ldots$$

where equality is attained if C is an ellipsoid. It is not known whether the ellipsoids are the only such bodies.

By choosing suitable polytopes in $\mathcal{P}^i_n(B)$ we see that there is a constant (which depends only on the volume of C) such that on the one hand we have the estimate

$$\delta^S(C,\mathcal{P}^i_n(C)) \leq \frac{\text{const}}{n^{2/(d-1)}} \text{ for } n = d+1,\ldots$$

The corresponding results for $\delta^S(C,\mathcal{Q}_n)$ where $\mathcal{Q}_n = \mathcal{P}^c_n, \mathcal{P}_n, \mathcal{P}^i_{(n)}, \mathcal{P}^c_{(n)}, \mathcal{P}_{(n)}$ follow from the analogous results for δ^H since, roughly speaking, the following inequality holds:

$$\delta^S(C,\mathcal{Q}_n) \leq \delta^H(C,\mathcal{Q}_n) \cdot \text{ area of bd}C.$$

See 3.3.1.

On the other hand the proof of a result of Gruber and Kenderov [1982] shows that for $C \in \mathcal{C} \cap \mathcal{C}^2$ there is a positive constant such that

$$\delta^S(C,\mathcal{Q}_n) \geq \frac{\text{const}}{n^{2/(d-1)}} \text{ for } n = d+1,\ldots, \text{ where } \mathcal{Q}_n = \mathcal{P}^i_n, \mathcal{P}^c_n, \mathcal{P}_n.$$

A slightly weaker lower estimate for $\mathcal{Q}_n = \mathcal{P}^i_{(n)}$ which holds for a wider class of convex bodies is due to Schütt [1991], see 5.1.2.

3.1.2. *Asymptotic Formula for* $\delta^S(C,\mathcal{P}^c_{(n)})$

Theorem 4. *(Gruber [1988,1991,1994]). Let $C \in \mathcal{C} \cap \mathcal{C}^2$ where $\kappa_C > 0$. There are positive constants* del_{d-1} *and* div_{d-1} *depending only on d, such that*

$$(1) \quad \delta^S(C,\mathcal{P}^i_n) \sim \frac{1}{2}\text{del}_{d-1}\Big(\int_{\text{bd}C} \kappa_C(x)^{1/(d+1)}d\sigma(x)\Big)^{(d+1)/(d-1)}\frac{1}{n^{2/(d-1)}} \text{ as } n \to \infty,$$

(2) $\delta^S(C,\mathcal{P}^c_{(n)}) \sim \frac{1}{2}\text{div}_{d-1}\big(\int\limits_{\text{bd}C} \kappa_C(x)^{1/(d+1)}d\sigma(x)\big)^{(d+1)/(d-1)}\frac{1}{n^{2/(d-1)}}$ *as* $n \to \infty$.

Here

$$\text{del}_1 = \frac{1}{6}, \text{del}_2 = \frac{1}{2\sqrt{3}}, \text{div}_1 = \frac{1}{12}, \text{div}_2 = \frac{5}{18\sqrt{3}}.$$

Also, σ denotes the ordinary surface area measure on bd$C$.

The integral in (1), resp. (2) is the *affine surface area* $S_a(C)$ of C. It is invariant with respect to volume-preserving affinities of $\mathbb{E}^d$ — as is δ^S. By the affine isoperimetric inequality,

$$S_a(C)^{d+1} \leq \kappa_d^2 d^{d+1} V(C)^{d-1}$$

of Blaschke [1923] and the precise discussion of the equality case due to Petty [1985] it follows that among all convex bodies $C \in \mathcal{C} \cap \mathcal{C}^2$ where $\kappa_C > 0$ and of given volume precisely the ellipsoids are asymptotically worst approximable.

How far can the assumptions in Theorem 4 be relaxed? We presume the assumption that $\kappa_C > 0$ may be omitted. Unfortunately our proof (as those of all other asymptotic formulae) relies heavily on this assumption. In the proof we have to compare the contributions of different points on bd$C$ (or rather of different small pieces of bd$C$) to $\delta^S(C,\mathcal{P}^i_n)$, say. This is achieved by comparing the corresponding osculating paraboloids which in turn requires that the paraboloids are non-degenerate, i.e. $\kappa_C > 0$. We doubt that $\mathcal{C}^2$ may be replaced by $\mathcal{C}^\alpha$ with a suitable $\alpha < 2$ while still retaining a simple asymptotic formulae. Definitely one may not take $\alpha = 1$: The convex polytopes in $\mathcal{C}$ are — trivially — very well approximated by $\mathcal{P}^i_n$ as $n \to \infty$. On the other hand Theorem 4 shows that the bodies $C \in \mathcal{C} \cap \mathcal{C}^2$ where $\kappa_C > 0$ are not very well approximated. Both the polytopes and the convex bodies of class $\mathcal{C}^2$ and with positive Gaussian curvature are dense in $\mathcal{C}$. Thus a result of Gruber [1983b] on the irregularity of approximation shows that for most convex bodies C (in the Baire category sense) the asymptotic behaviour of $\delta^S(C,\mathcal{P}^i_n), \delta^S(C,\mathcal{P}^c_n)$ as $n \to \infty$ is extremely irregular, cf. Gruber [1983b,1993b,c,1994] and Gruber and Kenderov [1982].

Outline of the proof for $\delta^S(C,\mathcal{P}^c_{(n)})$. The proof of the asymptotic formula (2) contains many technical details. Below we indicate the geometric ideas underlying a slightly different proof, omitting most details.

Let $\|\ \|$ denote the Euclidean norm on $\mathbb{E}^d$ and let S be the square $\{x \in \mathbb{E}^d : 0 \leq x_i \leq 1 \text{ for } i = 1,\dots,d-1, x_d = 0\}$.

Lemma 5. *For* $k \in \mathbb{N}$ *and* $p_1,\dots,p_k \in S$ *consider*

(3) $$\int\limits_S \min\{\|x-p_i\|^2 : i = 1,\dots,k\}dx$$

and let w_k *be the minimum of this quantity for* $p_1,\dots,p_k \in S$. *Then*

(4) $$\text{div}_{d-1} = \lim_{k\to\infty} w_k k^{2/(d-1)} \text{ exists and is positive.}$$

The function $x \to \min\{\|x - p_i\|^2 : i = 1, \ldots, k\}$ for $x \in S$ represents a mountain landscape consisting of k valleys with deepest points $p_1, \ldots, p_k$, respectively. Each valley is enclosed by mountains and the (orthogonal) projection of each valley into S is the Dirichlet–Voronoi cell of the deepest point (intersected with S); see e.g. Erdös, Gruber and Hammer [1989]. The integral (3) is the volume of the mountain landscape above S.

We now describe the proof of Lemma 5 using the interpretation just given. For $d = 3$ the Lemma is a special case of a moment lemma of Fejes Tóth [1953], p. 81.

By choosing $p_1, \ldots, p_k$ from a square grid we see that (for k of the form $k = l^{d-1}, l \in \mathbb{N}$ and then for all k)

$$w_k k^{2/(d-1)} \text{ is bounded above.}$$

If $p_1, \ldots, p_k$ are chosen such that the integral (3) attains its minimum w_k, we replace each valley after a suitable rotational symmetrization by a circular valley such that the area of its projection into $\mathbb{E}^{d-1}$ is the same as for the original valley, but where the volume of the enclosing mountains is less. The volume of the mountains of each circular valley may be expressed in terms of the area of the projection of the original valley into S. These areas sum up to 1. Thus a simple inequality for power means shows that

$$w_k k^{2/(d-1)} \text{ has a positive lower bound.}$$

Finally, from landscapes corresponding to a number k one may construct landscapes corresponding to $kl^{d-1}, l = 1, 2, \ldots$, by considering l^{d-1} homothetic images of the points $p_1, \ldots, p_k$. This relates w_k to $w_{kl^{d-1}}$ and finally leads to (4), concluding the proof of Lemma 5.

For the proof of the asymptotic formula for $\delta^S(C, \mathcal{P}^c_{(n)})$ let bd$C$ be endowed with the Riemannian metric $\varrho_{\mathrm{II}}(\cdot, \cdot)$ determined by the second fundamental form. Then, if $p, x \in \mathrm{bd}C$ are close to each other, $\frac{1}{2}\varrho_{\mathrm{II}}(p, x)^2$ is (approximately) the distance of x to the supporting hyperplane $H_C(p)$ of C at p. Since the directions of the exterior normals of bd$C$ at p and x almost coincide if p and x are close, $\frac{1}{2}\varrho_{\mathrm{II}}(p, x)^2$ is (approximately) the distance of x to $H_C(p)$ but measured along the normal of bd$C$ at x. Thus,

(5) if $P_n \in \mathcal{P}^c_{(n)}$ and if $p_1, \ldots, p_n$ are the points where the facets of P_n touch C and such that $p_1, \ldots, p_n$ are well distributed over bd$C$, then

$$\delta^S(C, P_n) \approx \int_{\mathrm{bd}C} \min\{\frac{1}{2}\varrho_{\mathrm{II}}(p_i, x)^2 : i = 1, \ldots, n\} d\sigma(x).$$

Similarly,

(6) if $p_1, \ldots, p_n \in \mathrm{bd}C$ such that $Q_n = H_C^+(p_1) \cap \ldots \cap H_C^+(p_n) \in \mathcal{P}^c_{(n)}$ and Q_n approximates C well, then

$$\delta^S(C, Q_n) \approx \int_{\mathrm{bd}C} \min\{\frac{1}{2}\varrho_{\mathrm{II}}(p_i, x)^2 : i = 1, \ldots, n\} d\sigma(x).$$

These propositions show that the best approximating polytopes $P_n \in \mathcal{P}^c_{(n)}$ and the point sets $\{p_1, \ldots, p_n\} \subset \text{bd}C$ for which the integral

$$\int_{\text{bd}C} \min\{\frac{1}{2}\varrho_{\text{II}}(p_i, x)^2 : i = 1, \ldots, n\} d\sigma(x)$$

is minimal cum grano salis correspond to each other. Now note that the Riemannian space $\langle \text{bd}C, \varrho_{\text{II}}(\cdot,\cdot)\rangle$ is locally almost Euclidean. Thus we may apply Lemma 5. Rather technical estimates then show that (5) leads to a lower estimate for $\delta^S(C, \mathcal{P}^c_{(n)})$ and (6) to an upper estimate which together yield the asymptotic formula (2).

3.1.3. *Supplementary Problems*
As in the planar case, also Theorem 4 gives rise to natural related problems:
(i) find asymptotic formulae for $\delta^S(C, \mathcal{P}_n), \delta^S(C, \mathcal{P}_{(n)})$,
(ii) find asymptotic expansions for $\delta^S(C, \mathcal{P}^i_n), \ldots$,
(iii) is it possible to replace polytopes by smooth convex bodies the boundaries of which consist of finitely many pieces of quadrics?

While these problems seem to be difficult we discuss in the next section related approximation problems which, in part, are more easily accessible.

3.2. DEVIATION OF QUERMASSINTEGRALS, IN PARTICULAR OF MEAN WIDTH

3.2.1. *The General Problem*
In Section 3.1 we considered asymptotic best approximation with respect to volume deviation. Volume is the first quermassintegral, cf. Schneider [1993]. This suggests the investigation of approximation problems using the deviation with respect to quermassintegrals as a measure of distance. In particular it would be of interest to consider "surface" and "mean width deviation". For $d = 2$ the latter notions coincide up to a multiplicative constant.

For $C \in \mathcal{C}$ let $W_0(C) = V(C), W_1(C) = \frac{1}{d}S(C)$ (surface area of C), $\ldots, W_{d-1}(C) = \frac{d\kappa_d}{2}W(C)$ (mean width of C) be the quermassintegrals of C. An elementary argument using a version of Steiner's formula for parallel bodies together with the corresponding result for δ^H yields the following result; see Glasauer and Gruber [1994+].

$$\delta^{W_k D}(C, \mathcal{P}^i_n) = \inf\{|W_k(C) - W_k(P)| : P \in \mathcal{P}^i_n\} \leq \frac{\text{const}}{n^{2/(d-1)}} \text{ for } n = d+1, \ldots,$$

where the constant depends only on C.

3.2.2. *Asymptotic Formula for $\delta^{WD}(C, \mathcal{P}^i_n)$*
Theorem 5. *(Glasauer and Gruber [1994+]). Let $C \in \mathcal{C} \cap \mathcal{C}^2$ where $\kappa_C > 0$. Then*

$$(7)\quad \delta^{WD}(C, \mathcal{P}^i_n) \sim \frac{\text{div}_{d-1}}{d\kappa_d}\Big(\int_{\text{bd}C} \kappa_C(x)^{d/(d+1)} d\sigma(x)\Big)^{(d+1)/(d-1)} \frac{1}{n^{2/(d-1)}} \text{ as } n \to \infty,$$

(8) $\delta^{WD}(C,\mathcal{P}^c_{(n)}) \sim \frac{\mathrm{del}_{d-1}}{d\kappa_d}\Big(\int\limits_{\mathrm{bd}C} \kappa_C(x)^{d/(d+1)}d\sigma(x)\Big)^{(d+1)/(d-1)}\frac{1}{n^{2/(d-1)}}$ *as* $n \to \infty$.

The case $d = 2$ has been settled before by Fejes Tóth [1948] and McClure and Vitale [1975]. A related problem (deviation of length) for curves in $\mathbb{E}^d (d \geq 2)$ was treated by Gleason [1979]. An extension of Gleason's result to curves in Riemannian spaces is due to Enomoto [1986].

Hölder's inequality and the property that the Gaussian curvature is constant precisely for balls show that among all convex bodies $C \in \mathcal{C} \cap \mathcal{C}^2$ where $\kappa_C > 0$ of given surface area the integral in (7), resp. (8) is maximal precisely for Euclidean balls.

About refinements of Theorem 5 the same can be said as for Theorem 4.

The Idea of the Proof for $\delta^{WD}(C, \mathcal{P}^i_n)$. We sketch a proof of (3) which slightly differs from that by Glasauer and Gruber [1994+].

Let bd$C$ be endowed with the Riemannian metric $\varrho_{\mathrm{II}}(\cdot,\cdot)$ corresponding to the second fundamental form. For $v, x \in \mathrm{bd}C$ close to each other, $\frac{1}{2}\varrho_{\mathrm{II}}(v,x)^2$ is (approximately) the distance of v and $H_C(x)$. Thus,

(9) if $P = \mathrm{conv}\{v_1, \ldots, v_n\} \in \mathcal{P}^i_n$ and $v_1, \ldots, v_n$ are well distributed over bd$C$, then

$$\delta^{WD}(C,P) = \frac{2}{d\kappa_d}\int\limits_{S^{d-1}} \mathrm{dist}(H(u),P)d\tau(u)$$
$$\approx \frac{2}{d\kappa_d}\int\limits_{\mathrm{bd}C} \min\{\frac{1}{2}\varrho_{\mathrm{II}}(v_i,x)^2 : i = 1,\ldots,n\}\kappa_C(x)d\sigma(x)$$

where $H(u)$ is the supporting hyperplane of C with exterior normal vector u and $d\tau$ is the ordinary surface area measure on S^{d-1}.

Note that $d\tau = \kappa_C d\sigma$. A similar argument shows:

(10) if $v_1, \ldots, v_n \in \mathrm{bd}C$ are such that $Q = \mathrm{conv}\{v_1, \ldots, v_n\} \in \mathcal{P}^i_n$ and Q_n approximates C well, then

$$\delta^{WD}(C,Q) \approx \frac{2}{d\kappa_d}\int\limits_{\mathrm{bd}C} \min\{\frac{1}{2}\varrho_{\mathrm{II}}(v_i,x)^2 : i = 1,\ldots,n\}\kappa_C(x)d\sigma(x).$$

(9) and (10) show that the best approximating polytopes $P_n \in \mathcal{P}^i_n$ (with respect to δ^{WD}) and the point sets $\{v_1, \ldots, v_n\} \subset \mathrm{bd}C$ for which the integral

$$\int\limits_{\mathrm{bd}C} \min\{\varrho_{\mathrm{II}}(v_i,x)^2 : i = 1,\ldots,n\}\kappa_C(x)d\sigma(x)$$

is minimal roughly correspond to each other.

Now take into account, that locally ϱ_{II} is a Euclidean norm and κ_C is constant. Hence one may use Lemma 5 in order to obtain from (9) and (10) lower and upper estimates for $\delta^{WD}(C, \mathcal{P}^i_n)$ which finally yield (7).

3.2.3. *The Open Cases*

The asymptotic formulae in Theorems 4 and 5 suggest that for $k = 1, \ldots, d-2$

$$\delta^{W_k D}(C, \mathcal{P}^i_n) \sim c_d\Big(\int\limits_{\mathrm{bd}C} \kappa_C(x)^{(k+1)/(d+1)} d\sigma(x)\Big)^{(d+1)/(d-1)} \frac{1}{n^{2/(d-1)}} \text{ as } n \to \infty,$$

where the constant c_d depends only on d, k and similarly for $\delta^{W_k D}(C, \mathcal{P}^c_{(n)})$.

There seem to be two main barriers for a proof of such formulae. First, it is difficult to relate the quermassintegrals for a convex body to those of an inscribed or circumscribed polytope. Second, how are the contributions to $\delta^{W_k D}(C, \mathcal{P}^i_n)$ or $\delta^{W_k D}(C, \mathcal{P}^c_{(n)})$ of parts of bd$C$ with different principal curvatures coupled? This is of importance for the application of tools such as Lemma 5. Both of these difficulties arise already in the interesting case of the surface deviation.

3.3. THE HAUSDORFF METRIC

Among all metrics it is the Hausdorff metric where the ideas underlying the proofs of the asymptotic formulae are most clearly visible. In part this is due to the fact that the tool of thinnest coverings of Euclidean space by balls can easily be transferred to Riemannian spaces.

3.3.1. *Earlier Results*

It was proved by Dudley [1974] and Bronštein and Ivanov [1975] (and independently by Betke and Wills [1979], unpublished) that for each convex body $C \in \mathcal{C}$ there is a constant such that

$$\delta^H(C, \mathcal{Q}_n) \leq \frac{\text{const}}{n^{2/(d-1)}} \text{ for } n = d+1, \ldots, \text{ and } \mathcal{Q}_n = \mathcal{P}^i_n, \mathcal{P}^c_{(n)}. \tag{11}$$

These results together with elementary arguments and using polarity yield that the estimate (11) holds also for $\mathcal{Q}_n = \mathcal{P}^c_n, \mathcal{P}_n, \mathcal{P}^i_{(n)}, \mathcal{P}_{(n)}$. From a paper of Schneider and Wieacker [1981] one may deduce that for $C \in \mathcal{C} \cap C^2$, there is a positive constant such that

$$\delta^H(C, \mathcal{P}^i_n) \geq \frac{\text{const}}{n^{2/(d-1)}} \text{ for } n = d+1, \ldots$$

3.3.2. *Asymptotic Formulae for* $\delta^H(C, \mathcal{P}^i_n)$

Theorem 6. *(Schneider [1981,1987], Gruber [1993c]). Let* $C \in \mathcal{C} \cap C^2$ *where* $\kappa_C > 0$. *Then*

$$\begin{aligned} &\delta^H(C, \mathcal{P}^i_n), \delta^H(C, \mathcal{P}^c_{(n)}) \\ &\sim \frac{1}{2}\Big(\frac{\vartheta_{d-1}}{\kappa_{d-1}} \int\limits_{\mathrm{bd}C} \kappa_C(x)^{1/2} d\sigma(x)\Big)^{2/(d-1)} \frac{1}{n^{2/(d-1)}} \text{ as } n \to \infty, \end{aligned} \tag{12}$$

Here ϑ_k is the density of the thinnest covering of $\mathbb{E}^k$ with Euclidean balls of equal radii. $\vartheta_1 = 1, \vartheta_2 = 2\pi/\sqrt{27}$ (Kerschner [1938]); no other values are known.

Since for $C \in \mathcal{C} \cap \mathcal{C}^1$

$$\begin{aligned}\delta^H(C,\mathcal{P}_n^i) &\sim \delta^H(C,\mathcal{P}_n^c) \sim 2\delta^H(C,\mathcal{P}_n),\\ \delta^H(C,\mathcal{P}_{(n)}^i) &\sim \delta^H(C,\mathcal{P}_{(n)}^c) \sim 2\delta^H(C,\mathcal{P}_{(n)}),\end{aligned}$$

as $n \to \infty$, Theorem 6 yields additional asymptotic formulae.

The integral in (12) among all convex bodies $C \in \mathcal{C} \cap \mathcal{C}^2$ where $\kappa_C > 0$ and of given surface area is maximal precisely for the balls.

Concerning refinements of Theorem 6 see the analogous remarks in 3.1.2.

The Idea of the Proof for $\delta^H(C,\mathcal{P}_n^i)$. By the *density* of a system of finitely many sets covering a given set we mean the sum of the areas of these sets divided by the area of the given set. A ball in a Riemannian space of dimension $d-1$ will be called a *Riemannian disc.*

Since a Riemannian space is locally (almost) Euclidean, the following result holds.

Lemma 6. *Let $\varrho(\cdot,\cdot)$ be a Riemannian metric on* bd$C$. *For $n = 1, 2, \dots$, let δ_n be the minimum density of a covering of* bd$C$ *by n Riemannian discs of equal radii. Then*

$$\delta_n \to \vartheta_{d-1} \text{ as } n \to \infty.$$

As in 3.1.2 and 3.2.2, let bd$C$ be endowed with the Riemannian metric $\varrho_{\mathrm{II}}(\cdot,\cdot)$, corresponding to the second fundamental form. Since ϱ_{II} measures the distance from supporting hyperplanes (see 3.1.2 or 3.2.2), we obtain the following: If $P = \operatorname{conv}\{v_1,\dots,v_n\} \in \mathcal{P}_n^i$ and $v_1,\dots,v_n$ are well distributed over bd$C$, then

$$\begin{aligned}\delta^H(C,P) &= \max_{x\in \mathrm{bd}C} \operatorname{dist}(P, H_C(x)) = \max_{x\in \mathrm{bd}C}\min_{i=1,\dots,n} \operatorname{dist}(v_i, H_C(x))\\ &\approx \frac{1}{2}\max_{x\in \mathrm{bd}C}\min_{i=1,\dots,n} \varrho_{\mathrm{II}}(v_i,x)^2.\end{aligned}$$

In other words,

(13) if $P = \operatorname{conv}\{v_1,\dots,v_n\} \in \mathcal{P}_n^i$ and $v_1,\dots,v_n$ are well distributed over bd$C$, then the Riemannian discs of radius $(2\delta^H(C,P))^{1/2}$ and centres at $v_1,\dots,v_n$ cover bd$C$.

Similarly,

(14) if $v_1,\dots,v_n$ are well distributed over bd$C$ and $\varrho > 0$ is such that the Riemannian discs of radius ϱ and centres at $v_1,\dots,v_n$ cover bd$C$, then $Q = \operatorname{conv}\{v_1,\dots,v_n\} \in \mathcal{P}_n^i$ and $\delta^H(C,Q) \le \frac{1}{2}\varrho^2$.

By (13) and (14) best approximating polytopes $P_n \in \mathcal{P}_n^i$ and thinnest coverings of bd$C$ by n Riemannian discs of equal radii are closely related. Thus, if $P_n = \operatorname{conv}\{v_1,\dots,v_n\} \in \mathcal{P}_n^i$ is best approximating, then the Riemannian discs with centres at $v_1,\dots,v_n$ and radius $(2\delta^H(C,P_n))^{1/2} = (2\delta^H(C,\mathcal{P}_n^i))^{1/2}$ form (almost) a thinnest covering of bd$C$. By Lemma 6 this implies that

$$\text{(15)} \qquad \frac{n(2\delta^H(C,\mathcal{P}_n^i))^{(d-1)/2}\kappa_{d-1}}{\operatorname{area}(\mathrm{bd}C)} \to \vartheta_{d-1} \text{ as } n \to \infty$$

where the areas used are those corresponding to our Riemannian metric ϱ_{II}. Thus

(16) $$\text{area}(\text{bd}C) = \int\limits_{\text{bd}C} \kappa_C(x)^{1/2} d\sigma(x).$$

The asymptotic formula for $\delta^H(C, \mathcal{P}_n^i)$ is an immediate consequence of (15) and (16).

3.4. BANACH–MAZUR DISTANCE AND SCHNEIDER'S DISTANCE

In the following we present two further approximation results which are based on a suitable choice of a Riemannian metric and on Lemma 6.

3.4.1. *Asymptotic Formula for* $\delta^{BM}(C, \mathcal{P}_{on})$

Let $\mathcal{C}_o$ denote the subspace of $\mathcal{C}$ consisting of all convex bodies which are symmetric with respect to the origin o. For $n = d, d+1, \dots$ let $\mathcal{P}_{on} = \mathcal{C}_o \cap \mathcal{P}_{2n}$ and $\mathcal{P}_{o(n)} = \mathcal{C}_o \cap \mathcal{P}_{(2n)}$. The *Banach–Mazur distance* δ^{BM} on $\mathcal{C}_o$ is defined by

$$\delta^{BM}(C, D) = \inf\{\lambda > 1 : C \subset l(D) \subset \lambda C, l : \mathbb{E}^d \to \mathbb{E}^d \text{ linear}\} \text{ for } C, D \in \mathcal{C}_o.$$

If $\mathcal{C}_o$ is interpreted as the set of unit balls of norms on $\mathbb{E}^d$, the well-known Banach–Mazur distance for (finite-dimensional) normed spaces results. δ^{BM} is invariant with respect to linear transforms,

(17) $$\delta^{BM}(C, D) = \delta^{BM}(lC, mD) \text{ for } C, D \in \mathcal{C}_o \text{ and } l, m : \mathbb{E}^d \to \mathbb{E}^d \text{ linear, onto.}$$

In particular, δ^{BM} does not distinguish between isometric norms. Being the "natural" notion of distance for spaces of norms, δ^{BM} plays a prominent rôle in (finite-dimensional) functional analysis, see e.g. Tomczak–Jaegermann (1988) or Lindenstrauss and Milman (1993).

The following result answers a question of Kashin; it may be interpreted in terms of approximation of a norm by piecewise linear norms.

Theorem 7. *(Gruber [1993c]). Let* $C \in \mathcal{C}_o \cap \mathcal{C}^2$ *where* $\kappa_C > 0$. *Then*

(18) $$\delta^{BM}(C, \mathcal{P}_{on}), \delta^{BM}(C, \mathcal{P}_{o(n)}) \sim 1 + \frac{1}{2}\Big(\frac{\vartheta_{d-1}}{\kappa_{d-1}} \int\limits_{\text{bd}C} \Big(\frac{\kappa_C(x)}{h_C(n_C(x))^{d-1}}\Big)^{1/2} d\sigma(x)\Big)^{2/(d-1)} \frac{1}{n^{2/(d-1)}}$$

as $n \to \infty$.

Here h_C is the *support function* of C,

$$h_C(u) = \max\{u \cdot x : x \in C\} \text{ for } u \in S^{d-1}$$

and $n_C(x)$ denotes the exterior normal unit vector of C at $x \in \text{bd}C$.

The Cauchy–Schwarz inequality and the inequality of Blaschke–Santaló show that the integral in (18) is maximal precisely for ellipsoids among all convex bodies $C \in \mathcal{C}_o \cap \mathcal{C}^2$ where $\kappa_C > 0$.

For a discussion of possible refinements of Theorem 7 we again refer to the analogous remarks after Theorem 4 in 3.1.2.

Considering the invariance property (17) of δ^{BM} a candidate for a Riemannian metric on bd$C$ which might be helpful for the proof of Theorem 7 is the Riemannian metric induced by the fundamental form of central affine differential geometry. It turns out that actually this is the right choice. A further tool is Lemma 6.

3.4.2. *Asymptotic Formula for* $\delta^{SCH}(C, \mathcal{P}_n^i)$
Schneider [1986] proposed the following measure of deviation of a convex body C from an inscribed convex polytope P : $\delta^{SCH}(C, P)$ is the maximal volume of a cap of C determined by a hyperplane containing a facet of P.

Theorem 8. *(Schneider [1986]). Let $C \in \mathcal{C} \cap \mathcal{C}^3$ where $\kappa_C > 0$. Then*

$$\delta^{SCH}(C, \mathcal{P}_{(n)}^i) \sim \frac{\kappa_{d-1}}{d+1}\Big(\frac{\vartheta_{d-1}}{\kappa_{d-1}} \int\limits_{\mathrm{bd}C} \kappa_C(x)^{1/(d+1)} d\sigma(x)\Big)^{(d+1)/(d-1)} \frac{1}{n^{(d+1)/(d-1)}} \tag{19}$$

as $n \to \infty$.

Since $\delta^{SCH}(C, \mathcal{P}_{(n)}^i)$ is invariant with respect to volume preserving affinities, it is not surprising that the integral on the right hand side of (19) is the affine surface area. In spite of this the idea underlying the proof of Theorem 8 is related to that of Theorems 6, 7 where δ^H and δ^{BM} are treated and does not follow the line of proof of Theorems 4, 5 which concern δ^S and δ^{MW}.

About the worst approximable convex bodies and refinements the same can be said as for δ^S, see the remarks after Theorem 4 in 3.1.2.

The Idea of the Proof for $\delta^{SCH}(C, \mathcal{P}_{(n)}^i)$. The appropriate Riemannian metric $\varrho(\cdot,\cdot)$ on bd$C$ is the one which has the property that small caps of C may (approximately) be represented in the form

$$\mathrm{conv}\{x \in \mathrm{bd}C : \varrho(c, x) \le \varrho\} \quad (\varrho > 0 \text{ fixed}) \tag{20}$$

and such that the volume of such a cap is (approximately) ϱ^{d+1}.

For $P \in \mathcal{P}_{(n)}^i$ we may determine $\delta^{SCH}(C, P)$ in the following way: For each facet F of P consider the point $c \in \mathrm{bd}C$ where the exterior normal unit vector of C is the same as the exterior normal unit vector of F and consider the cap of C determined by the facet F. Represent this cap in the form (20). Thus we obtain n points on bd$C$, say $c_1, \dots, c_n$ and n radii, say $\varrho_1, \dots, \varrho_n$. Clearly, the Riemannian discs with centres at $c_1, \dots, c_n$ and radii $\varrho_1, \dots, \varrho_n$, respectively, cover bd$C$ and $\delta^{SCH}(C, P)$ is (approximately) ϱ^{d+1} where $\varrho = \max\{\varrho_1, \dots, \varrho_n\}$.

From this (and a sort of converse of it) we obtain the following: $\delta^{SCH}(C, \mathcal{P}_{(n)}^i)$ is ϱ^{d+1} where $\varrho > 0$ is the smallest radius such that bd$C$ can be covered by n suitable Riemannian discs of radius ϱ. By Lemma 6,

$$\frac{n\varrho^{d-1}\kappa_{d-1}}{\mathrm{area(bd}C)} \sim \vartheta_{d-1} \text{ or } \varrho^{d+1} \sim \Big(\frac{\vartheta_{d-1}}{\kappa_{d-1}} \mathrm{area(bd}C)\Big)^{(d+1)/(d-1)} \frac{1}{n^{(d+1)/(d-1)}} \text{ as } n \to \infty,$$

where area means area in the sense of the Riemannian metric $\varrho(\cdot,\cdot)$. This gives (19).

4. Effective and Stepwise Approximation

4.1. THE GENERAL PROBLEM

Given some notion of distance δ on $\mathcal{C}$ and a convex body $C \in \mathcal{C}$ we ask for an explicit construction or an algorithm with the following property: for $n = d+1, \ldots$, it permits to find polytopes $P_n \in \mathcal{Q}_n$ where $\mathcal{Q}_n$ is one of the polytope classes $\mathcal{P}^i_n, \ldots, \mathcal{P}_{(n)}$ such that

$$\delta(C, P_n) \text{ is close to } \delta(C, \mathcal{Q}_n).$$

Only for very special convex bodies, mainly in $\mathbb{E}^2$, one may be able to construct explicitly polytopes $P_n \in \mathcal{Q}_n$ such that

$$\delta(C, P_n) = \delta(C, \mathcal{Q}_n);$$

but see the dynamical systems approach of Kenderov and Kirov [1993] for $\delta^H(C, \mathcal{P}_n)$ $(d = 2)$.

In some cases in two and three dimensions it is possible to specify sequences of polytopes $P_n \in \mathcal{Q}_n$ such that

$$(1) \qquad \delta(C, P_n) \sim \delta(C, \mathcal{Q}_n) \text{ as } n \to \infty.$$

Such sequences are called *asymptotically best approximating.*

For all planar convex bodies and all sufficiently smooth convex bodies in higher dimensions there are explicit constructions of sequences of polytopes $P_n \in \mathcal{Q}_n$ such that

$$(2) \qquad \delta(C, P_n) \leq \frac{\text{const}}{n^{2/(d-1)}} \text{ for } n = d + 1, \ldots$$

Here const is a positive constant; in case $d = 2$ it is possible to determine its numerical value. In part such constructions are step-by-step methods which are cheap from the computational viewpoint.

Finally, for any convex bodies in $\mathbb{E}^3$ Sonnevend [1983] has given a construction for sequences of polytopes $P_n \in \mathcal{Q}_n$ which satisfy inequalities of the form (2) but with larger right hand side.

4.2. EFFECTIVE ASYMPTOTIC BEST APPROXIMATION FOR $d = 2, 3$

4.2.1. *The Case* $d = 2$

Let δ be one of the metrics $\delta^S, \delta^{MW}, \delta^H, \delta^p$ (L_p-metric for support functions) and choose $C \in \mathcal{C}(\mathbb{E}^2) \cap \mathcal{C}^2$ where $\kappa_C > 0$. McClure and Vitale [1975] proposed two methods of constructing asymptotically best approximating sequences of polygons.

Special Cases of the Empirical Distribution Method. Refining a remark of Fejes Tóth [1953], p. 43, Ludwig [1993] showed that for $\delta = \delta^S, \delta^H$ and $\mathcal{Q}_n = \mathcal{P}^i_n, \mathcal{P}^c_n$ and suitable notions of arclengths on bd$C$ (for δ^S it is the affine arclength and for δ^H a different one; see 2.1.4 and 2.2.3) the following hold. For best approximating polygons $P_n \in \mathcal{Q}_n$ the vertices of P_n, resp. the points where the edges of P_n touch C are almost equally spaced along bd$C$ in the sense of the corresponding arclengths. Conversely, it is easy to show that for polygons $Q_n \in \mathcal{Q}_n$ with equally spaced vertices, resp. touching points, (1) holds. Actually Ludwig proved much more: such polygons Q_n even satisfy the precise asymptotic formulae of Theorems 1, 3.

McClure and Vitale [1975] (independently of Ludwig, of course,) proved that for all notions of distance considered here there are suitable notions of arclength such that for polygons $Q_n \in \mathcal{Q}_n^i$ with equally spaced vertices (1) holds.

The Balance of Errors Method. Given a best approximating polygon $P_n \in \mathcal{Q}_n$ of C with respect to $\delta = \delta^S, \delta^{MW}, \delta^H, \delta^p$, we may consider for each edge of P_n the "local deviation" from C. It is plausible that these "local errors" are approximately equal. (For $\delta = \delta^H$ and sufficiently large n they are, in fact, equal.)

McClure and Vitale [1975] showed that, if polygons $Q_n \in \mathcal{Q}_n$ are chosen such that all "local errors" are equal, then they form an asymptotically best approximating sequence.

In contrast to the empirical distribution method, this method seems to be difficult to apply directly.

In addition to these methods geometric constructions of asymptotically best approximating sequences of polygons have been proposed; compare the references in Gruber [1983a,1993a].

4.2.2. *The Case $d = 3$*

The results on asymptotic best approximation in $\mathbb{E}^d$ presented in Chapter 3 depend on properties of Dirichlet–Voronoi and Delone tilings in $\mathbb{E}^k (k = d - 1)$ and on thinnest coverings of $\mathbb{E}^k$ with Euclidean balls of equal radii. For $k = 2$ the corresponding constants and the extremal configurations are known through work of Fejes Tóth [1953] (see Gruber [1991]), Gruber [1988] and Kershner [1939]. A close inspection of the proofs of Gruber [1988,1991,1993c,1994], Schneider [1981,1986,1987] and Glasauer and Gruber [1994+] then shows that one may construct explicitly for $C \in \mathcal{C}(\mathbb{E}^3) \cap \mathcal{C}^2$ $(\kappa_C > 0)$ and $\delta^S, \delta^{WD}, \delta^H, \delta^{BM}, \delta^{SCH}$ asymptotically best approximating sequences of polytopes $Q_n \in \mathcal{P}_n^i$, resp. $\in \mathcal{P}_{(n)}^c$. While this is possible in principle, it does not seem to be of practical relevance.

4.3. EFFECTIVE STEP-BY-STEP APPROXIMATION OF OPTIMAL ORDER

4.3.1. *The Origin of the Idea*

To the best of our knowledge, the first step-by-step approximation methods for so-called Sandwich approximation of convex functions on $[0,1]^d$ were proposed by Sonnevend [1983] $(d = 2)$ and Fruhwirth, Burkard and Rote [1989] and Burkard, Hamacher and Rote [1991] $(d = 1)$; see Gruber [1993a] for a short survey. For $d = 1$ the order of these methods is optimal $(O(1/n^2))$ and the constants can be specified explicitly. Unfortunately they require much computation.

The idea for the step-by-step results that will be presented in the following was taken from the theory of uniform distribution, a branch of number theory. We give a short description.

A basic notion in this theory is that of a *uniformly distributed sequence* in $[0,1]$. This is a sequence $\omega_1, \omega_2, \ldots, \in [0,1]$ such that each subinterval of $[0,1]$ has the right share of elements of the sequence, or equivalently: if $f : [0,1] \to \mathbb{R}$ is continuous, then

$$\int_0^1 f(x)dx = \lim_{n\to\infty} \frac{1}{n} \sum_{i=1}^{n} f(\omega_i). \tag{3}$$

Thus a uniformly distributed sequence gives rise to a step-by-step integration formula. Of course, much more can be said in addition to (3); see Hlawka [1984] or Kuipers and Niederreiter [1974].

4.3.2. *The Problem of Step-By-Step Approximation*

Let $C \in \mathcal{C}$. For $x \in \text{bd}C$ we denote by $H_C^+(x)$ a supporting halfspace of C the boundary of which contains x. Let δ be a metric on $\mathcal{C}$.

We will consider two problems. First, given a sequence $x_1, x_2, \ldots \in \text{bd}C$, how well do the polytopes

$$\text{conv}\{x_1, \ldots, x_n\} \text{ and } H_C^+(x_1) \cap \ldots \cap H_C^+(x_n)$$

approximate C as $n \to \infty$? Second, what is the best rate of convergence that can be achieved by an appropriate choice of the sequence?

In order to state our result we need the following: Let M be a totally bounded metric space with metric δ and let $x_1, x_2, \ldots$ be a sequence in M. The *dispersion* $d_n(x_i), n = 1, 2, \ldots$, of this sequence (x_i) is defined as follows:

(4) $d_n(x_i)$ is the smallest radius $\varrho > 0$ such that the balls with radius ϱ and centres at $x_1, \ldots, x_n$ cover M.

Niederreiter [1984] constructed explicitly a sequence in $[0, 1]^k \subset \mathbb{E}^k$ with the property that its dispersion is bounded above by

$$\frac{\text{const}}{n^{1/k}} \text{ for } n = 1, 2, \ldots \tag{5}$$

4.3.3. *The Hausdorff Metric*

Theorem 9. *(Gruber [1993c]). Let $C \in \mathcal{C} \cap \mathcal{C}^2$ where $\kappa_C > 0$ and let* bd$C$ *be endowed by the Riemannian metric $\varrho_{\text{II}}(\cdot, \cdot)$ induced by the second fundamental form. Then*

(i) *for any sequence $x_1, x_2, \ldots \in \text{bd}C$ with dispersion $d_n(x_i) \to 0$ as $n \to \infty$,*

$$\begin{aligned} &\delta^H(C, \text{conv}\{x_1, \ldots, x_n\}), \delta^H(C, H_C^+(x_1) \cap \ldots \cap H_C^+(x_n)) \\ &\qquad \sim \frac{d_n(x_i)^2}{2} \text{ as } n \to \infty, \end{aligned} \tag{6}$$

(ii) *there exists a sequence $y_1, y_2, \ldots \in \text{bd}C$ which can be described explicitly and a constant depending on C, such that*

$$\begin{aligned} &\delta^H(C, \text{conv}\{y_1, \ldots, y_n\}), \delta^H(C, H_C^+(y_1) \cap \ldots \cap H_C^+(y_n)) \\ &\qquad \le \frac{\text{const}}{n^{2/(d-1)}} \text{ for } n = d+1, \ldots \end{aligned} \tag{7}$$

Thus the concept of dispersion with respect to the metric ϱ_{II} and approximation with respect to δ^H match well. (ii) shows that, surprisingly, a suitable step-by-step approximation method is of the same order as best approximation. It is worth noting that the construction of the sequence $y_1, y_2, \ldots$ in (ii) is rather easy and does not require much computation.

Outline of the Proof for $\delta^H(C, \mathrm{conv}\{\ \})$. In the proof of Theorem 6 in 3.3.2 the following was shown (compare (13), (14)):

$$\text{if } P = \mathrm{conv}\{x_1, \ldots, x_n\} \in \mathcal{P}_n^i \text{ then } \delta^H(C, P) \sim \frac{d_n(x_i)^2}{2}. \tag{8}$$

This yields (i). For the proof of (ii) we have to construct a sequence (y_i) on bd$C$ with dispersion

$$d_n(y_i) \leq \frac{\mathrm{const}}{n^{1/(d-1)}} \text{ for } n = 1, 2, \ldots \tag{9}$$

To achieve this cover bd$C$ by finitely many images of the unit square $[0,1]^{d-1}$ where the mappings, say $f_1, \ldots, f_m$, are non expansive. Let (u_i) be the sequence in $[0,1]^{d-1}$ specified by Niederreiter [1984]; compare (5). Then

$$y_1 = f_1(u_1), \ldots, y_m = f_m(u_1), y_{m+1} = f_1(u_2), \ldots, y_{2m} = f_m(u_2), \ldots$$

is a sequence in bd$C$ satisfying (9). Now apply (8).

4.3.4. *Other Metrics*
δ^{SCH}: It is easy to see that in this case there is a result analogous to Theorem 8.
δ^{BM} : The formulae (6) and (7) for the convex hull case have to be replaced by

$$1 + \mathrm{const}\, d_n(\pm x_i)^2 \leq \delta^{BM}(C, \mathrm{conv}\{\pm x_1, \ldots, \pm x_n\}) \leq 1 + \mathrm{const}\, d_n(\pm x_i)^2,$$
$$\delta^{BM}(C, \mathrm{conv}\{\pm y_1, \ldots, \pm y_n\}) \leq 1 + \frac{\mathrm{const}}{n^{2/(d-1)}} \text{ for } n = d, d+1, \ldots$$

δ^S : Instead of (6) and (7) the following inequalities hold; where we state only the case of the convex hull.

$$\mathrm{const}\, d_n(x_i)^{d+1} \leq \delta^S(C, \mathrm{conv}\{x_1, \ldots, x_n\}) \leq \mathrm{const}\, d_n(x_i)^2,$$
$$\delta^{BM}(C, \mathrm{conv}\{y_1, \ldots, y_n\}) \leq \frac{\mathrm{const}}{n^{2/(d-1)}} \text{ for } n = d+1, \ldots$$

5. A Short Survey of Recent Approximation Results

The following report supplements the review of Gruber [1993a]. Some results from other areas are cited since, possibly, they may induce related investigations within convexity

5.1. APPROXIMATION WITH RESPECT TO METRICS

5.1.1. *Surface and Perimeter Deviation*
For $C \in \mathcal{C}$ and $P \in \mathcal{P}_n$ let $\delta^1(C, P)$ denote the L_1-distance (of the support functions) of C and P.

Florian [1993+] showed that for $C \in \mathcal{C}(\mathbb{E}^2)$,

$$\delta^1(C, \mathcal{P}_n) \leq L(C)(1 - \frac{2n}{\pi} \arcsin(\frac{1}{2} \sin \frac{\pi}{n})) \text{ for } n = 3, 4, \ldots$$

Equality holds if C is a circular disc and Florian conjectures that it holds only in this case. (For $\mathcal{P}_n^i$ a similar result is due to Schneider [1971], who also proved a slightly weaker result for $\mathcal{P}_n^c$.)

For general d Müller [1990] gave a result on the approximation of Euclidean balls B by convex polytopes the vertices of which are chosen randomly in bd$B$.

5.1.2. *Symmetric Difference Metric*

A result of Schütt [1991] says that for any convex body $C \in \mathcal{C}$,

$$\delta^s(C, \mathcal{P}^i_{(n)}) \geq \frac{\text{const}}{n^{2/(d-1)}(\ln n)^2} \text{ for } n = d+1, d+2, \ldots,$$

where the constant depends on a notion of affine surface area of C due to Schütt and Werner. Compare Theorem 4 in 3.1.2. While this notion of affine surface area is defined for all convex bodies, it is nonzero only for a meagre set of convex bodies; see Gruber [1993b].

A good deal of the more recent results on random polytopes may be interpreted as results on volume approximation of convex bodies. We present some typical results. For related earlier results see the surveys of Schneider [1988] and Affentranger [1992].

Extending a result of Wieacker [1978] for balls, Bárány [1992] proved for convex bodies $C \in \mathcal{C} \cap \mathcal{C}^3$ where $\kappa_C > 0$ and $V(C) = 1$ the following asymptotic formula. If $E(C, n)$ denotes the mathematical expectation of $1 - V(\text{conv}\{x_1, \ldots, x_n\})$ where the n points $x_1, \ldots, x_n$ are chosen independently and uniformly in C, then

$$E(C, n) \sim \text{const} \int_{\text{bd}C} \kappa_C(x)^{1/(d+1)} d\sigma(x) \frac{1}{n^{2/(d+1)}} \text{ as } n \to \infty.$$

The constant depends only on d. Similar results hold also for the other quermass-integrals. Clearly, Bárány's result may be formulated also as follows: $E(C, n)n^{2/(d+1)}$ converges to a constant times the affine surface area. A striking recent result of Schütt [1993] shows that in this form the asymptotic formula of Bárány holds for all convex bodies C, where the affine surface area is replaced by the generalized affine surface area introduced by Schütt and Werner.

For convex polytopes P with $V(P) = 1$, Bárány and Buchta [1993+] proved that

$$E(C, n) \sim \text{const} \frac{(\log n)^{d-1}}{n} \text{ as } n \to \infty.$$

The constant depends on the combinatorial structure of P and, in principle, can be computed explicitly.

If the points $x_1, \ldots, x_n$ are chosen in bd$C$, the corresponding result is known only in case C is a ball. It is surprising to note that in this case the random approximation result of Müller [1990] is of order $O(1/n^{2/(d-1)})$, i.e. of the same order as asymptotic best approximation; compare 3.1.

Dyer, Füredi and McDiarmid [1990] investigated the expectation of the volume of the convex hull of $n = n(d)$ randomly chosen vertices of the unit cube in $\mathbb{E}^d$ as $d \to \infty$.

5.2. APPROXIMATION PROBLEMS DEALING WITH VOLUME AND HOMOTHETIES

5.2.1. *Special Inscribed and Circumscribed Convex Bodies*

Pelczynski and Szarek [1991] considered for convex bodies $C \in \mathcal{C}_0$ parallelepipeds $P \supset C$ of minimum volume and proved the inequality

$$V(P) \leq \frac{2^{3d/2}}{\kappa_d (d+1)^{d/2}} \binom{d(d+1)/2}{d}^{1/2} V(C),$$

thus improving an earlier estimate of Babenko [1988]. For $d = 2$ they give an alternative proof of a result of Petty [1955], also obtained by Babenko [1988] and which says that

$$A(P) \leq \frac{4}{3} A(C).$$

For a related result see Lassak [1991].

Given $C \in \mathcal{C}$, let c be the centre of an inscribed ellipsoid E of maximum volume. Further, let C^+, C^- be the convex bodies into which a given hyperplane through c dissects C. Then, if F is an ellipsoid of maximum volume contained in, say, C^+ a result of Khachiyan [1990] shows that

$$V(F) \leq 0{\cdot}844 V(E).$$

This refines an earlier result of Tarasov, Khachiyan and Èrlikh [1988]. The optimal constant in the inequality is conjectured to be $e^{1/2}/2$.

Svistunov [1991] discusses the "cost" of the determination of maximum inscribed and minimum circumscribed ellipsoids of a polytope. Goffin [1984] and Grötschel, Lovász and Schrijver [1988] specified polynomial time algorithms to calculate for a convex body C homothetic ellipsoids E, F which do not differ too much and such that

$$E \subset C \subset F.$$

(By a well-known theorem of John [1948] the best possible homothety ratio which applies to all convex bodies C is d.) For further results see Ball [1992] and Juhnke [1993a,b].

For planar convex bodies C several results on homothetic inscribed and circumscribed special convex bodies were given in recent years. In part they can be interpreted in terms of the Banach–Mazur distance. See e.g. Lassak [1992] (C centrally symmetric; triangles) and Schwarzkopf et.al. [1990] (general C; rectangles).

5.2.2. *Subspaces*

Ball [1991] showed that for $C \in \mathcal{C}_0$ there is a $(d-1)$-dimensional subspace H of $\mathbb{E}^d$ such that for the $(d-1)$-dimensional volume of $C \cap H$ the following inequality holds, where the constant is absolute:

$$v(C \cap H)^{1/d} \geq \text{const} V(C)^{(d-1)/d}.$$

5.2.3. *Calculation of Volumes*

The calculation of the volume of a convex body is difficult from the point of view of computation, more precisely, it is $\#P$ hard as shown by Dyer and Frieze [1988]. See also the discussion by Lovász and Simonovits [1990] and Lovász [1992]. To overcome this difficulty efficient *randomized* polynomial time algorithms have been proposed; compare Dyer, Frieze and Kannan [1989] and Dyer and Frieze [1991].

5.3. MISCELLANEOUS RESULTS ON APPROXIMATION OF CONVEX BODIES

5.3.1. *Geometric Transformations*

Many of the standard transformations of convex bodies (e.g. Steiner symmetrizations) have the following property: Given a convex body C, then there are suitable sequences of (iterations of) such transformations, say $s_1, s_2, \ldots$, such that the sequence

$$s_1C, s_2C, \ldots$$

converges in some sense to a Euclidean ball or another special convex body. Recent examples of such results are due to Gregorac [1992] (weighted p-mean in the sense of Firey of C and its polar body C^* where it is assumed that $o \in \text{int}C$) and Florian [1993] (Blaschke symmetrization).

5.3.2. *Zonoids*

A *zonotope* Z in $\mathbb{E}^d$ is convex body which may be represented in the form

$$Z = S_1 + \ldots + S_k (= \{x_1 + \ldots + x_k : x_i \in S_i\}),$$

where $S_1, \ldots, S_k$ are line segments. A convex body which can be represented as a limit (in the sense of δ^H or δ^S) of zonotopes is called a *zonoid.*

There are many results on approximation of zonoids by zonotopes. Extending a result of Wagner [1993] ($d \leq 6$), Bourgain and Lindenstrauss [1993] gave a result on the approximation of the unit ball in $\mathbb{E}^d$ (a zonoid) by zonotopes which are sums of line-segments of equal lengths.

5.3.3. *Lattice Point Results*

A result of Balog and Bárány [1991] says that for any convex body $C \in \mathcal{C}(\mathbb{E}^2) \cap \mathcal{C}^3$ with positive curvature there are constants $\alpha, \beta > 0$ such that

$$\alpha\lambda^{2/3} \leq \# \text{ vertices of conv}(\lambda C \cap Z^2) \leq \beta\lambda^{2/3} \text{ for all } \lambda \in \mathbb{R}^+.$$

By a result of Andrews [1963] the upper estimate can be extended to all dimensions; $2/3$ has to be replaced by $d(d-1)/(d+1)$.

Other results on lattice points in convex bodies which are close to our subject are due to Bárány, Howe and Lovász [1992], Kosygin and Sinai [1992] and Nowak [1991].

5.4. CONVEX FUNCTIONS

In the following we present a small selection of approximation results dealing with convex functions.

5.4.1. *Approximation of Convex Functions by Convex Polynomials*
See Yu [1989] and Lu [1990].

5.4.2. *Integration of convex functions*
Approximation of convex bodies with respect to δ^S gives rise to results on numerical integration of convex functions; see e.g. the survey of Gruber [1993a].

Forster and Petras [1990] showed that the error of the Gauss quadrature formula when applied to convex functions on $[0,1]$, say, is of order $O(1/n^2)$.

Du and Ko [1989] discuss the computational complexity of indefinite integration of polynomially computable convex functions on $[0,1]$.

5.4.3. *Approximation by convex functions*
Let $f : [0,1] \to \mathbb{R}$ be continuous. Then for $1 \leq p \leq \infty$ there are best L_p-approximations of f by continuous convex functions on $[0,1]$. For $1 \leq p < \infty$ these are unique, say f_p, but this is not necessarily true for $p = \infty$. On any closed interval $I \subset (0,1)$ the functions f_p converge uniformly to f_1 as $p \to 1$. There is a best L_∞-approximating convex function f_∞ of f such that again on any closed interval $I \subset (0,1)$ the functions f_p converge uniformly to f_∞ as $p \to +\infty$. Part of these results can be extended to measurable functions on $[0,1]$ and to continuous functions on $[0,1]^d$. See e.g. Huotari and Zwick [1989], Legg and Townsend [1989,1990] and Huotari, Legg and Townsend [1991].

There are similar results where instead of convex functions *n-convex functions* are considered. These are real functions on an interval in $\mathbb{R}$ for which the nth divided differences are non-negative. 1-convex functions are monotone increasing and 2-convex functions are convex.

Acknowledgements

For their helpful remarks I am obliged to Dr. M. Ludwig and Professors P. McMullen and F. J. Schnitzer.

References

Affentranger, F.: 1992, Approximación aleatoria de cuerpos convexos, *Publ. Sec. Mat. Univ. Autònoma Barcelona*, **36**, pp. 85–102.

Andrews, G.E.: 1963, A lower bound for the volume of strictly convex bodies with many boundary lattice points, *Trans. Amer. Math. Soc.* **106**, pp. 270–279.

Babenko, I.K.: 1988, Asymptotic volume of tori and the geometry of convex bodies, *Mat. Zametki* **44**, pp. 177–190.

Ball, K.M.: 1991, Normed spaces with a weak Gordon–Lewis property, In: *Functional analysis*, 36–47, *Lecture Notes in Math.* **1470**, Springer, Berlin.

Ball, K.M.: 1992, Ellipsoids of maximal volume in convex bodies, *Geom. Dedicata* **41**, pp. 241–250.

Balog, A., Bárány, I.: 1991, On the convex hull of integer points in a disc, In: *Discrete and computational geometry*, DIMACS Ser. **6**, pp. 39–44, Amer. Math. Soc., Providence RI.

Bárány, I.: 1992, Random polytopes in smooth convex bodies, *Mathematika* **39**, pp. 81–92.

Bárány, I., Buchta, Chr.: 1993, Random polytopes in a convex polytope, independence of shape and concentration of vertices, *Math. Ann.*, To appear.

Bárány, I., Howe, R., Lovász, L.: 1992, On integer points in polyhedra: a lower bound, *Combinatorica* **12**, 135–142.

Blaschke, W.: 1917, ber affine Geometrie III: Eine Minimumeigenschaft der Ellipse, *Leipziger Ber.* **69**, pp. 3–12, *Gesammelte Werke* **3**, pp. 217–226, Thales Verlag, Essen 1985.

Blaschke, W.: 1923, *Vorlesungen ber Differentialgeometrie II*, Springer, Berlin.

Bourgain, J., Lindenstrauss, J.: 1993, Approximating the ball by a Minkowski sum of segments with equal length, *Discrete Comput. Geom.* **9**, pp. 131–144.

Bronštein, E.M., Ivanov, L.D.: 1975, The approximation of convex sets by polyhedra, *Sibirsk. Mat. Zh.* **16** (1975), pp. 1110–1112, *Siberian Math. J.* **16**, pp. 852–853.

Buchta, Chr.: 1985, Zufllige Polyeder — eine bersicht, In: *Lecture Notes in Math.* **1114**, pp. 1–13, Springer, Berlin.

Burkard, R.E., Hamacher, H.W., Rote, G.: 1991, Sandwich approximation of univariate convex functions and application to separable convex programming, *Naval Res. Logist.* **38**, pp. 911–924.

Du, Ding Zhu, Ko, Ker-I: 1989, Computational complexity of integration and differentiation of convex functions, *Systems Sci. Math. Sci.* **2**, pp. 70–79.

Dudley, R.M.: 1974, Metric entropy of some classes of sets with differentiable boundaries, *J. Approx. Theory* **10**, pp. 227–236, Corr., ibid. **26**, (1979), pp. 192–193.

Dyer, M., Frieze, A.M.: 1988, On the complexity of computing the volume of a polyhedron *SIAM J. Comput.* **17**, pp. 967–974.

Dyer, M., Frieze, A.M.: 1991, Computing the volume of convex bodies: a case where randomness provably helps, In: *Probabilistic combinatorics and its applications*, pp. 123–169, *Proc. Sympos. Appl. Math.* **44** Amer. Math. Soc., Providence RI.

Dyer, M., Frieze, A.M., Kannan, R.: 1991, A random polynomial time algorithm for approximating the volume of convex bodies, *J. Assoc. Comput. Math.* **38**, pp. 1-17.

Dyer, M., Füredi, Z., McDiarmid, C.: 1990, Random volumes in the n-cube, In: *Polyhedral combinatorics*, pp. 33–38, DIMACS Ser. **1**, Amer. Math. Soc., Providence RI.

Enomoto, K.: 1986, Approximation of length of curves in Riemannian manifolds by geodesic polygons, *Yokohama Math. J.* **34**, pp. 53–58.

Erdös, P., Gruber, P.M., Hammer, J.: 1989, *Lattice points*, Longman Scientific, Harlow, Essex.

Fejes Tóth, L.: 1948, Approximation by polygons and polyhedra, *Bull. Amer. Math. Soc.* **54**, pp. 431–438.

Fejes Tóth, L.: 1953, *Lagerungen in der Ebene, auf der Kugel und im Raum*, Springer, Berlin, 2nd ed., 1972.

Florian, A.: 1993+, On the perimeter deviation of a convex disc from a polygon, In print.

Forster, K.-J., Petras, K.: 1990, On a problem proposed by H. Brass concerning the remainder term in quadrature for convex functions, *Numer. Math.* **57**, pp. 763–777.

Fruhwirth, B., Burkard, R.E., Rote, G.: 1989, Approximation of convex curves with applications to the minimum cost flow problem, *European J. Oper. Res.* **42**, pp. 326–338.

Glasauer, S., Gruber, P.M.: 1994+, Asymptotic estimates for best and stepwise approximation of convex bodies III, In preparation.

Gleason, A.M.: 1979, A curvature formula, *Amer. J. Math.* **101**, 86–93.

Goffin, J.L.: 1984, Variable metric relaxation methods II: the ellipsoid method, *Math. Programming* **30**, pp. 147–162.

Grötschel, M., Lóvasz, L., Schrijver, A.: 1988, *Geometric algorithms and combinatorial optimization*, Springer, Berlin.

Gregorac, R.J.: 1992, Iterated means of convex bodies, *Monatsh. Math.* **113**, pp. 189–198.

Gruber, P.M.: 1983a, Approximation of convex bodies, In: *Convexity and its applications*, pp. 131–162, Birkhuser, Basel.

Gruber, P.M.: 1983b, In most cases approximation is irregular, *Rend. Sem. Mat. Univers. Politecn. Torino* **41**, pp. 19–33.

Gruber, P.M.: 1988, Volume approximation of convex bodies by inscribed polytopes, *Math. Ann.* **281**, pp. 229–245.

Gruber, P.M.: 1991, Volume approximation of convex bodies by circumscribed polytopes, In: *Applied Geometry and discrete Mathematics. The Victor Klee Festschrift*, pp. 309–317, DIMACS Ser. **4**, Amer. Math. Soc., Providence RI.

Gruber, P.M.: 1993a, Aspects of approximation of convex bodies, In: *Handbook of convex geometry* **A**, pp. 319–345, Elsevier, North-Holland, Amsterdam.

Gruber, P.M.: 1993b, Baire categories in convexity, In: *Handbook of convex geometry* **B**, pp. 1327–1346, Elsevier, North-Holland, Amsterdam.

Gruber, P.M.: 1993c, Asymptotic estimates for best and stepwise approximation of convex bodies I, *Forum Math.* **5**, pp. 281–297.

Gruber, P.M.: 1994, Asymptotic estimates for best and stepwise approximation of convex bodies II, *Forum Math.*, to appear.

Gruber, P.M., Kenderov, P.: 1982, Approximation of convex bodies by polytopes, *Rend. Circ. Mat. Palermo* (2) **31**, pp. 195–225.

Gruber, P.M., Lekkerkerker, C.G.: 1987, *Geometry of numbers*, 2nd ed., North-Holland, Amsterdam.

Hlawka, E.: 1984, *The theory of uniform distribution*, AB Academic Publ., Berkhamsted.

Huotari, R., Legg, D., Townsend, D.W.: 1991, Best L_1-approximation by functions in several variables, In: *Progress in approximation theory*, pp. 475–481, Academic Press, Boston, MA.

Huotari, R., Zwick, D.: 1989, Approximation in the mean by convex functions, *Numer. Funct. Anal. Optimization* **10**, pp. 489–498.

John, F.: 1948, Extremum problems with inequalities as subsidiary conditions, in: *Studies and Essays presented to R. Courant*, pp. 187–204, Interscience, New York.

Juhnke, F.: 1993a, Embedded maximal ellipsoids and semi-infinite optimization, in preparation.

Juhnke, F.: 1993b, Polarity of embedded and circumscribed ellipsoids, in preparation.

Kaltenbach, F.J.: 1990, *Asymptotisches Verhalten zufälliger Polyeder*, Thesis, Univ. Freiburg.

Kenderov, P.S., Kirov, N.K.: 1993, A dynamical systems approach to the polygonal approximation of plane convex compacts, *J. Approximation Theory* **74**, pp. 1–15.

Khachiyan, L.G.: 1990, An inequality for the volume of inscribed ellipsoids, *Discrete Comput. Geom.* **5**, pp. 219–222.

Kosygin, D.V., Sinai, Ya.G.: 1992, On the number of lattice points inside a random domain, In: *From phase trasition to chaos*, pp. 547–566, World Sci. Publ., River Edge NJ.

Kuipers, L., Niederreiter, H.: 1974, *Uniform distribution of sequences*, J. Wiley, New York.

Lassak, M.: 1989, Approximation of plane convex bodies by centrally symmetric bodies, *J. London Math. Soc.* (2) **40**, pp. 369–377.

Lassak, M.: 1991, Approximation of convex bodies by parallelotopes, *Bull. Pol. Acad. Sci. Math.* **39**, pp. 219–223.

Lassak, M.: 1992, Approximation of convex bodies by triangles, *Proc. Amer. Math. Soc.* **115**, pp. 207–210.

Legg, D.A., Townsend, D.W.: 1989, The Polya algorithm for convex approximation, *J. Math. Anal. Appl.* **141**, pp. 431–441.

Legg, D.A., Townsend, D.W.: 1990, The Polya 1-property for convex approximation, *Approximation Theory Appl.* **6**, pp. 9–17.

Lindenstrauss, J., Milman, V.D.: 1993, The local theory of normed spaces and its applications to convexity, In: *Handbook of convex geometry* **B**, pp. 1149–1220, Elsevier, North-Holland, Amsterdam.

Lovász, L.: 1991, Geometric algorithms and algorithmic geometry, In: *Proc. Int. Congr. Math., Kyoto 1990*, **1**, pp. 139–154.

Lovász, L.: 1990, How to compute the volume, In: *Jahresber. d. DMV, Jubiläumstagung, 100 Jahre DMV, Bremen 1990*, pp. 138–151, (1992).

Lovász, L., Simonovits, M.: 1990, The mixing rate of Markov chains, an isoperimetric inequality, and computing volumes, In: *31st Annual Sympos. on Foundations of Computer Sci.* I, II, pp. 346–354, IEEE Comput. Soc. Press, Los Alamos CA.

Lu, Xu Guang: 1990, Convex approximation by multivariate polynomials, *Math. Numer. Sinica* **12**, pp 186–193, *Chinese J. Numer. Math. Appl.* **12**, pp. 59–69.

Ludwig, M.: 1992, *Asymptotische Approximation konvexer Krper*, Ph.D. Thesis, Techn. Univ. Vienna.

Macbeath, A.M.: 1951, An extremal property of the hypersphere, *Proc. Cambridge Philos. Soc.* **47**, pp. 245–247.

McClure, D.E., Vitale, R.A.: 1975, Polygonal approximation of plane convex bodies, *J. Math. Anal. Appl.* **51**, pp. 326–358.

Müller, J.S.: 1990, Approximation of a ball by random polytopes, *J. Approx. Theory* **63**, pp. 198–209.
Müller, J.S.: 1992, Step by step approximation of plane convex bodies, *Arch. Math.* **58**, pp. 606–610.
Niederreiter, H.: 1984, On a measure of denseness for sequences, In: *Topics in classical number theory*, pp. 1163–1208, *Colloq. Math. Soc. Janós Bolyai* **34**, North-Holland, Amsterdam.
Nowak, W.G.: 1991, On the lattice rest of a convex body III, *Czech. Math. J.* **41**, 359–367.
Pelczyński, A., Szarek, S.J.: 1991, On parallelepipeds of minimal volume containing a convex symmetric body in R^n, *Math. Proc. Cambridge Philos. Soc.* **109**, pp. 125–148.
Petty, C.M.: 1955, On the geometry of the Minkowski plane, *Riv. Mat. Univ. Parma* **6**, pp. 269–292.
Petty, C.M.: 1985, Affine isoperimetric problems, *Ann. New York Acad. Sci.* **440**, pp. 113–127.
Pisier, G.: 1989, *The volume of convex bodies and Banach space geometry*, Cambridge Univ. Press, Cambridge.
Popov, V.A.: 1968, Approximation of convex figures, *C. R. Acad. Bulgare Sci.* **21**, pp. 993–995.
Popov, V.A.: 1970, Approximation of convex sets, *Bulgar. Akad. Nauk. Otdel. Mat. Fiz. Nauk. Izv. Mat. Inst.* **11**, pp. 67–80.
Salkowski, E.: 1934, *Affine Differentialgeometrie*, de Gruyter, Berlin – Leipzig.
Sas, E.: 1939, ber eine Extremaleigenschaft der Ellipsen, *Composito Math.* **6**, pp. 468–470.
Schneider, R.: 1971, Zwei Extremalaufgaben fr konvexe Bereiche, *Acta Math. Acad. Sci. Hungar.* **22**, pp. 379–383.
Schneider, R.: 1981, Zur optimalen Approximation konvexer Hyperflchen durch Polyeder, *Math. Ann.* **256**, pp. 289–301.
Schneider, R.: 1986, Affine-invariant approximation by convex polytopes, *Studia Sci. Math. Hungar.* **21**, pp. 401–408.
Schneider, R.: 1987, Polyhedral approximation of smooth convex bodies, *J. Math. Anal. Appl.* **128**, pp. 470–474.
Schneider, R.: 1988, Random approximation of convex sets, *J. Microscopy* **151**, pp. 211–227.
Schneider, R.: 1993, *Convex bodies: the Brunn–Minkowski theory*, Cambridge Univ. Press.
Schneider, R., Wieacker, J.A.: 1981, Approximation of convex bodies by polytopes, *Bull. London Math. Soc.* **13**, pp. 149–156.
Schütt,C.: 1991, The convex floating body and polyhedral approximation, Israel J. Math. **73**, pp. 65–77.
Schwarzkopf, O., Fuchs, U., Rote, G., Welzl, E.: 1990, Approximation of convex figures by pairs of rectangles, In: STACS 90, pp. 240–249, *Lecture Notes in Comput. Sci.* **415**, Springer, Berlin.

Sonnevend, G.: 1983, An optimal sequential algorithm for the uniform approximation of convex functions on $[0,1]^2$, In: *Mathematical control theory*, 501–514, Banach Center Publ. **14**, PWN, Warsaw 1985, *Appl. Math. Optim.* **10**, pp. 127–142.

Svistunov, S.V.: 1991, Ellipsoidal upper and lower approximation of a given convex polytope, *Vestsi Akad. Navuk BSSR Ser. Fiz.-Mat. Navuk*, pp. 26–31, 124.

Tarasov, S.P., Khachiyan, L.G., Èrlikh, I.I.: 1988, The method of inscribed ellipsoids, *Soviet Math. Dokl.* **37**, pp. 226–230.

Tomczak-Jaegermann, N.: 1988, *Banach–Mazur Distances and Finite-Dimensional Operator Ideals*, Longman, Harlow, Essex.

Vitale, R.A.: 1985, L_p metrics for compact convex sets, *J. Approx. Theory* **45**, pp. 280–287.

Wagner, G.: 1993, On a new method for constructing good point sets on spheres, *Discrete Comput. Geom.* **9**, pp. 111–129.

Wieacker, J.-A.: 1978, *Einige Probleme der polyedrischen Approximation*, Diplomarbeit, Univ. Freiburg.

Yu, Xiang-Ming: 1989, Convex approximation of certain classes of convex functions, In: *Approximation theory VI, Proc. 6th Int. Sympos., (College Station, TX*, 1989) **2**, pp. 687–688.

SOME ASPECTS OF THE COMBINATORIAL THEORY OF CONVEX POLYTOPES

GIL KALAI
Institute of Mathematics
Hebrew University
Jerusalem, 91904, Israel

Abstract. We start with a theorem of Perles on the k-skeleton, $Skel_k(P)$ (faces of dimension $\leq k$) of d-polytopes P with $d+b$ vertices for large d. The theorem says that for fixed b and d, if d is sufficiently large, then $Skel_k(P)$ is the k-skeleton of a pyramid over a $(d-1)$-dimensional polytope. Therefore the number of combinatorially distinct k-skeleta of d-polytopes with $d+b$ vertices is bounded by a function of k and b alone. Next we replace b (the number of vertices minus the dimension) by related but deeper invariants of P, the g-numbers. For a d-polytope P there are $[d/2]$ invariants $g_1(P), g_2(P), ..., g_{[d/2]}(P)$ which are of great importance in the combinatorial theory of polytopes. We study polytopes for which g_k is small and move to related and slightly related problems.

Key words: Convex polytopes, skeleton, simplicial sphere, simplicial manifold, f-vector, g-theorem, ranked atomic lattices, stress, rigidity, sunflower, lower bound theorem, elementary polytopes

1. Introduction

1.1. Overview

In this paper, we will discuss several combinatorial problems concerning the combinatorial structure of polytopes. For a d-polytope P, the number of k-faces is denoted by $f_k(P)$. The vector $(f_0(P), f_1(P), \ldots f_{d-1}(P))$ is called the f-vector of P. The same definitions will apply to more general combinatorial objects considered below. The k-th skeleton $Skel_k(P)$ of a d-polytope P is the set of all faces of P of dimension $\leq k$. $Skel_1(P)$ is called the *graph* of P and is denoted by $G(P)$.

A simple basic fact is that for every d-dimensional polytope P

$$f_0(P) \geq d+1. \tag{1}$$

Equality holds if and only if P is a simplex. An important part of convex polytope theory is the study of polytopes with "few vertices", namely polytopes with a bounded difference between the number of vertices and the dimension. The following theorem of Perles is part of the theory of polytopes with "few vertices" and it will play a central role in this paper.

Theorem 1.1 (Perles, 1970) *Let $f(d,k,b)$ be the number of combinatorial types of k-skeleta of d-polytopes with $d+b+1$ vertices. Then for fixed b and k, $f(d,k,b)$ is bounded.*

T. Bisztriczky et al. (eds.), POLYTOPES: Abstract, Convex and Computational, 205–229.

A proof of Perles' theorem is given in Section 2. (The proof relies on only section 1.4 from the Introduction.) The proof given here is somewhat different from Perles' original proof. It relies, like the original proof, on the important concept of missing faces. The proof here uses the famous sunflower (Delta-system) theorem of Erdős and Rado.

A construction which increases by one both the dimension and the number of vertices is the forming of a pyramid over a polytope. Perles theorem asserts that for fixed b and d, if d is sufficiently large, then $Skel_k(P)$ is the k-skeleton of a pyramid over a $(d-1)$-dimensional polytope. In contrast, note that the number of combinatorial types of d-polytopes with $d+3$ vertices is bounded *below* by an exponential function of d, see [22].

Another theorem which is basic to the discussion in the second part of this paper is the lower bound theorem which was conjectured by Brückner in 1909 and was proved by Barnette [5] in 1970.

Theorem 1.2 *For every simplicial d-polytope P,*

$$f_1(P) \geq df_0(P) - \binom{d+1}{2}. \tag{2}$$

Equality is obtained by stacked polytopes, namely polytopes built by gluing simplices along facets.

For a d-polytope P, there are $[d/2]$ invariants $g_1(P), g_2(P), \ldots, g_{[d/2]}(P)$ which are of great importance in the combinatorial theory of polytopes. $g_1(P)$ is just the difference between the number of vertices of P and $d+1$. For simplicial polytopes, $g_2(P)$ is the difference between the left hand side and the right hand side of the lower bound relation (2).

In analogy with the theory of polytopes with "few vertices", we discuss in Sections 3 and 4 the combinatorial properties of polytopes with a bounded value of g_k for some fixed k. The nonnegativity of $g_k(P)$ is in general a deep fact (and for $k > 2$ it is not even known for general polytopes which cannot be realized by vertices with rational coordinates.) And we try to use the methods originally applied to prove the nonnegativity of $g_k(P)$ to study those polytopes for which $g_k(P)$ is small.

Section 3 deals with simplicial polytopes and Section 4 deals with general polytopes. In both cases the case $k = 2$ is substantially simpler than the general case. We will use in this discussion the notion of stresses and the connection between stresses and the g_k's as developed in Carl Lee's paper [34]. Our discussion in Section 4 is strongly related to the first section in Margaret Bayer's paper [8] and also to some topics in Richard Stanley's paper [42].

The paper is written in a somewhat ununiform style. The discussion in Section 2 is self-contained and elementary. In Sections 3-4 while technically the paper is still mostly self-contained, some prior familiarity with the notions of h-vectors, the lower bound theorem, the g-theorem and the algebraic tools which play a role in their study would be very useful. There are many problems and conjectures which are quoted and stated throughout the paper.

1.2. A START

Theorem 1.3 *Every d-dimensional polytope has at least $d+1$ vertices.*

Proof 1 (geometric-algebraic) : This follows at once from the fact that the vertices of a d-dimensional polytope affinely span a d-dimensional space.

Proof 2 (combinatorial) By induction: Let P be a d-polytope and let F be a facet of P. By the induction hypothesis F has at least d vertices. There must be a vertex in P not in F; therefore, P has at least $d+1$ vertices.

The combinatorial proof has the advantage that it applies to much more general combinatorial objects (ranked relatively-complemented lattices). The geometric proof show that $g_1(P)$ is the dimension of the space of affine relations among the vertices of P, and suggests a study polytopes with small value of g_1 by looking on the space of affine relations among vertices. This is the starting point of a very useful theory of "Gale diagrams" see [22] Ch. 6.

Both proofs show that equality holds if and only if P is a simplex.

The combinatorial proof easily extends to prove the inequality

$$g_1[r](P) =: f_r(P) - \binom{d+1}{r+1} \geq 0. \tag{3}$$

Indeed, given a d-polytope P and a facet F of P, every $(r-1)$-face G of F is included in an r-face H_G of P such that H_G itself is not contained in F. It follows that $H_G \cap F = G$, and therefore $G \to H_G$ is a one–to–one map from $(r-1)$-faces of F to r-faces of P which are not contained in F. Thus, by an induction hypothesis, there are at least $\binom{d-1}{r}$ r-faces of P contained in F, and at least $\binom{d-1}{r-1}$ r-faces of P which are not contained in F.

One of the interesting facts about the combinatorial theory of convex polytopes is that often algebraic arguments are needed. In some cases one needs a suitable mixture of algebraic and combinatorial arguments. We will see this in various places in this paper.

Remark: Relation 3 also has an algebraic interpretation. Each r-face of P determines an r-dimensional flat in R^d and thus, also a vector in the exterior $(r+1)$-power of R^{d+1}. The vectors corresponding to all r-faces linearly span this exterior power.

1.3. Polytopes, simplicial complexes, simplicial manifolds, polyhedral complexes and ranked atomic lattices

The set of faces of a polytope P, denoted by $L(P)$ is a ranked atomic lattice. $L(P)$ is called the *face lattice* of P. (A lattice L is ranked if for every element $x \in L$ all maximal chains of elements which are smaller than x have the same size. This size is called the rank of x. An *atom* is an element of rank 1, and L is atomic if every join-irreducible element is an atom.) For example, the face lattice of a simplex is a Boolean lattice. We say that P and Q are combinatorially isomorphic if there is an order preserving bijection between P and Q. In most parts of this paper, we will not distinguish between combinatorially isomorphic polytopes and we will also, by abuse of notation, not distinguish between a polytope P and its face lattice $L(P)$. We say that Q is dual to P and write $Q = P^*$, if there is an order *reversing* bijection between $L(P)$ and $L(Q)$. Every polytope has a dual given by the polar construction (see [22] Ch. 3).

A meet-semilattice is a poset with the meet operation. Every finite meet-semilattice becomes a lattice by adding to it a maximal element. A polyhedral complex is a meet-semilattice in which every lower interval is combinatorially isomorphic to a face lattice of a polytope. A simplicial complex is a meet-semilattice in which every lower interval is a Boolean lattice. To every polyhedral complex K, there is an associated topological space denoted by $|K|$.

Intervals in face lattices of polytopes are also face lattices of polytopes. If $L(Q)$ is combinatorially isomorphic to an interval of $L(P)$, we say that Q is a *quotient* of P. If F is a face of P, the interval $[F, P]$ in $L(P)$ is the face lattice of a polytope denoted by P/F. For every meet-semilattice L, we will use the notation L/F to denote the set of all elements of L which are $\geq$ than F. L/F is called the *link* of F in L. Let K be a simplicial complex and let F be a face of K. The *star* of F in K, denoted by $st(F, K)$, is the simplicial complex spanned by all the faces containing F. Note that if v is a vertex of K then $st(v, K)$ is a cone over the link of v in F. Clearly, a link of a face in a polyhedral complex is itself a polyhedral complex and a link of a face in a simplicial complex is a simplicial complex.

A simplicial polytope P is a convex polytope all whose (proper) faces are simplices. The set of faces of P is a simplicial complex, denoted by $B(P)$, and it is called the *boundary complex of* P. If P is a simplicial polytope and F is a face of P then P/F is also a simplicial polytope. The boundary complex of P/F is the link of the face F in the boundary complex of P. A *simplicial d-sphere* is a simplicial complex K such that $|K|$ is homeomorphic to the d-dimensional sphere S^d. Clearly the boundary complex of every simplicial d-polytope is a simplicial $(d-1)$-sphere, but the converse is far from being true. However many results on simplicial polytopes extend (or are believed to extend) to arbitrary simplicial spheres.

A ranked atomic lattice L is *relatively complemented* if every interval in L is atomic. It is sufficient to require that every interval of rank 2 is atomic or, in other words, that if $x > y$ are elements in L and x does not cover y, then there are at least two elements of L strictly between x and y. (See [11].) Clearly, the face lattice of every polytope is relatively complemented.

1.4. Empty faces

Let K be a simplicial complex. An *empty simplex* S of K is a minimal non-face of K, that is, S is a subset of the vertices of K, $S \notin K$ but every proper subset of S is in K. Empty simplices are called in [3, 40] *missing faces*. We want to distinguish between empty faces of various types and therefore we use a slightly different terminology.

Lemma 1.4 *The set of empty simplices of a simplicial complex K determine the complex.*

Proof: A set of vertices of K is a face if and only if it does not contain an empty simplex.

Problem 1 *Let $m_i(K)$ denotes the number of empty simplices of K of size $i+1$. Characterize the vectors $(m_1(K), m_2(K), \cdots m_d(K))$ which arise from simplicial d-polytopes.*

Let K be a polyhedral complex and let U be a subset of its vertices. The induced subcomplex of K on U, denoted by $K[U]$, is the set of all faces in K whose vertices are in U. An *empty face* of K is an induced polyhedral subcomplex of K which is homeomorphic to a polyhedral sphere. An empty 2-dimensional face is called an *empty polygon.*

For the proof of Perles' theorem, we need only the much simpler concept of empty pyramid. An *empty pyramid* of K is an induced subcomplex of K which consists of all the proper faces of a pyramid over a face of K.

1.5. h-VECTORS, g-VECTORS AND THE g-THEOREM

Let $d > 0$ be a fixed integer. Given a sequence $f = (f_0, f_1, \ldots, f_{d-1})$ of nonnegative integers, put $f_{-1} = 1$ and define $h[f] = (h_0, h_1, \ldots, h_d)$ by the relation

$$\sum_{k=0}^{d} h_k x^{d-k} = \sum_{k=0}^{d} f_{k-1}(x-1)^{d-k}. \tag{4}$$

If $f = f(K)$ is the f-vector of a $(d-1)$-dimensional simplicial complex K then $h[f] = h(K)$ is called the *h-vector* of K. For the case where K is the boundary complex of a simplicial sphere, the g-vector $g(K) = (g_0, g_1, \ldots, g_{[d/2]})$ associated with K is defined by $g_i = h_i - h_{i-1}$. Thus, $g_0 = 1$, $g_1 = f_0 - (d+1)$, $g_2 = f_1 - df_0 + \binom{d+1}{2}$ and $g_3 = f_2 - (d-1)f_1 + \binom{d}{2}f_0 + \binom{d+1}{3}$ and so on.

In 1970, P. McMullen [36] proposed a complete characterization of f-vectors of boundary complexes of simplicial d-dimensional polytopes. McMullen's conjecture was settled in 1980. L. Billera and C. Lee [10] proved the sufficiency part of the conjecture and R. Stanley [48] proved the necessity part. Stanley's proof relies on deep algebraic machinery including the hard Lefschetz theorem for toric varieties. Recently, McMullen [37] found a self-contained proof of the necessity part of the g-theorem. It is conjectured that the g-theorem applies to arbitrary simplicial spheres.

For positive integers $n \geq k > 0$, there is a unique expression of n of the form

$$n = \binom{a_k}{k} + \binom{a_{k-1}}{k-1} + \ldots + \binom{a_i}{i}, \tag{5}$$

where $a_k > a_{k-1} > \ldots > a_i \geq i > 0$. This given, define

$$\partial^k(n) = \binom{a_{k-1}-1}{k-1} + \binom{a_{k-1}-1}{k-2} + \ldots + \binom{a_i - 1}{i-1}. \tag{6}$$

Theorem 1.5 (g-theorem) *For a vector $h = (h_0, h_1, \ldots, h_d)$ of nonnegative integers the following conditions are equivalent:*
(i) h is the h-vector of some simplicial d-polytope.
(ii) h satisfies the following conditions
(a) $h_k = h_{d-k}$ for $k = 0, 1, \ldots, [\frac{d}{2}]$
Put $g_k = h_k - h_{k-1}$.
(b) g_0=1 and $g_k \geq 0$, $k = 1, 2, , \ldots, [\frac{d}{2}]$.
(c) $\partial^k(g_{k+1}) \leq g_k, k < [\frac{d}{2}]$

The relations of part *ii(a)* are the well-known Dehn-Sommerville relations. They hold for arbitrary simplicial spheres and even for arbitrary Eulerian simplicial complexes [33, 42]. Part *ii(b)* consists of linear inequalities called the generalized lower bound inequalities proposed by McMullen and Walkup in [35].

Both Stanley's original proof and McMullen's new proof of the necessity of the g-theorem give a (completely different) proof of the same algebraic statement. This algebraic statement can be expresses in terms of the Stanley-Reisner ring [49], or as a property of certain stresses on simplicial polytopes (and also in terms of algebraic shifting [31]). For a thorough explanation the reader is referred to Carl Lee's paper in this volume [34].

The g-theorem demonstrates the importance of the g-numbers to the combinatorial theory of simplicial polytopes (and spheres). It is natural to ask how combinatorial properties of P are reflected by its g-numbers.

A far reaching extension of the h-vector (and g-vector) for general polytopes was given by Stanley [45], see [8, 26]. We will discuss this in Section 4.

The following conjecture was suggested by Kalai, Kleinschmidt and Lee. If true, this conjecture gives a sharp form of Theorems 2.7 and 3.8 below.

Conjecture 2 *For all simplicial d-polytopes with prescribed h-vector $h = (h_0, h_1, \ldots, h_d)$, the number of i-dimensional missing simplices is maximized by the Billera-Lee polytopes $P_{BL}(h)$.*

$BP(h)$ is the polytope constructed by Billera and Lee [10] in their proof of the sufficiency part of the g-theorem. It is quite possible that the conjecture applies also to general polytopes.

2. Polytopes with few vertices and Perles' skeleton theorem

2.1. Monotonicity properties of g_1

Lemma 2.1 *For every face F of a polytope P, $g_1(P/F) + g_1(F) \leq g_1(P)$. In particular, $g_1(F) \leq g_1(P)$ and $g_1(P/F) \leq g_1(P)$.*

Proof: Let $k = dimF$ and note that $dim(P/F) = d - k - 1$. $f_0(P/F)$ is the number of $(k+1)$-faces G which contain F. Choose a vertex v_G in G which is not a vertex of F. Clearly if H and G are two different $(k+1)$-faces which contain F then $v_G \neq v_H$. Therefore $f_0(P/F) \leq f_0(P) - f_0(F)$. So $g_1(F) + g_1(P/F) = f_0(F) - k - 1 + f_0(P/F) - (d-k) \leq f_0(P) - d - 1 = g_1(P)$.

Lemma 2.2 *[9] Put $\nu(P) = Max\{f_0(P) - f_0(F) - 1 : F \text{ is a facet of } P\}$. Then for every facet F of P, $\nu(P) \geq \nu(F)$.*

Proof: Let F be a facet of P and let G be a facet of F. G is the intersection of F with another facet F' of P. Thus $V(P)\backslash V(F') \supset V(F)\backslash V(G)$. Therefore $f_0(P) - f_0(F') \geq f_0(F) - f_0(G)$ and the assertion follows.

Note that $\nu(P) = 0$ if and only if P is a simplex.

Lemma 2.3 *Every d-polytope with $d+b$ vertices contain a $(d-b+1)$-face which is a simplex.*

Proof: Choose a sequence $P \supset F_{d-1} \supset F_{d-1} \supset \cdots \supset F_{d-b} = G$ such that $\nu(F_i) = f_0(F_i) - f_0(F_{i-1}) - 1$. If G is not a simplex then $\nu(G) > 0$ and therefore by the previous Lemma, $\nu(F_i) > 0$ for every $i > d - b + 1$. Therefore $g_1(P) \geq g_1(G) + b - 1 \geq b$. A contradiction.

2.2. The simplicial case of Perles' theorem

Definition: A collection $\{A_1, A_2, \cdots, A_t\}$ of sets is a ***sunflower*** if every element which belongs to two or more of the sets belongs to all the sets. In other words, let $X = \cap_{i=1}^{t} A_i$ then for every $i \neq j$, $A_i \cap A_j = X$.

Lemma 2.4 *Let P be a simplicial d-polytope, and assume that P contains b disjoint empty simplices, then P has at least $d + b$ vertices.*

Proof: Let F be a facet of P. Clearly every empty simplex of P must contain a vertex not in F, but $V(P) \backslash V(F) = f_0(P) - d = b$. (Here $V(F)$ denotes the set of vertices of F.)

Lemma 2.5 *Let P be a simplicial d-polytope, and assume that P contains a sunflower of size b of empty simplices, then P has at least $d + b$ vertices.*

Proof: First note that if S is an empty simplex in P and $A \subset S$ then $S \backslash A$ is an empty simplex in P/A. Now, if P contain a sunflower $\{A_1, A_2, \cdots A_b\}$ of empty simplices and $\cap_{i=1}^{b} A_i = R$ then $A_1 \backslash R$, $A_2 \backslash R, \ldots, A_b \backslash R$ are b disjoint empty simplices in $Q = P/R$. Therefore by Lemma 2.5, $f_0(Q) - \dim Q \geq b$, and by Lemma 2.1, $f_0(P) - d \geq b$.

Lemma 2.6 (Erdős-Rado sunflower lemma [18]) *Let F be a collection of n-sets which contains no sunflower of size b then $|F| \leq m(n, b) = (b-1)^n \cdot n!$.*

Proof By induction on n. Let F be a collection of n-sets without a sunflower of size b, and let G be a maximal subcollection of pairwise disjoint sets. Put $A = \cup G$. Then $|G| < b$, $|A| \leq n(b-1)$ and every set in F contains an element from A. For each $a \in A$, the family $F(a) = \{S \backslash \{a\} : S \in F, a \in S\}$ is a family of $(n-1)$-sets without a sunflower of size b. Using the induction hypothesis, we get $|F| \leq n(b-1) \cdot (b-1)^{n-1}(n-1)! = m(n, b)$.

Let me mention an old and still very interesting conjecture of Erdős and Rado. Let $f(n, b)$ be the maximum size of a family of n-sets without a sunflower of size b.

Conjecture 3 (Erdős and Rado) *For a fixed b $f(n, b) \leq C(b)^n$, where $C(b)$ is a function of only b.*

As a corollary to the results we proved in this section we obtain:

Theorem 2.7 *Let P be a d-polytope with $d + b$ vertices. Then the total number of empty simplices of dimension $\leq k$ is bounded by a function of b and k.*

Proof of Perles' theorem, the simplicial case:

We want to bound the combinatorial types of k-skeleta of simplicial d-polytopes P with $d+b$ vertices. The k-skeleton of P is determined by the set of empty simplices of P of dimension $\leq k$. The number of empty simplices of dimension $\leq k$ is bounded by $m(k+1, b+1) = b^{k+1} \cdot (k+1)!$. Therefore, the number of all vertices of these empty faces is bounded by $(k+1) \cdot m(k+1, b+1)$ and the number of isomorphism types of the family of empty simplices is at most $\sum_{i=1}^{k+1} \binom{(k+1)m(k+1,b+1)}{i}$. This is roughly $\exp((k+1)^2(\log(k+1) + \log b - 1)$.

2.3. Perles theorem - the general case

Lemma 2.8 *A d-polytope P with $d+b$ vertices has at most 2b disjoint empty pyramids.*

Proof: It follows from Lemma 2.3 that P must have a $(d-b-1)$-face S which is a simplex. Every empty pyramid (or empty face) must contain vertices outside S. The lemma follows from the fact that $|V(P)\backslash V(S)| = 2b$.

Lemma 2.9 *Every collection of more than $(b-1)^r \cdot n^r$ r-faces, each having at most n vertices, contains a sunflower of size b.*

Proof: The proof follows the inductive proof of the sunflower lemma. Let F be a collection of faces of dimension r (or less), each having n vertices or less without a sunflower of size b. Let G be a maximal subcollection of pairwise disjoint faces. Let A be the set of vertices of all faces in G. Then $|G| < b$, $|A| \leq n(b-1)$ and every set in F contains an element from A. For each $a \in A$ the family $F(a) = \{S/\{a\} : S \in F, a \in S\}$ is a family of $(r-1)$-faces with at most $(n-1)$- vertices without a sunflower of size b. Using an induction hypothesis, we get $|F| \leq n(b-1) \cdot (b-1)^{r-1} n^{r-1}$.

Proof of Perles' theorem (end):[1] For a polyhedral complex K define the *kernel* of P, $Ker(P)$, to be the union of the sets of vertices of all empty pyramids in K. Clearly the combinatorial type of $Ker(K)$ and the number of vertices of K determine the combinatorial type of K. Namely the set of faces of K is precisely $F * T$ where F is a face of K and T is any subset of vertices which are disjoint from $Ker(K)$. $F * T$ is the $|T|$-fold pyramid with basis F.

Lemma 2.9 implies that the kernel of the k-th skeleton of a d-polytope with $d+b$ vertices has at most $(2b)^{r-1}(b+k))^r$ vertices. Therefore the number of isomorphism types of k-skeleta of d-polytopes with $d+b$ vertices is bounded by the number of k-dimensional polyhedral complexes with $(2b)^{r-1}(b+k))^r$ vertices.

2.4. The scope of Perles' theorem

As easily seen, the proof of Perles theorem for simplicial polytopes given above, applies to arbitrary pure simplicial complexes. The proof of the general case applies to a large class of ranked atomic lattices. Perles observed that his proof (and this applies to the proof given here) applies to arbitrary ranked atomic relatively complemented lattices. He went further to define an even larger class of lattices, the class of pyramidally perfect lattices, for which his proof applies. For an element x in an atomic lattice L, $J(x)$ denotes the set of atoms below x. An atom a is *pyramidal* with respect to $x \in L$ if $a \not< x$ and $J(x \vee a) = J(x) \cup \{a\}$. A ranked atomic lattice is called *pyramidally perfect* if, whenever a is pyramidal w.r.t. x, it is also pyramidal w.r.t. every y, where $y < x$,

3. Simplicial polytopes with small value of g_k

3.1. Overview

In this section we discuss simplicial polytopes with a small value of g_k. The situation is simpler for g_2 and more involved for higher k's.

[1] This part of the proof is taken from [40] without changes

The nonnegativity of g_2 can be proved by purely combinatorial methods as well as by the rigidity theory of frameworks. Both approaches apply to a very general class of simplicial complexes, the class of pseudomanifolds. The rigidity theoretic interpretation of g_2 gives much information on the structure of simplicial polytopes (and simplicial manifolds) with small values of g_2. This is described below in Section 3.3. The proofs of the necessity of the g-theorem (both Stanley's original proof and McMullen's recent proof) deduce the theorem from a certain crucial algebraic fact. This gives an interpretation of g_k which is closely related to the rigidity theoretic interpretation of g_2, see [34, 37, 31], and allows the extension of some of the results to simplicial polytopes with small values of g_k.

In Section 3.2 we state a conjecture giving a complete description of g-vectors of sequences of simplicial polytopes which converge to smooth bodies. Like the g-theorem, the conjecture consists of a linear part and a nonlinear part. The linear part of the conjecture may be doable by improving the methods and results described here. In Section 3.3 we describe the main tool we use, the notion of stresses. This is a very quick outline of some facts from Carl Lee's paper [34]. In Section 3.4 we state the lower bound inequalities, and in Section 3.5 we describe the structure of the proof that $g_2 = 0$ only for stacked polytopes. In Section 3.6 we describe some partial information on polytopes with vanishing g_k. In Section 3.7 we extend Perles theorem to simplicial polytopes with bounded g_k. In Section 3.8 we study in more detail the case $k = 2$. It turns out that every simplicial polytope with a small value of g_2 can be obtained by gluing together "small" pieces. In Section 3.8 we diverge to describe finer invariants of simplicial polytopes which give much more information than the g-numbers.

3.2. g-NUMBERS OF SIMPLICIAL POLYTOPES WHICH CONVERGE TO A SMOOTH BODY

We state two conjectures on the behavior of g-numbers of simplicial polytopes which converge to a smooth body. The first conjecture falls into our study of polytopes with bounded values of g_k. It is trivial for $k = 1$ and follows from the result of Section 3.8 for $k = 2$. The second conjecture calls for a similar study of polytopes for which $g_k^{<k>} - g_{k+1}$ is bounded.

Conjecture 4 ([24]) *Let k and d be positive integers, $d \geq 2k$. Let P_n be a sequence of d-polytopes which converge to a smooth body K. Then*

$$\lim_{n\to\infty} g_k(P_n) \to \infty. \tag{7}$$

Conjecture 5 *Let P_n be a sequence of d-polytopes which converge to a smooth body K. Then for $k < [d/2]$,*

$$\lim_{n\to\infty} (g_k(P_n) - \partial^k(g_{k+1}(P_n))) \to \infty. \tag{8}$$

If Conjectures 4 and 5 are true then they give a complete description of sequences of g-vectors which come from a sequence of simplicial polytopes converging to a smooth body K. (Note: the description is independent from K.)

If P_n is a sequence of polytopes which converges to a convex body K, and Q_n is any sequence of polytopes, then one can glue a projective copy of Q_n to one facet of P_n and the resulting sequence of polytopes will also converges to K.

If $\mathcal{G}_d$ is the set of g-vectors of simplicial d-polytopes, and $B_d(r)$ is the set of vectors $(a_1, \ldots, a_{[d/2]})$ such that $|a_i| \leq r$ for every i, then, if Conjectures 4 and 5 are true, a sequence of g-vectors $\{g^n\}$ is the sequence of g-vectors of simplicial polytopes converging to a smooth convex body K if and only if, for every r, there is N_r such that $g^n + B_d(r) \subset \mathcal{G}_d$ for every $n > N_r$.

Remarks: 1. *Connections between metrics on the sphere and combinatorial invariants*

It is possible to formulate similar questions in a purely combinatorial way for sequences of simplicial spheres (and even simplicial manifolds). Every triangulation of the sphere induces a metric and it is possible to consider limits of such metrics (see [21]) as the number of vertices tends to infinity. It is natural to study Conjectures 4 and 5 in this more general context, and more generally to study the following

Problem 6 *Given a metric on the $(d-1)$-sphere and sequence of simplicial spheres whose induced metrics converges to this metric, find relations between the g-numbers (and other combinatorial properties) of the simplicial spheres in the sequence and the geometric properties of the limiting metric.*

2. *Separation properties of $G(P^*)$.*

Let K be a pure $(d-1)$-dimensional simplicial complex. The dual graph of K, denoted by $G^*(K)$ is defined as follows: the vertices of $G^*(K)$ are the facets ($(d-1)$-faces) of K and two vertices F and G are adjacent if $F \cap G$ is a $(d-2)$-dimensional face. If K is the boundary complex of a simplicial polytope P then $G^*(K)$ is just the graph of the dual polytope P^*.

Let $f(r)$ be a function of the integer number r and consider the class $\mathcal{P}_f$ of all simplicial polytopes P (or even simplicial spheres) with the following property: whenever r vertices are deleted from $G(P_n^*)$ then the remaining graph has a connected component having all, except at most $f(r)$, vertices. It is plausible (and this would imply Conjecture 4) that for every function $f(r)$ and for all polytopes in $\mathcal{P}_f$, the value of $g_k(P)$ $(k < [d/2])$ is bounded. See [31] for some results on separation properties of graphs of special types of simple polytops.

3.3. Stresses

This is a very quick outline of some important ingredients of Carl Lee's paper [34]. See also the papers of Tay, White and Whiteley [50] on skeletal rigidity. The reader should also consult Lee's paper for the relations between stresses and the Stanley-Reisner's ring, and [31] for the connection with algebraic shifting.

Let K be a simplicial complex embedded into R^d. (By "embedded" we mean only that the vertices are embedded in such a way that the vertices of faces are affinely independent.) A k-stress (which is an abbreviation here for an affine k-stress since we will not consider linear stresses) is defined as follows. For every $(k-1)$-face F, choose a point $u_F \in F$. If G is a k-face containing a $(k-1)$-face F, let $v(F,G)$ denote the vertex of G not in F. A *k-stress* is an assignment of weights w_G to the k-faces G of K such that for every $(k-1)$-face F,

$$\sum \{w_G(v(F,G) - u_F) : G \supset F\} \in Aff(F). \tag{9}$$

Here, $Aff(F)$ is the affine span of the face F. Let S^a_k denote the space of k-stresses of K.

Let $A_k(K)$ be the space of all assignment of weights w_G to the k-faces G of K. Now consider the map T_k which assigns to every $w \in A_k(K)$ weights on $(k-1)$-faces F as follows: the weight of a $(k-1)$-face F is $\sum\{w_G(v(F,G) - u_F) : G \supset F\}$ considered as a vector in the quotient space $R^d/Aff(F)$. (The weights are vectors of dimensions $d-k+1$.) The space of k-stresses is precisely the kernel of T_k. There is a complementary notion of k-rigidity, (or skeletal k-rigidity) which is of importance here. Roughly speaking, K is k-rigid if the image of T_k is "as large as possible". What we will need is that if K is k-rigid and K' is obtained from K by adding just one k-face then K' has a non-zero k-stress.

All these concepts become classic for $k = 1$. 1-rigidity is called infinitesimal rigidity. 1-stress is the classical notion of a stress of a framework. 0-stresses are just affine relations among the vertices, and 0-rigid just means that the vertices of K affinely span R^d.

The following basic fact connecting stresses with the g-vector, follows from the two known proofs of the necessity part of the g-theorem. (In Stanley's proof, it is proved only for rational polytopes.)

Theorem 3.1 *Let P be a simplicial d-polytope and let $k < [(d+1)/2]$. Then $g_k(P) = dim S^a_k(P)$.*

This is equivalent to the fact that $Skel_k(P)$ is k-rigid.

An important fact about stresses is that they behave nicely under "forming a cone". Let K be a simplicial k-dimensional complex and consider a generic embedding of K in R^d. Consider also a generic embedding of a cone over K in R^{d+1}. Then the embedding of K is k-stress free iff the embedding of the cone is k-stress free. See [50]. (This is related to the fact that the operation of "forming a cone" commutes with algebraic shifting.)

Remark: Stresses can be regarded as analogs for Gale transforms which are one of the most useful tools in the study of polytopes with few vertices. However, the extension of the basic property of Gale transform is not yet known:

Conjecture 7 ([24]) *Let P_1 and P_2 be two simplicial d-polytopes and let ϕ be a bijection from $V(P)$ to $V(Q)$ such that ϕ is a combinatorial isomorphism from $Skel_k(P)$ to $Skel_k(Q)$ and, moreover, the map induced by ϕ gives an isomorphism between the space of k-stresses of P and the space of k-stresses of Q. Then ϕ induces a combinatorial isomorphism between P and Q.*

In other words, is the k-skeleton plus the vector space of k-stresses determine uniquely the combinatorial type of simplicial polytopes?

Note that for $k = [d/2]$, the space of stresses is trivial, but indeed an important theorem of Perles asserts that for two simplicial d-polytopes, every combinatorial isomorphism between $Skel_{[d/2]}(P)$ and $Skel_{[d/2]}(Q)$ can be extended to a combinatorial isomorphism between P and Q. Also, as we shall see later, if for $k < [d/2]$, the space of k-stresses of a simplicial d-polytope is trivial (that is, $g_k(P) = 0$) then P has no missing faces of dimension greater than k and smaller than $d-k$, and again, the k-skeleta determine the combinatorial structure of the polytope.

3.4. g_2 AND THE LOWER BOUND INEQUALITIES

A simplicial d-polytope is stacked if it can be obtained by gluing d-simplices along facets. Every stacked polytope with n vertices is obtained from a stacked polytope with $n-1$ vertices by adding a vertex, beyond exactly one facet. P is stacked if and only if it can be triangulated without introducing faces of dimension smaller than $d-1$. The boundary complex of a stacked polytope is called a stacked sphere.

It is easy to see that the f-vector of a stacked d-polytope is determined by the number of vertices. Let $\phi_k(n,d)$ denote the number of k-faces of a stacked d-polytope with n vertices. Thus, $\phi_k(n,d) = \binom{d}{k}n - \binom{d+1}{k+1}k$, for $1 \le k \le d-2$, and $\phi_{d-1}(n,d) = (d-1)n - (d+1)(d-2)$.

The lower bound inequalities assert that for every simplicial d-polytope P with n vertices, $f_k(P) \ge \phi_k(n,d)$. The case $k=1$ of this inequality is just the nonnegativity of g_2. There is an inductive way to deduce the lower bound inequalities from the nonnegativity of g_2, see [5, 35, 25]. However, this inductive argument does not apply to certain generalization of the lower bound inequalities; such as, those for centrally symmetric polytopes and for general polytopes. Thus, it may be useful to find a direct interpretation of $g_2[r] = f_k(P) - \phi_k(n,d)$ as the dimension of some vector space.

3.5. SIMPLICIAL POLYTOPE WITH VANISHING g_2

Theorem 3.2 *For $d > 4$, the following conditions are equivalent: (1) P is stacked (2) $P/\{v\}$ is stacked for every vertex v (3) P has no empty simplices of dimension r, for $1 < r < d-1$.*

The proof of this theorem is given in [25]. It applies to arbitrary simplicial $(d-1)$-manifolds which are simply connected. For non-simply connected manifolds M, conditions 2 and 3 remain valid and are equivalent to the fact that M is obtained from the boundary of stacked polytope by additional operations of "handle forming", via identifying the vertices of two disjoint facets and deleting the facet.

Theorem 3.3 *For $d \ge 4$, the following are equivalent: (1) P is stacked (2) P has no empty faces (of any kind) of dimension r, for $1 < r < d-1$ (3) P has no empty simplices of dimension r, for $1 < r < d-1$, and no empty polygons.*

The crucial point behind this theorem is the situation for a simplicial 3-polytope. A simplicial 3-polytope is stacked iff it has no missing polygons other than triangles. While the two theorems above are purely combinatorial, rigidity arguments are needed to prove the following

Theorem 3.4 *For $d > 3$, if $g_2 = 0$ then*
(1) P has no empty faces of dimension r, $1 < r < d-1$, (2) $g_2(P/v) = 0$ for every vertex v, and (3) P has no empty polygons.

It follows from the theorems quoted above that if $g_2(P) = 0$ then P is a stacked polytope. This result applies to arbitrary simplicial manifolds (and pseudomanifolds).

Remark: There is an interesting issue which is related to the preceding theorems. Consider a simplicial manifold K and assume that all links of vertices K/v (which are simplicial spheres) have certain combinatorial properties. What does this imply about the topology of K? If all links are stacked spheres then for dimension > 3, this implies severe restrictions on the topology of K, in particular, if K is simply connected then K is a sphere (the 3-dimensional case is open).

3.6. SIMPLICIAL POLYTOPES WITH VANISHING g_k

We do not have a good understanding of simplicial polytopes P with $g_k(P) = 0$. McMullen and Walkup conjectured that every such polytope is k-stacked, namely, it can be triangulated without introducing faces of dimension $< d - k$. Unlike stacked (1-stacked) polytopes which are well understood, k-stacked polytopes are themselves quite mysterious. Parts of the discussion concerning the vanishing of g_2 extend to higher k's but other parts are still not known (but perhaps doable).

Proposition 3.5 *For $d \geq 2k + 3$, the following are equivalent: (1) P is k-stacked (2) P/v is k-stacked for every vertex v.*

Proposition 3.6 *For $d > 2k + 1$, if $g_k(P) = 0$ then (1) P has no empty simplices of dimension r, $k \leq r \leq d - k$, and (2) $g_k(P/v) = 0$ for every vertex v.*

Proof: (1) Assume that S is an empty k-simplex. Now, the vertex figure P/v is k-rigid and therefore $st(v, P)$ (a cone over it) is k-rigid. $R = S\backslash v$ is a $(k-1)$-face in P which is not in $st(v, P)$. Therefore $st(v, P) \cup R$ has a non-zero stress and since $st(v, P) \cup R \subset P$, P has a nonzero k-stress, and $g_k(P) > 0$. If S is an empty simplex of size $k + i$, choose $V \subset S$, $|V| = i$ and a vertex $v \in V$. Apply the same argument for P/V inside $P/(V\backslash\{v\})$.

Part (2) follows at once from the cone property for k-stresses. In fact,

Lemma 3.7 $g_k(P/v) \leq g_k(P)$.

Proof: $g_k(P/v)$ is the dimension of the space of k-stresses of P/v w.r.t to embedding in R^{d-1} and, therefore, $g_k(P/v)$ is the dimension of the space of k-stresses of $st(v, P)$ w.r.t. embedding in R^d.

Part (2) of Proposition 3.6 also follows from the identity

$$\sum_v g_k(P/v) = (d - k + 1)g_k(P) + (k + 1)g_{k+1}(P),$$

and the nonnegativity of $g_{k+1}(P)$.

Conjecture 8 *For $d \geq 2k$, the following are equivalent: (1) P is k-stacked (2) P has no empty faces (of any kind) of dimension r, for $k < r < d - k$. For $d \geq 2k$, these two conditions are equivalent to (3) $g_k(P) = 0$.*

Remark: The k-skeleton of every d-polytope contains a subdivision of the k-skeleton of a d-simplex. For simplicial polytopes, the nonvanishing of g_k also seems related to the existence of a subdivision of the k-skeleton of a $(d+1)$-simplex. Indeed, the nonvanishing of g_2 for a simplicial polytope P is equivalent to the fact that the graph of P contains a refinement of K_5 [25]. For $k > 2$, the results of Stanley [47] seem relevant.

3.7. Analog of Perles' theorem for simplicial polytopes with small value of g_k

Theorem 3.8 *For positive integers $k \geq 1, r \geq k$ and $b \geq 0$, there is a function $b_k(r, b)$ with the following property: if P is a simplicial d-polytope with $g_k(P) \leq b$ then P has at most $b_k(r, b)$ empty r-simplices.*

Proof (sketch)

By the sunflower theorem, it is enough to prove that there is no sunflower of empty r-simplices of size $b+1$. We have seen that an empty r-simplex, $d-k > r > k$ is responsible for a k-stress, so it is enough to show that in case of a sunflower $\{S_1, S_2, \ldots, S_{b+1}\}$, we get a contribution of at least $b+1$ to the space of stresses. Let R be the intersection of the S_i's. If $|R| > r-k$ then the situation is very easy; choose $V \subset R$ such that $|V| = r-k$ and a vertex $v \in V$. Let $K = P/(V \backslash \{v\})$. K/v is k-rigid in dimension $d - |V|$ and therefore, $st(v, K)$ is k-rigid in dimension $d - |V| + 1$, and adding to it the $b+1$ k-faces $S_1 \backslash V, S_2 \backslash V \ldots, S_{b+1} \backslash V$ creates a k-stress space of dimension at least $b+1$. The case where $|R|$ is smaller is slightly more complicated and we omit the details.

Corollary 3.9 *There exists a function $u_k(r, b)$ with the following property: if K is the k-th skeleton of a simplicial d-polytope P with $g_k(P) < b$ (note: g_k can be read from the k-skeleton) then there are only $u_k(r, b)$ possibilities for the r-skeleton of P.*

Note that the number of 1-skeleta of stacked d-polytopes with n vertices is exponential in n.

It seems that Theorem 3.8 applies for general empty faces and not only for empty simplices. What we need to do is, given an empty face in dimension $r, k < r < d-k$, find a nonzero k-stress such that for disjoint empty faces, one gets linearly independent k-stresses. It looks as if an appropriate Meyer-Vietoris type statement for k-stresses is needed. Proving this may be helpful also in verifying Conjecture 4.

3.8. Simplicial polytopes with small value of g_2

A simplicial polytope P is *prime* if it does not contain an empty $(d-1)$-simplex. If P is not prime then P can be obtained by gluing together along facets of prime simplicial polytopes. We write $P = P_1 \# P_2 \# \cdots \# P_k$ for the description of P as the union of prime simplicial polytopes. It is easy to see that $g_2(P) = \sum g_2(P_i)$. The following theorem shows that if $g_2(P)$ is small then P is obtained by gluing together many small pieces. (Clearly most of these pieces must be simplices.)

Theorem 3.10 *There is a function $u(d, b)$ such that if P is a prime simplicial d-polytope and $g_2(P) \leq b$ then $g_1(P) \leq u(d, b)$.*

Proof: (sketch) For $d > 4$, we know that the number of empty simplices of P of dimension r, $1 < r < d-1$, is at most $X = X(d, b)$. The number of edges of P is bounded by a linear function of the number of vertices. Therefore (by Turán's theorem) if the number of vertices is large then $G(P)$ contains an independent set U, of vertices, of size $> X$. For $u \in U$, if $P/\{u\}$ (= the link of u in P) is not stacked and also if $P/\{u\}$ contains a empty $d-2$ simplex which is not a face of P, we get a non-zero stress s_u in $st(u, K)$. Moreover, s_u has nonzero weight on some edge containing u. Since U is an independent set of vertices, this edge is not included in the star of any other vertex of U. Therefore, all the stresses $\{s_u : u \in U\}$ are linearly independent. A contradiction.

It follows that for some $u \in U$, $st(u, P)$ does not contain a non-zero stress. Therefore either $P/\{u\}$ is a simplex and thus an empty $(d-1)$-simplex or $P/\{u\}$ is stacked, and contains an empty $(d-2)$-simplex R which is not an empty simplex in P. In this case, $R \cup \{u\}$ is an empty $(d-1)$-simplex in P. The proof for $d = 4$ is more involved and we will not include it here.

Remarks: 1. It is quite possible that $u(d, b)$ is actually independent from d, and it would be interesting to determine its best possible value. I do not know of examples of prime simplicial polytopes with $g_1 \leq g_2 - 1$.

2. The proof of Theorem 3.10 applies in much more general contexts. The proof as described above applies for simplicial spheres and even for simplicial manifolds. Every simply connected prime simplicial manifold can written as a connected sum of prime (simply connected) simplicial manifolds. For arbitrary manifolds, one has to add another operation - that of "handle forming" via an identification of two facets. Every simplicial manifold with small value of g_2 can be obtained from small "prime" pieces by the operations of connected sum and handle forming. (Each handle increases the value of g_2 by $\binom{d+1}{2}$.) It follows that for every b, there are only finitely many d-manifolds which have a triangulation K such that $g_2(K) \leq b$.

3. Even more generally, Theorem 3.10 applies to all pseudomanifolds such that the link of every face of codimension ≥ 1 is connected and the link of every link of dimension ≥ 2 is simply connected.

3.9. DIVERSION: FINER INVARIANTS

An order ideal of monomials I is a collection of monomials in variables (say) $x_1, x_2, \ldots$ such that $1 \in I$ and if $m \in I$ and m' divides m then $m' \in I$. We will denote by $f_k(I)$, the number of monomials of I of degree k, and call the vector $(f_0(I), f_1(I), \ldots)$, the f-vector of I. (Note: the indices here are shifted by 1.) An order ideal of monomials is *shifted* if for every monomial m in S, if x_j has positive degree in m and $i < j$ then $m \cdot x_i \cdot x_j^{-1} \in I$.

A sequence of integers $(m_0, m_1, \ldots)$ is an M-vector if $m_0 = 1$ and $0 \leq \partial^k(m_{k+1}) \leq m_k$ for every $k \geq 1$. An old theorem of Macaulay asserts that M-vectors are precisely the possible f-vectors of an order ideal of monomials. Every M-vector can be realized (usually in many ways) as the f-vector of a shifted order ideal of monomials.

For a simplicial d-polytope (and probably for every simplicial $(d-1)$-sphere), the vector $(1, g_1, \ldots, g_{[\frac{d}{2}]})$ is an M - vector. The proof of the necessity part of the g-theorem actually associates to every simplicial polytope P, a shifted order ideal of monomials $S(P)$ such that $g_i(K)$ is the number of monomials of degree i in S. (See [31].) $S(P)$ can be regarded as a delicate invariant of P. It is conjectured that the same algebraic construction applies to arbitrary simplicial spheres.

Conversely, for every shifted order ideal of monomial S, there is a construction of a simplicial sphere $K(S)$, see [28]. It is conceivable but not known that $S(K(S)) = S$.

It would be interesting to study the structural properties of P as a function of $S(P)$, in a manner similar to the approach of this paper. Here is a far-reaching extensions of Conjectures 4 and 2.

Let $M(d)$ denote the set of all monomials of degree $\leq [d/2]$ on the countable set of variables $x_1, x_2, \ldots, x_n, \ldots$.

Conjecture 9 *Let P_n be a sequence of simplicial d-polytopes which converges to a smooth body K. Then $\cup S(P_n) = M(d)$.*

Conjecture 10 *For simplicial spheres K with $S(K) = S$, the vector of empty simplices is maximized for the complex $K(S)$.*

Remarks: 1. It has been known for quite a long time that there are simplicial (and polyhedral) spheres that cannot be realized as the boundary complex of a simplicial polytope. In high dimensions, there is a striking gap between the number of combinatorial types of simplicial polytopes and the number of combinatorial types of simplicial spheres. See [20, 28]. However, most of the results mentioned here for simplicial polytopes are either known or conjectured to be known for simplicial spheres. While face numbers are probably too weak to distinguish simplicial polytopes from arbitrary triangulations of spheres, it is possible that the finer invariant $S(K)$ will contain some useful parameters for this problem.

2. It is interesting to note that neither the g-vectors nor the finer invariant $S(P)$ can distinguish between different neighborly polytopes. Indeed P is neighborly iff $S(P)$ is the ideal of all monomials of degree $\leq d/2$ in $n - d$ variables. The combinatorial structure of neighborly polytopes (even in dimension 4) is a rich topic, and it seems that completely different invariants are needed for their study.

4. General Polytopes

4.1. Overview

In this section we consider general polytopes. In this case even the definition of $g_k(P)$ is quite subtle. We describe the definition in Section 4.2. More details can be found in Bayer's paper [8]. Section 4.3 is devoted to $g_2(P)$. We describe the rigidity theoretic meaning of $g_2(P)$, and describe some facts on the remarkable class of polytopes with vanishing g_2. The nonnegativity of the g_k's implies many linear inequalities for flag numbers of polytopes. The possibility to use the large amount of complicated data given by such inequalities to prove basic, and easy to state properties of polytopes, is discussed in Section 4.4. We describe there results of Meisinger, who developed the automatic polytope theorem prover FLAGTOOL. In Section 4.5, we make some conjectures about additional linear inequalities for flag numbers of polytopes. In Section 4.6, we discuss special classes of polytopes, and in Section 4.7, we ask how to generalize the notions of h- and g-numbers.

4.2. g_k for general polytopes and flag numbers

Intersection homology theory has led to deep and mysterious extensions of g-numbers from simplicial polytopes to general polytopes.

The definition (which can be found also in [8, 42]) is as follows. For a polytope P, denote by P_k the set of k-faces of P.

Define by induction two polynomials

$$h_P(x) = \sum_{k=0}^{d} h_k x^{d-k}, g_P(x) = \sum_{k=0}^{[d/2]} g_k x^{d-k},$$

such that: (a) $g_k = h_k - h_{k-1}$, (b) if P is the empty polytope or a 0-polytope P then $h_P = g_P = 1$, and

$$h_P(x) = \sum_{k=0}^{d} (x-1)^{d-k} \sum \{g_F(x) : x \in P_k\}.$$

Thus $g_1(P) = f_0(P) - d - 1$ and

$$g_2(P) = f_1(P) + \sum \{f_0(F) - 3 : F \in P_2\} - df_0(P) + \binom{d+1}{2}.$$

The value of g_2 for general polytopes has also a rigidity theoretic meaning. In this case, however, the nonnegativity of g_2 is still open for more general objects such as polyhedral spheres and manifolds.

The higher g numbers for general polytopes are quite mysterious, and at present, their nonnegativity is known only for polytopes with rational vertices. Goresky and MacPherson (unpublished) developed a concrete way to describe $g_k(P)$ as certain hyperhomology groups based directly on the geometry of the polytope. This concrete description (which they proved only for rational polytopes,) may shed some light on their geometric meaning. McMullen's recent new proof of the necessity part of the g-theorem also gives some hope for an elementary interpretation of the g-numbers for general polytopes, and a new proof for their nonnegativity. McMullen's proof also contains a relatively easy reduction from simple polytopes to rational simple polytopes, and there is hope that this part, at least, can be extended to general polytopes.

Stanley [45] conjectured that the g-vector is an M-vector for every polytope. It is not even known that $g_k(P) = 0$ implies that $g_{k+1}(P) = 0$. Stanley also indicated the extreme combinatorial generality for the g-numbers as defined in this section: namely, for regular cell decomposition of (homology) spheres whose faces form a lattice.

For a d-polytope P and a subset $S = \{i_1, \ldots, i_k\} \subset \{0, 1, \ldots, d-1\}$, the flag number $f_S(P)$ is the number of chains $F_1 \subset F_2 \subset \cdots \subset F_k$ of faces of P such that $dimF_j = i_j$. (The same definition applies to ranked lattices.) For simplicial polytopes, the flag numbers are determined by the face numbers, but for general polytopes, flag numbers seem to be the "correct" invariants. A remarkable theorem of Bayer and Billera asserts that the affine dimension of the space of flag numbers of d-polytopes is $c_d - 1$, where c_d is the d-th Fibonacci number. The g-numbers are linear combination of flag numbers, and this is explained in detail in [8].

4.3. g_2 FOR GENERAL POLYTOPES AND ELEMENTARY POLYTOPES

Let P be a d-polytope. A framework based on P is a graph, embedded in R^d, which is obtained by triagulating all the 2-faces of P by polygons. Let f_1^+ be the number of edges in such a framework. $g_2(P) = f_1^+(P) - df_0(P) + \binom{d+1}{2}$. Whiteley [51] proved by a clever inductive argument starting with the case $d = 3$ (which was proved by Alexandrov) that every such framework is infinitesimally rigid (1-rigid). This implies that $g_2(P)$ is the dimension of the space of stresses of a framework based on P and therefore, $g_2(P)$ is nonnegative for every d-polytope.

A polytope P is *elementary* if $g_2(P) = 0$.

Theorem 4.1 *Let P be a d-polytope. Then $g_2(P) \geq g_2(F) + g_1(F)g_1(P/F) + g_2(P/F)$*

Proof: (Rough sketch) Let F be a k-face of P. Choose a vertex in each $(k+1)$-face G containing F, which is not in F. On all these vertices, form a graph H whose edges correspond to edges of a framework of P/F. Put $G = G(F)$. Consider now $G * F$, the join of G and H (the graph $H \cup G$: the union all edges between a vertex in H and a vertex in G,) as a framework in R^d. It is possible to move from $G(P)$ to a framework containing $G * H$ by successively applying the following operation: move from a framework A to a framework A', by adding an edge e_1 and deleting an edge e_2 when $A \cup A'$ contains a minimal stress containing both e_1 and e_2.

Theorem 4.2 *If $g_2(P) = 0$ then $g_2(P^*) = 0$.*

Proof: (sketch) Every d-polytope for $d \leq 3$ is elementary, so let $d \geq 4$. For $d = 4$, it is easy to verify that $g_2(P) = g_2(P^*)$. For $d > 4$, the Theorem follows by induction using the following:

Theorem 4.3 *For a d-polytope P, $d > 4$, the following conditions are equivalent:*

(a) For every proper face F of P, (1) $g_2(F) = 0$, (2) $g_2(P/F) = 0$ and (3) either F is a simplex or P/F is a simplex.

(b) $g_2(P) = 0$.

Proof: (sketch) (b) implies (a) by Theorem 4.1. (a) implies (b) for rational polytopes by the following identity (which can easily be verified by expanding both sides in terms of flag numbers):

$$3\sum\{g_2(F) : F \in P_4\} + 2\sum\{g_1(F) \cdot g_1(P/F) : F \in P_2\} \qquad (10)$$
$$+\sum\{g_2(P/v) : v \in P_0\} = (d-1)g_2(P) + 3g_3(P).$$

(A very rough sketch of the proof for general polytopes: follow the second proof of Theorem 9.3 in [25].)

Definition: An *abstract polytope* is a ranked lattice L such that

(a) every interval of length two has four elements, and

(b) the following connectivity property holds: for every two elements x and y of rank k in L, there is a sequence of elements of rank k, $x = x_1, x_2, \ldots, x_t = y$, such that $rank(x_i \wedge x_{i+1}) = k - 1$ and $x_i \leq x \vee y$, for every i.

This notion (which is purely combinatorial) clearly includes as special cases, simplicial and polyhedral manifolds and pseudomanifolds.

Conjecture 11 *The inequality $g_2 \geq 0$ holds for arbitrary abstract polytopes L. Equality holds only for face lattices of polytopes. Moreover, every elementary polytope can be realized with rational coordinates.*

It is conjectured that elementary polytopes have many of the pleasant properties of 3-polytopes. Here is one question in this direction.

Conjecture 12 *The graph of every elementary d-polytope is $d+1$ colorable. Moreover, every graph obtained by triangulating (with no new vertices) the 2-faces of an elementary d-polytope is $(d+1)$-colorable.*

We also conjecture that the main theorems of this section extend to higher g_k's.

Conjecture 13 ([26]) *(1) $g_k(P) = 0$ if and only if $g_k(P^*) = 0$.*
(2) For every face F, $g_k(P) \geq \sum_{i=0}^{k} g_i(F) g_{k-i}(P/F)$.

4.4. Diversion: quotients, faces and Meisinger's FLAGTOOL

The reader may have noticed that the inequalities $g_k \geq 0$ for general polytopes are rather complicated, and it may be asked to what extent are these relations (even if they will be proved completely) relevant to basic combinatorial properties of polytopes. As described in Bayer's paper [8] (see also [26]), a few basic linear inequalities for flag numbers of polytope imply, by convolutions, a large number of other inequalities. Günter Meisinger developed a computerized system called FLAGTOOL whose aim is to try to prove automatically theorems on polytopes using the large amount of (known and conjectured) inequalities for face numbers. The following three conjectures were (among others) some targets for FLAGTOOL.

Conjecture 14 (Perles) *For every integer $k > 0$, there exists $f(k)$ so that every d-polytope, $d \geq f(k)$, has a k-dimensional quotient which is a simplex.*

Conjecture 15 *For every integer $k > 0$, there exist integers $n(k)$ and $d(k)$ so that every d-polytope, $d \geq d(k)$, has a k-dimensional face with at most $n(k)$ vertices.*

It can be conjectured that $n(k)$ can be chosen to be 2^k and that the following stronger conjecture holds.

Conjecture 16 ([30]) *For every integer $k > 0$, there exists $d(k)$ so that every d-polytope, $d \geq d(k)$, has a k-dimensional face which is either a simplex or is combinatorially isomorphic to a cube.*

These conjectures are valid for $k = 2$. It follows easily from Euler's theorem that every polytope in 3-space has a triangular face or its dual has such a face. It also follows from Euler's theorem that every polytope in 3-space has a face with at most five vertices, and in [30], it is proved that every d-polytope, $d \geq 5$, has a face with at most four vertices.

The hope (which was fulfilled) was that FLAGTOOL will automatically prove some of these conjectures in low dimensions and moreover, (this was not fulfilled yet) it will give some insight into what is involved in a proof for arbitrary dimension.

FLAGTOOL proved automatically the following partial results and supported results to the conjectures above (among many other results).

Theorem 4.4 (Meisinger [39]) *1. Every rational d-polytope, $d \geq 9$, has a 3-face with at most 150 vertices.*

2. Every d-polytope, $d \geq 9$, has a 3-dimensional quotient which is a simplex.

3. Every d-polytope, $d \geq 7$, has a triangle as the quotient of 1-face in a 4-face.

4. Every 7-polytope has a 3-face with at most 17 vertices or its dual has such a face.

5. Every 5-polytope has a 3-quotient with at most 8 vertices, and every 7-polytope has a 4-quotient with at most 16 vertices.

4.5. Fake f-vectors and more linear inequalities

Every linear combination of face numbers which is nonnegative for all simplicial polytopes is a linear combination with nonnegative coefficients of $g_0, g_2, \ldots, g_{[d/2]}$. In [26], the author conjectured that the nonnegativity of $g_1, \ldots, g_{[d/2]}$ give, by convolutions (see [8]), all linear inequalities among flag numbers of polytopes. Meisinger [39] showed that this is false and, in fact, if we write $g_1[r](P) = f_r(P) - \binom{d+1}{r}$ then while $g_1[r]$ is nonnegative for every d-polytope, this inequality does *not* follow from the nonnegativity of the g_i's. So the problem of finding all linear inequalities for flag numbers of polytopes is wide open. We give in this section some conjectures about new inequalities of this type.

We suspect that the lower bound inequalities, which for simplicial polytopes are consequences of the nonnegativity of g_2, correspond to independent inequalities for general polytopes. We state now what seems to be the right "analogs" of the lower bound inequalities for general polytopes, and present a general conjecture which corresponds to the generalized lower bound inequalities.

Let $g_k^r(P) = \sum\{g_k(F) : F \in P_k\}$. Thus $g_0^r(P) = f_r(P)$. Recall that $\phi_k(n,d)$ is the number of k-faces of stacked d-polytopes with n vertices. Here is an extension of the lower bound inequalities for general polytopes. There is some hope that a certain rigidity type argument may be useful for a proof.

Conjecture 17 *Let P be a d-polytope (and more generally an abstract polytope). Then for $k < d-1$,*

$$f_k(P) + g_1^k(P) + g_1^{k+1}(P) \geq \phi_k(n,d),$$

and for $k = d-1$,

$$f_{d-1}(P) + g_1^{d-1}(P) \geq \phi_{d-1}(n,d).$$

Equality holds if and only if P is an elementary polytope.

Note that the case $k = 1$ is just the nonnegativity of $g_2(P)$.

We describe now a more general conjecture. Let P be a simplicial polytope with $g_k(P) = 0$. For the class of all such polytopes, all face numbers are determined by $f_0(P), \ldots, f_{k-1}(P)$. Define $\alpha_k(r,i)$ such that, for simplicial d-polytopes with vanishing g_k, we have $f_r = \sum_{i=0}^{k} \alpha_k(r,i) f_{i-1}$. ($\alpha_k(r,i)$ is determined uniquely.) For arbitrary simplicial d-polytopes P, one gets the inequalities

$$f_r \geq \sum_{i=0}^{k} \alpha_k(r,i) f_{i-1}. \tag{11}$$

In the simplicial case, these inequalities do not contribute anything new. They follow from the nonnegativity of the g-numbers. Moreover, similar to the case for the lower bound inequalities, they also follow from the nonnegativity of the g_i's, $i < r$, for the polytope and its quotients.

Remark: In order to get the $\alpha_k(r,i)$ explicitly, first expand f_r in terms of the h_i's, by the defining relations. Next, use the relations $h_i \geq h_k$ for $i, d-k \geq i \geq k$, and $h_i = h_{d-i}$ for $i > d-k$. This gives an inequality $f_r \geq \sum_{i=0}^{k} \beta_k(r,i) h_i$. Finally, expand the h_i's back in terms of the f_i's to obtain $f_r \geq \sum_{i=0}^{k} \alpha_k(r,i) f_{i-1}$.

We will consider now general polytopes:

Definition: Let P be an arbitrary d-polytope, and let $h(P) = (h_0, h_1, ..., h_d)$ be the h-vector of P. Define the *fake f-vector* of P by $\sum \hat{f}_{k-1}(P)(x-1)^{d-k} = \sum h_k x^{d-k}$.

Explicitly, one gets

$$\hat{f}_k = \sum_{i=0}^{k} \sum_{r=0}^{i} \binom{i}{r} g_i^{k+r}.$$

Note that the "fake number of edges" $\hat{f}_1$, is the number of edges in a framework based on P (denoted before by $f_1^+(P)$).

Now define a truncated version of the fake face number:

$$\hat{f}_k[m] = \sum_{i=0}^{m} \sum_{r=0}^{i} \binom{i}{r} g_i^{k+r}.$$

Conjecture 18

$$\hat{f}_r[k] \geq \sum_{i=0}^{k} \alpha_k(r,i) \hat{f}_i.$$

Remark: Another class of conjectures, for linear inequalities of flag numbers of polytopes, was suggested by Stanley. Let P be an Eulerian poset of rank n. The cd-index [42] of P associates for every word w, in noncommuting variables c and d such that the number of c's plus twice the number of d's is n, a certain linear combination of flag numbers of P denoted by $\Phi_P(w)$. Stanley conjectured that for P's, which are face-lattices of $(n-1)$-dimensional polytopes, the value of this linear combination of flag numbers is minimized precisely when P is a Boolean algebra (that is, the face lattice of a simplex).

4.6. Centrally symmetric polytopes, cubical polytopes, kupitopes and other classes of polytopes

It is of interest to study the combinatorial structure of polytopes in special classes of polytopes. We decribe here a few such classes.

– The class of polytopes, which were studied the most, are the class of centrally symmetric polytopes. There are known lower bound theorems for simplicial centrally symmetric polytopes [46]. But the proofs are non elementary and do not extend to more general structures. For general centrally symmetric polytopes, there are some partial results [19]. But even the simple question, are there always at least 3^d proper faces, is open [29]. For more information, see [8].

- Kupitopes are polytopes with no triangular 2-faces. Kupitz studied this class of polytopes and conjectured that the number of r-faces is at least the number of r-faces of the cube. It was quite a while before Blind and Blind [16] proved Kupitz' conjecture. It seem plausible that in analogy with inequality $g_2(P) \geq 0$ for general polytopes,

  Conjecture 19 *For every d-kupitope P,*

  $$f_1 + 1/2 \sum \{f_0(F) - 4 : F \in P_2\} \geq (d+1)/2 f_0 - 2^{d-1}. \qquad (12)$$

  In particular, for every cubical d-polytope,

  $$f_1 \geq (d+1)/2 f_0 - 2^{d-1}.$$

  Some variants of rigidity theory may be helpful here. This is part of a general concept of h-numbers for cubical polytopes and kupitopes introduced by Adin [2].
- Polytopes without r-faces which are simplices. This may be useful for the study of Conjectures 15 and 16.
- A class $\mathcal{M}$ of polytopes which are of interest is the class of polytopes defined by system of linear inequalities, each of which has the form $x_i \leq a x_j + b$. In the context of linear programming, these classes were studied by Megiddo [38] and others. But it seems that their combinatorial structure was not studied. Faces of polytopes in $\mathcal{M}$ are also in $\mathcal{M}$.
- Balanced d-polytopes of type $(k_1, k_2, \ldots, k_t)$ are simplicial polytopes whose ertices can be colored with t colors such that each facet contains exactly k_i vertices of colors i. Of particular interest are balanced polytopes of type $(1, 1, \ldots, 1)$, which are called completely balanced. Duals of completely balanced polytopes are precisely the simple polytopes with 2-chromatic graphs, or in other words, precisely the simple polytopes all whose 2-faces have even sides. For an extension of h-vector theory to this setting, see [43].
- Charney and Davis considered simplicial complexes with no empty simplices of dimension greater than 1, and called them flag complexes. They made exciting conjectures concerning face numbers of flag polytopes and spheres. (see [17]).
- Another class of polytopes which are of interest are polytopes with the property that every k-face has at most Ck facets.

4.7. h-VECTORS FOR MORE EXOTIC STRUCTURES

As we have seen, h-vectors and g-vectors play a crucial role in the study of polytopes and related combinatorial structures. It was suggested that these concepts can be extended to much more general classes of combinatorial objects. The extension from simplicial polytopes to general polytopes is instructive. What is needed is to add extra terms measuring the amount by which the faces are not simplices.

One direction would be to define h-vectors for arbitrary simplicial manifolds and even pseudomanifolds. For manifolds, one can expect that the "correcting terms" will be in terms of the Betti numbers. (See [25].) For pseudomanifolds, we can expect some terms of Betti numbers of links of faces.

Another direction proposed by Bjorner [12] is to give a definition of arbitrary regular cell decomposition of spheres. The definition and properties of h-vectors of general polytopes are expected to apply to regular cell decomposition of spheres whose faces form a lattice. For regular cell decomposition of spheres without the lattice property, one expects some correction terms for the non lattice property, but so far no one has been able to come up with a reasonable definition even for h_1. Such h-vectors should include as special cases, the Kazhdan-Lusztig polynomials.

h-vectors play, in the combinatorial theory of structures considered here, a role similar to the role of zeta functions in number theory. (This is not a totally artificial analogy since in some special cases, the generating function of the h numbers *is* a zeta function of some variety.) In simple cases, the definition is obvious but proving the basic properties is hard. In more general cases, the main challenge is to find the right definition.

References

1. Adin, R., *Combinatorial structure of simplicial complexes with symmetry*, Ph.D Thesis, Jerusalem, 1991.
2. Adin, R., *Cubical h-vectors*, preprint.
3. Altshuler, A. and Perles, M., *Quotient polytopes of cyclic polytopes, Part I: Structure and characterization*, Isr. J. Math., 36(1980), 97-125.
4. Barany, I. and Lovasz, L., *Borsuk's theorem and the number of facets of centrally symmetric polytopes*, Acta Math. Acad. Sci. Hung. 40(1982), 323-329.
5. Barnette, D., *A proof of the lower bound conjecture for convex polytopes*, Pac. J. Math. 46 (1971), 349-354.
6. Bayer, M. and Billera, L. J., *Generalized Dehn-Sommerville relation for polytopes, spheres and Eulerian partially ordered sets*, Invent. Math. 79 (1985), 143-157.
7. Bayer, M. and Klapper, A., *A new index for polytopes*, Discrete Comp. Geometry 6(1991), 33-47.
8. Bayer, M., *Face numbers and subdivisions of convex polytopes*, THIS VOLUME pp. 155-171.
9. Bezdek, A., Bezdek, K., Makai, E. and McMullen, P., *Facets with fewest vertices*, Monastach Math. 109(1990), 89-96.
10. Billera, L. J. and Lee, C. W., *A proof of the sufficiency of McMullen's conditions for f-vectors of simplicial convex polytopes*, J. Combin. Th. Ser. A 31 (1981), 237-255.
11. Björner, A., *On complements in lattices of finite length*, Discrete Math. 36(1981), 325-326.
12. Björner, A., *Face numbers of complexes and polytopes*, Proc. International Congress of Math., Berkeley, 1986.
13. Björner, A. and Kalai, G., *Extended Euler Poincaré relations*, Acta Math. 161 (1988), 279-303.
14. Björner A. and Kalai, G., *On f-vectors and homology, Combinatorial Mathematics* (G.S. Bloom, R.L. Graham and J. Malkevitch, eds.) Proc. NY Academy of Science 555(1989), 63-80.

15. Björner, A. and Kalai, G., *Extended Euler Poincaré relations for regular cell complexes*, in: Applied Geometry and Discrete Mathematics, The Klee Festschrift, DIMACS Series in Discrete Mathematics and Computer Science 4(1991).
16. Blind, G. and Blind, R., *Convex polytopes without triangular faces*, Isr. J. Math. 71(1990), 129-134.
17. Charney, R. and Davis, M., *The Euler characteristics of nonpositively curved, piecewise linear Euclidean manifolds*, preprint.
18. Erdős, P. and Rado, R., *Intersection theorems for system of sets*, J. London Math. Soc. 35(1960), 85-90.
19. Figiel, T., Lindenstrauss, J. and Milman, V., *The dimension of almost spherical sections of convex bodies*, Acta Math. 139(1977), 53-94.
20. Goodman, J. and Pollack, R., *There are asymptotically far fewer polytopes than we thought*, Bull. Amer. Math. Soc. 14(1986), 127-129.
21. Gromov, M., *Structures métriques pour les variétés Riemanniens*, Paris, CEDEX, 1981.
22. Grünbaum, B., *Convex Polytopes*, Wiley Interscience, London, 1967.
23. Kalai, G., *A characterization of f-vectors of families of convex sets in R^d, Part I: Necessity of Eckhoff's conditions*, Israel J. Math. 48 (1984), 175-195.
24. Kalai, G., *Open Problems: Collected open problems* Oberwolfach Convexity meeting, 1986.
25. Kalai, G., *Rigidity and the lower bound theorem I*, Invent. Math. 88(1987), 125-151.
26. Kalai, G., *A new basis of polytopes*, J. Comb. Th. (Ser. A), 49(1988), 191-209.
27. Kalai, G., *A simple way to tell a simple polytope from its graph*, J. Comb. Th. (Ser. A) 49(1988), 381-383.
28. Kalai, G., *Many triangulated spheres*, Discrete and Computational Geometry, 3(1988), 1-14.
29. Kalai, G., *On the number of faces of centrally-symmetric convex polytopes* (research problem), Graphs and Combinatorics, 5 (1989), 389-391.
30. Kalai, G., *On low-dimensional faces that high-dimensional polytopes must have*, Combinatorica 10(1990), 271-280.
31. Kalai, G., *The diameter of graphs of convex polytopes and f-vector theory*, in: Applied Geometry and Discrete Mathematics, The Klee Festschrift, DIMACS Series in Discrete Mathematics and Computer Science 4(1991), 387-411.
32. Kalai, G., *Algebraic shifting*, to appear in Proc. Jerusalem Combinatorics Conference.
33. Klee, V., *A combinatorial analogue of Poincaré's duality theorem*, Can. J. Math. 16(1964), 517-531.
34. Lee, C., *Generalized stresses and motions*, THIS VOLUME pp. 249-271.
35. McMullen, P. and Walkup, D., *A generalized lower bound conjecture for convex polytopes*, Mathematica 18 (1971), 264-273.
36. McMullen, P., *The number of faces of simplicial polytopes*, Israel J. Math. 9(1971), 559-570.
37. McMullen, P., *On simple Polytopes*, Inven. Math. 113(1993), 419-444.

38. Megiddo, N., *Toward a genuinely polynomial algorithm for linear programming*, SIAM J. Comp. 12(1983), 347-353.
39. Meisinger, G., *Flag numbers and quotients of convex polytopes*, Ph.D Thesis, Passau University, 1993.
40. Perles, M., *Truncation of atomic lattices*, unpublished manuscript, around 1970.
41. Stanley, R., *A monotonicity property of h-vectors and h*-vectors*, European J. Combin. 14 (1993), 251-258.
42. Stanley, R., *A Survey of Eulerian posets*, THIS VOLUME, pp. 301-333.
43. Stanley, R., *Balanced Cohen-Macaulay complexes*, Trans. Amer. Math. Soc., 249, (1979), 139-157.
44. Stanley, R., *Flag vectors and the CD-index*, Math. Z., to appear.
45. Stanley, R., *Generalized h-vectors, intersection cohomology of toric varieties, and related results*, in Commutative Algebra and Combinatorics, (M. Nagata and H. Matsumura, eds.), Advanced Studies in Pure Mathematics 11, Kinokuniya, Tokyo, and North-Holland, Amsterdam/New York, 1987, pp. 187-213.
46. Stanley, R., *On the number of faces of centrally symmetric simplicial polytopes*, Graphs and Combinatorics 3(1987), 55-66.
47. Stanley, R., *Subdivisions and local h-vectors*, J. Amer. Math. Soc., 5(1992), 805-851.
48. Stanley, R., *The number of faces of simplicial convex polytopes*, Adv. Math. 35(1980), 236-238.
49. Stanley, R., *The upper bound conjecture and Cohen-Macaulay ring*, Stud. Appl. Math. 54(1975), 135-142.
50. Tay, T-S., White, N. and Whiteley, W., *Skeletal rigidity of cell complexes*, to appear.
51. Whiteley, W., *Infinitesimal rigid polyhedra I. Statics of frameworks*, Trans. Amer. Math. Soc. 285(1984), 431-465.
52. Ziegler, G., *Lectures on Polytopes*, Springer, to appear.

ON VOLUMES OF NON-EUCLIDEAN POLYTOPES

RUTH KELLERHALS*
Max-Planck-Institut für Mathematik,
Gottfried-Claren-Strasse 26,
D-53225 Bonn,
Germany.

Abstract. The problem of calculating volumes of non-Euclidean polytopes was already studied by Lobachevsky and Schläfli and is of actual interest, in particular, in connection with the structure of volume spectra of hyperbolic space forms. A brief summary of these results and an introduction to the volume problem for hyperbolic polytopes are given. The covolumes of some hyperbolic five-dimensional manifolds are computed; these are built upon hyperbolic crystals of simple combinatorial metrical type, and their volumes are all commensurable with Riemann's zeta function at three.

Key words: Hyperbolic geometry, volumes of hyperbolic polytopes, polylogarithms

1. Introduction

For Euclidean polytopes, volumes are comparatively easy to compute; the volume of a simplex is - up to a dimension factor - the positively taken determinant of the matrix of its spanning vectors. This is not the case anymore for non-Euclidean polytopes in so far as the computation of their volumes involves transcendental functions such as polylogarithms.
Nevertheless, Lobachevsky, around 1836 (cf. [L, 1898]), solved completely the volume problem for hyperbolic polyhedra by expressing the volume of a 3-dimensional orthoscheme in terms of its angles, making use of a function related to Euler's Dilogarithm (see **Appendix**). In the spherical case, Schläfli, in 1852 (cf. [S, 1950]), studied intensively the normalized volume functional f_n on arbitrary n-simplices and, especially, on orthoschemes, and derived a variety of fundamental properties; for example, he proved a nice and important formula for the volume differential df_n, a reduction principle for even dimensional simplices and, finally, he computed explicitly volumes of Coxeter orthoschemes.
Around 1935 (cf. [C, 1935]), Coxeter combined results of Lobachevsky and Schläfli inspiring Böhm, Müller and others to generalize methods in order to study the volume problem in higher dimensions (cf. [K, 1992]).
In recent years, much interest arose again in volume computations for hyperbolic polytopes through the work of Jørgensen and Thurston. Volumes of hyperbolic n-dimensional space forms, which are actually topological invariants, are distributed according to the dimension n quite differently (cf. [K, 1991; pp. 328-331]). If $\mathbf{Vol}_n := \{\, \mathrm{vol}_n(M) \mid M = H^n/\Gamma\,,\, \Gamma < \mathrm{Iso}(H^n) \text{ discrete} \,\}$ denotes the volume spectrum for hyperbolic n-space forms, then Jørgensen and Thurston showed that $\mathbf{Vol}_3$

* Partially supported by the Swiss National Science Foundation

T. Bisztriczky et al. (eds.), POLYTOPES: Abstract, Convex and Computational, 231–239.

is a closed, non-discrete, well-ordered subset of $\mathbf{R}_+$ of order type ω^ω, with smallest element v, say. An open problem is to find a compact manifold M with $\mathrm{vol}_3(M) = v$. For $n \neq 3$, Wang proved that $\mathbf{Vol}_n \subset \mathbf{R}_+$ is discrete (for $n = 2m$, this is already contained in the Theorem of Gauss-Bonnet).
The goal of this paper is to study $\mathbf{Vol}_5$, only to some modest extent, however! We derive explicit volume formulae for hyperbolic 5-simplices in order to compute covolumes of certain arithmetic discrete reflection groups acting on $\overline{H^5}$.

Acknowledgements

This paper was written during a stay at MSRI in Berkeley (U.S.A.). I wish to thank the Institute for its hospitality.

2. Polytopes and their schemes

2.1. Some Generalities

Let $X \subset Y$ be either the Euclidean n-space, the n-sphere or the hyperbolic n-space with ambient space Y endowed with scalar product $<,>_Y$; in the hyperbolic case, we usually work in the model $Y = E^{n,1}$ of Lorentz space of signature $(n,1)$.
Denote by $P \subset X$ an acute-angled (that is, all dihedral angles different from $\frac{\pi}{2}$ are strictly less than $\frac{\pi}{2}$) convex polytope bounded by finitely many hyperplanes H_i with outer unit normal vectors $e_i \in Y$, $i \in I$, and Gram matrix $G(P) = (<e_i, e_j>_Y)_{i,j \in I}$. Many of the combinatorial and metrical properties of P can be read off from $G(P)$ (cf. [V, 1993; pp. 104-118]); for example, whether, in extended hyperbolic space $\overline{H^n}$, P is compact or of finite volume, and how many of its vertices are finite or not. However, if many of the dihedral angles $\alpha_{jk} = \angle(H_j, H_k)$ of P are right angles, it is most convenient to describe P by a *scheme* or *weighted graph* $\Sigma(P)$. To each hyperplane H_i (or normal vector e_i) corresponds a node n_i of $\Sigma(P)$, and if two hyperplanes H_j, H_k are not orthogonal, their nodes n_j, n_k are joined by a weighted edge; for example, if H_j, H_k intersect at angle α_{jk}, then n_j, n_k are connected by an edge with weight α_{jk}. In the case of Coxeter polytopes (all dihedral angles are of the form $\frac{\pi}{p}$, $p \in \mathbf{N}$, $p \geq 2$), we use the standard notation.
$\Sigma(P)$ is called *elliptic, parabolic* or *hyperbolic*, if P is spherical, euclidean or hyperbolic; this is equivalent to saying that $G(P)$ is positive definite, positive semidefinite of rank n or of signature $(n,1)$.
The simplest n-dimensional polytopes are described by linear and cyclic schemes of order $n+1$, that is, having $n+1$ nodes. In the sequel, we shall study such simplices in $X = \overline{H^n}$.

2.2. Simplices with linear schemes

Linear hyperbolic schemes of order $n+1$ describe *n-orthoschemes* $R \subset \overline{H^n}$ determined by n acute angles $\alpha_i = \angle(H_{i-1}, H_i)$, $1 \leq i \leq n$, uniquely up to motions:

$$\Sigma(R) \quad : \quad \circ\overset{\alpha_1}{\text{———}}\circ\text{———}\cdots\text{———}\circ\overset{\alpha_n}{\text{———}}\circ \quad .$$

Let P_i denote the vertex of R opposite to H_i. At most P_0 and P_n may be points at

infinity (at all other vertices one always finds right planar angles); hence, the scheme $\Sigma(R)$ contains one or two parabolic subschemes of order n, and R is called simply or doubly asymptotic.
Orthoschemes are the most elementary objects with respect to many properties; for example, they generate the scissors congruence groups. In particular, for odd dimensions $n \geq 3$, the scissors congruence groups of H^n and $\overline{H^n}$ are isomorphic, and the latter are generated by doubly asymptotic orthoschemes (cf. [K, 1991; pp. 331-332]).
All Coxeter orthoschemes are classified. The list of hyperbolic Coxeter orthoschemes ends in dimension five where we find precisely the following examples

σ_1 :  ,

σ_2 :  ,

σ_3 :  .

Notice that σ_1 is simply asymptotic while σ_2, σ_3 are doubly asymptotic.

2.3. Simplices with cyclic schemes

Cyclic hyperbolic schemes of order $n+1$ are simplices $Q \subset \overline{H^n}$ bounded by hyperplanes $H_0(=: H_{n+1}), H_1, \ldots, H_n$ with $n+1$ acute dihedral angles $\alpha_i = \angle(H_{i-1}, H_i)$, $1 \leq i \leq n+1$, determining Q uniquely up to isometry:

$\Sigma(Q)$: 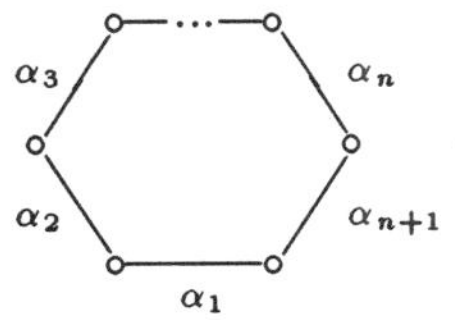

.

Q has vertex figures which are orthoschemes, but all its face schemes are cyclic. Moreover, all vertices of Q may lie at infinity. We say that Q is $k-$asymptotic if k of its $n+1$ vertices are infinite points.
The Coxeter simplices with cyclic graphs also exist just up to dimension five. In the top dimension we are left with

ρ_1 : 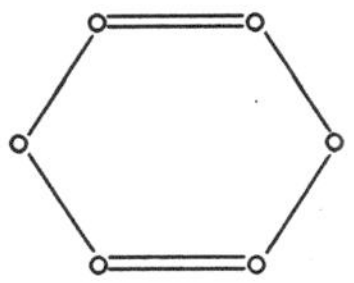 ,

ρ_2 : 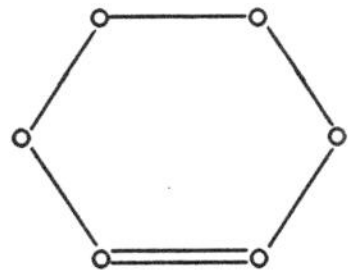 .

Notice that ρ_1 is 6– and ρ_2 is 2–asymptotic.

Another family of hyperbolic n-simplices $T \subset \overline{H^n}$ is described by the opposite, namely by star-shaped schemes of order $n+1$, determined by n acute angles uniquely up to isometry. There are very few Coxeter simplices among them; the only remaining example in dimensions at least 5 is the 5-asymptotic crystal of dimension five,

τ :

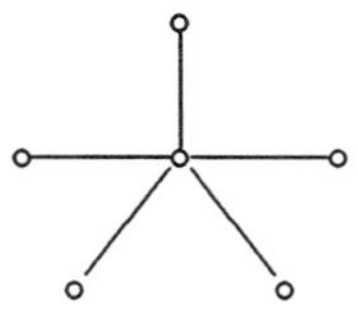

.

2.4. Dissection relations

On the scissors congruence level, there are the relations

$$\sigma_3 = 2\sigma_2 = 10\sigma_1 \quad ,$$

$$\tau = 24\sigma_2 \quad ,$$

arising from different dissections of the totally asymptotic regular cross-polytope in $\overline{H^5}$.

3. Volume computations

3.1. Schläfli's volume differential

The main tool for volume computations is Schläfli's formula expressing the volume differential in terms of infinitesimal angle perturbations. Its hyperbolic analogue is as follows:

THEOREM 1.

Let $S \subset H^n$, $n \geq 2$, denote a family of n-simplices with dihedral angles α_{jk} attached to the codimension two faces S_{jk}, $0 \leq j, k \leq n$, $j \neq k$. Then

$$d\,\mathrm{vol}_n(S) = \frac{1}{1-n} \sum_{j \neq k} \mathrm{vol}_{n-2}(S_{jk})\, d\,\alpha_{jk} \quad , \quad \mathrm{vol}_0(S) := 1 \quad . \qquad (*)$$

In the planar case, the integration of Schläfli's differential $(*)$ leads immediately to the well-known defect formula for hyperbolic triangles. Moreover, the inductive character of Schläfli's formula shows that there is a basic difference in dealing with volumes of polytopes of even or odd dimensions.

3.2. The reduction principle

For even dimensions $2m$, the successive integration of $(*)$ leads to a reduction property for the volumes of $2m$-dimensional polytopes P discovered first by Schläfli (cf. [K, 1991; pp. 310-311]); the volume of P is reducible to a sum of volumes, of certain of its lower (odd) dimensional vertex figures, multiplied by some coefficients. Which particular vertex figures of P and what coefficients appear in the reduction formula depend on the combinatorial and metrical properties of P. For example, for simplices with non-right dihedral angles, one has to sum over *all* (odd dimensional) vertex figures, and the coefficients are the tangent numbers.

3.3. The odd dimensional case

Therefore, it remains to consider the volume problem for non-Euclidean polytopes of odd dimensions. In the three-dimensional case, Lobachevsky derived a closed formula for the volume of a hyperbolic 3-orthoscheme R in terms of its angles $\alpha_1, \alpha_2, \alpha_3$; for this he introduced a transcendental function related to Euler's Dilogarithm, which nowadays is called the Lobachevsky-function:

$$\text{Л}(\alpha) = \frac{1}{2}\,\text{Im}\,(\text{Li}_2(e^{2i\alpha})) = -\int_0^\alpha \log|2\sin t|\,dt \quad .$$

The volume formula is as follows (cf. [L, 1898]):

$$\text{vol}_3(R) = \frac{1}{4}\{\text{Л}(\alpha_1+\theta) - \text{Л}(\alpha_1-\theta) + \text{Л}(\frac{\pi}{2}+\alpha_2-\theta) + \text{Л}(\frac{\pi}{2}-\alpha_2-\theta) + $$
$$+\text{Л}(\alpha_3+\theta) - \text{Л}(\alpha_3-\theta) + 2\,\text{Л}(\frac{\pi}{2}-\theta)\} \qquad (1)$$

where $0 \leq \theta \leq \frac{\pi}{2}$ is such that

$$\tan^2\theta = \frac{|\det G(R)|}{\cos^2\alpha_1\cos^2\alpha_3} = \frac{\cos^2\alpha_2 - \sin^2\alpha_1\sin^2\alpha_3}{\cos^2\alpha_1\cos^2\alpha_3} \quad .$$

In particular, for a doubly asymptotic orthoscheme

$$\Sigma(R) \quad : \quad \circ\overset{\alpha}{\text{———}}\circ\overset{\alpha'}{\text{———}}\circ\overset{\alpha}{\text{———}}\circ \quad , \quad \alpha' := \frac{\pi}{2} - \alpha \quad ,$$

one obtains $\text{vol}_3(R) = \frac{1}{2}\,\text{Л}(\alpha)$. Moreover, by dissection of the 4−asymptotic simplex Q with graph (see 2.3.)

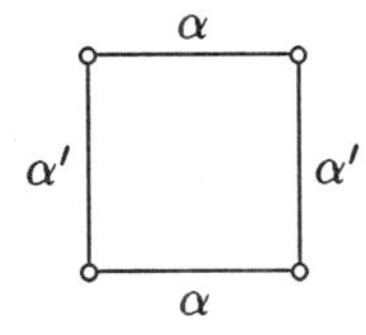

into orthoschemes, we derive the formula

$$\text{vol}_3(Q) = \text{Л}(\alpha) + \text{Л}(\alpha') \qquad (2)$$

3.4. Volumes of five-dimensional simplices

By means of Schläfli's differential (∗) and Lobachevsky's formula (1) we can proceed to study the volume problem for hyperbolic simplices of dimension five. Again, it is sufficient to consider doubly asymptotic orthoschemes R (see 2.2.)

$$\Sigma(R) \quad : \quad \circ\overset{\alpha_1}{\text{———}}\circ\overset{\alpha_2}{\text{———}}\circ\overset{\alpha_3}{\text{———}}\circ\overset{\alpha_4}{\text{———}}\circ\overset{\alpha_5}{\text{———}}\circ \quad ,$$

having therefore the parabolic subschemes

$$\circ\overset{\alpha_1}{\text{———}}\circ\overset{\alpha_2}{\text{———}}\circ\overset{\alpha_3}{\text{———}}\circ\overset{\alpha_4}{\text{———}}\circ \qquad \text{and} \qquad \circ\overset{\alpha_2}{\text{———}}\circ\overset{\alpha_3}{\text{———}}\circ\overset{\alpha_4}{\text{———}}\circ\overset{\alpha_5}{\text{———}}\circ \quad .$$

This parabolicity property is equivalent to the following set of relations:

$$\tan\alpha_1 \cot\alpha_4 = \cot\alpha_2 \tan\alpha_5 = \tan\alpha_3 \cot\alpha_6 = \tan\Theta =: \lambda$$

where $0 < \alpha_6 < \frac{\pi}{2}$ is such that the scheme

$$\circ\overset{\alpha_3}{\text{———}}\circ\overset{\alpha_4}{\text{———}}\circ\overset{\alpha_5}{\text{———}}\circ\overset{\alpha_6}{\text{———}}\circ$$

is parabolic, and $0 \le \Theta \le \frac{\pi}{2}$ is given by

$$\tan^2\Theta = \frac{|\det G(R)|}{\cos^2\alpha_1 \cos^2\alpha_3 \cos^2\alpha_5} \quad ,$$

analogously to the three dimensional case.

The case $\lambda = 1$, that is, $\alpha_1 = \alpha_4 =: \alpha\,,\, \alpha_2 = \alpha_5 =: \beta\,,\, \alpha_3 = \alpha_6 =: \gamma$ with $\cos^2\alpha + \cos^2\beta + \cos^2\gamma = 1$, can be treated easily. The integration of the volume differential (∗) for R leads, in terms of the Trilobachevsky function (see **Appendix**), to the formula (cf. [K, 1992; Theorem, p.659])

THEOREM 2.

Let R denote the doubly asymptotic 5-orthoscheme given by

$$\Sigma(R) \quad : \quad \circ\overset{\alpha}{\text{———}}\circ\overset{\beta}{\text{———}}\circ\overset{\gamma}{\text{———}}\circ\overset{\alpha}{\text{———}}\circ\overset{\beta}{\text{———}}\circ \quad \text{with} \quad \cos^2\alpha + \cos^2\beta + \cos^2\gamma = 1 \quad .$$

Then,

$$\mathrm{vol}_5(R) = \frac{1}{4}\{\mathrm{JI}_3(\alpha) + \mathrm{JI}_3(\beta) - \frac{1}{2}\mathrm{JI}_3(\frac{\pi}{2} - \gamma)\} - \frac{1}{16}\{\mathrm{JI}_3(\frac{\pi}{2} + \alpha + \beta) + \\ + \mathrm{JI}_3(\frac{\pi}{2} - \alpha + \beta)\} + \frac{3}{64}\zeta(3) \quad . \tag{3}$$

Therefore,

$$\mathrm{vol}_5(\sigma_2) = \frac{7}{9216}\zeta(3) \quad , \quad \mathrm{vol}_5(\sigma_3) = \frac{7}{4608}\zeta(3) \quad ,$$

and, finally, by 2.4.,

$$\mathrm{vol}_5(\sigma_1) = \frac{1}{5}\,\mathrm{vol}_5(\sigma_2) = \frac{1}{10}\,\mathrm{vol}_5(\sigma_3) = \frac{7}{46080}\,\zeta(3) \quad , \quad \text{as well as}$$

$$\mathrm{vol}_5(\tau) = 24\,\mathrm{vol}_5(\sigma_2) = \frac{7}{384}\,\zeta(3) \quad .$$

The general case $\lambda \neq 1$ is much more complicated and will be treated elsewhere.

To compute the volume of a general acute-angled simplex, one can always dissect it into orthoschemes by drawing successively perpendiculars to faces of lower dimensions, starting from a vertex, say (see 2.2.). Sometimes, however, it is easier to go the direct way using the preceding integration method. This is how we proceed for hyperbolic cycles of order 6,

$\Sigma(Q)$:

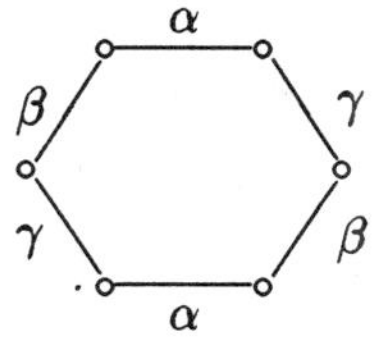

, $\cos^2\alpha + \cos^2\beta + \cos^2\gamma = 1$,

describing totally asymptotic simplices $Q \subset \overline{H^5}$ (see 2.3.). Since the codimension two faces F_ω , $\omega \in \{\alpha, \beta, \gamma\}$, of Q are described by schemes

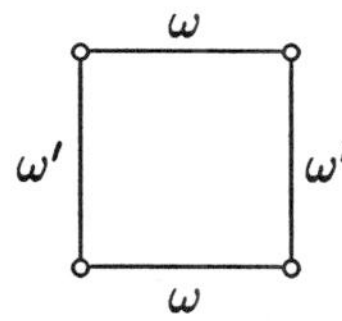

,

whose volumes are given by (2), we obtain, by integrating the volume differential (*), the following result:

THEOREM 3.
Let Q denote a totally asymptotic 5-simplex given by

$\Sigma(Q)$:

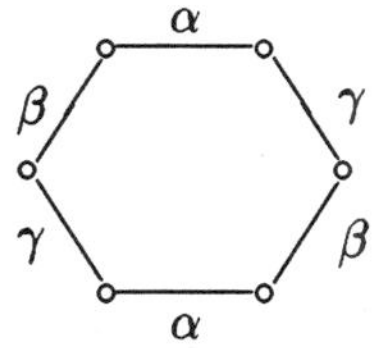

, $\cos^2\alpha + \cos^2\beta + \cos^2\gamma = 1$.

Then,

$$\mathrm{vol}_5(Q) = \frac{1}{2}\,\{\mathrm{JI}_3(\alpha) + \mathrm{JI}_3(\beta) + \mathrm{JI}_3(\gamma) - \mathrm{JI}_3(\frac{\pi}{2} - \alpha) - $$
$$- \mathrm{JI}_3(\frac{\pi}{2} - \beta) - \mathrm{JI}_3(\frac{\pi}{2} - \gamma)\} + \frac{7}{32}\,\zeta(3) \quad . \qquad (4)$$

In particular, for the Coxeter simplex ρ_1, we find that $\mathrm{vol}_5(\rho_1) = \frac{7}{288}\,\zeta(3)$.

3.5. Some information about Vol_5

The hyperbolic Coxeter simplices that we considered are fundamental domains of discrete reflection groups acting on $\overline{H^5}$. By a result of Borel (cf. [K, 1992; p.662]), each of these groups contains a torsion-free subgroup of finite index so that the quotient of $\overline{H^5}$ by this subgroup is a cusped hyperbolic manifold of dimension five. The spaces arising from the Coxeter simplices $\sigma_1, \sigma_2, \sigma_3, \rho_1, \tau$ have volumes which are certain rational multiples of $\zeta(3)$ (see 3.4.). In fact, τ describes an *arithmetic* reflection group, and the schemes $\sigma_1, \sigma_2, \sigma_3, \rho_1$ are precisely the *arithmetic* groups among the discrete reflection groups with linear and cyclic diagrams of order 6; since their fundamental polytopes are noncompact, these arithmetic groups are defined over the ground field $\mathbf{Q}$ (cf. [V, 1993; pp.217-222 and 226-228] for definitions and the arithmeticity criterion for hyperbolic reflection groups).

Appendix

The classical polylogarithms $\mathrm{Li}_n(z)$, $z \in \mathbf{C}$, $|z| \leq 1$, are functions defined iteratively from the ordinary logarithm

$$\mathrm{Li}_1(z) := -\log(1-z) \qquad \text{by}$$

$$\mathrm{Li}_n(z) := \int_0^z \frac{\mathrm{Li}_{n-1}(t)}{t}\, dt \quad , \quad n \geq 2 \quad ,$$

or, by

$$\mathrm{Li}_n(z) = \sum_{r=1}^{\infty} \frac{z^r}{r^n} \quad ,$$

and satisfy $\mathrm{Li}_n(1) = \zeta(n)$, as well as

$$\frac{1}{k^{n-1}} \mathrm{Li}_n(z^k) = \mathrm{Li}_n(z) + \mathrm{Li}_n(\omega z) + \cdots + \mathrm{Li}_n(\omega^{k-1} z) \quad ,$$

where $\omega = e^{2\pi i/k}$, $k \geq 1$.

The Lobachevsky function, related to the Dilogarithm by

$$\text{Л}(\alpha) = \frac{1}{2} \mathrm{Im}\,(\mathrm{Li}_2(e^{2i\alpha})) = -\int_0^{\alpha} \log|2\sin t|\, dt \quad ,$$

admits a geometrical interpretation as hyperbolic 3-volume (see 3.3.). It can be generalized to higher order according to

$$\text{Л}_{2m}(\alpha) = \frac{1}{2^{2m-1}} \mathrm{Im}(\mathrm{Li}_{2m}(e^{2i\alpha})) \quad ,$$

$$\mathrm{JI}_{2m+1}(\alpha) = \frac{1}{2^{2m}} \mathrm{Re}(\mathrm{Li}_{2m+1}(e^{2i\alpha})) \quad ,$$

with the properties

$$\mathrm{JI}_{2m}(\alpha) = \int_0^\alpha \mathrm{JI}_{2m-1}(t)dt \quad , \quad \mathrm{JI}_{2m+1}(\alpha) = \frac{1}{2^{2m}} \zeta(2m+1) - \int_0^\alpha \mathrm{JI}_{2m}(t)dt \quad .$$

Moreover, $\mathrm{JI}_m(\alpha)$ is π−periodic, even (odd) for m odd (even) and distributes according to

$$\frac{1}{k^{m-1}} \mathrm{JI}_m(k\alpha) = \sum_{r=0}^{k-1} \mathrm{JI}_m(\alpha + \frac{r\pi}{k}) \quad .$$

Therefore, its values at crystallographic entries $\frac{\pi}{n}$, $n = 2, 3, 4, 6$, are commensurable with $\zeta(m) = \mathrm{Li}_m(1)$. For example, for $m = 3$, we find

$$\mathrm{JI}_3(\frac{\pi}{2}) = -\frac{3}{16}\zeta(3) \quad , \quad \mathrm{JI}_3(\frac{\pi}{3}) = -\frac{1}{9}\zeta(3) \quad ,$$

$$\mathrm{JI}_3(\frac{\pi}{4}) = -\frac{3}{128}\zeta(3) \quad , \quad \mathrm{JI}_3(\frac{\pi}{6}) = \frac{1}{12}\zeta(3) \quad .$$

References

[C, 1935] Coxeter, H.S.M. (1935): 'The functions of Lobatschefsky and Schläfli', *Quart. J. Math. Oxford* **6**, 13–29.

[K, 1991] Kellerhals, R. (1991): 'The Dilogarithm and volumes of hyperbolic polytopes', in: *Structural Properties of Polylogarithms*, Leonard Lewin, Editor, AMS Mathematical Surveys and Monographs **37**.

[K, 1992] Kellerhals, R. (1992): 'On the volumes of hyperbolic 5-orthoschemes and the Trilogarithm', *Comment. Math. Helvetici* **67**, 648–663.

[L, 1898] Lobachevskij, N.I. (1898): *Zwei geometrische Abhandlungen*, Teubner, Leipzig.

[S, 1950] Schläfli, L. (1950): 'Theorie der vielfachen Kontinuität', in: *Gesammelte Mathematische Abhandlungen* Vol.no. **1**, Birkhäuser, Basel.

[V, 1993] Vinberg, E.B. (Ed.) (1993). *Geometry II*, Encyclopaedia of Mathematical Sciences Vol. 29, Springer, Berlin, Heidelberg, New York.

MANIFOLDS IN THE SKELETONS OF CONVEX POLYTOPES, TIGHTNESS, AND GENERALIZED HEAWOOD INEQUALITIES

WOLFGANG KÜHNEL
Fachbereich Mathematik
Universität Duisburg
47048 Duisburg
Germany

Abstract. If a 2-dimensional manifold in the 2-dimensional skeleton of a convex d-polytope P contains the 1-skeleton of P then d is bounded in terms of the genus of the surface: this is essentially Heawood's inequality. In this paper we prove a higher dimensional analogue about $2k$-dimensional manifolds containing the k-skeleton of a simplicial convex polytope. Related conjectures are formulated for tight polyhedral submanifolds and generalized Heawood inequalities, including an Upper Bound Conjecture for combinatorial manifolds.

A *polyhedral* n-*manifold* M^n is a finite cell complex whose i-dimensional cells are convex i-polytopes, such that the intersection of any two cells is either empty or a lower dimensional cell, and such that the vertex figure of each vertex is a polyhedral $(n-1)$-sphere. The latter condition guarantees that a typical neighbourhood of a vertex is a topological ball. A *polyhedral submanifold* of a convex d-polytope P is by definition a subcomplex of the boundary complex of P which is a polyhedral manifold.

Examples of 2-dimensional submanifolds of polytopes are the following:

1. Triangulated surfaces in the 2-skeleton of a higher dimensional simplex, in particular such with a complete edge graph [Ri2] ,
2. Coxeter's regular skew polyhedra $\{4, d \mid 4^{[d/2]-1}\}$ [Cox] , regarded as surfaces of type $\{4, d\}$ in the 2-skeleton of the d-cube containing the 1-skeleton of the d-cube.

These examples were of importance also for the investigation of tight submanifolds of Euclidean space. Tightness is a generalization of convexity — for a definition in general and basic facts compare [Kui1] or [Kui2] . In the special case of 2-manifolds M without boundary, we can define $M \hookrightarrow E^d$ to be *tight* if and only if every hyperplane cuts it into at most two pieces (Banchoff's *Two-Piece-Property*). For smooth tight surfaces in E^d, the substantial codimension is always at most 3. In the polyhedral case there are tight surfaces with arbitrarily high codimension. These are just the examples 1 and 2 [Ba1], [Ba2] according to the following lemma:

Lemma 1: *(i) If $M^2 \hookrightarrow E^d$ is a tight polyhedral surface then M contains the 1-skeleton of the convex hull $\mathcal{H}M$ of M.*

T. Bisztriczky et al. (eds.), POLYTOPES: Abstract, Convex and Computational, 241–247.

(ii) Moreover, if M^2 is a subcomplex of the boundary of its convex hull then the converse is also true:

$$M \ \ tight \quad \Leftrightarrow \quad Sk_1(\mathcal{H}M) \subseteq M\ .$$

The proof is more or less obvious from the definition. Observe that the 1-skeleton of a convex polytope certainly has the Two-Piece-Property. This is preserved if we add two-dimensional faces.

A subcomplex of the boundary complex of a polytope P may be called k-*Hamiltonian* if it contains the k-dimensional skeleton $Sk_k(P)$. In particular, Lemma 1 says that any 1-Hamiltonian 2-dimensional submanifold of a convex polytope is tight.

Theorem 1 [Kü4]: *Let M^2 be a 2-dimensional submanifold of a convex d-polytope P which is 1-Hamiltonian. Then the following holds:*

(i) $\binom{d-2}{2} \leq 3(2-\chi(M))$.

(ii) For $d \geq 4$, equality in (i) holds if and only if P is a simplex.

We shall not repeat the proof here, but just remark that (i) is essentially *Heawood's inequality*

$$\binom{d-2}{2} \leq 3(2-\chi) \quad \Leftrightarrow \quad d+1 \leq \frac{1}{2}\left(7+\sqrt{49-24\chi}\right)$$

where the integer part of the right hand side of the last inequality is known as the *Heawood colouring number* [Ri2] . Theorem 1 remains true under the weaker assumption that $M \hookrightarrow E^d$ is a tight polyhedral surface, not contained in any hyperplane [Ba1] , [Ba2] .

For the discussion of higher dimensional submanifolds, we remark that a k-Hamiltonian $2k$-dimensional submanifold of a convex polytope is necessarily $(k-1)$-connected in the sense of homotopy theory. On the other hand, for such a manifold, the tightness condition is easy to formulate: a $(k-1)$-connected $2k$-manifold embedded in E^d is called *tight* if every hyperplane cuts it into at most two pieces such that each piece is again $(k-1)$-connected.

Lemma 1 remains true for $(k-1)$-connected $2k$-manifolds of we just replace the 1-skeleton by the k-skeleton. In particular, any k-Hamiltonian $2k$-submanifold of a convex polytope is tight. Higher dimensional examples in the cube have been studied in the earlier paper [KS] . A particular consequence is that, for arbitrary $d \geq 2k+1$, there is a tight $(k-1)$-connected polyhedral $2k$-manifold in E^d.

Theorem 2: *Let M^{2k} be a $2k$-dimensional submanifold of a simplicial convex d-polytope P which is k-Hamiltonian. Then the following holds:*

(i) $\binom{d-k-1}{k+1} \leq (-1)^k\binom{2k+1}{k+1}(\chi(M)-2)$.

(ii) For $d \geq 2k+2$, equality in (i) holds if and only if P is a simplex.

We suggest calling the inequality in (i) a *generalized Heawood inequality.* Note that by assumption M is $(k-1)$-connected, and thus the right hand side of (i) is nonnegative:

$$(-1)^k(\chi(M)-2) = rk\ H_k(M;\mathbb{Z})\ .$$

In the case of equality in (ii), the submanifold must be a $(k+1)$-neighbourly triangulation. Examples exist in dimension $2k = 2, 4, 8$; see example 1 for $k = 1$, [KB] or [KL] for $k = 2$, and [BK2] for $k = 4$.

The assumption that P is simplicial is more of a technical nature. We conjecture that Theorem 2 is true for arbitrary convex polytopes. Note that (ii) does not hold for $d = 2k+1$ because the boundary of any d-polytope is an example of such a case.

Proof: The idea is to compare the h-vector of P with the h-vector of M. Recall that the f-vector $(f_{-1}, f_0, f_1, \ldots)$ consists of the numbers f_i of i-dimensional simplices, where formally $f_{-1} := 1$. We write $f(P)$ for the f-vector of P, $f(M)$ for the f-vector of M. By assumption, $f_i(M) = f_i(P)$ for $i = -1, 0, \ldots, k$. The h-vector $(h_0, h_1, \ldots)$ is defined by

$$h_j(P) = \sum_{i=-1}^{j-1} (-1)^{j-i-1} \binom{d-i-1}{j-i-1} f_i(P)$$

and

$$h_j(M) = \sum_{i=-1}^{j-1} (-1)^{j-i-1} \binom{2k-i}{j-i-1} f_i(M) \, .$$

The *Dehn-Sommerville equations* [Kl1] say that

$$h_j(P) - h_{d-j}(P) = 0 \qquad \text{for} \quad 0 \le j \le \frac{1}{2}(d-1)$$

and

$$h_j(M) - h_{2k+1-j}(M) = (-1)^{2k+1-j} \binom{2k+1}{j} (\chi(M) - 2) \quad \text{for} \quad 0 \le j \le k \, .$$

In particular,

$$h_{k+1}(M) - h_k(M) = (-1)^k \binom{2k+1}{k+1} (\chi(M) - 2) \, . \tag{1}$$

The most important ingredient of our proof is the *Generalized Lower Bound Theorem* [MW] , [St] :

$$h_{j+1}(P) - h_j(P) \ge 0 \qquad \text{for} \quad 0 \le j \le \frac{1}{2}(d-1) \, .$$

Another way of expressing this is

$$f_j(P) \ge \sum_{i=-1}^{j-1} (-1)^{j-i-1} \binom{d-i}{j-i} \cdot f_i(P) \, .$$

In order to prove the inequality in (i), we start with the equation (1) and then put in successively the inequalities of the Generalized Lower Bound Theorem for $j = k, k-1, \ldots, 0$. At each step, we get certain new coefficients $c^i_{j,d}$ for the f_i :

$$(-1)^k \binom{2k+1}{k+1} (\chi(M) - 2) \; = \; h_{k+1}(M) - h_k(M)$$

$$= \sum_{i=-1}^{k} (-1)^{k-i} \binom{2k+1-i}{k+1} \cdot f_i$$

$$\geq \sum_{i=-1}^{k-1} (-1)^{k-i-1} c_{k-1,d}^{i} \cdot f_i$$

$$\vdots$$

$$\geq \sum_{i=-1}^{j} (-1)^{j-i} c_{j,d}^{i} \cdot f_i$$

$$\vdots$$

$$\geq c_{1,d}^{1} \cdot f_1 - c_{1,d}^{0} \cdot f_0 + c_{1,d}^{-1}$$

$$\geq c_{0,d}^{0} \cdot f_0 - c_{0,d}^{-1}$$

$$\geq c_{0,d}^{0} \cdot (d+1) - c_{0,d}^{-1}$$

$$= c_{-1,d}^{-1}.$$

We still have to justify these inequalities by showing that all the coefficients $c_{j,d}^{j}$ are nonnegative.

The coefficients $c_{j,d}^{i}$ obey the following recursion formula:

$$c_{k,d}^{i} = \binom{2k+1-i}{k+1}$$

and

$$c_{j-1,d}^{i} = c_{j,d}^{j} \cdot \binom{d-i}{j-i} - c_{j,d}^{i} \qquad \text{for} \quad i < j\ .$$

By induction, we show that $\quad c_{i,d}^{i} = \binom{d-k-i-2}{k-i} \geq 0 \quad$ for $\quad i = -1, 0, \ldots, k$.

This is trivial for $i = k$ because $c_{k,d}^{k} = 1$.

Now we assume that the assertion holds for $j = i+1, i+2, \ldots, k$. By repeated applications of the recursion formula, we obtain

$$\begin{aligned} c_{i,d}^{i} &= c_{i+1,d}^{i+1} \binom{d-i}{1} - c_{i+1,d}^{i} \\ &= c_{i+1,d}^{i+1} \binom{d-i}{1} - c_{i+2,d}^{i+2} \binom{d-i}{2} + c_{i+2,d}^{i} \\ &\ \vdots \\ &= \sum_{j=i+1}^{k} (-1)^{j-i-1} c_{j,d}^{j} \binom{d-i}{j-i} + (-1)^{k-i} \binom{2k+1-i}{k+1} \end{aligned}$$

$$= \sum_{j=i+1}^{k} (-1)^{j-i-1} \binom{d-k-j-2}{k-j} \binom{d-i}{j-i} + (-1)^{k-i} \binom{2k+1-i}{k+1}$$

$$= \binom{d-k-i-2}{k-1} + \sum_{j=i}^{k} (-1)^{j-i-1} \binom{d-k-j-2}{k-j} \binom{d-i}{j-i} + (-1)^{k-i} \binom{2k+1-i}{k+1}$$

$$= \binom{d-k-i-2}{k-1} + (-1)^{k-i-1} \sum_{j=i}^{k} \binom{-d+2k+1}{k-j} \binom{d-i}{j-i} + (-1)^{k-i} \binom{2k+1-i}{k+1}$$

$$= \binom{d-k-i-2}{k-1} + (-1)^{k-i-1} \binom{2k-i+1}{k-i} + (-1)^{k-i} \binom{2k+1-i}{k+1} \qquad (2)$$

$$= \binom{d-k-i-2}{k-i}.$$

For the equation (2), see [Rn, 1.3] or [Kl1] .

Now part (i) of Theorem 2 follows from $c_{-1,d}^{-1} = \binom{d-k-1}{k+1}$.

For part (ii), we look at the inequality if it becomes an equality at each step. In particular, the last step

$$c_{0,d}^{0} \cdot f_0 - c_{0,d}^{-1} \geq c_{0,d}^{0} \cdot (d+1) - c_{0,d}^{-1}$$

tells us that equality implies $f_0 = d+1$ if $c_{0,d}^{0} = \binom{d-k-2}{k} > 0$.

This is true for $d \geq 2k+2$. Observe that this argument breaks down if $d = 2k+1$.

Conjecture A: *The assertion of Theorem 2 holds under the assumption that M^{2k} is $(k-1)$-connected and that $M \hookrightarrow E^d$ is a tight polyhedral embedding not lying in a hyperplane.*

Conjecture A is true for $k = 1$ and the bound is essentially sharp; see [Ba1] , [Kü1] . For arbitrary k, Theorem 2 is a special case of Conjecture A.

Conjecture B: *The assertion of Theorem 2 holds under the assumption that M^{2k} is $(k-1)$-connected and admits an embedding of the* k-*skeleton of the* d-*dimensional simplex.*

Again this is true for $k = 1$ by the discussion of the genus of the complete graph [Ri2] . Compare the theorem of van Kampen and Flores which says that the sphere S^{2k} does not admit an embedding of the k-skeleton of the d-simplex if $d \geq 2k+2$ [Grü, 11.1] . The complex projective plane does admit an embedding of the 2-skeleton of the 8-dimensional simplex [KB] .

Conjecture C: *For any triangulation of a manifold M^{2k} with* n *vertices, the following inequality holds:*

$$\binom{n-k-2}{k+1} \geq (-1)^k \binom{2k+1}{k+1} (\chi(M) - 2)$$

with equality if and only if the triangulation is (k+1)*-neighbourly; that is, if* $f_k(M) = \binom{n}{k+1}$.

Conjecture C is a weak form of an *Upper Bound Conjecture for combinatorial manifolds;* cf. [Kl2] . It is a consequence of a more general conjecture made by G. Kalai. Conjecture C is true if

- $k = 1$ or $k = 2$ ([Ri1] , [JR] , [Kü3]),
- M is a sphere (trivial),
- $n \leq 3k + 3$ ([BK1]),
- M has the homology of $S^j \times S^{2k-j}$, $j < k$ ([BK1]),
- M is a manifold like a projective plane in the sense of [EK] ([BK1]),
- $n > k^2 + 4k + 2$ (this holds by the same argument as in the case of the classical UBC, [Grü]).

In particular, any triangulation of a K3-surface must have at least 16 vertices. (There is a 16-vertex triangulation of the Kummer variety with 16 nodes in [Kü2].) Furthermore, any triangulation of the Cayley projective plane must have at least 27 vertices. It does not seem to be known whether these bounds are attained or not.

References

[Ba1] Banchoff, T.F., *Tightly-embedded 2-dimensional polyhedra manifolds*, Amer. J. Math. **87**, pp. 462–472, (1965).

[Ba2] Banchoff, T.F., *Tight polyhedral Klein bottles, projective planes, and Möbius bands*, Math. Ann., **207**, pp. 233–243, (1974).

[BK1] Brehm, U. and Kühnel, W., *Combinatorial manifolds with few vertices*, Topology, **26**, pp. 465–473, (1987).

[BK2] Brehm, U. and Kühnel, W., 15-*vertex triangulations of an*8-*manifold*, Math. Ann., **294**, 167–193, (1992).

[Cox] Coxeter, H.S.M., *Regular skew polyhedra in three and four dimensions and their topological analogues*, Proc. London Math. Soc., Ser. 2, **43**, pp. 33–62, (1937).

[EK] Eells, J. and Kuiper, N.H., *Manifolds which are like projective planes*, Publ. Math. I.H.E.S., **14**, pp. 181–222, (1962).

[Grü] Grünbaum, B., *Convex Polytopes*, Interscience Publ., New York, 1967.

[JR] Jungerman, M. and Ringel, G., *Minimal triangulations on orientable surfaces*, Acta Math., **145**, pp. 121–154, (1980).

[Kl1] Klee, V., *A combinatorial analogue of Poincaré's theorem*, Can. J. Math., **16**, pp. 517–531. (1964).

[Kl2] Klee, V., *On the number of vertices of a convex polytope*, Can. J. Math., **16**, pp. 701–720 (1964).

[Kü] Kühnel, W., *Tight and 0-tight polyhedral embeddings of surfaces*, Invent. Math., **58**, pp. 161–177, (1980).

[Kü2] Kühnel, W., *Minimal triangulations of Kummer varieties*, Abh. Math. Sem. Univ. Hamburg, **57**, pp. 7–20, (1987).

[Kü3] Kühnel, W., *Triangulations of manifolds with few vertices*, Advances in Diff. Geom. and Topology (F. Tricerri, ed.), pp. 59–114, World Scientific, 1990.

[Kü4] Kühnel, W., *Hamiltonian surfaces in polytopes*, Proc. Conf. Intuitive Geom., J. Bolyai Soc., Szeged, 1991.

[KB] Kühnel, W. and Banchoff, T.F., *The 9-vertex complex projective plane*, The Math. Intelligencer, **5-3**, pp. 11–22,(1983).

[KL] Kühnel, W. and Lassmann, G., *The unique 3-neighbourly 4-manifold with few vertices*, J. Combin. Th. (A), **35**, pp. 173–184, (1983).

[KS] Kühnel, W. and Schulz, Ch., *Submanifolds of the cube*, Appl. Geom. and Discr. Math., The Victor Klee Festschrift (P. Gritzmann amd B. Sturmfels, eds.), pp. 423–432, DIMACS Ser. in Discr. Math. and Theor. Comp. Sci. Vol.4, AMS, 1991.

[Kui1] Kuiper, N. H., *Tight embeddings and maps. Submanifolds of geometrical class three in E^N*, The Chern Symposium, Berkeley 1979, pp. 97–145, Springer, 1980.

[Kui2] Kuiper, N. H., *Geometry in total absolute curvature theory*, Perspectives in Math., Anniversary of Oberwolfach, pp. 377–392, Birkhäuser, 1984.

[MW] McMullen, P. and Walkup, D.W., *A generalized lower bound conjecture for simplicial polytopes*, Mathematika, **18**, pp. 264–273, (1971).

[Ri1] Ringel, G., *Wie man die geschlossenen nichtorientierbaren Flächen in möglichst wenig Dreiecke zerlegen kann*, Math. Ann., **130**, pp. 317-326, (1955).

[Ri2] Ringel, G., *Map Color Theorem*, Springer, 1974 (Grundlehren Band 209).

[Rn] Riordan, R., *Combinatorial Identities*, Wiley and Sons, New York, 1968.

[St] Stanley, R.P., *The number of faces of a simplicial convex polytope*, Adv. in Math., **35**, pp. 236–238, (1980).

GENERALIZED STRESS AND MOTIONS

In Memory of Paul Filliman

CARL W. LEE*
Department of Mathematics
University of Kentucky
Lexington, KY 40506, U.S.A.

Abstract. In 1987 Kalai presented a new proof of the Lower Bound Theorem for simplicial convex d-polytopes by linking the problem to results in rigidity and stress. He suggested that if higher-dimensional analogues of stress and rigidity were developed, they might lead to other combinatorial results on polytopes, and in particular another proof of the g-Theorem. Here we discuss such a generalization of stress and its relationship to face rings, h-vectors, shellings, bistellar operations, spheres, and simplicial polytopes. In particular, stress plays a role in McMullen's recent new geometric proof of the g-Theorem using his polytope algebra.

Key words: bistellar operations, convex polytope, face ring, h-vector, infinitesimal motion, p.l.-sphere, rigidity, shelling, stress.

1. Introduction

In 1987 Kalai [8] presented a new proof of the Lower Bound Theorem for simplicial convex d-polytopes by linking the problem to results in rigidity and stress. He suggested that if higher-dimensional analogues of stress and rigidity were developed, they might lead to other combinatorial results on polytopes, and in particular another proof of the g-Theorem. A proposal for such a generalization of stress was introduced in [10]. Here we provide details, discussing the relationship to face rings, h-vectors, shellings, bistellar operations, spheres, and simplicial polytopes. In particular, stress plays a role in McMullen's [11] recent new geometric proof of the g-Theorem using his polytope algebra.

2. Infinitesimal Rigidity and Stress

We first offer some background on rigidity and stress. See, for example, Kalai [8], Roth [14], and Whiteley [19] for more details and references. Begin by considering a graph $G = (V, E)$, where $V = \{1, \ldots, n\}$. Suppose that we make a structure by choosing a point $v_i \in \mathbf{R}^d$ for each vertex of the graph, and placing bars connecting

* Supported in part by NSF grants DMS-8504050 and DMS-8802933, by NSA grant MDA904-89-H-2038, by the Mittag-Leffler Institute, and by DIMACS (Center for Discrete Mathematics and Theoretical Computer Science), a National Science Foundation Science and Technology Center, NSF-STC88-09648.

T. Bisztriczky et al. (eds.), POLYTOPES: Abstract, Convex and Computational, 249–271.

pairs of points corresponding to edges. An *infinitesimal motion* of the vertices is a set of vectors $\overline{v}_1, \ldots, \overline{v}_n \in \mathbf{R}^d$ such that $d(||(v_i + t\overline{v}_i) - (v_j + t\overline{v}_j)||^2)/dt = 0$ when $t = 0$ for all bars v_iv_j. Equivalently, $(v_i - v_j)^T(\overline{v}_i - \overline{v}_j) = 0$ for all edges, or the projections of $\overline{v}_i$ and $\overline{v}_j$ onto the affine span of $\{v_i, v_j\}$ agree. For example, we could choose a single vector $u \in \mathbf{R}^d$ and set $\overline{v}_i = u$ for all vertices v_i. This would be a trivial motion in the sense that it could be extended to all of $\mathbf{R}^d$. That is to say, we can define an *infinitesimal motion of* $\mathbf{R}^d$ to be a choice of vector $\overline{v}$ for each point $v \in \mathbf{R}^d$ such that $(v - w)^T(\overline{v} - \overline{w}) = 0$ for all pairs v, w of points. Then we say that an infinitesimal motion of a structure is *trivial* if it is the restriction of an infinitesimal motion of $\mathbf{R}^d$. If a structure admits only trivial infinitesimal motions, we say it is *infinitesimally rigid.*

Now an infinitesimal motion of $\mathbf{R}^d$ is uniquely determined by its restriction to the vertices of any geometric $(d-1)$-simplex, and conversely, any infinitesimal motion of a structure consisting of the vertices and edges of a geometric $(d-1)$-simplex can be extended to $\mathbf{R}^d$. So a geometric $(d-1)$-simplex is infinitesimally rigid, and it is not hard to see that the dimension of the space of infinitesimal motions of such a simplex is $\binom{d+1}{2}$. Hence we conclude that this is also the dimension of the space of trivial infinitesimal motions. Thus a structure is infinitesimally rigid if and only if the dimension of its infinitesimal motion space is $\binom{d+1}{2}$.

The fact that a motion of an infinitesimally rigid structure is determined by the motions on d affinely independent vertices allows us to conclude that the union of two infinitesimally rigid structures in $\mathbf{R}^d$ sharing d affinely independent vertices is infinitesimally rigid.

The space of infinitesimally rigid motions of a structure is the nullspace of a certain *rigidity matrix* R. The rows of R are indexed by the edges v_iv_j, and the columns of R occur in n groups of d columns, one group for each vertex of R. The row vector of length d in row v_iv_j, group v_k, will be

$$\begin{cases} O^T & \text{if } k \neq i, j, \\ (v_i - v_j)^T & \text{if } k = i, \\ (v_j - v_i)^T & \text{if } k = j. \end{cases}$$

So a structure is infinitesimally rigid if and only if the dimension of the nullspace of R is $\binom{d+1}{2}$.

It is also useful to consider the left nullspace of R, elements of which are an assignment of numbers λ_{ij} to edges v_iv_j such that

$$\sum_{\{j : v_iv_j \in E\}} \lambda_{ij}(v_j - v_i) = O \tag{1}$$

holds for every vertex v_i. Such a vector of numbers is called a *stress*, and the vector space of all stresses of a structure is its *stress space.*

Dehn [2] proved the following:

Theorem 1 (Dehn) *The edge skeleton of a simplicial convex 3-polytope P is infinitesimally rigid.*

This is proved by first showing:

Theorem 2 (Dehn) *A simplicial convex 3-polytope P admits only the trivial stress in which all $\lambda_{ij} = 0$.*

PROOF. The proof we give here is a slight modification of that of Roth [14], which in turn uses some techniques of Cauchy. Suppose there is a non-trivial stress. Label each edge $v_i v_j \in E$ with the sign $(+, -, 0)$ of λ_{ij}. Suppose there is a vertex v such that all edges incident to it are labeled 0. Then delete v and take the convex hull of the remaining vertices. The resulting polytope cannot be two-dimensional, because it is clear that there can be no non-trivial stress on the edges of a single polygon. So the polytope is three-dimensional. If it is not simplicial, triangulate the non-triangular faces arbitrarily, labeling the new edges 0. Repeat this procedure until you have a simplicial 3-polytope Q (possibly with some coplanar faces) such that every vertex is incident to at least one nonzero edge. Note that every nonzero edge of Q is an edge of the original polytope P.

Now in each corner of each face (which is a triangle) of Q place the label 0 if the two edges meeting there are of the same sign, 1 if they are of opposite sign, and 1/2 if one is zero and the other nonzero.

Claim 1. The sum of the corner labels at each vertex v is at least four. First, because v is a vertex of P, the nonzero edges of P incident to v cannot all have the same sign. Consider now the cyclic changes in signs of just the nonzero edges of P incident to v. If there were only two changes in sign, the positive edges could be separated from the negative edges by a plane passing through v, since no three edges incident to v in P are coplanar. So there must be at least four changes in sign. The claim for the corner labels in Q now follows easily.

Claim 2. the sum of the three corner labels for each face is at most two. Just check all the possibilities of the edge and corner labels for a single triangle.

Now consider the sum S of all the corner labels of Q. By Claim 1 the sum is at least $4f_0$, where f_0 is the number of vertices of Q. By Claim 2 the sum is at most $2f_2$, where f_2 is the number of faces of Q. But Euler's relation and $3f_2 = 2f_1$ imply that $f_2 = 2f_0 - 4$. So $4f_0 \leq S \leq 4f_0 - 8$ yields a contradiction. □

PROOF OF THEOREM 1. Because P is simplicial, $f_1 = 3f_0 - 6$. So R has $f_1 = 3f_0 - 6$ rows and $3f_0$ columns. We need to show that the dimension of the nullspace of R is six, so we need to show that R has full row rank. But this is equivalent to there being no nontrivial stresses, which we have done. □

Whiteley [19] extended Theorem 1 to arbitrary $d > 3$:

Theorem 3 (Whiteley) *For $d > 3$, the edge skeleton of a simplicial convex d-polytope P is infinitesimally rigid.*

Using induction on d, he explained why the edge skeleton of clstar v, the closed star of v, is infinitesimally rigid for each vertex v of P. Then the rigidity of the entire edge skeleton of P results from the fact that the closed stars of two adjacent vertices share a $(d-1)$-simplex.

Regarding the matrix R for an arbitrary simplicial convex d-polytope, $d \geq 3$, the dimension of its nullspace is $\binom{d+1}{2}$. Hence its rank is $df_0 - \binom{d+1}{2}$. So the dimension of the stress space is $f_1 - df_0 + \binom{d+1}{2}$. In particular, this integer, usually now denoted $g_2(P)$ or g_2, is nonnegative. The Lower Bound Theorem follows from the nonnegativity of g_2, however, and this is Kalai's [8] striking proof. In fact, Kalai used this method to prove:

Theorem 4 *For all convex d-polytopes P, $d \geq 3$,*

$$f_{02} - 3f_2 + f_1 - df_0 + \binom{d+1}{2} \geq 0,$$

where f_{02} is the number of incidences of vertices with 2-faces.

3. McMullen's Conditions

On the other hand, the nonnegativity of g_2 for simplicial convex d-polytopes ($d \geq 3$) is a consequence of McMullen's conditions [17], which we will describe in this section.

Let Δ be a simplicial $(d-1)$-complex on the set $\{1,\ldots,n\}$ (its *vertices*). The *f-vector* of Δ is $f = (f_0, f_1, \ldots, f_{d-1})$, where f_j is the number of faces of Δ of dimension j (cardinality $j+1$). Taking $f_{-1} = 1$, the *h-vector* of Δ is $h = (h_0, \ldots, h_d)$ where

$$h_k = \sum_{j=0}^{k} (-1)^{j-k} \binom{d-j}{d-k} f_{j-1}, \quad k = 0, \ldots, d. \tag{2}$$

These relations are invertible:

$$f_j = \sum_{k=0}^{j+1} \binom{d-k}{d-j-1} h_k, \quad j = -1, \ldots, d-1.$$

Define also $g_0 = h_0 = 1$ and $g_k = h_k - h_{k-1}$, $k = 1, \ldots, \lfloor d/2 \rfloor$.

The *Stanley-Reisner ring* or *face ring* of Δ over $\mathbf{R}$ is $A = \mathbf{R}[x_1, \ldots, x_n]/I_\Delta$, where I_Δ is the ideal generated by all square-free monomials $x_{i_1} \cdots x_{i_s}$ such that $\{i_1, \ldots, i_s\} \notin \Delta$. The ring A inherits the grading by degree, $A = A_0 \oplus A_1 \oplus A_2 \oplus \cdots$. Stanley [15, 16] observed:

Theorem 5 (Stanley) *Let A be the face ring of a simplicial $(d-1)$-complex Δ. Then A is Cohen-Macaulay if and only if there exist $\theta_1, \ldots, \theta_d \in A_1$ such that $\dim B_k = h_k$, $k = 0, \ldots, d$, where $B = B_0 \oplus \cdots \oplus B_d = A/(\theta_1, \ldots, \theta_d)$. In this case the θ_j can be chosen generically (that is, with algebraically independent coefficients over $\mathbf{Q}$).*

If the above situation holds we say that Δ is *Cohen-Macaulay*. Reisner [13] derived a homological characterization of Cohen-Macaulay complexes. In particular, shellable simplicial complexes and simplicial balls and spheres (and hence boundary complexes of simplicial polytopes) are Cohen-Macaulay. To see the effect of this condition on the h-vector, we need another definition. For positive integers a and k, a can be expressed uniquely in the form

$$a = \binom{a_k}{k} + \binom{a_{k-1}}{k-1} + \cdots + \binom{a_\ell}{\ell},$$

where $a_k > a_{k-1} > \cdots > a_\ell \geq \ell \geq 1$. Using this, set

$$a^{\langle k\rangle} = \binom{a_k+1}{k+1} + \binom{a_{k-1}+1}{k} + \cdots + \binom{a_\ell+1}{\ell+1}.$$

Define also $0^{\langle k\rangle} = 0$. Stanley [15] proved:

Theorem 6 (Stanley) *Let Δ be a simplicial $(d-1)$-complex. If Δ is Cohen-Macaulay, then the h-vector is nonnegative and $h_{k+1} \leq h_k^{\langle k\rangle}$, $k = 1, \ldots, d-1$.*

Using a connection between the face ring of a simplicial convex polytope and the cohomology of an associated projective toric variety, Stanley [17] showed that the Hard Lefschetz Theorem implies:

Theorem 7 (Stanley) *Suppose that Δ is the boundary complex of a simplicial d-polytope, and that A is its face ring. Then, for some choice of $\theta_1, \ldots, \theta_d \in A_1$, there exists $\omega \in B_1$ such that multiplication by ω^{d-2k} is a bijection between B_k and B_{d-k}, $k = 0, \ldots, \lfloor d/2\rfloor$, where $B = B_0 \oplus \cdots \oplus B_d = A/(\theta_1, \ldots, \theta_d)$. In particular, multiplication by ω is an injection from B_k into B_{k+1}, $k = 0, \ldots, \lfloor d/2\rfloor - 1$. As a consequence, $g_k = \dim C_k$, $k = 0, \ldots, \lfloor d/2\rfloor$, where $C = C_0 \oplus \cdots \oplus C_{\lfloor d/2\rfloor} = B/(\omega)$. Hence*

1. *$h_k = h_{d-k}$, $k = 0, \ldots, d$ (the* Dehn-Sommerville Relations*),*
2. *$g_k \geq 0$, $k = 0, \ldots, \lfloor d/2\rfloor$ (the* Generalized Lower-Bound Inequalities*), and*
3. *$g_{k+1} \leq g_k^{\langle k\rangle}$, $k = 1, \ldots, \lfloor d/2\rfloor - 1$.*

The above three conditions are *McMullen's conditions* and characterize h-vectors of simplicial convex d-polytopes (the sufficiency was established by Billera and Lee [1]). This characterization is also known as the *g-Theorem.*

4. k-Stress

In this section we offer a generalization to the classical stress space of Section 2 that is motivated by the Stanley-Reisner ring. The original idea arose when contemplating Kalai's algebraic shifting technique [7].

First we give some notation. For $x = (x_1, \ldots, x_n)$, and for $r = (r_1, \ldots, r_n) \in \mathbf{Z}_+^n$, by x^r we mean $x_1^{r_1} \cdots x_n^{r_n}$. Define also $\operatorname{supp} x^r = \{i : r_i \neq 0\}$ (the *support* of x^r), $r! = r_1! \cdots r_n!$, and $|r| = r_1 + \cdots + r_n$.

Let Δ be a simplicial complex (not necessarily of dimension $d-1$) on the set $\{1, \ldots, n\}$, and let $v_1, \ldots, v_n \in \mathbf{R}^d$. Define M to be the $d \times n$ matrix with columns $v_1, \ldots, v_n$, and $\overline{M}$ to be the $(d+1) \times n$ matrix obtained from M by appending a final row of 1's.

For each $k = 0, 1, 2, \ldots$, a *linear k-stress* on Δ (with respect to $v_1, \ldots, v_n$) is a polynomial of the form

$$b(x) = \sum_{r:|r|=k} b_r \frac{x^r}{r!}$$

that satisfies

$$b_r = 0 \text{ if } \operatorname{supp} x^r \notin \Delta, \tag{3}$$

and

$$M\nabla b = O. \tag{4}$$

This last condition is equivalent to

$$\sum_{i=1}^{n}(\frac{\partial b}{\partial x_i})v_i = O,$$

where the left-hand side is to be regarded as a polynomial with vector coefficients, or

$$\sum_{i=1}^{n}(\frac{\partial b}{\partial x_i})v_{ij} = 0,\ j = 1, \ldots, d,$$

where $v_i = (v_{i1}, \ldots, v_{id})^T$, or

$$\sum_{i=1}^{n} b_{s+e_i} v_i = O \tag{5}$$

for every $s \in \mathbf{Z}_+^n$ such that $|s| = k - 1$, where e_i the ith unit vector in $\mathbf{R}^n$. That is to say, we have a linear relation on the vectors v_i for every such s. The collection of all linear k-stresses forms a vector space, which we will denote S_k^ℓ. (In [10] we used the notation $\overline{B}_k$.)

An *affine k-stress* on Δ (with respect to $v_1, \ldots, v_n$) is a linear k-stress that satisfies the additional condition

$$e^T \nabla b = 0,$$

where e denotes the vector $(1, \ldots, 1)^T$. Equivalently,

$$\sum_{i=1}^{n} \frac{\partial b}{\partial x_i} = 0,$$

or

$$\sum_{i=1}^{n} b_{s+e_i} = 0 \tag{6}$$

for every $s \in \mathbf{Z}_+^n$ such that $|s| = k - 1$; that is, we have an affine relation on the vectors v_i for every such s. Thus

$$\overline{M}\nabla b = O.$$

Clearly $b(x)$ is an affine k-stress with respect to $v_1, \ldots, v_n$ if and only if it is a linear k-stress with respect to $\overline{v}_1, \ldots, \overline{v}_n$, where

$$\overline{v}_i = \begin{bmatrix} v_i \\ 1 \end{bmatrix},\ i = 1, \ldots, n.$$

The collection of all affine k-stresses forms a subspace of S_k^ℓ, which we will denote S_k^a. (In [10] we used the notation $\overline{C}_k$.)

For $c \in \mathbf{R}^n$, define the function σ_c on the space of linear k-stresses by

$$\sigma_c(b) = c^T \nabla b = \sum_{i=1}^{n} c_i \frac{\partial b}{\partial x_i}$$

for any linear k-stress $b(x)$. In particular, define

$$\omega(b) = \sigma_e(b) = \sum_{i=1}^{n} \frac{\partial b}{\partial x_i}.$$

Theorem 8 *Let Δ be any simplicial complex with n vertices, and let $v_1, \ldots, v_n \in \mathbf{R}^d$. Then for $k = 1, 2, 3, \ldots$, the function σ_c maps S_k^ℓ into S_{k-1}^ℓ, and for $k = 0, 1, 2, \ldots$, the kernel of ω restricted to S_k^ℓ is S_k^a.*

PROOF. The statement about the kernel of ω follows immediately from the definition of S_k^a. Suppose that $b \in S_k^\ell$ for some $k = 1, 2, 3, \ldots$. For $r \in \mathbf{Z}_+^n$ such that $|r| = k-1$, the coefficient of $\frac{x^r}{r!}$ in $\sigma_c(b)$ is $\sum_{i=1}^n c_i b_{r+e_i}$. Suppose that $\operatorname{supp} x^r \notin \Delta$. Then $\operatorname{supp} x^{r+e_i} \notin \Delta$ for $i = 1, \ldots, n$. Hence $b_{r+e_i} = 0$, $i = 1, \ldots, n$, and so $\sigma_c(b)$ satisfies condition (3). Also, $M\nabla(c^T \nabla b) = M[(\nabla^2 b)c] = [\nabla(M\nabla b)]c = O$ since $M\nabla b = O$, so $\sigma_c(b)$ satisfies condition (4). □

Theorem 9 *Let Δ be any simplicial complex with n vertices, and let $v_1, \ldots, v_n \in \mathbf{R}^d$. Then*

1. *$S_0^\ell = S_0^a = \mathbf{R}$.*
2. *S_1^ℓ is isomorphic to the space of all linear relations on the vectors v_i.*
3. *S_1^a is isomorphic to the space of all affine relations on the vectors v_i.*
4. *S_2^a is isomorphic to the classical stress space, under the correspondence $\lambda_{ij} = b_{e_i+e_j}$.*

PROOF. The first three parts are trivial. For the fourth, assume $b \in S_2^a$. Let $\lambda_{ij} = b_{e_i+e_j}$ for all $i, j = 1, \ldots, n$. Note that $\lambda_{ij} = \lambda_{ji}$ and that $\lambda_{ij} = 0$ if $\{i, j\}$ is not an edge. From conditions (5) and (6) we see that for all $j = 1, \ldots, n$,

$$\begin{aligned} O &= \sum_{i=1}^{n} \lambda_{ij} v_i \\ &= \sum_{i:i\neq j} \lambda_{ij} v_i + \lambda_{jj} v_j \\ &= \sum_{i:i\neq j} \lambda_{ij} v_i + \sum_{i:i\neq j} (-\lambda_{ij}) v_j \\ &= \sum_{i:\{i,j\}\in E} \lambda_{ij}(v_i - v_j), \end{aligned}$$

where E denotes the edges of Δ. Hence the λ_{ij} satisfy condition (1).

Conversely, suppose we are given λ_{ij} for $\{i,j\} \in E$ that satisfy condition (1). For $j = 1, \ldots, n$ define

$$b_{2e_j} = -\sum_{i:\{i,j\}\in E} \lambda_{ij},$$

and for $i \neq j$ define

$$b_{e_i+e_j} = \begin{cases} \lambda_{ij} & \text{if } \{i,j\} \in E, \\ 0 & \text{otherwise.} \end{cases}$$

The above argument then reverses to show that these coefficients determine an affine 2-stress. □

Example 1 Let Δ be the boundary complex of a d-simplex in $\mathbf{R}^d$ and take $v_1, \ldots, v_{d+1} \in \mathbf{R}^d$ to be the vertices of the simplex. Assume that no proper subset of the vertices is linearly dependent. Then there exist nonzero c_i such that

$$\sum_{i=1}^{d+1} c_i v_i = O,$$

and all linear relations on the v_i are multiples of this one. Then for all $k = 0, \ldots, d$, S_k^ℓ is one-dimensional and is spanned by

$$\sum_{r:|r|=k} c^r \frac{x^r}{r!}.$$

For we can see that

$$\sum_{i=1}^{d+1} c^{s+e_i} v_i = c^s \sum_{i=1}^{d+1} c_i v_i = O$$

for all $s \in \mathbf{Z}_+^{d+1}$ such that $|s| = k-1$. Note that c^r is nonzero for all r. On the other hand, $\dim S_k^\ell = 0$ for all $k > d$, $\dim S_0^a = 1$, and $\dim S_k^a = 0$ for all $k \geq 1$, since the v_i are affinely independent and so $\sum_{i=1}^{d+1} c_i \neq 0$.

Example 2 Let Δ be the boundary complex of the standard octahedron in $\mathbf{R}^3$, with

$$\begin{aligned} v_1 &= (+1,0,0)^T \\ v_2 &= (-1,0,0)^T \\ v_3 &= (0,+1,0)^T \\ v_4 &= (0,-1,0)^T \\ v_5 &= (0,0,+1)^T \\ v_6 &= (0,0,-1)^T \end{aligned}$$

Then it can be checked that

1. $S_0^\ell = \mathbf{R}$.
2. S_1^ℓ is three dimensional and has a basis $\{x_1+x_2, x_3+x_4, x_5+x_6\}$.
3. S_2^ℓ is three dimensional and has a basis $\{x_1x_3+x_1x_4+x_2x_3+x_2x_4, x_1x_5+x_1x_6+x_2x_5+x_2x_6, x_3x_5+x_3x_6+x_4x_5+x_4x_6\}$.
4. S_3^ℓ is one-dimensional and has a basis $\{x_1x_3x_5+x_1x_3x_6+x_1x_4x_5+x_1x_4x_6+x_2x_3x_5+x_2x_3x_6+x_2x_4x_5+x_2x_4x_6\}$.

5. $S_k^\ell = \{0\}$ if $k > 3$.
6. $S_0^a = \mathbf{R}$.
7. S_1^a is two-dimensional and has a basis $\{x_1 + x_2 - x_3 - x_4, x_1 + x_2 - x_5 - x_6\}$.
8. $S_k^a = \{0\}$ if $k > 1$.

5. Relationship to the Face Ring

The definition of generalized stress follows somewhat naturally from the face ring. For suppose that Δ is a simplicial complex (not necessarily of dimension $d-1$) with n vertices $\{1, \ldots, n\}$, and let $R = \mathbf{R}[x_1, \ldots, x_n] = R_0 \oplus R_1 \oplus R_2 \oplus \cdots$ be the ring of polynomials, graded by degree. If we are given $\theta_1, \ldots, \theta_d \in R_1$, we are interested in the dimension of B_k (as a vector space over $\mathbf{R}$), where $B = B_0 \oplus B_1 \oplus B_2 \oplus \cdots$ equals R factored out by the ideal $J = J_0 \oplus J_1 \oplus J_2 \oplus \cdots$ generated by I_Δ and $\theta_1, \ldots, \theta_d$. Using the inner product $\langle \sum_{r:|r|=k} a_r x^r, \sum_{r:|r|=k} b_r x^r \rangle = \sum_{r:|r|=k} a_r b_r$ on R_k, write $R_k = J_k \oplus J_k^\perp$. Now $\sum_{r:|r|=k} b_r x^r$ is in $J_k^\perp$ if and only if it is orthogonal to

1. all monomials of the form $x^s x^q$ where x^q is square-free, $\operatorname{supp} x^q \notin \Delta$, and $|s| + |q| = k$; and
2. all polynomials of the form $x^s \theta_j$, where $|s| = k - 1$.

Writing $\theta_j = \sum_{i=1}^n v_{ij} x_i$, $j = 1, \ldots, d$ and defining $v_i = (v_{i1}, \ldots, v_{id})^T$, $i = 1, \ldots, n$, the first condition above is equivalent to condition (3) and the second condition is equivalent to condition (5). Hence $\sum_{r:|r|=k} b_r x^r \in J_k^\perp$ if and only if $\sum_{r:|r|=k} b_r \frac{x^r}{r!} \in S_k^\ell$. Recalling that an affine stress with respect to $v_1, \ldots, v_n$ is a linear stress with respect to $\overline{v}_1, \ldots, \overline{v}_n$, we have:

Theorem 10 *Suppose that Δ is a simplicial complex (not necessarily of dimension $d-1$) with n vertices. Let A be its face ring, and assume that we have $\theta_1, \ldots, \theta_d \in A_1$ and $v_1, \ldots, v_n \in \mathbf{R}^d$ such that $\theta_j = \sum_{i=1}^n v_{ij} x_i$, $j = 1, \ldots, d$ and $v_i = (v_{i1}, \ldots, v_{id})^T$, $i = 1, \ldots, n$. Let $A = A_0 \oplus A_1 \oplus A_2 \oplus \cdots = R/I_\Delta$, $B = B_0 \oplus B_1 \oplus B_2 \oplus \cdots = A/(\theta_1, \ldots, \theta_d)$, and $C = C_0 \oplus C_1 \oplus C_2 \oplus \cdots = B/(x_1 + \cdots + x_n)$. Then*

1. *Regardless of whether or not Δ is Cohen-Macaulay, $\dim B_k = \dim S_k^\ell$, $k = 0, \ldots, d$, and $\dim C_k = \dim S_k^a$, $k = 0, \ldots, d$.*
2. *Let $J = J_0 \oplus J_1 \oplus J_2 \oplus \cdots$ be the ideal of R generated by I_Δ and $\theta_1, \ldots, \theta_d$, and $J' = J_0' \oplus J_1' \oplus J_2' \oplus \cdots$ be the ideal of R generated by J and $x_1 + \cdots + x_n$. Then $b(x) = \sum_{r:|r|=k} b_r \frac{x^r}{r!}$ is a linear (respectively, affine) k-stress if and only if*

$$\sum_{r:|r|=k} a_r b_r = 0$$

for all $a(x) = \sum_{r:|r|=k} a_r x^r$ in J_k (respectively, J_k').

Corollary 11 *Let Δ be any simplicial $(d-1)$-complex with n vertices.*

1. *Δ is Cohen-Macaulay if and only if there exist $v_1, \ldots, v_n \in \mathbf{R}^d$ such that $\dim S_k^\ell = h_k$, $k = 1, \ldots, d$. In this case, the v_i can be chosen generically (that is, with algebraically independent components).*
2. *Suppose that Δ is in fact a simplicial $(d-1)$-sphere. If $\dim S_k^a = g_k$, $k = 0, \ldots, \lfloor d/2 \rfloor$, then its h-vector satisfies McMullen's conditions.*

PROOF. This follows from the above result and Theorems 5 and 7. □

6. Formulas for Coefficients

In this section we explain why, under suitable conditions on the v_i, the coefficients of the square-free monomials of a linear stress uniquely determine the coefficients of the non-square-free monomials. We then characterize the former coefficients. For a simplicial complex Δ with n vertices, and for $v_1, \ldots, v_n \in \mathbf{R}^d$, we say that the v_i are in *linearly general position with respect to* Δ if $\{v_{i_1}, \ldots, v_{i_s}\}$ is linearly independent for every face $\{i_1, \ldots, i_s\}$ of Δ.

Theorem 12 *Let Δ be any simplicial complex with n vertices, and let $v_1, \ldots, v_n \in \mathbf{R}^d$ be in linearly general position with respect to Δ. If $b(x)$ is a linear stress, then the coefficients of the non-square-free monomials in $b(x)$ are linear combinations of the coefficients of the square-free monomials and hence are uniquely determined by them.*

PROOF. Let $b(x) \in S_k^\ell$. We will use reverse induction on $\ell = \mathrm{card}(\mathrm{supp}\, x^r)$. The result is trivially true if $\ell = k$, so assume that the result is true for some ℓ such that $1 \le \ell \le k$, and suppose that $\mathrm{card}(\mathrm{supp}\, x^r) = \ell - 1$ where $\mathrm{supp}\, x^r \in \Delta$. Choose j such that $r_j > 1$ and let $s = r - e_j$. Condition (5) implies

$$\sum_{i=1}^{n} b_{s+e_i} v_i = O.$$

But, by the induction hypothesis, the coefficients b_{s+e_i} are linear combinations of the coefficients of the square-free monomials when $r_i = 0$, since $\mathrm{card}(\mathrm{supp}\, x^{s+e_i}) = \ell$ in this case. This leaves the $\ell - 1$ coefficients b_{s+e_i} for $i \in \mathrm{supp}\, x^r$ to be uniquely determined, since the corresponding v_i are linearly independent by assumption. In particular, $b_{s+e_j} = b_r$ is a linear combination of the coefficients of the square-free monomials. □

Therefore, if you are given the coefficients of the square-free coefficients of a k-stress, you can use conditions (3) and (5) to find the other coefficients systematically.

Corollary 13 *Let Δ be any simplicial complex with n vertices, and let $v_1, \ldots, v_n \in \mathbf{R}^d$ be chosen in linearly general position with respect to Δ. Then $\dim S_k^\ell = 0$ for all $k > \dim \Delta + 1$.*

PROOF. If $k > \dim \Delta + 1$ then there are no faces of cardinality k, so all coefficients of square-free monomials of a linear k-stress must be zero. □

The next theorem provides an explicit formula for the coefficients of the non-square-free monomials in terms of the coefficients of the square-free monomials. For $G = \{i_1, \ldots, i_s\} \in \Delta$, define $\mathrm{conv}\, G$ (with respect to $v_1, \ldots, v_n$) to be $\mathrm{conv}\{v_{i_1}, \ldots, v_{i_s}\}$. We similarly define $\mathrm{aff}\, G$ and $\mathrm{span}\, G$. We will sometimes abuse notation and write b_G and x^G for b_r and x^r, respectively, where $r_i = 1$ if $i \in G$ and $r_i = 0$ if $i \notin G$. We will also use the notation $G + i$ for $G \cup \{i\}$ and $G - i$ for $G \setminus \{i\}$. Fixing an ordering of the elements of G and assuming that $s \le d$, define

$$[G] = \det \begin{bmatrix} v_{i_1 1} & v_{i_2 1} & \cdots & v_{i_s 1} \\ v_{i_1 2} & v_{i_2 2} & \cdots & v_{i_s 2} \\ \vdots & \vdots & \ddots & \vdots \\ v_{i_1 s} & v_{i_2 s} & \cdots & v_{i_s s} \end{bmatrix},$$

a subdeterminant of M. Note that only the first s rows of M are used. If $i \in G$, we compute $[G-i]$ using the ordering induced by G and multiply by $+1$ (respectively, -1) if i is in an odd (respectively, even) position with respect to this ordering, and we compute $[G-i+j]$ by replacing the column corresponding to v_i with the column corresponding to v_j.

Theorem 14 *Let Δ be a simplicial complex on n vertices of dimension at most $d-1$, and let $v_1, \ldots, v_n$ be chosen generically in $\mathbf{R}^d$. Suppose that $b(x)$ is a linear k-stress for some $1 \leq k \leq d$. Suppose that $r \in \mathbf{Z}_+^n$ such that $|r| = k$ and $S = \operatorname{supp} x^r \in \Delta$. Then*

$$b_r = \sum_{(k-1)\text{-faces } F \text{ containing } S} b_F \frac{\prod_{i \in S} [F-i]^{r_i - 1}}{\prod_{i \in F \setminus S} [F-i]}.$$

PROOF. We will use reverse induction on $\ell = \operatorname{card} S$. The formula is trivially true when $\ell = k$, so assume that the formula for b_r is true whenever $\operatorname{card}(\operatorname{supp} x^r) = \ell+1$, for some ℓ such that $1 \leq \ell < d$. Suppose that the support S of x^r has cardinality ℓ. Write $S = \{i_1, \ldots, i_\ell\}$ where $i_1 < \cdots < i_\ell$. Since x^r is not square-free, there must be some m for which $r_m > 1$. Let M_ℓ be the submatrix of M consisting of the first ℓ rows of M, and let B be the submatrix of M_ℓ determined by the members of S. Multiplying the mth row of $B^{-1} M_\ell (x_1, \ldots, x_n)^T$ by $x^{r - e_{i_m}}$ yields a member of J_k, namely

$$x^r + \sum_{j \in \operatorname{lk} S} \frac{[S-m+j]}{[S]} x^{r - e_m + e_j},$$

Note that each monomial on the right-hand side has support of cardinality $\ell+1$. By the induction hypothesis, the orthogonality condition (2) of Theorem 10, and some

Grassman-Plücker relations, we compute

$$\begin{aligned}
b_r &= -\sum_{j\in \operatorname{lk} S} \frac{[S-m+j]}{[S]} b_{r-e_m+e_j} \\
&= -\sum_{j\in \operatorname{lk} S} \frac{[S-m+j]}{[S]} \sum_{(k-1)\text{-faces } F \text{ containing } S+j} b_F \frac{\prod_{i\in S+j} [F-i]^{(r-e_m+e_j)_i-1}}{\prod_{i\in F\setminus(S+j)} [F-i]} \\
&= -\sum_{(k-1)\text{-faces } F \text{ containing } S} \sum_{j\in F\setminus S} b_F \frac{[S-m+j]\prod_{i\in S}[F-i]^{(r-e_m)_i-1}}{[S]\prod_{i\in F\setminus(S+j)} [F-i]} \\
&= -\sum_{(k-1)\text{-faces } F \text{ containing } S} b_F \frac{\prod_{i\in S}[F-i]^{(r-e_m)_i-1}}{[S]} \sum_{j\in F\setminus S} \frac{[S-m+j]}{\prod_{i\in F\setminus(S+j)} [F-i]} \\
&= -\sum_{(k-1)\text{-faces } F \text{ containing } S} b_F \frac{\prod_{i\in S}[F-i]^{(r-e_m)_i-1}}{[S]\prod_{i\in F\setminus S} [F-i]} \sum_{j\in F\setminus S} [S-m+j][F-j] \\
&= \sum_{(k-1)\text{-faces } F \text{ containing } S} b_F \frac{\prod_{i\in S}[F-i]^{(r-e_m)_i-1}}{[S]\prod_{i\in F\setminus S} [F-i]} [S][F-m] \\
&= \sum_{(k-1)\text{-faces } F \text{ containing } S} b_F \frac{\prod_{i\in S}[F-i]^{r_i-1}}{\prod_{i\in F\setminus S} [F-i]}. \square
\end{aligned}$$

The formula is not symmetric with respect to permutations of the coordinates of the v_i, but can be made so by averaging over all permutations, for example. Since we know that the coefficients of the square-free monomials determine all of the others, it would be nice to characterize them somehow geometrically.

Theorem 15 *Let Δ be any simplicial complex with n vertices, and let $v_1, \ldots, v_n \in \mathbf{R}^d$. Let $b(x)$ be a linear (respectively, affine) k-stress, $k \geq 1$. Choose any face F of Δ of cardinality $k-1$ and any point v in* span F *(respectively* aff F*). Then*

$$v + \sum_{i\in \operatorname{lk} F} b_{F+i}(v_i - v)$$

lies in span F *(respectively,* aff F*). Equivalently, if w_i is the vector joining the projection of v_i onto* span F *(respectively,* aff F*) to v_i, then*

$$\sum_{i \in \operatorname{lk} F} b_{F+i} w_i = O.$$

Proof. Suppose that $v \in \operatorname{span} F$. Then, using condition (5),

$$\begin{aligned} v + \sum_{i \in \operatorname{lk} F} b_{F+i}(v_i - v) &= v + \sum_{i \in \operatorname{lk} F} b_{F+i} v_i - \sum_{i \in \operatorname{lk} F} b_{F+i} v \\ &= v - \sum_{i \in F} b_{F+i} v_i - \sum_{i \in \operatorname{lk} F} b_{F+i} v \end{aligned}$$

which is in span F (abusing notation slightly in the penultimate sum). If b is an affine stress, then by condition (6) the sum of the coefficients in the above expression is

$$1 - \sum_{i \in F} b_{F+i} - \sum_{i \in \operatorname{lk} F} b_{F+i} = 1.$$

So we have an element of aff F. □

Note that, for a linear k-stress, w_i is the altitude vector for the point v_i in the simplex $\operatorname{conv}(\{O\} \cup F)$, and, for an affine k-stress, w_i is the altitude vector for the point v_i in the simplex conv F. So affine k-stress is a natural generalization of classical stress (affine 2-stress), and is equivalent to the proposed generalization of Kalai (personal communication).

Example 3 Let Δ be the boundary complex of a simplicial d-polytope in $\mathbf{R}^d$, $d \geq 1$, and take the v_i to be its vertices. Then the above theorem implies that $\dim S_d^a = 0$. For take any $b(x) \in S_d^a$ and consider any subfacet F (i.e., of cardinality $d-1$). Then there are precisely two facets containing F, and hence only two altitude vectors w_i with respect to aff F, where $i \in \operatorname{lk} F$. By convexity these two vectors are not collinear and we must have

$$\sum_{i \in \operatorname{lk} F} b_{F+i} w_i = O,$$

from which it follows that $b_{F+i} = 0$ for $i \in \operatorname{lk} F$. Thus all the coefficients of the square-free monomials of $b(x)$ are zero, and hence all of the coefficients of $b(x)$ must also be zero.

The previous theorem provides necessary conditions for the coefficients of the square-free terms. But Filliman [3] and Tay-White-Whiteley [18] have shown that they are also sufficient. So we could just as well define linear or affine k-stress using the conditions provided by the previous theorem, and perhaps this would be more natural.

7. Infinitesimal k-Motions

What is the generalization of infinitesimal motions? Consider an affine k-stress on a simplicial complex Δ with respect to $\{v_1, \ldots, v_n\} \subset \mathbf{R}^d$. That is, for each $(k-1)$-face F, we have a number b_F such that for every $(k-2)$-face G,

$$\sum_{i \in \operatorname{lk} G} b_{G+i} w_i = O,$$

where w_i is the altitude vector of v_i in the simplex $\operatorname{conv}(G+i)$. Consider the matrix R with rows indexed by $(k-1)$-faces F and columns occurring in groups of d columns, one group for each $(k-2)$-face G. The row vector of length d in row F, group G, is

$$\begin{cases} O^T & \text{if } G \not\subset F, \\ w_{(G,F)}^T & \text{if } G \subset F, \end{cases}$$

where $w_{(G,F)}$ is the altitude vector of the simplex $\operatorname{conv} F$ with respect to $\operatorname{conv} G$. So the left nullspace of R is S_k^a. The other nullspace of R suggests a definition of infinitesimal $(k-1)$-motion, namely, an assignment of a vector $\overline{v}_G \in \mathbf{R}^d$ to each $(k-2)$-face G, such that for every $(k-1)$-face F,

$$\sum_{i \in F} w_i \cdot \overline{v}_i = 0.$$

In the above expression, we use the notation w_i for $w_{(F \setminus \{i\}, F)}$ and $\overline{v}_i$ for $\overline{v}_{F \setminus \{i\}}$.

Write G_i for $F \setminus \{i\}$ and let u_i be the unit outer normal vector of $\operatorname{conv} G_i$ with respect to $\operatorname{conv} F$ in $\operatorname{aff} F$. Then

$$\sum_{i \in F} u_i \|w_i\| \cdot \overline{v}_i = 0.$$

Dividing by $\operatorname{vol}_{k-1}(F)$ implies

$$\sum_{i \in F} u_i \cdot \frac{\overline{v}_i}{\operatorname{vol}_{k-2}(G_i)} = 0$$

and hence

$$\sum_{i \in F} u_i \operatorname{vol}_{k-2}(G_i) \cdot \overline{m}_i = 0,$$

where $\overline{m}_i = \overline{v}_i / \operatorname{vol}_{k-2}^2(G_i)$.

A better definition of infinitesimal $(k-1)$-motion might be a choice of vector $\overline{m}_G \in \mathbf{R}^d$ for each $(k-2)$-face G, such that the above expression is satisfied for each $(k-1)$-face F. Since $u_i \cdot m_i = u_i \cdot \overline{m}_i$ for the projection m_i of $\overline{m}_i$ onto $\operatorname{aff} F$, we also have

$$\sum_{i \in F} u_i \operatorname{vol}_{k-2}(G_i) \cdot m_i = 0.$$

It can be shown that this is equivalent to the existence of a vector $m_F \in \mathbf{R}^d$ for each $(k-1)$-face F such that $m_F \cdot u_i = m_i \cdot u_i$ for each $i \in F$. Notice in this case that

$$\sum_{i\in F} u_i \mathrm{vol}_{k-2}(G_i)\cdot m_i$$

$$=\sum_{i\in F} u_i \mathrm{vol}_{k-2}(G_i)\cdot m_F$$

$$=(\sum_{i\in F} u_i \mathrm{vol}_{k-2}(G_i))\cdot m_F$$

$$=O\cdot m_F$$

$$=0$$

by Minkowski's theorem.

For a real number t, let $F(t)$ be the $(k-1)$ simplex determined by translating $\mathrm{aff}\, G_i$ by the vector tm_i. Then our definition is also equivalent to

$$\frac{d}{dt}\mathrm{vol}_{k-1}^2(F(t))=0$$

when $t=0$. This was also observed by Filliman [3]. Equivalently, $F(1)$ is congruent to F.

So generalized stress leads to a fairly natural generalization of infinitesimal motions (2-motions). See Tay, White and Whiteley [18] for a deeper study of the relationship between generalized stress and skeletal rigidity of cell complexes.

8. Simplicial Spheres

Let Δ be a simplicial $(d-1)$-sphere (or connected $(d-1)$-pseudo-manifold) with n vertices, and choose $v_1,\ldots,v_n\in\mathbf{R}^d$ in generically. As noted before, Δ is Cohen-Macaulay. Also, Euler's relation implies that $h_d=1$. So $\dim S_d^\ell=1$, hence up to scalar multiple there is only one linear d-stress $b(x)$. By Theorem 12, it suffices to determine the square-free coefficients of b. Choose a consistent orientation of all the facets of Δ and use this to induce an ordering of the elements of each facet. Let G be a subfacet of Δ, and F_1,F_2 the two facets containing G. Examining the conditions on b_{F_1} and b_{F_2} in terms of their altitudes w_1,w_2 with respect to $\mathrm{span}G$, one can readily see that $[F_1]b_{F_1}=[F_2]b_{F_2}$. So we may without loss of generality assume that $b_F=[F]^{-1}$ for every facet F. The coefficients of the non-square-free monomials can then be determined.

In Section 10 we will see the geometrical significance of the canonical linear d-stress in the case that Δ is the boundary complex of a simplicial d-polytope.

For a subset $S=\{i_1,\ldots,i_s\}$ of $\{1,\ldots,n\}$, define the function τ_S on the space of linear stresses by

$$\tau_S(b)=\frac{\partial^s b}{\partial x_{i_1}\cdots\partial x_{i_s}}$$

In particular, write

$$\tau_i(b)=\frac{\partial b}{\partial x_i}.$$

Suppose that Δ is a simplicial $(d-1)$-sphere on $\{1,\ldots,n\}$. Let $G \subseteq \{1,\ldots,n\}$ be a face of Δ of cardinality s. Then $\operatorname{clstar} G$, the closed star of G in Δ, is a simplicial $(d-1)$-ball (hence Cohen-Macaulay), that is the join of a simplicial $(d-s-1)$-sphere to G. In such a case it is known that $h_{d-s}(\operatorname{clstar} G) = 1$. The next theorem shows how to obtain a canonical linear $(d-s)$-stress for $\operatorname{clstar} G$ from the canonical linear d-stress of Δ.

Theorem 16 *Let Δ be a simplicial $(d-1)$-complex with n vertices, and let $v_1,\ldots,v_n$ be chosen arbitrarily. Suppose that $b(x)$ is a linear d-stress and that G is a face of Δ of cardinality s. Then $\tau_G(b)$ is a linear $(d-s)$-stress supported on* $\operatorname{clstar} G$. *In particular, if Δ is a simplicial $(d-1)$-sphere, $v_1,\ldots,v_n \in \mathbf{R}^d$ are chosen generically, and $b(x)$ is the nonzero canonical linear d-stress, then $\tau_G(b)$ is a nonzero linear $(d-s)$-stress supported on* $\operatorname{clstar} G$.

PROOF. That $\tau_G(b)$ is a linear $(d-s)$-stress follows from the fact that $\tau_i = \sigma_{e_i}$ and $\tau_G = \tau_{i_1}\cdots\tau_{i_s}$, where $G = \{i_1,\ldots,i_s\}$. If the coefficient of x^r is nonzero in $\tau_G(b)$, then the coefficient of $x_{i_1}\cdots x_{i_s}x^r$ must be nonzero in $b(x)$, and hence $\operatorname{supp} x^r \in \operatorname{clstar} G$. Now suppose that Δ is a simplicial sphere, the v_i are chosen generically, and $b(x)$ is the canonical linear d-stress. Let F be a facet of Δ that contains G. Then the coefficient of $x^{F\setminus G}$ in $\tau_G(b)$ equals the coefficient of x^F in b, which is nonzero. So $\tau_G(b)$ is not the zero polynomial. □

Let us consider the case that Δ is a shellable simplicial $(d-1)$-sphere Δ, and $F_1,\ldots,F_m$ constitutes a shelling order of the facets. It is known that as each facet F_k is added, precisely one component, say h_s, of the h-vector increases by one, the remaining components remaining unchanged. For generic $v_1,\ldots,v_n$, this implies that the dimension of S_s^ℓ increases by one, while the dimensions of the other linear stress spaces remain unchanged. When F_k is added, the closed star of precisely one face G_k of cardinality $d-s$ is completed. The linear s-stress $\tau_G(b)$ now becomes a member of S_s^ℓ, where $b(x)$ is the canonical linear d-stress. Thus if Δ is shellable, we can use the shelling to derive a basis for S_s^ℓ.

Theorem 17 *If Δ is a shellable simplicial $(d-1)$-sphere with n vertices, $v_1,\ldots,v_n \in \mathbf{R}^d$ are chosen generically, and $F_1,\ldots,F_m$, $G_1,\ldots,G_m$ are as above, then $\{\tau_{G_k}(b) :$* $\operatorname{card} G_k = d-s\}$ *is a basis for S_s^ℓ. Hence the collection $\{\tau_G(b) : G$ is a face of Δ of cardinality $d-s\}$ spans S_s^ℓ.*

By working directly with the face ring, Kind and Kleinschmidt [9] found an inductive proof that arbitrary shellable simplicial complexes are Cohen-Macaulay. From the above ideas one can construct another inductive proof of this result using linear stress.

9. Bistellar Operations and P.L.-Spheres

Suppose that Δ is a simplicial $(d-1)$-complex whose vertices are contained in $\{1,\ldots,n\}$. Assume that F and G are disjoint subsets of $\{1,\ldots,n\}$ of cardinality k and ℓ, respectively, $(1 \le k,\ell \le d)$ such that $k+\ell = d+1$, $F \in \Delta$, $G \notin \Delta$, and $\operatorname{lk} F = \partial G = \{G' : G'$ is a proper subset of $G\}$, the boundary of G. Then we say that the simplicial complex $\Delta' = (\Delta \setminus F) \cup (G \cdot \partial F)$ is the result of a *bistellar operation* on Δ. That is, we remove from Δ all faces containing F, and then introduce all sets of the form $G \cup F'$ where $F' \in \partial F$. Notice that the faces of Δ' which are not in Δ are precisely those faces of Δ' which contain G, and the faces of Δ' which are not in Δ are precisely the faces of Δ which contain F.

Theorem 18 *If Δ and Δ' are as above and $v_1, \ldots, v_n \in \mathbf{R}^d$ are generic, then*

$$\dim S_s^\ell(\Delta') = \begin{cases} \dim S_s^\ell(\Delta) + 1 & \text{if } k > \ell \text{ and } \ell \le s \le d - \ell, \\ \dim S_s^\ell(\Delta) - 1 & \text{if } k < \ell \text{ and } k \le s \le d - k, \\ \dim S_s^\ell(\Delta) & \text{otherwise.} \end{cases} \quad (7)$$

PROOF. Let $\Delta'' = \Delta \cup (G \cdot \partial F)$. Note that Δ'' also equals $\Delta' \cup (F \cdot \partial G)$. It suffices to show that

$$\dim S_s^\ell(\Delta'') = \begin{cases} \dim S_s^\ell(\Delta) & \text{if } s = 0, \ldots, \ell - 1, \\ \dim S_s^\ell(\Delta) + 1 & \text{if } s = \ell, \ldots, d, \end{cases}$$

and

$$\dim S_s^\ell(\Delta'') = \begin{cases} \dim S_s^\ell(\Delta') & \text{if } s = 0, \ldots, k - 1, \\ \dim S_s^\ell(\Delta') + 1 & \text{if } s = k, \ldots, d. \end{cases}$$

Since Δ and Δ'' share the same faces of cardinality s when $s = 0, \ldots, \ell - 1$, it is clear that $S_s^\ell(\Delta) = S_s^\ell(\Delta'')$ for these values of s. So assume that $\ell \le s \le d$. Define $S = F \cup G$, a subset of cardinality $d+1$, all of whose proper faces are in Δ''. Let $c(x)$ be the nonzero canonical linear s-stress obtained from the essentially unique linear relation on the set $\{v_i : i \in S\}$, as in Example 1. Since all faces of Δ are also in Δ'', we see that every linear s-stress on Δ is also a linear s-stress on Δ''. Now let $b(x)$ be any linear s-stress on Δ'' that is not a linear s-stress on Δ. Thus b_r is nonzero for some r such that $\operatorname{supp} x^r \in \operatorname{openstar} G$, where by $\operatorname{openstar} G$ we mean the set of all faces of Δ'' that contain G. We will show that that there is a nonzero $t \in \mathbf{R}$ such that $b_r = tc^r$ whenever $\operatorname{supp} x^r \in \operatorname{openstar} G$, and hence that $b(x) - tc(x)$ is a linear s-stress on Δ. This is clearly true if $s = \ell$ since G is the only face of cardinality ℓ in $\operatorname{openstar} G$, so assume that $\ell + 1 \le s \le d$. Choose any r such that b_r is nonzero and $\operatorname{supp} x^r \in \operatorname{openstar} G$. Since $s > \ell$, there exists j such that $\operatorname{supp} x^{r-e_j} \in \operatorname{openstar} G$. Condition 5 implies that $\sum_{i=1}^n b_{r-e_j+e_i} v_i = O$. But this sum involves only the $d+1$ vectors v_i such that $i \in S$, so the coefficients must be multiples of the coefficients of $c(x)$. Repeating this procedure and using induction on the cardinality of $\operatorname{supp} b_r$ establishes the desired result. Therefore $\dim S_s^\ell(\Delta'') = \dim S_s^\ell(\Delta) + 1$ for $s = \ell, \cdots, d$. The proof relating $\dim S_s^\ell(\Delta'')$ and $\dim S_s^\ell(\Delta')$ is analogous. □

We remark that the changes in the dimensions of the linear stress spaces under a bistellar operation remain valid whether or not Δ is Cohen-Macaulay. But the case in which Δ is Cohen-Macaulay, and in particular a p.l.-sphere, is of special interest. Pachner [12] showed that every (simplicial) p.l.-sphere of dimension $d - 1$ can be obtained from the boundary of a d-simplex by a sequence of bistellar operations. This fact, together with the above result, can be used to obtain a new proof that p.l.-spheres are Cohen-Macaulay.

Corollary 19 *If Δ is a simplicial piecewise linear $(d-1)$-sphere, then Δ is Cohen-Macaulay.*

PROOF. Choose $v_1, \ldots, v_n \in \mathbf{R}^d$ generically. The boundary of a simplex is Cohen-Macaulay by part (1) of Corollary 11 and Example 1, since the dimensions of the linear stress spaces agree with the components of the h-vector. It is easy to see that for any simplicial complex, the components h_s of the h-vector increase and decrease in exactly the same manner as the dimensions $\dim S_s^\ell$ change in (7). So if Δ is obtained from the boundary of a simplex by a sequence of bistellar operations, it follows that $h_i(\Delta) = \dim S_s^\ell(\Delta)$ for all s. The result now follows from Corollary 11 and Pachner's theorem. □

10. Simplicial Convex Polytopes

In this section, we will assume that Δ is the boundary complex of some simplicial d-polytope P containing the origin in its interior. In discussing the stress spaces of Δ, take $v_1, \ldots, v_n \in \mathbf{R}^d$ to be the actual vertices of P. Kind and Kleinschmidt's shelling proof shows that this suffices to ensure that $\dim S_i^\ell = h_i$, $i = 0, \ldots, d$.

For $x \in \mathbf{R}^n$, consider the polytope $Q(x) = \{y \in \mathbf{R}^d : y^T v_i \leq x_i,\ i = 1, \ldots, n\}$. Of course, $Q(e)$ is the polar P^* of P. Since P^* is simple, for values of x_i near 1 the combinatorial structure of $Q(x)$ agrees with that of P^*. It is known that the volume of $Q(x)$ as a function of the x_i is a homogeneous polynomial $V(x) = \sum_{r:|r|=d} b_r \frac{x^r}{r!}$ of degree d and $b_r = 0$ whenever $\operatorname{supp} x^r \notin P$.

Theorem 20 *Let P be as above. Then the canonical d-stress $b(x)$ described in Theorem 14 is precisely $V(x)$.*

PROOF. For every $u \in \mathbf{R}^d$, $Q(x_1, \ldots, x_n) + u = Q(x_1 + u^T v_1, \ldots, x_n + u^T v_n)$. So $V(x_1, \ldots, x_n) - V(x_1 + u^T v_1, \ldots, x_n + u^T v_n) = 0$. Fix r such that $|r| = d - 1$. Then

$$\begin{aligned} O &= \frac{\partial^{d-1}}{(\partial x_1)^{r_1} \cdots (\partial x_n)^{r_n}} [V(x_1, \ldots, x_n) - V(x_1 + u^T v_1, \ldots, x_n + u^T v_n)] \\ &= \sum_{i=1}^n b_{r+e_i} x_i - \sum_{i=1}^n b_{r+e_i}(x_i + u^T v_i) \\ &= \sum_{i=1}^n b_{r+e_i} u^T v_i \\ &= u^T \left(\sum_{i=1}^n b_{r+e_i} v_i \right). \end{aligned}$$

But this is true for every u, so that $\sum_{i=1}^n b_{r+e_i} v_i = O$ and $V(x)$ is a linear d-stress. That $V(x)$ is the same as the canonical linear d-stress follows from the fact that $b_F = \frac{1}{[F]}$ for every facet F of P. □

Note that the above proof is analogous to the proof of Minkowski's Theorem [6] that $\sum_{i=1}^n \operatorname{vol}_{d-1}(F_i) \frac{v_i}{\|v_i\|} = O$ where $\operatorname{vol}_{d-1}(F_i)$ is the $(d-1)$-volume of the facet F_i of P^* corresponding to v_i. In fact, the relationship is much closer, as we shall soon see.

McMullen's conditions would be a consequence of the bijectivity of ω^{d-2i} : $S^\ell_{d-i} \to S^\ell_i$, $i = 0, \ldots, \lfloor d/2 \rfloor$, or more weakly, of the surjectivity of $\omega : S^\ell_i \to S^\ell_{i-1}$, $i = 1, \ldots, \lfloor d/2 \rfloor$. We will explore some special cases of these conjectures.

What is the geometrical interpretation of the canonical linear i-stresses $\omega^{d-i}(V(x))$, $i = 1, \ldots, d$? Let $W(x) = V(x_1 + 1, \ldots, x_n + 1)$. Then for small x, $W(x)$ is the volume of a polytope near P^*. Write $W(x) = \sum_{i=0}^{d} W_i(x)$, where each $W_i(x)$ is a homogeneous polynomial of degree i. Of course, $W_0(x)$ is the volume of P^*, $W_d(x) = V(x)$, and it is easy to see that $W_1(x) = \sum_{i=1}^{n} \frac{\mathrm{vol}_{d-1}(F_i)}{\|v_i\|} x_i$.

Theorem 21 *Let P be as above. Then for $i = 0, \ldots, d$, $\omega^{d-i}(V(x)) = (d-i)!W_i(x)$.*

Proof. We calculate the contribution of $b_r \frac{x^r}{r!}$ in $V(x)$ to the coefficient of x^s in $W_i(x)$, where $x^s | x^r$. Expanding

$$b_r \frac{(x_1 + 1)^{r_1} \cdots (x_n + 1)^{r_n}}{r_1! \cdots r_n!}$$

we see that the contribution is

$$b_r \frac{\binom{r_1}{s_1} \cdots \binom{r_n}{s_n}}{r_1! \cdots r_n!} = \frac{b_r}{s_1!(r_1 - s_1)! \cdots s_n!(r_n - s_n)!}.$$

On the other hand, the contribution of

$$b_r \frac{x^r}{r!} = b_r \frac{x_1^{r_1} \cdots x_n^{r_n}}{r_1! \cdots r_n!}$$

in $V(x)$ to the coefficient of x^s in $\omega^{d-i}(V(x))$, where $i = d - |s|$, is

$$b_r \frac{\binom{d-i}{r_1 - s_1, \ldots, r_n - s_n}}{r_1! \cdots r_n!} = \frac{(d-i)! b_r}{s_1!(r_1 - s_1)! \cdots s_n!(r_n - s_n)!}. \quad \square$$

Corollary 22 *Let P be as above.*

1. *The canonical linear 0-stress $\omega^d(V(x))$ equals $d!\mathrm{vol}(P^*)$.*
2. *The canonical linear 1-stress $\omega^{d-1}(V(x))$ equals $(d-1)! \sum_{i=1}^{n} \frac{\mathrm{vol}_{d-1}(F_i)}{\|v_i\|} x_i$. That is, the canonical linear combination of the v_i induced by ω is (up to scalar multiple) the same as that induced by Minkowski's Theorem.*

We remark that the coefficient of the square-free term of W_i corresponding to the $(i-1)$-face F of P equals

$$\frac{\mathrm{vol}_{d-i}(F^*)}{i!\mathrm{vol}_i(\mathrm{conv}(\{O\} \cup \{v_i : v_i \in F\}))}, \tag{8}$$

where F^* is the face of P^* corresponding to F. See also Filliman [5].

Theorem 23 *Let P be as above. Then $\omega^d : S^\ell_d \to S^\ell_0$ is a bijection. Further, if $d \geq 3$, then $\omega^{d-2} : S^\ell_{d-1} \to S^\ell_1$ is a bijection.*

PROOF. The first statement is true by part (1) of the previous corollary simply because P^* has positive (and hence nonzero) volume. From the Dehn-Sommerville Relations we know that $\dim S^\ell_{d-1} = h_{d-1} = h_1 = \dim S^\ell_1$. So it suffices to show that $\omega^{d-2} : S^\ell_{d-1} \to S^\ell_1$ is a surjection. From Theorem 17, we know that $\{\tau_1(V(x)), \ldots, \tau_n(V(x))\}$ spans S^ℓ_{d-1}. We need to show that $\{\omega^{d-2}\tau_1(V(x)), \ldots, \omega^{d-2}\tau_n(V(x))\}$ spans S^ℓ_1. Since $\dim S^\ell_1 = h_1 = n - d$, it is sufficient to demonstrate that the given subset of S^ℓ_1 has rank $n-d$. But since ω and τ_i commute, our set equals $(d-2)!\{\tau_1(W_2(x)), \ldots, \tau_n(W_2(x))\}$. It is straightforward to check that $\sum_{i=1}^n x_i\tau_i(W_2(x)) = 2W_2(x)$, and the result now follows from the Brunn-Minkowski Theorem [6], which implies that the rank of the Hessian of $W(x)$ is $n-d$. □

The above theorem implies that $\omega : S^\ell_2 \to S^\ell_1$ is s surjection when $d \geq 3$. Hence $\dim S^a_2 = h_2 - h_1 = g_2$ and Theorem 3 is a corollary.

Consider the ring of all differential operators with constant coefficients in the variables $x_1, \ldots, x_n$. Factor out the ideal of operators that annihilate the polynomial $V(x)$. Khovanskii (personal communication) observed that the resulting ring R is isomorphic to the cohomology ring of the projective toric variety associated with the polytope. This implies that R is isomorphic to $B = A/(\theta_1, \ldots, \theta_d)$, where the coefficients of the θ_i are given by the coefficients of the vertices of P. Indeed, we can see this latter fact directly. Clearly $\tau_S(V(x))$ equals 0 for any subset $S \notin \Delta$. But the invariance of the polynomial $V(x)$ under translation (see the proof of Theorem 20) implies that

$$\sum_{i=1}^n v_{ij}\frac{\partial V(x)}{\partial x_i} = 0$$

for each $j = 1, \ldots, d$. Finally, Theorem 17 implies that the image of $V(x)$ under the homogeneous differential operators of degree k spans S^ℓ_{d-k}, and hence has dimension $h_{d-k} = h_k$. Using Theorem 5, this suffices to prove that R is isomorphic to B under the map $\frac{\partial}{\partial x_i} \to x_i$.

11. Simplicial 3-Polytopes

We will summarize the consequences of the previous section in the case that $d = 3$. Suppose that P is a simplicial convex 3-polytope containing the origin in its interior with vertices $v_1, \ldots, v_n$, and that P^* is its polar. Then the canonical linear 0-stress is 3! times the volume of P^*, the canonical linear 1-stress is equivalent to the linear relation induced by Minkowski's Theorem, the canonical linear 2-stress is the classical Maxwell stress (shown by Filliman [4]), and the canonical linear 3-stress is the volume polynomial $V(x)$. Hence, $S^\ell_0 = \mathbf{R}$, S^ℓ_1 has dimension $n-3$ and is isomorphic to the space of all linear relations on the v_i, S^ℓ_2 has dimension $n-3$, is spanned by the $\tau_i V(x)$, and is isomorphic to S^ℓ_1, and S^ℓ_3 is spanned by $V(x)$. Also, $S^a_0 = \mathbf{R}$, S^a_1 has dimension $n-4$ and is isomorphic to the space of all affine relations on the v_i, and both S^a_2 and S^a_3 are trivial. That $\omega^3 : S^\ell_3 \to S^\ell_0$ is a bijection is equivalent to P^* having nonzero volume, and that $\omega : S^\ell_2 \to S^\ell_1$ is a bijection is equivalent to Dehn's Theorem (Theorem 2) and is a consequence of the Brunn-Minkowski Theorem. So in dimension 3 we see a striking confluence of geometric results.

12. The Polytope Algebra

McMullen [11] recently found a new proof of the g-Theorem for simplicial convex d-polytopes using the powerful tool of his polytope algebra. In particular, his arguments demonstrate that ω^{d-2i} of the previous section is bijective and hence $\dim S_k^a = g_k(P)$, $k = 0, \ldots, \lfloor d/2 \rfloor$. In the process, he discovered new geometric inequalities (the Hodge-Riemann-Minkowski inequalities), generalized the Aleksandrov-Fenchel Inequality, and elucidated the geometric role of the Hard Lefschetz Theorem.

Use a generator $[P]$ for each convex polytope in $\mathbf{R}^d$ to generate an abelian group, and take $0 = [\emptyset]$. Impose the relations $[P \cup Q] + [P \cap Q] = [P] + [Q]$ for all polytopes P and Q whose union is also a polytope, and $[P + t] = [P]$ for all polytopes P and translation vectors t. Define multiplication by $[P] \cdot [Q] = [P + Q]$, extending by linearity. The dilatation operator is defined by $\Delta(\lambda)[P] = [\lambda P]$ for polytopes P and real numbers λ. The result is the polytope algebra Π.

Now fix a simple d-polytope $P \subset \mathbf{R}^d$ and consider $\Pi(P)$, the subalgebra generated by the classes of the summands of P. McMullen shows that there exists a grading of $\Pi(P)$ such that kth weight space has dimension $h_k(P^*)$ and that $\Pi(P)$ admits a Lefschetz decomposition.

In so doing he defines a k-weight on P to be a choice of number $\omega(F)$ for each k-face F of P, which satisfies the Minkowski relations

$$\sum_{F_j \subset G} \omega(F_j) u_j = O,$$

where the sum runs over all the k-faces of a $(k+1)$-face G of P, with u_j the unit outer normal vector to G at F_j with respect to aff G.

Given simplicial convex polytope Q containing O in its interior, with linear stresses defined with respect to its vertices $v_1, \ldots, v_n$, an isomorphism between linear $(d-k)$-stresses for Q and the k-weights for Q^* is given by

$$\omega(F^*) = (d-k)! b_F \operatorname{vol}_{d-k}(\operatorname{conv}(\{O\} \cup \{v_i : v_i \in F\})).$$

In the above, F^* is the face of Q^* corresponding to F under duality. In particular, the canonical linear $(d-k)$-stresses for Q simply become the relative volumes of the dual faces — see (8).

Refer to Filliman [4] for other connections between stress and the Aleksandrov-Fenchel Inequality.

13. Acknowledgments

The author's definition of k-stress was motivated by Kalai's results on algebraic shifting and on stress. Most of the results in Sections 4, 5, 6, 8, and 9 were developed in 1987–88, but the basis for the formulas for the coefficients actually rests in some calculations begun during the author's stay at the Mathematical Institute, Ruhr University in Bochum, 1984–85, which was supported by a fellowship from the Alexander von Humboldt Foundation. Jonathan Fine's suggestion in Oberwolfach, 1989 that the volumes of the faces of the dual simple polytope should play an important role led to the material in Sections 10 and 11, mostly discovered during the author's stay at DIMACS, Rutgers University in 1989–90. The definition of infinitesimal motions (Section 7) resulted from conversations with White during a visit to the Mittag-Leffler Institute in early 1992, and has subsequently been evolving in dialogue with Sue Foege, Robert Hebble, Stewart Tung, Neil White, and Walter Whiteley.

The author is grateful for stimulating conversations with a number of individuals, including Louis Billera, Bob Connelly, Paul Filliman, Jonathan Fine, Sue Foege, Robert Hebble, A. Khovanskii, Peter Kleinschmidt, Peter McMullen, Rolf Schneider, Stewart Tung, Neil White, and Walter Whiteley.

References

1. Billera, L. J. and Lee, C. W. (1981), 'A proof of the sufficiency of McMullen's conditions for f-vectors of simplicial convex polytopes', *J. Combin. Theory Ser. A* **31**, 237–255.
2. Dehn, M. (1916), 'Über die Starrheit konvexer Polyeder', *Math. Ann.* **77**, 466–473.
3. Filliman, P. (1991), 'Face numbers of pl-spheres', manuscript.
4. Filliman, P. (1992), 'Rigidity and the Alexandrov-Fenchel inequality', *Monats. Math.* **113**, 1–22.
5. Filliman, P. (1992), 'The volume of duals and sections of polytopes', *Mathematika* **39**, 67–80.
6. Grünbaum, B. (1967), *Convex Polytopes*, Interscience Publishers, New York.
7. Kalai, G. (1984), 'Characterization of f-vectors of families of convex sets in $\mathbf{R}^d$. Part I: Necessity of Eckhoff's conditions', *Israel J. Math.* **48**, 175–195.
8. Kalai, G. (1987), 'Rigidity and the lower bound theorem. I', *Invent. Math.* **88**, 125–151.
9. Kind, B. and Kleinschmidt, P. (1979), 'Schälbare Cohen-Macauley-Komplexe und ihre Parametrisierung', *Math. Z.* **167**, 173–179.
10. Lee, C. W. (1990), 'Some recent results on convex polytopes', *Contemp. Math.* **114**, 3–19.
11. McMullen, P. (1993), 'On simple polytopes', *Invent. Math.* **113**, 419–444.
12. Pachner, U. (1990), 'Shellings of simplicial balls and p.l. manifolds with boundary', *Discrete Math.* **81**, 37–47.
13. Reisner, G. (1976), 'Cohen-Macaulay quotients of polynomial rings', *Adv. in Math.* **21**, 30–49.
14. Roth, B. (1981), 'Rigid and flexible frameworks', *Amer. Math. Monthly* **88**, 6–21.
15. Stanley, R. P. (1975), 'The upper bound conjecture and Cohen-Macaulay rings', *Stud. Appl. Math.* **54**, 135–142.

16. Stanley, R. P. (1978), 'Hilbert functions of graded algebras', *Adv. in Math.* **28**, 57–83.
17. Stanley, R.P. (1980), 'The number of faces of a simplicial convex polytope', *Adv. in Math.* **35**, 236–238.
18. Tay, T.-S., White, N., and Whiteley, W., (1992), 'Skeletal rigidity of cell complexes', to appear.
19. Whiteley, W. (1984), 'Infinitesimally rigid polyhedra. I. Statics of frameworks', *Trans. Amer. Math. Soc.* **285**, 431–465.

POLYTOPES AND BRUNN-MINKOWSKI THEORY

ROLF SCHNEIDER
Mathematisches Institut, Albert-Ludwigs-Universität
Albertstr. 23b, D-79104 Freiburg, Germany

Abstract. We consider several questions from the Brunn-Minkowski theory of convex bodies, concentrating on problems that can be treated successfully if restricted to polytopes. Topics under investigation are a variant of indecomposability with respect to Minkowski addition, addition and decomposition problems involving intermediate area measures, determination by volumes of projections, special representations of mixed volumes, and the equality problem for the Aleksandrov-Fenchel inequality.

Key words: Polytope, Brunn-Minkowski theory, Minkowski addition, summand, indecomposability, area measure, intermediate area measure, Blaschke sum, Aleksandrov-Fenchel-Jessen theorem, projection, projection function, Baire category, mixed volume, Aleksandrov-Fenchel inequality

1. Introduction

By the Brunn-Minkowski theory one understands that part of the classical geometry of convex bodies which is concerned with Minkowski addition, volume, surface area, and with the outcome of combining these notions. Central objects of the theory are mixed volumes, mixed surface area measures, quermassintegrals, and related uniqueness, existence and extremal problems. Polytopes play a particular role in Brunn-Minkowski theory, under different aspects. On one hand they are a useful tool, since often results are conveniently proved for polytopes first and then extended to general convex bodies by approximation. On the other hand, it may happen that for polytopes, or for special classes of polytopes, one can prove results that either do not hold for general convex bodies or are still out of reach for these. In the following, we shall treat some new results of this type. In some cases, the results obtained for polytopes will allow us to conclude that most convex bodies, in the Baire category sense, have the property in question.

The Brunn-Minkowski theory of convex bodies is the subject of a recent book by the author (Schneider, 1993). In the following, this book will be referred to as CB. The present article can be considered as a collection of supplements to, and extensions of, several sections of that book, with a special view to polytopes. We shall also quote and comment on some related and more recent literature.

Section 2 is devoted to Minkowski addition alone, that is, volume or other notions from Brunn-Minkowski theory do not play a role in this section. We treat a variant of the notion of indecomposability with respect to Minkowski addition. The subject of Section 3 are the lower order area measures of convex bodies and in particular of polytopes. This section contains refinements and strengthenings of various former results on addition and decomposition problems connected with intermediate area measures, and on the determination of convex bodies from projection functions.

T. Bisztriczky et al. (eds.), POLYTOPES: Abstract, Convex and Computational, 273–299.

In Section 4, devoted to mixed volumes, we first generalize a representation, due to Betke, for mixed volumes of polytopes, and then extend a result of Ewald and Tondorf, concerning equality in the Aleksandrov-Fenchel inequality in a special case involving polytopes.

2. Indecomposable Pairs of Polytopes

For convex bodies (nonempty, compact, convex subsets) K, M in n-dimensional euclidean vector space $\mathbb{E}^n$, the *Minkowski sum* is defined by $K + M := \{x + y : x \in K,\ y \in M\}$. The convex body M is a *summand* of the convex body K if $K = M + L$ with a suitable convex body L. The body K is called *indecomposable* if every summand of K is homothetic to K, that is, of the form $\lambda K + t$ with $\lambda \geq 0$ and $t \in \mathbb{E}^n$. The indecomposable bodies in $\mathbb{E}^2$ are precisely the segments and the triangles, but for $n \geq 3$, most convex bodies (in the sense of Baire category) in $\mathbb{E}^n$ are indecomposable. We refer to Section 3.2 of CB for proofs and references. Note 8 of that section gives hints to results on summands and decomposability of polytopes.

If K and M are indecomposable convex bodies, their sum $K + M$ may well have non-trivial summands, that is, other than sums of homothets of K and M. For example, the sum of a triangle and its image under reflection in a point is a centrally symmetric hexagon, which is the sum of three segments, and hence the sum of two parallelograms. In the following we say that the pair (K, M) of convex bodies in $\mathbb{E}^n$ is *indecomposable* if K and M are indecomposable and every summand of the sum $K + M$ is of the form $\lambda K + \mu M + t$ with $\lambda,\ \mu \geq 0$ and $t \in \mathbb{E}^n$. It is easy to see (using CB, Th. 3.2.11 and pp. 144 - 145) that in $\mathbb{E}^2$ a pair (K, M) is indecomposable in precisely the following cases: (a) K and M are segments, (b) K and M are triangles having together at most four different outward unit normal vectors at edges, (c) the pair (K, M) consists of a triangle and a segment which is parallel to an edge of the triangle. In $\mathbb{E}^n$, examples of indecomposable convex bodies K, M for which the pair (K, M) is not indecomposable can be obtained as follows. Let K, M be indecomposable convex bodies (for example, simplicial polytopes) such that $K \cup M$ is convex. By CB (3.1.4) we have $K + M = (K \cup M) + (K \cap M)$. If neither of the bodies K, M contains the other, this yields the required example. On the other hand, Theorem 1.1 below will show that quite general pairs of polytopes are indecomposable. From this we can deduce that for $n \geq 3$ most pairs (in the sense of Baire category) of convex bodies in $\mathbb{E}^n$ are indecomposable; see Theorem 1.2.

2.1. Preliminaries

Most of the terminology and the results used in the following are found in CB. The subsequent investigation can be considered as a continuation of Section 3.2 of that book.

We denote by $S^{n-1} := \{x \in \mathbb{E}^n : ||x|| = 1\}$ the unit sphere and by $B^n := \{x \in \mathbb{E}^n : ||x|| \leq 1\}$ the unit ball of $\mathbb{E}^n$. The space of convex bodies in $\mathbb{E}^n$ is denoted by $\mathcal{K}^n$ and is, as usual, equipped with the Hausdorff metric δ. It contains the set of simplicial polytopes as a dense subset. By $\mathcal{K}_0^n \subset \mathcal{K}^n$ we denote the set of convex bodies with interior points.

For $K \in \mathcal{K}^n$ and $u \in S^{n-1}$ we denote by $F(K,u)$ the support set of K with exterior normal vector u. We shall often make use of the relation

$$F(K+M,u) = F(K,u) + F(M,u)$$

for K, $M \in \mathcal{K}^n$ (see CB, Th. 1.7.5).

For a set $F \subset \mathbb{E}^n$ we denote by $L(F)$ the linear subspace of $\mathbb{E}^n$ that arises by translation of the affine hull of F. Let $P, Q \subset \mathbb{E}^n$ be convex polytopes. The polytopes P and Q are said to be *in general relative position* if for any two nonempty faces F of P and G of Q lying in parallel supporting hyperplanes of P and Q, respectively, we have

$$L(F) \cap L(G) = \{0\}.$$

Let P and Q be two polytopes. Then the set of rotations ρ in the rotation group $SO(n)$ of $\mathbb{E}^n$ for which P and ρQ are not in general relative position has Haar measure zero in $SO(n)$. This follows from CB, Lemma 4.5.1, and implies the following lemma.

Lemma 2.1. *If $P, Q \subset \mathbb{E}^n$ are convex polytopes, then every neighbourhood of the identity in $SO(n)$ contains a rotation ρ such that P and ρQ are in general relative position.*

2.2. Indecomposable Pairs of Polytopes

First we find indecomposable pairs of a very special type.

Lemma 2.2. *If $P_1 \subset L_1$ and $P_2 \subset L_2$ are indecomposable polytopes in complementary linear subspaces L_1, L_2 of $\mathbb{E}^n$, then the pair (P_1, P_2) is indecomposable.*

Proof. We may assume, without loss of generality, that $P_1 + P_2$ is n-dimensional. Let K be a summand of $P_1 + P_2$, say $P_1 + P_2 = K + M$ with $K, M \in \mathcal{K}^n$. Then K and M are polytopes (see CB, p. 143). First we assume that K is of dimension n. Let $u \in S^{n-1}$ be an outer normal vector of a facet of K. Then

$$F(K,u) + F(M,u) = F(P_1,u) + F(P_2,u).$$

If $u \notin L_1^\perp \cup L_2^\perp$, then

$$n-1 = \dim(F(P_1,u) + F(P_2,u)) = \dim F(P_1,u) + \dim F(P_2,u) < n-1,$$

a contradiction. It follows that $u \in L_1^\perp$ or $u \in L_2^\perp$ for each normal vector of a facet of K. Since K is the intersection of the supporting halfspaces determined by its facets, we deduce that $K = K_1 + K_2$ with polytopes $K_1 \subset L_1$ and $K_2 \subset L_2$.

If K is of dimension less than n, then

$$P_1 + P_2 = (K + \lambda M) + (1-\lambda)M$$

for $0 < \lambda \leq 1$, where $K + \lambda M$ is n-dimensional and thus of the form $K + \lambda M = K_1(\lambda) + K_2(\lambda)$ with $K_1(\lambda) \subset L_1$ and $K_2(\lambda) \subset L_2$. With $\lambda \to 0$ we obtain a representation $K = K_1 + K_2$ with $K_1 \subset L_1$, $K_2 \subset L_2$.

Analogously, we show that $M = M_1 + M_2$ with $M_1 \subset L_1$ and $M_2 \subset L_2$. It follows that $P_1 = K_1 + M_1$. Since P_1 is indecomposable, K_1 is homothetic to P_1. Similarly, K_2 is homothetic to P_2, and thus $K = \lambda_1 P_1 + \lambda_2 P_2 + t$ with $\lambda_1, \lambda_2 \geq 0$ and $t \in \mathbb{E}^n$. Since K was an arbitrary summand of $P_1 + P_2$, the pair (P_1, P_2) is indecomposable. ∎

The following theorem yields more general examples of indecomposable pairs of polytopes.

Theorem 2.1. *Let $n \geq 3$, and let $P, Q \subset \mathbb{E}^n$ be n-dimensional polytopes in general relative position. If all the two-dimensional faces of P and Q are triangles, then the pair (P, Q) is indecomposable.*

Proof. First we note that all faces of P and Q, including P and Q themselves, are indecomposable (CB, Corollary 3.2.13). Suppose that K is a summand of $P + Q$, thus $P + Q = K + M$ with suitable polytopes $K, M \in \mathcal{K}^n$. Let F be a face of $P + Q$ and choose a vector $u \in S^{n-1}$ for which $F = F(P + Q, u)$. Then

$$F(K, u) + F(M, u) = F = F(P, u) + F(Q, u).$$

Since P and Q are in general relative position, the sum $F(P, u) + F(Q, u)$ is direct. From Lemma 2.2 and the indecomposability of $F(P, u)$ and $F(Q, u)$ we deduce

$$F(K, u) = \lambda_F F(P, u) + \mu_F F(Q, u) + t_F \tag{1}$$

with $\lambda_F, \mu_F \geq 0$ and $t_F \in \mathbb{E}^n$. Here λ_F, μ_F, t_F do, in fact, depend only on F and not on the choice of u, since the same is true for the faces $F(P, u)$ and $F(Q, u)$ (see CB, pp. 143 - 144).

As above, let $F = F(P+Q, u) = F(P, u) + F(Q, u)$ be a face of $P+Q$. If $F(Q, u)$ is one–pointed, we call F a *red* face; if $F(P, u)$ is one–pointed, we call F a *blue* face. If, in particular, F is an edge of $P + Q$, then not both $F(P, u)$ and $F(Q, u)$ can be (non-degenerate) edges, since P and Q are in general relative position. Hence, each edge of $P + Q$ is either red or blue. Returning to a general face F, let G be a face of F. Choose $v \in S^{n-1}$ with $G = F(P + Q, v)$. Then $F(P, v)$ is a face of $F(P, u)$ (CB, p. 144). In particular, if F is a red face, then G is a red face. For arbitrary F and G as before,

$$F(K, v) = \lambda_G F(P, v) + \mu_G F(Q, v) + t_G \tag{2}$$

by (1). Since $F(K, v)$ is a face of $F(K, u)$, we deduce from (1) that

$$\begin{aligned} F(K, v) &= F(F(K, u), v) \\ &= \lambda_F F(F(P, u), v) + \mu_F F(F(Q, u), v) + t'_F \\ &= \lambda_F F(P, v) + \mu_F F(Q, v) + t'_F. \end{aligned}$$

Since $\lambda_F F(P, v) + \mu_F F(Q, v)$ is a direct sum, we can conclude from (2) that $\lambda_G = \lambda_F$ if $\dim F(P, v) \geq 1$. In particular, this holds if G contains a red edge.

If $F = F(P, u)$ is a facet of P, then $F(Q, u)$ is of dimension zero, since P and Q are in general relative position; say $F(Q, u) = \{q_F\}$. Then $F' := F + q_F$ is a facet of $P + Q$, and we define $\lambda_F := \lambda_{F'}$.

Now let $F = F(P,u)$ and $G = F(P,v)$, where $u,v \in S^{n-1}$, be two facets of P having an $(n-2)$-face H in common. For each vector $w \in \operatorname{relint} \operatorname{pos}\{u,v\}$ the face $F(Q,w)$ of Q either is of dimension zero or is an edge not parallel to H, since $F(P,w) = H$ and P and Q are in general relative position. Let $u_1,\dots,u_m \in S^{n-1}$ be the vectors in $\operatorname{relint} \operatorname{pos}\{u,v\}$ for which $F(Q,u_i) = E_i$ is an edge of Q. If $u_1,\dots,u_m$ are suitably ordered, the edges $E_1,\dots,E_m$, in this order, form a polygonal path connecting the vertices q_F and q_G of Q. (To see this, let Π be the orthogonal projection onto the linear subspace spanned by u and v. The vertices Πq_F and Πq_G of the polygon ΠQ can be connected by a chain of edges with normal vectors in $\operatorname{relint} \operatorname{pos}\{u,v\}$. The pre-images of the edges (vertices) of this chain are edges (vertices) of Q.) In the sequence

$$F + q_F, H + E_1, \dots, H + E_m, G + q_G$$

of facets of $P+Q$ each two consecutive ones have a red $(n-2)$-face in common. It follows that $\lambda_F = \lambda_G$. Since any two facets of P can be connected by a chain of facets of which any two consecutive ones intersect in an $(n-2)$-face, we deduce that $\lambda_F = \lambda$, say, for each facet F of P.

Now let E be a red edge of $P+Q$ and choose $v \in S^{n-1}$ with $E = F(P+Q,v)$. The edge $F(P,v)$ of P belongs to some facet F of P, say $F = F(P,u)$ with $u \in S^{n-1}$. Consider the faces of the polytope $P+Q$ of the form $F(P+Q,w)$ with $w \in \operatorname{relint} \operatorname{pos}\{u,v\}$. They can be ordered to form a chain connecting $F + q_F$ and E and such that any two consecutive faces have a common red edge. (To see this, we extend the argument used above. Let Π be the orthogonal projection onto the linear subspace spanned by u and v. The vertices Πq_F and Πq_E, where $\{q_E\} = F(Q,v)$, can be connected by a chain of edges with normal vectors in $\operatorname{relint} \operatorname{pos}\{u,v\}$. Let w_1, w_2 be the normal vectors of two consecutive edges in this chain and let $w \in \operatorname{relint} \operatorname{pos}\{w_1,w_2\}$. Let x be a vertex of $F(Q,w)$; then x is a common vertex of $F(Q,w_1)$ and $F(Q,w_2)$. Since the sum $F(P,w) + F(Q,w)$ is direct, $F(P,w) + \{x\}$ is a face of $F(P,w) + F(Q,w)$ and thus a face of $P+Q$. This face is common to $F(P+Q,w_1)$ and $F(P+Q,w_2)$, and it is a red edge.) It follows that $\lambda_E = \lambda$.

In a similar way one shows the existence of a number μ so that $\mu_E = \mu$ for all blue edges E of $P+Q$.

Let x,y be two distinct vertices of K. We can choose vectors $u,v \in S^{n-1}$ so that $F(K,u) = \{x\}$ and $F(K,v) = \{y\}$, $F(P,u) = \{x_P\}$ and $F(P,v) = \{y_P\}$ for vertices x_P, y_P of P, $F(Q,u) = \{x_Q\}$ and $F(Q,v) = \{y_Q\}$ for vertices x_Q, y_Q of Q, finally $u \neq -v$. Let Π denote the orthogonal projection onto the linear subspace spanned by u and v. Since u and v can be chosen from open sets on S^{n-1}, we can assume, without loss of generality, that $P+Q$ has a sharp shadow boundary under the projection Π. Let $u_1,\dots,u_k$ be those unit vectors in $\operatorname{pos}\{u,v\}$ for which $F(\Pi(P+Q),u_i)$ is an edge of $\Pi(P+Q)$. For each $i \in \{1,\dots,k\}$, either $F(\Pi P,u_i)$ is an edge of ΠP or $F(\Pi Q,u_i)$ is an edge of ΠQ, but not both, since $F(P+Q,u_i)$ is an edge and P and Q are in general relative position. All edges of ΠP and ΠQ with outer normal vector in $\operatorname{pos}\{u,v\}$ are obtained in this way, since $\Pi P + \Pi Q = \Pi(P+Q)$. We may assume that $F(\Pi P,u_i)$ is an edge of ΠP for $i = 1,\dots,m$, and $F(\Pi Q,u_i)$ is an edge of ΠQ for $i = m+1,\dots,k$.

The edges $F(P, u_i)$, $i = 1, \ldots, m$, form a simple polygonal path connecting x_P and y_P, and the edges $F(Q, u_i)$, $i = m+1, \ldots, k$, form a simple polygonal path connecting x_Q and y_Q. Similarly, the edges among the faces $F(K, w)$ with $w \in \operatorname{pos}\{u, v\}$ connect x and y, and each edge of this form is parallel to either a red or a blue edge of $P + Q$. Hence, if $\bar{e}_i$ is a vector parallel to the edge $F(P + Q, u_i)$, of the same length and suitably oriented, then

$$\begin{aligned} y_P - x_P &= \sum_{i=1}^{m} \bar{e}_i, \\ y_Q - x_Q &= \sum_{i=m+1}^{k} \bar{e}_i, \\ y - x &= \sum_{i=1}^{m} \lambda \bar{e}_i + \sum_{i=m+1}^{k} \mu \bar{e}_i \\ &= \lambda(y_P - x_P) + \mu(y_Q - x_Q). \end{aligned}$$

After suitable translations, we may assume that $x = x_P = x_Q = 0$. Then

$$y = \lambda y_P + \mu y_Q \in \lambda P + \mu Q$$

for arbitrary vertices y of K. On the other hand, every vertex of $\lambda P + \mu Q$ is obtained in this way. It follows that $K = \lambda P + \mu Q$. Since K was an arbitrary summand of $P + Q$, this shows that the pair (P, Q) is indecomposable and thus completes the proof of Theorem 2.1. ∎

2.3. Application to General Convex Bodies

Theorem 2.1 can be used to show that, for $n \geq 3$, most pairs of convex bodies in $\mathbb{E}^n$ are indecomposable. Here 'most' is understood in the sense of Baire category (for explanations, see, e.g., the introduction and the notes of Section 2.6 in (Schneider, 1993)). The product space $\mathcal{K}^n \times \mathcal{K}^n$, endowed with the metric defined by $\delta((K, M), (\bar{K}, \bar{M})) = \delta(K, \bar{K}) + \delta(M, \bar{M})$, is a complete metric space and hence a Baire space. Theorem 2.2 shows that the complement of the set of indecomposable pairs in $\mathcal{K}^n \times \mathcal{K}^n$ $(n \geq 3)$ is meagre.

Theorem 2.2. *For $n \geq 3$, the set of indecomposable pairs of convex bodies in $\mathcal{K}^n \times \mathcal{K}^n$ is a dense G_δ-set.*

Proof. Let $\mathcal{A}$ denote the set of indecomposable pairs in $\mathcal{K}^n \times \mathcal{K}^n$. If $(K, M) \in \mathcal{K}^n \times \mathcal{K}^n$ and a number $\epsilon > 0$ are given, there exist simplicial polytopes P, Q such that $\delta(K, P) < \epsilon/3$ and $\delta(M, Q) < \epsilon/3$. By Lemma 2.1, every neighbourhood of the identity in the rotation group $SO(n)$ contains a rotation ρ such that P and ρQ are in general relative position. Since the operation of the rotation group on $\mathcal{K}^n$ is continuous, ρ can be chosen so that $\delta(Q, \rho Q) < \epsilon/3$. Thus $\delta((K, M), (P, \rho Q)) < \epsilon$, and by Theorem 2.1 the pair $(P, \rho Q)$ is indecomposable. Thus $\mathcal{A}$ is dense in $\mathcal{K}^n \times \mathcal{K}^n$.

For $k \in \mathbb{N}$ we now define the set

$$\mathcal{A}_k := \{(K, M) \in \mathcal{K}^n \times \mathcal{K}^n : K + M \text{ has a summand } A \text{ with}$$

$$\delta(A, \lambda K + \mu M + t) \geq 1/k \text{ for all } \lambda, \mu \in [0,1] \text{ and } t \in \mathbb{E}^n\}$$

We show that $\mathcal{A}_k$ is closed. For this, let $((K_i, M_i))_{i\in\mathbb{N}}$ be a sequence in $\mathcal{A}_k$ converging to $(K, M) \in \mathcal{K}^n \times \mathcal{K}^n$. For each $i \in \mathbb{N}$ there exist convex bodies $A_i, B_i \in \mathcal{K}^n$ such that

$$K_i + M_i = A_i + B_i, \qquad \delta(A_i, \lambda K_i + \mu M_i + t) \geq 1/k \tag{3}$$

for all $\lambda, \mu \in [0,1]$ and all $t \in \mathbb{E}^n$. By changing B_i appropriately, we may assume that each A_i has Steiner point 0. Then the sequence $(A_i)_{i\in\mathbb{N}}$ is bounded. By the Blaschke selection theorem, it has a subsequence converging to some convex body A. We may assume (changing the notation) that the sequence $(A_i)_{i\in\mathbb{N}}$ converges to A. From $K_i \to K$, $M_i \to M$ and $A_i \to A$ for $i \to \infty$ we get $B_i \to B$ for some convex body B (e.g., using CB, Th. 1.8.12). Then $K + M = A + B$, hence A is a summand of $K + M$. For arbitrary $\lambda, \mu \in [0,1]$ and $t \in \mathbb{E}^n$ the inequality (3) holds, hence $\delta(A, \lambda K + \mu M + t) \geq 1/k$ and thus $(K, M) \in \mathcal{A}_k$. Thus $\mathcal{A}_k$ is closed.

We want to show that

$$(\mathcal{K}^n \times \mathcal{K}^n) \setminus \bigcup_{k\in\mathbb{N}} \mathcal{A}_k = \mathcal{A}. \tag{4}$$

Let $(K, M) \in (\mathcal{K}^n \times \mathcal{K}^n) \setminus \bigcup_{k\in\mathbb{N}} \mathcal{A}_k$. Suppose that $A \in \mathcal{K}^n$ is a summand of $K + M$. Let $k \in \mathbb{N}$. Since $(K, M) \notin \mathcal{A}_k$, there exist $\lambda_k, \mu_k \in [0,1]$ and $t_k \in \mathbb{E}^n$ such that $\delta(A, \lambda_k K + \mu_k M + t_k) < 1/k$. Selecting suitable subsequences and changing the notation, we may assume that $\lambda_k \to \lambda$, $\mu_k \to \mu$, $t_k \to t$ for $k \to \infty$. Then $\delta(A, \lambda K + \mu M + t) = 0$ and hence $A = \lambda K + \mu M + t$. We have proved that (K, M) is indecomposable and thus $(K, M) \in \mathcal{A}$.

Let $(K, M) \in \mathcal{A}$ and suppose that (K, M) is not an element of the left side of (4). Then $(K, M) \in \mathcal{A}_k$ for some $k \in \mathbb{N}$, hence there exist $A, B \in \mathcal{K}^n$ satisfying

$$K + M = A + B, \tag{5}$$

$$\delta(A, \lambda K + \mu M + t) \geq 1/k \tag{6}$$

for all $\lambda, \mu \in [0,1]$ and all $t \in \mathbb{E}^n$. Since (K, M) is indecomposable, we must have $A = \alpha K + \beta M + z$ with suitable $\alpha, \beta \geq 0$ and $z \in \mathbb{E}^n$. Here necessarily $\alpha > 1$ or $\beta > 1$ by (6), say $\alpha = 1 + \tau$ with $\tau > 0$. Then $\beta < 1$, and (5) yields $(1-\beta)M = \tau K + B + z$; hence $\tau/(1-\beta)K$ is a summand of M. Since M is indecomposable, M and K are homothetic; thus A is homothetic to K. But then A, being a summand of $K + M$, has also a representation $A = \lambda K + \mu M + t$ with $\lambda, \mu \in [0,1]$ and $t \in \mathbb{E}^n$, which contradicts (6). Thus (4) holds, which completes the proof of Theorem 2.2. ■

3. Surface Area Measures of Lower Order

Background material for the following is found in CB, Sections 7.1 and 7.2. For the purpose of introduction, we recall Minkowski's existence theorem and the notion of area measures of lower order.

Let $K \in \mathcal{K}^n$ be a convex body. For a Borel set ω in the unit sphere S^{n-1}, one denotes by $S_{n-1}(K, \omega)$ the surface area ($= (n-1)$-dimensional Hausdorff measure) of the set of boundary points of K at which K has an outer unit normal vector belonging to ω. Then $S_{n-1}(K,\cdot)$ is a finite measure on S^{n-1}, called the (surface) area measure of K. If φ is the area measure of a convex body, then

$$\int_{S^{n-1}} u\,d\varphi(u) = 0. \tag{7}$$

Vice versa, if φ is a finite (Borel) measure on S^{n-1} satisfying (7) and not concentrated on any great subsphere of S^{n-1}, then there exists an n-dimensional convex body K for which $S_{n-1}(K,\cdot) = \varphi$. The body K is uniquely determined up to a translation. This is (the general version of) Minkowski's existence and uniqueness theorem (proof and references are given in CB, Section 7.1).

The area measure S_{n-1} generates a series of further measures by means of the Steiner–type formula

$$S_{n-1}(K+\rho B^n,\cdot) = \sum_{i=0}^{n-1} \rho^{n-1-i}\binom{n-1}{i} S_i(K,\cdot),$$

where $B^n := \{x \in \mathbb{E}^n : \|x\| \le 1\}$ is the unit ball of $\mathbb{E}^n$ (see CB, Section 4.2). One calls $S_i(K,\cdot)$ the *area measure of order* i (or ith area measure) of the convex body K. In order that a finite measure φ on S^{n-1} be the ith area measure of some convex body, for given $i \in \{1,\dots,n-1\}$, condition (7) is still necessary, but for $i < n-1$ no longer sufficient. One knows necessary and sufficient conditions for $i = 1$ (as quoted in CB, 4.3), but not for $1 < i < n-1$. The uniqueness assertion, however, does extend: if $i \in \{1,\dots,n-1\}$ and if $K, L \in \mathcal{K}^n$ are of dimension at least $i+1$ and satisfy

$$S_i(K,\cdot) = S_i(L,\cdot),$$

then K and L are translates of each other. This result is known as the Aleksandrov–Fenchel–Jessen theorem (CB, Corollary 7.2.5).

The general theme of the following considerations is, vaguely speaking, the feeling that for a measure φ on S^{n-1} (satisfying (7)) it is exceptional to be an ith area measure, if $1 \le i < n-1$. For polytopes, this can be made concrete in several ways, and from the results for polytopes one can deduce results for most convex bodies, in the sense of Baire category.

3.1. Preliminaries

In the following, we shall exploit the fact that the area measures of polytopes have a very special structure. Let $P \in \mathcal{K}^n$ be a polytope and F a proper face of P. By $N(P,F)$ we denote the normal cone of P at its face F (see CB, p. 72), and we put

$$\nu(P,F) := N(P,F) \cap S^{n-1}.$$

Thus

$$\nu(P,F) = \sigma(P, \operatorname{relint} F),$$

where σ denotes the spherical image in the sense of CB, p. 77, and relint F is the relative interior of F. The ith area measure of P is given by

$$S_i(P,\omega) = \binom{n-1}{i}^{-1} \sum_{F \in \mathcal{F}_i(P)} V_i(F) \sigma^{(n-1-i)}(\nu(P,F) \cap \omega) \tag{8}$$

for Borel sets $\omega \subset S^{n-1}$; here V_i denotes the i-dimensional volume of the i-face F of P and $\sigma^{(j)}$ is j-dimensional spherical Lebesgue measure (for (8), see CB, (4.2.18) together with (4.2.11)).

From (8) it follows, in particular, that the support of the measure $S_i(P,\cdot)$ (i.e., the complement of the largest open subset of S^{n-1} on which the measure vanishes) is given by

$$\operatorname{supp} S_i(P,\cdot) = \sigma_{n-i-1}(P), \tag{9}$$

where

$$\sigma_{n-i-1}(P) := \bigcup_{F \in \mathcal{F}_i(P)} \nu(P,F) \tag{10}$$

is the *spherical* $(n-i-1)$-*skeleton* of P. It is the union of finitely many $(n-i-1)$-dimensional spherical polytopes.

3.2. Addition and Decomposition Problems

If $K, L \in \mathcal{K}^n$ are convex bodies, then there exists a convex body $M \in \mathcal{K}^n$ for which

$$S_1(K,\cdot) + S_1(L,\cdot) = S_1(M,\cdot),$$

namely the Minkowski sum of K and L. Similarly, if $K, L \in \mathcal{K}^n$ are n-dimensional convex bodies, there exists a convex body $M \in \mathcal{K}^n$, called the Blaschke sum of K and L (and determined only up to a translation), for which

$$S_{n-1}(K,\cdot) + S_{n-1}(L,\cdot) = S_{n-1}(M,\cdot).$$

On the other hand, for $1 < i < n-1$ there are convex bodies $K, L \in \mathcal{K}^n$ for which $S_i(K,\cdot) + S_i(L,\cdot)$ is not the ith area measure of a convex body. This was proved independently by (Fedotov, 1979) and by (Goodey *et al.*, 1980). The special examples given by these authors (pairs of simplices and of parallelepipeds, respectively) can now be generalized considerably. From this one can deduce (cf. Theorem 3.5) that for $1 < i < n-1$ the sum of two ith area measures is typically not an ith area measure.

Theorem 3.1. *Let $P, Q \subset \mathbb{E}^n$ be n-polytopes in general relative position. If $1 < i < n-1$, then $S_i(P,\cdot) + S_i(Q,\cdot)$ is not the i-th area measure of a convex body.*

Proof. We shall use (9). Suppose that $P, Q \subset \mathbb{E}^n$ are n-polytopes in general relative position and that $K \subset \mathbb{E}^n$ is a convex body for which

$$S_i(K,\cdot) = S_i(P,\cdot) + S_i(Q,\cdot) \tag{11}$$

for some $i \in \{2, \ldots, n-2\}$. As shown in (Goodey *et al.*, 1980), K must be a polytope.

Suppose first that

$$\sigma_{n-i-1}(P) \cap \sigma_{n-i-1}(Q) = \emptyset.$$

Then $\sigma_{n-i-1}(P) \cup \sigma_{n-i-1}(Q)$ is not connected. By (11), this set is the support of $S_i(K, \cdot)$, hence equal to $\sigma_{n-i-1}(K)$. But for $i < n-1$ this set is connected, a contradiction. (The connectedness of $\operatorname{supp} S_i(K, \cdot)$ for $i < n-1$ was already used by (Fedotov, 1982).)

We can, therefore, assume in the following that there exists a vector

$$u \in \sigma_{n-i-1}(P) \cap \sigma_{n-i-1}(Q). \tag{12}$$

In this case, we have

$$i \leq \frac{n-1}{2}. \tag{13}$$

In fact, if $2i > n-1$, then no vector $v \in S^{n-1}$ has the property that both faces $F(P, v)$ and $F(Q, v)$ are of dimension $\geq i$, since P and Q are in general relative position. Hence, in that case, $\sigma_{n-i-1}(P) \cap \sigma_{n-i-1}(Q) = \emptyset$.

From (12) it follows that $u \in N(P, F) \cap N(Q, G)$ for some i-face F of P and some i-face G of Q. Suppose that

$$\operatorname{relint} N(P, F) \cap \operatorname{relint} N(Q, G) = \emptyset. \tag{14}$$

By CB, Th. 1.3.8, there exists an $(n-1)$-dimensional linear subspace H of $\mathbb{E}^n$ that properly separates the cones $N(P, F)$ and $N(Q, G)$. Then $u \in H$. The intersection $N(P, F) \cap H$ is of positive dimension and hence is the normal cone $N(P, F')$ of a proper face F' of P containing F. Similarly, $N(Q, G) \cap H$ is the normal cone $N(Q, G')$ of a proper face G' of Q containing G. Then $L(F')^\perp \subset H$ and $L(G')^\perp \subset H$, hence $H^\perp \subset L(F') \cap L(G')$ (for the definition of $L(F)$, see subsection 2.1). Since u is a normal vector for F' and for G', this contradicts the assumption that P and Q are in general relative position. Thus (14) was false and we can, therefore, assume in the following that there exists a vector

$$u \in \operatorname{relint} N(P, F) \cap \operatorname{relint} N(Q, G). \tag{15}$$

If the cones $N(P, F)$ and $N(Q, G)$ do not span $\mathbb{E}^n$, then there is a linear subspace H of dimension $n-1$ containing them, hence $H^\perp \subset L(F) \cap L(G)$. Again this contradicts the fact that P and Q are in general relative position. Thus $N(P, F)$ and $N(Q, G)$ span $\mathbb{E}^n$, and it follows that

$$\dim\,[N(P, F) \cap N(Q, G)] = n - 2i. \tag{16}$$

The polytope K satisfies

$$\sigma_{n-i-1}(K) = \sigma_{n-i-1}(P) \cup \sigma_{n-i-1}(Q). \tag{17}$$

By (15), we can choose a neighbourhood U of u in $\mathbb{E}^n$ such that

$$U \cap N(P, \bar{F}) = \emptyset, \qquad U \cap N(Q, \bar{G}) = \emptyset$$

for all i-faces $\bar{F} \neq F$ of P and all i-faces $\bar{G} \neq G$ of Q. Then, in view of (17),

$$\sigma_{n-i-1}(K) \cap U = [\nu(P,F) \cup \nu(Q,G)] \cap U. \tag{18}$$

Let $Z := F(K,u)$. By (18), u is contained in the normal cones of at least two i-faces of K, where the normal cones have an intersection of dimension $n-2i$, by (16). We deduce that $\dim Z \geq 2i$. We choose an $(i+1)$-face X of Z and then two i-faces F', F'' of X such that

$$\operatorname{lin} N(K,F') \neq \operatorname{lin} N(K,F''). \tag{19}$$

We have

$$\nu(K,F') \cap U \subset \sigma_{n-i-1}(K),$$

hence by (18)

$$\nu(K,F') \cap U \subset \nu(P,F) \cup \nu(Q,G),$$

similarly

$$\nu(K,F'') \cap U \subset \nu(P,F) \cup \nu(Q,G).$$

Since $\nu(K,F') \cap U$ is of dimension $n-i-1$, it must be contained either in $\nu(P,F)$ or in $\nu(Q,G)$, say

$$\nu(K,F') \cap U \subset \nu(P,F).$$

Similarly, $\nu(K,F'') \cap U$ is either contained in $\nu(P,F)$ or in $\nu(Q,G)$. The first case is impossible by (19), and the second case is impossible since then

$$\nu(K,X) \cap U \subset \nu(K,F') \cap U \subset \nu(P,F),$$
$$\nu(K,X) \cap U \subset \nu(K,F'') \cap U \subset \nu(Q,G),$$

which would imply

$$\dim N(K,X) \leq \dim [N(P,F) \cap N(Q,G)] = n-2i,$$

thus $i+1 = \dim X \geq 2i$ and hence $i=1$, a contradiction. This completes the proof of Theorem 3.1. ■

The fact that ith area measures of polytopes have small supports of special structure has the consequence that such a measure can dominate another ith area measure only if the latter is of a similarly special type. This leads to some results which at first sight may seem surprising. The first of these, from which others can be deduced, may be viewed as an improvement of the Aleksandrov-Fenchel-Jessen theorem for special polytopes.

Theorem 3.2. *Let $1 \leq i \leq n-1$, and let $P \in \mathcal{K}^n$ be an n-polytope with the property that all its $(i+1)$-faces are simplices. If K is a convex body of dimension* at least *$i+1$ satisfying*

$$S_i(K,\cdot) \leq S_i(P,\cdot), \tag{20}$$

then K is homothetic to P.

Proof. If (20) holds, then

$$\text{supp } S_i(K,\cdot) \subset \text{ supp } S_i(P,\cdot) = \sigma_{n-i-1}(P),$$

hence K is a polytope (see (Goodey *et al.*, 1980)), and thus

$$\sigma_{n-i-1}(K) \subset \sigma_{n-i-1}(P). \tag{21}$$

If $i = n-1$, then P is by assumption an n-simplex. Its area measure $S_{n-1}(P,\cdot)$ is concentrated in $n+1$ points, hence (21) implies the same for K. It is then clear that K is a simplex homothetic to P. We may, therefore, assume in the following that $i < n-1$.

Let F be an i-face of P. With F we associate a number $\tau(F)$ in the following way. If

$$\sigma_{n-i-1}(K) \cap \operatorname{relint} \nu(P,F) = \emptyset, \tag{22}$$

we put $\tau(F) = 0$. Suppose that (22) does not hold, hence there exists an i-face G of K such that

$$\nu(K,G) \cap \operatorname{relint} \nu(P,F) \neq \emptyset.$$

Let

$$A := S^{n-1} \setminus \bigcup_{F' \in \mathcal{F}_i(P)\setminus\{F\}} \nu(P,F'),$$

then A is an open subset of S^{n-1} and

$$A \cap \sigma_{n-i-1}(P) = \operatorname{relint} \nu(P,F).$$

Since $\nu(K,G) \subset \sigma_{n-i-1}(P)$ by (21), it follows that

$$\nu(K,G) \cap A \subset \operatorname{relint} \nu(P,F).$$

Suppose this inclusion were strict. Then $\operatorname{relint} \nu(P,F)$ contains a relative boundary point u of $\nu(K,G)$. The support set $F(K,u)$ is a face of K of dimension $> i$. This face contains an i-face G' of K such that $\nu(K,G') \cap A$ is not contained in $\operatorname{relint} \nu(P,F)$, a contradiction. Thus

$$\nu(K,G) \cap A = \operatorname{relint} \nu(P,F). \tag{23}$$

In particular, the face G is uniquely determined, and we can define

$$\tau(F) := \frac{V_i(G)}{V_i(F)}.$$

We assert that (23) can be strengthened to

$$\nu(K,G) = \nu(P,F). \tag{24}$$

Suppose (24) were false. Since $\operatorname{relint} \nu(P,F) \subset \nu(K,G)$ by (23), there must be a point u in the relative boundary of $\nu(P,F)$ that is not a relative boundary point of $\nu(K,G)$. Such a point u can, in fact, be chosen in the relative interior of a (spherical)

$(n-i-2)$-face φ of $\nu(P,F)$. The face $S := F(P,u)$ satisfies $\nu(P,S) = \varphi$ and hence is an $(i+1)$-face, by assumption a simplex. It has, therefore, $i+2$ faces of dimension i, say F_1 $(= F), F_2, \dots, F_{i+2}$. Since $u \in \operatorname{relint} \nu(P,S)$, we can choose a neighbourhood U of u in S^{n-1} that does not meet $\nu(P,F')$ for an i-face F' of P different from $F_1, \dots, F_{i+2}$. Since $\nu(K,G) \subset \sigma_{n-i-1}(P)$, we deduce that

$$U \cap \nu(K,G) \subset \bigcup_{r=1}^{i+2} \nu(P,F_r).$$

Since S is a simplex, the linear subspaces $\operatorname{lin} N(P,F_r)$, $r = 1,\dots,i+2$, are pairwise different, hence $U \cap \nu(K,G) \subset \nu(P,F)$. Thus the point u is a relative boundary point of $\nu(K,G)$. This contradiction shows that (24) holds.

Now let S be an arbitrary $(i+1)$-face of P. The simplex S has $i+2$ faces $F_1, \dots, F_{i+2}$ of dimension i. Choose

$$u \in \operatorname{relint} \nu(P,S). \tag{25}$$

Suppose, say, that $\tau(F_1) > 0$. Then there is an i-face G_1 of K such that

$$\nu(K,G_1) = \nu(P,F_1). \tag{26}$$

Since u is a relative boundary point of $\nu(K,G_1)$, the face $F(K,u)$ of K is of dimension greater than i. From (25) and (26) we deduce that $F(K,u)$ is an $(i+1)$-face of K. It has at least $i+2$ faces $G_1,\dots,G_{i+2}$ of dimension i. Since $u \in \nu(K,G_r) \subset \sigma_{n-i-1}(P)$ by (21), we can assume, after renumbering, that

$$\nu(K,G_r) = \nu(P,F_r) \tag{27}$$

for $r = 1,\dots,i+2$. The face $F(K,u)$ cannot have further i-faces and hence is a simplex. By (27), the affine hull of this simplex is parallel to the affine hull of S, and the affine hull of G_r is parallel to the affine hull of F_r $(r = 1,\dots,i+2)$. Hence, $F(K,u)$ is homothetic to S. This implies that $V_i(G_r)/V_i(F_r)$ is independent of r and thus shows that $\tau(F_1) = \ldots = \tau(F_{i+2})$. This has been proved under the assumption that at least one $\tau(F_r)$ is positive. If this assumption is not satisfied, then $\tau(F_1) = \ldots = \tau(F_{i+2}) = 0$. Since any two $(i+1)$-faces of P can be connected by a chain of $(i+1)$-faces such that any two consecutive ones have a common i-face, we conclude that $\tau(F)$ does not depend on F, say $\tau(F) = \tau$ for all i-faces F of P. Since K does have i-faces, the case $\tau = 0$ is impossible.

Let $\omega \subset S^{n-1}$ be a Borel set. Then

$$S_i(P,\omega) = \binom{n-1}{i}^{-1} \sum_{F \in \mathcal{F}_i(P)} V_i(F)\sigma^{(n-i-1)}(\nu(P,F) \cap \omega).$$

Let $F \in \mathcal{F}_i(P)$. Then there exists a unique face $G \in \mathcal{F}_i(K)$ such that

$$\nu(P,F) = \nu(K,G),$$

and we have $V_i(G) = \tau V_i(F)$. Every i-face of K corresponds in this way to some i-face of P. It follows that $S_i(K,\cdot) = \tau S_i(P,\cdot) = S_i(\tau^{1/i}P,\cdot)$. By the Aleksandrov-Fenchel-Jessen theorem, K and $\tau^{1/i}P$ are translates, hence K and P are homothetic.

■

In Theorem 3.1 we have described a general situation where the sum of two ith area measures is not an ith area measure. The next theorem exhibits a class of ith area measures which cannot be decomposed into a sum of two other ith area measures except in a trivial way.

Theorem 3.3. *Let $1 \leq i < n-1$. Let $P \in \mathcal{K}^n$ be an n-polytope with the property that all its $(i+1)$-faces are simplices. If $K, M \in \mathcal{K}_0^n$ are convex bodies satisfying*

$$S_i(K,\cdot) + S_i(M,\cdot) = S_i(P,\cdot), \tag{28}$$

then K and M are homothetic to P.

Proof. If (28) holds, then $S_i(K,\cdot) \leq S_i(P,\cdot)$, hence K is homothetic to P by Theorem 3.2. In a similar way, M is homothetic to P. ■

The case $i = 1$ of Theorem 3.3 is well known, since condition (28) for $i = 1$ is equivalent to $P = K + M + t$ with suitable $t \in \mathbb{E}^n$. Thus Theorem 3.3 can be considered as an extension of the known indecomposability of polytopes with triangular faces (first shown by (Shephard, 1963)). For $i = n-1$ and P not a simplex, however, decomposition in the sense of (28) is often possible in a non-trivial way, see (Firey *et al.*, 1964).

3.3. Projections

From the preceding results one can also deduce information on the determination of convex bodies from projection functions. For a convex body $K \in \mathcal{K}^n$ and for $E \in \mathcal{L}_i^n$, the Grassmannian of i-dimensional linear subspaces of $\mathbb{E}^n$, we denote by $V_i(K|E)$ the i-dimensional volume of the orthogonal projection $K|E$ of K onto E. The function $V_i(K|\cdot)$, defined on $\mathcal{L}_i^n$, is called the ith *projection function* of K $(1 \leq i \leq n-1)$. A connection between projection functions and area measures is given in the following lemma. Here $K^* := \{-x : x \in K\}$ denotes the image of K under reflection in the origin.

Lemma 3.1. *Let $i \in \{1, \ldots, n-1\}$ and let $K, M \in \mathcal{K}^n$ be convex bodies of dimension at least $i+1$. If*

$$V_i(K|E) = V_i(M|E) \qquad \text{for all } E \in \mathcal{L}_i^n, \tag{29}$$

then

$$S_i(K,\cdot) + S_i(K^*,\cdot) = S_i(M,\cdot) + S_i(M^*,\cdot). \tag{30}$$

An immediate consequence of this lemma and of the Aleksandrov-Fenchel-Jessen theorem is Aleksandrov's classical result: if $K, M \in \mathcal{K}^n$ are centrally symmetric and of dimension at least $i+1$, then (29) implies that K and M are translates of each other. We refer to (Goodey *et al.*, 1993+) for an investigation of projection functions related to this result and to Lemma 3.1. As already observed there, sharper results may be possible for polytopes, and we continue this line of research here. In particular, we show that if in Aleksandrov's theorem the body M is a special polytope (and $i < n-1$), then for K we can drop the assumption of central symmetry.

Theorem 3.4. *Let $1 \leq i < n-1$, and let $P \in \mathcal{K}_0^n$ be a centrally symmetric polytope such that all the $(i+1)$-faces of P are simplices. If $K \in \mathcal{K}^n$ is a convex body with*

$$V_i(K|\cdot) = V_i(P|\cdot) \qquad \text{on } \mathcal{L}_i^n, \tag{31}$$

then K is a translate of P.

Proof. By Lemma 3.1, assumption (31) yields

$$S_i(K,\cdot) + S_i(K^*,\cdot) = 2S_i(P,\cdot),$$

hence $S_i(K,\cdot) \leq S_i(2^{1/i}P,\cdot)$. It follows from (31) that K cannot be of dimension at most i. By Theorem 3.2, K is homothetic to P, by (31) it must then be a translate of P. ■

It is conceivable that Theorem 3.4 can be extended to suitable polytopes without a centre of symmetry, where the conclusion would have to state that K is a translate of either P or P^*. In (Goodey *et al.*, 1993+) the following was shown. If P is the direct sum of a triangle and an $(n-2)$-dimensional parallelepiped, then any convex body K satisfying (31) for some $i \in \{2,\ldots,n-2\}$ is a translate of P or P^*.

We mention another direction in which part of Aleksandrov's theorem can be improved in the case of polytopes. By an *equator zone* we understand any subset of the sphere S^{n-1} of the form $\{x \in S^{n-1} : |\langle x,v\rangle| < \epsilon\}$ with $v \in S^{n-1}$ and $\epsilon > 0$. For $u \in S^{n-1}$ we write

$$E_u := \{x \in \mathbb{E}^n : \langle x,u\rangle = 0\}.$$

Theorem 3.5. *Let $P \in \mathcal{K}_0^n$ be a centrally symmetric polytope, and let $Z \subset S^{n-1}$ be an equator zone. If $K \in \mathcal{K}^n$ is a centrally symmetric convex body satisfying*

$$V_{n-1}(K|E_u) = V_{n-1}(P|E_u) \qquad \text{for } u \in Z,$$

then K is a translate of P.

This is a special case of a result proved in (Schneider, 1970).

We want to mention briefly an unsolved problem related to Lemma 3.1. Let $i \in \{1,\ldots,n-2\}$ and let $K, M \in \mathcal{K}^n$ be convex bodies of dimension at least $i+1$. If K and M satisfy

$$V_i(K|E) = V_i(M|E) \qquad \text{for all } E \in \mathcal{L}_i^n, \tag{32}$$

then they also satisfy

$$V_i(K|H) = V_i(M|H) \qquad \text{for all } H \in \mathcal{L}_{n-1}^n. \tag{33}$$

(and hence (30); see Theorem 2.1 in (Goodey *et al.*, 1993+) for these implications). Here $V_i(K|H)$ denotes the ith intrinsic volume of the orthogonal projection of K onto the hyperplane H. The question is whether conversely (33) implies (32). This is known to be true for $i = n$, but unknown for $i > 1$. It was recently proved by (Goodey *et al.*, 1993+) for the case where K or M is a polytope.

3.4. Applications to General Convex Bodies

In some cases, the properties established above for polytopes can be used to show that most convex bodies have the property in question. Again, 'most' is understood in the Baire category sense.

Theorem 3.6. *Let $1 < i < n-1$. For most pairs (K, M) of convex bodies in $\mathcal{K}_0^n \times \mathcal{K}_0^n$, the sum $S_i(K,\cdot) + S_i(M,\cdot)$ is not the i-th area measure of a convex body.*

Proof. Let $\mathcal{A}$ denote the set of pairs (K, M) in $\mathcal{K}_0^n \times \mathcal{K}_0^n$ for which $S_i(K,\cdot)+S_i(M,\cdot)$ is not an i-th area measure. It follows from Lemma 1.1 that the set of pairs of polytopes in general relative position is dense in $\mathcal{K}_0^n \times \mathcal{K}_0^n$, hence Theorem 3.1 implies that $\mathcal{A}$ is dense in $\mathcal{K}_0^n \times \mathcal{K}_0^n$.

For $k \in \mathbb{N}$, we denote by $\mathcal{A}_k$ the set of pairs $(K, M) \in \mathcal{K}_0^n \times \mathcal{K}_0^n$ for which there exists a convex body L such that

$$S_i(L,\cdot) = S_i(K,\cdot) + S_i(M,\cdot)$$

and

$$\delta(L, K) \leq k,$$

where δ denotes the Hausdorff metric. We show that $\mathcal{A}_k$ is closed. Let $((K_j, M_j))_{j\in\mathbb{N}}$ be a sequence in $\mathcal{A}_k$ converging to a pair $(K, M) \in \mathcal{K}_0^n \times \mathcal{K}_0^n$. For each $j \in \mathbb{N}$ we can choose a convex body L_j satisfying

$$S_i(L_j,\cdot) = S_i(K_j,\cdot) + S_i(M_j,\cdot) \tag{34}$$

and

$$\delta(L_j, K_j) \leq k. \tag{35}$$

From $K_j \to K$ and (35) it follows that the sequence $(L_j)_{j\in\mathbb{N}}$ is bounded. By the Blaschke selection theorem it contains a convergent subsequence. Changing the notation, we may assume that the sequence $(L_j)_{j\in\mathbb{N}}$ itself converges to a convex body L. From (34) and the weak continuity of the ith area measure we deduce that $S_i(L,\cdot) = S_i(K,\cdot)+S_i(M,\cdot)$, and from (35) we get $\delta(L, K) \leq k$. Thus $(K, M) \in \mathcal{A}_k$. This shows that $\mathcal{A}_k$ is closed. Evidently

$$\mathcal{A} = \mathcal{K}_0^n \times \mathcal{K}_0^n \setminus \bigcup_{k\in\mathbb{N}} \mathcal{A}_k,$$

hence $\mathcal{A}$ is a dense G_δ set in $\mathcal{K}_0^n \times \mathcal{K}_0^n$. This completes the proof of Theorem 3.6. ∎

The following result shows in a strong sense that there is no analogue of Minkowski or Blaschke symmetrization involving intermediate area measures.

Theorem 3.7. *Let $1 < i < n-1$. For most convex bodies K in $\mathcal{K}_0^n$ the measure $S_i(K,\cdot) + S_i(K^*,\cdot)$ is not the i-th area measure of a convex body.*

Proof. Let P be an n-polytope with the property that P and P^* are in general relative position. According to Theorem 3.1, $S_i(P,\cdot) + S_i(P^*,\cdot)$ is not the ith area measure of a convex body.

One can argue as before as soon as it is shown that the set of polytopes P with the property above is dense in $\mathcal{K}^n$. This can be seen as follows. We start with a simplicial n-polytope P and a given number $\epsilon > 0$. Let $x_1, \ldots, x_k$ be the vertices of P. For $j = 1, \ldots, k$, we can choose a ball B_j with centre x_j so that, for arbitrary points $y_j \in B_j$ ($j = 1, \ldots, k$), the polytope $Q := \mathrm{conv}\{y_1, \ldots, y_k\}$ is n-dimensional and simplicial and satisfies $\delta(P, Q) < \epsilon$. Let $y_1, \ldots, y_k$ be chosen in this way and suppose that $Q := \mathrm{conv}\,\{y_1, \ldots, y_k\}$ and Q^* are not in general relative position. Then there are disjoint subsets $I, J \subset \{1, \ldots, k\}$ such that $\mathrm{conv}\,\{y_i : i \in I\}$ and $\mathrm{conv}\,\{-y_j : j \in J\}$ are (simplicial) faces of Q lying in two different parallel supporting hyperplanes of Q and such that their sum is not a direct one. Assuming, without loss of generality, that $I = \{1, \ldots, p\}$ and $J = \{p+1, \ldots, p+q\}$ with $2 \leq p, q \leq n$, we deduce that the matrix with column vectors $y_2 - y_1, \ldots, y_p - y_1, y_{p+2} - y_{p+1}, \ldots, y_{p+q} - y_{p+1}$ (with respect to some basis) has rank less than $p + q - 2$. Hence the k-tuple $(y_1, \ldots, y_k)$ belongs to the union of certain finitely many subsets of measure zero in $(\mathbb{E}^n)^k$. Therefore, the ϵ-neighbourhood of P contains polytopes Q for which Q and Q^* are in general relative position. ∎

The following theorem contains a (surprising) strengthening of the Aleksandrov-Fenchel-Jessen theorem for most convex bodies.

Theorem 3.8. *Let $1 \leq i < n - 1$. For most convex bodies K in $\mathbb{E}^n$, the inequality*

$$S_i(L, \cdot) \leq S_i(K, \cdot)$$

for a convex body L of dimension at least $i + 1$ implies that L is homothetic to K.

Proof. Let $\mathcal{A}$ denote the set of convex bodies K in $\mathbb{E}^n$ having the property in question. By Theorem 3.2, $\mathcal{A}$ is dense in $\mathcal{K}^n$.

For $k \in \mathbb{N}$, let $\mathcal{A}_k$ denote the set of convex bodies $K \in \mathcal{K}^n$ for which there exists a convex body $L \in \mathcal{K}^n$ containing an $(i + 1)$-ball of radius $1/k$ and satisfying

$$S_i(L, \cdot) \leq S_i(K, \cdot),$$

$$\delta(L, K) \leq k$$

and

$$\delta(L, K') \geq 1/k$$

for all homothets K' of K. Then

$$\mathcal{A} = \mathcal{K}^n \setminus \bigcup_{k \in \mathbb{N}} \mathcal{A}_k.$$

For the proof, let $K \in \mathcal{A}$. Suppose that there exists a number $k \in \mathbb{N}$ for which $K \in \mathcal{A}_k$. Then there is a body $L \in \mathcal{K}^n$ such that $\dim L \geq i + 1$, $S_i(L, \cdot) \leq S_i(K, \cdot)$, and $\delta(L, K') \geq 1/k$ for all homothets K' of K. In particular, L is not homothetic to K, hence $K \notin \mathcal{A}$, a contradiction. Thus $K \in \mathcal{K}^n \setminus \bigcup_{k \in \mathbb{N}} \mathcal{A}_k$.

Vice versa, let $K \in \mathcal{K}^n \setminus \bigcup_{k \in \mathbb{N}} \mathcal{A}_k$. Let $L \in \mathcal{K}^n$ be a body such that $\dim L \geq i + 1$ and $S_i(L, \cdot) \leq S_i(K, \cdot)$. There exists a number k_0 such that L contains an $(i+1)$-ball of radius $1/k$ and $\delta(L, K) \leq k$ for all $k \geq k_0$. Let $k \in \mathbb{N}$, $k \geq k_0$. Since $K \notin \mathcal{A}_k$, we can choose a homothet K_k of K for which $\delta(L, K_k) < 1/k$. Since $\delta(L, K_k) \to 0$ for $k \to \infty$, the sequence $(K_k)_{k \geq k_0}$ converges to L. On the other hand, each K_k and hence L is homothetic to K. This shows that $K \in \mathcal{A}$.

To prove that $\mathcal{A}_k$ is closed, let $(K_j)_{j \in \mathbb{N}}$ be a sequence in $\mathcal{A}_k$ converging to a convex body K. For $j \in \mathbb{N}$, there exists a body $L_j \in \mathcal{K}^n$ such that L_j contains an $(i+1)$-ball B_j of radius $1/k$, $S_i(L_j, \cdot) \leq S_i(K_j, \cdot)$, $\delta(L_j, K_j) \leq k$ and $\delta(L_j, K_j') \geq 1/k$ for all homothets K_j' of K_j. From $\delta(L_j, K_j) \leq k$ and the convergence $K_j \to K$ it follows that the sequence $(L_j)_{j \in \mathbb{N}}$ is bounded. Choosing a suitable subsequence and changing the notation, we may assume that $L_j \to L$ and $B_j \to B$ $(j \to \infty)$ for convex bodies $L, B \in \mathcal{K}^n$. Then B is an $(i+1)$-ball of radius $1/k$ contained in L. From the weak continuity of S_i and the continuity of δ we conclude that $S_i(L, \cdot) \leq S_i(K, \cdot)$ and $\delta(L, K) \leq k$. Let K' be a homothet of K, say $K' = \lambda K + t$ with $\lambda \geq 0$ and $t \in \mathbb{E}^n$. Then $\lambda K_j + t$ is a homothet of K_j, hence $\delta(L_j, \lambda K_j + t) \geq 1/k$, from which we deduce that $\delta(L, K') \geq 1/k$. Thus $K \in \mathcal{A}_k$, which completes the proof of Theorem 3.8. ∎

The final results of this section concern the determination of convex bodies from projection functions. Let $i \in \{1, \ldots, n-1\}$. According to Aleksandrov's theorem, a centrally symmetric convex body $K \in \mathcal{K}_0^n$ is uniquely determined, up to translation, among all centrally symmetric convex bodies, by its ith projection function. However, there may exist non-symmetric convex bodies which have the same ith projection function. For example, (Firey, 1970) has constructed non-spherical convex bodies of revolution with constant ith projection function. For $i = n-1$, a recent result of (Gardner *et al.*, 1993+) has clarified the situation. If $K \in \mathcal{K}_0^n$ is not a parallelepiped, then there exist convex bodies with the same $(n-1)$-st projection function as K which are not translates of K or K^*. On the other hand, the following holds true.

Theorem 3.9. *Let $1 \leq i < n-1$. Most centrally symmetric convex bodies K in $\mathcal{K}^n$ have the property that any convex body with the same i-th projection function as K is a translate of K.*

Proof. It is sufficient to consider the space $\mathcal{K}_c^n$ of convex bodies in $\mathcal{K}^n$ which are centrally symmetric with respect to the origin. Let $\mathcal{A}$ be the set of all bodies $K \in \mathcal{K}_c^n$ having the property of the theorem. By Theorem 3.4, $\mathcal{A}$ is dense in $\mathcal{K}_c^n$. For $k \in \mathbb{N}$ we define

$$\begin{aligned} \mathcal{A}_k := \{ & K \in \mathcal{K}_c^n : \exists\, L \in \mathcal{K}^n \text{ such that } V_i(L|\cdot) = V_i(K|\cdot), \\ & \delta(L, K) \leq k \text{ and } \delta(L, K+t) \geq 1/k \text{ for all } t \in \mathbb{E}^n \}. \end{aligned}$$

Then it is easy to see that $\mathcal{A} = \mathcal{K}_c^n \setminus \bigcup_{k \in \mathbb{N}} \mathcal{A}_k$ and that each $\mathcal{A}_k$ is closed. Hence, $\mathcal{A}$ is a dense G_δ set in $\mathcal{K}_c^n$, from which the assertion follows. ∎

The case $i = 1$ of Theorem 3.9 is known. Let $K \in \mathcal{K}^n$ be a centrally symmetric convex body and suppose that the body $L \in \mathcal{K}^n$ has the same first projection function as K. Then $h(L, u) + h(L, -u) = h(K, u) + h(K, -u)$ for all $u \in S^{n-1}$ and hence $L + L^* = 2K + t$ with $t \in \mathbb{E}^n$. If this equality is only possible if L is homothetic to K, then K is said to be irreducible. It is known that most centrally symmetric convex bodies in $\mathbb{E}^n$ ($n \geq 3$) are irreducible (see (Shephard, 1966) and, in particular, (Yost, 1991), Proposition 20.

Also in the following theorem, 'most' cannot be replaced by 'all', as is well known.

Theorem 3.10. *Let $Z \subset S^{n-1}$ be an equator zone. Most centrally symmetric convex bodies K in $\mathcal{K}^n$ have the property that every centrally symmetric convex body L satisfying*

$$V_{n-1}(K|E_u) = V_{n-1}(L|E_u) \qquad \text{for all } u \in Z$$

is a translate of K.

Proof. Let $\mathcal{A} \subset \mathcal{K}_c^n$ be the set of those bodies for which

$$V_{n-1}(L|E_u) = V_{n-1}(K|E_u) \qquad \text{for all } u \in Z$$

for some $L \in \mathcal{K}_c^n$ implies that $L = K$. By Theorem 3.5, $\mathcal{A}$ is dense in $\mathcal{K}_c^n$. For $k \in \mathbb{N}$ we define

$$\mathcal{A}_k := \{K \in \mathcal{K}_c^n : \exists\, L \in \mathcal{K}_c^n \text{ such that } V_i(L|\cdot) = V_i(K|\cdot) \text{ on } Z,\ \ 1/k \leq \delta(L,K) \leq k\}$$

and then complete the proof as above. ■

The result that the equality (33) implies (32) has so far not been extended from polytopes to general convex bodies, except for $i = n-2$. The latter was done in (Goodey *et al.*, 1993+), essentially by proving the inequality

$$\begin{aligned} &|V_{n-2}(K|E) - V_{n-2}(M|E)| \\ \leq\ &\frac{n-1}{4\pi}\|S_{n-2}(K,\cdot) + S_{n-2}(K^*,\cdot) - S_{n-2}(M,\cdot) - S_{n-2}(M^*,\cdot)\|_{TV} \end{aligned}$$

for $E \in \mathcal{L}_{n-2}^n$, where $\|\cdot\|_{TV}$ denotes the total variation of a signed measure. This inequality, first proved for polytopes K and M, extends to general convex bodies by approximation. As an example for the result obtained in this way, we mention the four-dimensional case. If $K \subset \mathbb{E}^4$ is a convex body with the property that all its three-dimensional orthogonal projections have the same surface area, then all two-dimensional orthogonal projections of K have the same area.

4. Mixed Volumes

4.1. Mixed Volumes of Polytopes

The purpose of this subsection is the derivation of a special formula for representing the mixed volume of polytopes. Let $P_1, \ldots, P_k \subset \mathbb{E}^n$ be polytopes. Applying, e.g., the lifting theorem of Walkup and Wets to the affine map

$$P_1 \times \ldots \times P_k \to P_1 + \ldots + P_k, \quad (x_1, \ldots, x_k) \mapsto x_1 + \ldots + x_k,$$

one finds a representation

$$V_n(P_1 + \ldots + P_k) = \sum V_n(F_1 + \ldots + F_k), \tag{36}$$

where the sum extends over certain k-tuples of faces $F_1 \in \mathcal{F}(P_1), \dots, F_k \in \mathcal{F}(P_k)$ ($\mathcal{F}$ denotes the set of faces). This approach was mentioned, in a different context, in (McMullen *et al.*, 1983), p. 200; a similar idea appears in (Pedersen *et al.*, 1992+), after (2.3). By homogeneity considerations, one deduces from (36) formulas of the type

$$\binom{n}{r_1 \dots r_k} V(P_1[r_1], \dots, P_k[r_k]) = \sum V_n(F_1 + \dots + F_k), \tag{37}$$

where $V(P_1[r_1], \dots, P_k[r_k])$ denotes the mixed volume with the argument P_i repeated r_i times. The summation now extends over certain k-tuples of faces $F_1 \in \mathcal{F}_{r_1}(P_1), \dots, F_k \in \mathcal{F}_{r_k}(P_k)$. Exactly over which faces one has to sum, is hard to tell from the approaches mentioned above. For the case $k = 2$, an explicit summation rule, depending on a vectorial parameter, was found by (Betke, 1992). It appears that his method of proof does not extend to $k > 2$. By a different procedure, we shall now generalize Betke's result to arbitrary $k \geq 2$. The result is given in Theorem 4.1.

In Euclidean space $\mathbb{E}^p$ of dimension $p > n$ let L be an n-dimensional linear subspace and let $\Pi : \mathbb{E}^p \to L$ denote the orthogonal projection. For a vector $v \in L^\perp \setminus \{0\}$ we define $L_v := L + \operatorname{lin}\{v\}$ and $L_v^+ := L + \{v\}$. By $\Pi_v : \mathbb{E}^p \to L_v$ we denote the orthogonal projection onto L_v.

Let $P \subset \mathbb{E}^p$ be a polytope of dimension at least n. Let $v \in L^\perp \setminus \{0\}$ and suppose, first, that there is a vector $u \in L_v^+$ such that $\dim F(P, u) > n$. Writing $F(P, u) = F$ and $u = v + x$ with $x \in L$, we have $v + x \in N(P, F)$ and hence

$$v \in L^\perp \cap \operatorname{lin}(N(P, F) \cup L).$$

The right-hand side is of dimension less than $p - n$, since $\dim N(P, F) < p - n$. Hence, all vectors $v \in L^\perp \setminus \{0\}$, except those in a certain finite union of proper linear subspaces, have the property that

$$\dim F(P, u) \leq n \qquad \text{for all } u \in L_v^+. \tag{38}$$

We choose v so that this holds. Then every n-face F' of the polytope $\Pi_v P$ with a normal vector $u \in L_v^+$ is the projection of an n-face F of P such that

$$N(P, F) \cap L_v^+ \neq \emptyset.$$

The projection Π maps F' into an n-polytope in L. The n-faces F' of $\Pi_v P$ with a normal vector $u \in L_v^+$ are precisely the facets, relative to L_v, of the polytope $\Pi_v P$ having an outer normal vector u' with $\langle u', v \rangle > 0$. Hence, all such $\Pi F' = \Pi F$ cover ΠP and have pairwise disjoint relative interiors. It follows that

$$V_n(\Pi P) = \sum_{*(v)} V_n(\Pi F), \tag{39}$$

where the sum $\sum_{*(v)}$ extends over the n-faces F of P with

$$N(P, F) \cap L_v^+ \neq \emptyset. \tag{40}$$

The foregoing will now be applied to the case where $k \geq 2$ and $\mathbb{E}^p$ is identified with $\mathbb{E}^n \times \dots \times \mathbb{E}^n$ (k factors). We define

$$L := \{(x, \dots, x) \in (\mathbb{E}^n)^k : x \in \mathbb{E}^n\}$$

and

$$\Pi(x_1,\ldots,x_k) := \frac{1}{k}(x_1+\ldots+x_k,\ldots,x_1+\ldots+x_k)$$

for $(x_1,\ldots,x_k)\in(\mathbb{E}^n)^k$. Then $\Pi:(\mathbb{E}^n)^k\to L$ is the orthogonal projection. Let $P_1,\ldots,P_k\subset\mathbb{E}^n$ be polytopes. We apply (39), with a suitable vector v, to the polytope $P=P_1\times\ldots\times P_k$. Since every face of P is of the form $F_1\times\ldots\times F_k$, where F_i is a face of P_i $(i=1,\ldots,k)$, we get

$$V_n(\Pi(P_1\times\ldots\times P_k))=\sum_{*(v)}V_n(\Pi(F_1\times\ldots\times F_k))$$

with the appropriate specification for summation. The addition map

$$(x_1,\ldots,x_k)\mapsto x_1+\ldots+x_k$$

from $(\mathbb{E}^n)^k$ to $\mathbb{E}^n$ is the composition of Π, the dilatation by k, and the projection from $(\mathbb{E}^n)^k$ onto the first factor. It follows that

$$V_n(P_1+\ldots+P_k)=\sum_{*(v)}V_n(F_1+\ldots+F_k). \tag{41}$$

Here the summation has to be done in the following way. First we have to choose the vector v in such a way that

$$v=(v_1,\ldots,v_k)\in L^{\perp}\backslash\{0\} \tag{42}$$

and

$$\dim F(P,u)\le n \qquad \text{for all } u\in L_v^+. \tag{43}$$

Condition (42) is satisfied if and only if

$$v_1+\ldots+v_k=0, \qquad (v_1,\ldots,v_k)\neq(0,\ldots,0). \tag{44}$$

If $u=(u_1,\ldots,u_k)$ with $u_i\in\mathbb{E}^n$, then $u\in L_v^+$ if and only if

$$u_i=v_i+x \qquad \text{for } i=1,\ldots,k \text{ with some } x\in\mathbb{E}^n. \tag{45}$$

Now

$$F(P_1\times\ldots\times P_k,(u_1,\ldots,u_k))=F(P_1,u_1)\times\ldots\times F(P_k,u_k),$$

with the convention that $F(P_i,0)=P_i$. Thus condition (43) is equivalent to

$$\dim F(P_1,u_1)+\ldots+\dim F(P_k,u_k)\le n \qquad \text{for } (u_1,\ldots,u_k)\in L_v^+. \tag{46}$$

In other words, the vectors $v_1,\ldots,v_k\in\mathbb{E}^n$ have to be chosen so that (44) holds and that

$$\dim F(P_1,v_1+x)+\ldots+\dim F(P_k,v_k+x)\le n \qquad \text{for all } x\in\mathbb{E}^n. \tag{47}$$

Condition (47) can also be written in the following form. If F_i is a face of P_i for $i=1,\ldots,k$ and if

$$\dim F_1+\ldots+\dim F_k>n, \tag{48}$$

then

$$\bigcap_{i=1}^{k} [\operatorname{relint} N(P_i, F_i) - v_i] = \emptyset. \tag{49}$$

When $v = (v_1, \ldots, v_k)$ has been chosen in this way, then the summation $\sum_{*(v)}$ in (41) extends over all k–tuples $(F_1, \ldots, F_k)$ where F_i is a face of P_i $(i = 1, \ldots, k)$ such that $F_1 \times \ldots \times F_k$ is an n–face of $P_1 \times \ldots \times P_k$, equivalently

$$\dim F_1 + \ldots + \dim F_k = n, \tag{50}$$

and

$$N(P_1 \times \ldots \times P_k, F_1 \times \ldots \times F_k) \cap L_v^+ \neq \emptyset. \tag{51}$$

Condition (51) is equivalent to the existence of a vector $u = (u_1, \ldots, u_k) \in (\mathbb{E}^n)^k$ such that

$$(u_1, \ldots, u_k) \in N(P_1 \times \ldots \times P_k, F_1 \times \ldots \times F_k) \tag{52}$$

and

$$(u_1, \ldots, u_k) = (v_1, \ldots, v_k) + (x, \ldots, x) \quad \text{with some } x \in \mathbb{E}^n. \tag{53}$$

Relation (52) holds if and only if

$$u_i \in N(P_i, F_i) \qquad \text{for } i = 1, \ldots, k. \tag{54}$$

The existence of $(u_1, \ldots, u_k)$ satisfying (53) and (54) is equivalent to

$$\bigcap_{i=1}^{k} [N(P_i, F_i) - v_i] \neq \emptyset. \tag{55}$$

Hence, the summation in (41) extends over all $(F_1, \ldots, F_k)$ such that F_i is a face of P_i and (50) and (55) are satisfied.

Now we replace P_i by $\lambda_i P_i$ with $\lambda_i > 0$, then (41) yields

$$V_n(\lambda_1 P_1 + \ldots + \lambda_k P_k) = \sum_{*(v)} V_n(\lambda_1 F_1 + \ldots + \lambda_k F_k).$$

Since $N(\lambda_i P_i, \lambda_i F_i) = N(P_i, F_i)$, the admissible choices for $v_1, \ldots, v_k$ and the summation rule do not depend on $\lambda_1, \ldots, \lambda_k$. Each k–tuple $(F_1, \ldots, F_k)$ which is relevant for the summation satisfies (50) and $\dim(F_1 + \ldots + F_k) = n$, hence $\lambda_1 F_1 + \ldots + \lambda_k F_k$ is a direct sum and therefore

$$V_n(\lambda_1 F_1 + \ldots + \lambda_k F_k) = \lambda_1^{\dim F_1} \cdots \lambda_k^{\dim F_k} V_n(F_1 + \ldots + F_k).$$

This gives

$$V_n(\lambda_1 P_1 + \ldots + \lambda_k P_k) = \sum_{\substack{r_1, \ldots, r_k = 0 \\ r_1 + \cdots + r_k = n}}^{n} \lambda_1^{r_1} \cdots \lambda_k^{r_k} \sum_{*} V_n(F_1 + \ldots + F_k),$$

where now the inner summation $\sum_*$ extends over $F_i \in \mathcal{F}_{r_i}(P_i)$ $(i = 1, \ldots, k)$ satisfying (55). On the other hand

$$V_n(\lambda_1 P_1 + \ldots + \lambda_k P_k) = \sum_{r_1,\ldots,r_k=0}^{n} \binom{n}{r_1 \ldots r_k} \lambda_1^{r_1} \cdots \lambda_k^{r_k} V(P_1[r_1], \ldots, P_k[r_k]).$$

We compare the coefficients and collect what we have obtained.

Theorem 4.1. *Let $P_1, \ldots, P_k \in \mathcal{K}^n$ be polytopes $(k \geq 2)$. Choose vectors $v_1, \ldots, v_k \in \mathbb{E}^n$ such that*

$$v_1 + \ldots + v_k = 0, \qquad (v_1, \ldots, v_k) \neq (0, \ldots, 0)$$

and

$$\bigcap_{i=1}^{k} [\operatorname{relint} N(P_i, F_i) - v_i] = \emptyset$$

whenever $F_i \in \mathcal{F}(P_i)$ and $\dim F_1 + \ldots + \dim F_k > n$. If $r_1, \ldots, r_k \in \{0, \ldots, n-1\}$ and $r_1 + \ldots + r_k = n$, then

$$\binom{n}{r_1 \ldots r_k} V(P_1[r_1], \ldots, P_k[r_k]) = \sum_{*(v)} V_n(F_1 + \ldots + F_k), \tag{56}$$

where the summation extends over the k–tuples $(F_1, \ldots, F_k)$ of faces $F_i \in \mathcal{F}_{r_i}(P_i), i = 1, \ldots, k$, for which $\dim(F_1 + \ldots + F_k) = n$ and

$$\bigcap_{i=1}^{k} [N(P_i, F_i) - v_i] \neq \emptyset. \tag{57}$$

We add four remarks. (1) As shown in the proof, the admissible k–tuples $(v_1, \ldots, v_k)$ satisfying the assumptions of the theorem can be chosen from a dense open subset of the space

$$\{(x_1, \ldots, x_k) \in (\mathbb{E}^n)^k : x_1 + \ldots + x_k = 0\}.$$

(2) The volume appearing on the right side of (56) can also be written in the form

$$V_n(F_1 + \ldots + F_k) = [F_1, \ldots, F_k] V_{r_1}(F_1) \cdots V_{r_k}(F_k).$$

Here $[F_1, \ldots, F_k]$ denotes the volume of the parallelepiped that is the sum of unit cubes in the affine hulls of $F_1, \ldots, F_k$, respectively.

(3) For $k = 2$, Theorem 4.1 reduces to the result of (Betke, 1992), because in that case $v_1 + v_2 = 0$, so that (57) is equivalent to

$$v_1 \in N(P_1, F_1) - N(P_2, F_2).$$

(4) Taking $P_1 = \ldots = P_k = P$ in Theorem 4.1, we obtain various formulas for computing the volume of a polytope P in terms of volumes of parallelepipeds formed by lower dimensional faces.

4.2. THE ALEKSANDROV–FENCHEL INEQUALITY

The Aleksandrov–Fenchel inequality states that

$$V(K,L,\mathcal{C})^2 \geq V(K,K,\mathcal{C})V(L,L,\mathcal{C}) \tag{58}$$

for convex bodies $K, L \in \mathcal{K}^n$ and any $(n-2)$–tuple $\mathcal{C} = (C_1, \ldots, C_{n-2})$ of convex bodies. For a proof (due to Aleksandrov) and for references, we refer to Section 6.3 of CB. An attempt of (Ewald, 1990) to give a new short proof of (58) is open to doubt. Generalizations of the Aleksandrov–Fenchel inequality for a simple polytope P were recently discovered by (McMullen, 1993); they are formulated in terms of the subalgebra of McMullen's polytope algebra generated by the classes of summands of P.

It is an open problem to characterize the cases of equality in (58). A conjecture can be found in CB, Section 6.6. We give here an equivalent formulation of it for the case where $C_1, \ldots, C_{n-2}$ are polytopes.

Let $P_1, \ldots, P_{n-2} \in \mathcal{K}^n$ be polytopes. A vector $u \in S^{n-1}$ is called $(P_1, \ldots, P_{n-2})$–*extreme* if each face $F(P_i, u)$ contains a segment S_i $(i = 1, \ldots, n-2)$ such that $\dim(S_1 + \ldots + S_{n-2}) = n-2$. A supporting hyperplane of a convex body K is $(P_1, \ldots, P_{n-2})$–*extreme* if its outer unit normal vector is $(P_1, \ldots, P_{n-2})$–extreme.

Conjecture. If $K, L \in \mathcal{K}^n$ are convex bodies and $P_1, \ldots, P_{n-2} \in \mathcal{K}^n$ are n–dimensional polytopes, then equality in (58) for $\mathcal{C} = (P_1, \ldots, P_{n-2})$ holds if and only if suitable homothets of K and L have the same $(P_1, \ldots, P_{n-2})$–extreme supporting hyperplanes.

For the case where $P_1, \ldots, P_{n-2}$ are strongly isomorphic simple polytopes, this is proved in CB, Theorem 6.6.20 (see also (Schneider, 1991+), which was written earlier). If $P_1, \ldots, P_{n-2}$ are not n–dimensional, the conjecture has to be modified. For the case where $P_1, \ldots, P_{n-2}$ lie in parallel hyperplanes, such a modification was found by (Ewald *et al.*, 1990+) and proved by them if K and L are polytopes. We generalize their result to the case of arbitrary convex bodies K and L and at the same time give an easier proof.

If we are interested in the equality case in (58), we can assume that

$$V(K,K,\mathcal{C}) = V(L,L,\mathcal{C}) \neq 0. \tag{59}$$

In fact, if the right side of (58) is zero, then one can use CB, Theorem 5.1.7, to discuss the equality cases in (58). Otherwise, (59) can be achieved by replacing K by a suitable homothet.

Theorem 4.2. *Let $K, L \in \mathcal{K}^n$ be convex bodies and let $P_1, \ldots, P_{n-2} \in \mathcal{K}^n$ be polytopes lying in parallel hyperplanes with unit normal vector w. Then, under the condition* (59), *equality in* (58) *for $\mathcal{C} = (P_1, \ldots, P_{n-2})$ holds if and only if the bodies $K + F(L,w)$ and $L + F(K,w)$ have the same $(P_1, \ldots, P_{n-2})$–extreme supporting hyperplanes.*

Proof. We may assume that $P_1, \ldots, P_{n-2}$ lie in the linear subspace

$$H_{w,0} := \{x \in \mathbb{E}^n : \langle x, w \rangle = 0\}.$$

Let $P := P_1 + \ldots + P_{n-2}$ and let $F = F(P, v)$ with $v \in H_{w,0}$ be a face of P. We say that F is an *essential* face of P if there exist segments $S_i \subset F(P_i, v)$ $(i = 1, \ldots, n-2)$ such that $\dim(S_1 + \ldots + S_{n-2}) = n-2$. Thus every essential face of P is an $(n-2)$–face of P, but in general not vice versa. From $V(K, K, P_1, \ldots, P_{n-2}) \neq 0$ one can deduce that P has at least one essential face (cf. CB, Theorem 5.1.7 and its proof). We denote by $F_1, \ldots, F_m$ the essential faces of P and by $v_k \in H_{w,0}$ the outer unit normal vector of P at F_k $(k = 1, \ldots, m)$. Further we define

$$\sigma_k := \{u \in S^{n-1} : F(P, u) = F_k\} \cup \{w, -w\}$$

for $k = 1, \ldots, m$. If $m > 1$, then each σ_k is a half great circle, and if $m = 1$, then σ_1 is a great circle. The vectors in $\bigcup_{k=1}^{m} \sigma_k$ are precisely the $(P_1, \ldots, P_{n-2})$–extreme unit vectors.

Now suppose that (58) holds with equality. We may assume that the support functions $h(K, \cdot)$ and $h(L, \cdot)$ of K and L satisfy

$$h(K, w) = h(L, w) = 0, \tag{60}$$

since this can be achieved by applying suitable translations to K and L (whereby $K + F(L, w)$ and $L + F(K, w)$ undergo the same translation).

We can proceed as in the proof of Theorem 6.6.20 in CB, as long as no special assumptions on the polytopes $P_1, \ldots, P_{n-2}$ are involved. Thus we obtain the equality

$$S(K, \mathcal{C}, \cdot) = S(L, \mathcal{C}, \cdot) \tag{61}$$

for the mixed area measures, which implies the existence of a vector $t_k \in \mathbb{E}^n$ such that

$$h(K, u) - h(L, u) = \langle t_k, u \rangle \qquad \text{for } u \in \sigma_k \tag{62}$$

and (cf. CB, p. 374 to 375, line 3)

$$h(F(K, w), v_k) - h(F(L, w), v_k) = \langle t_k, v_k \rangle \tag{63}$$

for $k = 1, \ldots, m$.

We put $\bar{K} := K + F(L, w)$ and $\bar{L} := L + F(K, w)$. Let $u \in \sigma_k$; then u is a linear combination of w and v_k. Since $F(K, w) \subset H_{w,0}$ by (60), we have

$$h(F(K, w), u) = h(F(K, w), v_k)\langle u, v_k \rangle,$$

and similarly for L instead of K. From (62) and (63) we derive for $u \in \sigma_k$

$$\begin{aligned}
& h(\bar{K}, u) - h(\bar{L}, u) \\
= \ & h(K, u) - h(L, u) - [h(F(K, w), u) - h(F(L, w), u)] \\
= \ & \langle t_k, u \rangle - \langle t_k, v_k \rangle \langle u, v_k \rangle \\
= \ & \langle t_k, u - \langle u, v_k \rangle v_k \rangle \\
= \ & \langle t_k, w \rangle \langle u, w \rangle.
\end{aligned}$$

By (60) and (62) we have $\langle t_k, w \rangle = 0$, hence

$$h(\bar{K}, u) = h(\bar{L}, u) \qquad \text{for } u \in \sigma_k. \tag{64}$$

Hence, $\bar{K}$ and $\bar{L}$ have the same $(P_1, \ldots, P_{n-2})$–extreme supporting hyperplanes. This shows that the conditions of the theorem are necessary for equality in (58).

Vice versa, we assume now that (64) is satisfied. Let $\omega \subset S^{n-1}$ be a Borel set. It has a disjoint decomposition $\omega = \omega_1 \cup \omega_2 \cup \omega_3$, where $\omega_1 \subset \bigcup_{k=1}^m (\operatorname{relint} \sigma_k)$, $\omega_2 \cap \sigma_k = \emptyset$, and $\omega_3 \subset \{w, -w\}$. From (64) we get

$$S(K, \mathcal{C}, \omega_1) = S(L, \mathcal{C}, \omega_1)$$

(see CB, p. 374, in particular (6.6.73)), and by CB (6.6.74) we have

$$S(K, \mathcal{C}, \omega_2) = 0 = S(L, \mathcal{C}, \omega_2).$$

Therefore, (59) and CB (5.1.18) yield

$$\begin{aligned} 0 &= V(K, K, \mathcal{C}) - V(K, L, \mathcal{C}) + V(L, K, \mathcal{C}) - V(L, L, \mathcal{C}) \\ &= \frac{1}{n} \int_{S^{n-1}} h(K, u)\, d[S(K, \mathcal{C}, u) - S(L, \mathcal{C}, u)] \\ &\quad + \frac{1}{n} \int_{S^{n-1}} h(L, u)\, d[S(K, \mathcal{C}, u) - S(L, \mathcal{C}, u)] \\ &= \frac{1}{n}[h(K, w) + h(L, w)][v(F(K, w), \mathcal{C}) - v(F(L, w), \mathcal{C})] \\ &\quad + \frac{1}{n}[h(K, -w) + h(L, -w)][v(F(K, -w), \mathcal{C}) - v(F(L, -w), \mathcal{C})]. \end{aligned}$$

Here v denotes the $(n-1)$–dimensional mixed volume of convex bodies in hyperplanes orthogonal to w, and we have applied CB (6.6.72). Using (60) and $h(K, -w) > 0$, $h(L, -w) > 0$ (as follows from $V(K, K, \mathcal{C}) > 0$, $V(L, L, \mathcal{C}) > 0$) we conclude that

$$v(F(K, -w), \mathcal{C}) - v(F(L, -w), \mathcal{C}) = 0.$$

In a similar way we now obtain

$$\begin{aligned} & V(L, K, \mathcal{C}) - V(L, L, \mathcal{C}) \\ &= \frac{1}{n} \int_{S^{n-1}} h(L, u)\, d[S(K, \mathcal{C}, u) - S(L, \mathcal{C}, u)] \\ &= \frac{1}{n} h(L, w)[v(F(K, w), \mathcal{C}) - v(F(L, w), \mathcal{C})] \\ &\quad + \frac{1}{n} h(L, -w)[v(F(K, -w), \mathcal{C}) - v(F(L, -w), \mathcal{C})] \\ &= 0. \end{aligned}$$

This together with (59) implies equality in (58). ∎

References

Betke, U. 1992, Mixed volumes of polytopes. *Arch. Math.* **58** (1992), 388–391.

Ewald, G. 1990, A short proof of Alexandrov-Fenchel's inequality. *Note di Matematica* **10** , Suppl. n. 1 (1990), 243–249.

Ewald, G. and Tondorf, E. 1990+, A contribution to equality in Alexandrov-Fenchel's inequality. Preprint 1990 (and lecture at Oberwolfach, July 1990).

Fedotov, V. P. 1979, A counterexample to a hypothesis of Firey. (Russian) *Mat. Zametki* **26** (1979), 269–275. English translation: *Math. Notes* **26** (1979), 626–629.

Fedotov, V. P. 1982, Polar representation of a convex compactum. (Russian) *Ukrain. Geom. Sb.* **25** (1982), 137 - 138.

Firey, W. J. 1970, Convex bodies of constant outer p-measure. *Mathematika* **17** (1970), 21–27.

Firey, W. J. and Grünbaum, B. 1964, Addition and decomposition of convex polytopes. *Israel J. Math.* **2** (1964), 91–100.

Gardner, R. J and Volčič, A. 1993+, Determination of convex bodies by their brightness functions. (to appear)

Goodey, P. R. and Schneider, R. 1980, On the intermediate area functions of convex bodies. *Math. Z.* **173** (1980), 185–194.

Goodey, P. R., Schneider, R. and Weil, W. 1993+, Projection functions on higher rank Grassmannians. In: *Geometric Aspects of Functional Analysis* (eds. J. Lindenstrauss & V. D. Milman), *Lecture Notes in Math.*, Springer, Heidelberg etc. (to appear)

McMullen, P. 1993, On simple polytopes. *Invent. Math.* **113** (1993), 419–444.

McMullen, P. and Schneider, R. 1983, Valuations on convex bodies. In: *Convexity and its Applications* (eds. P. M. Gruber & J. M. Wills), Birkhäuser, Basel 1983, pp. 170–247.

Pedersen, P. and Sturmfels, B. 1992+, Product formulas for resultants and Chow forms. *Math. Z.* (to appear)

Schneider, R. 1970, On the projections of a convex polytope. *Pacific J. Math.* **32** (1970), 799–803.

Schneider, R. 1993, *Convex Bodies: the Brunn–Minkowski Theory.* (Encyclopedia of Math. and its Appl., vol. 44) Cambridge University Press, Cambridge 1993.

Schneider, R. 1991+, Equality in the Aleksandrov-Fenchel inequality - present state and new results. In: Colloquia Math. Soc. János Bolyai **63** (Intuitive Geometry, Szeged 1991) (to appear)

Shephard, G. C. 1963, Decomposable convex polyhedra. *Mathematika* **10** (1963), 89–95.

Shephard, G. C. 1966, Reducible convex sets. *Mathematika* **13** (1966), 49–50.

Yost, D. 1991, Irreducible convex sets. *Mathematika* **38** (1991), 134–155.

A SURVEY OF EULERIAN POSETS

RICHARD P. STANLEY*
Department of Mathematics
Massachusetts Institute of Technology
Cambridge, MA 02139

Abstract. An *Eulerian poset* is a finite graded poset with $\hat{0}$ and $\hat{1}$ such that every interval of length at least one has the same number of elements of odd rank as of even rank. For instance, the face lattice of a convex polytope is Eulerian. We survey some numerical and polynomial invariants associated with an Eulerian poset P. The *flag f-vector* counts the number of chains of P whose elements have specified ranks. A convenient way to represent the flag f-vector is by a noncommutative polynomial $\Phi_P(c, d)$ called the *cd-index* of P. The problem of characterizing the flag f-vector of certain classes of Eulerian posets, notably those which are Cohen-Macaulay, is best approached in the context of the cd-index. For the special class of *simplicial* Eulerian posets (which include face lattices of simplicial polytopes and triangulations of spheres), much more can be said about the flag f-vector. A high point of this subject is the *g-theorem*, which characterizes the f-vectors of simplicial convex polytopes. In Section 4 we discuss the concept of the h-vector of a lower Eulerian poset and its connection with intersection homology theory. The notion of h-vector leads naturally to the theory of *acceptable functions* on a lower Eulerian poset and their connection with subdivisions and the Ehrhart polynomial.

Key words: Eulerian poset, f-vector, h-vector, flag f-vector, flag h-vector, cd-index, simplicial poset, polytope, subdivision, local h-vector, acceptable function, Ehrhart polynomial

1. Preliminaries.

In this paper we will survey some of the fascinating properties of a class of posets (partially ordered sets) called *Eulerian posets.* We will be concerned almost exclusively with certain numerical and polynomial invariants associated with Eulerian posets. For results of a more structural nature, see e.g. [54]. Basic poset notation and terminology may be found in [65]. All posets, CW complexes, simplicial complexes, etc., considered in this paper are always assumed to be *finite.* Let P be a finite graded poset of rank $n+1$ with $\hat{0}$ and $\hat{1}$. (Often the letter d is used in place of our n. We use n to avoid confusion with the variable d of the cd-index, discussed in the next section.) Let ρ denote the rank function and μ the Möbius function of P. Thus $\rho(\hat{0}) = 0$ and $\rho(\hat{1}) = n+1$. If $s \leq t$ in P then we write $\rho(s,t) = \rho(t) - \rho(s)$, the rank (length) of the interval $[s,t]$. We say that P is *Eulerian* if $\mu(s,t) = (-1)^{\rho(s,t)}$ for all $s \leq t$ in P. Equivalently [65, Exer. 3.69(a)], P is Eulerian if and only if every interval of rank at least one contains as many elements of even rank as of odd rank, i.e.,

$$\sum_{u\in[s,t]} (-1)^{\rho(u)} = 0, \text{ if } s < t \text{ in } P.$$

* Partially supported by NSF grant #DMS-9206374.

T. Bisztriczky et al. (eds.), POLYTOPES: Abstract, Convex and Computational, 301–333.

The prototypical example of an Eulerian poset is the face lattice $P_{\mathcal{P}}$ of a convex polytope $\mathcal{P}$. If $\dim(\mathcal{P}) = n$, then $P_{\mathcal{P}}$ has rank $n+1$. For instance, if $\mathcal{P}$ is an n-simplex, then $P_{\mathcal{P}}$ is isomorphic to the boolean algebra B_{n+1}. A more general example of an Eulerian poset is the following. Recall that a (finite) *regular CW complex* Γ consists of a finite collection of disjoint open cells σ_i in a Euclidean space such that each σ_i is homeomorphic to an open ball $\mathbb{B}^{n_i}$ of some dimension n_i, and such that the boundary $\partial\sigma_i$ is homeomorphic to a sphere $\mathbb{S}^{n_i-1}$ of dimension $n_i - 1$. From this one can show that each cell boundary $\partial\sigma_i$ is a union of cells of Γ. By convention the empty set $\emptyset$ is also a cell of Γ, unless $\Gamma = \emptyset$. We will henceforth identify Γ with its *face poset*, whose elements are the cells of Γ, partially ordered by inclusion of their closures. The *dimension* $\dim \Gamma$ of Γ is the maximum dimension of any cell. The *body* $|\Gamma|$ is defined by

$$|\Gamma| = \bigcup_{\sigma\in\Gamma} \sigma.$$

An equivalent definition of a regular CW complex is due to Björner [14]. An (abstract) *simplicial complex* Δ is a collection of sets F (called *faces*) such that if $F \in \Delta$ and $G \subset F$, then $G \in \Delta$. We sometimes identify Δ with its *face poset*, i.e., the set of faces of Δ partially ordered by inclusion. A nonempty poset is then a simplicial complex if and only if it is a meet-semilattice (and hence has a $\hat{0}$ since all posets considered here are finite), and every interval $[\hat{0}, t]$ is a boolean algebra. We assume familiarity with the notion of the *geometric realization* $|\Delta|$ of an abstract simplicial complex Δ. For us, $|\Delta|$ is only considered to be a topological space. Thus, for instance, if $|\Delta|$ is a sphere then we say that Δ is a *triangulation of a sphere*. Given any poset P, define the *order complex* $\Delta(P)$ of P to be the set of all chains of P, regarded as an abstract simplicial complex. We can now state Björner's result: A (finite) poset Γ with $\hat{0}$ is a regular CW-complex if and only if for all $t > \hat{0}$ the geometric realization $|\Delta(\hat{0}, t)|$ of the order complex $\Delta(\hat{0}, t)$ of the open interval $(\hat{0}, t)$ is a sphere. Note that a simplicial complex is a special case of a regular CW complex, since if $[\hat{0}, t]$ is a boolean algebra then $|\Delta(\hat{0}, t)|$ is a sphere.

It is now an elementary topological result that a regular CW sphere (i.e., a regular CW complex whose body is a sphere), with a $\hat{1}$ adjoined, is Eulerian. By slight abuse of notation, we will call such posets (after the $\hat{1}$ has been adjoined) also regular CW spheres. Such posets need not even be lattices, as shown by Figure 1. An Eulerian poset need not be a regular CW complex. For example, let P be the poset of Figure 2. Note that P is a regular CW complex but not a sphere. Let C_2 be a two-element chain. Then $P \times C_2$ is Eulerian but not a regular CW complex.

A fundamental combinatorial invariant of a graded poset P of rank $n+1$ is the number p_i of elements of rank i. We also write $f_i = p_{i-1}$ and call the vector $f(P) = (f_0, f_1, \ldots, f_{d-1})$ the *f-vector* of P. Note that $f_{-1} = 1$ unless P is empty. Though as we shall see the f-vector of an Eulerian poset (or indeed of any graded poset) has many interesting properties, often it is too crude an invariant to be useful. We need to count not just elements of a given rank, but rather chains (or *flags*) of element of specified ranks. To this end, if $n \in \mathbb{N}$ (the set of nonnegative integers), write $[n] = \{1, 2, \ldots, n\}$, so in particular $[0] = \emptyset$. Also write $S = \{a_1, a_2, \ldots, a_k\}_<$ to denote that $S = \{a_1, a_2, \ldots, a_k\} \subset \mathbb{R}$ and $a_1 < a_2 < \cdots < a_k$. If now $S =$

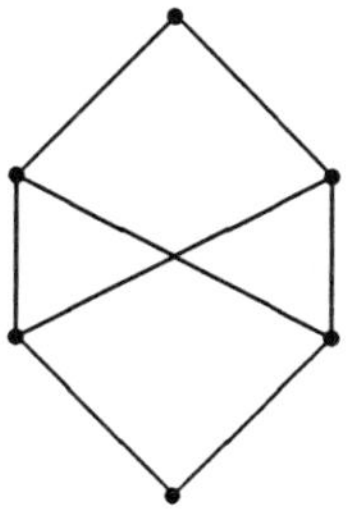

Fig. 1. An Eulerian nonlattice

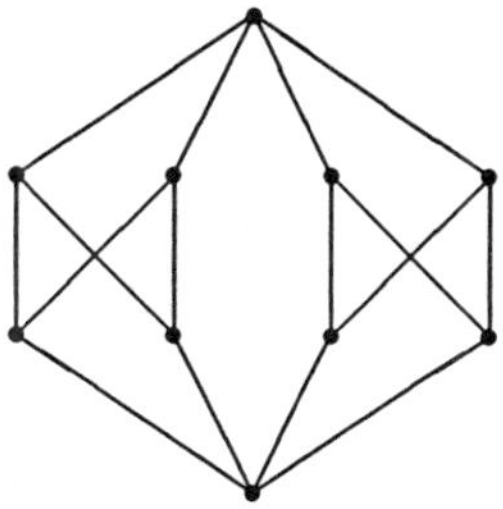

Fig. 2. A nonspherical Eulerian poset

$\{a_1, \ldots, a_k\}_< \subseteq [n]$, then define the *rank-selected subposet* P_S of P by

$$P_S = \{\hat{0}, \hat{1}\} \cup \{t \in P : \rho(t) \in S\}.$$

Now define $\alpha_P(S)$ to be the number of maximal chains of P_S. Equivalently, $\alpha_P(S)$ is the number of *S-flags* of P, i.e., flags (= chains) $\hat{0} < t_1 < \cdots < t_k < \hat{1}$ such that $S = \{\rho(t_1), \ldots, \rho(t_k)\}$. Thus for instance $\alpha_P(\emptyset) = 1$, $\alpha_P(\{i\})$ (abbreviated $\alpha_P(i)$) is the number $p_i = f_{i-1}$ of elements of P of rank i, and $\alpha_P([n])$ is the number of maximal chains of P. The function α_P is called the *flag f-vector* of P.

It is often the case that it is not the flag f-vector itself that is most natural to use, but rather a certain linear transformation of it defined as follows. For $S \subseteq [n]$ let

$$\beta_P(S) = \sum_{T \subseteq S} (-1)^{|S-T|} \alpha_P(T). \tag{1}$$

Equivalently (by the Principle of Inclusion-Exclusion),

$$\alpha_P(S) = \sum_{T \subseteq S} \beta_P(T). \tag{2}$$

The function β_P is called the *flag h-vector* of P. This function was first defined for distributive lattices in [56][57], and subsequently extended to other posets. It is an immediate consequence of "Philip Hall's theorem" [52, §3, Prop. 6] [65, Prop. 3.8.5] in the theory of Möbius functions that

$$\beta_P(S) = (-1)^{\#S+1}\mu_S(\hat{0}, \hat{1}), \tag{3}$$

where μ_S denotes the Möbius function of P_S (see [65, (34) on p. 131]).

We briefly explain the reason for the terminology "flag h-vector." Let Δ be any simplicial complex of dimension $n-1$, with f_i i-dimensional faces. The *h-vector* $(h_0, h_1, \ldots, h_n)$ of Δ is defined by

$$\sum_{i=0}^{n} f_{i-1}(x-1)^{n-i} = \sum_{i=0}^{n} h_i x^{n-i}. \tag{4}$$

The h-vector has long been known to be of fundamental importance; see for instance [62] and Section 3 below. If P is a poset with $\hat{0}$ and $\hat{1}$, define the *reduced* order complex $\tilde{\Delta}(P)$ of P to be the ordinary order complex $\Delta(P - \{\hat{0}, \hat{1}\})$ of $P - \{\hat{0}, \hat{1}\}$. If now P is graded of rank $n+1$, then it is easy to see that

$$f_{i-1}(\tilde{\Delta}(P)) = \sum_{\substack{S \subseteq [n] \\ \#S = i}} \alpha_P(S)$$

$$h_i(\tilde{\Delta}(P)) = \sum_{\substack{S \subseteq [n] \\ \#S = i}} \beta_P(S).$$

Thus the flag f-vector is a refinement, in a natural way, of the usual f-vector; and in exactly the same way the flag h-vector is a refinement of the usual h-vector.

We now discuss an important class of Eulerian posets. A poset P with $\hat{0}$ and $\hat{1}$ is said to be *Cohen-Macaulay* (over a fixed ground field K) if for every $s < t$ in P we have

$$\dim \tilde{H}_i(\Delta(s,t); K) = 0 \text{ if } i < \dim \Delta(s,t),$$

where $\tilde{H}_i(\Delta(s,t); K)$ denotes the ith reduced simplicial homology group (over K) of the order complex $\Delta(s,t)$ of the open interval (s,t). (The coefficient field K is fixed once and for all and is often suppressed from the notation.) In other words, every open interval has all its (reduced) homology only in the top possible dimension. It is easy to show from this definition that every Cohen-Macaulay poset is graded.

If f is a function whose domain consists of the set of intervals $[s,t]$ of a poset P, then we often abbreviate $f([s,t])$ as f_{st}. Let μ denote the Möbius function of P. Recall that if $s < t$ in P, then Philip Hall's theorem is equivalent to the formula

$$\mu_{st} = \tilde{\chi}(\Delta(s,t)),$$

where $\tilde{\chi}$ denotes the reduced Euler characteristic. It follows that if P is Cohen-Macaulay, then for any $s < t$ in P we have

$$\mu_{st} = (-1)^m \dim \tilde{H}_m(\Delta(s,t)), \tag{5}$$

where $m = \dim \Delta(s,t)$. (Because P is graded, we in fact have $m = \rho(s,t) - 2$.) In particular, if P is Cohen-Macaulay and Eulerian then

$$\dim \tilde{H}_m(\Delta(s,t)) = 1, \text{ for all } s < t \text{ in } P.$$

A poset P which is both Cohen-Macaulay and Eulerian is called a *Gorenstein** poset.

The following theorem summarizes two basic results in the theory of Cohen-Macaulay posets. For further information and references, see [16].

1.1 Theorem. *(a) Cohen-Macaulayness is a topological property, i.e., the question of whether a poset P (say with $\hat{0}$ and $\hat{1}$) is Cohen-Macaulay (over a fixed field K) depends only on the geometric realization $|\tilde{\Delta}(P)|$. In particular, P is Cohen-Macaulay if $|\tilde{\Delta}(P)|$ is a sphere.*

(b) If a poset P is Cohen-Macaulay, then so is any rank-selected subposet P_S.

1.2 Corollary. *If P is Cohen-Macaulay then $\beta_P(S) \geq 0$ for all S.*

Proof. Immediate from Theorem 1.1(b) and equations (3) and (5). □

A more general notion than Cohen-Macaulay poset that we will need later is a *Cohen-Macaulay simplicial complex* (over a field K). This is a simplicial complex Δ such that for every $F \in \Delta$ (including $F = \emptyset$), we have

$$\dim_K \tilde{H}_i(\mathrm{lk}_\Delta F; K) = 0, \text{ if } i < \dim(\mathrm{lk}_\Delta F).$$

Here $\tilde{H}_i(\mathrm{lk}_\Delta F; K)$ denotes the ith reduced homology group of $\mathrm{lk}_\Delta F$ with coefficient field K, and $\mathrm{lk}_\Delta F$ is the *link* of F in Δ, defined by

$$\mathrm{lk}_\Delta F = \{G \in \Delta : G \cap F = \emptyset \text{ and } G \cup F \in \Delta\}.$$

In particular, $\mathrm{lk}_\Delta \emptyset = \Delta$. Examples of Cohen-Macaulay simplicial complexes include triangulations of spheres and balls, and the independent sets of a matroid. It can be shown that a simplicial complex Δ is Cohen-Macaulay as a simplicial complex if and only it is Cohen-Macaulay as a poset (i.e., its face poset is a Cohen-Macaulay poset). Moreover, using commutative algebra techniques one can show that the h-vector of a Cohen-Macaulay complex (short for Cohen-Macaulay simplicial complex) is nonnegative. In fact, Cohen-Macaulay complexes seem to be the most general "natural" class of simplicial complexes whose h-vector is nonnegative.

2. The *cd*-Index.

In this section we will discuss what can be said about the flag f-vector (or flag h-vector) of a general Eulerian poset. First let us consider *linear* conditions on the flag f-vector. For instance, it is an immediate consequence of the definition of an Eulerian poset (say of rank $n+1$) that

$$f_0 - f_1 + f_2 - \cdots + (-1)^{n-1} f_{n-1} = 1 + (-1)^n.$$

This is a linear condition on the flag f-vector (since $f_i = \alpha(i+1)$). However, there are many additional linear conditions. We will mention a couple of special cases before moving to the general result.

Let P be a poset with $\hat{0}$ and $\hat{1}$. The *zeta polynomial* $Z(P, r)$ is defined for $r \in \mathbb{P}$ to be the number of multichains $\hat{0} = t_0 \leq t_1 \leq \cdots \leq t_r = \hat{1}$ of length r in P. It is easily seen that $Z(P, r)$ is a polynomial function of r whose degree is the rank of P. In particular, since $Z(P, r)$ is a polynomial it is defined for all integers (or even complex numbers) r. For these and other basic properties of the zeta polynomial see [65, Ch. 3.11]. For an Eulerian poset P of rank $n+1$, it is not hard to show [65, Prop. 3.14.1] that

$$Z(P,r) = (-1)^{n+1}Z(P,-r). \tag{6}$$

For instance, the face-lattices of the dodecahedron and icosahedron have zeta polynomial $5r^4 - 4r^2$. Equation (6) is equivalent to certain linear conditions on the flag f-vector of P, which turn out to be given by

$$h_i(\tilde{\Delta}(P)) = h_{n-i}(\tilde{\Delta}(P)).$$

In fact, there holds the much stronger result

$$\beta_P(S) = \beta_P(\bar{S}), \tag{7}$$

for all $S \subseteq [n]$, where $\bar{S} = [n] - S$. See e.g. [65, Cor. 3.14.6] for a proof.

Equation (7) yields 2^{n-1} independent linear relations satisfied by the flag f-vector of an Eulerian poset. But there are still others; for instance, every element of rank two in an Eulerian poset covers exactly two elements, from which it follows that $\alpha(1,2) = 2\alpha(2)$. This is a linear relation independent from (7). A complete set of linear relations was found by Bayer and Billera [6]. An elegant formulation of their result due to Fine (see [7]) will now be discussed. (Another proof appears in [70].) The idea is to work with certain noncommutative generating functions for the flag f and h-vectors. Fix a graded poset P of rank $n+1$ with $\hat{0}$ and $\hat{1}$. If $S \subseteq [n]$, then define a noncommutative monomial $u_S = u_1u_2\cdots u_n$ in the variables a and b by

$$u_i = \begin{cases} a, & \text{if } i \notin S \\ b, & \text{if } i \in S. \end{cases}$$

For instance, if $n = 6$ and $S = \{2,6\}$, then $u_S = abaaab$. Let

$$\Upsilon_P(a,b) = \sum_{S\subseteq[n]} \alpha_P(S)u_S \tag{8}$$

$$\Psi_P(a,b) = \sum_{S\subseteq[n]} \beta_P(S)u_S.$$

It is an immediate consequence of (1) or (2) that

$$\Psi_P(a,b) = \Upsilon_P(a-b,b) \tag{9}$$

$$\Upsilon_P(a,b) = \Psi_P(a+b,b).$$

For instance, if P is the boolean algebra B_3, then

$$\Upsilon_P(a,b) = a^2 + 3ba + 3ab + 6b^2$$

$$\Psi_P(a,b) = a^2 + 2ba + 2ab + b^2. \tag{10}$$

The result of Bayer and Billera, as formulated by Fine, is the following.

2.1 Theorem. *Let P be an Eulerian poset. Then $\Psi_P(a,b)$ can be written as a polynomial $\Phi_P(c,d)$ in $c = a + b$ and $d = ab + ba$.*

The polynomial $\Phi_P(c,d)$ is called the *cd–index* of P. The cd–index of an Eulerian poset is unique since $a+b$ and $ab+ba$ are algebraically independent (as noncommutative polynomials) over any field K. For instance, from (10) we see that $\Phi_{B_3}(c,d) = c^2 + d$. If we define $\deg(c) = 1$ and $\deg(d) = 2$, then clearly $\Phi_P(c,d)$ is homogeneous of degree n with integer coefficients. The cd-index is a very compact way of presenting the flag f-vector of an Eulerian poset. For instance, if P is the Bruhat order of the symmetric group $\mathcal{S}_4$ (as defined e.g. in [18]), then P is a fairly complicated poset of rank 6 with 24 elements and 168 maximal chains, yet

$$\Phi_P(c,d) = c^5 + dc^3 + 2cdc^2 + 2c^2dc + c^3d + 2d^2c + dcd + 2cd^2.$$

Note that since $a+b$ and $ab+ba$ are invariant under interchanging a and b, it follows from Proposition 2.1 that for an Eulerian poset P we have $\Psi_P(a,b) = \Psi_P(b,a)$. This formula is equivalent to (7). This symmetry condition is not enough to guarantee the existence of the cd–index. Indeed, a noncommutative polynomial $\Psi(a,b)$ satisfies $\Psi(a,b) = \Psi(b,a)$ if and only if $\Psi(a,b)$ is a polynomial (necessarily unique) in the variables $a+b, a^2+b^2, a^3+b^3, \ldots$. On the other hand, $\Psi(a,b) = \Phi(a+b, ab+ba)$ for some polynomial Φ if and only if $\Psi(a,b)$ is a polynomial just in the variables $a+b$ and $a^2+b^2 = (a+b)^2 - (ab+ba)$.

The existence of the cd-index imposes certain linear relations on the flag f-vector of an Eulerian poset, and it is not hard to show that there are no other relations. Indeed, Bayer and Billera [6] prove the stronger result that there are no additional relations even among flag f-vectors of face-lattices of convex polytopes. For arbitrary Eulerian posets the proof is easier and appears in [70, Prop. 1.5].

Even the cd-index $U_m(c,d)$ of the boolean algebra B_m is a subtle object. It turns out to be identical (as shown by Purtill [50]) to a polynomial studied earlier by Foata and Schützenberger [29], called a (noncommutative) *André polynomial.* Several related combinatorial interpretations of $U_m(c,d)$ were given by Foata and Schützenberger. We state here a slightly simpler interpretation due to S. Sundaram. We say that a sequence $\rho = \rho_1\rho_2\cdots\rho_m$ of integers *has no double descents* if there does not exist i for which $\rho_{i-1} > \rho_i > \rho_{i+1}$. A *simsun permutation* (named after Rodica Simion and Sheila Sundaram; see [71]) is a permutation π of $[m]$ such that for any $j \geq 0$, the word $\rho_1\rho_2\cdots\rho_{m-j}$ obtained from $\pi = \pi_1\pi_2\cdots\pi_m$ by removing $n, n-1, \ldots, n-j+1$ has no double descents. (In particular, the case $j = 0$ states that π has no double descents.) The number of simsun permutations of $[m]$ is the *Euler number* E_m, defined by

$$\sum_{m\geq 0} E_m \frac{x^m}{m!} = \tan x + \sec x.$$

E_m is also the number of *alternating permutations* of $[m]$, as defined e.g. in [65, pp. 148–149]. If π has no double descents, then define U_π to be the noncommutative monomial obtained from π by replacing each factor $\pi_i\pi_{i+1}$ with a d if $\pi_i > \pi_{i+1}$, and replacing each remaining term π_k with a c.

2.2 Proposition. *We have*

$$U_m(c,d) = \sum U_\pi,$$

summed over all simsun permutations of $[m]$.

For instance, the sixteen simsun permutations π of [4], together with the monomials U_π, are given by

1234	$cccc$	1324	cdc	3412	cdc	2143	dd
1243	ccd	1423	cdc	2134	dcc	3142	dd
1342	ccd	2314	cdc	3124	dcc	4132	dd
2341	ccd	2413	cdc	4123	dcc	4231	dd

Hence $U_4(c,d) = c^4 + 3c^2d + 5cdc + 3dc^3 + 4d^2$. One can also give a formula for the (noncommutative) exponential generating function $\sum_{m\geq 0} U_m(c,d)x^m/m!$; see [70, Cor. 1.9] for details.

Note that Proposition 2.2 gives a combinatorial interpretation to the coefficients of $U_m(c,d)$, thereby showing that they are nonnegative. We can ask whether there are similar results for other Eulerian posets. In general the *cd*-index of an Eulerian poset need not be nonnegative (i.e., have nonnegative coefficients). It is easy to see [70, Prop. 1.6] that for an Eulerian poset P, the numbers $\beta_P(S)$ are sums of certain coefficients of $\Phi_P(c,d)$. Hence a necessary condition for $\Phi_P(c,d)$ to be nonnegative (denoted $\Phi_P(c,d) \geq 0$) is that each $\beta_P(S)$ is nonnegative. This observation, together with Corollary 1.2, suggests the following conjecture, which is one of the most intriguing open problems in the theory of Eulerian posets.

2.3 Conjecture. *Let P be a Gorenstein* poset. Then* $\Phi_P(c,d) \geq 0$.

Conjecture 2.3 was first formulated by Fine for face lattices of convex polytopes, then for regular CW spheres by Bayer and Klapper [7, Conj. 5] and extended to its present form in [70, Conj. 2.1]. To prove Conjecture 2.3, presumably one would have to give either a combinatorial or algebraic (e.g., the dimension of a vector space) interpretation of the coefficients of $\Phi_P(c,d)$. It is easy to see that the sum $\Phi_P(1,1)$ of all the coefficients is given by $\Phi_P(c,d) = \beta_P(1,3,5,\ldots) = \beta_P(2,4,6,\ldots)$. Hence when $\beta_P(1,3,5,\ldots)$ has a natural combinatorial or algebraic interpretation, one can try to find a "refinement" into the coefficients of the *cd*-index. Only for very special posets called *lexicographically shellable* is a combinatorial interpretation known for $\beta_P(1,3,5,\ldots)$ (or more generally for any $\beta_P(S)$). For certain of these, most notably face lattices of *simplicial* polytopes, Purtill [50] was able to interpret the *cd*-index combinatorially and thereby prove nonnegativity. However, even though the face lattice of *any* convex polytope is lexicographically shellable, Purtill's methods have not yet been extended to them. Note that equations (3) and (5) give an interpretation of $\beta_P(S)$ as the dimension of a certain vector space (homology group). Thus a more promising line of attack on Conjecture 2.3 would be to find a refinement of the homology of $\tilde{\Delta}(P)$. The face ring techniques discussed in the next section can be used to interpret $\beta_P(S)$ as the dimension of a certain homogeneous part of a graded algebra [59], so one could also try to impose further structure (such as a finer gradation or filtration) on this algebra. Despite the plethora of techniques involving face rings which are now available, no progress has been made on the problem of finding a suitable refinement of the gradation.

Despite the lack of progress discussed in the previous paragraph, nevertheless some positive results related to Conjecture 2.3 have been obtained. One such result appears in the next section (Theorem 3.6). An additional result is given by the following theorem [70, Thm. 2.6].

2.4 Theorem. *Let P be the face poset of an S-shellable regular CW sphere Γ, with a $\hat{1}$ adjoined. Then $\Phi_P(c,d) \geq 0$.*

We will not give the rather technical definition of "S-shellability" here. It says roughly that there is a "nice" ordering of the facets of Γ. The recursive nature of the definition allows one to prove Theorem 2.4 by induction. The definition of S-shellability differs slightly from the more standard notion [14, §4] of shellability of a regular CW complex. However, the Bruggesser-Mani line shellings of a polytope [20] turn out to be S-shellable. Hence as a corollary to Theorem 2.2 we get a proof of the original conjecture of Fine that polytopes have nonnegative cd-index.

Conjecture 2.3 (and the weaker Theorem 2.4) give linear inequalities satisfied by the flag f-vector of certain Eulerian posets. It is natural to ask whether there are any additional inequalities independent from these. The answer is given by the next result, which explains the importance of Conjecture 2.3.

2.5 Theorem. *Conjecture 2.3, if true, gives all linear inequalities satisfied by flag f-vectors of Gorenstein* posets. Moreover, Theorem 2.4 gives (without any conjectures) all the linear inequalities satisfied by flag f-vectors of S-shellable regular CW spheres.*

Proof. (sketch) For any Eulerian poset P, the coefficient of c^n in $\Phi_P(c,d)$ is equal to one. Hence it suffices to show that for any cd-word $w \neq c^n$ of degree n (where $\deg c = 1$ and $\deg d = 2$), we can find an S-shellable CW sphere whose face poset with a $\hat{1}$ adjoined has cd-index for which the coefficient of w is an arbitrary factor larger than any of the other coefficients. Suppose $w = w_1 \cdots w_k$, where $w_i = c$ or d. Fix $m \geq 3$. Define posets T_i for $1 \leq i \leq k$ by $T_i = B_2$ (the boolean algebra of rank two) if $w_i = c$, and T_i is the face lattice of a convex m-gon if $w_i = d$. Given two Eulerian posets P and Q, define their *join* $P*Q$ to be the poset obtained by removing the $\hat{1}$ of P and $\hat{0}$ of Q, and putting every element of $Q - \hat{0}$ above every element of $P - \hat{1}$. It is easy to see that $P*Q$ is Eulerian. Now define $P_{w,m} = T_1 * T_2 * \cdots * T_k$. It is easy to show that $P_{w,m}$ is S-shellable (and a regular CW sphere), and that

$$\Phi_{P_{w,m}}(c,d) = w(c, c^2 + d),$$

i.e., substitute $c^2 + d$ for d in w. The coefficient of w in $w(c, c^2+d)$ is m^t, where t is the number of d's in w. The coefficient of any other cd-word in $w(c, c^2+d)$ is m^s for some $s < t$. Hence the proof follows by letting $m \to \infty$. □

Unfortunately the posets $P_{w,m}$ are not face lattices of polytopes (indeed, they are not even lattices), so Theorem 2.5 does not (necessarily) give all the linear inequalities satisfied by flag f-vectors of convex polytopes. In fact, one can show that flag f-vectors of polytopes satisfy linear inequalities not implied by the nonnegativity of the cd-index. An example of such an inequality, due to Kalai [37, Thm. 7.1], is

$$3f_2(\mathcal{P}) \geq 2f_1(\mathcal{P}) + 2f_3(\mathcal{P}),$$

valid for every 5-polytope $\mathcal{P}$ (where $f(\mathcal{P}) = (f_0, f_1, f_2, f_3, f_4)$ is the usual f-vector of $\mathcal{P}$).

In general, in trying to understand f-vectors, flag f-vectors, etc., of certain posets, geometric complexes, etc., there are three successively stronger levels of results: (a) linear equalities, (b) linear inequalities, (c) complete characterization. For flag f-vectors of Gorenstein* posets, we have just discussed the first two levels. Namely, (a) is completely solved, while (b) has a conjectured solution which has been proved to be necessary in some significant special cases. At present very little progress has been made toward a complete characterization, so this remains an interesting area of research. In the next section we will see some further examples of this hierarchy of results, including a difficult complete characterization (Theorem 3.9).

The next result, proved in [59, Cor. 4.5], does give some nonlinear information about flag f-vectors. However, it is a rather weak result when applied to Gorenstein* posets, since it holds for a much wider class of objects ("completely balanced Cohen-Macaulay simplicial complexes").

2.6 Theorem. *Let P be a Gorenstein* poset of rank $n+1$. Then there exists a simplicial complex Δ of dimension $n-1$ on a vertex set V and a map $\eta : V \to [n]$ satisfying the following conditions:*

(a) The restriction of η to any face F of Δ is injective.

(b) For any $S \subseteq [n]$, the number of faces $F \in \Delta$ for which $\eta(F) = S$ is equal to $\beta_P(S)$.

A numerical form of Theorem 2.6 can be gleaned from the results of [15] and [30]. (The statement in [59] after Corollary 4.5 that the converse to Corollary 4.5 is false is incorrect.)

Conjecture 2.3 and Theorems 2.4 and 2.5 suggest looking also at *inhomogeneous* linear equalities and inequalities satisfied by the coefficients of the cd-index, but in general they give only trivial additional information. For instance, the only inhomogeneous linear equality or inequality satisfied by flag f-vectors of Gorenstein* posets independent of Conjecture 2.3 is that the coefficient of c^n in $\Phi_P(c,d)$ is equal to one. However, additional hypotheses lead to more interesting possibilities. For instance, suppose that one particular coefficient of $\Phi_P(c,d)$ is specified. What can be said about the minimum value of the other coefficients? This question has yet to be investigated. We also venture the following conjecture.

2.7 Conjecture. *Let P be a Gorenstein* lattice of rank $n+1$. Then $\Phi_P(c,d) \geq U_{n+1}(c,d)$. In other words, among all Gorenstein* lattices of rank $n+1$, the boolean algebra B_{n+1} minimizes all the coefficients of the cd-index.*

Some of the inequalities implied by Conjecture 2.3 have the particularly simple form $\beta_P(T) \leq \beta_P(S)$. These are summarized by the next result. To this end, given $S \subseteq [n]$ define $\omega(S) \subseteq [n-1]$ by the condition $i \in \omega(S)$ if and only if exactly one of i and $i+1$ belongs to S. For instance, if $n = 8$ and $S = \{2,4,5,8\}$ then $\omega(S) = \{1,2,3,5,7\}$. If $n \geq 9$ then $\omega(S) = \{1,2,3,5,7,8\}$.

2.8 Proposition. *Let P be Eulerian with $\Phi_P(c,d) \geq 0$. Then $\beta_P(T) \leq \beta_P(S)$ whenever $\omega(T) \leq \omega(S)$. Moreover, if S and T are subsets of $[n]$ such that $\beta_P(T) \leq \beta_P(S)$ for every Gorenstein* poset (or even face poset of an S-shellable regular CW complex, with a $\hat{1}$ adjoined) P of rank $n+1$, then $\omega(T) \subseteq \omega(S)$.*

2.9 Corollary. *If $\Phi_P(c,d) \geq 0$ as above, then $\beta_P(S)$ is maximized for $S = \{1,3,5,\ldots\} \cap [n]$ and $S = \{2,4,6,\ldots\} \cap [n]$ (and possibly other values of S, depending on the poset P).*

There is an application of the cd-index to an intriguing conjecture of Charney and Davis [24, Conj. D]. The conjecture is a kind of combinatorial analogue of a well-known conjecture of Hopf that the Euler characteristic of a $2n$-dimensional closed Riemannian manifold M of nonpositive sectional curvature satisfies $(-1)^n\chi(M) \geq 0$. To state the Charney-Davis conjecture, define a ***flag complex*** to be a simplicial complex for which every minimal set of vertices which do not form a face (sometimes called a "missing face") has two elements. Charney and Davis conjecture that the h-vector $(h_0, \ldots, h_{2m})$ of a $(2m-1)$-dimensional Gorenstein* flag complex satisfies

$$(-1)^m(h_0 - h_1 + h_2 - \cdots + h_{2m}) \geq 0.$$

Now the order complex of any poset is a flag complex. When specialized to Gorenstein* posets P, it was observed by E. Babson that the Charney-Davis conjecture is equivalent to the statement that the coefficient of d^m in $\Phi_P(c,d)$ is nonnegative. Hence the Charney-Davis conjecture for the special case of order complexes follows from Conjecture 2.3. Moreover, it follows from Theorem 2.4 that the Charney-Davis conjecture holds for face-lattices of convex polytopes (or even S-shellable regular CW spheres), as noted by Charney and Davis in Section 7 of their paper.

3. Simplicial Eulerian Posets.

In this section we discuss some Eulerian posets for which much more can be said about the flag f-vector. We define a poset P with $\hat{0}$ to be *simplicial* if the interval $[\hat{0}, t]$ is a boolean algebra for all $t \in P$. For instance, the face poset P of a simplicial complex Δ is simplicial. One should think of an arbitrary simplicial poset as a kind of "generalized simplicial complex" in which the intersection of two faces can be any subcomplex of their boundaries, and not just a single face. By slight abuse of terminology, we say that an *Eulerian* poset P is simplicial if $P - \{\hat{1}\}$ is simplicial (as defined above). Thus if Δ is a triangulation of a sphere (regarded as a poset), then $\Delta \cup \{\hat{1}\}$ is Eulerian and simplicial.

Suppose that P is simplicial and Eulerian of rank $n+1$, with f_i elements of rank $i+1$. If t is an element of P of rank $m < n+1$, then since the interval $[\hat{0}, t]$ is a boolean algebra we see that the number of chains $\hat{0} = t_0 < t_1 < \cdots < t_k = t$ such that $\rho(t_i) = a_i$ is given by the multinomial coefficient $\binom{m}{a_1, a_2 - a_1, \ldots, m - a_{k-1}}$. Hence if $S = \{a_1, a_2, \ldots, a_k\}$, then there follows

$$\alpha_P(S) = \binom{m}{a_1, a_2 - a_1, \ldots, m - a_{k-1}} f_{m-1}.$$

As a consequence, the flag f-vector is completely determined by the f-vector. Thus we may restrict our attention just to the f-vector when dealing with the flag f-vector of simplicial (Eulerian) posets. Just as we defined the h-vector of a simplicial complex in (4), so we can define the h-vector $(h_0, h_1, \ldots h_n)$ of any simplicial Eulerian poset P of rank $n+1$. More precisely, let $\tilde{P} = P - \{\hat{1}\}$, and define $h(\tilde{P}) = (h_0, h_1, \ldots, h_n)$ by

$$\sum_{i=0}^{n} f_{i-1}(x-1)^{n-i} = \sum_{i=0}^{n} h_i x^{n-i}, \tag{11}$$

where $\tilde{P}$ has f_{i-1} elements of rank i. The reason we define $(h_0, h_1, \ldots, h_n)$ to be the h-vector of $\tilde{P}$ rather than P is that in the next section we will define the h-vector of a class of posets (called *lower Eulerian*) which includes both P and $\tilde{P}$, such that the definition agrees with (11) for the case $\tilde{P}$. (Of course we could define $h(\tilde{P})$ for *any* Eulerian poset P by (11). In fact, one could define the h-vector of any *graded* poset P by (11), but this turns out to be a not very useful definition when P is not simplicial. The "correct" definition of $h(\tilde{P})$ for an arbitrary Eulerian poset P is quite subtle and will be discussed in the next section. There is no "correct" definition known of the h-vector of an arbitrary (graded) poset.)

Our main concern here will be with the following classes of simplicial Eulerian posets: (a) Gorenstein*, (b) face lattices of simplicial polytopes, and (c) Gorenstein* lattices. For the first case, the f-vector (or h-vector) has been completely characterized for, loosely speaking, 3/4 of the posets. More precisely, we have the following result [67, §4].

3.1 Theorem. *Let* $\mathbf{h} = (h_0, h_1, \ldots, h_n) \in \mathbb{Z}^{n+1}$. *Suppose that either* n *is odd, or that* n *is even and* $h_{n/2}$ *is even. Then the following two conditions are equivalent.*

(a) $\mathbf{h}$ *is the* h*-vector of a simplicial Gorenstein* poset of rank* $n+1$.

(b) $h_0=1$, $h_i \geq 0$, *and* $h_i = h_{n-i}$ *for all* i.

The first step in the proof of Theorem 3.1 is the following result.

3.2 Theorem. *Let* P *be a Gorenstein* simplicial poset (or more generally a Cohen-Macaulay simplicial poset) with* h*-vector* $(h_0, h_1, \ldots, h_n)$. *Then* $h_i \geq 0$ *for all* i.

Proof (sketch). The idea is to associate a commutative ring A_P with P and use techniques from commutative algebra. Hence we first review some concepts from commutative algebra. Let K be an arbitrary ground field. A *finitely-generated graded* K*-algebra* is a commutative ring R which is an algebra over the field K, together with a vector space direct sum decomposition $R = R_0 \oplus R_1 \oplus \cdots$ satisfying the conditions:

- $R_0 = K$ (so that the identity element 1 of K is the identity element of R)
- $R_i R_j \subseteq R_{i+j}$. We then say that elements of R_i are *homogeneous* of degree i.
- R is finitely-generated as a K-algebra.

Henceforth we will call a finitely-generated graded K-algebra R simply a *graded algebra*. If the graded algebra R is generated as a K-algebra by elements of degree one, then we say that R is *standard*. If R is integral over the subalgebra $K[R_1]$ of R generated by elements of degree one (equivalently, R is a finitely-generated $K[R_1]$-module), then we say that R is *semistandard*. Clearly standard graded algebras are semistandard.

The *Krull dimension* $\dim R$ of a graded K-algebra R is the maximum number of elements of R which are algebraically independent over K. (See e.g. [1, Thm. 11.14] for some equivalent definitions of Krull dimension.) The *Noether normalization lemma* asserts that if $d = \dim R$, then we can find d homogeneous elements $\theta_1, \ldots, \theta_d$ of R of positive degree such that the quotient ring $R/(\theta_1, \ldots, \theta_d)$ has Krull dimension zero (equivalently, is a finite-dimensional vector space). The elements $\theta_1, \ldots, \theta_d$ are called a *homogeneous system of parameters* (h.s.o.p.). If each θ_i has degree one, then we call $\theta_1, \ldots, \theta_d$ a *linear system of parameters* (l.s.o.p.). If the ground field K is infinite, then the existence of an l.s.o.p. is equivalent to R being semistandard.

The hypothesis that R is finitely-generated guarantees that each homogeneous component R_i is a finite-dimensional vector space. The *Hilbert function* $H(R,\cdot)$ is defined for nonnegative integers i by $H(R,i) = \dim_K(R_i)$. It is the most natural invariant which measures the "size" of a graded algebra R. The *Hilbert series* of R is the formal power series

$$F(R,x) = \sum_{i\geq 0} H(R,i)x^i.$$

A well-known result of Hilbert (see e.g. [58, eqn. (1)]) asserts that when R is semistandard, then

$$F(R,x) = \frac{P(x)}{(1-x)^d}, \tag{12}$$

where $P(x) \in \mathbb{Z}[x]$, $P(1) \neq 0$, and $d = \dim R$. It follows from standard results about generating functions (e.g., [65, Prop. 4.2.2 and Cor. 4.3.1]) that $H(R,i)$ is a polynomial in i of degree $d-1$ for sufficiently large i. This polynomial is called the *Hilbert polynomial* of R. The Hilbert polynomial agrees with the Hilbert function for all $i \geq 0$ if and only if $\deg P(x) < d$. If $P(x) = h_0 + h_1x + \cdots + h_sx^s$ for some $s \geq 0$, then we write $h(R) = (h_0, h_1, ..., h_s)$ and call $h(R)$ the *h-vector* of R. Similarly $P(x)$ is called the *h-polynomial* of R. The h-vector and h-polynomial are only defined when R is semistandard. Note that trailing 0's in the h-vector are irrelevant, e.g, $(1,2,1)$ and $(1,2,1,0,0)$ are considered to be the same h-vector.

From now on let us assume that R is semistandard and K is infinite, so R has an l.s.o.p. We now come to the crucial concept of a *Cohen-Macaulay ring*. We will be content with an "operational" definition of what it means for R to be Cohen-Macaulay. It is not difficult to show that this definition is equivalent to the more customary definition involving regular sequences. For further information see e.g. [21][36][43][58][62][63]. Namely, we say that R is a Cohen-Macaulay ring if for some (equivalently, every) l.s.o.p. $\theta_1,\ldots,\theta_d$, the quotient ring $S = S_0 \oplus S_1 \oplus \cdots := R/(\theta_1,\ldots,\theta_d)$ (where the grading is inherited from R) satisfies $H(S,i) = h_i(R)$ for all i. Equivalently (by (12)) the Hilbert series $F(S,x)$ of S is given by

$$F(S,x) = (1-x)^d F(R,x). \tag{13}$$

It is an immediate consequence of the above definition that $h(R) \geq 0$ if R is Cohen-Macaulay.

Example. Let $R = K[u,v]/(uv)$ with $\deg u = \deg v = 1$. Let $\theta_1 = u+v$. Then θ_1 is an l.s.o.p. for R, and $S := R/\theta_1 R$ has a K-basis $\{1,u\}$. Hence $H(S,x) = 1+x$. Moreover, a K-basis for R is given by $\{1,u,v,u^2,v^2,\ldots\}$, so

$$F(R,x) = 1 + \frac{2x}{1-x} = \frac{1+x}{1-x}.$$

Hence $F(S,x) = (1-x)F(R,x)$, so R is Cohen-Macaulay. On the other hand, we leave it to the reader to verify that the ring $K[u,v,w]/(uv,uw)$ is not Cohen-Macaulay.

We are now ready to discuss the ring A_P which we associate with a simplicial poset P. First let us consider the case when P is a meet-semilattice, so that P is the face lattice of a simplicial complex Δ. In this case A_P turns out to be the *face ring* (also called the *Stanley-Reisner ring*) $K[\Delta]$ of Δ, where K is an arbitrary (infinite) ground field as before. Let us review the definition of $K[\Delta]$. Let the vertices of Δ be $x_1, \ldots, x_m$, and let $K[x_1, \ldots, x_m]$ denote the polynomial ring over K in the variables (indeterminates) $x_1, \ldots, x_m$. Set $K[\Delta] = K[x_1, \ldots, x_m]/I_\Delta$, where I_Δ is the ideal generated by all monomials (which we may assume are squarefree) $x_{i_1}x_{i_2}\cdots x_{i_k}$ such that $\{x_{i_1}, x_{i_2}, \ldots, x_{i_k}\} \notin \Delta$. We wish to extend this definition to A_P, for any simplicial poset P. Suppose that the elements of P are $\hat{0} = y_0, y_1, \ldots, y_p$. Define I_P to be the ideal of the polynomial ring $K[y_0, y_1, \ldots, y_p]$ generated by the following elements:

(R_1) $y_i y_j$, if y_i and y_j have no common upper bound in P.

(R_2) $y_i y_j - (y_i \wedge y_j)(\sum_z z)$, where z ranges over all *minimal* upper bounds of y_i and y_j, if y_i and y_j have a common upper bound in P.

(R_3) $y_0 - 1$.

NOTE. It is clear that the greatest lower bound $y_i \wedge y_j$ exists whenever y_i and y_j have an upper bound z in P, since the interval $[\hat{0}, z]$ is a boolean algebra (and therefore a lattice). Thus (R_2) is well-defined.

It is easy to see that when P is the face poset of a simplicial complex Δ then $A_P \cong K[\Delta]$, the face ring of Δ. Hence A_P is a generalization of the face ring. We can give the polynomial ring $K[y_0, y_1, \ldots, y_p]$ the structure of a graded K-algebra by defining the degree of $y \in P$ to be its rank $\rho(y)$. The relations (R_1)–(R_3) are homogeneous with respect to this grading, so we obtain a grading $A_P = (A_P)_0 \oplus (A_P)_1 \oplus \cdots$ of A_P. An important property of A_P [67, Lemma 3.9] is that it is semistandard (with respect to the grading just defined).

A simple computation [67, Prop. 3.8] shows that the Hilbert series of A_P is given by

$$F(A_P, x) = \frac{h_0 + h_1 x + \cdots + h_n x^n}{(1-x)^n}, \tag{14}$$

where $(h_0, h_1, \ldots, h_n)$ is the h-vector of P. In other words $h(P) = h(A_P)$. This is the means by which the h-vector enters into the structure of A_P.

A famous theorem of G. Reisner [21, Cor. 5.3.9][36, Ch. VI][51][62, Cor. 4.2] characterizes those simplicial complexes Δ for which the face ring $K[\Delta]$ is Cohen-Macaulay.

3.3 Theorem (Reisner's theorem). *The face ring $K[\Delta]$ of a simplicial complex Δ is Cohen-Macaulay if and only if Δ is a Cohen-Macaulay simplicial complex (over K). In particular, the face ring $K[\Delta(P)]$ of the order complex of a poset P is Cohen-Macaulay if and only if P is a Cohen-Macaulay poset.*

Theorem 3.3 of course explains the terminology "Cohen-Macaulay complex" and "Cohen-Macaulay poset." We will assume the validity of Theorem 3.3 in this survey. A detailed proof involves concepts from homological algebra which require some work for the uninitiated to absorb. The extension of Reisner's theorem to simplicial posets P is given by the next result [67, Cor. 3.5].

3.4 Theorem. *Let P be a simplicial poset. Then A_P is Cohen-Macaulay if and only P is Cohen-Macaulay.*

Proof (brief sketch). The only part to concern us here is the "if" part. By Reisner's theorem the face ring $K[\Delta(P)]$ of the order complex of P is Cohen-Macaulay if (and only if) P is Cohen-Macaulay. One then shows that A_P is an "algebra with straightening law" (ASL) on P. This means that A_P in a sense is "nicely approximated" by $K[\Delta(P)]$. In particular, the property that $K[\Delta(P)]$ is Cohen-Macaulay is transferred to A_P, so A_P is Cohen-Macaulay when P is Cohen-Macaulay. □

Note that the proof of Theorem 3.2 is now immediate. If P is Cohen-Macaulay then A_P is a Cohen-Macaulay ring; and hence by (13) the h-vector of P is nonnegative. □

Theorem 3.2 can be used to prove Conjecture 2.3 for Eulerian simplicial posets (generalizing the special case of simplicial polytopes due to Purtill [50], so now we have two generalizations of this result, viz., Theorems 2.4 and 3.6). We simply state the relevant result [70, Thm. 3.3] here.

3.5 Theorem. *Let n be a positive integer. Then there exist homogeneous polynomials $\check{\Phi}_1, \ldots, \check{\Phi}_{n-1}$ of degree n in the variables c, d (with $\deg c = 1$, $\deg d = 2$) with nonnegative coefficients such that for every simplicial Eulerian poset P of rank $n+1$ with h-vector $(h_0, h_1, \ldots, h_n)$, we have*

$$\Phi_P(c,d) = \sum_{i=0}^{n-1} h_i \check{\Phi}_i(c,d).$$

Putting together Theorems 3.2 and 3.5 yields the following result.

3.6 Theorem. *Let P be a Gorenstein* simplicial poset. Then $\Phi_P(c,d) \geq 0$.*

The next ingredient we need for our 3/4-characterization of f-vectors of Gorenstein* simplicial posets is the following result, known as the *Dehn-Sommerville equations* for Eulerian simplicial posets.

3.7 Theorem. *Let P be a simplicial Eulerian poset of rank $n+1$ with h-vector $(h_0, h_1, \ldots, h_n)$. Then $h_i = h_{n-i}$ for all i.*

Theorem 3.7 goes back to Dehn, who proved it in 1905 for simplicial 4-polytopes and 5-polytopes and suggested that it be extended to higher dimensions. Such an extension was carried out by Sommerville in 1927, who proved it for arbitrary simplicial polytopes. A generalization to what is essentially the special case of Theorem 3.7 when P is a meet semilattice (i.e., the face poset of a simplicial complex) was proved by Klee in 1964. Numerous generalizations and extensions have been subsequently given. A result equivalent to Theorem 3.7 itself appears in [65, Thm. 3.14.9]. A very general version of the Dehn-Sommerville equations may be found in [68, Lemma 6.4].

Theorems 3.2 and 3.7 show that (a) implies (b) in Theorem 3.1. To show the converse, it is necessary to construct simplicial Gorenstein* posets with suitable h-vectors. This is a fairly easy inductive construction whose details appear in [67, Lemmas 4.1 and 4.2]. With this the proof of Theorem 3.1 is complete. □

It is an intriguing problem to fill in the gap inherent in Theorem 3.1, i.e., to characterize h-vectors of simplicial Gorenstein* posets P of odd rank $n+1$ for which $h_{n/2}$ is also odd. Given that P has odd rank, the condition that $h_{n/2}$ is odd is equivalent to the condition that P has an odd number of facets (elements of rank n). This is because $h_i = h_{n-i}$ and $h_0 + h_1 + \cdots + h_n = f_{n-1}$, the number of facets. We collect together from [67] and [27, Ch. 2] some partial results related to this problem.

3.8 Proposition.

(a) *Suppose* $h_0 = 1$, $h_i = h_{n-i}$, *and* $h_i > 0$ *for all* i. *Then there exists a simplicial Gorenstein* poset* P *with* $h(P) = (h_0, h_1, \ldots, h_n)$.

(b) *There is no Eulerian simplicial poset* P *with an odd number of facets (equivalently,* P *has odd rank* $n+1$ *and* $h_{n/2}$ *is odd) and* $h_1(P) = 0$.

(c) *There is no Eulerian simplicial poset* P *of rank* $n+1$ *with an odd number of facets such that*

$$h_0(P) = h_1(P) = \cdots = h_i(P) = 1, \quad h_{i+1}(P) = 0,$$

where $i \geq 1$ *and* $i + 1 < n$.

(d) *There is no Eulerian simplicial poset* P *of rank* $n+1$ *with an odd number of facets such that* $\sum_{i=0}^{n} h_i(P) \leq n$.

(e) *There is no Eulerian simplicial poset* P *with an odd number of facets such that* $h_1(P) = 2$, $h_2(P) = 0$.

(f) *There is no Gorenstein* simplicial poset* P *with an odd number of facets such that* $h_1(P) = 3$, $h_2(P) = 0$.

The following question in particular remains open concerning simplicial Gorenstein* posets P. If P has odd rank $n+1$ and $h_i = 0$ for some $1 \leq i \leq n-1$, then is $h_{n/2}$ even? If the answer is positive then Theorem 3.1 and Proposition 3.8(a) would give a complete solution to the problem of characterizing h-vectors of simplicial Gorenstein* posets. Let us finally remark that the problem of characterizing h-vectors of simplicial Gorenstein* posets bears some similarities to the problem of characterizing the Betti numbers of compact manifolds (see [23, Thm. 2.2]).

We now consider one of the most interesting classes of simplicial Gorenstein* posets, viz., the face lattices $L_{\mathcal{P}}$ of simplicial convex polytopes $\mathcal{P}$ (i.e., convex polytopes all of whose facets are simplices). For general information about convex polytopes see e.g. [19][35][47][73]. According to Theorems 3.2 and 3.7 the h-vector of $\mathcal{P}$ satisfies $h_0 = 1$, $h_i \geq 0$, and $h_i = h_{n-i}$, where $n = \dim \mathcal{P}$. In 1971 McMullen and Walkup [48] published the conjecture that the h-vector was *unimodal*, i.e., weakly increases and then decreases. In view of the Dehn-Sommerville equations ($h_i = h_{n-i}$) this is equivalent to the inequalities

$$h_0 \leq h_1 \leq \cdots \leq h_{\lfloor n/2 \rfloor}. \tag{15}$$

The inequalites (15) were called the *Generalized Lower Bound Conjecture* (GLBC), because they implied an earlier *Lower Bound Conjecture* (proved by Barnette [2][3] in 1971 and 1973 by geometric reasoning). McMullen and Walkup [48] showed that the GLBC, if true, gave the all the homogeneous linear inequalities satisfied by the f-vector of a simplicial convex polytope (except for $h_0 \geq 0$, which is trivial since $h_0 = 1$). Also, the only additional *inhomogeneous* linear equality or inequality is given by the trivial $h_0 = 1$.

In 1970 in East Lansing, Michigan, McMullen made an inspired conjecture, based on very little evidence. Namely, he conjectured a complete characterization of the h-vector of a simplicial polytope. This conjecture was published in 1971 [44] and became known as *McMullen's g-conjecture* (because of the use of the notation g_{i-1} to denote the crucial quantity $h_i - h_{i-1}$; it is now customary to use g_i for this quantity). It was later proved by Billera-Lee and Stanley (as discussed below), and so is now called the *g-theorem*. To state this result, we use the fact (easily proved by induction) that given positive integers h and i, there is a unique way to write

$$h = \binom{n_i}{i} + \binom{n_{i-1}}{i-1} + \cdots + \binom{n_j}{j}, \tag{16}$$

where $n_i > n_{i-1} > \cdots > n_j \geq j \geq 1$. For a nice discussion of the significance of this representation, see [34, Section 8]. Now define

$$h^{\langle i \rangle} = \binom{n_i+1}{i+1} + \binom{n_{i-1}+1}{i} + \cdots + \binom{n_j+1}{j+1},$$

and set $0^{\langle i \rangle} = 0$. The number $h^{\langle i \rangle}$ is sometimes called the *ith upper pseudopower of h*. Call a vector $(h_0, h_1, \ldots, h_d) \in \mathbb{Z}^{d+1}$ an *M-vector* (after F. S. Macaulay, for a reason soon to be made clear) if $h_0 = 1$ and $0 \leq h_{i+1} \leq h_i^{\langle i \rangle}$ for $1 \leq i \leq d-1$. We can now state the remarkable g-theorem first conjectured by McMullen.

3.9 The g-Theorem. *A vector $(h_0, \ldots, h_d) \in \mathbb{Z}^{d+1}$ is the h-vector of some simplicial d-polytope $\mathcal{P}$ if and only if $h_i = h_{d-i}$ and $(h_0, h_1 - h_0, h_2 - h_1, \ldots, h_{\lfloor d/2 \rfloor} - h_{\lfloor d/2 \rfloor - 1})$ is an M-vector.*

The sufficiency of McMullen's g-theorem (i.e., the "if" part of the theorem) was first proved by Billera and Lee in 1979 (see [12] and [13]) and the necessity soon afterwards by this writer. The proof of Billera and Lee involved a clever inductive construction. The proof of necessity, on the other hand, required deep techniques from algebraic geometry. The first step is an algebraic interpretation of M-vectors.

3.10 Theorem. *A vector $(k_0, k_1, \ldots, k_s) \in \mathbb{Z}^{s+1}$ is an M-vector if and only if there exists a standard graded algebra $R = R_0 \oplus R_1 \oplus \cdots \oplus R_s$ satisfying $H(R, i) = h_i$ for $0 \leq i \leq s$.*

The difficult part of the proof of Theorem 3.10 is due to F. S. Macaulay [40]. His intricate argument has subsequently been simplified [55][72][25], though it is still not an easy result. The purely numerical form of the theorem as stated here first appeared in [58].

To prove the necessity of the g-theorem (i.e., the "only if" part), it follows from Theorem 3.10 that we need to find a standard graded algebra R whose Hilbert function satisfies $H(R, i) = h_i - h_{i-1}$ for $1 \leq i \leq \lfloor n/2 \rfloor$, where $(h_0, h_1, \ldots, h_n)$ is the h-vector of the simplicial polytope $\mathcal{P}$. Reisner's theorem implies that the face ring $K[\mathcal{P}]$ of the boundary complex of $\mathcal{P}$ is Cohen-Macaulay. Hence if $\theta_1, \ldots, \theta_n$ is any l.s.o.p., then the quotient ring $S = K[\mathcal{P}]/(\theta_1, \ldots, \theta_n)$ satisfies $H(S, i) = h_i$. This is not what we want, but it is close. Suppose we could find an element $\omega \in S_1$ which is a non-zerodivisor on S_i for $i < \lfloor n/2 \rfloor$. Let $R = S/\omega S$. Then the condition on ω implies that for $i < \lfloor n/2 \rfloor$,

$$\begin{aligned} \dim_K R_i &= \dim_K S_i - \dim_K \omega S_{i-1} \\ &= \dim_K S_i - \dim_K S_{i-1} \\ &= h_i - h_{i-1}. \end{aligned}$$

Hence R is the desired ring, and the necessity of the g-theorem would be proved.

To show the existence of the critical element ω (called a *Lefschetz element*), assume without loss of generality that the simplicial n-polytope $\mathcal{P}$ is embedded in

$\mathbb{R}^n$ so that the origin is in the interior and the vertices have rational coordinates. (Not every convex polytope is *rational*, i.e., can be embedded in $\mathbb{R}^n$ so that the vertices have rational coordinates. However, all simplicial polytopes are rational since small perturbations of their vertices do not change their combinatorial type. For further information, see [35, p. 95] [17, Fig. 8.4.1][73, Example 6.21].) It is then possible to associate with $\mathcal{P}$ an algebraic variety $X_{\mathcal{P}}$, which is a special case of a *toric variety*. In general there is an intimate connection between convex polytopes and related geometric objects on the one hand, and toric varieties on the other. For further information on these fascinating connections, see for example [26][28][31][49]. The connection between $\mathcal{P}$ and $X_{\mathcal{P}}$ which is of concern to us here is the following: The (singular) cohomology ring (say with real coefficients) $H^*(X_{\mathcal{P}};\mathbb{R})$ of $X_{\mathcal{P}}$ is isomorphic to $\mathbb{R}[\mathcal{P}]/(\theta_1,\ldots,\theta_n)$, where $\theta_1,\ldots,\theta_n$ is a certain l.s.o.p. of $\mathbb{R}[\mathcal{P}]$ (which depends on how $\mathcal{P}$ is embedded in $\mathbb{R}^n$).

The varieties $X_{\mathcal{P}}$ are projective and have very simple singularities (called "finite quotient singularities"). From this it can be shown that they satisfy the *hard Lefschetz theorem* (so called because Lefschetz stated two theorems about the cohomology of algebraic varieties, one of which is much more difficult to prove than the other). The hard Lefschetz theorem implies the existence of the desired Lefschetz element ω, completing the proof of the necessity of the g-theorem.

For a more detailed exposition of the above proof, see [10][28][31][49][64]. The reference [64] contains in particular some historical comments on the proof (p. 221). Let us just add here an additional historical remark which does not appear in the previous references. The crucial and most difficult result used in the proof is the hard Lefschetz theorem for the varieties $X_{\mathcal{P}}$. In the original published proof [61] of the necessity of the g-theorem, a proof of the hard Lefschetz theorem for $X_{\mathcal{P}}$ was attributed to Steenbrink in 1976. However, it turned out that the proof of Steenbrink was incorrect. The first correct proof was not given until 1983 by M. Saito [53], using the deep machinery of filtered $\mathcal{D}$-modules. Fortunately I did not become aware of the incorrectness of Steenbrink's proof until after Saito had published his correct proof!

After the g-theorem was proved by the methods indicated above, it became a challenge to try to simplify the proof. The use of ring theory seems inevitable because of the simple algebraic interpretation of M-vectors. One would like, however, to show the existence of a Lefschetz element ω without using the theory of toric varieties. Recently exactly such a proof was found by McMullen [46], using his important concept of the *polytope algebra* [45]. McMullen's proof is a high point in the geometric tradition founded by Minkowski. The polytope algebra is a fascinating object which will undoubtedly receive intense further study.

A natural question at this point is the following: Does the g-theorem hold for a wider class of geometric objects than simplicial polytopes? Three classes of objects to consider in this context are (a) PL spheres (which may be defined as triangulations of spheres such that the link of every vertex is also a sphere), (b) triangulations of spheres, and (c) Gorenstein* lattices. Each of the three classes is more general than the preceding. For all three classes it remains open whether the g-theorem is valid. A promising approach toward proving the g-theorem for case (b) (and possibly even (c)) was found by Kalai, based on his theory of algebraic shifting and some work on embeddability of simplicial complexes due to Sarkaria. However, this approach has not yet led to a proof of the g-theorem for spheres. The problem of extending the g-theorem to spheres is perhaps the most outstanding open problem in the subject of Eulerian posets. About the only significant generalization of the g-theorem to date is an extension of the GLBC to triangulations of $(n-2)$-dimensional spheres which are subcomplexes of boundary complexes of simplicial n-polytopes. This result was first proved by Kalai [38, §8] using algebraic shifting, and was given a somewhat simpler proof based on face rings in [69, Cor. 2.4].

4. The h-Vector of an Eulerian Poset.

In the previous section we defined the h-vector of a simplicial Eulerian poset P using the simple formula (11). (CAVEAT: Do not confuse this h-vector with the h-vector of the order complex of P.) We mentioned that extending this definition in the obvious way to an arbitrary Eulerian poset does not seem interesting. Instead there is a subtle generalization motivated by topology and algebraic geometry. Recall from our discussion of McMullen's g-theorem that if $\mathcal{P}$ is a simplicial n-polytope (embedded in $\mathbb{R}^n$ with rational vertices and origin in the interior), then the cohomology ring $H^*(X_\mathcal{P};\mathbb{R})$ of the toric variety $X_\mathcal{P}$ is isomorphic to $S := \mathbb{R}[\mathcal{P}]/(\theta_1,\ldots,\theta_n)$ for a certain l.s.o.p $\theta_1,\ldots,\theta_n$. The usual grading of $H^*(X_\mathcal{P};\mathbb{R})$ is such that $H^{2i}(X_\mathcal{P};\mathbb{R})$ corresponds to S_i (so that $H^{2i+1}(X_\mathcal{P};\mathbb{R})=0$ for all i). Thus

$$\dim_\mathbb{R} H^{2i}(X_\mathcal{P};\mathbb{R}) = h_i(\mathcal{P}). \tag{17}$$

Now a toric variety $X_\mathcal{P}$ can be defined for *any* rational polytope $\mathcal{P}$. (It need not be simplicial.) We could try to define an h-vector for $\mathcal{P}$ *via* equation (17). Unfortunately this naive definition is not satisfactory. For one thing, it is not determined solely by the combinatorial type of $\mathcal{P}$ but rather depends on how $\mathcal{P}$ is embedded into $\mathbb{R}^n$. There is, however, a homology theory which is better behaved, viz., the *intersection homology* (or cohomology) theory of Goresky and MacPherson (see e.g. [32][33][42]). Let $IH^i(X_\mathcal{P};\mathbb{R})$ denote the i-th real middle-perversity intersection cohomology group of the toric variety $X_\mathcal{P}$, and let

$$h_i = h_i(\mathcal{P}) = \dim_\mathbb{R} IH^{2i}(X_\mathcal{P};\mathbb{R}).$$

One can show the following facts:

(IH$_1$) The odd-degree groups $IH^{2i+1}(X_\mathcal{P};\mathbb{R})$ are all 0.

(IH$_2$) h_i depends only on the combinatorial type of the rational polytope $\mathcal{P}$, and not on its embedding.

(IH$_3$) $h_i = h_{n-i}$ for all i (where $n = \dim\mathcal{P}$), and $h_i = 0$ for $i > n$. This result is a consequence of *Poincaré duality* for intersection homology.

(IH$_4$) $1 = h_0 \le h_1 \le \cdots \le h_{\lfloor n/2\rfloor}$. This is a consequence of the difficult *hard Lefschetz theorem* for intersection homology due to Beilinson, Bernstein and Deligne [8] (see [42, §1.6.2]).

Since $h_i(\mathcal{P})$ depends only on the combinatorial type of $\mathcal{P}$, we would like a combinatorial formula or rule for computing $h_i(\mathcal{P})$. Such a result was given independently by Khovanskiĭ and MacPherson (and perhaps others), and was first published in [66]. This rule turns out to make sense for any Eulerian poset P, so we use it to *define* the h-vector of P (more precisely, of $P - \{\hat{1}\}$). The definition depends on the following result.

4.1 Theorem. *Let P be an Eulerian poset of rank $n+1$ with rank function ρ. Then there exists a unique function $h : P \to \mathbb{Z}[x]$ satisfying the conditions (where we write h_t for the value of h at the point $t \in P$):*

(a) For all $t > \hat{0}$ in P we have

$$\deg h_t(x) \leq \lfloor \frac{1}{2}(\rho(t) - 1) \rfloor.$$

(b) For all $t \in P$, we have

$$\sum_{s \leq t} h_s(x)(x-1)^{\rho(s,t)} = x^{\rho(t)} h_t(1/x). \qquad (18)$$

For instance, when the interval $[\hat{0}, t]$ is a boolean algebra, then the binomial theorem implies that $h_t(x) = 1$.

If P is Eulerian, then the polynomial $h(P, x) := h_{\hat{1}}(x)$ is called the *h-polynomial* of P or *g-polynomial* of $P - \{\hat{1}\}$. We want, however, to define the h-polynomial of $P - \{\hat{1}\}$. More generally, define a (finite) poset P to be *lower Eulerian* if P has a $\hat{0}$ and every interval $[\hat{0}, t]$ is Eulerian. Note that for such a poset, the polynomial $h_t(x)$ of Theorem 4.1 is still defined for each $t \in P$, viz., $h_t(x) = h([\hat{0}, t], x)$. Thus for any lower Eulerian poset P of rank n, we can define the *h-polynomial* of P by the formula

$$x^n h(P, 1/x) = \sum_{t \in P} h_t(x)(x-1)^{n-\rho(t)}. \qquad (19)$$

Moreover, if $h_P(x) = h_0 + h_1 x + \cdots + h_n x^n$, then we define $(h_0, h_1, \ldots, h_n)$ to be the *h-vector* of P.

The definition of the h-vector of a lower Eulerian poset is not easy to understand, and many of its properties remain mysterious. Let us point out a few salient facts concerning the above definition.

- Suppose that P is simplicial (and hence lower Eulerian) of rank n. Since $h(B_k, x) = 1$ for any boolean algebra B_k, equation (19) reduces to

$$\begin{aligned} x^n h(P, 1/x) &= \sum_{t \in P} (x-1)^{n-\rho(t)} \\ &= \sum_{i=0}^{n} f_{i-1}(x-1)^{n-i}, \end{aligned}$$

  where P has f_{i-1} elements of rank i. Comparing with (11) shows that the the two definitions we have given of the h-vector (or h-polynomial) of P agree.
- We mentioned that the original motivation for the definition of $h(P, x)$ came from the intersection homology of toric varieties. Specifically, if $\mathcal{P}$ is a *rational*

n-dimensional polytope, then let P be its face poset with $\hat{1}$ removed. Thus P is lower Eulerian. It is then the case that $h_i(P) = \dim_{\mathbb{R}} IH^{2i}(X_{\mathcal{P}}; \mathbb{R})$. In particular, $h_i(P) \geq 0$. Note that (IH$_3$) gives $h_i = h_{n-i}$ (a generalization of the Dehn-Sommerville equations), while (IH$_4$) yields $1 = h_0 \leq h_1 \leq \cdots \leq h_{\lfloor n/2 \rfloor}$.

- It is natural to ask to what extent the three results $h_i = h_{n-i}$, $h_i \geq 0$, and $h_0 \leq h_1 \leq \cdots \leq h_{\lfloor n/2 \rfloor}$ hold for lower Eulerian posets more general than face posets of rational polytopes (with $\hat{1}$ removed). So far as the Dehn-Sommerville equations are concerned, the following result can be obtained by a formal computation [66, Thm. 2.4] [65, Thm. 3.14.9][68, Lemma 6.4].

4.2 Theorem. *Let P be an Eulerian poset of rank $n+1$ with $\hat{1}$ removed (so P has rank n). Then $h_i(P) = h_{n-i}(P)$ for all i.*

- Suppose that P is Eulerian of rank $n+1$, and let $\tilde{P} = P - \{\hat{1}\}$. Let $h(\tilde{P}, x) = h_0 + h_1 x + \cdots + h_n x^n$. Thus by Theorem 4.2 we have $h_i = h_{n-i}$. It is a simple consequence of the definition of $h(P, x)$ that

$$h_i(P) = \begin{cases} h_i(\tilde{P}) - h_{i-1}(\tilde{P}), & \text{if } i \leq \lfloor n/2 \rfloor \\ 0, & \text{otherwise.} \end{cases} \tag{20}$$

In particular, knowing either of $h(P, x)$ or $h(\tilde{P}, x)$ determines the other.

- Let us consider the question of the nonnegativity and unimodality of the h-vector of an arbitrary lower Eulerian poset. The unimodality question seems natural only in the presence of the Dehn-Sommerville equations $h_i = h_{n-i}$. The most general "natural" class of posets with this property consists of (by Theorem 4.2) Eulerian posets with $\hat{1}$ removed. Now note that by (20), we have $h(P, x) \geq 0$ if and only if $h(\tilde{P}, x)$ is unimodal. Hence the unimodality question is subsumed by the nonnegativity question. Moreover, since the most general "natural" class of simplicial complexes with nonnegative h-vector are the Cohen-Macaulay ones, we should restrict our attention to Cohen-Macaulay lower Eulerian posets.

It is false that $h_i(P) \geq 0$ for all Cohen-Macaulay lower Eulerian posets. For instance, the Cohen-Macaulay Eulerian poset P given by Figure 1 satisfies $h(P, x) = 1 - x$, since

$$(x-1)^3 + 2(x-1)^2 + 2(x-1) + (1-x) = x^3(1 - \frac{1}{x}).$$

Moreover, the Cohen-Macaulay lower Eulerian poset P (which is an Eulerian poset with $\hat{1}$ removed) given by Figure 3 satisfies

$$h(P, x) = x^3 h(P, 1/x) = (x-1)^3 + 2(x-1)^2 + 2(x-1) + 2(1-x) = x^3 - x^2 - x + 1.$$

Although not all Cohen-Macaulay lower Eulerian posets have nonnegative h-vector, we do have the following conjecture [66, Conj. 4.2(b)].

4.3 Conjecture. *Let P be a Cohen-Macaulay lower Eulerian meet-semilattice. Then $h_i(P) \geq 0$ for all i.*

Conjecture 4.3 is open even for face lattices of *nonrational* polytopes, with or without $\hat{1}$ removed. Conjecture 4.3 is one of the outstanding open problems in the theory of Eulerian posets.

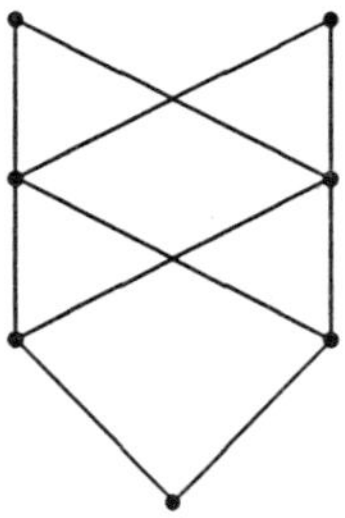

Fig. 3. A lower Eulerian poset with nonpositive h-vector

- It is not difficult to show that for any lower Eulerian poset P, the h-polynomial $h(P,x)$ can be computed from the flag f-vector of P [5, Prop. 6.1]. In fact, if P is Eulerian then each $h_i(P)$ or $h_i(\tilde{P})$ is a *unique* linear combination of the coefficients of the cd-index $\Phi_P(c,d)$. More precisely, for each n there is a unique linear transformation φ from the space of all cd-polynomials of degree n (with deg $c = 1$ and deg $d = 2$ as usual) with rational coefficients to the space $\mathbb{Q}^{n+1}$ which for every Eulerian poset P of rank $n+1$ satisfies $\varphi(\Phi_P(c,d)) = h(\tilde{P})$. However, this expression for $h_i(P)$ or $h_i(\tilde{P})$ in terms of the cd-index seems quite messy and not of much use. For example, writing $[w]$ for the coefficient of the cd-word w in $\Phi_P(c,d)$, we have:

$$\begin{aligned} h_0(P) &= [c^n] \\ h_1(P) &= -(n-2)[c^n] + [dc^{n-2}] \\ h_2(P) &= \frac{1}{2}(n-1)(n-4)[c^n] - (n-4)[dc^{n-2}] - [c^2dc^{n-4}] + [d^2c^{n-4}]. \end{aligned}$$

5. Acceptable Functions on a Lower Eulerian Poset.

The condition (18) satisfied by the h-vector of an Eulerian (or even lower Eulerian) poset is quite interesting for its own sake. Thus given a lower Eulerian poset P, define a function $f : P \to \mathbb{Q}[x]$ (other fields could be used instead of $\mathbb{Q}$, but there is no reason to do so here) to be *acceptable* if for all $t \in P$ we have

$$\sum_{s \le t} f_s(x)(x-1)^{\rho(s,t)} = x^{\rho(t)} f_t(1/x). \tag{21}$$

Note that if f is acceptable then deg $f_t \le \rho(t)$ for all $t \in P$, since otherwise the right-hand side of (21) would not be a polynomial. Let $\mathcal{A}(P)$ denote the set of all acceptable functions $f : P \to \mathbb{Q}[x]$. Clearly $\mathcal{A}(P)$ is a vector space over $\mathbb{Q}$; moreover, since deg $f_t \le \rho(t)$ it follows that

$$\dim \mathcal{A}(P) \le \sum_{t \in P} (\rho(t) + 1).$$

The next theorem gives a more precise result.

5.1 Theorem. *Suppose P is lower Eulerian. Then*

$$\dim \mathcal{A}(P) = \sum_{t \in P} \lfloor \frac{1}{2}\rho(t) + 1 \rfloor.$$

Proof (sketch). Let $t \in P$. Suppose that we have defined f_s for all $s < t$ so that equation (21) holds (with t replaced by s). We claim that the set of all polynomials f_t satisfying (21) is a vector space of dimension $\lfloor \frac{1}{2}\rho(t) + 1 \rfloor$, from which the proof will follow. The key point is that the polynomial $k_t(x)$ given by

$$k_t(x) = \sum_{s<t} f_s(x)(x-1)^{\rho(s,t)}$$

is *antisymmetric*, i.e., $k_t(1/x) = -k_t(x)$. This is proved by a formal argument, using the fact that $\mu(u,v) = (-1)^{\rho(u,v)}$ in a lower Eulerian poset P. For complete details (in a more general setting), see [68, Lemma 6.4]. Now from (21) we have

$$k_t(x) = x^{\rho(t)} f_t(1/x) - f_t(x). \tag{22}$$

Since both $k_t(x)$ and the right-hand side of (22) are antisymmetric, we are free to choose the first $\lfloor \frac{1}{2}\rho(t)+1 \rfloor$ coefficients of f_t arbitrarily, and the remaining coefficients are then uniquely determined. From this the proof follows. (Note that if $k_t(x)$ were not antisymmetric, then there would be no way to choose $f_t(x)$, so it would not be possible to build up f "greedily" as we have done.) □

Theorem 5.1 describes the structure of the space of all acceptable functions on P. We would now like to give some interesting examples of acceptable functions, in addition to the h-polynomial. We will give two classes of examples, the first dealing with subdivisions and the second with Ehrhart polynomials.

The definition of a subdivision of an Eulerian poset is motivated by the usual notion in topology of a subdivision of a simplicial complex (or of more general geometric objects). A *geometric simplex* is the convex hull in some Euclidean space $\mathbb{R}^m$ of a set of affinely independent points. A *geometric simplicial complex* is a collection Σ of finitely many geometric simplices contained in $\mathbb{R}^m$ such that (a) if $\alpha \in \Sigma$ and τ is a face of α, then $\tau \in \Sigma$, and (b) if $\alpha, \tau \in \Sigma$, then $\alpha \cap \tau$ is a common face (possibly empty) of both α and τ. A *geometric simplicial subdivision* (or *geometric triangulation*, often called just a *triangulation*) of Σ is a geometric simplicial complex Σ' such that

$$\bigcup_{\alpha \in \Sigma} \alpha = \bigcup_{\alpha' \in \Sigma'} \alpha',$$

and such that every face of Σ' is contained in a face of Σ. Now let Δ and Δ' be the face posets of Σ and Σ', respectively. We may regard Δ and Δ' as *abstract* simplicial complexes. How can we abstract the notion that Σ' is a subdivision of Σ in the most general useful way to define what it means for Δ' to be a subdivision of Δ? There turn out to be many possible definitions, depending on how much of the geometric essence of the notion of subdivision we wish to preserve. In all cases, we will have a map $\sigma : \Sigma' \to \Sigma$ satisfying certain properties. One should think of $\sigma(t)$ as the smallest face of Δ which contains t (called the *carrier* of t). Thus the motivation for the following definition should be clear. We say that Δ' (or more precisely, the pair (Δ', σ)) is a *topological subdivision* of Δ if (a) for every face F of Δ, the subset $\sigma^{-1}(2^F)$ of Δ' (where 2^F denotes the subcomplex of Δ consisting of F and all its faces) is a subcomplex of Δ' whose geometric realization $|\sigma^{-1}(2^F)|$ is homeomorphic to a ball of dimension $\#F - 1$, and (b) $\sigma^{-1}(F)$ consists of the interior faces of the ball $\sigma^{-1}(2^F)$. We have called this type of subdivision *topological* because it depends on the topological notion of *ball* to define it. As such it is a difficult definition to work with, because there is no useful combinatorial or algebraic characterization of simplicial complexes whose geometric realization is a ball. In fact, it is known that the question of deciding whether a simplicial complex is a ball is undecidable! What are the properties of balls that are essential here? This depends on what kind of applications we have in mind. If one is only interested in showing that certain functions are acceptable, then the key property we need is the following result. For a proof of a more general result, see [62, Ch. II, Cor. 7.2]. A result equivalent to a generalization of Theorem 5.2 to manifolds with boundary was earlier proved by Macdonald [41, Thm. 2.1]. Let Δ be a simplicial complex whose geometric realization is an $(n-1)$-dimensional ball. Recall that the h-polynomial $h(\Delta, x) = h_0 + h_1 x + \cdots + h_n x^n$ of Δ is defined by

$$x^n h(\Delta, 1/x) = \sum_{F \in \Delta} (x-1)^{n-\#F}.$$

Define the *interior h-polynomial* $h(\mathrm{int}(\Delta), x)$ by

$$x^n h(\mathrm{int}(\Delta), 1/x) = \sum_{F \in \mathrm{int}(\Delta)} (x-1)^{n-\#F},$$

where $\mathrm{int}(\Delta)$ denotes the set of interior faces (faces not contained in the boundary) of Δ.

5.2 Theorem. *With Δ as above, we have*

$$h(\mathrm{int}(\Delta), x) = x^n h(\Delta, 1/x).$$

Now let P and P' be lower Eulerian posets, and let $\sigma : P' \to P$ be a map satisfying:

(S$_1$) For all $t \in P'$, $\rho(t) \le \rho(\sigma(t))$ (where ρ denotes the rank function of P).

(S$_2$) For every $u \in P$, let $P_{\le u} = \{v \in P : v \le u\}$ (the principal order ideal generated by u). Then $P'_u := \sigma^{-1}(P_{\le u})$ is an order ideal of P', i.e., if $t \in P'$ and $s \le t$ then $s \in P'$.

Let $h(P', x)$ denote the h-polynomial of P', as defined by equation (19). Define the *interior* $\mathrm{int}(P'_u)$ of the poset P'_u by $\mathrm{int}(P'_u) = \sigma^{-1}(u)$. Finally define the h-polynomial $h(\mathrm{int}(P'_u), x)$ by

$$x^{\rho(u)} h(\mathrm{int}(P'_u), 1/x) = \sum_{t \in \mathrm{int}(P'_u)} h_t(x)(x-1)^{\rho(u)-\rho(t)}.$$

Guided by Theorem 5.2, we now say that P' (or more accurately, the pair (P', σ)) is a *formal subdivision* of P if (S_1) and (S_2) are satisfied, and if the following condition is also satisfied:

(S_3) For all $u \in P$ we have

$$x^{\rho(u)} h(P'_u, 1/x) = h(\mathrm{int}(P'_u), x). \tag{23}$$

In particular, topological subdivisions of simplicial complexes (or even of regular CW complexes, with an obvious extension of the definition of topological subdivision) are formal subdivisions. The main theorem on formal subdivisions is the following [68, Thm. 7.5].

5.3 Theorem. *Let $\sigma : P' \to P$ be a formal subdivision of the lower Eulerian poset P. Let $f : P \to \mathbb{Z}[x]$ be defined by $f_u(x) = h(P'_u, x)$. Then f is acceptable.*

Theorem 5.3 shows that every formal subdivision of an Eulerian poset P "induces" an acceptable function f on P. For topological triangulations Δ' of a simplicial complex Δ, the value f_t of f on a face u of Δ is just the usual h-polynomial of the restriction of Δ' to Δ. For a formal subdivision $\sigma : P' \to P$ of an *Eulerian* (rather than lower Eulerian) poset P, we can define a highly interesting polynomial $\ell(\sigma, x)$, called the *local h-polynomial* of σ, by the formula

$$\ell(\sigma, x) = \sum_{u \in P} (-1)^{\rho(u,\hat{1})} h(P'_u, x) h([u, \hat{1}]^*, x). \tag{24}$$

Here $[u, \hat{1}]^*$ denotes the dual of the interval $[u, \hat{1}]$. One fundamental property of $\ell(\sigma, x)$ is the following symmetry result [68, Cor. 7.7], which is proved as a formal consequence of equation (23).

5.4 Theorem. *Let $\sigma : P' \to P$ be a formal subdivision of the Eulerian poset P of rank $n + 1$. Then*

$$x^n \ell(\sigma, 1/x) = \ell(\sigma, x). \tag{25}$$

In the more intuitive case of simplicial complexes, P is a boolean algebra (the face lattice of a simplex 2^V), so $h([u, \hat{1}]^*, x) = 1$ for all $u \in P$. Moreover P' is just a simplicial complex Γ, and we often write $\ell_V(\Gamma, x)$ for $\ell(\sigma, x)$ (though $\ell_V(\Gamma, x)$ actually depends on σ [68, Ex. 2.3(e)]). Equation (24) takes the simple form

$$\ell_V(\Gamma, x) = \sum_{u \in P} (-1)^{\rho(u,\hat{1})} h(P'_u, x), \tag{26}$$

i.e., the alternating sum of the h-polynomials of the restrictions of Γ to the faces of the simplex 2^V.

We do not have the space here to go into a detailed discussion of the local h-polynomial $\ell_V(\Gamma, x)$, but let us mention a few highlights. We need to introduce two further types of subdivisions. Define a topological subdivision $\sigma : \Delta' \to \Delta$, where Δ' and Δ are simplicial complexes, to be *quasi-geometric* if no face t of Δ' has the following property: There is a face F of Δ such that $\dim(F) < \dim(t)$ and every vertex x of t satisfies $\sigma(x) \leq F$. In other words, no face t of Δ' has all its vertices on a face of Δ of lower dimension than t.

Note that geometric subdivisions (more accurately, triangulations of abstract simplicial complexes which can be geometrically realized by a geometric triangulation) are quasi-geometric, since the vertices of a geometric simplex are affinely independent. It can be shown that there exist quasi-geometric subdivisions which are not geometric [68, p. 814]. The simplest example of a topological subdivision which is not quasi-geometric is the following. Let Δ be a triangle with vertices a, b, c. Add a new vertex d on the line ab and let Δ' have facets abc and abd. This subdivision is not quasi-geometric since the two-dimensional face abd of Δ' has all its vertices on the one-dimension face ab of Δ.

The last type of subdivision which we consider (though there are still others of interest, such as homological subdivisions and homotopical subdivisions) is the following. A geometric triangulation of a simplicial complex is called *regular* if (speaking somewhat informally) its restriction to each face (simplex) can be realized as the projection of a strictly convex polyhedral surface. For a more detailed definition, see [15][39][68, Def. 5.1]. An abstract triangulation of a simplicial complex is *regular* if it has a regular geometric realization. It can be shown that not every geometric subdivision of a geometric simplicial complex is regular [15, Fig. 1][39, Fig. 2(b)].

We now have all the definitions necessary to summarize some basic properties of the local h-polynomial $\ell_V(\Gamma, V)$ of a triangulation Γ of the simplex 2^V.

First, as a special case of the symmetry result (25), we have

$$x^n \ell_V(\Gamma, 1/x) = \ell_V(\Gamma, x), \tag{27}$$

where $\#V = n$.

Now let Δ be a pure (i.e., all maximal faces have the same dimension) $(n-1)$-dimensional simplicial complex, and let Δ' be a topological triangulation. Then the h-polynomial $h(\Delta', x)$ is given by the following result [68, Thm. 3.2].

5.5 Theorem. *With notation as above, we have*

$$h(\Delta', x) = \sum_{F \in \Delta} \ell_F(\Delta'_F, x) h(\mathrm{lk}_\Delta F, x), \tag{28}$$

where Δ'_F denotes the restriction of Δ' to the face F of Δ.

Equation (28) is a fundamental result for reducing questions about f-vectors of triangulations to properties of local h-vectors.

A much deeper result on local h-vectors than equation (27) and Theorem 5.5 is the following.

5.6 Theorem. *Let Γ be a quasi-geometric triangulation of the simplex 2^V. Then $\ell_V(\Gamma, x) \geq 0$ (i.e., every coefficient is nonnegative).*

Proof (brief sketch). The proof requires commutative and homological algebra. The idea is as follows. Let $K[\Gamma]$ be the face ring of Γ, and let $V = \{x_1, \ldots, x_n\}$. An l.s.o.p. $\theta_1, \ldots, \theta_n$ is called *special* if for each i, we have that θ_i is a linear combination of vertices of Γ whose carrier contains x_i. It is not hard to show that if Γ is a topological triangulation of 2^V, then a special l.s.o.p. exists if and only if Γ is quasi-geometric [68, Cor. 4.4]. Supposing that $\theta_1, \ldots, \theta_n$ is special, let Ω be the ideal of $K[\Gamma]$ generated by all monomials whose support does not lie on the boundary of Γ. Let $L_V[\Gamma]$ be the image of Ω in the quotient $K[\Gamma]/(\theta_1, \ldots, \theta_n)$. Then $L_V[\Gamma]$ is a graded $K[\Gamma]$-module, say $L_V[\Gamma] = L_0 \oplus L_1 \oplus \cdots$. One shows by a homological argument [68, Thm. 4.6] that the Hilbert series $F(L_V[\Gamma], x) := \sum_i (\dim_K L_i)x^i$ is just $\ell_V(\Gamma, x)$, and the proof follows. □

The conclusion of Theorem 5.6 fails for topological triangulations. For example, let $V = \{a, b, c, d\}$. Let Γ be the triangulation of 2^V obtained by adding a vertex e inside face abc, and letting the facets of Γ be $abcd$ and $abce$. Then $\ell_V(\Gamma, x) = -x^2$. This example was found by Clara Chan.

The next result gives an important application of Theorem 5.6 (in fact, the original *raison d'etre* for the creation of local h-vectors).

5.7 Corollary. *Let Δ be a Cohen-Macaulay simplicial complex and Δ' a quasi-geometric triangulation. Then $h(\Delta') \geq h(\Delta)$. In other words, the h-vector of a Cohen-Macaulay complex increases under quasi-geometric triangulation.*

Proof. Consider equation (28). By Theorem 5.6 each polynomial $\ell_F(\Delta'_F, x)$ is nonnegative. By definition of a Cohen-Macaulay complex Δ, each link $\mathrm{lk}_\Delta F$ is Cohen-Macaulay. We mentioned in Section 1 the result that Cohen-Macaulay complexes have nonnegative h-vectors. Hence each polynomial $h(\mathrm{lk}_\Delta F, x)$ appearing in (28) is nonnegative. Moreover, the term in the sum on the right-hand side of (28) indexed by $F = \emptyset$ is just $h(\Delta, x)$. From this the proof is immediate. □

Corollary 5.7 is false if Δ is not assumed to be Cohen-Macaulay (but is assumed to be pure) [68, Ex. 4.12]. It is not known whether the h-vector of a Cohen-Macaulay complex increases under *topological* triangulation. This is an intriguing open problem in the theory of Cohen-Macaulay complexes (due to G. Kalai and this writer), which we state as an explicit conjecture.

5.8 Conjecture. *Let Δ' be a topological triangulation of a Cohen-Macaulay simplicial complex. Then*

$$h(\Delta') \geq h(\Delta).$$

We next consider the situation for regular triangulations.

5.9 Theorem. *Let Γ be a regular triangulation of the simplex 2^V. Then $\ell_V(\Gamma, x)$ is unimodal, i.e. (using Theorem 5.4), if $\ell_V(\Gamma, x) = \ell_0 + \ell_1 x + \cdots + \ell_n x^n$, then $\ell_0 \leq l_1 \leq \cdots \leq \ell_{\lfloor n/2 \rfloor}$.*

The proof of Theorem 5.9 is very deep; it requires the hard Lefschetz theorem for the decomposition theorem of intersection homology (though the proof is quite easy if one is willing to accept all this machinery). For further details see [68, Thm. 5.2]. There is an application of Theorem 5.9 analogous to how Corollary 5.7 follows from Theorem 5.6. The easy proof is omitted.

5.10 Corollary. *Let Δ be a pure simplicial complex such that for every face $F \in \Delta$ the h-vector of $\mathrm{lk}_\Delta F$ is symmetric ($h_i = h_{d-\#F-i}$ for all i) and unimodal ($h_0 \leq h_1 \leq \cdots \leq h_{\lfloor \frac{1}{2}(d-\#F) \rfloor}$, assuming symmetry). Let Δ' be a regular triangulation of Δ. Then*

$$h_i(\Delta) - h_{i-1}(\Delta) \leq h_i(\Delta') - h_{i-1}(\Delta'), \quad 0 \leq i \leq \lfloor d/2 \rfloor.$$

Note that by Theorem 3.7 the boundary complex of a simplicial convex polytope satisfies the hypothesis of the above corollary. At the end of Section 3 some conjectured extensions are discussed. There is an intriguing conjectured generalization of Theorem 5.9, analogous to the conjectured generalization of the g-theorem for simplicial polytopes to the case of spheres.

5.11 Conjecture. *Theorem 5.9 continues to hold when Γ is a geometric (or even quasi-geometric) subdivision of 2^V.*

Note that the example $\ell_V(\Gamma, x) = -x^2$ of Clara Chan mentioned after Theorem 5.6 shows that Theorem 5.9 certainly cannot be extended to topological subdivisions.

There are many additional interesting open problems associated with local h-vectors. For instance, can one find a "nice" characterization of quasi-geometric (or geometric or regular) triangulations Γ of 2^V for which $\ell_V(\Gamma, x) = 0$? For further information see [68, pp. 821–823]. Can one characterize (or at least obtain significant new information on) the h-vectors of suitable classes of triangulations (e.g., topological, quasi-geometric, geometric, regular) Γ of 2^V? Even more strongly, can one characterize the functions f defined on subsets W of V by $f(W) = \ell_W(\Gamma_W, x)$, where Γ is a (suitably restricted) triangulation of 2^V? For some work on this problem, see [22].

We mentioned earlier that we would like to discuss two classes of acceptable functions. The first class, just discussed, deals with subdivisions. We now come to the second class, which is concerned with Ehrhart polynomials. There is by now a vast literature on Ehrhart polynomials, but we will be content with a few remarks which show the connection with acceptability. For an introduction to Ehrhart polynomials see [65, pp. 235–241] or [36, Part 2]. Our account below is taken from [68, Ex. 7.13].

Let L be a lattice in $\mathbb{R}^N$. (One can take $L = \mathbb{Z}^N$ without significant loss of generality.) Let $\mathcal{P}$ be an *L-polytope*, i.e., a convex polytope with vertices in L. (Much of what we say below can be generalized to L-polyhedral complexes, but for simplicity we only consider polytopes.) As usual we partially order the faces of $\mathcal{P}$ by inclusion; this makes $\mathcal{P}$ into an Eulerian poset. Given an integer $n > 0$ and a d-dimensional L-polytope $\mathcal{P} \subset \mathbb{R}^N$, define

$$i(\mathcal{P}, n) = \#(n\mathcal{P} \cap L),$$

where $n\mathcal{P} = \{n\alpha : \alpha \in \mathcal{P}\}$. Also set $i(\mathcal{P}, 0) = 1$. It is known that $i(\mathcal{P}, n)$ is a polynomial function of n of degree d, called the *Ehrhart polynomial* of $\mathcal{P}$. It follows that if we set

$$\omega(\mathcal{P}, x) = (1 - x)^{d+1} \sum_{n \geq 0} i(\mathcal{P}, n) x^n, \tag{29}$$

then $\omega(\mathcal{P}, x)$ is a polynomial in x of degree $\leq d$. Since every face F of $\mathcal{P}$ is also an L-polytope, we may regard ω as a function on $\mathcal{P}$, i.e., the value of ω at F is just $\omega(F, x)$. It is noted in [66, p. 201] (for the lattice $L = \mathbb{Z}^N$) that ω is an acceptable function on $\mathcal{P}$. This result is equivalent to Ehrhart's "law of reciprocity," which states that

$$(-1)^d i(\mathcal{P}, -n) = \#(\text{int}(\mathcal{P}) \cap L),$$

where $\text{int}(\mathcal{P})$ denotes the relative interior of $\mathcal{P}$.

Now assume that $\mathcal{S}$ is a d-dimensional L-simplex in $\mathbb{R}^N$. The "Ehrhart analogue" of the local h-vector of a triangulation of a simplex is given by $\ell(\omega, x)$. We will use the notation $\ell^*(\mathcal{S}, x)$ instead of $\ell(\omega, x)$, where the asterisk * indicates that we are working in the context of lattice points and Ehrhart polynomials. Thus we have explicitly the formula

$$\ell^*(\mathcal{S}, x) = \sum_F \omega(F, x)(-1)^{d-\dim(F)},$$

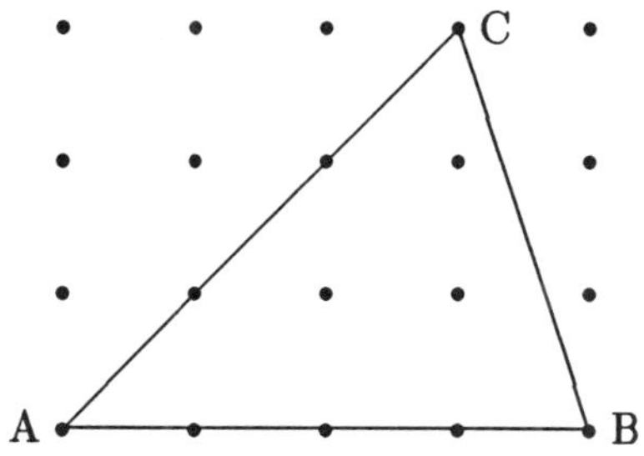

Fig. 4. A lattice triangle

where F runs over all faces of $\mathcal{S}$. Note a fundamental difference between ℓ and ℓ^*: $\ell_V(\Gamma, x)$ depends on the choice Γ of triangulation of 2^V, while $\ell^*(\mathcal{S}, x)$ depends only on $\mathcal{S}$ (but with the structure of an L-simplex, not just an abstract simplex). By Theorem 5.4 we have the symmetry formula

$$x^{d+1}\ell^*(\mathcal{S}, 1/x) = \ell^*(\mathcal{S}, x).$$

Moreover, Betke and McMullen [9] (using a different notation and viewpoint) give a geometric interpretation of the coefficients of $\ell^*(\mathcal{S}, x)$ which show that they are nonnegative (a considerably easier result than the analogous Theorem 5.6 for triangulations). From this and the lattice analogue of Theorem 5.5 it follows that for any L-polytope $\mathcal{P}$ the polynomial $\omega(\mathcal{P}, x)$ has nonnegative coefficients. Another proof of this fact based on shelling was given in [60]. An algebraic approach to ℓ^* analogous to the theory needed to prove Theorem 5.6 was given by Batyrev [4, Def. 9.1ff]. For further information and references, see [68, Ex. 7.13].

We conclude with an example illustrating the definition of $\ell^*(\mathcal{S}, x)$. Let $\mathcal{S} = ABC$ be the triangle with vertices $A = (0,0)$, $B = (4,0)$, and $C = (3,3)$, as shown in Figure 4. Then $\sum_{n \geq 0} i(ABC, n)x^n = (1 + 8x + 3x^2)(1 - x)^{-3}$, so $\omega(ABC, x) = 1 + 8x + 3x^2$. Similarly $\omega(AB, x) = 1 + 3x$, $\omega(AC, x) = 1 + 2x$, $\omega(BC, x) = 1$, $\omega(A, x) = \omega(B, x) = \omega(C, x) = \omega(\emptyset, x) = 1$. Hence

$$\begin{aligned} \ell^*(ABC, x) &= 1 + 8x + 3x^2 - (1 + 3x) - (1 + 2x) - 1 + 1 + 1 + 1 - 1 \\ &= 3x + 3x^2. \end{aligned}$$

(For a lattice triangle $\mathcal{S}$ it is in fact not difficult to see that $\ell^*(\mathcal{S}, x) = kx + kx^2$, where k is the number of interior lattice points of $\mathcal{S}$. For higher dimensional simplices the situation is considerably more complicated.) Note that $x^3\ell^*(ABC, 1/x) = \ell^*(x)$, in accordance with Theorem 5.4.

References

1. Atiyah, M. F., and Macdonald, I.G.: 1969, *Introduction to Commutative Algebra*, Addison-Wesley, Reading, Massachusetts.
2. Barnette, D. W.: 1971, The minimum number of vertices of a simple polytope, *Israel J. Math.* **9**, pp. 121–125.

3. Barnette, D. W.: 1973, A proof of the lower bound conjecture for convex polytopes, *Pacific J. Math.* **46**, pp. 349–354.
4. Batyrev, V. V.: 1993, Variations of the mixed Hodge structure of affine hypersurfaces in algebraic tori, *Duke Mathematics J.* **69**, pp. 349–409.
5. Bayer, M.: 1986, The generalized Dehn-Sommerville equations revisited, Technical Report, Northeastern University.
6. Bayer, M. M., and Billera, L. J.: 1985, Generalized Dehn-Sommerville relations for polytopes, spheres and Eulerian partially ordered sets, *Invent. math.* **79**, pp. 143–157.
7. Bayer, M. M., and Klapper, A.: 1991, A new index for polytopes, *Discrete Comput. Geom.* **6**, pp. 33–47.
8. Beilinson, A., Bernstein, J., and Deligne, P.: 1983, Faisceaux Pervers, *Astérisque* **100**.
9. Betke, U., and McMullen, P.: 1985, Lattice points in lattice polytopes, *Monatsh. Math.* **99**, pp. 253–265.
10. Billera, L. J.: 1983, Polyhedral theory and commutative algebra, *Mathematical Programming: The State of the Art* (A. Bachem, M. Grötschel, and B. Korte, eds.), Springer-Verlag, New York/Berlin, pp. 57–77.
11. Billera, L. J., Filliman, P., and Sturmfels, B.: 1990, Constructions and complexity of secondary polytopes, *Advances in Math.* **83**, pp. 155–179.
12. Billera, L. J., and Lee, C. W.: 1980, Sufficiency of McMullen's condition for f-vectors of simplicial polytopes, *Bull. Amer. Math. Soc.* **2**, pp. 181–185.
13. Billera, L. J., and Lee, C. W.: 1981, A proof of the sufficiency of McMullen's conditions for f-vectors of simplicial polytopes, *J. Combinatorial Theory (A)* **31**, pp. 237–255.
14. Björner, A.: 1984, Posets, regular CW complexes and Bruhat order, *Europ. J. Combinatorics* **5**, pp. 7–16.
15. Björner, A., Frankl, P., and Stanley, R.: 1987, The number of faces of balanced Cohen-Macaulay complexes and a generalized Macaulay theorem, *Combinatorica* **7**, pp. 23–34.
16. Björner, A., Garsia, A., and Stanley, R.: 1982, An introduction to the theory of Cohen-Macaulay partially ordered sets, *Ordered Sets* (I. Rival, ed.), Reidel, Dordrecht/Boston/London, pp. 583–615.
17. Björner, A., Las Vergnas, M., Sturmfels, B., White, N., and Ziegler, G. M.: 1993, *Oriented Matroids*, Encyclopedia of Mathematics **46**, Cambridge University Press, Cambridge.
18. Björner, A., and Wachs, M.: 1982, Bruhat order of Coxeter groups and shellability, *Advances in Math.* **43**, pp. 87–100.
19. Brøndsted, A.: 1983, *An Introduction to Convex Polytopes*, Graduate Texts in Mathematics **90**, Springer-Verlag, New York/Berlin.
20. Bruggesser, M., and Mani, P.: 1971, Shellable decompostions of cells and spheres, *Math. Scand.* **29**, pp. 197–205.
21. Bruns, W., and Herzog, J.: 1993, *Cohen-Macaulay Rings*, Cambridge University Press, Cambridge.
22. Chan, C.: On subdivisions of simplicial complexes: Characterizing local h-vectors, preprint.

23. Chan, C., Jungreis, D., and Stong, R.: 1993, Buchsbaum and Eulerian complexes, preprint.
24. Charney, R., and Davis, M.: The Euler characteristic of a nonpositively curved, piecewise linear Euclidean manifold, preprint.
25. Clements, G., and Lindström, B.: 1969, A generalization of a combinatorial theorem of Macaulay, *J. Combinatorial Theory* **7**, pp. 230–238.
26. Danilov, V. I.: 1978, The geometry of toric varieties, *Russian Math. Surveys* **33**, pp. 97–154.
27. Duval, A. M.: 1991, *Simplicial posets: f-vectors and free resolutions*, Ph.D. thesis, M.I.T.
28. Ewald, G.: in preparation, *Combinatorial Convexity and Algebraic Geometry*, Graduate Texts in Mathematics, Springer-Verlag, New York/Berlin.
29. Foata, D., and Schützenberger, M.: 1973, Nombres d'Euler et permutations alternantes, *A Survey of Combinatorial Theory* (J. N. Srivastava, et al., eds.), North-Holland, Amsterdam, pp. 173–187.
30. Frankl, P., Furedi, Z., and Kalai, G.: 1988, Shadows of colored complexes, *Math. Scand.* **63**, pp. 169–178.
31. Fulton, W. F.: 1993, *Introduction to Toric Varieties*, Annals of Mathematics Studies, no. 131, Princeton University Press, Princeton, N.J.
32. Goresky, M., and MacPherson, R.: 1983, Intersection homology theory, *Topology* **19**, pp. 135–162.
33. Goresky, M., and MacPherson, R.: 1983, Intersection homology theory II, *Invent. math.* **71**, pp. 77–129.
34. Greene, C., and Kleitman, D. J.: 1978, Proof techniques in the theory of finite sets, *Studies in Combinatorics* (G.-C. Rota, ed.), Mathematical Association of America, pp. 22–79.
35. Grünbaum, B.: 1967, *Convex Polytopes*, John Wiley & Sons, New York, N.Y.
36. Hibi, T.: 1992, *Algebraic Combinatorics on Convex Polytopes*, Carslaw, Glebe, Australia.
37. Kalai, G.: 1988, A new basis for polytopes, *J. Combinatorial Theory (A)* **49**, pp. 191–209.
38. Kalai, G.: 1991, The diameter of graphs of convex polytopes and f-vector theory, *Applied Geometry and Discrete Mathematics — The Victor Klee Festschrift* (P. Gritzmann and B. Sturmfels, eds.), DIMACS Series in Discrete Mathematics and Theoretical Computer Science **4**, Amer. Math. Soc., pp. 387–411.
39. Lee, C. W.: 1991, Regular triangulations of convex polytopes, *Applied Geometry and Discrete Mathematics — the Victor Klee Festschrift* (P. Gritzmann and B. Sturmfels, eds.), DIMACS Series in Discrete Mathematics and Theoretical Computer Science **4**, Amer. Math. Soc., pp. 443–456.
40. Macaulay, F. S.: 1927, Some properties of enumeration in the theory of modular systems, *Proc. London Math. Soc.* **26**, pp. 531–555.
41. Macdonald, I. G.: 1971, Polynomials associated with finite cell complexes, *J. London Math. Soc. (2)* **4**, pp. 181–192.
42. MacPherson, R.: 1990, Intersection homology and perverse sheaves, preprint.
43. Matsumura, H.: 1986, *Commutative Ring Theory*, Cambridge University Press, Cambridge.

44. McMullen, P.: 1971, The numbers of faces of simplicial polytopes, *Israel J. Math.* **9**, pp. 559–570.
45. McMullen, P.: 1989, The polytope algebra, *Advances in Math.* **87**, pp. 76–130.
46. McMullen, P.: 1993, On simple polytopes, *Invent. math.* **113**, pp. 419–444.
47. McMullen, P., and Shephard, G. C.: 1971, *Convex Polytopes and the Upper Bound Conjecture*, Cambridge University Press, Cambridge.
48. McMullen, P., and Walkup, D.: 1971, A generalized lower-bound conjecture for simplicial polytopes, *Mathematika* **18**, pp. 264–273.
49. Oda, T.: 1988, *Convex Bodies and Algebraic Geometry (An Introduction to the Theory of Toric Varieties*, Springer-Verlag, Berlin/New York.
50. Purtill, M.: 1993, André permutations, lexicographic shellability and the *cd*-index of a convex polytope, *Trans. Amer. Math. Soc.* **338**, pp. 77–104.
51. Reisner, G.: 1976, Cohen-Macaulay quotients of polynomial rings, *Advances in Math.* **21**, pp. 30–49.
52. Rota, G.-C.: 1964, On the foundations of combinatorial theory I. Theory of Möbius functions, *Z. Wahrscheinlichkeitstheorie* **2**, pp. 340–368.
53. Saito, M.: 1983, Hodge filtrations on Gauss-Manin systems. II, *Proc. Japan Acad., Ser. A* **59**, pp. 37–40.
54. Santhi, V. K.: 1992, *Topics in Commutative Algebra*, Ph.D. thesis, Madurai Kamaraj University.
55. Sperner, E.: 1930, Über einen kombinatorischen Satz von Macaulay und seine Anwendung auf die Theorie der Polynomideale, *Abh. Math. Sem. Univ. Hamburg* **7**, pp. 149–163.
56. Stanley, R.: 1971, *Ordered Structures and Partitions*, Ph.D. thesis, Harvard University.
57. Stanley, R.: 1972, Ordered structures and partitions, *Mem. Amer. Math. Soc.*, no. 119.
58. Stanley, R.: 1978, Hilbert functions of graded algebras, *Advances in Math.* **28**, pp. 57–83.
59. Stanley, R.: 1979, Balanced Cohen-Macaulay complexes, *Trans. Amer. Math. Soc.* **249**, pp. 139–157.
60. Stanley, R.: 1980, Decompositions of rational convex polytopes, *Annals of Discrete Math.* **6**, pp. 333–342.
61. Stanley, R.: 1980, The number of faces of a simplicial convex polytope, *Advances in Math.* **35**, pp. 236–238.
62. Stanley, R.: 1983, *Combinatorics and Commutative Algebra*, Birkhäuser, Boston/Basel/Stuttgart.
63. Stanley, R.: 1984, An introduction to combinatorial commutative algebra, *Enumeration and Design* (D. M. Jackson and S. A. Vanstone, eds.), Academic Press, Toronto/Orlando, pp. 3–18.
64. Stanley, R.: 1985, The number of faces of simplicial polytopes and spheres, *Discrete Geometry and Convexity* (J. E. Goodman, et al., eds.), *Ann. New York Acad. Sci.*, vol. 440, pp. 212-223.
65. Stanley, R.: 1986, *Enumerative Combinatorics*, vol. 1, Wadsworth and Brooks/Cole, Pacific Grove, CA.

66. Stanley, R.: 1987, Generalized h-vectors, intersection cohomology of toric varieties, and related results, *Commutative Algebra and Combinatorics* (M. Nagata and H. Matsumura, eds.), *Advances Studies in Pure Mathematics* **11**, Kinokuniya, Tokyo, and North-Holland, Amsterdam/New York, pp. 187–213.
67. Stanley, R.: 1991, f-vectors and h-vectors of simplicial posets, *J. Pure Appl. Alg.* **71**, pp. 319–331.
68. Stanley, R.: 1992, Subdivisions and local h-vectors, *J. Amer. Math. Soc.* **5**, pp. 805–851.
69. Stanley, R.: 1993, A monotonicity property of h-vectors and h^*-vectors, *European J. Combinatorics* **14**, pp. 251–258.
70. Stanley, R.: to appear, Flag f-vectors and the cd-index, *Math. Z.*
71. Sundaram, S.: The homology of partitions with an even number of blocks, preprint.
72. Whipple, F.: 1928, On a theorem due to F. S. Macaulay, *J. London Math. Soc.* **8**, pp. 431–437.
73. Ziegler, G. M.: 1993, *Lectures on Polytopes*, Technical Report TR 93-6, Konrad-Zuse-Zentrum für Informationstechnik Berlin.

ON RECENT PROGRESS IN COMPUTATIONAL SYNTHETIC GEOMETRY

JÜRGEN BOKOWSKI
Fachbereich Mathematik
TH Darmstadt
Schloßgartenstraße 7
D-64289 Darmstadt
Germany

Abstract. *Computational Synthetic Geometry* deals with methods for realizing abstract geometric objects in concrete vector spaces. A general method how to accomplish realizations can be subdivided into three major problem steps. Recent results representing typical ways to tackle these problems deal; for example, with pseudoline arrangements, immersions of 2-manifolds, equifacetted simple tilings of 3-spheres, non-polytopal uniform matroid polytopes, neighborly spatial polyhedra, or the cube problem of Las Vergnas.

Key words: neighborly polyhedra, pseudoline arrangements, oriented matroids, matroid polytopes, triangulations, polyhedral subdivisions.

1. Introduction

Computational Synthetic Geometry deals with methods for realizing abstract geometric objects in concrete vector spaces. We consider a first example class of problems. For a given face lattice of a questionable d-polytope, we search for a corresponding d-polytope P; that is, the abstract geometric object in this case is a $(d-1)$-sphere or just a $(d-1)$-manifold for which we might not even know the sphere property. We try to find coordinates of the vertices in euclidean d-space or a proof that such coordinates do not exist. There are many other classes of problems which are of a similar nature.

The notion of computational synthetic geometry appeared for the first time in contributions to an INRIA Workshop in Sophia-Antipolis, France 1987, on Computer-Aided Geometric Reasoning. It was also chosen as the title of the Ph.D. Thesis of Sturmfels [56], and it was the title of volume 1355 of the Springer Lecture Notes Series, 1989, [22]. The latter contains contributions about the algorithmic Steinitz problem, projective incidence theorems, arrangements of lines and pseudolines, diophantine problems in combinatorial geometry, the embedding of triangulated manifolds, and related aspects.

The present survey describes additional problem classes in computational synthetic geometry and results which have not been covered in previous surveys. We discuss for instance how the matroid information can determine the oriented matroid structure in special cases, and how equifacetted simple tilings of 3-spheres occur when we search for non-polytopal uniform matroid polytopes. The reader will find additional aspects in [22], in sections of the research monograph [11] or

T. Bisztriczky et al. (eds.), POLYTOPES: Abstract, Convex and Computational, 335–358.

in survey articles of the handbook, by Gruber and Wills (eds.) [31]. The articles about polyhedral manifolds by Brehm and Wills [23], about oriented matroids by Bokowski [12], about characterizing convex sets by Mani–Levitska [44], and about tilings by Schulte [52], are closely related to some aspects of this article.

We start by describing a general program in the next section which subdivides many problems in computational synthetic geometry into three major problem steps. The cube problem of Las Vergnas will serve to illustrate the step between the matroid and the oriented matroid or chirotope structure (Problem 2). The fact that pseudoline arrangements and oriented matroids or chirotopes are essentially isomorphic, tells us that when we discuss new decisions for the strechability of pseudoline arrangements, it concerns the third step in our general subdivision of problems. We discuss convex matroid polytopes, general spatial polyhedra, and immersions of manifolds as problem classes in which all three problem steps occur.

The solution of many problems depends on an effective way of handling oriented matroids. We mention implementations for finding oriented matroids with prescribed properties and for finding coordinates for chirotopes. We assume some familarity of the reader with the matroid and oriented matroid structure.

The motivation for studying the matroid structure can be carried over literally from that of groups: *The mathematical concept of matroids unifies many apparently disparate ideas. It is an abstraction of essential mathematical content from particular situations. Abstract matroid theory is the study of this essential content. There are several advantages to working at this level of generality. First, any result obtained at this level may be applied to many different situations, and so the result does not have to be worked out or re-discovered in each particular context. Furthermore, it is often easier to discover facts when working at this abstract level since one has shorn away details which, though perhaps pertinent at some level of analysis, are irrelevant to the broad picture. Of course, to work effectively in the abstract one has to develop some intuition at this level. Although some people can develop this intuition by working only with abstract concepts, most people need to combine such work with the detailed study of particular examples, in order to built up an effective understanding*, compare [37]. The examples provided in this article can also be seen to serve this purpose.

2. The General Program

$$\textbf{Abstract Object} \overset{Problem1}{\longrightarrow} \text{admissible matroid} \overset{Problem2}{\longrightarrow} \text{admissible chirotope} \overset{Problem3}{\longrightarrow} \textbf{Concrete Object}$$

There are many problems in computational synthetic geometry in which the emphasis lies on problems with a restricted problem size. In particular, when we look for minimal point configurations serving as counter–examples, we can expect limited input data; for example, a possible counter example for the Hirsch conjecture, compare the article of Kleinschmidt within this volume. There are also problems in which the search for an infinite class of samples starts with investigating their beginning sequence with computational support.

We subdivide our given problem in computational synthetic geometry into three major problem steps each of which is far from being trivial in the general case. We refer to this subdivision as a *program.* It has already been applied in a wide class of problems in computational synthetic geometry. The main suggestion is to insert intermediate combinatorial structures between the abstract object (for example, the face lattice of a doubted d-polytope) and the concrete object (for example, the d-polytope, either presented as its set of vertices or by its set of supporting hyperplanes): an admissible matroid and an admissible oriented matroid.

In many instances, this basic idea has been applied successfully. Considering the way in which the smallest uniform non-polytopal matroid polytope was found by Bokowski in 1978, cf. [18], it can be viewed as a typical result of pursuing this general idea. In fact, most problems in this article fit into this general framework.

The *first problem step* in the program (transition from the abstract object to an admissible matroid) depends very much on the given problem, and some examples presented below show how one can proceed.

The *second problem step* (transition from the admissible matroid to an admissible oriented matroid or chirotope) describes a difficult problem in oriented matroid theory. Although each choice of axioms for defining a matroid and each choice for defining an oriented matroid provides implicitly a method to check whether a matroid can be oriented, it is not at all clear how to proceed in order to decide within a reasonable time whether the matroid can be oriented. Some conditions are known in which matroids are orientable but in general we are far from having an easy characterization of matroids which are orientable. We come back to this problem in the next section.

Depending on the problem, it might well be that this second step does not occur. This is for example the case when we consider simplicial polytopes and the oriented matroids are uniform, in other words, the underlying matroid structure is trivial. It can also happen that we use additional information from the abstract object when we determine all admissible chirotopes corresponding to our given problem.

There are problems when only the first two steps are involved (the determination of pseudoline arrangements with prescribed properties can be considered as an example). The search for matroid polytopes with given properties belongs to another class of problems in which the *third problem step* (transition from the admissible oriented matroid or chirotope to a concrete geometrical object) is not neccessarily involved. In those cases we have a finite problem; in principle, all admissible matroids and oriented matroids (=chirotopes) can be generated inductively. There is only a finite number of samples for a given dimension d and a given number n of elements. When symmetries are involved or other additional conditions require subsets only, we can expect fast decisions for our problems, once we have managed the difficulties of turning the corresponding algorithms into applicable implementations. We mention implementations for finding oriented matroids with prescribed properties and for finding coordinates for chirotopes later on.

The third problem step in our program (transition from the admissible oriented matroid to the concrete geometric object) is again a major problem in oriented matroid theory. The recent research monograph [11] contains a whole chapter about this subject. Dress pointed out that from real algebraic geometry (the real version of Hilbert's Nullstellensatz) comes a satisfactory theoretical result. Sturmfels worked out the idea of Dress in [56], compare Theorem 8.5.2. [11]. But the result from real algebraic geometry also shows that a general algorithm leading to solutions and decisions in all cases is not easy to implement and, moreover, it is not applicable in actual interesting cases.

We add aspects of the third problem step later on.

3. The Cube Problem of Las Vergnas

We start with a section in which the interaction of a concrete point set with its matroid structure and its oriented matroid structure plays a decisive role. As an introduction we consider this problem to be of much help for understanding the whole article, especially when the reader has to recall some concepts from matroid theory. We do not repeat here the basic concepts from oriented matroid theory. The example we deal with is that of a cube, and the corresponding problem serves us as a representative for the problem of finding all *oriented* matroids for a given matroid. This concerns our second problem step in the program of Section 2.

For the second problem step, there are no easy answers available. In one extreme case, in the uniform case, we have, by starting with the trivial matroid structure, the problem of finding all uniform oriented matroids with given rank and given number of elements. In the other extreme case, we mention [58] as a contribution where we find matroids which can not be oriented at all. In some cases, we have a satisfactional answer. Las Vergnas, Roudneff, and Salaün have shown that the reorientation classes in some cases; for example, for the 24–cell and for the 4–cube, are unique. The general case for the d-cube (Las Vergnas has conjectured that the uniqueness holds in the general case as well) remains open. But some progress can be seen in [14]. It is known, and was mentioned for instance by Victor Klee, that certain problems which we can easily formulate can turn out to be very hard, e.g. *find all triangulations of a d-cube* or *determine the largest d-simplex in a d-cube.* The problem of Las Vergnas is of a similar nature, and in fact, it is related to the former to some extend. The Las Vergnas cube problem reads as follows, compare [41]:

Problem: Given the matroid of the d–cube. Does it have a unique reorientation class?

In other words: when two oriented matroids are given such that they each have as underlying matroid that of the d–cube, are they equal up to a reorientation, are they reorientation equivalent? In general, there are many reorientation classes for a given matroid. But in special cases like in the case of a cube, in the 24–cell example, or in other cases with high symmetries or many dependencies, this is different. When studying this problem, we are forced to distinguish carefully the notion of a matroid from that of an oriented matroid, both on an abstract level and in the concrete case when the coordinates of the cube are given.

3.1. The 4–cube

We study the rank 5 case, the 4-cube. The proof in this case shows the typical arguments in this context, cf. [14]. We have the reorientation class of the 4–cube C_4, represented by its Folkman–Lawrence representation with unoriented topological

3–planes. In order to define them, it suffices to present all rank 2 contractions. It is known that they determine uniquely the reorientation class, cf. [26]. We will discuss how the underlying matroid defines the reorientation class in this case.

The word *matroid* comes from matrix but in matroid theory the abstract dependences play the essential role, concrete matrices are rather seldom. In this case they guide us when looking at the matroid structure and at the oriented matroid structure of the cube at the same time. We have the abstract notion in mind, and we use only the appropriate information (sometimes the matroid alone, sometimes the oriented matroid as a representative for the reorientation class) but coordinates or familiar pictures of the cube help us understanding the symmetry of the cube.

We assume the vertices of the 4–cube given as rows of a matrix M, the i'th row representing homogeneous coordinates $(1, x_1, x_2, x_3, x_4)$ of vertex number i of the cube. We get the signed bases of the corresponding rank 5 oriented matroid $\mathcal{M} = \chi(C_4)$ of the 4–cube C_4 by calculating the signs of all determinants of (5×5)–submatrices of M. The map assigning to each formal determinant (bracket) the absolute value of the signs of these determinants defines the underlying matroid $\underline{\mathcal{M}}$ of the 4–cube.

Contracting the oriented matroid of the cube at vertex number 1 yields $\chi(C_4)/\{1\}$, an oriented matroid of rank 4 with 15 elements, the combinatorial vertex figure of the cube. Because of the symmetry of the cube, it suffices to consider just one of all the 2^4 vertex figures (contractions). The following first matrix $M/\{1\}$ with the heading $\chi(C_4)/\{1\}$ defines the labels of all vertices of our 4-cube.

We assume (induction hypothesis) that the assertion of Las Vergnas' cube–conjecture has been proven for the 3–cube, which we leave as an exercise.

Our proof proceeds as follows. We consider all rank 2 matroid contractions of the matroid $\mathcal{M}$ of the 4–cube. By using the lower dimensional assertion of the theorem for the 3–cube, we are going to show that each rank 2 matroid contraction of the matroid of the 4–cube induces the same ordering of the vertices of the cube in the corresponding rank 2 Folkman Lawrence representation (with unoriented elements) as that ordering of the 4–cube.

We have to use only two rank 2 contractions because of the cube symmetry and because other cases are settled by induction easily. We are referring to a particular rank 2 contraction of the matroid of the 4–cube $\mathcal{M}/\{i,j,k\}$ simply by writing down the corresponding triple (i,j,k). It turns out that all the rank 2 contractions that we have to consider are minors of a single rank 3 contraction $\mathcal{M}/\{i,j\}$. We are going to determine all our relevant minors.

Contracting a point out of the set $\{2,3,4,5\}$, say 2, i.e. contracting two points of the 4–cube with distance 1, leaves us with the vertex figure of a 3–cube, with each element counted twice. In other words, subsequent rank 2 contractions are contractions of a 3-cube. The induction argument settles this case.

Contracting two opposite points in the 4–cube; for example, 1 and F; that is, contracting two points with distance $\sqrt{4}$, yields the same case as before, we get the vertex figure of a 3–cube, each vertex counted twice. This can be seen immediately by constructing a matrix representation of $\chi(C_4)/\{1, F\}$. We obtain such a matrix representation by subtracting the last column of $M/\{1\}$ from all the others and by deleting the last column. We have the vertex figure of a 3–cube again as in the (distance 1)–case. Whenever a pair of opposite vertices occurs in the contraction of the 4–cube, our problem degenerates to that of a lower dimensional cube. We conclude that we can assume that in the 3–tuples of vertices i, j, k, used to get all rank 2 contractions of $\chi(C_4)$, pairs of vertices with distance 1 or with distance $\sqrt{4}$ do not occur.

This implies that among our three points (i, j, k) we have at least one pair of vertices with distance $\sqrt{2}$. Without loss of generality, we choose these vertices to be 1 and 6, i.e. we contract these two points with distance $\sqrt{2}$ first. We have the corresponding rank 3 contraction $\chi(C_4)/\{1,6\}$. We get a matrix representation by adding in $M/\{1\}$ to the first column the negative second one and by deleting the second column. We have identified the new line vectors when they were equal up to a nonzero factor.

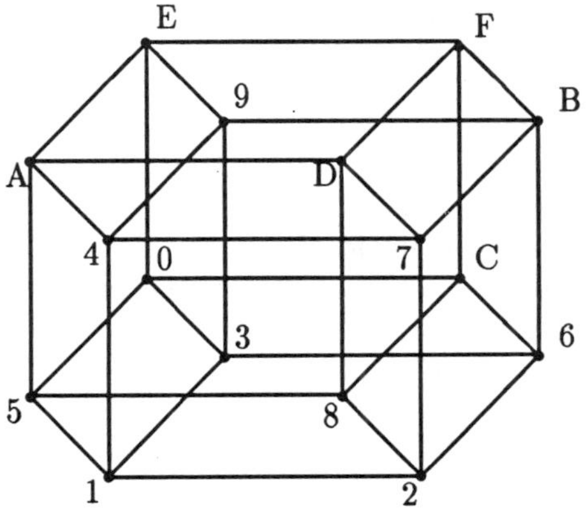

$\chi(C_4)/\{1\}$

vertex	x_1	x_2	x_3	x_4
1	0	0	0	0
2	1	0	0	0
3	0	1	0	0
4	0	0	1	0
5	0	0	0	1
6	1	1	0	0
7	1	0	1	0
8	1	0	0	1
9	0	1	1	0
0	0	1	0	1
A	0	0	1	1
B	1	1	1	0
C	1	1	0	1
D	1	0	1	1
E	0	1	1	1
F	1	1	1	1

$\chi(C_4)/\{1,6\}$

vertex	x_1	x_2	x_3	case
1,6	0	0	0	
2,3	1	0	0	distance 1
4,B	0	1	0	distance 1
5,C	0	0	1	distance 1
7	1	1	0	First case
8	1	0	1	First case
A,F	0	1	1	distance 2
D	1	1	1	Second case
9	-1	1	0	First case
0	-1	0	1	First case
E	-1	1	1	Second case

What are all combinatorially different rank 2 contractions? Vertices 7, 8, 9, 0 are equivalent under the cube symmetry yielding the first case. Vertices D and E are equivalent under the cube symmetry yielding the second case.

$\chi(C_4)/\{1,6,7\}$		
vertex	x_1	x_2
1,6,7	0	0
2,3,4,9,B	1	0
5,C,D	0	1
8	1	1
A,F,0	-1	1
E	-2	1

$\chi(C_4)/\{1,6,D\}$		
vertex	x_1	x_2
1,6,D	0	0
2,3,A,F,E	1	0
4,8,B	0	1
5,7,C	1	1
9	-1	1
0	2	1

3.1.1. *The case* $\chi(C_4)/\{1,6,7\}$

The first rank 2 contraction along $1,6,7$ has 5 elements occuring in the cyclic order

1. $\{2,3,4,9,B\}$, 2. $\{8\}$, 3. $\{5,C,D\}$, 4. $\{0,A,F\}$, 5. $\{E\}$.

We have to prove that this cyclic order has to occur by using the matroid information alone. We think first in geometric terms. When we rotate a 3-dimensional hyperplane around the affine 2-plane which is determined by the points $1,6,7$, we get the above cyclic sequence. This rotating 3-dimensional hyperplane leads to the following sequence.

1.	the 3-cube	$\{1,6,7\}$	$\{2,3,4,9,B\}$,
2.	the tetrahedron	$\{1,6,7\}$	$\{8\}$,
3.	the prism	$\{1,6,7\}$	$\{5,C,D\}$,
4.	the octahedron	$\{1,6,7\}$	$\{0,A,F\}$,
5.	and the tetrahedron	$\{1,6,7\}$	$\{E\}$.

The rotating 3-dimensional hyperplane induces e.g. a rotation of a 2-dimensional plane around $\{1,7\}$, in the affine hull of the facet $\{1,2,4,7,5,8,A,D\}$. We delete the other vertices $3,6,9,B,E,F,0,C$ in the 4-cube to obtain the facet, the 3-cube. In the contraction of this 3-cube along 1 and 7, we find: the rotating 2-dimensional plane leads to the following sequence.

1.	the subfacet	$\{1,7\}$	$\{2,4\}$,
2.	the triangle	$\{1,7\}$	$\{8\}$,
3.	the square	$\{1,7\}$	$\{5,D\}$,
4.	the triangle	$\{1,7\}$	$\{A\}$.

This sequence is uniquely determined by induction. It must be compatible with our above rank 2 contraction of the 4-cube; that is, we obtain the same sequence when we first contract $1,6,7$ and when we later delete the vertices $3,6,9,B,E,F,0,C$. This is in fact the case.

What about the position of point E in the above rank 2 contraction of the 4-cube? We find a similar argument for this. The rotating 3-dimensional hyperplane induces a rotation of a 2-dimensional plane around $\{1,6\}$, in the affine hull of the combinatorial cube with vertices $\{1,2,3,6,A,D,E,F\}$. We delete the other vertices $4,7,9,B,5,8,0,C$ in the 4-cube to obtain this combinatorial 3-cube. In the contraction of this 3-cube along 1 and 6, we find: the rotating 2-dimensional plane cuts off in the following circular sequence

1.	the subfacet	$\{1,6\}$	$\{2,3\}$,
2.	the triangle	$\{1,6\}$	$\{D\}$,
3.	the square	$\{1,6\}$	$\{A,F\}$,
4.	and the triangle	$\{1,6\}$	$\{E\}$.

This sequence is uniquely determined by induction. It must be compatible with our above rank 2 contraction of the 4-cube, i.e. we have to obtain the same sequence when we contract first $1,6,7$ and when we delete later the vertices $4,7,9,B,5,8,0,C$. This is in fact the case.

Both sequences with 4 elements each define the sequence of all 5 elements uniquely.

We conclude this decisive first case by analysing the proof structure once more on an abstract level. We have seen that the reorientation class property of our oriented matroid rank 2 contraction of the 4-cube, stored in $\chi(C_4)/\{1,6,7\}$, can be determined by two matroid rank 2 contractions of 3-cubes. On one hand, the reorientation class of $\chi(C_4)/\{1,6,7\}\setminus\{E\}$ is isomorphic to the reorientation class of $\chi(C_a)/\{1,7\}$ with C_a being the 3-cube with vertices $\{1,2,4,7,5,8,A,D\}$. On the other hand, the reorientation class of $\chi(C_4)/\{1,6,7\}\setminus\{8\}$ is isomorphic to the reorientation class of $\chi(C_b)/\{1,6\}$ with C_b being the 3-cube with vertices $\{1,2,3,6,A,D,E,F\}$.

3.1.2. *The case* $\chi(C_4)/\{1,6,D\}$

The second rank 2 contraction along $1,6,D$ has 5 elements occuring in the cyclic order

1. $\{2,3,A,E,F\}$, 2. $\{0\}$, 3. $\{5,7,C\}$, 4. $\{4,8,B\}$, 5. $\{9\}$.

It can be shown in precisely the same way that this ordering is unique. The axis of rotation and the 3-cubes involved are $\{1,6\}$, and $\{1,2,3,6,4,7,9,B\}$ on one hand and $\{1,6\}$ and $\{1,2,3,6,5,8,0,C\}$ on the other. From the former case, it is clear how the abstract argument works.

Up to reorientation, there is precisely one orientation for the matroid of the 4-cube.

3.2. The 5-cube

The proof in the 5-cube case in [14] proceeds as follows. In order to cut down the number of cases, the first step is a reduction argument. Some symmetry of (the oriented matroid of) the cube and as a consequence, some symmetry of the matroid of the cube, is used together with the induction hypothesis with respect to the rank. This reduction yields in the rank 6 case that we have to consider not more than one rank 4 contraction, afterwards not more than four rank 3 cases, and finally not more than 16 rank 2 subcases.

With the help of the matrix of all coordinates, we can construct explicitly in each subcase its ((oriented) matroid) rank 2 contraction of the 5-cube. In the same way we construct the ((oriented) matroid) rank 2 contractions of certain subsets of vertices of the 5-cube which have as underlying matroid structure that of a lower-dimensional cube (with dimension less than 5). In the latter case we know by the induction hypothesis the ordering of the unoriented elements in the oriented matroid rank 2 contractions. By deleting certain elements in a subcase matroid rank 2 contraction, we see after the deletion that such a rank 2 matroid equals a rank 2 matroid which can be obtained by contracting an oriented matroid isomorphic to that of a lower-dimensional cube. This implies that we know by induction the ordering of the unoriented elements for its corresponding oriented matroid of rank 2.

These orderings (within the minors) after the deletion of an oriented matroid rank 2 contraction of the rank 6 oriented matroid, which has as its underlying matroid that of the 5-cube, have to be compatible when there is a part in which there is some overlap. It turns out that this observation determines uniquely the global ordering in each oriented matroid rank 2 contraction of the rank 6 oriented matroid which has as its underlying matroid that of the 5 cube.

A subset of vertices of the 5-cube has, as underlying matroid structure, that of a lower-dimensional cube when it is a translate of the vertex set of the Minkowski sum of a set of main diagonals of 1-, 2-, or 3-faces of the 5-cube, provided these diagonals lie in pairwise orthogonal subspaces. In particular, a face is a translate of the Minkowski sum of a set of main diagonals of 1-faces of the 5-cube.

4. Coordinates for Chirotopes

We consider again a generic problem in computational synthetic geometry. After the (inductive) generation of admissible matroids and admissible oriented matroids (=chirotopes) for our problem, we are left with the problem of finding coordinates for the oriented matroid.

Before we discuss this problem, here are some remarks about the implementation for the inductive generation of chirotopes. These implementations require a thorough knowledge of the oriented matroid axiom structure. The author worked on these implementations and applied them to various problems. There are at least three pairwise independent implementations for finding oriented matroids inductively. There is a first FORTRAN program (Anheuser, Bokowski) available in the uniform case in rank 4. A more general C-program based on the same ideas was implemented in arbitrary rank by Guedes de Oliveira. Another algorithm (implemented by Peter Schuchert) for generating recursively all uniform oriented matroids is based on the circuit–axioms of Las Vergnas (the abstract RADON partitions) of an oriented matroid.

When the number of vertices exceeds about 10 in the rank 4 case, the complexity of the algorithm of generating all admissible oriented matroids has turned out to be very bad. Therefore, in general, it is difficult to find decisions within a reasonable CPU–time of whether a given complex has a corresponding admissible oriented matroid unless additional assumptions can be made. Depending on the particular problem, there are many posssibilities of reducing the CPU–time. There are for instance symmetry assumptions on the oriented matroid, assumptions on the number of extreme points, assumptions on subdivisions on the oriented matroid, forbidden Radon–partitions, and so on.

For methods of obtaining coordinates, we find in the literature the solvability sequence method, general reduction techniques, combinatorial reduction techniques, cf. [46], and algebraic solving techniques. An interactive method has been successfull in the rank 3 case [20]. From the theory of real algebraic geometry, we know that we cannot expect a good behavior of algorithms in all cases. Nevertheless, we find that many problems can be tackled, where the size is small enough and in which heuristic arguments can be used effectively.

A method due to Bokowski which solved remaining difficult cases in the 10 point classification of all rank 3 oriented matroids, cf. [15], was first implemented by Klaus-Peter Pock and in a refined version by Jens Scharnbacher, cf. [17] in the higher rank case. The main idea is a process of generating realizations of a sequence of minors of the oriented matroid, starting from a simplex and adding a new point in each step. Whenever this is possible, the algorithm finds and determines the next point (LP–problem). In cases where this is not possible, a suitable change in the foregoing points is performed such that the region for the next point becomes non-empty. Although this is only a heuristic method, the program was very effective in many instances, and many decisions, which seemed difficult in the past, were obtained in a few seconds. As an example, the Roudneff example from [43] was shown to be realizable within 6 seconds CPU-time.

Whenever there are some hints that the oriented matroid is not realizable, there is a test for confirming that. A method due to Bokowski (it was first mentioned in Cologne 1986, and was also mentioned in the talk of the author at the INRIA Workshop in Sophia-Antipolis, France 1987) was tested in [16], under the notion bi-quadratic final polynomial. The problem of determining an oriented matroid which is non–realizable and which does not have a bi-quadratic final polynomial is still open.

5. Pseudoline Arrangements

For the first time, we find questions about line configurations in a wider context named pseudolines; that is, with axiomatisized intersection properties in [42].

Levi (1926) emphasized explicitly that the careful analysis of many proofs of assertions about line arrangements shows that these proofs do not depend on the metrical properties of the lines: they are purely combinatorial in nature, they are invariant under the transformation group of homeomorphisms in the projective plane. This was the beginning of studying pseudoline arrangements.

Meanwhile, we have additional motivations for studying line arrangements and pseudoline arrangements in [33], in applications for algebraic geometry, cf. [9], or even more general as an example class for oriented matroids with all its applications, cf. [11]. We have these general applications in mind when we study particular examples. The methods studied in a particular case are applicable also in the general case.

A pseudoline arrangement in the projective plane is a set of closed topological curves with the property that any two of them have precisely one point in common where they intersect. We stress that it is the equivalence class of homeomorphic pseudoline arrangements that we are interested in. The equivalence classes of homeomorphic pseudoline arrangements are combinatorial objects which can be stored in terms of their intersection properties. A particular pseudoline arrangement plays the role of a representative for the whole equivalence class.

The following problem arises in many problems in a natural way. Can we find such a representative of a class of homeomorphisms of a pseudoline arrangement which is a line arrangement? This is referred to as the problem of stretchability of a pseudoline arrangement. We are going to consider as an example a class of problems in which this question about stretchability has been asked several times, cf. [35], [36]. It can be seen in the following more general context.

Among the topological invariants of unoriented pseudoline arrangements and more general among the topological invariants of unoriented pseudosphere arrangements (=reorientation classes of oriented matroids), the number of triangular cells (or more general: the number of simplicial cells, also called mutations) has been studied extensively, see [11]. The upper and the lower bounds for these numbers have been studied for a fixed number of pseudolines, or pseudo-hyperspheres, respectively. See [43], [35] for recent contributions in this area.

The conjecture of Las Vergnas about the minimal number of mutations is still unsolved and remains a challenging problem in this area. In [33] we find contributions to the problem of finding classes of pseudoline arrangements in which the number of triangles is maximal. Harborth and Roudneff found infinite classes of pseudoline arrangements with a maximal number of triangles [35], [47].

In this context Harborth has asked in [35], see also [36], whether the following pseudoline arrangement $\mathcal{A}$ with 28 elements is stretchable. The picture below shows the projective plane with the understanding that the outer circle represents the line at infinity.

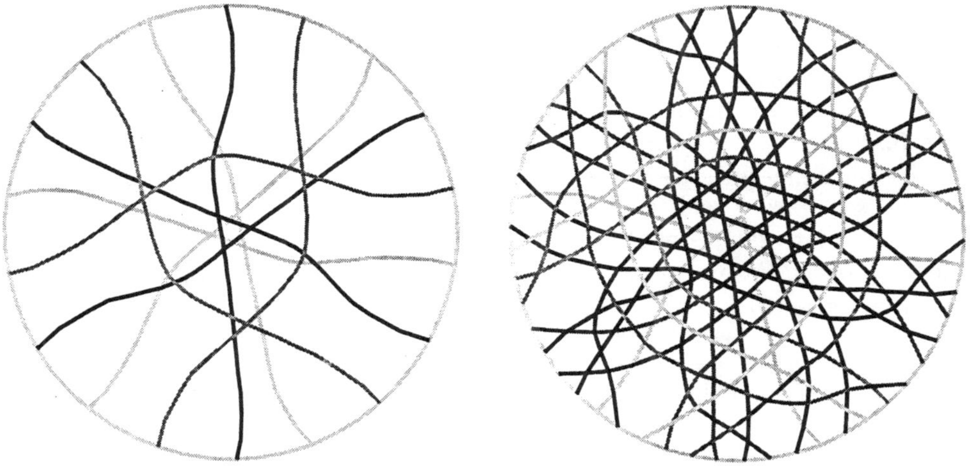

Fig. 1. Pseudoline Arrangement of Harborth with 28 Elements and a Maximal Number of Triangles Obtained from the Unique Arrangement with 10 Elements and a Maximal Number of Triangles

In Harborth's example $\mathcal{A}$, the number $p_3(\mathcal{A})$ of triangular regions attains the maximal value $p_3(\mathcal{A}) = \frac{n(n-1)}{3}$ for $n = 28$, i.e. $p_3 = 252$. This example has been used for some time to test algorithms for deciding the strechability problem in a more general framework. Therefore, instead of providing just one solution related to this particular problem, the methods involved can be of much help in other problems as well. The embedding of the stretchability problem into real algebraic geometry has shown that we cannot expect a general algorithm for solving the stretchability problem in the most general case, cf. [11] Chapter 8 on realizability. On the other hand, not only the example of Harborth shows that there is a need for having partial anwers for large problem classes. So far, the state of the art was to solve examples with up to 10 elements.

A classification of all pseudoline arrangements with 10 elements has been achieved in [15]. The same article provides also an overview about all projective incidence theorems with up to 10 lines including Pappus' and Desargues' theorem. The rank 4 investigation of neighborly pseudomanifolds with nine and ten points led to problems with limited complexity.

With regard to these results, Harborth's example with 28 elements was challenging. In order to find computational solutions and because of the symmetrical shape, this example was used to test methods for deciding the stretchability of pseudoline arrangements, or the realizability problem of oriented matroids.

We are going to describe why this example is not stretchable. A detailed discussion can be found in [20]. The guiding idea is a transition from the combinatorial information of the pseudoline arrangement to a sign structure of formal 3×3–determinants which are available because we assume the stretchability. These real determinants fulfil many wellknown conditions, named Graßmann–Plücker relations. The signs of these determinants (signed bases of our oriented matroid) are known, and by omitting certain terms, we obtain many inequalities of the form $|det_i| \times |det_j| < |det_k| \times |det_l|$ which must hold simultaneously.

By taking the logarithm log $|det_m|$ of their absolute values $|det_m|$ as new variables, we get a large linear programming problem in the new variables within the appropriate dimension. It turns out that the admissible region of the LP-problem in the case of the above Harborth problem is empty. This implies that the strechability assumption is false. In the Harborth problem we get a linear programming problem with 5209 inequalities and 1098 variables. After preparatory steps in Darmstadt thanks to P.Schuchert, this was later solved in Berlin (thanks to the program *cplex*) applied by G.Ziegler and R. Borndörfer within 264.24 seconds on a Sun Spark 10. This particular example with 28 elements occurs as a minor whenever we perform two steps of the Harborth construction for finding pseudoline arrangements $\mathcal{A}$ with the property $p_3(\mathcal{A}) = \frac{n(n-1)}{3}$. This leads to the result in [20].

Theorem (Bokowski and Strempel 93): Let H be the recursively constructable infinite class of pseudoline arrangements $\mathcal{A}_n$ with n elements with the property that the maximal possible number $\frac{n(n-1)}{3}$ of mutations is attained and which is obtained by starting with any of the 5 known stretchable examples of this kind. Only a finite number of pseudoline arrangements of H is stretchable.

The decisive element for this assertion is the above example of a pseudoline arrangements with 28 elements first published by Harborth in [34].

6. Matroid Polytopes

Frequent investigations of polytopes deal with their topological invariants, such as the f-vector or the combinatorial type of the polytope. According to Felix Klein's Erlanger Program, it is natural to use already the topological framework for describing the objects under consideration. Matroid polytopes provide an answer to this general idea.

6.1. On the Steinitz problem

For many years there was an active area of research in convexity around the following question, named Steinitz problem after to a proposal of Ewald. Characterize, among all combinatorial (d-1)-spheres, the d-polytopal ones, see for example Question 6.1 in [44] and the papers cited there.

A contribution of Sturmfels published in [57] stopped this investigation for several years. The result appeared under the title: *boundary complexes of convex polytopes cannot be characterized locally.* It has been quoted in several articles of the Handbook of Convex Geometry, for instance in [12], [10], but in the meantime G. Ziegler has discovered an error in this paper which is unfixable. The chirotope on the "outer simplices" of a polytope is not determined by the face lattice. The assertions in Theorem 1.4 in that paper and Theorem 1.8 in the same paper (Nonrational polytope in dimension 6) are open problems.

Nevertheless, the hope of finding a proof that polytopal spheres cannot be characterized locally among all spheres remains. Support comes from corresponding results in matroid theory and oriented matroid theory, see [12], Theorem 4.1 and Theorem 4.2. One natural approach to tackle the problem might lie in constructing infinite series of say uniform non-polytopal matroid polytopes with (for simplicity reasons) isomorphic vertex figures such that when deleting a point, the corresponding sphere is polytopal. Do there exist such matroid polytopes except the two examples we know so far? But first of all, what is a matroid polytope?

A convex $(d-1)$-polytope in euclidean $(d-1)$-space with n facets can be viewed as an arrangement of n oriented hyperplanes. We assume the polytope to lie on the positive side of each hyperplane, and we consider the natural embedding into projective space with projective hyperplanes. We obtain matroid polytopes when the oriented hyperplanes are replaced with oriented topological hyperplanes which have intersection properties like ordinary hyperplanes. For technical details see [11]. There is an abstract convexity structure for matroid polytopes due to Las Vergnas, cf. [40]. A matroid polytope (having the Folkman Lawrence representation of an oriented matroid in mind) can be viewed as a topological d-ball which is a topological d-cell within an (oriented) topological hyperplane arrangement in projective d-space.

Let us consider a small example of an interesting matroid polytope in order to explain some more aspects. In dimension 2 all matroid polytopes are n-gons, and therefore they are trivial. In dimension 3 each matroid polytope (even a non-realizable one) has a face lattice which can also be represented (theorem of Steinitz) by a realizable polytope. In other words: the 3-dimensional case is not very interesting either.

There is only one uniform matroid polytope in dimension 4 with fewer than 10 elements, cf. [7], [18], which has only a small degree of symmetry compared with the following example depicted in color. In this sense it is a unique small example which should be investigated when studying matroid polytopes.

We consider the 4–cell in the Folkman Lawrence representation showing the matroid polytope with all its (combinatorially equal) 3-dimensional facets, its 2-dimensional subfacets and its edges and vertices. This face lattice (polar to the matroid polytope) is a simple equifacetted 3-sphere.

There are no more than two minor minimal non-polytopal uniform matroid polytopes known with combinatorially equivalent vertex figures, cf. [19]. The colored picture describes one of them.

The combinatorial description of the 3-sphere was given by Altshuler, cf. [3]. The uniqueness of the corresponding matroid polytope was obtained implicitely by Bokowski and can be confirmed by using a theorem of Shemer [53]. The boundary structure can not be that of an ordinary polytope, this has been proven in [13]. The representation used here in the colored picture is due to computational support developed by J. Bokowski, M. Gebhard, J. Richter-Gebert and P. Schuchert.

On the main diagonal, we have 10 Schlegel diagrams representing the 3-dimensional (topological) facets of the topological 4-cell, the decisive tope in the Folkman Lawrence representation. To each 3-dimensional facet defining its corresponding topological 3-dimensional hyperplane, there corresponds one basic color. The 2-dimensional topological subfacets lie in corresponding intersections of two topological hyperplanes. The color of each subfacet and the whole 2-dimensional projective plane containing the subfacet is composed of the two basic colors of the two defining topological 3-dimensional hyperplanes. Corresponding to each subfacet (marked by a red dot) of a facet, the whole topological projective plane can be seen. One row in the scheme corresponds to all subfacets of the 3-facet in this row. The pseudolines depicted in each projective plane indicate the intersection of the projective topological plane with an additional 3-dimensional topological hyperplane. The orientations of all topological hyperplanes are defined by the adjacent topological 4-cell, an abstract convex polytope with a face lattice anti-isomorphic to the Las Vergnas face lattice of our matroid polytope which is by definition this set of 10 oriented topological hyperplanes. The fact that this is indeed a matroid polytope was known and is implicitly contained in [13]. In the oriented matroid language, we can say that we have depicted all rank 2 contractions of our matroid polytope.

The isomorphic vertex figures of uniform matroid polytopes lead to equifacetted simple topological 3-spheres, to tilings of the 3-sphere. Such non-polytopal examples are interesting in view of the Steinitz problem. In [19] we find another new example of a non-polytopal equifacetted tiling of the 3-sphere with 10 elements and it is shown that there are only two such examples with up to 10 elements. The new example was found by starting with the final polynomial structure of our colored example. Here is the short proof that this can not be polytopal. We use from the symmetry of the matroid polytope of order 20 only the cyclic subgroup defined by the permutation $(1,3,5,7,9)(2,4,6,8,0)$. We assume that we have determined

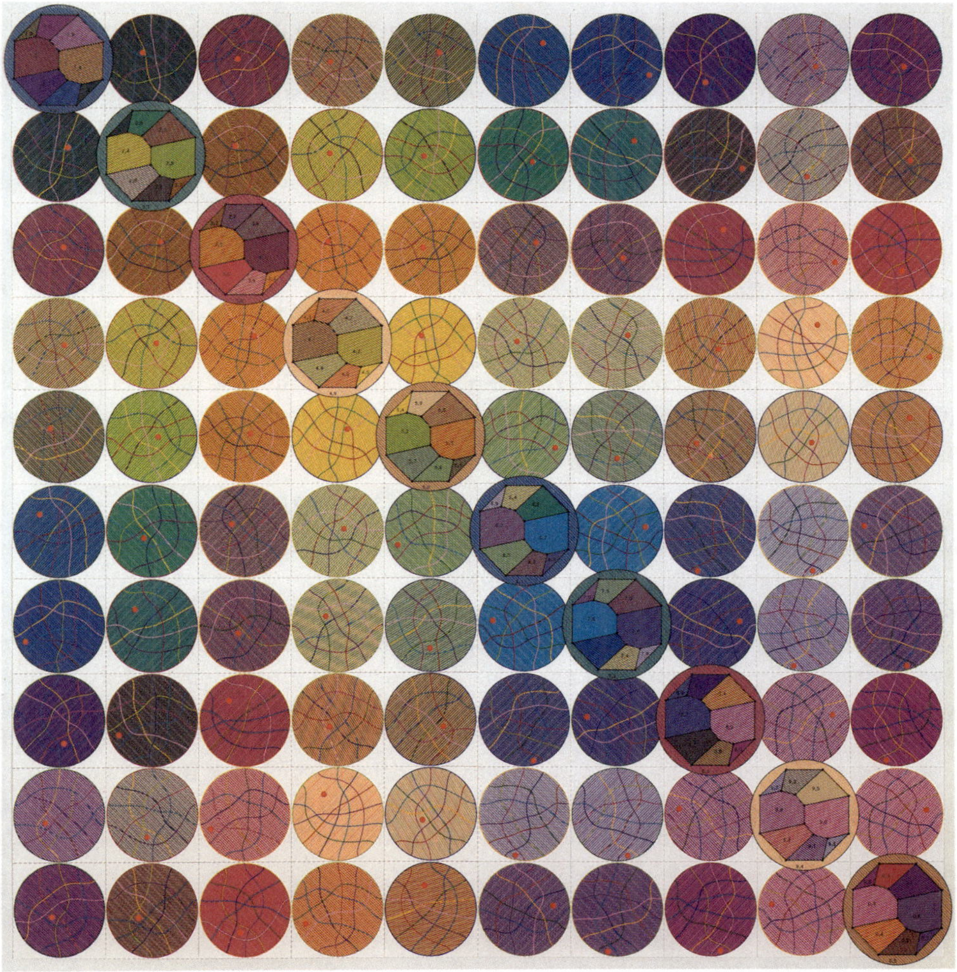

Fig. 2. Folkman Lawrence Representation of Altshuler's 3-Sphere M^{10}_{425}. It is a Uniform Non-Polytopal Matroid Polytope [13]

A Simple Equifacetted Tiling of the 3-Sphere with Symmetry $Z_5 \times Z_4$

already all orientations of the bases of the oriented matroid. The sphere is known to be rigid and uniform; that is, all signs of bases are determined and they are different from zero. We sum the vanishing Grassmann-Plücker polynomials $\{246|7308\}$ and $\{260|1384\}$ to obtain in terms of brackets by definition of the polynomials

$$0 = +[24673][24608] - [24670][24638] + [24678][24630]$$
$$+[26013][26084] - [26018][26034] + [26014][26038]$$

which can be written because of the symmetry and by using the alternating rules for determinants as

$$[24673][24608] - [24670][24638] + [24678][24630]$$
$$+[26013][26084] - [82674][26034] + [48236][60472] = 0.$$

But the last equation can be reduced to $[24680]\,([23467] - [12360]) = 0$ on the one hand, and on the other hand we find opposite signs for the brackets [23467] and [12360] from the oriented matroid property. This is a contradiction to polytopality.

The 3-facet in the Folkman Lawrence representation of the new matroid polytope [19] is a cube with truncated opposite vertices. The symmetry group is $Z_5 \times Z_4$ as in the above example.

6.2. Bi-cyclic polytopes and equifacetted 3-spheres

Whereas in higher dimensions G. Kalai has shown in [38] that there are much more spheres than boundary complexes of polytopes, a corresponding assertion for 3-spheres is missing. A large and interesting class of realizable equifacetted simple 3-spheres can be obtained as polar polytopes of the so called bi-cyclic 4-polytopes investigated by Smilansky in [54] and [55]. They occured also independently from a combinatorial point of view in the investigation of equifacetted tilings of the 3-sphere in [19].

In order to describe this class, or rather a subclass, we choose integers n q (e.g. with greatest common divisor $gcd(n,q)$ not greater than 2), and we take the convex hull $P_{n,q}$ of the n points in 4-space which lie equidistributed on the 3-torus $S^1 \times S^1$ as follows. The i'th point has coordinates

$$(\cos\frac{i2\pi}{n}, \sin\frac{i2\pi}{n}, \cos\frac{iq2\pi}{n}, \sin\frac{iq2\pi}{n}).$$

The polar convex body $P^*_{n,q}$ of $P_{n,q}$ has as its boundary structure an equifacetted 3-sphere. $P_{n,2}$ and $P_{2k-1,k}$ are cyclic polytopes in R^4. $P^*_{8,3}$ is equivalent to the 4-cube with Schläfli symbol $\{4,3,3\}$ and $P^*_{10,3}$ has relations to the Petrie-Coxeter polytope with symbol $\{6,\frac{6}{1,2}\}$. $P^*_{11,3}$ seems not to be well known (neither paper of Smilansky was cited in the handbook of convexity). The polytope $P^*_{11,3}$ provides a simple equifacetted tiling of the 3 sphere with 11 tiles. We depict the coordinates as projections onto two totally orthogonal 2-dimensional planes. We obtain corresponding labels of the vertices in both projections via the facet structure which is depicted in bold lines. In other words the two drawings provide us with the coordinates in 4-space. Note the similarity to the picture of the Petrie-Coxeter polytope with symbol

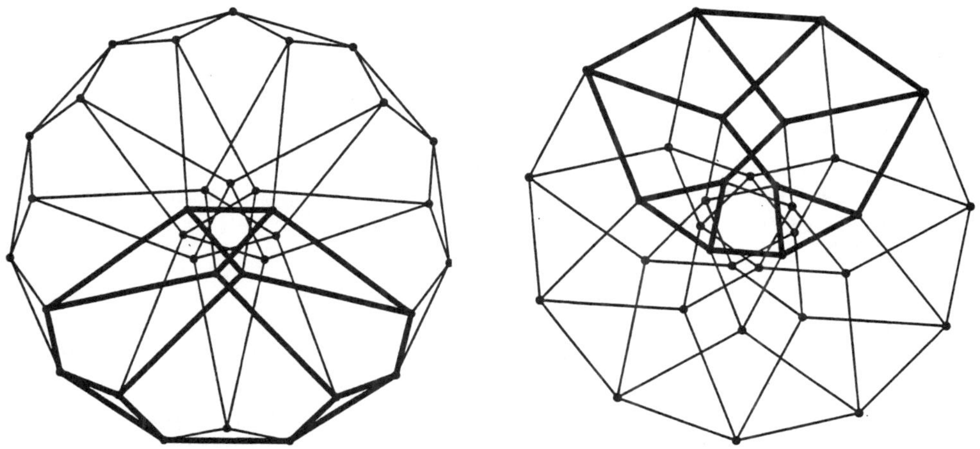

Fig. 3. $P^*_{11,3}$: Complementary Projections of a Simple Equifacetted 3-Sphere with 11 Facets and 33 Vertices.

$\{6, \frac{6}{1,2}\}$ in [27]. In Coxeter's article we find one projection, the corresponding other one would be equal.

Are there additional equifacetted simple polytopes in dimension 4 which have not been mentioned above and which are different from the 120 cell? An interesting facet type of an equifacetted simple 3-manifold connects the new matroid polytope (the new non-polytopal uniform matroid polytope with 10 vertices mentioned above) and the 120 cell. Here are the three corresponding Schlegel diagrams of the 3-dimensional facets in these cases.

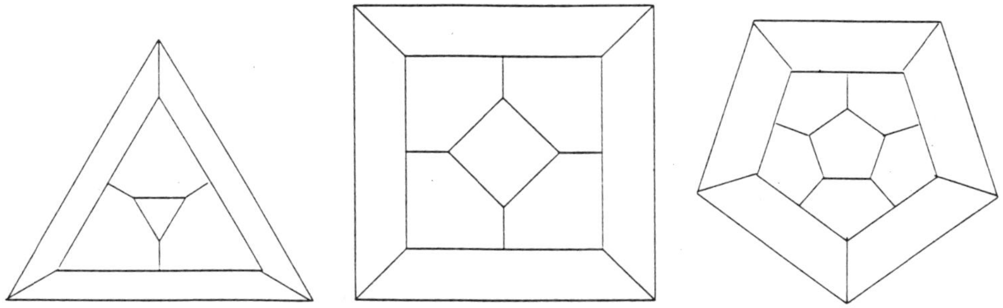

Fig. 4. Schlegel Diagrams of Facets of Equifacetted 3-Manifolds Number of Facets: 10, 26, 120.

Are there additional non-polytopal uniform matroid polytopes in rank 5 which are not extensions of the three known examples with 9 and 10 elements? A more detailed investigation of these questions can be found in [19] and in [51].

7. Spatial Polyhedra and Immersions

7.1. Polyhedra with a minimal number of vertices and maximal symmetry

Several solved and unsolved problems under the headline *polyhedra with minimal number of vertices and maximal symmetry* fit into the class of problems in computational synthetic geometry. In particular, those problems with small number of vertices and symmetries can be solved automatically. For details we refer the reader to the survey of U. Brehm and J. M. Wills in [23].

7.2. Neighborly spatial polyhedra

Cyclic polytopes and more general neighborly polytopes in higher dimensions have been studied extensively, e.g. because of their extreme f-vector properties. In the spatial case, we get interesting aspects when the manifold structure is weakened a bit. Neighborly spatial polyhedra or polyhedra without diagonals have been studied in [4] and [5]. A complete classification into realizable and non-realizable polyhedra has been obtained in the 9 vertex case, cf. [4]. In this case all samples lead to a polyhedral realization. In the 10 point case we know all strongly connected realizations.

Theorem (Altshuler, Bokowski, Schuchert 1992): There are altogether 32 orientable neighborly 2–pseudomanifolds with 10 vertices. 29 of them are strongly–connected and three are non–strongly–connected.

In this class of combinatorial polyhedra we find a first non–strongly–connected non-realizable abstract object.

Theorem (Altshuler, Bokowski, Schuchert 1992): There is at least one not strongly connected orientable neighborly pseudomanifold with 10 vertices which cannot be realized in 3–space.

We find also polyhedra which have pinched spheres as their boundary.

Theorem (Altshuler, Bokowski, Schuchert 1992): There are precisely four neighborly pinched 2–spheres with 10 vertices each of which is realizable in 3–space.

We have depicted one example: a Schlegel diagram of a 3-polytope. By identifying points with equal labels, we obtain a neighborly pinched sphere.

A 3–dimensional triangulation of the objects on different stages of abstraction (abstract complex, oriented matroid, concrete polyhedron) played a decisive role as an algorithmic tool for these findings. The combinatorial objects in these cases are block designs $S_2(2,3,n)$.

7.3. 2-manifolds in the 2-skeleton of 4-polytopes

A survey about contributions to the chosen topic *2-manifolds in the 2-skeleton of 4-polytopes* can be found in the article of Brehm and Wills in [23].

One method for finding a realization of a combinatorially presented 2-manifold M in R^3 is to determine a 4-polytope P first, such that the 2-skeleton of it contains M as a subcomplex. Any Schlegel diagram of P provides a realization of M in 3-space. The search for P can be performed in an automatic fashion. We can construct (all) pairs of combinatorial 3-manifolds having M as its boundary. After gluing together the pairs of 3-manifolds along their boundary, one can check the matroid polytope

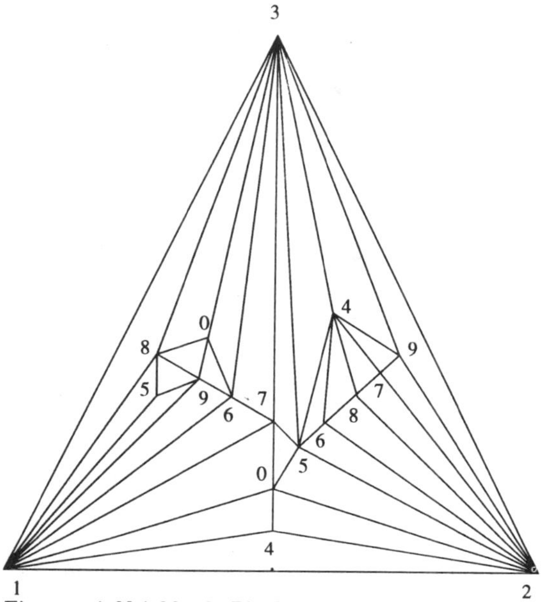

Fig. 5. A Neighborly Pinched Sphere with 10 Vertices

property automatically. The existence of a matroid polytope has been confirmed if and only if we have found an admissible oriented matroid with extreme points. In the affirmative case, the methods of Section 4 can be appended to find coordinates for the extreme points of the matroid polytope.

This method has been applied in [21] to determine realizations for all tori among the finite class of 21 unshrinkable tori as a first step to tackle the open Steinitz problem for tori via embeddings in 2-skeletons of 4-polytopes. Here is one result of this type:

Torus with 8 vertices with its list of 16 triangles

124	152	143	135	263	238	274	256
287	348	375	367	465	458	476	578

The resulting polytope has the facets:

3541	5241	3854	8735	2645	7328	7263	6247
2351	2431	2843	3652	4782	7548	5763	4765

One can try to find in the same way 4-polytopes such that the 2-skeleton contains as a subcomplex a neighborly 2-manifold with 12 vertices. It turns out that this is not possible. In none of altogether 59 combinatorially different cases of such 2-manifolds does there exist such a corresponding 4-polytope. This result has been obtained by Altshuler, Bokowski and Schuchert in [6].

7.4. Immersions

In his Ph.D.-thesis [25], Cervone found vertex-minimal immersions of simplicial surfaces into three-space. This topic fits to the series of problems in computational synthetic geometry. The oriented matroids which are involved (especially when a symmetry is assumed) can be generated automatically. In this case the forbidden Radon partitions in the generation of all admissible oriented matroids are those which correspond to the local forbidden intersections.

Here is how the generation of David Cervone's Klein-bottle can work. With the labels (A=1, B=2, C=3, D=4, E=5, F=6, G=7, H=8, I=9) we have the input files for a corresponding program as follows.

1. The oriented matroid as a signlist is simply a positive sign + representing a tetrahedron.

2. The list of triangles of the immersion determined by Cervone is the following

1 2 4	1 2 9	1 3 4	1 3 7	1 6 7	1 6 8	1 8 9	2 3 5	2 3 9
2 4 5	3 4 9	3 5 7	4 5 9	5 6 8	5 6 9	5 7 8	6 7 9	7 8 9

3. We require e.g. a symmetry given by the permutation (1,5)(2,6)(3,7)(4,8)(9). Now the program can be started to obtain as the first admissible oriented matroid of the immersed Klein bottle the following one. We list the signed bases indirectly by listing the hyperline sequences, compare [12].

1 :	1	2	3	9	7	6	-8	5	4
2 :	1	3	2	-6	8	-7	-4	-9	-5
3 :	2	3	1	6	9	5	4	7	-8
4 :	1	4	2	5	9	3	7	-8	6
5 :	2	4	1	8	-7	-3	-9	-6	-5
6 :	3	4	1	9	8	-7	5	2	-6
7 :	1	5	2	-4	3	9	7	-8	6
8 :	2	5	1	8	-7	-3	-9	-6	4
9 :	3	5	1	8	-7	-4	2	-9	-6
10 :	4	5	1	-2	6	9	3	7	-8
11 :	1	6	2	9	7	3	-4	-5	8
12 :	2	6	1	-7	-9	-3	5	4	-8
13 :	3	6	1	2	5	9	4	7	-8
14 :	4	6	1	8	-7	-3	-9	-2	-5
15 :	5	6	1	8	-7	-9	-3	-2	4
16 :	1	7	2	9	-6	8	3	-4	-5
17 :	2	7	1	6	5	4	-3	-8	-9
18 :	3	7	1	8	9	4	5	2	-6

19 :	4	7	1	6	-2	-5	-3	-9	-8
20 :	5	7	1	6	-2	4	-3	-8	-9
21 :	6	7	1	8	3	-4	-5	-2	-9
22 :	1	8	2	-6	5	4	-3	-7	-9
23 :	2	8	1	6	9	7	3	-4	-5
24 :	3	8	1	6	-2	-5	-4	-9	-7
25 :	4	8	1	7	9	3	5	2	-6
26 :	5	8	1	9	7	3	-4	2	-6
27 :	6	8	1	5	4	-3	-7	-9	-2
28 :	7	8	1	6	-2	-5	-9	-4	3
29 :	1	9	2	-7	-6	8	-4	3	-5
30 :	2	9	1	6	-3	5	4	-8	7
31 :	3	9	1	-4	8	-7	2	5	-6
32 :	4	9	1	6	-2	-5	7	-8	3
33 :	5	9	1	6	-3	-2	4	-8	7
34 :	6	9	1	8	-4	3	-5	-2	7
35 :	7	9	1	-2	-5	8	-4	3	-6
36 :	8	9	1	-3	4	-7	5	2	-6

Now the search for coordinates as described in Section 4 can be appended. Cervone in [25] has chosen a direct way for finding coordinates.

References

1. Altshuler, A.: 1971, Polyhedral realization in R^3 of triangulations of the torus and 2–manifolds in cyclic 4-polytopes, *Discrete Math.*, 1, pp. 211–235.
2. Altshuler, A.: 1976, A peculiar triangulation of the 3-sphere, of the torus and 2–manifolds in cyclic 4-polytopes, *Proc. Am. Math. Soc.* **54** pp. 449–452.
3. Altshuler, A.: 1977, Neighborly 4-polytopes and neighborly combinatorial 3-manifolds with ten vertices, *Can.J.Math.*, **29**, pp. 400–420.
4. Altshuler, A., Bokowski, J. and Schuchert, P.: 1992, Spatial polyhedra without diagonals, *Israel J. Math.*, in print.
5. Altshuler, A., Bokowski, J. and Schuchert, P.: 1993, Sphere systems and neighborly spatial polyhedra with 10 vertices, *Palermo proceedings.*
6. Altshuler, A., Bokowski, J. and Schuchert, P.: 1993, The orientable neighborly 2-manifolds with 12 vertices can not be embedded in the 2-skeleton of 4-polytopes, *manuscript.*

7. Altshuler, A., Bokowski, J. and Steinberg, L.: 1980, The classification of simplicial 3-spheres with nine vertices into polytopes and nonpolytopes, *Discrete Math., 31,* pp. 115–124.
8. Altshuler, A. and Brehm, U.: 1992 Neighborly 2–manifolds with few vertices, *Discrete Comput. Geom.*, 8, pp. 93–104.
9. Barthel, G., Hirzebruch, F. and Höfer, T.: 1987, Geradenkonfigurationen und Algebraische Flächen, *Vieweg Verlag.*
10. Bayer, M. M. and Lee, C. W.: 1993, Combinatorial aspects of convex polytopes, *Handbook of Convex Geometry, Gruber P. and Wills J.M., (eds.), Elsvier, North-Holland, Netherland.*
11. Björner, A., Las Vergnas, M., Sturmfels, B., White, N. and Ziegler, G.M.: 1993, Oriented matroids, *Cambridge University Press, Cambridge.*
12. Bokowski, J.: 1993, Oriented matroids, *Handbook of Convex Geometry, Gruber P. and Wills J.M., (eds.), Elsvier, North-Holland, Netherland.*
13. Bokowski, J. and Garms, K.: 1987, Altshuler's sphere M^{10}_{425} is not polytopal, *European J. Comb.,* **8,** pp. 227–229.
14. Bokowski, J., Guedes de Oliveira, A. and Veloso da Costa, A.: 1993, On the cube problem of Las Vergnas, *manuscript.*
15. Bokowski, J., Laffaille, G. and Richter, J.: 1991, Classification of nonstrechable pseudoline arrangements and related properties, *manuscript.*
16. Bokowski, J. and Richter, J.: 1990, On the finding of final polynomials, *Europ. J. Combinatorics*, 11, pp. 21–34.
17. Bokowski, J. and Scharnbacher, J.: 1993, On the finding of realizations of oriented matroids, *manuscript.*
18. Bokowski, J. and Schuchert, P.: 1993, Altshuler's sphere M^{9}_{963} revisited, *Preprint.*
19. Bokowski, J. and Schuchert, P.: 1993, Equifacetted 3-spheres as topes of nonpolytopal matroid polytopes, *Preprint.*
20. Bokowski, J. and Strempel, T. K.: 1993, Pseudogeraden Arrangements mit maximaler Dreiecksanzahl, *manuscript.*
21. Bokowski, J. and Strempel, T. K.: 1993, Tori in the 2-skeleton of 4-polytopes, *manuscript.*
22. Bokowski, J. and Sturmfels, B.: 1989, Computational synthetic geometry, *Lecture Notes in Mathematics*, 1355, Springer, Heidelberg.
23. Brehm, U. and Wills, J.M.: 1993, Polyhedral manifolds, *Handbook of Convex Geometry, Gruber P. and Wills J.M., (eds.), Elsvier, North-Holland, Netherland.*
24. Cervone, D. P., Vertex-minimal immersions of simplicial surfaces into three-space, *Ph.D. Thesis, Brown University, Rhode Island.*
25. Cervone, D. P., Vertex-minimal immersions of the Klein bottle, *Geometriae Dedicata,* to appear.
26. Cordovil, R., Guedes de Oliveira, A. and Fukuda, K.: 1991, On the cocircuit-graph of an oriented matroid, *preprint,* 11 pages.
27. Coxeter, H.S.M.: 1937, Regular skew polyhedra in three and four dimensions and their topological analogues, *Proc. London Math. Soc., Ser. 2* **43**.

28. Folkman, J. and Lawrence, J.: 1978, Oriented matroids, *J. Comb. Theory, Ser. B,* **25** pp. 199–236.
29. Füredi, Z. and Palasti, I.: 1984, Arrangements of lines with a large number of triangles, *Proc.Am.Math.Soc.,* **92,2** pp. 561–566.
30. Ganter B., Mathon R. and Rosa A.: 1977, A complete census of (10, 3, 2)–block designs and of Mendelsohn triple systems of order ten. I. Mendelsohn triple systems without repeated blocks, *Proc. Seventh Manitoba Conf. Numerical Math. and Computing,* pp. 383–398.
31. Gruber, P. and Wills, J. M.: 1993, *Handbook of Convex Geometry, Gruber P. and Wills J.M., (eds.), Elsvier, North-Holland, Netherland.*
32. Grünbaum B.: 1967, Convex Polytopes, *Interscience Publ.,* London.
33. Grünbaum B.: 1972, Arrangements and Spreads, *Regional Conf., Vol. 10 (Amer. Math. Soc., Providence, RI).*
34. Harborth, H.: 1981, Two-colorings of simple arrangements, *Coll. Math. Soc. Janos Bolyai* 37, pp. 371–378.
35. Harborth, H.: 1985, Some simple arrangements of pseudolines with a maximum number of triangles, *Disc. Geom. Conv., N.Y. Acad. Sc.* 440, pp. 30–31.
36. Harborth, H.: 1988, Problem about an simple arrangement of pseudolines, *Math. Semesterberichte.*
37. Humphreys, J.F. and Prest, M.Y.: 1989, Numbers, groups and codes, *Cambridge University Press.*
38. Kalai, G.: 1988, Many triangulated spheres, *Discrete Comput. Geom.,* **3**, pp. 1–14.
39. Las Vergnas, M.: 1975, Matroïdes orientables, *C.R. Acad. Sci. Paris, Ser. A,***280** pp. 61–64.
40. Las Vergnas, M.: 1980, Convexity in oriented matroids, *J. Comb. Theory, Ser. B,***29,** pp. 231–243.
41. Las Vergnas, M., Roudneff, J.-P. and Salaün, I.: 1991, Regular polytopes and oriented matroids, *preprint.*
42. Levi, F.: 1926, Die Teilung der projektiven Ebene durch Gerade und Pseudogerade, *Ber. Math.-Phys. Kl. Sächs. Akad. Wiss.,* **78,** pp. 256–267.
43. Ljubić, D., Roudneff, J.-P. and Sturmfels, B.: 1989, Arrangements of lines and pseudolines without adjacent triangles, *J.Comb.Theor.Ser.A.,* 50, pp. 24–32.
44. Mani–Levitska, P.: 1993, Characterizations of convex sets, *Handbook of Convex Geometry, Gruber P. and Wills J.M., (eds.), Elsvier, North-Holland, Netherland.*
45. Richter–Gebert, J.: 1992, Oriented matroids with few mutations, *Discrete Comp. Geom.* **10** 251–269.
46. Richter–Gebert, J.: 1992, On the realizability problem of combinatorial geometries – Decision methods, *Ph.D. Thesis, University Darmstadt.*
47. Roudneff, J.-P.: 1986, On the number of triangles in simple arrangements of pseudolines in the real projective plane, *Disc. Math.,* 60, pp. 234–251.
48. Roudneff, J.-P.: 1988, Reconstruction of the orientation class of an oriented matroid, *Europ. J. Combinatorics,* **9** pp. 423–429.
49. Ruppert J. and Seidel R.: 1992, On the difficulty of triangulating three-dimensional nonconvex polyhedra, *Discrete Comput. Geom.,* **7** pp. 227–253.

50. Salaün, I.: 1988, Deux problèmes de gèometrie combinatroire, *Thèse de troisième Cycle, Université Paris 6.*
51. Schuchert, P.: 1993, *Ph.D. Thesis, University Darmstadt,* in preparation.
52. Schulte, E.: 1993, Tilings, *Handbook of Convex Geometry, Gruber P. and Wills J.M., (eds.), Elsvier, North-Holland, Netherland.*
53. Shemer, I.: 1982, Neighborly polytopes, *Israel J. Math.,* **43** pp. 291–314.
54. Smilansky, Z.: 1985, Convex hulls of generalized moment curves, *Israel J. Math.,* **52,** pp. 115–128.
55. Smilansky, Z.: 1990, Bi-cyclic 4-polytopes, *Israel J. Math.,* **70,1** pp. 82–92.
56. Sturmfels, B.: 1987, Computational synthetic geometry, *Ph.D. Thesis, University of Washington, Seattle.*
57. Sturmfels, B.: 1987, Boundary complexes of convex polytopes cannot be characterized locally, *J. London Math. Soc.* (2) 35, pp. 314–326.
58. Ziegler, G. M.: 1991, Some minimal non-orientable matroids of rank 3, *Geometriae Dedicata* **38**, pp. 365–371.

THE RIDGE GRAPH OF THE METRIC POLYTOPE AND SOME RELATIVES

ANTOINE DEZA
Tokyo Institute of Technology, Dept. of Information Sciences, Tokyo, Japan, and Université de Paris-Sud, Orsay, France

and

MICHEL DEZA
CNRS-LIENS, Ecole Normale Supérieure, Paris, France

Abstract. The metric polytope is a $\binom{n}{2}$-dimensional convex polytope defined by its $4\binom{n}{3}$ facets. The vertices of the metric polytope are known only up to $n = 6$, for $n = 7$ they number more than 60 000. The study of the metric polytope and its relatives (the metric cone, the cut polytope and the cut cone) is mainly motivated by their application to the maximum cut and multicommodity flow feasibility problems. We characterize the ridge graph of the metric polytope, i.e. the edge graph of its dual, and, as corollary, obtain that the diameter of the dual metric polytope is 2. For $n \geq 5$, the edge graph of the metric polytope restricted to its integral vertices called cuts, and to some $\{\frac{1}{3}, \frac{2}{3}\}$-valued vertices called anticuts, is, besides the clique on the cuts, the bipartite double of the complement of the folded n-cube. We also give similar results for the metric cone, the cut polytope and the cut cone.

Key words: Metric polytope, cut polytope, ridge graph, edge graph.

1. Introduction

We first recall the definitions of the *metric polytope* $MetP_n$ and its relatives, the *metric cone* Met_n, the *cut polytope* $CutP_n$ and the *cut cone* Cut_n. Then we present some applications to well known optimization problems and some combinatorial and geometric properties of those four polyhedra.

For all 3-sets $\{i, j, k\} \subset \{1, \ldots, n\}$, we consider the following inequalities:

$$x_{ij} - x_{ik} - x_{jk} \leq 0 \tag{1}$$
$$x_{ij} + x_{ik} + x_{jk} \leq 2. \tag{2}$$

The inequalities (1) define the metric cone Met_n and the metric polytope $MetP_n$ is obtained by bounding Met_n by the inequalities (2). The $3\binom{n}{3}$ facets defined by the inequalities (1), which can be seen as triangle inequalities for distance x_{ij} on $\{1, 2, \ldots, n\}$, are called *homogeneous triangle facets* and are denoted by $Tr_{ij,k}$. The $\binom{n}{3}$ facets defined by the inequalities (2) are called *non-homogeneous triangle facets* and are denoted by Tr_{ijk}.

Given a subset S of $V_n = \{1, 2, \ldots, n\}$, the cut determined by S consists of the pairs (i, j) of elements of V_n such that exactly one of i, j is in S. $\delta(S)$ denotes both the cut and its incidence vector in $\mathbb{R}^{\binom{n}{2}}$, i.e. $\delta(S)_{ij} = 1$ if exactly one of i, j is in S and 0 otherwise for $1 \leq i < j \leq n$. By abuse of language, we use the term cut for

T. Bisztriczky et al. (eds.), POLYTOPES: Abstract, Convex and Computational, 359–372.

both the cut itself and its incidence vector, so $\delta(S)_{ij}$ are considered as coordinates of a point in $\mathbb{R}^{\binom{n}{2}}$. The cut polytope $CutP_n$ is the convex hull of all 2^{n-1} cuts, and the cut cone Cut_n is the conic hull of all $2^{n-1}-1$ nonzero cuts.

We have $CutP_n \subseteq MetP_n$ and $Cut_n \subseteq Met_n$ with equality only for $n \leq 4$. Any facet of the metric polytope contains a facet of the cut polytope and the vertices of the cut polytope are vertices of the metric polytope, in fact the cuts are precisely the integral vertices of the metric polytope. Actually the metric polytope $MetP_n$ wraps the cut polytope $CutP_n$ very tightly since, in addition to the vertices, all edges and 2-faces of $CutP_n$ are also faces of $MetP_n$ [12]. There is a $1-1$ correspondence between the elements of the metric cone Met_n and all the semi-metrics on n points, and the elements of the cut cone Cut_n correspond precisely to the semi-metrics on n points that are isometrically embeddable into some l_1^m, see[2], it is easy to see that $m \leq \binom{n}{2}$. Those polyhedra were considered by many authors, see for instance [1, 3, 6, 9, 10, 11, 12, 13, 16, 17] and references there.

One of the motivations for the study of these polyhedra comes from their applications in combinatorial optimization, see for instance [11]. Given a graph $G=(V_n,E)$ and nonnegative weights w_e, $e \in E$, assigned to its edges, the max-cut problem consists in finding a cut $\delta(S)$ whose weight $\sum_{e\in\delta(S)} w_e$ is as large as possible. By setting $w_e = 0$ if e is not an edge of G, we can consider the complete graph on V_n. Then the max-cut problem can be stated as a linear programming problem over the cut polytope $CutP_n$ as follows:

$$\begin{cases} \max \;\; w^T \cdot x \\ x \in CutP_n. \end{cases}$$

Since $MetP_n$ is a relaxation of $CutP_n$, optimizing $w^T \cdot x$ over the metric polytope instead of the cut polytope provides an upper bound for the max-cut problem [3].

With E the set of edges of the complete graph on V_n, an instance of the multicommodity flow problem is given by two nonnegative vectors indexed by E: a capacity $c(e)$ and a requirement $r(e)$ for each $e \in E$. Let $U = \{e \in E : r(e) > 0\}$. If T denotes the subset of V_n spanned by the edges in U, then we say that the graph $G=(T,U)$ denotes the *support* of r. For each edge $e=(s,t)$ in the support of r, we seek a flow of $r(e)$ units between s and t in the complete graph. The sum of all flows along any edge $e' \in E$ must not exceed $c(e')$. If such a flow exists, we call c, r *feasible*. A necessary and sufficient condition for feasibility is given by the Japanese theorem [15]: a pair c,r is feasible if and only if $(c-r)^T x \geq 0$ is valid over Met_n. For example, $Tr_{ij,k}$ can be seen as an elementary solvable flow problem with $c(ij) = r(ik) = r(jk) = 1$ and $c(e) = r(e) = 0$ otherwise, so the inequalities (1) correspond to $(c-r)^T x \geq 0$ for $x \in Met_n$. Therefore, the metric cone Met_n is the dual cone to the cone of feasible multicommodity flow problems.

The metric polytope $MetP_n$ and the cut polytope $CutP_n$ share the same symmetry group induced by permutations on $V_n = \{1,\ldots,n\}$ and switching reflections [9, 16]. This group is isomorphic to $Aut(\square_n)$, see Remark 3.4 below. Given a cut $\delta(S)$, the switching reflection $r_{\delta(S)}$ is defined by $y = r_{\delta(S)}(x)$ where $y_{ij} = 1 - x_{ij}$ if $(i,j) \in \delta(S)$ and $y_{ij} = x_{ij}$ otherwise. These symmetries, which preserve adjacency, are widely used in the study of $MetP_n$ and its relatives.

$CutP_3$ and $MetP_3$ are combinatorially equivalent to the tetrahedron and $CutP_4$ and $MetP_4$ are combinatorially equivalent to the 6-dimensional cyclic polytope with 8 vertices. More generally the cut polytope is a 3-neighbourly polytope [12]. Any two cuts are adjacent both on $CutP_n$ [3] and on $MetP_n$ [18]; in other words $MetP_n$ is quasi-integral in terms of [19], i.e. the edge graph of the convex hull of its integral vertices, the edge graph of the cut polytope, is an induced subgraph of the edge graph of the metric polytope itself.

The paper is organized as follows. In Section 2 we characterize the ridge graph of the metric polytope $MetP_n$ and the metric cone Met_n. In Section 3, respectively Section 4, we give a partial result on the edge graph of the metric polytope $MetP_n$ and the ridge graph of the cut polytope $CutP_n$. Section 5 contains the proofs of Lemma 2.1 and Theorem 2.2. A general reference for the graph theory used in this paper is [4].

2. Ridge Graph of the Metric Polytope and the Metric Cone

2.1. Ridge graph of the Metric Polytope

The ridge graph G_n of the metric polytope $MetP_n$ is the edge graph of its dual. The nodes of G_n are the $4\binom{n}{3}$ triangle facets of $MetP_n$ and two facets are adjacent if and only if their intersection is a ridge, i.e. a face of codimension 2. We first determine the various intersections of two facets of $MetP_n$ by the following lemma:

Lemma 2.1 *For $n \geq 4$, the intersection of two facets of $MetP_n$ is either:*
(a) a face of codimension $n-1$, combinatorially equivalent to $MetP_{n-1}$, or
(b) a face of codimension 3, or
(c) a face of codimension 2, i.e. a ridge.

We call faces of type (a): *weak triangle faces.* From the proof of Lemma 2.1 given in Section 5, it is easy to check that all weak triangle faces belong to a facet of the cube $[0,1]^{\binom{n}{2}}$ and form one orbit of the symmetry group of $MetP_n$, i.e. are equivalent under permutation and switching. Since each weak triangle face is combinatorially equivalent to $MetP_{n-1}$, the metric polytope $MetP_n$ contains $2\binom{n}{2}$ copies of $MetP_{n-1}$. The same proof also shows that, for $n \geq 4$, two facets intersect in a ridge if and only if they are non-conflicting. Two facets are called *conflicting* if there exists a pair i,j such that the two facets have nonzero coordinates of distinct signs at the position i,j. For example, Tr_{123} and $Tr_{24,3}$ are conflicting at pair 2, 3. The notion of conflicting facets was introduced in [13, 16]. Using this property we are able to characterize G_n, the ridge graph of $MetP_n$. $G_3 = K_4$ and G_4 is the (4×4)-grid. For higher values of n, it is more convenient to consider $\bar{G}_n$, the complement of G_n, which has a smaller valency. For $i = 1,\ldots,n$, let $G_i \approx G$ be n isomorphic graphs and $\Gamma_i \approx \Gamma$ their isomorphic induced subgraphs, we call *bouquet of n graphs G with common Γ* the graph with vertex set $V = \cup_{i=1}^{n} V(G_i \setminus \Gamma_i) \cup V(\Gamma)$ and edge set $E = \cup_{i=1}^{n} E(G_i \setminus \Gamma_i) \cup \{(x,y) : x \in \Gamma, y \in G_i \setminus \Gamma_i, x \text{ and } y \text{ adjacent}\}$. For example, Fig. 1 and Fig. 2 represent a bouquet of 3 (3×3)-grids with common K_3 and a bouquet of 2 hexagons with common edge. With v denoting the number of nodes, k the valency of each node, λ the number of nodes adjacent to two adjacent nodes and μ the number of nodes adjacent to two non-adjacent nodes, we have:

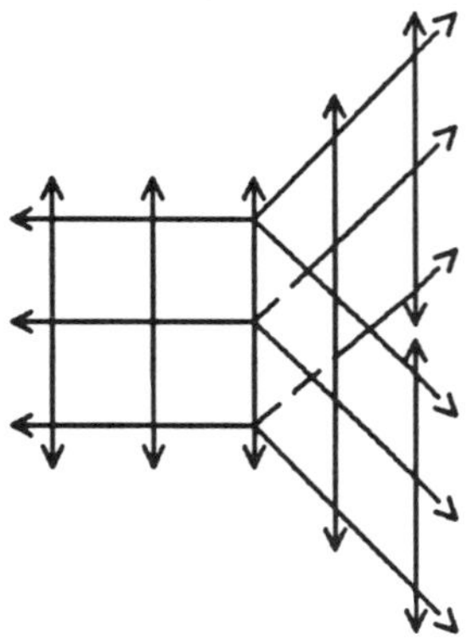

Fig. 1. The local graph of $\bar{G}_6$

Theorem 2.2 *For $n \geq 4$, $\bar{G}_n$ is locally the bouquet of $(n-3)$ (3×3)-grids with common K_3 having parameters: $v = 4\binom{n}{3}$, $k = 3(2n-5)$, $\lambda = 2(n-2)$ or 4, and $\mu = 6$, 4 for $n \geq 5$ or 0 for $n \geq 6$.*

Fig. 1 illustrates Theorem 2.2 for the case $n = 6$. From the parameters of $\bar{G}_n$ we can compute some parameters of G_n: the valency $v = \frac{2(n-3)(n-7)}{3}$ and $\mu = \frac{2(n-3)(n-13)}{3}$ or $\frac{2(n-3)(n-16)}{3} + 2$. This gives the two following corollaries:

Corollary 2.3 *For $n \geq 4$, the diameter of G_n is 2.*

PROOF. For $n \geq 4$, $\mu > 0$, i.e. any two non-adjacent nodes of G_n share a common neighbour. The diameter of G_3 is obviously 1. □

Corollary 2.4 *The metric polytope has exactly $\frac{16(n-7)\binom{n}{4}}{3}$ ridges.*

PROOF. The number of faces of codimension 2 of a polytope is half of the total valency of its ridge graph. Since we know the common valency of all $4\binom{n}{3}$ nodes of G_n, the result is a straightforward calculation. □

G_4, the ridge graph of $MetP_4$, is the (4×4)-grid $= L(K_{4,4}) = L(\square_4)$, the line graph of the folded 4-cube. G_4 is a strongly regular graph with parameters $v = 16$, $k = 6$, $\lambda = 2$ and $\mu = 2$. There exists only one other strongly regular graph with the same parameters, namely the Shrikhande graph [4]. For $\bar{G}_5$, the complement of the ridge graph of $MetP_5$, both λ and μ take their values in $\{4, 6\}$, providing first example of an interesting generalization of strongly regular graph: regular graph of diameter 2 with $\lambda, \mu \in \{a, b\}$ with $b > a > 0$.

Remark 2.5 *As a direct consequence of the proof of Lemma 2.1, we obtain that two facets F_1 and F_2 of $CutP_n$ contained in two facets F_1' and F_2' of $MetP_n$ are adjacent in $CutP_n$ if and only if F_1' and F_2' are adjacent in $MetP_n$. This implies that any ridge of the metric polytope contains a ridge of the cut polytope.*

Before presenting another interesting consequence of the proof of Lemma 2.1, we recall the notion of *0-lifting* which was considered in[10]. Let v be a vector of length $\binom{n}{2}$ and $v' = (v, 0 \ldots, 0)$ of length $\binom{n+1}{2}$, then the inequality $v' \cdot x \leq a$ defines a facet F' of $CutP_{n+1}$, called 0-lifting of F, if and only if the inequality $v \cdot x \leq a$ defines a facet F of $CutP_n$. We extend this notion to a ridge of $CutP_n$, $W = F_1 \cap F_2$, by defining the 0-lifting of W: $W' = F_1' \cap F_2'$, where F_1', respectively F_2', is the 0-lifting of F_1', respectively F_2'. For example, the $(n-3)$-times 0-lifting of a ridge of $MetP_3$, i.e. of an edge, is a weak triangle face of $MetP_n$. By a proof similar to the one used for the Lemma 2.1 and by direct checking for $n = 4$, we have:

Lemma 2.6 *For $n \geq 4$, the 0-lifting of ridge of $CutP_n$ is a ridge of $CutP_{n+1}$.*

In the next theorem, we give some additional characteristics of the ridge graph of the metric polytope. With $\omega(G_n)$, respectively $\alpha(G_n)$, the size of the largest clique, respectively co-clique, of G_n and $k(G_n)$ the number of maximal cliques of full rank $\binom{n}{2}$, where the rank of a clique is the rank of the set of the $\{-1, 0, 1\}$-valued vectors representing the coefficients of the triangle facets belonging to the clique, we have:

Theorem 2.7 *For $n \geq 5$, the intersection of the facets belonging to a maximal clique C of G_n is a $(\binom{n}{2} - rank(C))$-face containing a unique $\{\frac{1}{3}, \frac{2}{3}\}$-valued point of $MetP_n$ which is a vertex if and only if $rank(C) = \binom{n}{2}$. $k(G_n)$ equals the number of $\{\frac{1}{3}, \frac{2}{3}\}$-valued vertices of $MetP_n$, $\omega(G_n) = \binom{n}{3}$ and $\alpha(G_n) = 4$.*

PROOF The *support* of a triangle facet $Tr_{ij,k}$ or Tr_{ijk} is the 3-set $\{i, j, k\}$. Since the 4 triangle facets sharing a same support obviously form the largest clique of $\bar{G}_n$, we have $\alpha(G_n) = 4$. Let C be a maximal clique of G_n and f be the face of $MetP_n$ which is the intersection of the facets belonging to C. We show that any 2-set $\{i, j\}$ belongs to the support of a facet of C by exhibiting, otherwise, a new facet T non-conflicting with any facet F of C (which contradicts the maximality of C). For any $k \in \{1, \ldots, n\} \setminus \{i, j\}$, the coefficients F_{ik} and F_{jk} of F can be $(\leq 0, \leq 0)$, $(\leq 0, \geq 0)$, $(\geq 0, \leq 0)$ or $(\geq 0, \geq 0)$; the corresponding facet T is $Tr_{ij,k}, Tr_{jk,i}, Tr_{ik,j}$ or Tr_{ijk}. Then the point y defined by $y_{ij} = \frac{1}{3}$ if $F_{ij} = -1$ and $y_{ij} = \frac{2}{3}$ if $F_{ij} = 1$ is the unique $\{\frac{1}{3}, \frac{2}{3}\}$-valued point of $MetP_n$ in f. Since clearly $dim(f) = \binom{n}{2} - rank(C)$, f is a vertex v if and only if $rank(C) = \binom{n}{2}$, then, by unicity, $v = y$. One can easily check that a $\{\frac{1}{3}, \frac{2}{3}\}$-valued vertex can belong to at most 1 of the 4 triangle facets sharing a same support (it holds for any vertex v such that $0 < v_{ij} < 1$), and that any *anticut* $\hat{\delta}(S) = \frac{2}{3}(1, \ldots, 1)$-$\frac{1}{3}\delta(S)$, which is a vertex of $MetP_n$ for $n \geq 5$ (see Section 3), belongs to exactly $\binom{n}{3}$ triangle facets. It implies that $\omega(G_n) = \binom{n}{3}$. For example, the 10 non-homogeneous triangle facets of $MetP_5$ intersect on the vertex $y = \frac{2}{3}(1, \ldots, 1)$ while the 6 facets $Tr_{12,3}, Tr_{12,4}, Tr_{12,5}, Tr_{45,1}, Tr_{45,2}$ and $Tr_{45,3}$ intersect on a 4-face f of $MetP_5$ generated by the vertices $\delta(S)$, $S = \emptyset, \{1, 4\}, \{1, 5\}, \{2, 4\}$ and $\{2, 5\}$. The unique $\{\frac{1}{3}, \frac{2}{3}\}$-valued point of f is $v = \frac{1}{6}(\delta(1, 4) + \delta(1, 5) + \delta(2, 4) + \delta(2, 5) + 2\delta(\emptyset))$, so $3v$ is the path metric of K_5 with the edges 1, 2 and 4, 5 deleted. □

Remark 2.8 *A vertex of the metric polytope belongs to at most $3\binom{n}{3}$ triangle facets, i.e. to 3/4 of the total number of facets of the metric polytope, and this value is attained only by the cuts. On the other hand, a facet of the cut polytope contains at most $3 \cdot 2^{n-3}$ cuts, i.e. 3/4 of the total number of vertices of the cut polytope, and this value is attained by the triangle facets, see [7].*

2.2. Ridge graph of the Metric Cone

The ridge graph G'_n of the metric cone Met_n can be easily deduced from G_n since, for $n \geq 4$, two facets of the metric cone intersect in a ridge if and only if they are non-conflicting. We have:

Theorem 2.9 *For $n \geq 4$, $\bar{G}'_n$ is locally the bouquet of $(n-3)$ hexagons with common edge having parameters: $v = 3\binom{n}{3}$, $k = 2(2n-5)$, $\lambda = n-2$ or 2, and $\mu = 4$, 3 or 2 or 0 for $n \geq 5$.*

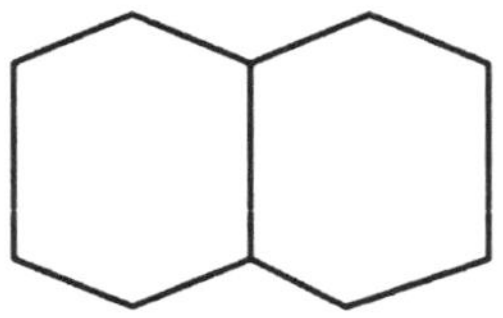

Fig. 2. The local graph of $\bar{G}'_5$ is the graph of the molecule of naphthalene

Fig. 2 illustrates Theorem 2.9 for the case $n = 5$, $\bar{G}'_5$ is locally the graph of the molecule of naphthalene. As for G_n, we can compute some parameters of G'_n: the valency $v = \frac{(n-3)(n-6)}{2}$ and $\mu = \frac{(n-3)(n-12)}{2}$ or $\frac{(n-3)(n-14)}{2} + 1$. This gives the two following corollaries:

Corollary 2.10 *For $n \geq 4$, the diameter of G'_n is 2.*

Corollary 2.11 *The metric cone has exactly $3(n^2-6)\binom{n}{4}$ ridges.*

G'_4, the ridge graph of Met_4, is the (4×3)-grid, i.e. $L(K_{4,3})$. G'_4 is a co-edge regular graph with parameters $v = 12$, $k = 5$, $\lambda = 2$ or 1 and $\mu = 2$.

In the same way as for the metric polytope and using a result of [13], we have: for $n \geq 5$, $k(G'_n)$, the number of maximal cliques of rank $\binom{n}{2} - 1$ of the ridge graph of the metric cone, is equal to the number of the $\{1,2\}$-valued, up to multiple, extreme rays of the metric cone.

3. Edge Graph of the Metric Polytope

While the edge graph of the cut polytope $CutP_n$ is known to be a clique, even to find all the vertices of the metric polytope $MetP_n$ seems to be hopeless. Therefore, we first consider the restriction of Γ_n, the edge graph of $MetP_n$, to the 2^{n-1} cuts $\delta(S)$ and 2^{n-1} *anticuts* $\hat{\delta}(S) = \frac{2}{3}(1,\ldots,1)$-$\frac{1}{3}\delta(S)$ which are vertices of $MetP_n$ for $n \geq 5$. The cuts, respectively the anticuts, belong to the same orbit, i.e are equivalent under permutation and switching. We recall the definitions of the *folded n-cube* and the *bipartite double* of a graph, see[4]. The folded n-cube is the graph whose vertices are the partitions of V_n into two subsets, two partitions being adjacent when their common refinement contains a set of size one. The bipartite double of a graph G is the graph whose vertices are the symbols γ^+, γ^- $(\gamma \in G)$ and whose edges are the 2-sets $\{\gamma^\sigma, \delta^\tau\}$, with γ and δ adjacent in G and $\sigma \neq \tau$. We have:

Theorem 3.1 *For $n \geq 5$, the restriction of the edge graph of the metric polytope to the cuts and anticuts, is, besides the clique on the cuts, the bipartite double of the complement of the folded n-cube.*

PROOF. The cuts, respectively the anticuts, are known to form a clique [18], and a co-clique [16]. To find the adjacency relations between the cuts and the anticuts, we use the following result proved in [1]. An anticut $\hat{\delta}(S)$ lies on an extreme ray of the metric cone Met_n if and only if $1 < |S| < n-1$. In other words, $\delta(\emptyset)$ and $\hat{\delta}(S)$ are adjacent if and only if $\min(|S|, |\bar{S}|) > 1$. Using the fact that switching preserves adjacency and that $r_{\delta(T)}(\delta(S)) = \delta(S \Delta T)$, it implies that $\delta(S')$ and $\hat{\delta}(S)$ are adjacent if and only if $\min(|S \Delta S'|, |\bar{S} \Delta S'|) > 1$, i.e. $\delta(S')$ and $\hat{\delta}(S)$, seen as vertices of the folded n-cube, are not adjacent. □

Corollary 3.2 *The diameter of the restriction of the edge graph of the metric polytope to the cuts and anticuts is 2.*

PROOF. Since the cuts, respectively the anticuts, form a clique and a co-clique, we shall exhibit a cut $\delta(T)$ adjacent to both anticuts $\hat{\delta}(S)$ and $\hat{\delta}(S')$ for any pair S, S'. Without loss of generality we can assume that $|S| \leq \frac{n}{2}$. Let $S' = \{1,2\}$, if $|S| > 1$, $\delta(T) = \delta(\emptyset)$ is obviously adjacent to both $\hat{\delta}(S)$ and $\hat{\delta}(S')$. If $|S| = 1$, $\delta(T) = \delta(I)$ with $I = \{i\}$ and $i \notin S \cup S'$. For $|S| = 0$, $\delta(T) = \delta(J)$, with $J = \{i, j\}$ and $i \notin S'$ and $j \in S'$, is adjacent to both $\hat{\delta}(S)$ and $\hat{\delta}(S')$. Then, using permutations and switching, we can exhibit a cut $\delta(T)$ adjacent to any pair of anticuts $\hat{\delta}(S)$ and $\hat{\delta}(S')$. □

The 32 vertices of $MetP_5$ are exactly the 16 cuts and the 16 anticuts, moreover, the complement of the folded 5-cube is isomorphic to the Clebsch graph [4], i.e. the skeleton $\frac{1}{2}H(5,2)$ of the 5-dimensional half-cube. Therefore, Γ_5 is, besides the clique K_{16} on the cuts, the bipartite double of the Clebsch graph.

Corollary 3.3 *$MetP_5$ has exactly 280 edges and its diameter is 2.*

PROOF. Since the valency of the Clebsch graph is 10, adjacency relations between the cuts and anticuts create $10 \times 16 = 160$ edges. The clique on the cuts creates $\binom{16}{2} = 120$ edges, and the co-clique on the anticuts none. Corollary 3.2 implies $\delta(\Gamma_5) = 2$. □

Remark 3.4 *[9] The symmetry group of the metric polytope $MetP_n$ and the cut polytope $CutP_n$ is isomorphic to the automorphism group of the folded n-cube, i.e. $Is(MetP_n) = Is(CutP_n) \approx Aut(\square_n)$.*

We consider for $n \geq 6$, in addition to the orbits of cuts and anticuts, the orbit formed by the $2^{n-1}\binom{n}{2}$ vertices called *trivial extension* of anticuts of $MetP_{n-1}$. The notion of trivial extension was introduced in [13, 16]. Given $S \subset V_n = \{1, 2, \ldots, n\}$ and a pair i, j with $1 \leq i < j \leq n$, the trivial extension of anticut $V(ij, S)$ determined by S and the pair i, j is given by:

$$V_{kl}(ij, S) = \hat{\delta}_{kl}(S) \qquad \text{for } 1 \leq k < l \leq n \text{ and } l \neq i$$
$$V_{il}(ij, S) = \delta_{ij}(S) + (-1)^{\delta_{ij}(S)} V_{jl}(ij, S) \qquad \text{for } l \notin \{i, j\},$$
$$V_{ij}(ij, S) = \delta_{ij}(S).$$

For $n \geq 6$, the $2^{n-1}\binom{n}{2}$ trivial extensions of anticuts are vertices of the metric polytope. $V_{k,l}(ij, S)$ for $i \notin \{k, l\}$, the projection of $V(ij, S)$, is the anticut $\hat{\delta}(S \backslash \{i\})$ of $MetP_{n-1}$. The adjacency relations induced by trivial extensions of anticuts were given in [14]. Without loss of generality we can choose S' such as $|S \Delta S'| \leq |S \Delta \bar{S}'|$, otherwise we take $\bar{S}'$. Then the adjacency relations induced by the trivial extensions of anticuts are:

(a) $V(ij, S)$ and $V(i'j', S')$ are adjacent if and only if $|\{i, j\} \cap \{i', j'\}| = 1$ and $S \Delta S' \subset \{i, j\} \cap \{i', j'\}$.
(b) $V(ij, S)$ and $\delta(S')$ are adjacent if and only if, either $|S \Delta S'| > 2$, or $|S \Delta S'| = 2$ and $\{i, j\} \subset (\overline{S \Delta S'})$.
(c) $V(ij, S)$ and $\hat{\delta}(S')$ are adjacent if and only if $|S \Delta S'| \leq 1$ and $S \Delta S' \subset \{i, j\}$.

This leads to a generalization of Theorem 3.1, where $M_{ij} = \{V(ij, S) : S \subset V_n\}$ and $M_S = \{V(ij, S) : 1 \leq i < j \leq n\}$:

Theorem 3.5 *For $n \geq 6$, besides the clique on the cuts and the bipartite double of the complement of the folded n-cube on the cuts and anticuts, the restriction of the edge graph of the metric polytope to the cuts, anticuts and trivial extensions of anticuts, is given by:*
(1) $M_{ij} = \bar{K}_{2^{n-1}}$.
(2) $M_S = L(K_n)$, the line graph of the complete graph.
(3) Adjacencies between M_S and $M_{S'}$ exist if and only if $\delta(S)$ and $\delta(S')$ are adjacent on the folded n-cube and, in this case, they form the bipartite double of $L(K_n)$.
(4) $V(ij, S)$ is adjacent to 3 anticuts, namely $\hat{\delta}(S \Delta \{i\})$, $\hat{\delta}(S \Delta \{j\})$ and $\hat{\delta}(S)$, and to $2^{n-1} - 3n + 2$ cuts.

Corollary 3.6 *For $n \geq 6$, the diameter of the restriction of the edge graph of the metric polytope to the cuts, anticuts and trivial extensions of anticuts is 2.*

PROOF. Since the diameter of the restriction of edge graph of the metric polytope to the cuts and anticuts is 2, we shall find a common neighbour to any pair of trivial extensions $V(ij, S)$ and $V(i'j', S')$. For $n \geq 7$, $2^{n-1} - 3n + 2 > 2^{n-2}$, i.e. the number of cuts adjacent to a given trivial extension $V(ij, S)$ is more than half the number of cuts. This implies that any pair of non-adjacent trivial extensions of anticuts is adjacent to a common cut; by same reasons any pair of non-adjacent anticut and trivial extension of anticut shares a common neighbour. For $n = 6$, one can directly check it. □

Corollary 3.7 *$MetP_6$ has exactly* 14 256 *edges and its diameter is* 2.

PROOF. The 544 vertices of $MetP_6$ are exactly the 32 cuts, 32 anticuts and 480 trivial extensions of anticuts. Since each trivial extension of anticut is adjacent to $4(n-2)$ trivial extensions, and to $2^{n-1} - 3n + 5$ cuts and anticuts, they create 12 960 edges. The valency of the complement of the folded 6-cube being 25, the cuts and anticuts create 1 296 edges. Corollary 3.6 gives $\delta(\Gamma_6) = 2$. □

Remark 3.8 *It was conjectured in [16] that the integral vertices of the metric polytope, i.e. the cuts, form a dominating clique in its edge graph Γ_n, i.e. that every vertex of the metric polytope is adjacent to a cut. It would imply that the diameter of the metric polytope satisfies $\delta(\Gamma_n) \leq 3$.*

4. Ridge Graph of the Cut Polytope

As for the vertices of the metric polytope $MetP_n$, the determination of all the facets of the cut polytope $CutP_n$ seems to be hopeless. For n odd, we consider the restriction of Ω_n, the ridge graph of the cut polytope, to the $4\binom{n}{3}$ triangle facets, which are all equivalent under permutations and switching, and the 2^{n-1} facets of the orbit of the equicut facet, where, for n odd, the equicut facet is defined by the following inequality:

$$\sum_{1 \leq i < j \leq n} x_{ij} \leq \frac{n^2-1}{4}.$$

All cuts contained in the equicut facet are equicuts, i.e. cuts $\delta(S)$ such that $|S| = \lfloor \frac{n}{2} \rfloor$ or $|S| = \lceil \frac{n}{2} \rceil$. We have:

Lemma 4.1 *For odd $n \geq 5$, the facets of the orbit of the equicut facet form the co-clique $\bar{K}_{2^{n-1}}$ in the ridge graph of the cut polytope.*

PROOF. Let F and F' be two facets of the orbit of the equicut facet, since switching preserves adjacency, we can assume that F is the equicut facet. Let $\delta(S)$ be the nonzero cut such that F' is the switching of F by $\delta(S)$, and let W_S, respectively $W_{\bar{S}}$, be the projection of $W = F \cap F'$ on S, respectively $\bar{S}$. To prove that F and F' are not adjacent in the ridge graph of the cut polytope, we shall show that $codim(W) > 2$. We have: $dim(W) \leq dim(W_S) + dim(W_{\bar{S}}) + |S| \cdot |\bar{S}|$, then, since W_S is in $\mathbb{R}^{\binom{|S|}{2}}$ and $W_{\bar{S}}$ is in in $\mathbb{R}^{\binom{|\bar{S}|}{2}}$, the last inequality can be rewritten: $codim(W) \geq codim(W_S) + codim(W_{\bar{S}})$. By construction W_S, respectively $W_{\bar{S}}$, is the face E_S defined by the inequality: $\sum_{i,j \in S, i<j} x_{ij} \leq \lfloor \frac{|S|}{2} \rfloor \cdot \lceil \frac{|S|}{2} \rceil$, respectively the face $E_{\bar{S}}$ defined by the inequality: $\sum_{i,j \in \bar{S}, i<j} x_{ij} \leq \lfloor \frac{|\bar{S}|}{2} \rfloor \cdot \lceil \frac{|\bar{S}|}{2} \rceil$. Thus we have: $codim(W) \geq codim(E_S) + codim(E_{\bar{S}})$. A result of [5] states that $codim(E_S) = 1$ if $|S|$ is odd and $|S|$ otherwise, then assuming that $|S|$ is even (otherwise we consider $\bar{S}$), we obtain $codim(W) \geq |S| + 1 > 2$ since $S \neq \emptyset$. □

Theorem 4.2 *For odd $n \geq 7$, besides G_n on the triangle facets, and the co-clique on the orbit of the equicut facet, the restriction of the ridge graph of the cut polytope to the triangle facets and the orbit of the equicut facet is the complete bipartite graph $K_{k,l}$ with $k = 4\binom{n}{3}$ and $l = 2^{n-1}$, between the k triangle facets and the l facets switching equivalent to the equicut facet.*

PROOF. We use the following result proved in [8]. Given $v \in R^{\binom{m}{2}}$, if the inequality $v \cdot x \leq 0$ defines a facet of Cut_m, then the inequality $v' \cdot x \leq 0$ defines a facet of the equicut polytope EP_n for all odd $n \geq 2m+1$, where $v' = (v, 0 \ldots, 0) \in R^{(\ )}$ and EP_n is the convex hull of all the equicuts of V_n. We apply this result for $m = 3$ to a triangle facet of Cut_3 defined by the inequality $v \cdot x \leq 0$, then any facet defined by $(v, 0 \ldots, 0) \cdot x \leq 0$, i.e. any homogeneous triangle facet of Cut_n, is a facet of EP_n for odd $n \geq 7$. In other words, any homogeneous triangle facet of Cut_n intersects with the equicut facet on a ridge for odd $n \geq 7$. Then, similarly, one can check that all triangle facets are ridge-adjacent to the equicut facet and, by switching, ridge-adjacent to all facets of the orbit of the equicut facet. □

The 56 facets of $CutP_5$ are exactly the 40 triangle facets and the 16 facets switching equivalent to the equicut facet. One can directly check that a pair of facets of $CutP_5$ are ridge-adjacent if and only if they are non-conflicting. It implies that the facets switching equivalent to the equicut facet form a co-clique and that each of these 16 facets is ridge-adjacent to 10 triangle facets. For example, the equicut facet is ridge-adjacent to the 10 non-homogeneous triangle facets.

Corollary 4.3 *The diameter of Ω_5 is* 2.

PROOF. Since the diameter of G_5 is 2, we shall exhibit a facet adjacent to any pair of facets of the orbit of the equicut facet. Let F denote the equicut facet and $F_{\delta(S)}$ denote the facet obtained by the switching of F by the nonzero cut $\delta(S)$. If $|S| = 1$, respectively $|S| = 2$, $Tr_{\bar{S}\setminus\{i\}}$ for $i \in \bar{S}$, respectively $Tr_{\bar{S}}$, are adjacent to F and $F_{\delta(S)}$. Then, by switching, any pair of facets of the orbit of F has 1 or 4 common neighbours. □

Corollary 4.4 *$CutP_5$ has exactly* 640 *ridges.*

PROOF. Ridges of $CutP_5$ which are not among the 480 ridges of $MetP_5$, see Corollary 2.4, arise from the 160 adjacency relations between the orbit of the triangle facets and the orbit of the equicut facet. □

Remark 4.5 *With $f_i(P)$ denoting the number of i-faces of a polytope P, we have:*

$$f_i(MetP_5) + f_{9-i}(MetP_5) = f_i(CutP_5) + f_{9-i}(CutP_5) \qquad \textit{for } i = 0 \textit{ or } 1.$$

Corollary 4.6 *For odd $n \geq 5$, the diameter of the restriction of the edge graph of the cut polytope to the triangle facets and the orbit of the equicut facet is* 2.

PROOF. Since the diameter of G_n is 2, it is a direct consequence of Theorem 4.2 and Corollary 4.3. □

Conjecture 4.7 *The triangle facets form a dominating set in the ridge graph of the cut polytope, i.e. every facet of the cut polytope is adjacent to a triangle facet. Since the diameter of the restriction of Ω_n to the triangle facets is* 2, *it would imply that the diameter of the dual cut polytope satisfies $\delta(\Omega_n) \leq 4$. This conjecture holds for $n \leq 7$, moreover, $\delta(\Omega_6) = 3$, see[7].*

Remark 4.8 *One can check that, with $d(G)$ the dominating number of a graph G, i.e. the size of the smallest dominating set in G, we have, for the ridge graph of the cut polytope, $d(\Omega_3) = 1$, $d(\Omega_4) = 4$ and $d(\Omega_5) = 10$ (take the* 10 *non-homogeneous facets), and for the ridge graph of the metric polytope $d(\Gamma_3) = 1$, $d(\Gamma_4) = 1$ and $d(\Gamma_5) = 6$ (take the cuts $\delta(\{i\}), i = 1, \ldots, 5$ and the cut $\delta(\{1,2\})$).*

5. Proofs

5.1. Proof of Lemma 2.1

The proof of Lemma 2.1 is based on a case by case analysis of the different intersections of pairs of triangle facets. The common support σ of two triangle facets $Tr_{ij,k}$ (or Tr_{ijk}) and $Tr_{i'j',k'}$ (or $Tr_{i'j'k'}$) is defined by $\sigma = \{i,j,k\} \cup \{i',j',k'\}$. Then we consider the 4 cases corresponding to the 4 possible values taken by σ.

Case $\sigma = 3$
We first recall the definition of trivial extension of a vertex of $MetP_{n-1}$ given in [16], which generalizes the trivial extension of anticuts of $MetP_{n-1}$ given in Section 3. For $\delta = 0$ or 1, we consider the map ϕ_δ from $R^{\binom{n-1}{2}}$ to $R^{\binom{n}{2}}$ defined as follows. For $v \in MetP_{n-1}$, the metric polytope on the $(n-1)$ points of $\{1,\ldots,n\}\backslash\{i\}$,

$$\begin{aligned} &\phi_\delta(v)_{kl} = v_{kl} && \text{for } 1 \leq k < l \leq n \text{ and } l \neq i \\ &\phi_\delta(v)_{il} = \delta + (-1)^\delta v_{jl} && \text{for } l \notin \{i,j\}, \\ &\phi_\delta(v)_{ij} = \delta \end{aligned}$$

$\phi_\delta(v)$ is called a trivial extension of v. Since permutations and switching reflections preserve the adjacency, without loss of generality, we can consider the following intersection of a pair of triangle facets: $F = Tr_{12,3} \cap Tr_{13,2}$. We have:

$$v \in F = Tr_{12,3} \cap Tr_{13,2}$$

$$\Rightarrow \begin{cases} v_{12} - v_{13} - v_{23} \leq 0 \\ v_{13} - v_{12} - v_{23} \leq 0 \end{cases}$$

$$\Rightarrow \begin{cases} v_{23} = 0 \\ v_{13} = v_{12} \end{cases}$$

using $0 \leq v_{ij} \leq 1$ for $1 \leq i < j \leq n$

$$\Rightarrow \begin{cases} v_{23} = 0 \\ v_{i3} = v_{i2} \text{ for } i \notin \{2,3\} \end{cases}$$

$$\Rightarrow \begin{cases} v \in \phi_0(MetP_{n-1}) \\ v \in (T_{2i,3} \cap Tr_{3i,2}) \text{ for } i \notin \{2,3\} \end{cases}$$

On the other hand, we obviously have: $\phi_0(MetP_{n-1}) \subset (Tr_{12,3} \cap Tr_{13,2})$. Then, since the map ϕ_δ preserves the faces of $MetP_{n-1}$ as well as their dimension [16], $F = \phi_0(MetP_{n-1})$ implies that F, which is called a weak triangle face, is a face of dimension $\binom{n-1}{2}$ combinatorially equivalent to $MetP_{n-1}$.

Case $\sigma = 4$
We first recall the definitions of the intersection vector $\pi(S)$ and the characteristic vector $\chi(S)$. For $S \subset V_n$, $\pi(S) \in \{0,1\}^{\binom{n+1}{2}}$ is defined by: $\pi(S)_{ij} = 1$ if $\{i,j\} \subset S$

and 0 otherwise, and $\chi(S) \in \{0,1\}^n$ is defined by: $\chi(S)_i = 1$ if $\{i\} \subset S$ and 0 otherwise. We also recall that the linear independence of the characteristic vectors $\chi(S_i)$ for $i = 1, \ldots, n$ implies the linear independence of the cuts $\delta(S_i)$ for $i = 1, \ldots, n$, and that linear independence of the intersection vectors $\pi(S_i)$ for $i = 1, \ldots, n$ is equivalent to the linear independence of the cuts $\delta(S_i)$ for $i = 1, \ldots, n$.

We consider the intersection of a pair of non-conflicting triangle facets, for example: $F = Tr_{23,1} \cap Tr_{23,4}$. Since F is the intersection of 2 facets, its codimension, $codim(F)$, is at least 2. We show that $codim(F)$ is exactly equal to 2 by exhibiting, besides the cut $\delta(\emptyset)$, a family of $\binom{n}{2}-2$ linearly independent cuts of the metric polytope $MetP_n$ belonging to F. For $n = 4$, the vertices of F are $\delta(\emptyset), \delta(2), \delta(3), \delta(1,2)$ and $\delta(1,3)$, and one can easily check that the last 4 cuts are linearly independent. Then, assume that F contains $\binom{n}{2}-2$ linearly independent nonzero cuts of $MetP_n$: $\delta(S_i)$ for $i = 1, \ldots, \binom{n}{2}-2$. Then, in addition to the $\binom{n}{2}-2$ following nonzero cuts of $MetP_{n+1}$: $\delta(S_i)$ for $i = 1, \ldots, \binom{n}{2}-2$, we consider the n cuts of $MetP_{n+1}$: $\delta(S_i \cup \{n+1\})$ for $i = 1, \ldots, n$. To check the linear independence of these $\binom{n}{2}-2+n = \binom{n+1}{2}-2$ cuts of $MetP_{n+1}$, we consider the matrix I_{n+1} of their associated intersection vectors. With the last $(n+1)$ columns corresponding to the coordinates $\delta_{i,n+1}$ for $i = 1, \ldots, n+1$ and the last n rows corresponding to $\pi(S_i \cup \{n+1\})$ for $i = 1, \ldots, n$, we have:

$$I_{n+1} = \begin{pmatrix} I_n & 0 \\ A & X_n \end{pmatrix}$$

where I_n is the matrix of the intersection vectors associated to the cuts of the metric polytope $MetP_n$: $\delta(S_i)$ for $i = 1, \ldots, n$ and X_n the matrix of the characteristic vectors associated to the cuts of $MetP_{n+1}$: $\delta(S_i \cup \{n+1\})$ for $i = 1, \ldots, n$. Since $rank(I_n) = \binom{n}{2}-2$ and $rank(X_n) = n$, we have $rank(I_{n+1}) = \binom{n}{2}-2+n = \binom{n+1}{2}-2$. So, for $n \geq 4$ the codimension of the intersection of a pair of non-conflicting triangle facets is 2.

Then, we consider the intersection of a pair of conflicting triangle facets, for example: $F = Tr_{23,1} \cap Tr_{24,3}$. Exactly in the same way as we did for the non-conflicting case, we can show that the codimension of F cannot increase with n. Since for $n = 4$, the vertices of F are $\delta(\emptyset), \delta(2), \delta(4)$ and $\delta(1,2)$, the codimension of F is 3. It implies that for $n \geq 4$, $codim(F) \leq 3$. One can easily check that $F = Tr_{23,1} \cap Tr_{24,3} = Tr_{24,1} \cap Tr_{14,3}$, which implies that F, being an intersection of more than 3 facets, can not be a ridge, i.e. $codim(F) = 3$.

Case $\sigma = 5$ *or* 6

For $\sigma = 5$ or 6, any pair of triangle facets is non-conflicting. As for the case of a non-conflicting pair for $\sigma = 4$, we first determine the dimension of F for $n = 5$ or 6. It turns out that F is always a ridge. For example, $F = Tr_{23,1} \cap Tr_{34,5}$ contains the cuts: $\delta(\emptyset)$, $\delta(2), \delta(3), \delta(4), \delta(1,2), \delta(1,3), \delta(2,4), \delta(4,5)$ and $\delta(3,5)$ and one can check that the last 8 cuts are linearly independent. Then, using once again the fact that the codimension of F cannot increase with n, we obtain that, for $n \geq 5$ or 6, F is a ridge.

Remark 5.1 *From the proof of Lemma 2.1, one can easily check that a weak triangle face contains* 2^{n-2} *cuts, a face resulting from the intersection of a pair of conflicting facets with* $\sigma = 4$ *contains* 2^{n-2} *cuts, and that a ridge contains either* $5 \cdot 2^{n-4}$ *cuts for* $\sigma = 4$*, or* $9 \cdot 2^{n-5}$ *for* $\sigma = 5$ *or* 6.

5.2. Proof of Theorem 2.2

Theorem 2.2 is a direct consequence of the fact that, for $n \geq 4$, a pair of triangle facets are not ridge-adjacent if and only if they are conflicting. For example, the 3 neighbours of the facet Tr_{123} in $\bar{G}_n$, the complement of the ridge graph of the metric polytope, corresponding to the case $\sigma = 3$, i.e. $Tr_{12,3}, Tr_{13,2}$ and $Tr_{23,1}$ form the common K_3 of a bouquet of $(n-3)$ (3×3)-grids. The $6(n-3)$ other neighbours, corresponding to the case $\sigma = 4$, belong to $(n-3)$ (3×3)-grids. For $i = 4, \ldots, n$, each grid is:

$$\begin{array}{ccc}
| & | & | \\
-Tr_{3i,2}- & Tr_{2i,3}- & Tr_{12,3}- \\
| & | & | \\
-Tr_{1i,3}- & Tr_{3i,1}- & Tr_{13,2}- \\
| & | & | \\
-Tr_{2i,1}- & Tr_{1i,2}- & Tr_{23,1}- \\
| & | & |
\end{array}$$

References

1. Avis D. (1980) 'On the extreme rays of the metric cone', *Canadian Journal of Mathematics XXXII* **1**, pp 126-144.
2. Assouad P. and Deza M. (1982) 'Metric subspaces of L^1', *Publications mathématiques d'Orsay* **3**.
3. Barahona F. and Mahjoub A.R. (1986) 'On the cut polytope', *Mathematical Programming* **36**, pp 157-173.
4. Brouwer A.E., Cohen A.M. and Neumaier A. (1989) *Distance-Regular Graphs*, Springer Verlag, Berlin.
5. Conforti M., Rao M.R. and Sassano A. (1990) 'The equipartition polytope I and II', *Mathematical Programming* **49**, pp 49-70 and 71-90.
6. Deza M. (1973) 'Matrices de formes quadratiques non négatives pour des arguments binaires', *Comptes rendus de l'Académie des Sciences de Paris* **277**, pp 873-875.
7. Deza A. and Deza M. (1993) 'On the skeleton of the dual cut polytope', Report LIENS 93-26, Ecole Normale Supérieure, Paris.
8. Deza M., Fukuda K. and Laurent M. (1993) 'The inequicut cone', *Discrete Mathematics* **119**, pp 21-48.

9. Deza M., Grishukhin V.P. and Laurent M. (1991) 'The symmetries of the cut polytope and of some relatives', in P. Gritzmann and B. Sturmfels (eds.), *Applied Geometry and Discrete Mathematics, the "Victor Klee Festschrift" DIMACS Series in Discrete Mathematics and Theoretical Computer Science* **4**, pp 205-220.
10. Deza M. and Laurent M. (1992) 'Facets for the cut cone I', *Mathematical Programming* **56 (2)**, pp 121-160.
11. Deza M. and Laurent M. (1992) 'Applications of cut polyhedra' Report LIENS 92-18, Ecole Normale Supérieure, Paris, *to appear in Journal of Computational and Applied Mathematics.*
12. Deza M., Laurent M. and Poljak S. (1992) 'The cut cone III: on the role of triangle facets', *Graphs and Combinatorics* **8**, pp 125-142.
13. Grishukhin V.P. (1992) 'Computing extreme rays of the metric cone for seven points', *European Journal of Combinatorics* **13**, pp 153-165.
14. Grishukhin V.P., private communication.
15. Iri M. (1970-71) 'On an extension of maximum-flow minimum-cut theorem to multicommodity flows', *Journal of the Operational Society of Japan* **13**, pp 129-135.
16. Laurent M. (1991) 'Graphic vertices of the metric polytope', Research report No. 91737-OR, Institut für Diskrete Mathematik, Universität Bonn.
17. Laurent M. and Poljak S. (1992) 'The metric polytope', in E. Balas, G. Cornuejols and R. Kannan (eds.), *Integer Programming and Combinatorial Optimization*, Carnegie Mellon University, GSIA, Pittsburgh, pp 274-286.
18. Padberg M. (1989) 'The boolean quadratic polytope: some characteristics, facets and relatives', *Mathematical Programming* **45**, pp 139-172.
19. Trubin V. (1969) 'On a method of solution of integer linear problems of a special kind', *Soviet Mathematics Doklady* **10**, pp 1544-1546.

ON THE COMPLEXITY OF SOME BASIC PROBLEMS IN COMPUTATIONAL CONVEXITY:

II. Volume and mixed volumes

PETER GRITZMANN*
Fb IV, Mathematik
Universität Trier
D-54286 Trier, Germany

and

VICTOR KLEE*
Department of Mathematics
University of Washington
Seattle, WA 98195, U.S.A.

Introduction

This paper is the second part of a broader survey of computational convexity, an area of mathematics that has crystallized around a variety of results, problems and applications involving interactions among convex geometry, mathematical programming and computer science. The first part [GrK94a] discussed containment problems. This second part is concerned with computing volumes and mixed volumes of convex polytopes and more general convex bodies. In order to keep the paper self-contained we repeat some aspects that have already been mentioned in [GrK94a]. However, this overlap is limited to Section 1. For further background material and references, see [GrK94a], and for other parts of the survey see [GrK94b] and [GrK94c].

Our section headings are as follows.

* Research of the authors was supported in part by the Deutsche Forschungsgemeinschaft and in part by a joint Max-Planck Research Award. Research of the second author was also supported in part by the National Science Foundation, U.S.A.

T. Bisztriczky et al. (eds.), POLYTOPES: Abstract, Convex and Computational, 373–466.

Acknowledgement. The authors gratefully acknowledge valuable comments and references from T. Burger, P. Goodey, A. Hufnagel, L. Lovász, E. Sachs and E. Weidner.

1. Preliminaries

1.1. What is Computational Convexity?

The subject of Computational Convexity draws its methods from discrete mathematics and convex geometry, and many of its problems from operations research, computer science, and other applied areas. In essence, it is the study of the computational and algorithmic aspects of high-dimensional convex bodies (especially polytopes), with a view to applying the knowledge gained to bodies that arise in other mathematical disciplines or in the mathematical modeling of problems from outside mathematics.

One of the requirements for turning a potential application into a real one is to have efficient algorithms for computing (or approximating or bounding) the functionals involved. The subject of computational convexity is centered on the search for such algorithms and on results concerning the intrinsic complexity of the functionals that may help to guide this search. Basic and typical problems deal with the efficient computation or approximation of geometric functionals such as the volume or the diameter of a polytope, or with the algorithmic reconstruction of a polytope from data concerning it, or with algorithmic versions of geometric theorems.

The name *Computational Convexity* is of recent origin, having first appeared in print in 1989 (see [GrK89]). However, results that retrospectively belong to this area go back a long way. In particular, many of the basic ideas of *Linear Programming* have an essentially geometric character and fit very well into our conception of Computational Convexity. The same is true of the subject of *Polyhedral Combinatorics* and of the *Algorithmic Theory of Polytopes and Convex Bodies*; see [GrK94a] for a brief survey of these areas and a list of references to related research papers, survey articles and books.

As opposed to the area that has come to be called *Computational Geometry* ([PrS85], [Ed87], [Ya90], [Ed93a], [Mu94], [Or94a]), the emphasis in Computational Convexity is on problems whose underlying structure is convexity in normed vector spaces of finite but generally *not* restricted dimension, rather than of fixed (low) dimension. This leads to much closer connections with the optimization problems that arise in a wide variety of disciplines.

In the study of Computational Convexity, the underlying model of computation is mainly the binary (Turing machine) model that is common in studies of computational complexity. This requirement is imposed by prospective applications, particularly in mathematical programming. For the study of algorithmic aspects of convex bodies the binary model is often augmented by additional devices called "oracles"; see Subsection 1.2 and [GrK94a]. Some cases of interest do involve other underlying models of computation, but the present paper focuses on aspects of computational convexity for which binary models seem most natural.

1.2. Presentations of polytopes and general convex bodies

The setting for everything in this paper is a finite-dimensional real vector space $\mathbb{R}^n$. In the present context, $\mathbb{R}^n$ may be assumed to carry the usual Euclidean norm, thus forming the Euclidean n-space $\mathbb{E}^n = (\mathbb{R}^n, \|\ \ \|_2)$.

As the terms are used here, a *body* in $\mathbb{R}^n$ is a compact convex set and a *polytope* is a body that has only finitely many extreme points. (These objects are usually called *convex bodies* and *convex polytopes*, but we often omit the adjectives in the interest of brevity.) We use the symbols $\mathcal{K}^n$ and $\mathcal{P}^n$ to denote respectively the family of all bodies in $\mathbb{R}^n$ and the family of all polytopes in $\mathbb{R}^n$.

A body or a polytope in $\mathbb{R}^n$ is *proper* if it is n-dimensional and hence has nonempty interior. From an algorithmic point of view, polytopes are dealt with much more easily than general bodies, because polytopes can be presented in a finite manner. However, even for a polytope P the precise manner of presentation must be specified, and the difficulty of answering basic questions about P can be greatly influenced by the manner of presentation.

A *$\mathcal{V}$-presentation* of a polytope P consists of positive integers n and m, and m points $v_1, \ldots, v_m$ in $\mathbb{R}^n$ such that $P = \text{conv}\{v_1, \ldots, v_m\}$. An *$\mathcal{H}$-presentation* of a polytope P consists of integers n and m with $m > n \geq 1$, a real $m \times n$ matrix A, and a vector $b \in \mathbb{R}^m$ such that $P = \{x \in \mathbb{R}^n : Ax \leq b\}$.

A *face* of a polytope P is P itself, the empty set, or the intersection of P with some supporting hyperplane. Faces of dimension i are called *i-faces* (with the convention that $\dim(\emptyset) = -1$). The 0-faces, 1-faces, and $(n-1)$-faces of an n-polytope P are respectively its *vertices*, *edges*, and *facets*. For $i = -1, 0, \ldots, n$, $\mathcal{F}_i(P)$ denotes the set of all i-faces of P and $f_i(P) = \text{card}(\mathcal{F}_i(P))$, the number of i-faces.

A $\mathcal{V}$- $\langle\mathcal{H}$-$\rangle$ presentation of P is *irredundant* if the omission of any of the points $v_1, \ldots, v_m$ $\langle$any of the inequalities in $Ax \leq b\rangle$ changes the polytope, reducing it in the first case and enlarging it in the second. In geometric terms, a $\mathcal{V}$-presentation is irredundant if each point v_i is a vertex of P, and when P is n-dimensional, an $\mathcal{H}$-presentation is irredundant if each inequality induces a facet of P.

Each polytope $P \subset \mathbb{R}^n$ admits a $\mathcal{V}$-presentation and also admits an $\mathcal{H}$-presentation, and we refer to [Dy83], [Sw85], [Se87], [AvF91], and [ChHJ92] for algorithms that convert one sort of presentation into the other. However, since P may have many more vertices than facets, and vice-versa [Mc70] (see also [Gr67], [McS71], [Br83]), it can happen that the minimum size of one sort of presentation is much larger than the minimum size of the other sort.

Our discussion here is based mainly on the *binary* or *Turing machine* model of computation [GaJ79], in which the *size of the input* is defined as the length of the binary encoding needed to present the input data to a Turing machine and the *time-complexity* of an algorithm is also defined in terms of the operations of a Turing machine. For this model, each computation involving polytopes begins with a *rational* $\mathcal{V}$- or $\mathcal{H}$-presentation of a *rational* polytope P. The presentation's being rational means that $v_1, \ldots, v_m \in \mathbb{Q}^n$ if P is $\mathcal{V}$-presented, or that the matrix A has rational entries and $b \in \mathbb{Q}^m$ if P is $\mathcal{H}$-presented; integer presentations are defined in a similar way. The *(binary) size* of a rational $\mathcal{V}$- or $\mathcal{H}$-presentation – usually denoted by L (for "length") – is the number of binary digits needed to encode the data of the presentation; see e.g. [GrLS88].

As was mentioned earlier, our main emphasis is on questions involving the dimension n as part of the input. However, we will also mention corresponding results for fixed dimensions. As a notational convention, we restrict the use of L to the case of varying dimension; further, whenever the dimension is regarded as fixed, we explicitly say so. Thus when no assumption on the dimension is stated, the dimension

n is always regarded as part of the input.

For algorithmic purposes it is usually not the rational polytope P as a *geometric* object that is relevant, but rather its *algebraic presentation*. We will speak of a *$\mathcal{V}$-polytope P* or of an *$\mathcal{H}$-polytope P* when a specific rational $\mathcal{V}$-presentation $(n, m; v_1, \ldots, v_m)$ or a specific rational $\mathcal{H}$-presentation $(n, m; A, b)$ is given. For most of the problems discussed here the focus is on polynomial-time computability or on various hardness results, and hence we may assume without loss of generality that presentations are irredundant. That is because, for a given $\mathcal{V}$- (or $\mathcal{H}$-) polytope P, linear programming can be used to produce, in polynomial time, an irredundant $\mathcal{V}$- (or $\mathcal{H}$-) presentation.

In order to further illuminate the algorithmic differences between $\mathcal{V}$- and $\mathcal{H}$-polytopes and the difficulties that may be expected in attempting to transfer an algorithmic approach from one sort of presentation to the other, we mention that LINIAL [Li86] has established the $\#\mathbb{P}$-completeness (a strong measure of difficulty) of each of the following two problems:

> Given a positive integer n and an n-dimensional $\mathcal{V}$-polytope P, determine the number of facets of P.
>
> Given a positive integer n and an n-dimensional $\mathcal{H}$-polytope P, determine the number of vertices of P.

Among important special classes of polytopes, the zonotopes are particularly interesting because they can be so compactly presented. A *zonotope* is the vector sum (Minkowski sum) of a finite number of line segments. When the segments are $S_1, \ldots, S_m$ and their centers are $c_1, \ldots, c_m$, we have

$$\sum_{i=1}^{m} S_i = \left(\sum_{i=1}^{m} c_i\right) + \sum_{i=1}^{m}(S_i - c_i)$$

where the point $\sum_{i=1}^{m} c_i$ is the center of symmetry of the set $\sum_{i=1}^{m} S_i$ and each segment $S_i - c_i$ is centered at the origin. Hence it is convenient to define an *$\mathcal{S}$-presentation* of a zonotope in $\mathbb{R}^n$ as a sequence $(c; z_1, \ldots, z_m)$ of points in $\mathbb{R}^n$, where c is the center and the z_i are the ends of segments centered at the origin 0, with one end listed for each segment. This sequence represents the zonotope

$$c + \sum_{i=1}^{m}[-1, 1]z_i = c + \left\{\sum_{i=1}^{m} \lambda_i z_i : |\lambda_i| \le 1 \text{ for all } i\right\}.$$

This zonotope is a polytope of dimension at most $\min\{m, n\}$, and it is a proper polytope if and only if there are n linearly independent points among the z_i. We speak of *integer* and *rational* $\mathcal{S}$-presentations, and define their sizes in the natural way. Sometimes we will also work with zonotopes whose relationship to the origin (and whose scaling) is different. Specifically, zonotopes of the form $\sum_{i=1}^{m}[0, 1]z_i$ are used quite frequently. To keep the notation simple, we refrain, however, from introducing an additional name for such a presentation.

Again, in our algorithmic model it is usually not the rational zonotope Z as a geometric object that is relevant, but rather its presentation; hence we speak of an *$\mathcal{S}$-zonotope Z* when a specific rational $\mathcal{S}$-presentation $(n, m; c; z_1, \ldots, z_m)$ is given. Although each zonotope is a polytope, in general neither the vertices nor the facets of a zonotope are readily accessible from an $\mathcal{S}$-presentation. In fact, for zonotopes generated by m segments in general position, both the number of facets and the number of vertices grow exponentially as m increases.

A zonotope $Z = c + \sum_{i=1}^{m}[-1,1]z_i$ is called a *parallelotope* when the points $z_1, \ldots, z_m$ are linearly independent. In contrast to the case of general zonotopes, the facial structure of a parallelotope is immediately accessible from an $\mathcal{S}$-presentation: The passage between a rational $\mathcal{S}$-presentation of a parallelotope P and a rational $\mathcal{H}$-presentation of P can be accomplished in polynomial time.

A different approach is required to deal with bodies K that are not polytopes, since an enumeration of all the extreme points of K or of K's polar is not possible. Sometimes K can be described explicitly in terms of an easily computable function. An example of this kind is the Euclidean unit ball $\mathbb{B}^n$ given in the form $\mathbb{B}^n = \{(\xi_1, \ldots, \xi_n)^T : \xi_1^2 + \ldots + \xi_n^2 \leq 1\}$. However, such a description is often not available. A convenient way to deal with the general situation is to assume that the body in question is given by an algorithm (called an *oracle*) that answers certain sorts of questions about the body. All information about the specific body must be obtained from the oracle, which functions as a "black box." In other words, while it is assumed that the oracle's answers are always correct, nothing is assumed about the manner in which it produces those answers. This oracular approach has been extensively studied and utilized for combinatorial optimization problems by GRÖTSCHEL, LOVÁSZ & SCHRIJVER [GrLS81], [GrLS88]. In order to describe some oracles that have figured prominently in their work, let us recall that for $\epsilon \geq 0$ the *outer parallel body* and the *inner parallel body* of a convex body K are given respectively by

$$K(\epsilon) = K + \epsilon\mathbb{B}^n = \bigcup_{b \in \epsilon\mathbb{B}^n} (K + b) \quad \text{and} \quad K(-\epsilon) = K \setminus \bigcup_{b \in \epsilon\mathbb{B}^n} ((\mathbb{R}^n \setminus K) + b).$$

The three most important oracles of [GrLS88] are the ones that solve the following problems for proper bodies K.

WEAK MEMBERSHIP PROBLEM. *Given $K \in \mathcal{K}^n$, $y \in \mathbb{Q}^n$, and a rational number $\epsilon > 0$, conclude with one of the following:*

assert that $y \in K(\epsilon)$;
assert that $y \notin K(-\epsilon)$.

WEAK SEPARATION PROBLEM. *Given $K \in \mathcal{K}^n$, $y \in \mathbb{Q}^n$, and a rational number $\epsilon > 0$, conclude with one of the following:*

assert that $y \in K(\epsilon)$;
find a vector $z \in \mathbb{Q}^n$ such that $\| z \|_\infty = 1$ and $z^T x < z^T y + \epsilon$ for every $x \in K(-\epsilon)$.

WEAK (LINEAR) OPTIMIZATION PROBLEM. *Given $K \in \mathcal{K}^n$, a vector $c \in \mathbb{Q}^n$, and a rational number $\epsilon > 0$, conclude with one of the following:*

find a vector $y \in \mathbb{Q}^n \cap K(\epsilon)$ such that $c^T x \leq c^T y + \epsilon$ for every $x \in K(-\epsilon)$;

assert that $K(-\epsilon) = \emptyset$.

If a proper body K is given by an algorithm that solves the weak membership problem, the weak separation problem, or the weak linear optimization problem, we say that K is described by a *weak membership oracle*, a *weak separation oracle* or a *weak (linear) optimization oracle*. The oracle is called *strong* if it solves the corresponding strong problem that is obtained by setting $\epsilon = 0$. A body K is called *circumscribed* or *well-bounded* if a positive rational number R or positive rational numbers r, R are given explicitly such that $K \subset R\mathbb{B}^n$ or, in addition, K contains a ball of radius r. If, further, we are given a vector $b \in \mathbb{Q}^n$ such that $b + r\mathbb{B}^n \subset K$, then K is called *centered*.

To place the weak linear optimization oracle in the perspective of classical convexity theory, recall Minkowski's useful functional approach to convex bodies by means of the *support function* $h : \mathcal{K}^n \times \mathbb{S}^{n-1} \to \mathbb{R}$, which is defined for $K \in \mathcal{K}^n$ and $u \in \mathbb{S}^{n-1} = \text{bd}\,\mathbb{B}^n$ by

$$h(K, u) = \max_{x \in K} \langle x, u \rangle.$$

Note that presenting a $c \in \mathbb{Q}^n$ as a call to a weak optimization oracle for K provides us with an approximation of $h(K, c/\|c\|_2)$, and, in addition, with a "weak support point" in this direction.

The above three problems are very closely related in the sense that when the classes of proper bodies are appropriately restricted to those that are circumscribed, well-bounded, or centered, and when input sizes are properly defined, an algorithm that solves any one of the problems in polynomial time can be used as a subroutine to solve the others in polynomial time also; see Theorem 1.2.1. The definition of input size involves the size of ϵ, the dimension of K, the given a priori information, and the input required by the oracle. Suppose that K is well-bounded with parameters r and R. Then the *input size* is defined as the sum of the following numbers:

size(ϵ);
size(K) $= n + \text{size}(r) + \text{size}(R)$;
an additional term size(b) when the oracle is centered with center b;
an additional term size(y) for the membership problem and the separation problem, and size(c) for the optimization problem.

The following theorem contains a list of the precise relationships among the three basic oracles for bodies (see [GrLS88]). The notation "$(\mathcal{A}; \textit{prop}) \to_\pi \mathcal{B}$" indicates the existence of an (oracle-) polynomial-time algorithm that solves problem $\mathcal{B}$ for every body that is given by the oracle $\mathcal{A}$ and has all the properties specified in *prop*. (*prop*$= \emptyset$ means that the statement holds for general bodies.)

1.2.1 (WEAK MEMBERSHIP; centered, well-bounded) $\to_\pi$ WEAK SEPARATION;
(WEAK MEMBERSHIP; centered, well-bounded) $\to_\pi$ WEAK OPTIMIZATION;
(WEAK SEPARATION; $\emptyset$) $\to_\pi$ WEAK MEMBERSHIP;
(WEAK SEPARATION; circumscribed) $\to_\pi$ WEAK OPTIMIZATION;
(WEAK OPTIMIZATION; $\emptyset$) $\to_\pi$ WEAK MEMBERSHIP;
(WEAK OPTIMIZATION; $\emptyset$) $\to_\pi$ WEAK SEPARATION.

We want to emphasize the following fact, for it implies that the "oracular" approach to convex bodies is in an important sense the most general sort of presentation introduced in this subsection. It also helps to clarify the way in which the formulation in terms of oracles leads to an efficient modular approach to the problems of computational convexity.

1.2.2 *There are polynomial-time algorithms which, accepting as input a proper $\mathcal{V}$-polytope or a proper $\mathcal{H}$-polytope P, or a proper $\mathcal{S}$-zonotope Z, produce membership, separation and optimization oracles for P and Z, and also compute lower bounds on P's and Z's inradius, upper bounds on P's and Z's circumradius, and "centers" b_P and b_Z for P and Z respectively.*

This implies that if an algorithm performs certain tasks for bodies given by some of the above (appropriately specified) oracles, then the same algorithm can also serve as a basis for procedures that perform these tasks for $\mathcal{V}$- or $\mathcal{H}$-polytopes and for $\mathcal{S}$-zonotopes. (On the other hand, it is possible to derive some lower bounds on the performance of approximate algorithms for the oracle model that do not carry over to the case of $\mathcal{V}$- or $\mathcal{H}$-polytopes or $\mathcal{S}$-zonotopes. Examples of this kind can be found in Subsection 6.3.)

Let us end this section by stating some basic algorithmic problems of volume computation. Other variants of these problems (including those asking for weak approximations) will be introduced later. Here are the problems.

VOLUME COMPUTATION

Instance: *A positive integer n, an $\mathcal{H}$-polytope $\langle$or a $\mathcal{V}$-polytope, or an $\mathcal{S}$-zonotope$\rangle$ P.*

Task: *Compute the volume $V(P)$ of P.*

WEAK VOLUME COMPUTATION

Instance: *A positive integer n, a body K in $\mathbb{R}^n$ that is given by a weak optimization oracle $\langle$or a weak membership oracle or a weak separation oracle; a rational vector $b \in \mathbb{R}^n$ and positive rationals r, R such that $b + r\mathbb{B}^n \subset K \subset R\mathbb{B}^n\rangle$; a positive rational λ.*

Task: *Compute a rational μ such that $|\mu - V(K)| \leq \lambda$.*

It should be emphasized that this survey concentrates on providing some idea of the principal methods that are available for computing or approximating volumes and mixed volumes, and sketches (or, much less frequently, details) of proofs are given only for the purpose of enhancing the intuitive understanding of these underlying ideas and concepts. To further research in this fruitful area of computational convexity we have formulated unsolved problems that seem especially natural or important, and in some cases of particular interest we have even included some "speculative" material, speculative in the sense that we show how certain procedures (which may not be available at present) could in principle be used to solve certain other problems efficiently. Finally, it should be mentioned that much of this survey is "qualitative" in the sense that the primary distinction in computational complexity is that between polynomial-time solvability on the one hand and $\mathbb{NP}$-hardness or

#ℙ-hardness on the other hand. We recognize that this classification is only a first step toward finding optimal algorithms, but we believe it to be a useful guide for the latter effort.

2. Foundations

In the present section, Subsection 2.1 fixes some terminology and Subsections 2.3–2.5 discuss the aspects of volume and mixed volumes that are most relevant to what follows. Subsection 2.2 represents a deviation (but a fascinating one) from our main line of discussion. Much of the material in 2.2 has been treated in books by BOLTYANSKII [Bo78] and WAGON [Wa85]. The material in 2.3–2.5 has been the subject of various books and survey articles, including the book by HADWIGER [Ha57], the survey by MCMULLEN & SCHNEIDER [McS83], and a recent handbook article by MCMULLEN [Mc93]. For this reason, and also because the present article is concerned mainly with algorithmic aspects, we will be rather brief in this section. More details can be found in the cited references.

2.1. BACKGROUND AND TERMINOLOGY

We could begin by simply noting that convex bodies are Lebesgue measurable, and that our term *volume* is synonymous to *Lebesgue measure*. However, when restricted to bodies and especially when restricted to polytopes, Lebesgue measure exhibits many properties that are of fundamental geometric significance. Further, these properties can in some cases be formulated in an "elementary" way – i.e., without recourse to limiting processes – and it turns out that some of the notation and terminology needed to describe the properties is also useful for algorithmic studies.

We speak of a *dissection* of an n-polytope P into n-polytopes $P_1, \ldots, P_k$ if

$$P = \bigcup_{i=1}^{k} P_i$$

$$\operatorname{int}(P_i \cap P_j) = \emptyset \qquad \text{for } i, j = 1, \ldots, k, i \neq j.$$

With respect to a subgroup G of the group of all affine automorphisms of $\mathbb{R}^n$, two polytopes $P, Q \subset \mathbb{R}^n$ are said to be *G-equidissectable* (or *equidissectable under G*) if (for some k) there exist dissections $P_1, \ldots, P_k$ of P and $Q_1, \ldots, Q_k$ of Q and elements $g_1, \ldots, g_k$ of G such that $P_i = g_i(Q_i)$ for all i.

A related notion is that of equicomplementability. Two polytopes $P, Q \subset \mathbb{R}^n$ are called *G-equicomplementable* if there are polytopes P_1, P_2 and Q_1, Q_2 such that P_2 is dissected into P and P_1, Q_2 is dissected into Q and Q_1, P_1 and Q_1 are G-equidissectable, and P_2 and Q_2 are G-equidissectable. HADWIGER [Ha57, p.47] showed that two polytopes are G-equidissectable if and only if they are G-equicomplementable.

Let $\mathcal{S}^n$ be a family of subsets of $\mathbb{R}^n$. A functional $\varphi : \mathcal{S}^n \to \mathbb{R}$ is called a *valuation* on $\mathcal{S}^n$ if

$$\varphi(S_1) + \varphi(S_2) = \varphi(S_1 \cup S_2) + \varphi(S_1 \cap S_2) \qquad \text{whenever } S_1, S_2, S_1 \cup S_2, S_1 \cap S_2 \in \mathcal{S}^n.$$

The families of principal interest to us here are $\mathcal{P}^n$ and $\mathcal{K}^n$. A valuation φ is called *G-invariant* if

$$\varphi(S) = \varphi(g(S)) \text{ for all } S \in \mathcal{S}^n \text{ and } g \in G,$$

simple if

$$\varphi(S) = 0 \text{ whenever } S \in \mathcal{S}^n \text{ and } S \text{ is contained in a hyperplane,}$$

and *monotone* if

$$\varphi(S_1) \leq \varphi(S_2) \text{ whenever } S_1, S_2 \in \mathcal{S}^n \text{ with } S_1 \subset S_2.$$

Obviously, if φ is a G-invariant simple valuation on $\mathcal{P}^n$ and P and Q are G-equidissectable (or G-equicomplementable) then $\varphi(P) = \varphi(Q)$. HADWIGER [Ha57] showed that this leads already to a characterization of G-equidissectability of polytopes.

2.1.1 *Two polytopes P and Q are G-equidissectable if and only if $\varphi(P) = \varphi(Q)$ for all G-invariant simple valuations on $\mathcal{P}^n$.*

2.2. "ELEMENTARY" APPROACHES TO VOLUME

The present subsection states some results on isometry-based elementary approaches to volume and contrasts them later with a result on an affinity-based approach.

The most famous result concerning equidissectability involves the group D of isometries. It is the following *Bolyai-Gerwien theorem* (see [Ge1833], [Bo78], [Wa85]).

2.2.1 *Two plane polygons are of equal area if and only if they are D-equidissectable.*

If one agrees that an a-by-b rectangle should have area ab, and also agrees that the area function should be a D-invariant simple valuation, it then follows from 2.2.1 that the area of any plane polygon P can be determined (at least in theory) by finding a rectangle R to which P is equidissectable. This provides a satisfyingly geometric theory of area that does not require any limiting considerations. Several refinements of the Bolyai-Gerwien theorem have been established. For example, rather than using the group of all isometries, it suffices to use translations and half-turns. Also, the pairs (P,Q) of polygons that are equidissectable under translations alone have been characterized by HADWIGER & GLUR [HaG51]. (See [Bo78] for this and other refinements.) Although the original proof of 2.2.1 was algorithmic in nature, there remain open questions concerning how rapidly, under various hypotheses, one can find an equidissection of two given polygons of equal area, and there are also open questions concerning the minimum number of pieces needed in certain equidissections. See [KoK94] for some of the algorithmic aspects, and see [Mo91, p.215] for a problem concerning minimum dissections.

The third problem of HILBERT [Hi00] asked, in effect, whether the Bolyai-Gerwien result extends to 3-polytopes. A negative answer was supplied by DEHN [De00], who showed that a regular tetrahedron and a cube are not equidissectable. His work led to the notion of a *Dehn invariant* of a 3-polytope P. Let $f : \mathbb{R} \to \mathbb{R}$ be an arbitrary additive function such that $f(\pi) = 0$ but f is not identically zero. (This implies that f is discontinuous.) For each such f, and for each 3-polytope P, let

$$f^*(P) = \sum_{i=1}^{k} \sigma_i f(\alpha_i),$$

where $\sigma_1, \ldots, \sigma_k$ are the lengths of the various edges of P and $\alpha_1, \ldots, \alpha_k$ are the radian measures of the corresponding dihedral angles. Then the number $f^*(P)$ is known as the *Dehn invariant* of P associated with the functional f. In the following result, the necessity was proved by DEHN [De00] and the sufficiency by SYDLER [Sy65] 65 years later.

2.2.2 *For two 3-polytopes P and Q to be equidissectable under the group of all isometries of $\mathbb{R}^3$, it is necessary and sufficient that $f^*(P) = f^*(Q)$ for each additive function f such that $f(\pi) = 0$.*

Dehn's necessary conditions (for equidissectability of proper polytopes in 3-space) were extended to n-space by HADWIGER (see [Bo78] and [Sa79] for references), and the sufficiency of the extended conditions was proved by JESSEN [Je68], [Je72] when $n = 4$. However, the case of $n \geq 5$ is still unsettled. See [Bo78] and [Sa79] for expositions of Jessen's proof, and see SAH [Sa79] for an account of algebraic studies that have been inspired by Hilbert's third problem.

A notion related to G-equidissectability is that of *G-equidecomposability*, where a *decomposition* of a set X is a way of expressing X as a union $X_1 \cup \ldots \cup X_k$ of a finite number of pairwise disjoint sets X_i. In contrast to the notion of a dissection, these sets X_i are not even permitted to have boundary points in common, and there is no restriction on the nature of the individual sets (they may even be nonmeasurable). Hence the study of equidecomposability is far from our algorithmic approach. Nevertheless, we feel that its principal results should at least be mentioned here because they are the most striking of all results related to volume.

Even though equidecomposability does not require measurability of the sets in the decomposition, in the following result these sets may be taken as open triangles together with nice portions of their boundaries.

2.2.3 *Two plane polygons are of equal area if and only if they are D-equidecomposable.*

Theorem 2.2.3 is due to TARSKI (see [BaT24]), and it led to the question as to which pairs of nonpolygonal plane bodies are D-equidecomposable. Although the sets in a decomposition need not be measurable, equidecomposability of two plane bodies does imply that the bodies are of equal area. That is a consequence of the following result.

2.2.4 *Lebesgue measure on the line or in the plane can be extended to a D-invariant simple monotone valuation μ that is defined and finite for all bounded sets.*

In fact, μ can also be required to multiply properly with respect to all similarity transformations of $\mathbb{R}^2$. See [Wa85] for references to proofs of 2.2.4, and for a discussion of further ramifications of the theorem.

LACZKOVICH [La90] sharpened 2.2.3 as follows.

2.2.5 *Any two plane polygons of equal area are equidecomposable under the group of translations.*

He also settled Tarski's old problem of "squaring the circle" by showing that a square and a circular disk of equal area are equidecomposable. There too, he needed only translations. (See [GaW89] for an expositon of the methods and theorems of [La90].) On the other hand, a theorem of DUBINS, HIRSCH & KARUSH [DuHK63] implies that a disk and a square cannot be "scissors congruent"; i.e., there is no equidissection (with respect to rigid motions) into pieces which, roughly speaking, could be cut out with a pair of scissors.

The relationship of equidecomposability to volume in $\mathbb{R}^n$ changes dramatically with the passage from $n = 2$ to $n = 3$. That is clear from the following result, which is known as the *Banach-Tarski paradox* [BaT24].

2.2.6 *If X and Y are bounded subsets of $\mathbb{R}^n$ (with $n \geq 3$), and each set has nonempty interior, then X and Y are D-equidecomposable.*

The essential difference between the cases $n \leq 2$ and $n \geq 3$ lies in the fact that for $n \leq 2$ the group of all isometries of $\mathbb{R}^n$ is *solvable* (a condition of near-commutativity), while for $n \geq 3$ it contains a free nonabelian subgroup and hence is not solvable. Thus, for example, while it is clear from 2.2.6 that 2.2.4 does not (as stated) extend to $\mathbb{R}^n$, it does extend when the group D of all isometries is replaced by the group of translations. See WAGON [Wa85] for a survey of the Banach-Tarski paradox and of several other results and problems related to the results stated in this subsection; see also [St79].

We see from 2.2.1 that an "elementary" theory of the area of plane polygons (i.e., a theory free of limiting processes) can be based on isometries and equidissectability, and we see from 2.2.3 that such a theory can also be based on isometries and equidecomposability. When the underlying group is the group D of isometries, both of these statements fail in $\mathbb{R}^n$ for each $n \geq 3$. Equidecomposability fails because (by 2.2.6) the associated equivalence class is far too large, having no connection with equality of volume. Equidissectability fails because, although this condition implies equality of volume, proper polytopes of equal volume can fail to be equidissectable (see 2.2.2). A pleasant contrast to these difficulties is provided by the following result (see [Mc93, p.966]), which is valid for all n. It is based on volume-preserving affinities rather than on isometries.

2.2.7 *Under the group of all volume-preserving affinites of $\mathbb{R}^n$, two n-polytopes are equidissectable if and only if they are of equal volume.*

For this result, as for 2.2.1, it seems that little is known about minimizing the number of pieces in an equidissection or about the computational complexity of finding an equidissection.

2.3. Characterizations of the volume

The volume function can be characterized as follows in terms of valuations.

2.3.1 *If φ is a translation-invariant, nonnegative simple valuation on $\mathcal{P}^n$ $\langle \mathcal{K}^n \rangle$, then there exists a nonnegative real α such that $\varphi = \alpha V$.*

2.3.2 *A translation-invariant valuation on $\mathcal{P}^n$ which is homogeneous of degree n is a constant multiple of the volume.*

2.3.3 *A continuous rigid-motion-invariant simple valuation on $\mathcal{K}^n$ is a constant multiple of the volume.*

2.3.4 *A nonnegative simple valuation on $\mathcal{P}^n$ $\langle \mathcal{K}^n \rangle$ which is invariant under all volume-preserving linear maps of $\mathbb{R}^n$ is a constant multiple of the volume.*

Proofs of 2.3.1, 2.3.2 and 2.3.3 can be found in the book of Hadwiger [Ha57], in Section 2.1.3 and on pages 79 and 221. Theorem 2.3.4 is also due to Hadwiger [Ha70]. See, in addition, the surveys [McS83] and [Mc93].

It is unknown whether, in 2.3.3, $\mathcal{K}^n$ can be replaced by $\mathcal{P}^n$.

2.4. Mixed volumes

The study of mixed volumes, the *Brunn-Minkowski theory*, forms the backbone of classical convexity theory and is also useful for applications in various other areas including combinatorics and algebraic geometry (see Section 9). Schneider [Sc93] has an excellent treatment of the theory that includes proofs for all the results presented in this section.

The starting point for the Brunn-Minkowski theory is the following famous theorem of Minkowski [Mi11] (see [BoF34], [Sa93], [Sc93]):

2.4.1 *Let $K_1, \dots, K_r$ be convex bodies in $\mathbb{R}^n$, and let $\xi_1, \dots, \xi_r$ be nonnegative reals. Then the function $V(\sum_{i=1}^r \xi_i K_i)$ is a homogeneous polynomial of degree n in the variables $\xi_1, \dots, \xi_r$, and can be written in the form*

$$V\left(\sum_{i=1}^r \xi_i K_i\right) = \sum_{i_1=1}^r \sum_{i_2=1}^r \cdots \sum_{i_n=1}^r \xi_{i_1}\xi_{i_2}\cdots\xi_{i_n} V(K_{i_1}, K_{i_2}, \dots, K_{i_n}),$$

where the coefficients $V(K_{i_1}, K_{i_2}, \dots, K_{i_n})$ are order-independent, i.e. invariant under permutations of their argument.

The coefficient $V(K_{i_1}, K_{i_2}, \dots, K_{i_n})$ is called the *mixed volume* of $K_{i_1}, K_{i_2}, \dots, K_{i_n}$. We will also use the term *mixed volume* for the functional

$$\begin{aligned} V : \overbrace{\mathcal{K}^n \times \ldots \times \mathcal{K}^n}^{n} &\to \mathbb{R} \\ K_1, \dots, K_n &\mapsto V(K_1, \dots, K_n) \end{aligned}$$

as well as for restrictions of this functional to certain subsets of $\mathcal{K}^n \times \ldots \times \mathcal{K}^n$.

Here are some of the most important properties of mixed volumes.

2.4.2 *Let $K_1, \ldots, K_n \in \mathcal{K}^n$. Then the mixed volumes have the following properties.*

(i) *Mixed volumes are nonnegative, monotone, multilinear, and continuous valuations.*

(ii) $V(K_1) = V(\overbrace{K_1, \ldots, K_1}^{n})$.

(iii) *If A is an affine transformation, then*

$$V(A(K_1), \ldots, A(K_n)) = |\det(A)| V(K_1, \ldots, K_n).$$

Property 2.4.2 (ii) shows that mixed volumes directly generalize the ordinary volume. This implies that in general, computing mixed volumes of polytopes is no easier than computing volumes of polytopes.

The multilinearity of the mixed volume is important for certain algorithmic approaches outlined in Subsection 7.2. It says that mixed volumes can again be expanded into mixed volumes, or, more explicitly, that for $C_1, \ldots, C_r \in \mathcal{K}^n$, $\xi_1, \ldots, \xi_r \in [0, \infty[$, $K = \sum_{i=1}^{r} \xi_i C_i$ and $k \in \mathbb{N}$, $k \leq n-1$,

$$V\,(\overbrace{K, \ldots, K}^{k}, K_{k+1}, \ldots, K_n) =$$

$$= \sum_{i_1=1}^{r} \sum_{i_2=1}^{r} \cdots \sum_{i_k=1}^{r} \xi_{i_1} \xi_{i_2} \cdots \xi_{i_k} V(C_{i_1}, \ldots, C_{i_k}, K_{k+1}, \ldots, K_n).$$

One of the most famous inequalities in convexity theory is the following, known as the *Aleksandrov-Fenchel inequality*, [Al37], [Al38], [Fe36].

2.4.3 *For $K_1, \ldots, K_n \in \mathcal{K}^n$, it is true that*

$$V(K_1, K_2, K_3, \ldots, K_n)^2 \geq V(K_1, K_1, K_3, \ldots, K_n)\; V(K_2, K_2, K_3, \ldots, K_n).$$

The cases for which equality holds in 2.4.3 have not been fully characterized; see [Sa93], [Sc93].

A famous consequence of Theorem 2.4.3 is the following *Brunn-Minkowski theorem* (see [Sa93], [Sc93]):

2.4.4 *If $K_0, K_1 \in \mathcal{K}^n$ and $\lambda \in [0,1]$, then*

$$V^{\frac{1}{n}}((1-\lambda)K_0 + \lambda K_1) \geq (1-\lambda)V^{\frac{1}{n}}(K_0) + \lambda V^{\frac{1}{n}}(K_1).$$

Let us close this subsection by introducing the *quermassintegrals* and the *intrinsic volumes* of a body K.

The quermassintegrals $W_0, \ldots, W_n$ are defined on $\mathcal{K}^n$ by

$$W_i(K) = V(\overbrace{K, \ldots, K}^{n-i}, \overbrace{\mathbb{B}^n, \ldots, \mathbb{B}^n}^{i}),$$

and the intrinsic volumes $V_0, \ldots, V_n$, are given by

$$\kappa_{n-i} V_i = \binom{n}{i} W_{n-i},$$

where κ_k denotes the k-volume of $\mathbb{B}^k$. (See [Ha57], [Mc75], [Mc77], [Mc93], [Sa93], [Sc93]). In particular, V_n is the volume V, V_{n-1} is half the surface area, and $V_0 = 1$. Note that the intrinsic volumes are *dimension-invariant* in the sense that $V_i(K)$ depends only on the body K and not on the dimension of the space in which K is embedded. This implies, in particular, that for an i-dimensional body K the ith intrinsic volume $V_i(K)$ is just the i-volume of K.

Note that for $\xi \geq 0$, the body $K + \xi\mathbb{B}^n$ is the outer parallel body (already introduced in Subsection 1.2) and the mixed volume expansion becomes the *Steiner-formula* [St1840]:

$$V(K + \xi\mathbb{B}^n) = \sum_{i=0}^{n} \binom{n}{i} W_i(K)\xi^i = \sum_{i=0}^{n} \kappa_{n-i} V_i(K)\xi^{n-i}.$$

2.5. Characterizations of Mixed Volumes

The following characterization is due to Firey [Fi76] for $k = 1$ and to McMullen [Mc90] for $k = n - 1$.

2.5.1 *Let $k = 1$ or $k = n - 1$. If φ is a monotone translation-invariant valuation on $\mathcal{K}^n$ and is homogeneous of degree k, then there exist bodies $K_{k+1}, \ldots, K_n$ such that*

$$\varphi(K) = V(\overbrace{K, \ldots, K}^{k}, K_{k+1}, \ldots, K_n).$$

The problem of extending 2.5.1 to general k is wide open, and examples of Goodey (private communication) show that the general situation is more complicated than that for $k = 1$ and $k = n - 1$. For instance, let X and Y be orthogonal 2-spaces in $\mathbb{E}^4$ and for each $K \in \mathcal{K}^n$ set $\varphi(K) = V_2(\Pi_X(K)) + V_2(\Pi_Y(K))$, where Π_X and Π_Y denote the orthogonal projections onto X and Y respectively. The valuation φ is monotone, translation-invariant and homogeneous of degree 2, but it cannot be expressed as a mixed volume. Goodey gives similar examples in arbitrary dimensions.

Let us close this section with the famous characterization theorems of Hadwiger [Ha57, Section 6.1.10], showing that the *quermassintegrals* or *intrinsic volumes* form a basis for a certain space of valuations. (Recall that the intrinsic volume V_i is a continuous valuation, invariant under rigid motions and homogeneous of degree i.)

2.5.2 *If φ is a continuous valuation that is invariant under rigid motions, then there are constants $\alpha_0, \ldots, \alpha_n$ such that*

$$\varphi(K) = \sum_{i=0}^{n} \alpha_i V_i(K) \qquad \textit{for all } K \in \mathcal{K}^n.$$

2.5.3 *If φ is a monotone valuation that is invariant under rigid motions, then there are nonnegative constants $\alpha_0, \ldots, \alpha_n$ such that*

$$\varphi(K) = \sum_{i=0}^{n} \alpha_i V_i(K) \qquad \text{for all } K \in \mathcal{K}^n.$$

The following example indicates the manner in which the characterization results 2.5.2 and 2.5.3 can be used to identify certain functionals as intrinsic volumes. For a body K in $\mathbb{R}^n$, and for $u \in \mathbb{S}^{n-1}$, the *breadth* of K in the direction u is defined as the number

$$b_u(K) = \max_{y \in K} \langle u, y \rangle - \min_{y \in K} \langle u, y \rangle;$$

this is just the distance between the two supporting hyperplanes of K that are orthogonal to the line $\mathbb{R}u$. The *width* of K is the minimum of the numbers $b_u(K)$. The *mean width* of K is obtained by integrating the function $b_u(K)$ over $u \in \mathbb{S}^{n-1}$ and then dividing the result by the $(n-1)$-measure of $\mathbb{S}^{n-1}$. As a consequence of 2.5.3 we see that up to a positive factor the mean width of K is just $V_1(K)$.

3. Deterministic methods for volume computation

The problem of computing polytope volume has been studied by many authors. The present section will summarize several of the basic ideas for deterministic volume computation. In addition to the papers that are mentioned below in connection with the various methods, the reader may be interested in the following papers that are not mentioned below: [AlS86], [BaS79], [CoH79], [Ka94], [Ko82], [La83], [LeR82a], [LeR82b], [ShH54], [Sp86].

3.1. Triangulation

If $v_0, \ldots, v_n$ are affinely independent points of $\mathbb{R}^n$, and $T = \operatorname{conv}\{v_0, \ldots, v_n\}$, then

$$V(T) = \frac{1}{n!} |\det(v_1 - v_0, \ldots, v_n - v_0)|.$$

Thus computing the volume of an n-simplex is equivalent to computing the determinant of an $n \times n$ matrix and can be handled very efficiently by means of *Gaussian elimination*. Other formulas for computing the volume of a simplex are stated in Subsection 3.6.

Since simplex volumes can be computed so easily, the most natural approach to the problem of computing the volume of a polytope P is to produce a dissection of P into simplices. Then compute the volumes of the individual simplices and add them up to find the volume of P. (This uses the fact that the volume is a simple valuation.)

In fact, we shall tell, for both $\mathcal{V}$-polytopes and $\mathcal{H}$-polytopes, how to produce a triangulation. As the term is used here, a *triangulation* is not merely a dissection into simplices, but it has the additional property that the intersection of any two simplices in the dissection is a face of each.

We will first outline an "incremental" algorithm that constructs a triangulation of a $\mathcal{V}$-polytope. The case of $\mathcal{H}$-polytopes is treated later in this subsection.

The problem of constructing triangulations of $\mathcal{V}$-polytopes is intimately related to the task of computing the *face-lattice* of the convex hull of a given finite point set, and this is a fundamental task in *computational geometry*; see EDELSBRUNNER [Ed87], [Ed93] and CHAZELLE [Ch93]. The *incremental* method, a paradigmatic procedure in computational geometry, uses the *beneath-beyond* approach that goes back to GRÜNBAUM [Gr67, p.78]. The basic strategy is to add one of the given points at a time and hence compute the convex hull step by step. This requires the use of an ordering of the input vectors.

The following algorithm is based on a convex hull algorithm due to SEIDEL [Se91], and can also be found in [Ed87] and [Ed93].

Let $v_1, \ldots, v_m \in \mathbb{Q}^n$ be given, and suppose that

(i) $\mathrm{aff}\{v_1, \ldots, v_{n+1}\} = \mathbb{R}^n$, and
(ii) a rational vector $z_0 \in \mathbb{R}^n$ is given such that $\langle v_1, z_0\rangle < \langle v_2, z_0\rangle < \ldots < \langle v_{m-1}, z_0\rangle < \langle v_m, z_0\rangle$.

Then the incremental algorithm proceeds as follows:

- Let $P_{n+1} = \mathrm{conv}\{v_1, \ldots, v_{n+1}\}$, set $\mathcal{T}_{n+1} = \{P_{n+1}\}$, and assume that for some $i \geq n+1$ a triangulation $\mathcal{T}_i$ of the convex hull P_i of $\{v_1, \ldots, v_i\}$ has already been constructed.
- Let $\mathcal{B}_i$ denote the induced triangulation of the boundary $\mathrm{bd}(P_i)$. (Note that P_{n+1} is a simplex, whence $\mathcal{B}_{n+1} = \mathcal{F}_{n-1}(P_{n+1})$.)
- Let $\hat{\mathcal{B}}_i$ denote the set of all $(n-1)$-simplices in $\mathcal{B}_i$ that are visible from v_{i+1} with respect to P_i, i.e. the affine hull of a simplex in $\mathcal{B}_i$ separates v_{i+1} from P_i.
- Finally, set $P_{i+1} = \mathrm{conv}(\{v_{i+1}\} \cup P_i)$ and $\mathcal{T}_{i+1} = \mathcal{T}_i \cup \{\mathrm{conv}(\{v_{i+1}\} \cup F) : F \in \hat{\mathcal{B}}_i\}$.

Before mentioning some complexity issues of the main algorithm, let us comment on the assumptions (i) and (ii). From a theoretical point of view, none of these assumptions constitutes any restriction of generality. However, since we are here interested in *algorithmic* questions, we need efficient computational procedures for satisfying the assumptions in order to conclude that they are not too restrictive for our purposes.

Using Gaussian elimination, we can determine a maximal affinely independent subset of $\{v_1, \ldots, v_m\}$ in polynomial time, and (possibly after reordering) we may assume that it consists of the first k vectors. If $k < n+1$ we may terminate the algorithm (in view of the fact that we are here interested in triangulations only as a tool for volume computation), or we may decide to continue in aff V. In any case, Assumption (i) is "algorithmically acceptable."

As to Assumption (ii), it is possible as follows to obtain such a hyperplane $H_0 = \{x : \langle x, z_0\rangle = 0\}$ through the origin with the property that no line determined by two of the vectors of V is parallel to H_0. For any pair (v_i, v_j) of different vectors of V, let $\delta_{ij} \in \{-1, 1\}$ and $y_{ij} = \delta_{ij}(v_i - v_j)$, where the choice of the sign δ_{ij} is such that the first nonzero coordinate of y_{ij} is positive. Then the positive hull of all such vectors y_{ij} is a pointed convex cone, and we can use linear programming to find (in polynomial time) a vector z_0 such that $\langle y_{ij}, z_0\rangle \geq 1$ for all such vectors y_{ij}. Clearly,

Assumption (ii) is then satisfied by z_0 after a suitable sorting of the inner-product values $\langle v_i, z_0 \rangle$, and hence is "algorithmically acceptable" as well.

(It is also possible to deal with Assumption (ii) by choosing any hyperplane and then "simulating" a perturbation of the input points [EdM90], an approach similar to the lexicographic rule of the simplex algorithm, see [Da63].)

Note that the ordering of $\{v_1, \dots, v_m\}$ implies that the segment $\text{conv}\{v_i, v_{i+1}\}$ meets P_i only in the point v_i, and hence v_i belongs to an $(n-1)$-simplex of $\mathcal{B}_i$ that is visible from v_{i+1}. This allows us in the main algorithm to find an element of $\hat{\mathcal{B}}_i$ by looking at all simplices of $\mathcal{B}_i$ that contain v_i; and then we proceed by looking at neighboring boundary simplices.

It is not hard to see that the running time of the above incremental algorithm is of the order $O(\pi(L)m^{\lfloor (n+1)/2 \rfloor})$, where π denotes a suitable polynomial in L. Observe that this bound is polynomial only in the case of *fixed* dimension; for general $\mathcal{V}$-polytopes, the number of simplices in a triangulation is indeed exponential in n since the number of facets may already be exponential in the dimension.

Let us mention in passing that triangulations with special properties are studied prominently in computational geometry, see [Ed87], [Ed93]. A particular class of triangulations that has received a lot of attention because of its wide range of applications is the class of *Delauney-triangulations* that are "dual" to the *Voronoi-diagrams.* Properties of triangulations that are important for practical application (for instance in "surface-design" in the automobile industry) include "good conditioning" in the sense that the ratio of a longest to a shortest edge of the triangulation is bounded above by a reasonably small constant.

For some structural properties of triangulations and dissections, and a related bibliography, see BAYER & LEE [BaL93].

For a given $\mathcal{H}$-polytope we could, of course, compute all vertices and then proceed just as before. However, we will outline an algorithmically different approach that is based on the fact that linear programming problems can be solved in polynomial time. It will allow us to derive an additional result in the case when the dimension is part of the input (and then, of course, rational $\mathcal{V}$-presentations and rational $\mathcal{H}$-presentations are no longer "polynomially equivalent").

Suppose that P is an $\mathcal{H}$-polytope given by the irredundant presentation $(n, m; A, b)$. A triangulation $\mathcal{T}(\mathcal{P})$ of P can be computed recursively as follows:

- Determine a vertex v of P. This can be done in polynomial time by an application of the ellipsoid algorithm or by interior-point methods.
- Determine the set $\mathcal{F}$ of (irredundant $\mathcal{H}$-presentations of) facets of P that do not contain v. This can again be done by linear progamming. (Note that from $(n, m; A, b)$ we can easily obtain $\mathcal{H}$-presentations for the facets in $\mathcal{F}$, and the subsequent redundancy tests require the solution of at most $O(m^2)$ suitable linear programs; hence this step can be done in time that is polynomial in the size of the original input.)

The same step is now repeated for the facets in $\mathcal{F}$ and so on, and the results are stored in a layered graph. The 0th node is the pair $(\emptyset, P)$ and the nodes of layer j are pairs S and F, where S is a set of j vertices and F is a face of dimension $n-j$. The recursive process stops with the $(n+1)$st layer. Then the respective faces are all empty, and the sets S contain the vertices of the simplices of the so constructed triangulation. (A close relative of this method appears in a paper by VON HOHENBALKEN [Vo81].)

A first observation confirms the result that for *fixed* n, the volume of an $\mathcal{H}$-polytope can be computed in polynomial time. (Note that this follows already from the triangulation routine for $\mathcal{V}$-polytopes that was outlined earlier, in conjunction with the fact that, when the dimension is fixed, a passage from a rational $\mathcal{V}$-presentation of a polytope to a rational $\mathcal{H}$-presentation of the same polytope can be carried out in polynomial time.)

3.1.1 *When the dimension n is fixed, the volume of $\mathcal{V}$-polytopes and $\mathcal{H}$-polytopes can be computed in polynomial time.*

Clearly, the above algorithm may require time that is exponential in n. However, for the case of $\mathcal{H}$-polytopes that are simplicial, the algorithm runs in polynomial time even when the dimension n is part of the input. To see this, observe first that the problem of deciding whether P has volume 0 can be solved by way of linear programming. So, suppose that P is n-dimensional. Further, note that the number of simplices of $\mathcal{T}(\mathcal{P})$ is bounded by m, the number of constraints in the given $\mathcal{H}$-presentation. We can then use linear programming to identify for each facet F of P the constraint hyperplanes of the given presentation that contain F's facets (which are $(n-2)$-dimensional faces of P), and then it is easy to derive irredundant $\mathcal{H}$-presentations for all simplices of $\mathcal{T}(\mathcal{P})$. For simplices, one presentation can be converted easily into the other, so we obtain, in time that is polynomial in L, rational $\mathcal{V}$-presentations of all simplices in $\mathcal{T}(\mathcal{P})$. As the final step we compute the volumes of these simplices and add them all up to obtain the volume of P. This result can be easily extended to "near-simplicial" $\mathcal{H}$-polytopes. To be more precise, let σ be a nonnegative integer and define the class $\mathcal{P}_{\mathcal{H}}(\sigma)$ by

$$\mathcal{P}_{\mathcal{H}}(\sigma) = \bigcup_{n\in\mathbb{N}} \{P \in \mathcal{P}^n : P \text{ is an } \mathcal{H}\text{-polytope, and } f_0(F) \leq n+1+\sigma \text{ for each facet } F \text{ of } P\}.$$

Then we obtain the following result:

3.1.2 *Let σ be a nonnegative integer constant. When restricted to $\mathcal{P}_{\mathcal{H}}(\sigma)$, the problem of volume computation can be solved in time that is bounded by a polynomial in L.*

Let us conclude with a result about the binary size of the volume of $\mathcal{V}$-polytopes. Clearly, when the dimension is fixed, the volume of a ($\mathcal{V}$- or $\mathcal{H}$-) polytope P can be computed in polynomial time and its size is therefore polynomial in the size of the input. It is not clear *a priori* whether this property of the volume persists when the dimension is part of the input. It is true that each vertex of P is rational of size that is bounded above by a polynomial in L, and that each simplex in a triangulation has volume of size that is again bounded by a polynomial in L. However, it is also true that there may be exponentially many simplices in any possible triangulation, and thus it is conceivable that $V(P)$ (the sum of all the simplex-volumes) may be of exponential size. (Remember that we are speaking here of the size or length of the volume as a binary number, and not of its magnitude as a real number!) As we will see in the next subsection, this may actually be the case for $\mathcal{H}$-polytopes. On the other hand, it is easy to see (by multiplying with the common denominator or simply with the product of all denominators of the rational entries $(v_1, \ldots, v_m)$), that the size of the volume of a $\mathcal{V}$-polytope is indeed bounded above by a polynomial in L.

3.1.3 *If P is a $\mathcal{V}$-polytope, then the binary size of $V(P)$ is bounded above by a polynomial in the size L of P's $\mathcal{V}$-presentation.*

3.2. Sweeping-plane formulas

Another approach that has become a standard tool for many algorithmic questions in geometry is the *sweeping-plane* technique. It goes back (at least) to HADWIGER [Ha55], who used it in the context of the Euler characteristic on the convex ring. It has been applied to volume computation by BIERI & NEF [BiN83], LAWRENCE [La91] and KHACHIYAN [Kh88], [Kh89], [Kh93].

The general idea is to "sweep" a hyperplane through a polytope P, keeping track of the changes that occur when the hyperplane sweeps through a vertex. Let us illustrate this idea for the problem of computing the volume of a triangle $T = \{v_0, v_1, v_2\}$ in the plane. Let $c \in \mathbb{R}^2$ be a rational vector such that

$$\langle c, v_0 \rangle < \langle c, v_1 \rangle < \langle c, v_2 \rangle.$$

For $\tau \in \mathbb{R}$, let

$$H(\tau) = \{x : \langle c, x \rangle \leq \tau\} \quad \text{and} \quad \varphi(\tau) = V(T \cap H(\tau)).$$

Clearly, $\varphi(\tau) = 0$ for $\tau \leq \langle c, v_0 \rangle$ and $\varphi(\tau) = V(T)$ for $\tau \geq \langle c, v_2 \rangle$. Now define the following three cones:

$$\begin{aligned} &C_0 = v_0 + \text{pos}\{v_1 - v_0, v_2 - v_0\}, \quad C_1 = v_1 + \text{pos}\{v_1 - v_0, v_2 - v_1\}, \\ &C_2 = v_2 + \text{pos}\{v_2 - v_0, v_2 - v_1\}. \end{aligned}$$

Note that $C_i \cap H(\tau)$ is bounded for each i, whence (as an easy case of the *inclusion-exclusion principle*)

$$\varphi(\tau) = V(C_0 \cap H(\tau)) - V(C_1 \cap H(\tau)) + V(C_2 \cap H(\tau)).$$

Further, for $i = 0, 1, 2$,

$$V(C_i \cap H(\tau)) = \begin{cases} 0 & \text{for } \tau \leq \langle c, v_i \rangle; \\ \gamma_i(\tau - \langle c, v_i \rangle)^2 & \text{for } \tau \geq \langle c, v_i \rangle, \end{cases}$$

where the γ_i are suitable (easily computable) constants. Hence for $\tau \geq \langle c, v_2 \rangle$ we obtain

$$V(T) = \sum_{i=0}^{2} (-1)^i \gamma_i (\tau - \langle c, v_i \rangle)^2.$$

Clearly, this approach can be generalized to arbitrary polytopes, and it yields a volume formula that does not explicitly involve triangulations. This formula was first derived by BIERI & NEF [BiN83] (even for more general bounded polyhedral sets); other proofs are due to LAWRENCE [La91] and FILLIMAN [Fi92]. We will give LAWRENCE's [La91] statement of the volume formula (under additional restrictions) since it is formulated in terms of the standard ingredients of the simplex tableau of linear programming. Later we will comment on some generalizations.

Suppose that $(n, m; A, b)$ is an irredundant $\mathcal{H}$-presentation of a *simple* polytope P. Recall that P's being simple means geometrically that each vertex of P is contained in precisely n facets. Let $b = (\beta_1, \ldots, \beta_m)^T$ and denote the row-vectors of A by $a_1^T, \ldots, a_m^T$. Let $M = \{1, \ldots, m\}$ and for each nonempty subset I of M, let A_I denote the submatrix of A of rows with indices in I and let b_I denote the corresponding right-hand side. For each vertex v of $P = \{x \in \mathbb{R}^n : Ax \leq b\}$ there is a subset $I = I_v \subset M$ of cardinality n such that $A_I v = b_I$ and $A_{M \setminus I} v \leq b_{M \setminus I}$. Since P is assumed to be simple and its $\mathcal{H}$-presentation to be irredundant, the set I_v is unique. Stated in the terminology of the (dual) simplex algorithm this means that the basic feasible solutions of $Ax \leq b$ are in one-to-one correspondence with the vertices of P, and hence the corresponding linear program is nondegenerate.

Let $c \in \mathbb{R}^n$ such that $\langle c, v_1 \rangle \neq \langle c, v_2 \rangle$ for any pair of vertices v_1, v_2 that form an edge of P, and set $H(\tau) = \{x \in \mathbb{R}^n : \langle c, x \rangle \leq \tau\}$ for $\tau \in \mathbb{R}$.

3.2.1 *If the polytope P is simple, and $(n, m; A, b)$ is an irredundant $\mathcal{H}$-presentation, then (with the above notation)*

$$V(P \cap H(\tau)) = \frac{(-1)^n}{n!} \sum_{v \in \mathcal{F}_0(P)} \frac{(\max\{0, \tau - \langle c, v \rangle\})^n}{\prod_{i=1}^n e_i^T A_{I_v}^{-1} c |\det(A_{I_v})|},$$

where $e_1, \ldots, e_n$ denote the standard unit vectors of $\mathbb{R}^n$.

Consequently,

$$V(P) = \frac{(-1)^n}{n!} \sum_{v \in \mathcal{F}_0(P)} \frac{(\tau - \langle c, v \rangle)^n}{\prod_{i=1}^n e_i^T A_{I_v}^{-1} c |\det(A_{I_v})|}$$

whenever $\tau \geq \max_{x \in P} \langle c, v \rangle$. It follows that as a polynomial in τ, the right-hand side of this formula is constant. Evaluation at $\tau = 0$ yields the following volume formula.

3.2.2 *If the polytope P is simple, and $(m, n; A, b)$ is an irredundant $\mathcal{H}$-presentation, then (with the above notation)*

$$V(P) = \frac{1}{n!} \sum_{v \in \mathcal{F}_0(P)} \frac{\langle c, v \rangle^n}{\prod_{i=1}^n e_i^T A_{I_v}^{-1} c |\det(A_{I_v})|}.$$

As was mentioned earlier, the ingredients of this volume formula are those which are computed in the (dual) simplex algorithm. More precisely, $\langle c, v \rangle$ is just the value of the objective function at the current basic feasible solution v, $\det(A_{I_v})$ is the determinant of the current basis, and $A_{I_v}^{-1}c$ is the vector of reduced costs, i.e. the (generally infeasible) dual point that belongs to v. Note that the signs that were present in our introductory example (which come from the inclusion-exclusion principle, or from Gram's relation) are now hidden in $A_{I_v}^{-1}c$; each dually infeasible component contributes a negative sign.

For practical computations, 3.2.2 has to be combined with some vertex enumeration technique. Its closeness to the simplex algorithm suggests the use of a *reverse search* method of AVIS & FUKUDA [AvF91], which is based on the simplex method with Bland's pivoting rule.

As it stands, the volume formula 3.2.2 does not involve triangulation. However, if we interpret it in a polar setting, it becomes clear that we are really dealing with the faces of the simplicial polytope P°. Accordingly, generalization to nonsimple polytopes involves polar triangulation. In fact, for general polytopes P, [BiN83] and [La91] suggest a "lexicographic rule" to move from one basis to another, and this is just a particular triangulation of P°; see also FILLIMAN [Fi92, Theorem 1].

As an application of formula 3.2.2, LAWRENCE [La91] derives the following negative result for the binary size of $V(P)$ for $\mathcal{H}$-polytopes P.

3.2.3 *The binary size of the volume of $\mathcal{H}$-polytopes is in general not bounded above by a polynomial in L.*

This result is in striking contrast to the case of $\mathcal{V}$-polytopes (3.1.3) and answers a question of DYER & FRIEZE [DyF88].

The example given in [La91] is a projective image of the standard cube. More precisely, let

$$C_n = [0,1]^n = \{x = (\xi_1, \dots, \xi_n)^T \in \mathbb{R}^n : 0 \le \xi_1, \dots, \xi_n \le 1\},$$

let

$$a = \frac{1}{2^n}(2^{n-1}, 2^{n-2}, \dots, 2^0)^T,$$

and consider the projective transformation π_a defined by

$$\pi_a(x) = \frac{x}{1 + \langle a, x \rangle}.$$

Then $P = \pi_a(C_n)$ is a polytope which is defined by the inequalities

$$\begin{aligned} \xi_i + \langle a, x \rangle &\le 1 \qquad i = 1, \dots, n \\ \xi_i &\ge 0 \qquad i = 1, \dots, n, \end{aligned}$$

and this is a rational $\mathcal{H}$-presentation of P of size that is polynomial in n. However,

$$V(P) = \frac{1}{n!} 2^{(n^2+3)/2} \sum_{i=2^n}^{2^{n+1}-1} (-1)^{\omega(i)-1} \frac{1}{i},$$

where $\omega(i)$ is the number of 1's in i's binary representation. Now, write this number as a rational β/γ in coprime representation (with $\beta, \gamma > 0$), and let p be a prime with $2^n < p < 2^{n+1}$. Note that p divides a denominator i of a summand if and only if $p = i$. Let $\tau = \prod q^k$, where the product extends over all primes q and q^k is the highest power of q that divides at least one of the integers between 2^n and $2^{n+1}-1$; certainly γ divides τ. Then

$$\sum_{i=2^n}^{2^{n+1}-1} (-1)^{\omega(i)-1}\frac{1}{i} = \sum_{i=2^n}^{2^{n+1}-1} (-1)^{\omega(i)-1}\frac{\tau/i}{\tau},$$

and p divides each τ/i with $p \neq i$ but does not divide τ/p. Thus the numerator

$$\sum_{i=2^n}^{2^{n+1}-1} (-1)^{\omega(i)-1}\frac{\tau}{i}$$

is not divisible by p and p is not factored out when producing the coprime representation β/γ. This implies that

$$\gamma \geq \prod\{p : p \text{ is a prime with } 2^n < p < 2^{n+1}\}.$$

By the prime number theorem (see e.g. [HaW68]) there are asymptotically $2^n/n$ such primes and thus γ is an integer of order 2^{2^n}. Hence the binary size of the denominator of $V(P)$ is not bounded by a polynomial in n.

3.3. Exponential integrals

Another possibility to compute the volume of a polytope P – at least if P belongs to some special classes of polytopes – is to study the *exponential integral*

$$\int_P e^{\langle c,x\rangle}\,dx,$$

where c is an arbitrary vector of $\mathbb{R}^n$. (Note that for $c = 0$, the above integral gives just the volume of P.) Exponential integrals satisfy certain relations, some of which are stated later, that make it possible to compute the integrals efficiently in some important cases.

Let us begin by stating formulas for the cube $C_n = [0,1]^n$ and for the regular $(n-1)$-simplex $T_n = C_n \cap H$, where $H = \{x \in \mathbb{R}^n : \xi_1 + \ldots + \xi_n = 1\}$, that is embedded in $\mathbb{R}^n$; see [Ba93a]. Let $c = (\gamma_1, \ldots, \gamma_n)^T$. Then in the first case we have

$$\int_{C_n} e^{\langle c,x\rangle}\,dx = \prod_{i=1}^n \alpha_i \qquad \text{where} \quad \alpha_i = \begin{cases} 1 & \text{if } \gamma_i = 0; \\ \frac{e^{\gamma_i}-1}{\gamma_i} & \text{else.} \end{cases}$$

In the second case, let μ denote the Lebesgue measure on H induced from $\mathbb{R}^n$. Then

$$\int_{T_n} e^{\langle c,x\rangle}\,d\mu = \sqrt{n}\sum_{i=1}^n e^{\gamma_i} \prod_{\substack{j=1 \\ j\neq i}}^n \frac{1}{\gamma_i - \gamma_j},$$

for all $c \in \mathbb{R}^n$ with pairwise distinct coordinates. This result is due to Podkorytov [Po80] and a different proof was given by Barvinok [Ba93a]. The following proposition stems from [Ba93a].

3.3.1 *Let C be an n-dimensional line-free polyhedral cone in $\mathbb{R}^n$. Further, let $z_1, \ldots, z_m$ be a minimal set of vectors that generate all extreme rays of C, and let $H_i = \{x : \langle z_i, x\rangle = 0\}$. Then the integral*

$$\hat{\sigma}_C(c) = \int_C e^{\langle c,x\rangle} dx$$

exists for all vectors c that are contained in the interior of C's polar C°, and the function $\hat{\sigma}_C$ is rational in c. Further, $\hat{\sigma}_C$ can be naturally extended to a rational function σ_C on $\mathbb{C}^n$, with singularities precisely in $H_1 \cup \ldots \cup H_m$.

The following theorem is the central result in this context. It is due to BRION [Br88] for rational polytopes, and was later extended to the form stated below by KHOVANSKII & PUHLIKOV (1989, unpublished) and BARVINOK [Ba91], [Ba93a].

3.3.2 *Let P be an n-polytope, and for each vertex v of P let C_v denote the cone $v + pos(P - v)$. Further, let σ_{C_v} be the functions defined in 3.3.1 (with C replaced by C_v). Then*

$$\int_P e^{\langle c,x\rangle} dx = \sum_{v\in\mathcal{F}_0(P)} \sigma_{C_v}(c),$$

whenever $c \in \mathbb{C}^n$ is nonsingular for all functions σ_{C_v}.

Note that in some sense Theorem 3.3.2 can be regarded as a generalization of the Gram-relation of LAWRENCE's [La91] approach.

The vector $c = 0$ that corresponds to volume computation is singular for all functionals σ_{C_v}; so we have to resort to computing the exponential integrals for nonzero vectors c with $0 < \|c\|_2 < \epsilon$ for some sufficiently small positive ϵ. Using such an approximation, BARVINOK [Ba93a] proves a theorem which, when combined with the fact that the volume of a given $\mathcal{V}$-polytope is polynomial in the size of the input, yields the following result.

3.3.3 *There is an algorithm which, for a given $\mathcal{V}$-polytope P, computes the volume of P in time $O(\pi(L)\beta(P))$, where π is a polynomial and*

$$\beta(P) = \sum_{v\in\mathcal{F}_0(P)} \binom{\hat{f}_1(v)}{n} \qquad \text{with } \hat{f}_1(v) = \text{card}(\{e \in \mathcal{F}_1(P) : v \in e\}).$$

As a corollary to this theorem we see that for "near-simple" $\mathcal{V}$-polytopes P the volume of P can be computed in polynomial time. To be more precise, let τ be a nonnegative integer and define the class $\mathcal{P}_{\mathcal{V}}(\tau)$ by

$$\mathcal{P}_{\mathcal{V}}(\tau) = \bigcup_{n\in\mathbb{N}} \{P \in \mathcal{P}^n : P \text{ is a } \mathcal{V}\text{-polytope, and } \hat{f}_1(v) \leq n + \tau \text{ for each vertex } v \text{ of } P.\}$$

Then we obtain the following result, which is the "dual" counterpart of Theorem 3.1.2:

3.3.4 *Let τ be a nonnegative integer constant. Then there is a polynomial-time algorithm which, for a given $P \in \mathcal{P}_{\mathcal{V}}(\tau)$, computes the volume of P.*

Let us point out that 3.3.4 can also be derived from 3.2.2. Note that the validity of 3.3.4 is based on the fact that the number of facets of a simple polytope is bounded above by its number of vertices. Since, on the other hand, the number of vertices may be exponential in the number of facets, a similar result is not likely to be true for near-simple $\mathcal{H}$-polytopes. In fact, as we will see in Subsection 5.1, the problem of computing the volume of the intersection of a cube with a rational halfspace is already $\#\mathbb{P}$-hard.

3.4. Numerical integration

It may be fair to say that the modern study of volume computations began with Kepler [Ke1615] who derived the first *cubature formula* for measuring the capacities of wine barrels (see [St69, pp.192–197]), and that it was the task of volume computation that motivated the general field of integration. Many efforts have been made in *numerical analysis* to devise efficient algorithms for computing or approximating integrals, and it seems very natural to browse through the fund of numerical analysis to see what kind of approaches to numerical integration may, when suitably specialized, lead to efficient methods for volume computation.

Of course, we do not attempt to give a full account of the methods of numerical integration; for general treatments of this subject see any standard monograph, e.g. Stroud [St71] or Davis & Rabinovitz [DaR84]. Here we want to concentrate on two main approaches to numerical integration, the (*degree d*) *integration formulas*, and the (*quasi-*) *Monte Carlo methods.*

Many of the approximate methods for integration of a functional f over a compact region B of $\mathbb{R}^n$ have the form

$$\int_B f(x)dx \sim \sum_{i=1}^{r} \alpha_i f(y_i),$$

where the points $y_1, \ldots, y_r \in \mathbb{R}^n$ are the *nodes* and the numbers $\alpha_1, \ldots, \alpha_r \in \mathbb{R}$ are the *coefficients* of the formula. Of course, the nodes and coefficients must not depend on f, and it is numerically desirable (to avoid annihilation) to have nonnegative coefficients. The integration formula is of *degree d* if it is exact for all multivariate polynomials f of degree at most d but inexact for some polynomial of degree $d+1$.

The theory of integration formulas for functions of one variable is well developed; subjects like the *Newton-Cotes formulas* or the *Gaussian quadrature formulas* are standard fare in every undergraduate course on this subject. However, already in dimension 2 the situation becomes significantly more complicated. One reason is that up to affine equivalence there is only one compact connected region in $\mathbb{R}^1$, while there are uncountably many such affinity classes in higher dimensions. Further, integration formulas for functions in one variable can be easily obtained by integration of interpolation polynomials, while for arbitrary point sets in higher dimensions, suitable interpolation is not always possible. Moreover, in contrast to the multivariate case, the theory of univariate orthogonal polynomials (which is of great use for constructing integration formulas in one variable) is simple and fairly well understood.

Suppose now that, given a region B in $\mathbb{R}^n$, we want to construct an integration formula of degree d. For $x = (\xi_1, \ldots, \xi_n)$ and $q = (\kappa_1, \ldots, \kappa_n) \in (\mathbb{N} \cup \{0\})^n$ let x^q denote the monomial

$$x^q = \xi_1^{\kappa_1} \cdot \xi_2^{\kappa_2} \cdot \ldots \cdot \xi_n^{\kappa_n}.$$

Further, let

$$S_{n,d} = \{q = (\kappa_1, \ldots, \kappa_n) \in (\mathbb{N} \cup \{0\})^n : \sum_{i=1}^{n} \kappa_i \leq d\}.$$

Note that there are

$$\binom{n+d}{d}$$

different multivariate monomials of degree at most d in $\mathbb{R}^n$. Thus, in order to obtain an integration formula of degree d we have to solve the system

$$\int_B x^q dx = \sum_{i=1}^{r} \alpha_i y_i^q, \qquad q \in S_{n,d},$$

of $\binom{n+d}{d}$ nonlinear equations in $r(d+1)$ variables.

It is quite easy to see (e.g. [DaR84, p.366]) that the system cannot be solved with fewer than

$$\binom{n + \lfloor d/2 \rfloor}{\lfloor d/2 \rfloor}$$

nodes. The following theorem of TCHAKALOFF [Tc57] shows on the positive side that the system is always solvable with

$$r_0 = \binom{n+d}{d}$$

nodes even under the additional constraints that all coefficients be positive.

3.4.1 *Let $B \subset \mathbb{R}^n$ be compact with positive volume. Then there exist nodes $y_1, \ldots, y_{r_0} \in B$ and positive coefficients $\alpha_1, \ldots, \alpha_{r_0}$ such that*

$$\int_B f(x) dx = \sum_{i=1}^{r_0} \alpha_i f(y_i),$$

whenever f is a multivariate polynomial in x of degree at most d.

It may be worthwhile in our context to point out that the most elegant proof of this theorem makes fundamental use of the theory of convex cones.

Note that the equation in the above nonlinear system that corresponds to the monomial x^0 of degree 0 is just

$$V(B) = \int_B dx = \int_B x^0 dx = \sum_{i=1}^{r} \alpha_i.$$

Hence 3.4.1 is just tautological when applied to volume computation for a body K in its formulation $B = K$, $f \equiv 1$.

There are other ways of formulating (WEAK) VOLUME COMPUTATION as a problem in integration. For instance, it is equally natural to write

$$V(K) = \int_C \chi_K(x)dx,$$

where χ_K denotes the characteristic function of the body K and C is a body with $K \subset C$ whose volume can be computed easily. Then the formula would not be tautological. However, since the quality of the approximation

$$\int_C \chi_K(x)dx \sim \sum_{i=1}^{r} \alpha_i \chi_K(y_i)$$

would depend on the error in approximating the (noncontinuous) function χ_K by polynomials of (preferably) small degree, this formula would not be of great practical use.

If mere continuity were the issue, we could use yet another formulation. Suppose that $0 \in \text{int } K$, and let γ_K denote the *gauge functional* of K; i.e. for $x \in \mathbb{R}^n$,

$$\gamma_K(x) = \min\{\lambda \geq 0 : x \in \lambda K\}.$$

Then

$$V(K) = \frac{1}{n!} \int_{\mathbb{R}^n} e^{-\gamma_K(x)} dx,$$

and hence we could get good approximations of $V(K)$ from the numerical value of the integral $\int_C e^{-\gamma_K(x)}dx$, where C is, say, a sufficiently large cube centered at the origin. By the *Stone-Weierstrass* theorem, any continuous function on C can be approximated uniformly on C by multivariate polynomials. However, in order to obtain sufficiently close approximation, the degrees of the polynomials must be very high.

There are many other ways in which one could try to utilize the rich fund of integration formulas (and their accompanying, sometimes very deep, theory of error bounds) for the apparently simpler task of volume computation. However, as we will see in Sections 5 and 6, there are some serious, apparently unavoidable obstacles to obtaining efficient deterministic algorithms for (WEAK) VOLUME COMPUTATION. With this in mind, it is natural to investigate techniques that use (or simulate) some kind of sampling. The general idea of the classical *Monte Carlo method* for numerical integration is to devise a *stochastic process* whose expected value is the integral under consideration, and then to estimate this expected value by sampling. More precisely, for approximating the integral $\int_B f(x)dx$ we choose, for a given integer r, random points $y_1, \ldots, y_r$ independently uniformly distributed in B. Then the integral is estimated by

$$\int_B f(x)dx \sim \frac{1}{r} V(B) \sum_{i=1}^{r} f(y_i).$$

The expression $\frac{1}{r}V(B)\sum_{i=1}^{r} f(y_i)$ is a random variable with expected value $\int_B f(x)dx$ and standard deviation

$$\frac{1}{\sqrt{r}}\left(V(B)\int_B f^2(x)dx - \left(\int_B f(x)dx\right)^2\right).$$

Since the standard deviation does not decrease very rapidly in r, and since for most regions B it does not seem possible to actually perform random sampling, most practical applications resort to sequences of points that are specifically tailored for integration. They are in fact *quasi-Monte Carlo methods.*

(Note that the latter two of the three mentioned methods of formulating volume computation as a problem of integration do at least avoid the obvious drawback of the first formulation – that we would have to know in advance the volume of the body under consideration.)

A natural approach to deterministic sampling uses the points of B that belong to the point lattice $\delta\mathbb{Z}^n$ for some parameter $\delta \in]0,1]$. The corresponding formula is then

$$\int_B f(x)dx \sim \frac{V(B)}{G_\delta(B)} \sum_{y\in B\cap\delta\mathbb{Z}^n} f(y),$$

where $G_\delta(B) = \text{card}(B \cap \delta\mathbb{Z}^n)$ is the *lattice-point enumerator* with respect to $\delta\mathbb{Z}^n$. Under assumptions on f that involve its variation it is possible to derive error estimates for such formulas; see [DaR84, p.352]. In the next subsection we will consider this lattice-point approach more closely in the context of (WEAK) VOLUME COMPUTATION.

Improved quasi-Monte Carlo methods can be obtained by "optimizing" the set of sampling points. The error estimates then rely explicitly on measures of equidistribution of the point set; see [St71, Sections 6.2 and 6.3], [DaR84, Section 5.5].

Let us point out in passing that the lattice-point sampling corresponds to a dissection of space into cubes with centers at the lattice points. Rather than choosing these centers as sampling points one can choose one or more random points in each cube; this leads to the method of *stratified sampling*, see [St71, p.209].

As we will see in Section 7, the general idea of random sampling (when appropriately elaborated, utilizing special properties of convex bodies) does indeed lead to a randomized polynomial-time algorithm for volume computation (and hence also for some special integration problems; see Subsection 9.3). In fact, after suitable transformations, DYER, FRIEZE & KANNAN [DyFK89], [DyFK91] construct an ascending sequence of bodies

$$\mathbb{B}^n = K_0 \subset K_1 \subset \ldots \subset K_k = K$$

such that the corresponding volume ratios are small, and they then use random walks on the lattice points inside K_i to generate random points from the uniform distribution over K_i that lead to an estimate for $V(K_{i-1})/V(K_i)$.

3.5. LATTICE POINT ENUMERATION

As was mentioned in the previous subsection, it is quite natural (though in general not optimal) to use the points of suitable lattices for sampling in a quasi-Monte Carlo approach to numerical integration. We want to consider the sampling with lattice points more closely now in the context of (WEAK) VOLUME COMPUTATION. Let $R \in \mathbb{N}$, and suppose that the body K is contained in $R\mathbb{B}^n$. Set for $\delta \in]0,1[$

$$k_\delta = \left\lceil \frac{R}{\delta} - \frac{1}{2} \right\rceil \quad \text{and} \quad B_\delta = \left(k_\delta + \frac{1}{2}\right)[-\delta,\delta]^n.$$

Note that

$$K \subset B_\delta, \quad G_\delta(B_\delta) = (2k_\delta + 1)^n \quad \text{and} \quad V(B_\delta) = (2k_\delta+1)^n\delta^n.$$

Then, when applied to $B = B_\delta$ and $f = \chi_K$, the corresponding quasi-Monte Carlo integral formula of Subsection 3.4 becomes

$$V(K) = \int_{B_\delta} \chi_K(x)dx \sim \frac{V(B_\delta)}{G_\delta(B_\delta)} G_\delta(K) = \delta^n G_\delta(K),$$

and this relates (WEAK) VOLUME COMPUTATION to the problem of counting lattice points. (See [GrW93] for a survey of lattice-point problems.) Now, we have the trivial upper bound

$$\delta^n G_\delta(K) \le V(K + \delta[-1,1]^n) \le \sum_{i=0}^{n} \binom{n}{i} V(\overbrace{K,\ldots,K}^{n-i}, \overbrace{[-1,1]^n,\ldots,[-1,1]^n}^{i})\delta^i,$$

and using the monotonicity of mixed volumes we obtain

$$\begin{aligned} \delta^n G_\delta(K) &\le V(K) + \delta \sum_{i=1}^{n} \binom{n}{i} V(\overbrace{R[-1,1]^n,\ldots,R[-1,1]^n}^{n-i}, \overbrace{[-1,1]^n,\ldots,[-1,1]^n}^{i}) \\ &\le V(K) + \delta(2(R+1))^n. \end{aligned}$$

On the other hand, the inequality of BOKOWSKI, HADWIGER & WILLS [BoHW72] yields

$$V(K) - \delta n(2R)^{n-1} \le V(K) - \delta V_{n-1}(K) \le \delta^n G_\delta(K),$$

whence

$$|V(K) - \delta^n G_\delta(K)| \le \delta(3R)^n.$$

Thus if λ is a positive rational, and we set $\delta = \lambda/(3R)^n$, the volume of K is approximated by $\delta^n G_\delta(K)$ up to the additive error λ.

By results of DYER [Dy91] for $n \le 4$ (see also ZAMANSKII & CHERKASSKII [ZaC83], [ZaC85]) and of BARVINOK [Ba93b] in general (see also [DyK93]), the number of lattice points of an $\mathcal{H}$- or a $\mathcal{V}$-polytope can be computed in polynomial time when the dimension is *fixed*. Hence the above approach yields again Theorem 3.1.1 as a corollary.

Now, suppose there is an algorithm $\mathcal{A}$ which, accepting as input a pair of rationals $\epsilon, \delta \in]0,1[$ and a centered well-bounded body in $\mathbb{R}^n$ that is given by a weak membership oracle, produces a number g such that

$$|G_\delta(K) - g| \le G_\delta(K(\epsilon)) - G_\delta(K(-\epsilon)).$$

Suppose further that for *fixed* n the running time of $\mathcal{A}$ is polynomial in size(K), size(ϵ) and size(δ). Then we can use the algorithm $\mathcal{A}$ to solve WEAK VOLUME COMPUTATION in fixed dimensions. In fact, let b, r, R denote the parameters of a centered well-bounded body K in $\mathbb{R}^n$, and suppose (without loss of generality) that $r = 1$. Note first that

$$K \subset K(-\epsilon) + \epsilon R\mathbb{B}^n,$$

and hence

$$V(K(-\epsilon)) \geq V(K) - \epsilon \sum_{i=1}^{n} \binom{n}{i} V(\overbrace{R[-1,1]^n, \ldots, R[-1,1]^n}^{n-i}, \overbrace{[-1,1]^n, \ldots, [-1,1]^n}^{i}) R^i$$

$$\geq V(K) - \epsilon(4R)^n.$$

With $\hat{\delta} = \delta(3(R+1))^n$, this implies that

$$\begin{aligned} \delta^n |G_\delta(K) - g| &\leq \delta^n G_\delta(K(\epsilon)) - \delta^n G_\delta(K(-\epsilon)) \\ &\leq V(K(\epsilon)) + |\delta^n G_\delta(K(\epsilon)) - V(K(\epsilon))| - V(K(-\epsilon)) \\ &\qquad + |\delta^n G_\delta(K(-\epsilon)) - V(K(-\epsilon))| \\ &\leq V(K(\epsilon)) - V(K(-\epsilon)) + \delta(3(R+\epsilon))^n + \delta(3(R-\epsilon))^n \\ &\leq 2\hat{\delta} + 2\epsilon(4R)^n \end{aligned}$$

and hence

$$\begin{aligned} |V(K) - \delta^n g| &\leq |V(K) - \delta^n G_\delta(K)| + \delta^n |G_\delta(K) - g| \\ &\leq 3\hat{\delta} + 2\epsilon(4R)^n. \end{aligned}$$

If λ is now the error parameter of the given instance of WEAK VOLUME COMPUTATION, we choose

$$\epsilon \leq \frac{\lambda}{4(4R)^n} \qquad \text{and} \qquad \delta = \frac{\hat{\delta}}{(3(R+1))^n)} \leq \frac{\lambda}{6(3(R+1))^n},$$

and run algorithm $\mathcal{A}$. This proves the following result.

3.5.1 *When the dimension is fixed, there is an algorithm for* WEAK VOLUME COMPUTATION *that uses a polynomial number of arithmetic operations on rationals of polynomially bounded sizes and a polynomial number of calls to the hypothetical algorithm $\mathcal{A}$.*

Note that we can of course check in oracle-polynomial time for each point y of $B_\delta \cap \delta\mathbb{Z}^n$ whether y is weakly contained in K. More precisely, given $y \in B_\delta \cap \delta\mathbb{Z}^n$, and a rational number $\epsilon > 0$, the oracle for K asserts that $y \in K(\epsilon)$ or that $y \notin K(-\epsilon)$. Further, the number g of input points $y \in B_\delta \cap \delta\mathbb{Z}^n$ for which the oracle asserts $y \in K(\epsilon)$ would satisfy the above requirements. Unfortunately, while the number of lattice points in $R[-1,1]^n$ that we have to check is polynomial in R and $1/\delta$, it is not bounded by a polynomial in size(K) and size(λ) and hence is not polynomial in the size of the input. Thus this simple checking procedure does not yield a suitable algorithm $\mathcal{A}$.

At present, we don't know whether such a polynomial algorithm $\mathcal{A}$ for WEAK LATTICE POINT ENUMERATION exists, nor do we know the precise status of WEAK VOLUME COMPUTATION in fixed dimensions. The latter question is of course closely related to the question of devising suitable algorithms for approximating bodies by polytopes.

3.6. SPECIAL CONVEX BODIES

Since simplices are the most basic and elementary polytopes, formulas for volumes of simplices are of special interest. We begin this subsection with some formulas which supplement the basic determinantal formula given at the beginning of Subsection 3.1 and which are for some purposes more useful than that one. The following result expresses the volume of a j-dimensional simplex (short *j-simplex*) in $\mathbb{R}^n$ in terms of its edge-lengths.

3.6.1 *Suppose that S is a j-simplex in $\mathbb{R}^n$ with vertices $v_1, \dots, v_{j+1}$. Let $B = (\beta_{ik})$ denote the $(j+1) \times (j+1)$ matrix given by $\beta_{ik} = \|v_i - v_k\|_2^2$. Then*

$$2^j (j!)^2 V_j^2(S) = |\det(\hat{B})|,$$

where $\hat{B}$ is the $(j+2) \times (j+2)$ matrix obtained from B by bordering B with a top row $(0, 1, \dots, 1)$ and a left column $(0, 1, \dots, 1)^T$.

The determinant appearing in 3.6.1 is often called the *Cayley-Menger determinant.* See [Dö65, p.285] and [BlG43] for references to low-dimensional cases of 3.6.1 associated with the names of Euler, Lagrange, Cayley, and Sylvester, and see SOMMERVILLE [So29, p.125] and BLUMENTHAL [Bl53, p.98] for proofs of 3.6.1.

The next formula, a close relative of 3.6.1, expresses the volume of a suitably located simplex in terms of the *Gram matrix* of inner products of its vertices; see [GrKL94] and [Bl53] for proofs of 3.6.2 and a variant of it.

3.6.2 *Suppose that S is a j-simplex in $\mathbb{R}^n$ with $0 \in \operatorname{aff} S$, and A is the $(j+1) \times n$ matrix whose rows list the coordinates of the vertices of S. Then*

$$(j!)^2 V_j^2(S) = \det(M + AA^T),$$

where M is the $(j+1) \times (j+1)$ matrix whose entries are all 1. If the origin is a vertex of S then

$$(j!)^2 V_j^2(S) = \det(A_0 A_0^T),$$

where A_0 is formed from A by discarding A's zero row.

For an n-simplex S in $\mathbb{R}^n$, the following formula expresses the volume of S in terms of the coefficients that appear in the affine functionals defining the facets of S. For general n the formula is due to KLEBANER, SUDBURY AND WATTERSON [KlSW89].

3.6.3 *Suppose that an n-simplex S in $\mathbb{R}^n$ is bounded by the $n+1$ hyperplanes whose equations are*

$$\alpha_{i0} + \sum_{j=1}^{n} \alpha_{ij} x_j = 0 \qquad (i = 0, 1, \ldots, n),$$

and let A denote the $(n+1) \times (n+1)$ matrix with elements α_{ij}, $0 \leq i, j \leq n$. Then the volume of S is given by

$$V(S) = \frac{|\det(A)|^n}{n! \prod_{i=0}^{n} A_{i0}},$$

where A_{i0} is the cofactor of α_{i0} in A.

The paper [KlSW89] also contains two formulas giving the volume of a j-simplex S in $\mathbb{R}^n$ when $j < n$. One formula is in terms of the coefficients that appear in the affine functionals defining the affine hull of S and the facets of S. The other formula is in terms of the coordinates of the vertices of the simplex. Like the volume formulas in [LL90] and [Be92], it may be regarded as a higher-dimensional analogue of the Pythagorean theorem of plane geometry.

There are, of course, other classes of bodies or polytopes for which special volume formulas or special methods of computing volumes are known. Some of these can be found in the references listed at the start of Section 3. We do not discuss these here, but the case of zonotopes does seems worthy of special mention.

Let $(n, r; c; z_1, \ldots, z_r)$ be an $\mathcal{S}$-zonotope and set $Z = \sum_{i=1}^{r} [0,1] z_i$. Further, for $i = 1, \ldots, r$, let $K_i = [0,1] z_i$, i.e. the K_i's are all line segments. It follows from 2.4.1 that

$$V(Z) = V\left(\sum_{i=1}^{r} K_i\right) = \sum_{i_1=1}^{r} \sum_{i_2=1}^{r} \cdots \sum_{i_n=1}^{r} V(K_{i_1}, \ldots, K_{i_n}).$$

If the indices $i_1, \ldots, i_n$ are not pairwise distinct, the zonotope $K_{i_1} + \cdots + K_{i_n}$ has volume 0, whence $V(K_{i_1}, \ldots, K_{i_n}) = 0$ and it follows with

$$V(K_{i_1} + \ldots + K_{i_n}) = V(K_{i_1}, \ldots, K_{i_n}) = n!\, |\det(z_{i_1}, \ldots, z_{i_n})|$$

that

$$V(Z) = \sum_{1 \leq i_1 < i_2 < \cdots < i_n \leq r} |\det(z_{i_1}, \ldots, z_{i_n})|;$$

see also [Mo89], [St91] and [Sh74].

Note that this formula for the volume of a zonotope leads to a polynomial-time algorithm for fixed dimensions, and also for varying dimensions if the input is restricted to the class of all "near-parallelotopal" $\mathcal{S}$-zonotopes, where $r - n$ is bounded by an a priori constant.

3.6.4 *When the dimension n is fixed or when $r - n$ is fixed,* VOLUME COMPUTATION *can be solved in polynomial time for $\mathcal{S}$-zonotopes.*

In general the above volume formula involves exponentially many summands, and this feature of zonotope volume computation cannot be avoided (unless $\mathbb{P} = \mathbb{NP}$); see 5.1.7.

For combining two convex sets J and K to form a third, the three most important ways are those of intersection, vector addition, and joining – forming the sets $J \cap K$, $J + K$, and $\text{conv}(J \cup K)$ respectively. When the sets J and K are sufficiently "independent," there are useful formulas relating $V(J + K)$ and $V(\text{conv}(J \cup K))$ to $V(J)$ and $V(K)$.

For any two subsets X and Y of $\mathbb{E}^n$, we define the *distance*

$$\text{dist}(X, Y) = \inf\{\|x - y\|_2 : x \in X, y \in Y\}.$$

Now suppose that X and Y are both flats (affine subspaces), that there is a unique pair of points $x_0 \in X$ and $y_0 \in Y$ for which $\|x_0 - y_0\|_2 = \text{dist}(X, Y)$, and that the linear subspaces $X - x_0$ and $Y - y_0$ are mutually orthogonal. Then the flats X and Y are said to be *orthogonal* when $x_0 = y_0$ (so that $\text{dist}(X, Y) = 0$) and *skew-orthogonal* when $x_0 \neq y_0$ (so that $\text{dist}(X, Y) > 0$).

3.6.5 *Suppose that J is a j-dimensional body in $\mathbb{R}^n$ and K is a k-dimensional body in $\mathbb{R}^n$. Let*

$$P = J + K \qquad \text{and} \qquad Q = \text{conv}(J \cup K).$$

If the flats $\text{aff}\, J$ *and* $\text{aff}\, K$ *are orthogonal, then P is a $(j+k)$-dimensional body with*

$$V_{j+k}(P) = V_j(J) \cdot V_k(K).$$

If the flats $\text{aff}\, J$ *and* $\text{aff}\, K$ *are skew-orthogonal, then Q is a $(j + k + 1)$-dimensional body with*

$$V_{j+k+1}(Q) = \frac{j!k!}{(j+k+1)!} \,\text{dist}\,(\text{aff}\, F, \text{aff}\, G) \cdot V_j(J) \cdot V_k(K).$$

The first formula in 3.6.5 is just the standard one for the volume of a cartesian product. Suppose, in particular, that $k, n_1, \ldots, n_k$ are fixed positive integers. Then, for given ($\mathcal{H}$- or $\mathcal{V}$-) polytopes $P_1 \subset \mathbb{R}^{n_1}, \ldots, P_k \subset \mathbb{R}^{n_k}$ and $P = P_1 \times \ldots \times P_k$ we have

$$V(P) = V(P_1) \cdot \ldots \cdot V(P_k),$$

and hence by 3.1.1, $V(P)$ can be computed in polynomial time. The computation of $V(P_1), \ldots, V(P_k)$ is generally more efficient than direct computation of $V(P)$.

The second formula in 3.6.5 appears in [GrKL94] for the case in which J and K are both simplices, whence the general formula follows easily for polytopes by dissection and then for general bodies by approximation.

4. Deterministic methods for computing mixed volumes

4.1. Using a volume oracle

In this first subsection we will outline the most natural approach for computing mixed volumes, a method directly suggested by Theorem 2.4.1, which is based on a procedure for volume computation.

Let us consider an arbitrary procedure $\mathcal{B}$ (efficient or not) for Weak Volume Computation; so, suppose that $\mathcal{B}$ is an algorithm which, for a body K given by a weak optimization oracle and for a given positive rational λ, produces a rational number μ such that $|V(K) - \mu| \le \lambda$. In this subsection we show how such an algorithm can be used for the (weak) computation of mixed volumes.

Note first that when the number r of bodies $K_1, \ldots, K_r$ and the binary sizes of the positive rationals $\xi_1, \ldots, \xi_r$ are bounded by a polynomial in n, then the weak optimization oracles for $K_1, \ldots, K_r$ can be used to devise a weak optimization oracle for $\sum_{i=1}^r \xi_i K_i$. We will now try to gain information about mixed volumes by calling $\mathcal{B}$ for various such linear combinations of $K_1, \ldots, K_r$.

Let us begin with some remarks about the maximum number of different mixed volumes. By 2.4.1,

$$V\left(\sum_{i=1}^r \xi_i K_i\right) = \sum_{i_1=1}^r \sum_{i_2=1}^r \cdots \sum_{i_n=1}^r \xi_{i_1}\xi_{i_2}\cdots\xi_{i_n} V(K_{i_1}, K_{i_2}, \ldots, K_{i_n}).$$

Thus we have r^n coefficients $V(K_{i_1}, K_{i_2}, \ldots, K_{i_n})$. However, these coefficients are order-independent and hence only

$$\binom{n+r-1}{r-1}$$

of them can actually be distinct, for this is the number of different multivariate monomials of degree n in $\mathbb{R}^r$. It follows that *if r is fixed*, their number is polynomial in n, and *if n is fixed*, their number is polynomial in r. However, it also follows that in general, the task of computing *all* mixed volumes cannot be accomplished in polynomial time since the number of different mixed volumes may grow exponentially.

Now, for $x = (\xi_1, \ldots, \xi_r)$ and $q = (\kappa_1, \ldots, \kappa_r) \in (\mathbb{N} \cup \{0\})^r$, let x^q denote again the monomial

$$x^q = \xi_1^{\kappa_1} \cdot \xi_2^{\kappa_2} \cdot \ldots \cdot \xi_r^{\kappa_r}.$$

Further, let

$$Q_n = \left\{ q = (\kappa_1, \ldots, \kappa_r) \in (\mathbb{N} \cup \{0\})^r : \sum_{i=1}^r \kappa_i = n \right\},$$

and for $q \in Q_n$ let

$$c_q = \binom{n}{\kappa_1, \ldots, \kappa_r} V(\overbrace{K_1, \ldots, K_1}^{\kappa_1}, \ldots, \overbrace{K_r, \ldots, K_r}^{\kappa_r}).$$

Here, as usual,

$$\binom{n}{\kappa_1, \ldots, \kappa_r} = n! \cdot \prod_{i=1}^r \frac{1}{\alpha_i} \qquad \text{where} \quad \alpha_i = \begin{cases} 1 & \text{if } \kappa_i = 0; \\ \kappa_i! & \text{else} \end{cases}$$

is a *multinomial coefficient*. Setting

$$\pi(x) = V\left(\sum_{i=1}^r \xi_i K_i\right),$$

the mixed volume expression 2.4.1 of π reads

$$\pi(x) = \sum_{q \in Q_n} c_q x^q.$$

Assume that the elements of Q_n are ordered (for instance lexicographically),

$$Q_n = \{q_1, \ldots, q_k\}, \qquad \text{with} \quad k = \binom{n+r-1}{r-1}.$$

Then, a choice of k nonnegative rational row vectors $y_1, \ldots, y_k$ of $\mathbb{R}^r$ and evaluation of $\pi(x)$ at these interpolation points leads to the matrix equation

$$p = \begin{pmatrix} \pi(y_1) \\ \vdots \\ \pi(y_k) \end{pmatrix} = \begin{pmatrix} y_1^{q_1} & y_1^{q_2} & \cdots & y_1^{q_k} \\ \vdots & \vdots & & \vdots \\ y_k^{q_1} & y_k^{q_2} & \cdots & y_k^{q_k} \end{pmatrix} \begin{pmatrix} c_{q_1} \\ \vdots \\ c_{q_k} \end{pmatrix} = Yz.$$

Note that for $r = 2$ and $y_1 = (1, \eta_1), \ldots, y_{n+1} = (1, \eta_{n+1})$, Y is just a *Vandermonde-matrix* and hence is nonsingular whenever $\eta_1, \ldots, \eta_{n+1}$ are pairwise distinct. In this case $\hat{\pi} = \pi|_{\{1\}\times\mathbb{R}}$ can be expressed in terms of the standard *Lagrange interpolation polynomials*, and there is a considerable literature on how to choose the interpolation nodes and do the computation in an efficient and numerically stable way; see e.g. [BeZ65], [Sa'74], [Ri75], [Ri90], or [MiM85]; see also 7.2 for a more explicit description of the case $r = 2$ in terms of Lagrange polynomials.

Now suppose we have chosen $y_j = (\eta_{j,1}, \ldots, \eta_{j,r})$ for $j = 1, \ldots, k$ such that Y is nonsingular. Further, let $\mu_1, \ldots, \mu_k$ be the rationals produced by $\mathcal{B}$ when applied to the bodies $\sum_{i=1}^{r} \eta_{j,i} K_i$, respectively, whence

$$|\pi(y_j) - \mu_j| \leq \lambda \qquad \text{for } j = 1, \ldots, k.$$

Now, let

$$A = Y^{-1}, \quad \hat{p} = (\mu_1, \ldots, \mu_k)^T, \quad \text{and} \quad \hat{z} = A\hat{p}.$$

Then

$$||\hat{z} - z||_\infty = ||A(\hat{p} - p)||_\infty \leq ||A||\,||\hat{p} - p||_\infty \leq \lambda ||A||,$$

where $||A||$ is the matrix norm induced by $||\ \ ||_\infty$, i.e. the maximum of the ℓ_1 norms of the rows of A.

It can now be shown (see [St71, p.55], [ChY77], [Ol86]; cf. also 3.4.1) that the interpolation points $y_1, \ldots, y_k$ can be appropriately specified so as to yield the first case of the following result; the assertion for fixed n but varying r then follows in a standard way.

4.1.1 *Whenever r is fixed or n is fixed, there is an algorithm for (weakly) computing all mixed volumes of r bodies in $\mathbb{R}^n$ given by weak optimization oracles that uses a polynomial number of arithmetic operations on rationals of polynomially bounded sizes and a polynomial number of calls to the algorithm $\mathcal{B}$.*

Note that in order to compute one specific mixed volume by this method, we must essentially compute all of them. Further, 4.1.1 does not cover the case of varying n

and r. In particular, it is unclear whether there is an efficient way of computing, say, $V(K_1, \ldots, K_n)$. Theorem 5.2.2 below gives some indication that this might not be the case.

Finally note that we have used here a quite strong algorithm for volume approximation. We will use a weaker approximation routine in Subsection 7.2, and we will comment on the difference there. Different measures for approximation errors will be introduced in Subsection 6.1.

4.2. Polytopes

Theorems 3.1.1 and 4.1.1 together show that there is a polynomial-time algorithm for computing all mixed volumes of polytopes in fixed dimensions.

4.2.1 *When the dimension n is fixed, there is a polynomial-time algorithm which, given $r \in \mathbb{N}$ and ($\mathcal{V}$- or $\mathcal{H}$-) polytopes $P_1, \ldots, P_r$, computes all mixed volumes $V(P_{i_1}, \ldots, P_{i_n})$.*

This algorithm makes use of the fact that the given bodies are actually polytopes only in the subroutine for Volume Computation. It is, however, possible to express the mixed volumes of polytopes as the sum of volumes of polytopes formed as sums of faces, and hence devise an algorithm that makes much stronger use of the facial structure of the polytopes, is more combinatorial and therefore possibly numerically more stable.

For $r \geq 2$ let $P_1, \ldots, P_r$ be polytopes in $\mathbb{R}^n$. When applied to the mapping

$$\varphi : P_1 \times \ldots \times P_r \to P_1 + \ldots + P_r \qquad \text{defined by} \quad \varphi(x_1, \ldots, x_r) = x_1 + \ldots + x_r,$$

the lifting theorem of Walkup & Wets [WaW69] yields a dissection of $P_1 + \ldots + P_r$ into polytopes $F_1 + \ldots + F_r$, where $(F_1, \ldots, F_r)$ varies over suitable r-tuples of faces F_i of P_i; see [McS83], [PeS92], [HuS93]. Using the expansion of $V(P_1 + \ldots + P_r)$ into mixed volumes and comparing coefficients, this dissection can be used to obtain a representation for mixed volumes in terms of the volumes $V(F_1 + \ldots + F_r)$. An explicit formula of this kind was given by Betke [Be'92] for $r = 2$ and Schneider [Sc94] in the general case.

In order to state Schneider's result (in Theorem 4.2.2) precisely, we need to introduce some notation.

For a polytope P and a face F of P, let $N(P, F)$ denote the cone of outer normals of P at F. Further, let us call the vectors $v_1, \ldots, v_r \in \mathbb{R}^n$ *admissible* for $P_1, \ldots, P_r$ if

(i) there is an $i \in \{1, \ldots, r\}$ such that $v_i \neq 0$;
(ii) $\sum_{i=1}^{r} v_i = 0$;
(iii) $\bigcap_{i=1}^{r} (\operatorname{relint} N(P_i, F_i) - v_i) = \emptyset$ whenever, for $i = 1, \ldots, r$, F_i is a face of P_i and $\sum_{i=1}^{r} \dim F_i > n$.

Note that the third condition is invariant under a common translation of the vectors $v_1, \ldots, v_r$, whence (iii) is the only relevant condition for admissibility. Now suppose that $v_1, \ldots, v_r$ do not satisfy (iii). Then there exist faces $F_1, \ldots, F_r$ of $P_1, \ldots, P_r$, respectively, with $\sum_{i=1}^r \dim F_i > n$, and a vector $x_0 \in \mathbb{R}^n$ such that for each $i = 1, \ldots, r$ the appropriate hyperplane perpendicular to the vector $x_0 + v_i$ supports P_i in the face F_i. Hence $z = ((x_0+v_1)^T, \ldots, (x_0+v_r)^T)^T \in \mathbb{R}^{nr}$ supports the polytope $P = P_1 \times \ldots \times P_r$ of $\mathbb{R}^{nr}$ in its face $F = F_1 \times \ldots \times F_r$, whence $z \in \operatorname{relint} N(P, F)$. This implies in particular that $v + S \subset \operatorname{lin}(N(P,F))$, where $v = (v_1^T, \ldots, v_r^T)^T$ and S is the n-dimensional subspace of $\mathbb{R}^{nr}$ of vectors of the form $(x^T, \ldots, x^T)^T$ with $x \in \mathbb{R}^n$. Now consider the (linear) hyperplane arrangement $\mathcal{H}$ in $\mathbb{R}^{nr}$ that is formed by all hyperplanes that are orthogonal to an edge of P. The condition that $\sum_{i=1}^r \dim F_i > n$ implies that $\dim N(P,F) \leq (r-1)n - 1$, whence the n-dimensional affine subspace $v+S$ of $\mathbb{R}^{nr}$ meets a face of $\mathcal{H}$ of dimension $(r-1)n-1$; so, admissibility is just a general position condition that is "generically" satisfied.

In practice, to find vectors $v_1, \ldots, v_r$ that are admissible for $P_1, \ldots, P_r$ one would essentially choose $v_1, \ldots, v_r$ at random. In a deterministic approach one might construct the face-lattice of $\mathcal{H}$ (using the algorithm of EDELSBRUNNER, O'ROURKE & SEIDEL [EdOS86], [EdSS91]), then add S to each cell of dimension $(r-1)n-1$ (and if necessary add further lines to obtain hyperplanes), and then find an interior point of a full-dimensional cell of this new arrangement.

Let, as in the previous subsection,

$$Q_n = \left\{q = (\kappa_1, \ldots, \kappa_r) \in (\mathbb{N} \cup \{0\})^r : \sum_{i=1}^r \kappa_i = n\right\},$$

and, for $q \in Q_n$, let $\mathcal{F}_q$ denote the set of all r-tuples $(F_1, \ldots, F_r)$ of faces of $P_1, \ldots, P_r$, respectively, for which

$\dim F_i = \kappa_i$, for $i = 1, \ldots, r$;
$\dim(\sum_{i=1}^r F_i) = n$;
$\bigcap_{i=1}^r (N(P_i, F_i) - v_i) \neq \emptyset$.

Then SCHNEIDER [Sc94] proved the following representation theorem.

4.2.2 *Let $P_1, \ldots, P_r$ be polytopes of $\mathbb{R}^n$, let $v_1, \ldots, v_r \in \mathbb{R}^n$ be admissible for $P_1, \ldots, P_r$, and let $q = (\kappa_1, \ldots, \kappa_r) \in Q_n$. Then*

$$\binom{n}{\kappa_1, \ldots, \kappa_r} V(\overbrace{P_1, \ldots, P_1}^{\kappa_1}, \ldots, \overbrace{P_r, \ldots, P_r}^{\kappa_r}) = \sum_{\mathcal{F}_q} V(F_1 + \ldots + F_r).$$

Let us point out that, when $P_1, \ldots, P_r$ are $\mathcal{H}$-polytopes, for a given r-tuple $(F_1, \ldots, F_r)$ of ($\mathcal{V}$- or $\mathcal{H}$-presented) faces of $P_1, \ldots, P_r$, respectively, it can be checked in polynomial time (using Gaussian elimination and linear programming) whether $(F_1, \ldots, F_r) \in \mathcal{F}_q$. Note, further, that 4.2.2 can be used to prove 4.2.1, that in fixed dimension, mixed volumes of polytopes can be computed in polynomial time.

We close this section with a tractability result of [DyGH94] that holds even when the dimension is part of the input.

4.2.3 *There is a polynomial time algorithm for checking whether, for given $n, r \in \mathbb{N}$,*

($\mathcal{H}$- or $\mathcal{V}$-) polytpes $P_1, \ldots, P_r$ of $\mathbb{R}^n$, and $q = (\kappa_1, \ldots, \kappa_r) \in Q_n$,

$$V(\overbrace{P_1, \ldots, P_1}^{\kappa_1}, \ldots, \overbrace{P_r, \ldots, P_r}^{\kappa_r}) = 0.$$

This result does not seem to be striking, but it is not trivial. In fact, suppose for notational simplicity that $r = n$, and that the origin belongs to the relative interior of all polytopes. Then select for each i a basis A_i of $\text{lin}P_i$. It is easy to see that $V(P_1, \ldots, P_n) \neq 0$ if and only if there is a choice of $(a_1, \ldots, a_n) \in A_1 \times \ldots \times A_n$ such that $\det(a_1, \ldots, a_n) \neq 0$. Let $A = \bigcup_{i=1}^n A_i$, let $\mathcal{I}_L$ be the family of all linearly independent subsets of A, and let $\mathcal{I}_P$ denote the family of all subsets of A which meet each of the A_i in at most one element. Then the pairs $\mathcal{M}_L = (A, \mathcal{I}_L)$ and $\mathcal{M}_P = (A, \mathcal{I}_P)$ are *matroids*, called respectively a *linear matroid* and a *partition matroid.* Now $V(P_1, \ldots, P_n) \neq 0$ if and only if $\mathcal{M}_L$ and $\mathcal{M}_P$ have a common basis, and this can be detected in polynomial time by the *matroid intersection theorem* of EDMONDS [Ed70]; see also [GrLS88, Section 7.5].

4.3. POLYTOPES AND BODIES

This subsection discusses a specific formula due to MINKOWSKI [Mi11] for the mixed volume of a convex body and $n-1$ copies of a polytope; see also [BoF34] and [Sc93].

With $h : \mathcal{K}^n \times \mathbb{S}^{n-1} \to \mathbb{R}$ denoting as before the support function, Minkowski proved the following result.

4.3.1 *Let K be a body and P a polytope in $\mathbb{R}^n$, let $F_1, \ldots, F_m$ denote P's facets, and let $u_1, \ldots, u_m$ be the corresponding outer unit normals. Then*

$$V(K, \overbrace{P, \ldots, P}^{n-1}) = \frac{1}{n} \sum_{i=1}^{m} h(K, u_i) V_{n-1}(F_i).$$

Let us mention, as a side remark, that 4.3.1 can be applied to a polytope P of the form $P = P_1 + \ldots + P_{n-1}$; it then yields a similar representation for $V(K, P_1, \ldots, P_{n-1})$ (see [BoF34, p.42]).

Suppose now that P is an $\mathcal{H}$-polytope and that the given presentation is irredundant. This means, in particular, that (not necessarily unit) normal vectors of all facets are given. Further suppose that the volumes of the facets of P are known, and that K is given by a weak optimization oracle. Then formula 4.3.1 allows us to approximate $V(K, P, \ldots, P)$ with the aid of m calls to the optimization oracle.

For general polytopes, and when the dimension is part of the input, this is not particularly encouraging since the problem is only polynomially reduced to VOLUME COMPUTATION for the facets of P. If, however, P belongs to a class of polytopes for which the facet volumes can be obtained efficiently, or if we just consider all computations that involve only P as "preprocessing" (since we may want to compute $V(K, P, \ldots, P)$ for many different bodies K but fixed P), then 4.3.1 may even be algorithmically useful.

It may be worthwhile to point out that some of the problems disappear when different *data structures* are used. This is particularly apparent in connection with the algorithmic significance of 4.3.1. Indeed, recall that by a theorem of MINKOWSKI [Mi1897], [Mi03], a polytope is uniquely determined (up to translation) by its facet volumes and its facet normals. Hence, the "tractability statement" related to 4.3.1 says essentially that if we choose, as our data structure for polytopes, a *Minkowski presentation* – i.e., a list of facet volumes and the associated facet normals – then $V(K, P, \ldots, P)$ can be approximated in polynomial time for arbitrary bodies given by a weak optimization oracle. However, the problem of passing from a given $\mathcal{V}$- or $\mathcal{H}$-presentation to a Minkowski-presentation is algorithmically difficult (see 5.2) unless the dimenson is fixed. The same is true for the reverse transformation; see [GrH94].

4.4. SPECIAL CONVEX BODIES

There are other formulas and integral representations known for mixed volumes in general or for certain classes of bodies (see e.g. [BoF34], [Ha57], [BuZ88], [Sc93]) whose algorithmic significance seems, however, restricted to very particular cases. In the present subsection we will just mention two explicit formulas for the mixed volumes of a body and a ball or a parallelotope. We begin with the *intrinsic volumes of polytopes.*

As was noted already in Subsection 2.4, the expansion of $V(P+\xi\mathbb{B}^n)$ into mixed volumes leads to quermassintegrals or intrinsic volumes of a polytope P. For a face F of P let $\gamma(F,P)$ denote the *outer angle* of F at P (i.e. the fraction of space that is taken up by the cone of outer normals of P at some point that is relatively interior to F). Then MCMULLEN [Mc75] gave the following representation of the intrinsic volumes.

4.4.1 *For* $i = 0, \ldots, n,$

$$V_i(P) = \sum_{F \in \mathcal{F}_i(P)} \gamma(F,P) V_i(F).$$

Evaluation of this formula involves computing the volumes of all i-dimensional faces of P, and also of the $(n-i-1)$-dimensional (spherical) volumes of spherical polytopes that are obtained by intersecting the cones $N(P,F)$ of outer normals with $\mathbb{S}^{n-1}$. While the former is algorithmically easy only for small values of i (see 3.1.1), the latter is easy only for small values of $n-i$.

We mention in passing that HADWIGER [Ha75] has given the integral formula

$$\sum_{i=0}^{n} V_i(K) = \int_{\mathbb{R}^n} e^{-\pi \operatorname{dist}^2(K,x)} dx,$$

which is a useful tool for certain lattice-point problems. The same is true (see [GrW93]) for a formula that we are going to develop now; see [BuZ88, p.141] or [Sc93, p.294].

Let $a_1, \ldots, a_n \in \mathbb{R}^n$ such that $Z = \sum_{i=1}^n [0,1]a_i$ is a proper parallelotope, let $0 < k < n$, and let $K_1, \ldots, K_{n-k} \in \mathcal{K}^n$. Further, $\Pi_S K$ denotes again the orthogonal projection of a body K onto a linear subspace S. Then the multilinearity of the mixed volume implies that

$$V(\overbrace{Z,\ldots,Z}^{k},\ K_1,\ldots,K_{n-k})$$

$$= \sum_{i_1=1}^{n}\sum_{i_2=1}^{n}\cdots\sum_{i_k=1}^{n} V([0,1]a_{i_1},\ldots,[0,1]a_{i_k},K_1,\ldots,K_{n-k}).$$

Now let $S_{i_1,\ldots,i_k} = \operatorname{lin}\{a_{i_1},\ldots,a_{i_k}\}$, then

$$\binom{n}{k}\ V([0,1]a_{i_1},\ldots,[0,1]a_{i_k},K_1,\ldots,K_{n-k})$$

$$= \begin{cases} \alpha_{i_1,\ldots,i_k} & \text{if } \dim S_{i_1,\ldots,i_k} = k; \\ 0 & \text{otherwise,} \end{cases}$$

where

$$\alpha_{i_1,\ldots,i_k} = V_{S^{\perp}_{i_1,\ldots,i_k}}(\Pi_{S^{\perp}_{i_1,\ldots,i_k}}K_1,\ldots,\Pi_{S^{\perp}_{i_1,\ldots,i_k}}K_{n-k})\cdot V_{S_{i_1,\ldots,i_k}}([0,1]a_{i_1},\ldots,[0,1]a_{i_k});$$

the subscripts $S^{\perp}_{i_1,\ldots,i_k}$ and $S_{i_1,\ldots,i_k}$ indicate that the corresponding mixed volumes are taken with respect to the spaces $S^{\perp}_{i_1,\ldots,i_k}$ and $S_{i_1,\ldots,i_k}$, respectively. If we specialize this formula to the case $Z = C_n = [0,1]^n = \sum_{i=1}^{n}[0,1]e_i$ and $K_1 = \ldots = K_{n-k} = K$ we obtain the following result.

4.4.2 *For* $k = 0,\ldots,n$,

$$\binom{n}{k} V(\overbrace{C_n,\ldots,C_n}^{k},\overbrace{K,\ldots,K}^{n-k}) = \sum V_{n-k}(\Pi_S K),$$

where S ranges over all $(n-k)$-dimensional coordinate subspaces of $\mathbb{R}^n$.

Another useful specialization is obtained for $K_1 = \ldots = K_{n-k} = \mathbb{B}^n$; it leads to a simple formula for the intrinsic volumes of a parallelotope.

4.4.3 *Let $Z = \sum_{i=1}^{n}[0,1]a_i$ be a proper parallelotope. Then, for $k = 0,\ldots,n$,*

$$2^{n-k}V_k(Z) = \sum_{F\in\mathcal{F}_k(Z)} V_k(F).$$

Hence, in order to compute the intrinsic volumes of Z, we need only compute the k-volume of its *k-skeleton* $\bigcup_{F\in\mathcal{F}_k(Z)} F$. This can be done inductively. In fact, if for $j = 0,\ldots,k$ and $m = 1,\ldots,n$

$$S(j,m) = \sum_{F\in\mathcal{F}_j(\sum_{i=1}^{m}[0,1]a_i)} V_j(F),$$

then we have (with appropriate conventions in the "boundary cases")

$$S(j,m+1) = 2S(j,m) + S(j-1,m)\cdot\|a_{m+1} - A(A^TA)^{-1}A^Ta_{m+1}\|_2,$$

where A is the $n \times m$ matrix with column vectors $a_1, \ldots, a_m$. Thus we obtain the following result as a corollary of 4.4.3.

4.4.4 *The mixed volumes of an $\mathcal{S}$-parallelotope Z in $\mathbb{R}^n$ can be approximated up to an additive rational error $\epsilon > 0$ in time that is polynomial in n, in the size of Z's presentation, and in* size(ϵ).

5. Intractability results

5.1. Volume computations

In striking contrast to the "positive" results 3.1.1, 3.1.2, 3.3.4, 3.6.4, there are several strong intractability results for VOLUME COMPUTATION. Theorem 5.1.1 summarizes the former, and the latter appear in Theorems 5.1.3–5.1.5 and 5.1.7.

5.1.1 *The volume of a polytope P can be computed in polynomial time in the following cases:*

(i) *if the dimension is fixed and P is a $\mathcal{V}$- or $\mathcal{H}$-polytope or an $\mathcal{S}$-zonotope;*
(ii) *if the dimension is part of the input and P is a near-simple $\mathcal{V}$- or a near-simplicial $\mathcal{H}$-polytope or a near-parallelotopal $\mathcal{S}$-zonotope.*

By 5.1.1 (i) there is a polynomial-time algorithm for VOLUME COMPUTATION when the dimension is fixed. However, the methods of volume computation that we described in Section 3 all require exponential time when n is part of the input. Hence it is natural to wonder whether there is a more robustly polynomial-time procedure for volume computation.

To set the stage, let us begin with a negative result that was mentioned in a different setting in Theorem 3.2.3.

5.1.2 *There does not exist a polynomial-space algorithm for exact computation of the volume of $\mathcal{H}$-polytopes.*

Since each polynomial-time algorithm uses only a polynomial amount of space, Theorem 5.1.2 implies that there is no polynomial-time algorithm which, given a dimension n and an n-dimensional $\mathcal{H}$-polytope P, computes $V(P)$. Doesn't this result already show that volume computation is actually much harder than such $\mathbb{NP}$-complete problems as the TRAVELING SALESMAN PROBLEM? The answer is "Not really!" and we take a few sentences to explain why (to a reader who is less familiar with the relevant concepts of complexity theory).

In the realm of $\mathbb{P}$ and $\mathbb{NP}$, complexity theory usually deals with problems whose answer is "yes" or "no" since this corresponds to the results of a halting Turing machine computation. (When dealing with the class $\#\mathbb{P}$, the Turing machine is augmented by a device that counts accepting computations.) This means that when dealing with related complexity results, the proper formulation of VOLUME COMPUTATION is as follows.

VOLUME.

Instance: *A positive integer n, an $\mathcal{H}$-polytope (or a $\mathcal{V}$-polytope, or an $\mathcal{S}$-zonotope) P, a nonnegative rational ν.*

Question: *Is the volume of P bounded above by ν, i.e. is $V(P) \leq \nu$?*

In order to distinguish the different classes of input polytopes we will sometimes speak of the problems $\mathcal{H}$-VOLUME, $\mathcal{V}$-VOLUME and $\mathcal{S}$-VOLUME, respectively.

Note that the above problems could have equally well been phrased in terms of lower bounding $V(P)$. We use upper-bounding only to associate a "yes" answer with instances in which P is lower-dimensional, a special and easy case.

Suppose now that we had a polynomial-time routine for solving $\mathcal{H}$-VOLUME. Then, using binary search (with appropriately specified values of ν) we could approximate $V(P)$ with any polynomial-size accuracy in polynomial time. Hence, Theorem 5.1.2 does not rule out the possibility that $\mathcal{H}$-VOLUME is in $\mathbb{P}$ and that computing any number of polynomially many digits of $V(P)$ for $\mathcal{H}$-polytopes P is actually easy.

DYER & FRIEZE [DyF88] showed, however, that both $\mathcal{H}$-VOLUME and $\mathcal{V}$-VOLUME are $\#\mathbb{P}$-hard, and we are going to describe various hardness proofs that are all geometric in nature. (See also KHACHIYAN [Kh88], [Kh89], [Kh93]). We begin with $\mathcal{H}$-VOLUME.

Let us point out that, in the following, we are going to deal with hardness results which involve classes of $\mathcal{H}$-polytopes for which the volume is of polynomial size. Hence, a polynomial time method for VOLUME would, in fact, result in a polynomial-time algorithm for VOLUME COMPUTATION.

The first proof stated here for the $\mathbb{NP}$-hardness of the problem of computing the volume of certain simple $\mathcal{H}$-polytopes utilizes the sweeping-plane formula 3.2.1; it is due to KHACHIYAN [Kh88], [Kh93].

As in 3.2.1, let P be a simple $\mathcal{H}$-polytope with corresponding irredundant $\mathcal{H}$-presentation $(n, m; A, b)$. Further, let $c \in \mathbb{R}^n$ such that $\langle c, \ \rangle$ is not constant on any edge of P, and let $H(\tau) = \{x \in \mathbb{R}^n : \langle c, x\rangle \leq \tau\}$ for $\tau \in \mathbb{R}$. Then, by 3.2.1,

$$V(P \cap H(\tau)) = \frac{(-1)^n}{n!} \sum_{v \in \mathcal{F}_0(P)} \frac{(\max\{0, \tau - \langle c, v\rangle\})^n}{\prod_{i=1}^n e_i^T A_{I_v}^{-1} c |\det(A_{I_v})|}.$$

This implies that

$$\varphi(\tau) = V(P \cap H(\tau))$$

is a piecewise polynomial of degree at most n, and is $(n-1)$-times continuously differentiable. (This result has also been proved in the theory of splines (cf. DE BOOR & HÖLLIG [BoH82]), and it is also relevant to some problems in geometric tomography; see [GaG94].) Further, if $\tau_0 < \ldots < \tau_k$ are the (ordered) values of τ for which bd$H(\tau)$ contains a vertex of P, the nth derivative of φ is discontinuous at most at $\tau_0, \ldots, \tau_r$, and at these points the one-sided derivatives satisfy the equation

$$\frac{d^n\varphi(\tau_j+)}{d\tau^n} - \frac{d^n\varphi(\tau_j-)}{d\tau^n} = \sum_{v \in \mathcal{F}_0(P), \langle c, v\rangle = \tau} \frac{(-1)^{\delta(v)}}{|\prod_{i=1}^n e_i^T A_{I_v}^{-1} c \det(A_{I_v})|},$$

where $\delta(v) = \text{card}\{w \in \mathcal{F}_0(P) : \langle c, w\rangle < \langle c, v\rangle\}$.

Specifically, if $C_n = [0,1]^n$ and $c > 0$, then for each vertex v of C_n we have $\delta(v) = ||v||_1$. Let us assume that c has the following "constant-one" property that whenever two vertices v, w of C_n are such that $\langle c, v\rangle = \langle c, w\rangle$, then $||v||_1 = ||w||_1$. Then,

$$\text{there is a vector } v \in \{0,1\}^n \text{ with } \langle c, v\rangle = \kappa$$

$$\text{if and only if}$$

$$\frac{d^n\varphi(\kappa+)}{d\tau^n} - \frac{d^n\varphi(\kappa-)}{d\tau^n} \neq 0.$$

Now suppose that we could compute the volume of the intersection $\varphi(\tau) = C_n \cap H(\tau)$ in polynomial time. Since, unless $\text{bd}H(\tau)$ meets a vertex of C_n, $\varphi(\tau)$ is a polynomial in τ of degree at most n, we can check the differentiability condition in polynomial time. Hence the problem,

> Given $c \in \mathbb{N}^n$ with the constant-one property and $\kappa \in \mathbb{N}$, is there a 0-1-vector v with $\langle c, v\rangle = \kappa$?,

is transformed in polynomial time into the problem of computing the volume of an $\mathcal{H}$-polytope. But except for the additional property of c this is equivalent to SUBSET-SUM:

> Given positive integers n, $\gamma_1, \ldots, \gamma_n$ and a positive integer κ, is there a subset I of $\{1, \ldots, n\}$ such that $\sum_{i \in I} \gamma_i = \kappa$?

SUBSET-SUM is known to be $\mathbb{NP}$-complete, [Sa74] (see also [GaJ79]). On the other hand, any instance of SUBSET-SUM can easily be transformed to the n instances of the required restricted version that are obtained by replacing c by the vector $c+(||c||_1+1, \ldots, ||c||_1+1)^T$, and κ successively for $\theta = 1, \ldots, n$ by $\kappa+(||c||_1+1)\theta$. This shows that computing the volume of an $\mathcal{H}$-polytope is $\mathbb{NP}$-hard even for polytopes that are intersections of C_n with one additional (rational) halfspace.

DYER & FRIEZE [DyF88] actually proved $\#\mathbb{P}$-hardness of $\mathcal{H}$-VOLUME; see also [Kh89], [DyF91], [Kh93].

5.1.3 *The problem of computing the volume of the intersection of the unit cube with a rational halfspace is $\#\mathbb{P}$-hard.*

To prove 5.1.3, [DyF88] use a reduction of the following counting version of 0-1 KNAPSACK, a problem that is known to be $\#\mathbb{P}$-hard; see [GaJ79].

#(0-1 KNAPSACK).

Instance: *A positive integer n, a positive integer n-vector a, a positive integer β.*

Task: *Determine the cardinality of $\{v \in \{0,1\}^n : \langle a, v\rangle \leq \beta\}$.*

The polynomial-time reduction of #(0-1 KNAPSACK) to $\mathcal{H}$-VOLUME uses some ideas that are similar in spirit to the ideas exploited in the above $\mathbb{NP}$-hardness proof.

In particular, an inclusion-exclusion formula is used, and the volumes that will be computed are again values of a certain polynomial. There are, however, important differences, and it may be useful to sketch the explicit geometric construction underlying the reduction of 5.1.3.

Let $(n; a_0, \beta_0)$ be an instance of #(0-1 KNAPSACK). We may assume (by considering the instance $(n; 2a_0, 2\beta_0+1)$, if necessary) that $\{v \in \{0,1\}^n : \langle a_0, v\rangle = \beta_0\} = \emptyset$.

Now, let us define for each $v \in \{0,1\}^n$ the polytope

$$S_v = \{x \in \mathbb{R}^n : x \geq v,\ \langle a_0, x\rangle \leq \beta_0\}.$$

If S_v is full-dimensional, it is the simplex with vertices v and $(\beta_0 - \langle a_0, v\rangle)e_i/\alpha_i$, where α_i is the ith coordinate of a_0. Hence, by the standard determinant formula for the volume of a simplex,

$$V(S_v) = \frac{1}{n!}\prod_{i=1}^{n} \frac{\max\{0, \beta_0 - \langle a_0, v\rangle\}}{\alpha_i}.$$

Now, let $P = \{x \in C_n : \langle a_0, x\rangle \leq \beta_0\}$, and let $\mathbf{1} = (1, \ldots, 1)^T \in \mathbb{R}^n$. Then the inclusion-exclusion principle yields

$$\begin{aligned} V(P) &= \sum_{v\in\{0,1\}^n} (-1)^{\langle v,\mathbf{1}\rangle} V(S_v) \\ &= \frac{1}{n!}\left(\prod_{i=1}^{n}\frac{1}{\alpha_i}\right) \sum_{v\in\{0,1\}^n} (-1)^{\langle v,\mathbf{1}\rangle}(\max\{0, \beta_0 - \langle a_0, v\rangle\})^n. \end{aligned}$$

In the neighborhood $]\beta_0 - 1, \beta_0 + 1[$ of β_0, the function π_{a_0,β_0} defined by

$$\pi_{a_0,\beta_0}(\beta) = \sum_{v\in\{0,1\}^n} (-1)^{\langle v,\mathbf{1}\rangle}(\max\{0, \beta - \langle a_0, v\rangle\})^n$$

is a polynomial in β, and a procedure for volume computation would allow us to compute all coefficients of π_{a_0,β_0}. Note that the coefficient of β^n is just

$$\sum_{\substack{v \in \{0,1\}^n \\ \langle a_0, v\rangle \leq \beta_0}} (-1)^{\langle v,\mathbf{1}\rangle}.$$

Let us now compute the leading coefficients for various choices of a_0 and β_0; set for $k = 1, \ldots, n$

$$\mu = \langle a_0, \mathbf{1}\rangle + 1, \quad a_k = a_0 + \mu\mathbf{1}, \quad \text{and} \quad \beta_k = \beta_0 + \mu k.$$

We may assume in the following that $\langle a_0, \mathbf{1}\rangle > \beta_0$, since otherwise the original instance of #(0-1 KNAPSACK) is trivial.

Now let $v \in \{0,1\}^n$. Then $\langle a_k, v\rangle \leq \beta_k$ if and only if v satisfies one of the following two conditions:

(i) $\langle v, \mathbf{1}\rangle < k$;

(ii) $\langle v, \mathbf{1}\rangle = k$ and $\langle a_0, v\rangle \leq \beta_0$.

Hence the leading coefficient of π_{a_k,β_k} is

$$(-1)^k \mathrm{card}\{v \in \{0,1\}^n : \langle a_0, v\rangle \leq \beta_0,\ \langle v, \mathbf{1}\rangle = k\} + \sum_{i=0}^{k-1}(-1)^i \binom{n}{i}.$$

Since

$$\sum_{k=0}^{n} \mathrm{card}\{v \in \{0,1\}^n : \langle a_0, v\rangle \leq \beta_0,\ \langle v, \mathbf{1}\rangle = k\}$$

is actually the solution for the given instance $(n; a_0, \beta_0)$ of #(0-1-KNAPSACK), we see that a polynomial-time algorithm for volume computation would yield a polynomial-time algorithm for #(0-1-KNAPSACK).

Note that this and the previous hardness result involve, as part of the input, integers whose absolute values are not bounded by a polynomial in n. In fact, a result of KOZLOV [Ko86] shows that the volume of the intersection of the unit cube with a constant number of rational halfspaces can be computed in *pseudopolynomial time*. Thus it is natural to wonder whether the problem retains its hardness if we restrict all input data to numbers whose absolute values are bounded by a polynomial in n. It turns out that the problem of computing the volume of $\mathcal{H}$-polytopes is #$\mathbb{P}$-hard even in this strong sense. This follows from the two facts that the problem of computing the number of linear extensions of a given partially ordered set $\mathcal{O} = (\{1,\ldots,n\}, \prec)$ is #$\mathbb{P}$-complete, BRIGHTWELL & WINKLER [BrW91], and that this number is equal to $n!V(P_{\mathcal{O}})$, where the set

$$P_{\mathcal{O}} = \{x = (\xi_1,\ldots,\xi_n)^T \in [0,1]^n : \xi_i \leq \xi_j \Longleftrightarrow i \prec j\}$$

is the *order polytope* of $\mathcal{O}$; STANLEY [St86a]. In the following we will indicate the geometric essence of the latter result.

Let $N = \{1,\ldots,n\}$, and let $\mathcal{O} = (N, \prec)$ be an arbitrary poset. A *linear extension* of $\mathcal{O}$ is a total ordering of N that is compatible with $\prec$. A linear extension of $\mathcal{O}$ can be regarded as a permutation π of N (or, equally, as a vector $(\pi(1), \pi(2), \ldots, \pi(n))$) which has the property

$$i, j \in N \wedge i \prec j \quad \Longrightarrow \quad \pi^{-1}(i) < \pi^{-1}(j).$$

Let $E(\mathcal{O})$ denote the set of linear extensions of $\mathcal{O}$. Now consider for a given linear extension $\pi \in E(\mathcal{O})$ the polytope

$$T_\pi = \{x \in [0,1]^n : \xi_{\pi(1)} \leq \xi_{\pi(2)} \leq \cdots \leq \xi_{\pi(n)}\}.$$

Observe that T_π is a simplex, and that all the constraints that define the order polytope $P_{\mathcal{O}}$ are also constraints of T_π; hence $T_\pi \subset P_{\mathcal{O}}$. Further,

if π_1 and π_2 are different linear extensions of $\mathcal{O}$ then $\mathrm{int}(T_{\pi_1}) \cap \mathrm{int}(T_{\pi_2}) = \emptyset$,

and also

$$\bigcup_{\pi \in E(\mathcal{O})} T_\pi = P_{\mathcal{O}}.$$

Hence the simplices T_π, $\pi \in E(\mathcal{O})$ form a dissection of the order polytope $P_{\mathcal{O}}$. Finally note that all these simplices are congruent, and hence

$$V(T_\pi) = \frac{1}{n!}.$$

But this shows that each linear extension π of $\mathcal{O}$ contributes $1/(n!)$ to the volume of $P_{\mathcal{O}}$, and therefore

$$\operatorname{card}(E(\mathcal{O})) = n!V(P_{\mathcal{O}}).$$

Observe that the number of inequalities defining P is $O(n^2)$.

5.1.4 $\mathcal{H}$-VOLUME *is* $\#\mathbb{P}$*-hard in the strong sense.*

Let us now turn to $\mathcal{V}$-VOLUME, a problem that, in general, is slightly easier since the volume of $\mathcal{V}$-polytopes is of polynomial size; see 3.1.3. However, as DYER & FRIEZE [DyF88] show, it is not much easier; see also [Kh89], [Kh93].

5.1.5 *The problem of computing the volume of the convex hull of the regular* $\mathcal{V}$*-cross-polytope and an additional integer vector is* $\#\mathbb{P}$*-hard.*

The following proof is due to KHACHIYAN [Kh89]. Let $Q_n = \operatorname{conv}\{\pm e_1, \ldots, \pm e_n\}$, the regular cross-polytope, and for each $a \in \mathbb{Z}^n$ let $P_a = \operatorname{conv}(\{a\} \cup Q_n)$. Then P_a can be dissected into Q_n and the set $\mathcal{S}$ of all simplices $S_F = \operatorname{conv}(F \cup \{a\})$, where F is a facet of Q_n that is visible from a. Now, let $S_F \in \mathcal{S}$, and let $z \in \{-1,1\}^n$ be an outer normal to F. Then

$$V(S_F) = V(F) \cdot \operatorname{dist}(a, F) = V(F) \cdot \frac{\langle a, z\rangle - 1}{\sqrt{n}} = \frac{1}{n!}(\langle a, z\rangle - 1).$$

Therefore

$$n!V(P_a) = n!V(Q_n) + n!\sum_{S\in\mathcal{S}} V(S) = 2^n + \sum_{z\in\{-1,1\}^n} \max\{0, \langle a, z\rangle - 1\},$$

whence

$$n!\,(V(P_{a+e_1}) - 2V(P_a) + V(P_{a-e_1})) =$$

$$= \sum_{z\in\{-1,1\}^n} (\max\{0, \langle a, z\rangle - 2\} - 2\max\{0, \langle a, z\rangle - 1\} + \max\{0, \langle a, z\rangle\})$$

$$= \sum_{\substack{z \in \{-1,1\}^n \\ \langle a, z\rangle = 1}} 1.$$

This implies that if we could compute the volume of a $\mathcal{V}$-polytope in polynomial time, then we could also solve the following counting problem in polynomial time:

> Given $n \in \mathbb{N}$ and $a \in \mathbb{Z}^n$, determine the number of solutions $z \in \{-1,1\}^n$ of $\langle a, z\rangle = 1$.

However, this problem is closely related to #(0-1 KNAPSACK) and is in fact #$\mathbb{P}$-complete.

It is not known whether the problem of computing the volume of a $\mathcal{V}$-polytope is #$\mathbb{P}$-hard in the strong sense.

DYER & FRIEZE [DyF88] also show that the problem of computing the volume of a $\mathcal{V}$-polytope is #$\mathbb{P}$-easy in the following sense.

5.1.6 *Let Π be any #$\mathbb{P}$-complete problem. Then any oracle $\mathcal{O}_\Pi$ for solving Π can be used to produce an algorithm that runs in time that is oracle-polynomial in L and* size(ϵ) *for solving the following problem:*

> *Given $n \in \mathbb{N}$, a $\mathcal{V}$- or an $\mathcal{H}$-polytope P and a positive rational ϵ, compute a rational number μ such that $V(P) - \epsilon \le \mu \le V(P) + \epsilon$.*

It follows from 5.1.6 that for $\mathcal{V}$-polytopes, $\mathcal{O}_\Pi$ can be used to actually *compute* $V(P)$, while (due to Theorem 5.1.2) for $\mathcal{H}$-polytopes, $V(P)$ can only be approximated (yet in a very strong sense). Note, however, that (as remarked in [DyF88]) the question remains open as to whether there exist a *fixed constant* λ and a polynomial-time algorithm which, given $n \in \mathbb{N}$ and a $\mathcal{V}$- or an $\mathcal{H}$-polytope P, computes a rational number μ such that

$$(1-\lambda)V(P) \le \mu \le (1+\lambda)V(P).$$

See Subsection 6.3 for some related "negative" results in a different model of computation.

The final subject of this subsection is the complexity of volume computations for zonotopes.

The fact that $\mathcal{V}$- and $\mathcal{H}$-VOLUME is #$\mathbb{P}$-hard does not necessarily mean that the same is true for $\mathcal{S}$-zonotopes since, typically, zonotopes have a number of vertices and a number of facets that grow exponentially in the number of generating segments. Recall from 3.6 that we can express the volume of the zonotope $Z = \sum_{i=1}^{r}[0,1]z_i$ as a sum of determinants

$$V(Z) = \sum_{1 \le i_1 < i_2 < \cdots < i_n \le r} |\det(z_{i_1}, \ldots, z_{i_n})|.$$

Hence $\mathcal{S}$-VOLUME is equivalent to the following problem, SUM-OF-DETERMINANTS:

> Given positive integers n, r with $r \ge n$, and an integer $n \times r$ matrix A, compute $\sum |\det(B)|$, where the sum extends over all $n \times n$ submatrices of A.

Clearly, in fixed dimension, this problem can be solved in polynomial time (see 3.6.4), and, even when the dimension is part of the input, each summand can be computed in polynomial time. There are, however, exponentially many summands, and this fact accounts for the hardness of the problem. (See BEN ISRAEL [Be92] for a related but different notion of "volume" associated with the determinants of the $n \times n$ submatrices of an $n \times r$ matrix, and for the relevance of his notion to Moore-Penrose inverses of rectangular matrices.)

5.1.7 $\mathcal{S}$-VOLUME *is* #$\mathbb{P}$*-hard and also* #$\mathbb{P}$*-easy.*

Theorem 5.1.7 is due to [DyGH94]. Its hardness result is obtained by a reduction of #PARTITION, the #$\mathbb{P}$-complete task to

> determine for given $m \in \mathbb{N}$, and $\alpha_1, \ldots, \alpha_m \in \mathbb{N}$, the number of different subsets $I \subset M = \{1, \ldots, m\}$ such that $\sum_{i \in I} \alpha_i = \sum_{i \in M \setminus I} \alpha_i$.

It is not known whether $\mathcal{S}$-VOLUME is #$\mathbb{P}$-hard in the strong sense.

Let us mention in passing that the problem

> given positive integers n, r with $r \geq n$, an integer $n \times r$ matrix A, a positive integer λ, determine whether there exists an $n \times n$ submatrix of A such that $|\det(B)| \geq \lambda$,

is $\mathbb{NP}$-complete. This follows from the $\mathbb{NP}$-completeness of HAMILTONIAN CYCLE for directed graphs by a construction of PAPADIMITRIOU & YANNAKAKIS [PaY90]; see [GrKL94] for applications of this result to the problem of finding j-simplices of maximum volume in n-polytopes.

5.2. COMPUTING MIXED VOLUMES

Since volume computation is just a special case of computing mixed volumes, the hardness results of the previous subsection carry over:

5.2.1 *For each fixed* $k \in \mathbb{N}$, *and for each fixed sequence* $(q_n)_{n \in \mathbb{N}}$, *where each* q_n *is a* k*-tuple* $(\kappa_1, \ldots, \kappa_k)$ *of nonnegative integers with* $\sum_{i=1}^{k} \kappa_i = n$, *the following problem is* #$\mathbb{P}$*-hard:*

Instance: *A positive integer* n, $\mathcal{H}$- *(or* $\mathcal{V}$-*) polytopes (or* $\mathcal{S}$*-zonotopes)* $P_1, \ldots, P_k$ *of* $\mathbb{R}^n$

Task: *Determine the mixed volume* $V(\overbrace{P_1, \ldots, P_1}^{\kappa_1}, \ldots, \overbrace{P_k, \ldots, P_k}^{\kappa_k})$.

In the remainder of this subsection we will give some additional hardness results for mixed volumes that do not trivially depend on the hardness of volume computations. Let us start with the following extreme example of such a result, the hardness of computing mixed volumes of *boxes*, by which we mean rectangular parallelotopes with axis-aligned edges.

5.2.2 *The following problem is* #$\mathbb{P}$*-hard.*

Instance: *A positive integer* n; *for* $i, j = 1, \ldots, n$, *positive integers* $\alpha_{i,j}$.

Task: *Determine the mixed volume* $V(Z_1, \ldots, Z_n)$, *where* $Z_i = \sum_{j=1}^{n} [0, \alpha_{i,j}] e_j$ *for* $i = 1, \ldots, n$.

Note that this result, which is due to [DyGH94], is indeed of a different nature than 5.2.1. In fact, each of the Z_i is just a rectangular box, and so is $Z = \sum_{i=1}^{n} Z_i$.

Hence the volume $V(Z) = \prod_{j=1}^{n}(\sum_{i=1}^{n}\alpha_{i,j})$ can be computed very easily. Nevertheless, the mixed volume $V(Z_1,\ldots,Z_n)$ is hard to compute. This is in interesting contrast to the hardness result of 5.1.7, where the volume of a sum of segments is hard to compute even though each of their mixed volumes can be computed in polynomial time.

As was shown in [DyGH94], 5.2.2 can be extended to show that the #$\mathbb{P}$-hardness persists even if the integers $\alpha_{i,j}$ have only two different values α and β.

To sketch the reasoning for this result, let us compute $V(Z_1,\ldots,Z_n)$, where $Z_i = \sum_{j=1}^{n}[0,\alpha_{i,j}]e_j$.

Let $\xi_1,\ldots,\xi_n \geq 0$. Then

$$V\left(\sum_{i=1}^{n}\xi_i Z_i\right) = V\left(\sum_{j=1}^{n}\left[0,\sum_{i=1}^{n}\xi_i\alpha_{i,j}\right]e_j\right) = \prod_{j=1}^{n}\left(\sum_{i=1}^{n}\xi_i\alpha_{i,j}\right),$$

and a comparison of the coefficients of $\xi_1\cdot\xi_2\cdot\ldots\cdot\xi_n$ yields

$$V(Z_1,\ldots,Z_n) = \frac{1}{n!}\sum_{j_1=1}^{n}\cdots\sum_{j_n=1}^{n}\delta_{j_1,\ldots,j_n}\alpha_{1,j_1}\cdot\ldots\cdot\alpha_{n,j_n},$$

where

$$\delta_{j_1,\ldots,j_n} = \begin{cases} 1 & \text{if } \{j_1,\ldots,j_n\} \text{ is a permutation of } \{1,2,\ldots,n\}; \\ 0 & \text{otherwise.}\end{cases}$$

Thus

$$n!V(Z_1,\ldots,Z_n) = \operatorname{per}(A)$$

is the *permanent* of the matrix $A = (\alpha_{i,j})_{i,j=1,\ldots,n}$.

Now, VALIANT [Va77] has established the #$\mathbb{P}$-hardness of the problem of computing the permanent even for 0-1-matrices. (In fact, this problem is equivalent to counting the number of *perfect matchings* in a bipartite graph.) This gives already the hardness result 5.2.2. The sharpening, however, relies on an extension of Valiant's result since it requires α and β to be positive or, equivalently, the parallelotopes to be full-dimensional.

Note that by 4.1.1 (in conjunction with 1.2.2), the mixed volumes of boxes $Z_1,\ldots,Z_r$ can be computed in polynomial time if the number r of boxes is constant. (Recall that in 5.2.2 we had $r = n$.) However, this result relies in an essential way on the fact that each of the rectangular parallelotopes has axis-parallel edges. When this restriction is lifted, even the case $r = 2$ becomes hard [DyGH94].

5.2.3 *The following problem is #$\mathbb{P}$-hard.*

Instance: *Positive integers n and k with $k < n$, two n-tuples $v_1, \dots, v_n$ and $w_1, \dots, w_n$ of integer vectors which each form an orthogonal basis of $\mathbb{R}^n$.*

Task: *Compute the mixed volume*

$$V(\overbrace{Z_1, \dots, Z_1}^{k}, \overbrace{Z_2, \dots, Z_2}^{n-k}),$$

where $Z_1 = \sum_{j=1}^{n} [0,1] v_j$ and $Z_2 = \sum_{j=1}^{n} [0,1] w_j$.

6. Deterministic approximation of volumes and mixed volumes

6.1. Measures for approximation errors

Since it is algorithmically difficult to compute the volume of a given body (or polytope) K, it is of interest to approximate $V(K)$ from above or below. The same is true for mixed volumes.

In general, the approximation of a (nonnegative) functional ρ defined on a class of bodies involves, first, an a priori measure for the closeness of approximation.

Typical measures of the closeness of a number μ and the function value $\rho(K)$ for a given convex body K include the *absolute error*

$$|\mu - \rho(K)|$$

and the *relative error*

$$\left| \frac{\mu - \rho(K)}{\rho(K)} \right|.$$

Obviously, the results of Section 5 and the fact that the absolute error changes after scaling K indicate that the absolute error is not an adequate measure for our purposes. The relative error introduced above is adequate for "positive" results that involve a small positive rational error bound λ. However, the relative error is biased toward underestimation in the sense that $\mu = 0$ always produces the error 1. Since we are interested in a *symmetric relative error* measure we define for an arbitrary positive rational λ a (rational) λ-*approximation* of $\rho(K)$ to be a positive rational number μ such that

$$\frac{\rho(K)}{\mu} \le 1 + \lambda \qquad \text{and} \qquad \frac{\mu}{\rho(K)} \le 1 + \lambda.$$

Note that this criterion can also be stated as follows

$$-\frac{\lambda}{1+\lambda} \le \frac{\mu - \rho(K)}{\rho(K)} \le \lambda.$$

In the remainder of the section we will deal mainly with the following problem for a positive functional $\lambda : \mathbb{N} \to \mathbb{R}$ and with ρ representing the volume or some mixed volume.

λ-APPROXIMATION for ρ

Instance: *A positive integer n, a well-bounded body K given by a (strong or weak) separation oracle.*

Task: *Determine a positive rational μ such that*

$$-\frac{\lambda}{1+\lambda} \le \frac{\mu - \rho(K)}{\rho(K)} \le \lambda.$$

For abbreviation we will sometimes use the terms VOLUME APPROXIMATION and MIXED VOLUME APPROXIMATION for the task of solving λ-APPROXIMATION for the volume or for some mixed volumes, respectively.

6.2. UPPER BOUNDS

A quite general tool for obtaining estimates of functionals, even for arbitrary convex bodies, is suggested by a theorem of JOHN [Jo48]. (A strengthening of this result for symmetric bodies appeared in [Jo42].)

6.2.1 *For a body K in $\mathbb{R}^n$, let $a_0 \in \mathbb{R}^n$ and let A_0 be a linear transformation such that $E_0 = a_0 + A(\mathbb{B}^n)$ is the ellipsoid of maximum volume inscribed in K. Then*

$$a_0 + A_0(\mathbb{B}^n) \subset K \subset a_0 + nA_0(\mathbb{B}^n).$$

Any ellipsoid $E = a + A(\mathbb{B}^n)$ that satisfies the inclusion relation $a + A(\mathbb{B}^n) \subset K \subset a+nA(\mathbb{B}^n)$ is called a *Löwner-John ellipsoid* for K. Observe that the dilatation factor n in John's theorem is best possible for the simplex (and only for the simplex [Pa92]). See the book [Pi89] for additional results on contained and containing ellipsoids.

In order to obtain approximative algorithms, one needs of course an algorithmic version of Theorem 6.2.1, or at least a polynomial-time method for approximating the ellipsoid E_0 in 6.2.1 (and in this way obtaining *weak Löwner-John ellipsoids*). Such an algorithm was devised by GRÖTSCHEL, LOVÁSZ & SCHRIJVER [GrLS88], using the ellipsoid method of linear programming.

6.2.2 *There exists an oracle-polynomial-time algorithm which, for any well-bounded body K of $\mathbb{R}^n$ given by a weak separation oracle, finds a point a and a linear transformation A such that*

$$a + A(\mathbb{B}^n) \subset K \subset a + (n+1)\sqrt{n}A(\mathbb{B}^n).$$

Further, the dilatation factor $(n+1)\sqrt{n}$ can be replaced by $\sqrt{n(n+1)}$ when K is symmetric, by $(n+1)$ when K is an $\mathcal{H}$-polytope, and by $\sqrt{n+1}$ when K is a symmetric ($\mathcal{V}$- or $\mathcal{H}$)-polytope.

Since the volume of the ellipsoid $a + A(\mathbb{B}^n)$ can be easily computed, taking the geometric mean of the upper and lower bound in 6.2.2 gives a polynomial-time $(n+1)^{3n/4}$-approximation μ to $V(K)$.

TARASOV, KHACHIYAN & ERLICH [TaKE88] and KHACHIYAN & TODD [KhT93] give polynomial-time algorithms for approximating the ellipsoid of maximum volume that is contained in an $\mathcal{H}$-polytope. In particular, the following appears in [KhT93].

6.2.3 *For each rational $\gamma \in]0,1[$ there exists a polynomial-time algorithm which, given $n, m \in \mathbb{N}$, and $a_1, \ldots, a_m \in \mathbb{Q}^n$, produces an ellipsoid $E = a + A(\mathbb{B}^n)$ such that*

$$E \subset P = \{x \in \mathbb{R}^n : \langle a_i, x\rangle \leq 1, \text{ for } i = 1, \ldots, m\} \quad \text{and} \quad V(E) \geq \gamma \cdot V(E_0),$$

where E_0 is the ellipsoid of maximum volume contained in P. The running time of the algorithm is

$$O\left(m^{3.5} \log(mR)/(r\log(1/\gamma)) \log(nR)/(r\log(1/\gamma))\right),$$

where r and R are respectively a lower bound on P's inradius and an upper bound on P's circumradius.

Note that it can be determined in polynomial time whether a given $\mathcal{H}$-polytope has interior points, and, if it does, such a point b can be found in polynomial time. Then, if necessary, a translation about $-b$ and a suitable scaling will transform the given $\mathcal{H}$-polytope into one of the kind used in 6.2.3. Hence, the condition on the right-hand side of P's $\mathcal{H}$-presentation does not impose any severe restrictions. It is not known whether a result similar to 6.2.3 can also be obtained for $\mathcal{V}$-polytopes; see [KhT93, p.158].

Now note that, as shown in [TaKE88], an approximation of E_0 of the kind given in Theorem 6.2.3 leads to the following inclusion:

$$a + A(\mathbb{B}^n) \subset K \subset a + \frac{n(1 + 3\sqrt{1-\gamma})}{\gamma} A(\mathbb{B}^n),$$

and hence leads, for every $\alpha > 1$, to an $(\alpha n)^{n/2}$-approximation of $V(P)$ for $\mathcal{H}$-polytopes P.

A similar bound can also be derived for convex bodies that are given by an appropriate oracle. In particular, APPLEGATE & KANNAN [ApK90] give the following algorithmic Löwner-John-type result for parallelotopes.

6.2.4 *There exists an oracle-polynomial-time algorithm which, for any well-bounded body K of $\mathbb{R}^n$ given by a weak separation oracle, finds a point a and a linear transformation A such that*

$$a + A([-1,1]^n) \subset K \subset a + 2(n+1)A([-1,1]^n).$$

While this result has direct applications in the design of improved randomized algorithms for volume computation (see Subsection 8.1), the following result of BETKE & HENK [BeH93] gives a slightly better approximation error.

6.2.5 *There exists an oracle-polynomial-time algorithm which, for any body K of $\mathbb{R}^n$ given by a weak optimization oracle, and for every $\epsilon > 0$, finds rationals μ_1 and μ_2 such that*

$$\mu_1 \leq V(K) \leq \mu_2 \quad \textit{and} \quad \mu_2 \leq n!(1+\epsilon)^n \mu_1.$$

In fact, two calls to a strong optimization oracle for directions c_1 and $-c_1$ give two supporting halfspaces H_1^+, H_1^- and two optimizers z_1^+, z_1^-. This procedure is now repeated for the directions $\pm c_2$, with a c_2 orthogonal to $\operatorname{aff}\{z_1^+, z_1^-\}$, etc. After n steps one gets the parallelotope $P = \bigcap_{i=1}^n (H_i^+ \cap H_i^-) \supset K$ and the polytope $Q = \operatorname{conv}\{z_i^+, z_i^- : i = 1, \dots, n\} \subset K$, with $V(P)/V(Q) \leq n!$. The use of a weak rather than a strong optimization oracle brings in an additional factor $(1+\epsilon)^n$.

Let us now turn to the case of mixed volumes for some bodies $K_1, \dots, K_r$. There are two natural general approaches to this problem, namely to approximate the bodies $K_1, \dots, K_r$ by bodies $C_1, \dots, C_r$, respectively, and then to use the corresponding mixed volumes of $C_1, \dots, C_r$ as approximations, or to approximate $V(\sum_{i=1}^r \xi_i K_i)$ for appropriately chosen nonnegative parameters $\xi_1, \dots, \xi_r$, and then to use the techniques outlined in Subsection 4.1 to derive estimates for the mixed volumes of $K_1, \dots, K_r$. The remainder of this subsection will address both possibilities.

Note, first, that the Minkowski sum of two ellipsoids is in general no longer an ellipsoid. Hence a straightforward extension of the Löwner-John approach to mixed volumes fails because of the lack of an efficient algorithmic procedure for computing mixed volumes of ellipsoids. Also the approach of 6.2.4 is bound to fail for mixed volumes, for we have seen in Theorem 5.2.3 that computing mixed volumes of parallelotopes is #$\mathbb{P}$-hard. The general problem that we are facing here is that there don't seem to be rich enough classes of bodies (which could be used for approximating the given bodies $K_1, \dots, K_r$) for which mixed volumes can actually be computed, and this is closely related to the obvious lack of rich enough classes of bodies for which the volume of their Minkowski sums can actually be computed.

There is however one case where the mixed volumes can be (weakly) computed, and this is the case $r = 2$ where $C_1 = \mathbb{B}^n$ and C_2 is a parallelotope. Recall, in fact, that by Theorem 4.4.4 the intrinsic volumes of an $\mathcal{S}$-parallelotope can be approximated (with respect to arbitrarily small additive error) in polynomial time. Hence we can combine Theorems 6.2.2 and 6.2.4 as follows. First we construct an ellipsoid $E = a_1 + A_1(\mathbb{B}^n)$ and a parallelotope $Z = a_2 + A_2([-1,1]^n)$ such that

$$a_1 + A_1(\mathbb{B}^n) \subset K_1 \subset a_1 + (n+1)\sqrt{n}A_1(\mathbb{B}^n) \qquad \text{and}$$
$$a_2 + A_2([-1,1]^n) \subset K_2 \subset a_2 + 2(n+1)A_2([-1,1]^n).$$

Then, with

$$C_1' = a_1 + A_1(\mathbb{B}^n), \qquad C_1'' = a_1 + (n+1)\sqrt{n}A_1(\mathbb{B}^n),$$
$$C_2' = a_2 + A_2([-1,1])^n, \quad C_2'' = a_2 + 2(n+1)A_2([-1,1])^n,$$

we have

$$V(\overbrace{C_1', \dots, C_1'}^{k}, \overbrace{C_2', \dots, C_2'}^{n-k}) \leq V(\overbrace{K_1, \dots, K_1}^{k}, \overbrace{K_2, \dots, K_2}^{n-k})$$

$$\leq V(\overbrace{C_1'', \dots, C_1''}^{k}, \overbrace{C_2'', \dots, C_2''}^{n-k}).$$

By 2.4.2 (iii), application of the affine transformation $x \mapsto A_1^{-1}(x - a_1)$ changes the mixed volume only by the common factor $|\det(A_1)|^{-1}$, and this is irrelevant for relative approximation. But now we have arrived at an approximation by means of the intrinsic volumes of the parallelotope $Z = A_1^{-1}(a_2 + A_2([-1,1]^n) - a_1)$ which can, in fact, be (weakly) computed. Hence we can compute a lower bound μ ($= V(\overbrace{C_1', \ldots, C_1'}^{k}, \overbrace{C_2', \ldots, C_2'}^{n-k})$) such that

$$\mu \leq V(\overbrace{K_1, \ldots, K_1}^{k}, \overbrace{K_2, \ldots, K_2}^{n-k}) \leq \left((n+1)\sqrt{n}\right)^k \left(2(n+1)\right)^{n-k} \mu.$$

Taking the geometric mean of the lower and upper bound and, if necessary, interchanging the roles of K_1 and K_2, we obtain the following result; see [DyGH94].

6.2.6 *There is a polynomial-time algorithm for* $O(2^{n/4}n^{5n/8})$-APPROXIMATION *of all mixed volumes of any two well-bounded bodies* K_1 *and* K_2 *given by a weak separation oracle.*

The approximation error in 6.2.6 is only an upper bound for the precise value that we get from 6.2.2 and 6.2.4 with the outlined method; it is in fact

$$\lambda(n) = (n+1)^{n/2} \min\{n^{\frac{k}{4}} 2^{\frac{n-k}{2}}, n^{\frac{n-k}{4}} 2^{\frac{k}{2}}\}.$$

Note further that for $\mathcal{H}$-polytopes, 6.2.6 can be improved by using 6.2.3 rather than 6.2.2. However, we don't know of any result that extends 6.2.6 to the general case of n bodies. It is easy to obtain some approximation results that depend on auxiliary parameters such as the inradius or the circumradius of the specific bodies, but such results are much weaker than 6.2.6 which depends only on the dimension.

Another way of attempting to obtain, for some functional $\lambda : \mathbb{N} \to \mathbb{R}$, a λ-APPROXIMATION of certain mixed volumes, is to try to extend 4.1.1 to λ-APPROXIMATION.

Recall that 4.1.1 utilized the fact that an algorithm for approximating a polynomial with respect to the absolute error can be used to obtain approximations of the coefficients (again with respect to the absolute error). It turns out, however, that such a procedure does not exist with respect to the (symmetric) relative error. In fact, let us consider the following simple univariate example. Suppose that we want to estimate the middle coefficient α of a quadratic polynomial π with constant 1 and leading coefficient 1. In other words, we know that $\pi = \pi_\alpha = x^2 + \alpha x + 1$ for some α, and we want to find or approximate α. Now let $\epsilon > 0$, and suppose that $\eta_0, \ldots, \eta_k$ are nodes at which we want to approximately evaluate π in order to estimate α. We may further suppose that $\eta_0, \ldots, \eta_k > 0$ (for this is the only situation that is relevant in the context of MIXED VOLUME APPROXIMATION, and also, the construction can be easily adapted to the general case if desired). Now assume that the approximation oracle uses the exact values of $\pi_0 = 1 + x^2$ at $\eta_0, \ldots, \eta_k$ to produce estimates for $\pi(\eta_0), \ldots, \pi(\eta_k)$.

For $j = 0, \dots, k$ and each α with $0 \le \alpha \le \epsilon/(\max_{i=0,\dots,k} \eta_i)$, we have

$$\frac{\pi_0(\eta_j)}{\pi(\eta_j)} \le 1 \quad \text{and} \quad \frac{\pi(\eta_j)}{\pi_0(\eta_j)} \le 1 + \alpha\eta_j \le 1 + \epsilon.$$

Hence the approximation oracle produces estimates for the values of the polynomial with symmetric relative error bounded by ϵ. On the other hand, since α may be (at least) any coefficient between 0 and $\epsilon/(\max_{i=0,\dots,k} \eta_i)$, we cannot use the approximations of the function values to obtain any symmetric relative approximation for this coefficient with finite error bound.

The obstruction here is the lack of some kind of correlation between the various coefficients of π. However, with mixed volumes we are here in a special situation since we can use the *Aleksandrov-Fenchel inequality.* For two bodies K_1 and K_2, 2.4.3 reads as follows:

$$V(\overbrace{K_1, \dots, K_1}^{n-i}, \overbrace{K_2, \dots, K_2}^{i})^2$$

$$\ge V(\overbrace{K_1, \dots, K_1}^{n-i+1}, \overbrace{K_2, \dots, K_2}^{i-1}) V(\overbrace{K_1, \dots, K_1}^{n-i-1}, \overbrace{K_2, \dots, K_2}^{i+1}),$$

This implies that the sequence of coefficients $\tau_0, \dots, \tau_n$ is unimodal. Furthermore, in the special case of mixed volumes of two bodies an appropriate "scaling" can be utilized, [DyGH94].

6.2.7 *For any pair K_1, K_2 of well-bounded bodies given by a weak separation oracle and for any $k = 1, \dots, n$ one can construct in polynomial time an affine transformation α and a positive rational scaling factor λ such that the mixed volumes*

$$\tau_i = V(\overbrace{K_1', \dots, K_1'}^{n-i}, \overbrace{K_2', \dots, K_2',}^{i}) \qquad i = k-1, k,$$

of the transformed bodies $K_1' = \alpha(K_1)$ and $K_2' = \lambda\alpha(K_2)$ satisfy the inequality

$$1 \le \frac{\tau_{k-1}}{\tau_k} \le (n+1)^{8k}.$$

Note that the right-hand bound does depend only on n and k, and not on special properties or measures of the bodies K_1 and K_2.

These special properties of mixed volumes can be used to obtain approximation results, and they are crucial for the randomized algorithm described in Subsection 7.2. There are, however, still major obstacles to extending Theorem 4.1.1 to relative volume approximation, and we will deal with these problems in Subsection 7.2.

6.3. Lower bounds in the oracle model

It turns out that the above bounds for Volume Approximation are not too far away from the best one can achieve. Elekes [El86] showed that even if our bodies K are given by a strong separation oracle, a subexponential number of calls to the oracle does not suffice to obtain a polynomial approximation. His argument is based on the following observation. Suppose that $K \subset \mathbb{B}^n$, that for some $k \in \mathbb{N}$ the inputs to the oracle are points $y_1, \ldots, y_k \in \mathbb{B}^n$, and suppose further that all membership tests are affirmative (and hence we never get a separating hyperplane). Then, with $P = \mathrm{conv}\{y_1, \ldots, y_k\}$ we know that $P \subset K \subset \mathbb{B}^n$, but this is all the information that is available, and based on this information an approximation μ of $V(K)$ is determined by our approximation algorithm. This implies that

$$\max_{\{K: P \subset K \subset \mathbb{B}^n\}} \left\{ \frac{\mu}{V(K)}, \frac{V(K)}{\mu} \right\} \geq \sqrt{\frac{V(\mathbb{B}^n)}{V(P)}}.$$

Now, ELEKES [El86] shows that

$$P \subset \frac{1}{2} \bigcup_{i=1}^{k} (y_i + \mathbb{B}^n),$$

and this yields

$$\frac{V(\mathbb{B}^n)}{V(P)} \geq \frac{k}{2^n}.$$

BÁRÁNY & FÜREDI [BaF86] improve this result by proving the following theorem.

6.3.1 *Suppose that*

$$\lambda(n) < \left(\frac{n}{\log n}\right)^{n/2} - 1 \qquad \textit{for all } n \in \mathbb{N}.$$

Then there is no deterministic oracle-polynomial-time algorithm for λ-APPROXIMATION of the volume.

Now, it is clear by Theorem 2.4.2 (ii) that Theorem 6.3.1 carries over to MIXED VOLUME APPROXIMATION simply because it includes the case where all bodies are the same. It is very likely, though, that in more general situations the bound of 6.3.1 can be improved. In particular, the worst-case approximation error for $V(K_1, \ldots, K_n)$ (where the worst case is taken over all possible choices of $K_1, \ldots, K_n$) should be much worse than $(n/\log n)^{n/2} - 1$.

7. Randomized algorithms

7.1. APPROXIMATING THE VOLUME

As we have seen, volume computation and even volume approximation is hard when we restrict our algorithms to deterministic ones. The situation changes drastically if we allow randomized algorithms. In fact, DYER, FRIEZE & KANNAN [DyFK89] give a polynomial-time randomized algorithm for relative approximation of the volume of convex bodies that are given by appropriate oracles. The algorithm is a *random walk*, and its analysis is based on the notion of *rapidly mixing Markov chains*. We are going to describe the basic ideas of this approach, skipping however a lot of technical details, particularly those related to the stochastic analysis. For further details, background information, a sketch of the corresponding history and more references see the papers by DYER & FRIEZE [DyF91], KHACHIYAN [Kh93], LOVÁSZ [Lo92], [Lo94] and LOVÁSZ & SIMONOVITZ [LoS93].

EXPECTED VOLUME COMPUTATION.

Instance: *A positive integer n, a centered well-bounded body K in $\mathbb{R}^n$ given by a weak membership oracle, positive rationals β and ϵ.*

Task: *Determine a positive rational random variable μ such that*

$$\text{prob}\left\{\left|\frac{\mu}{V(K)} - 1\right| \leq \epsilon\right\} \geq 1 - \beta.$$

Note that in the above problem, the relative error measure is employed; see Subsection 6.1. This indicates already that we are aiming at "close approximation," and in fact, the main theorem of this section due to DYER, FRIEZE & KANNAN [DyFK89] is as follows.

7.1.1 *There is a randomized algorithm for* EXPECTED VOLUME COMPUTATION *which runs in time that is oracle-polynomial in n, $1/\epsilon$ and* $\log(1/\beta)$.

Before giving an (informal) description of the algorithm let us clarify that the existence of a polynomial-time randomized algorithm for volume computations does not contradict the negative results of Subsections 5.1 and 6.3. In fact, for a deterministic algorithm all that counts is what it produces as output, while for a nondeterministic algorithm what can *potentially* be produced is relevant. In fact, the results depend on the distribution of these potential outcomes rather than on the outcomes themselves. As will become clear, the randomized algorithm described below does have the potential to reach *exponentially* many points, and this is crucial for the polynomial running time.

Let us now describe the original algorithm for 7.1.1; some improvements will be outlined later in this subsection. The first step is a rounding procedure that utilizes (in conjunction with 1.2.1) the algorithmic version 6.1.2 [GrLS88] of JOHN's [Jo48] result. According to this version, there exists an oracle-polynomial-time algorithm which, for any well-bounded body K of $\mathbb{R}^n$ given by a weak separation oracle, finds a point a and a linear transformation A such that

$$a + A(\mathbb{B}^n) \subset K \subset a + (n+1)\sqrt{n}A(\mathbb{B}^n).$$

Hence,

$$\mathbb{B}^n \subset A^{-1}a - A^{-1}(K) \subset (n+1)\sqrt{n}\mathbb{B}^n.$$

This rounding procedure is a deterministic algorithm that uses $O(n^4(\text{size}(r)+\text{size}(R)))$ operations on numbers of size $O(n^2(\text{size}(r)+\text{size}(R)))$, where r, R are (as usual) the a priori bounds for K's inradius and circumradius; see [GrLS88, p.122]. Since

$$V(K) = |\det(A)| V\left(A^{-1}a + A^{-1}(K)\right),$$

we may, for the second step of the randomized algorithm, assume that

$$\mathbb{B}^n \subset K \subset (n+1)\sqrt{n}\mathbb{B}^n.$$

One could now try to estimate the ratio $V((n+1)\sqrt{n}\mathbb{B}^n)/V(K)$ by means of a randomized procedure. However, this ratio may be exponential, and this leads to a blowup of the complexity of the randomized approach outlined below. For this reason, the next step reduces the problem to a series of problems with suitably bounded volume ratios. Let

$$k = \left\lceil \frac{3}{2}(n+1)\log(n+1) \right\rceil, \qquad \text{and} \qquad K_i = K \cap \left(1+\frac{1}{n}\right)^i \mathbb{B}^n \quad \text{for } i = 0, \ldots, k.$$

Then

$$\mathbb{B}^n = K_0 \subset K_1 \subset \ldots \subset K_{k-1} \subset K_k = K \subset (n+1)\sqrt{n}\mathbb{B}^n,$$

and, more importantly, for $i = 1, \ldots, k$,

$$1 \leq \frac{V(K_i)}{V(K_{i-1})} \leq \left(1+\frac{1}{n}\right)^n < e.$$

Clearly,

$$V(K) = V(\mathbb{B}^n) \prod_{i=1}^{k} \frac{V(K_i)}{V(K_{i-1})},$$

whence it suffices to estimate each ratio $V(K_i)/V(K_{i-1})$ up to a relative error of order $\epsilon/(n \log n)$ with error probability of order $\beta/(n \log n)$.

Now, the main step of the algorithm of Dyer, Frieze & Kannan [DyFK89] is based on a method for sampling nearly uniformly from within K_i. It superimposes a chess-board grid of small cubes (say of edge length δ) on K_i (compare 3.4 and 3.5) and performs a random walk over the set $\mathcal{C}_i$ of cubes in this grid that intersect a suitable parallel body $K + \alpha\mathbb{B}^n$ where α is small. This walk is performed by moving through a facet with probability $1/f_{n-1}(C_n) = (2n)^{-1}$ if this move ends up in a cube of $\mathcal{C}_i$, and staying at the current cube if the move would lead outside of $\mathcal{C}_i$. The random walk gives a *Markov chain* which is irreducible (since the moves are connected), aperiodic and hence ergodic. But this implies that there is a unique stationary distribution, the limit distribution of the chain, which is easily seen to be a *uniform distribution*. Thus after a sufficiently large number of steps we can use the current cube in the random walk to sample nearly uniformly from $\mathcal{C}_i$. Having obtained such a uniformly sampled cube, it is determined whether it belongs to $\mathcal{C}_{i-1}$ or to $\mathcal{C}_i \setminus \mathcal{C}_{i-1}$.

Now note that if ν_i is the number of cubes in $\mathcal{C}_i$, then the number $\mu_i = \nu_i/\nu_{i-1}$ is an estimate for the volume ratio $V(K_i)/V(K_{i-1})$. It is this number μ_i that can now be "randomly approximated" using the above constructed approximation of a uniform sampling over $\mathcal{C}_i$. In fact, a cube C that is reached after sufficiently many steps in the random walk will lie in $\mathcal{C}_{i-1}$ with probability approximately $1/\mu_i$; hence by repeated sampling we can approximate this number closely.

This informal description of the randomized algorithm must of course be rigorously analyzed to determine its complexity. A main question is just how quickly the random walk approximates a "reasonably uniform" distribution. In their analysis, DYER, FRIEZE & KANNAN [DyFK89] use a result of SINCLAIR & JERRUM [SiJ89] that relates the speed of convergence to the *conductance* of the chain. With the aid of a geometric interpretation of this quantity and an isoperimetric inequality of BÉRAD, BESSON & GALLOT [BeBG85] on the volume of subsets of smooth Riemannian manifolds with positive curvature, it is shown in [DyFK89] that the Markov chain is, indeed, mixing rapidly enough to yield polynomiality. The following inequality (which is stronger than what was needed in [DyFK89]'s original proof) is taken from [DyF91]; see also [LoS90], [ApK90] and [LoS93].

7.1.2 *Let K be a convex body in $\mathbb{R}^n$, and let f be a real-valued log-concave function on* $\mathrm{int}(K)$. *Further, let $S_1, S_2 \subset K$ be measurable, $S = K \setminus (S_1 \cup S_2)$, and suppose that* $\mathrm{dist}(S_1, S_2) > 0$. *Then*

$$\min\left\{\int_{S_1} f(x)dx, \int_{S_2} f(x)dx\right\} \leq \frac{R_1(K)}{\mathrm{dist}(S_1, S_2)} \int_S f(x)dx,$$

where $R_1(K)$ is half of K's diameter.

A corollary which conveys the flavor of this inequality (and which is sufficient for the proof of polynomiality of the randomized volume-algorithm) says that if K is a convex body in $\mathbb{R}^n$, and S is a minimal surface that partitions K into two sets S_1, S_2, then

$$\min\{V(S_1), V(S_2)\} \leq R_1(K)A(S),$$

where $A(S)$ denotes the surface area of S. This formulation shows that 7.1.2 is an extension of the result that a body K is contained in any cylinder whose base is the projection of K on the hyperplane orthogonal to some direction u, and whose height in direction u is K's breadth in this direction.

[DyFK89]'s polynomial-time randomized algorithm for EXPECTED VOLUME COMPUTATION was subsequently improved in various papers, including [LoS90], [ApK90], [DyF91], [LoS93] and [KaLS94].

One key ingredient for improvements is 7.1.2, while another major improvement can be obtained by replacing the "rounding"

$$\mathbb{B}^n \subset K \subset (n+1)\sqrt{n}\mathbb{B}^n$$

by the "normalization"

$$[-1,1]^n \subset K \subset 2(n+1)[-1,1]^n;$$

APPLEGATE & KANNAN [ApK90], see 6.2.4. Another idea of [ApK90] that avoids difficulties caused by inherent "nonsmoothness" is to approximate the characteristic function of K by a smooth function; cf. 3.4.

LOVÁSZ & SIMONOVITS [LoS93] improve on these ideas, extend the theory of conductance and rapid mixing from the finite case to arbitrary Markov chains (so that now steps can be chosen uniformly from a ball with fixed radius about the current point), and replace the rounding phase by an "approximate sandwiching:"

an affine transformation α is produced such that 2/3 of the volume of $\mathbb{B}^n$ is contained in $\alpha(K)$ and 2/3 of the volume of $\alpha(K)$ is contained in $n\mathbb{B}^n$. In their extensive study, they achieve the following complexity bound for the second step (after the normalization) of

$$\mathrm{O}\left(\frac{1}{\epsilon^2}n^7\log^2(n)\log^3(\frac{1}{\epsilon})\log(\frac{1}{\beta})\right)$$

Very recently, KANNAN, LOVÁSZ & SIMONOVITS [KaLS94] gave a further substantial improvement; see [Lo94]. They achieve the currently best known bound where now n enters only in fifth power.

Let us close this subsection with a few remarks.

Sometimes it is possible to devise random walks not over a superimposed grid of cubes but over objects that are more closely related to the specific bodies. One natural example is the class of *order polytopes*. As we have seen in Subsection 5.1 (the discussion preceding Theorem 5.1.4), an order polytope can be dissected into simplices of the same volume which correspond to the linear extensions of the given order $\mathcal{O}$. This approach gives rise to a random walk over the linear extensions of $\mathcal{O}$ which, itself, has interesting applications; see KARZANOV & KHACHIYAN [KaK90], KHACHIYAN [Kh93] and LOVÁSZ [Lo94].

A second class of bodies that come with a natural dissection are the zonotopes. Zonotopes can be dissected into parallelotopes, and it is intriguing to try to use these parallelotopes instead of the cubes. Unfortunately, the volumes of the parallelotopes may in general vary exponentially, and hence a direct extension of the above approach will work only in very special cases. Thus it is unknown whether, for general zonotopes, there is a randomized algorithm for volume computation that is more efficient than randomized algorithms that work for arbitrary convex bodies.

The key step of the randomized volume-algorithms is to compute a *nearly uniform distribution* on a body K. DYER & FRIEZE [DyF91] show that the converse is also true: A polynomial number of calls to a volume approximator suffice to generate with high probability uniformly distributed points in K.

7.2. APPROXIMATING MIXED VOLUMES

Now that we have a randomized polynomial-time algorithm at hand for solving EXPECTED VOLUME COMPUTATION, it is natural to try to use it for devising a similar procedure for mixed volumes. This subsection will outline such an approach of [DyGH94].

Let us begin with the case of two centered well-bounded bodies K_1 and K_2 that are given by weak membership oracles. Let us consider the polynomial π given by

$$\pi(\xi) = V(K_1 + \xi K_2) = \sum_{i=1}^{n} \binom{n}{i} V(\overbrace{K_1,\ldots,K_1}^{n-i},\overbrace{K_2,\ldots,K_2}^{i})\xi^i.$$

We will sometimes use the abbreviation

$$\tau_i = V(\overbrace{K_1,\ldots,K_1}^{n-i},\overbrace{K_2,\ldots,K_2}^{i}) \quad \text{and} \quad \zeta_i = \binom{n}{i}\tau_i.$$

Then, following the approach of Subsection 4.1, our goal is to use the randomized volume algorithm to evaluate the polynomial $\pi(\xi) = \sum_{i=1}^{n} \zeta_i \xi^i$ at suitable nodes in order to obtain estimates for its coefficients ζ_i.

As we have already seen in Subsection 6.2, there is no general way to derive *relative* estimates for the coefficients of a polynomial from *relative* estimates of certain function values of π. However, we are here in a special situation in which we can use both 6.2.7 and the specialization to two bodies of the *Aleksandrov-Fenchel inequality*. As we have already mentioned at the end of Subsection 6.2, it turns out, though, that there are still major obstacles to extending Theorem 4.1.1 to randomized relative volume approximation, and before we state the results of [DyGH94] we want to point out what the additional problems are.

As we have seen in Subsection 4.1, computing the coefficients of a polynomial from some of its values can in principle be done by numerical differentiation. Let $\eta_0, \ldots, \eta_n$ be pairwise different *interpolation points*, and let for $j = 0, \ldots, n$

$$\ell_j(\xi) = \sum_{i=0}^{n} \beta_{ij} \xi^i$$

denote the jth *Lagrange interpolation polynomial* on the node set $Y = \{\eta_0, \ldots, \eta_n\}$. Recall that for $j, k = 0, \ldots, n$,

$$\ell_j(\xi_k) = \sum_{i=0}^{n} \beta_{ij} \xi_k^i = \delta_{jk} = \begin{cases} 1 & \text{for } j = k \\ 0 & \text{otherwise.} \end{cases}$$

Therefore

$$\pi(\xi) = \sum_{j=0}^{n} \pi(\xi_j) \ell_j(\xi) = \sum_{i=0}^{n} \left(\sum_{j=0}^{n} \beta_{ij} \pi(\xi_j) \right) \xi^i,$$

whence for each $i = 0, \ldots, n$,

$$\zeta_i = \sum_{j=0}^{n} \beta_{ij} \pi(\xi_j).$$

Now, suppose we have approximations $\mu_0, \ldots, \mu_n$ of the values $\pi(\xi_0), \ldots, \pi(\xi_n)$, respectively, with relative error bounded by some $\epsilon > 0$, and for $i = 0, \ldots, n$ we use

$$\hat{\zeta}_i = \sum_{j=0}^{n} \beta_{ij} \mu_j$$

as an estimate for ζ_i. Then it is easy to see that

$$|\hat{\zeta}_i - \zeta_i| \leq \epsilon \max_{k=0,\ldots,n} \pi(\xi_k) \sum_{j=0}^{n} |\beta_{ij}|,$$

and this bound is tight. This means, in order to bound the relative error of the approximation $\hat{\zeta}_k$ of ζ_k we need to be able to bound the right-hand side in terms of ζ_k. Unfortunately, as is pointed out in [DyGH94], $\max_{k=0,\dots,n} \pi(\xi_k) \sum_{j=0}^{n} |\beta_{ij}|$ grows exponentially, and that is why only a certain portion of the coefficients may become approximable by such an approach. (Recall that the randomized volume algorithm is polynomial in $1/\epsilon$ but exponential in $\log(1/\epsilon)$.) Hence we introduce a version of the problem that depends on an additional function $\psi : \mathbb{N}_0 \to \mathbb{N}_0$ with $\psi(n) \leq n$ for every $n \in \mathbb{N}_0$.

EXPECTED ψ-MIXED VOLUME COMPUTATION.

Instance: *A positive integer n, centered well-bounded bodies K_1 and K_2 in $\mathbb{R}^n$ given by weak membership oracles, positive rationals β and ϵ.*

Task: *Determine for each nonnegative integer i with $i \leq \psi(n)$ a positive rational random variable $\hat{\tau}_i$ such that*

$$\text{prob}\left\{\left|\frac{\hat{\tau}_i}{\tau_i} - 1\right| \leq \epsilon\right\} \geq 1 - \beta.$$

Then [DyGH94] prove the following theorem.

7.2.1 *Let $\psi : \mathbb{N}_0 \to \mathbb{N}_0$ with $\psi(n) \leq n$ for every $n \in \mathbb{N}_0$, and suppose that*

$$\psi(n) \log \psi(n) = \mathrm{o}(\log n).$$

Then there is a randomized algorithm for EXPECTED ψ-MIXED VOLUME COMPUTATION *which runs in time that is oracle-polynomial in n, $1/\epsilon$ and* $\log(1/\beta)$.

Observe that $\psi(n) = \lceil \log(n)/\log^2 \log(n) \rceil$ is a choice that satisfies the assumptions of Theorem 7.2.1.

The algorithm underlying 7.2.1 proceeds inductively, beginning with τ_0 which can be approximated by the volume procedure 6.1.1. Suppose that for some k all mixed volumes $\tau_0, \dots, \tau_{k-1}$ have already been approximated. As was mentioned earlier, the algorithm now uses the scaling described in 6.2.7 as preprocessing for the next step. This yields a rough estimate for τ_k. Then, using the volume algorithm again and choosing the nodes appropriately, approximations of $\pi(\xi)$ are computed. Next, a binary search procedure is used to improve the initial relative estimate of τ_k to within a constant error, and finally the last step achieves an approximation of τ_k to within a relative error ϵ, as desired. Of course, the interpolation points now depend on ϵ, and they are chosen in such a way that the higher order terms of π can be bounded appropriately so as to allow the use of only a small part of the coefficient matrix $B = (\beta_{ij})$. This makes it possible to keep the error small.

It may be worth mentioning that as compared to algorithms for EXPECTED VOLUME COMPUTATION, the complexity of the above algorithm is only marginally worse.

Let us point out explicitly that it is not known whether EXPECTED ψ-MIXED VOLUME COMPUTATION can be solved in polynomial time under assumptions on ψ that are less restrictive than those stated in Theorem 7.2.1. In particular, it is not known how to efficiently approximate $V(\overbrace{K_1, \dots, K_1}^{n}, \overbrace{K_2, \dots, K_2}^{n})$ for bodies in $\mathbb{R}^{2n}$.

On the positive side, it is possible to extend 7.2.1 to the case of more than two bodies and to show that there is a randomized polynomial time algorithm for computing

$$V(\overbrace{K_1,\ldots,K_1}^{i_1},\overbrace{K_2,\ldots,K_2}^{i_2},\ldots,\overbrace{K_{r-1},\ldots,K_{r-1}}^{i_{r-1}},\overbrace{K_r,\ldots,K_r}^{i_r}),$$

where $\sum_{j=1}^{r} i_j = n$ and $\sum_{j=1}^{r-1} i_j = \psi(n)$ with a function ψ as in 7.2.1. In fact, suppose we have a procedure for r sets. Then, utilizing the multilinearity of the mixed volume, we consider

$$\pi(\xi) = V(\overbrace{K_1,\ldots,K_1}^{i_1},\ldots,\overbrace{K_{r-1},\ldots,K_{r-1}}^{i_r-1},\overbrace{K_r+\xi K_{r+1},\ldots,K_r+\xi K_{r+1}}^{i_r+i_{r+1}}),$$

which can be estimated recursively for fixed ξ. On the other hand, π is a polynomial of degree $i_r + i_{r+1}$ in ξ, for which we wish to estimate the coefficient of $\xi^{i_{r+1}}$. The coefficients of π are themselves mixed volumes, and consequently satisfy the Alexandrov-Fenchel inequalities. Thus the approach above for two sets can be used with very little change. There is, however, one difficulty. We do not have a polynomial-time procedure for producing a "good" initial scaling of the sets, as we had with 6.2.7 for two bodies; and we leave as an open question whether such a procedure exists. Without such a scaling, one has to resort to the "well-boundedness" parameters r_k, R_k that come as bounds for the inradius and the circumradius of the bodies. Unfortunately, these parameters may be exponentially large, and this feeds into the recursion. However, [DyGH94] show that one can approximate the mixed volumes for any *fixed* r in polynomial time, where each of the first $r-1$ sets may be repeated up to $o(\log n/\log\log n)$ times. Further, if the ratios R_k/r_k are "quasi-polynomial" in n, i.e. of the form $O(2^{\gamma(\log(n))})$, where γ is a polynomial, we can approximate mixed volumes for any $r = o(\log n/\log\log n)$ in polynomial time. For larger ratio $\rho = \max_{k=1,\ldots,r} R_k/r_k$ we can approximate up to $r = o(\log n/\log\log \rho)$ in similar time.

Let us finally point out that, particularly in view of the applications stated in Subsections 9.6 – 9.9, it would be desirable to be able to extend the above results to the general case. Specifically, it would be useful to be able to compute $V(K_1,\ldots,K_n)$ by means of a randomized polynomial–time algorithm. It is not known whether such a procedure exists.

8. Miscellaneous

In the present section, we will mention some results that are closely related to volume computation.

8.1. Projections and sections

The problem of maximizing or minimizing the volumes of orthogonal projections of polytopes onto hyperplanes has received some attention in geometry because it is related to various illumination and optimization problems, see e.g. Martini [Ma85]. It has been treated from a computational viewpoint by McKenna & Seidel [McS85], whose algorithm finds a direction in which the orthogonal projection has maximum (or minimum) volume. Their algorithm is asymptotically optimal when the dimension is fixed.

The more general case of projections onto subspaces of arbitrary intermediate dimension is studied (for fixed and for variable dimensions) in [BuGK94a]. Let $\gamma : \mathbb{N} \rightarrow \mathbb{N}$ denote a functional with the property that $1 \leq \gamma(n) \leq n-1$ for each n. Then we have the following decision problems.

MAXIMUM γ-PROJECTION ⟨MINIMUM γ-PROJECTION⟩

Instance: *A positive integer n, an $\mathcal{H}$-polytope ⟨a $\mathcal{V}$-polytope, or an $\mathcal{S}$-zonotope⟩ P, a nonnegative rational μ.*

Question: *Is there a $\gamma(n)$-dimensional subspace S of $\mathbb{R}^n$ such that $V^2_{\gamma(n)}(\Pi_S P) \geq \mu$ ⟨$V^2_{\gamma(n)}(\Pi_S P) \leq \mu$⟩?*

Here, as before, $V_{\gamma(n)}(\Pi_S P)$ denotes the $\gamma(n)$-dimensional volume of the orthogonal projection $\Pi_S P$ of P on S.

Note that with the special choice $\gamma \equiv 1$, MAXIMUM γ-PROJECTION is the problem of lower bounding (the square of) a polytope P's *diameter*. This problem is easy for $\mathcal{V}$-polytopes; however, it is already $\mathbb{NP}$-complete for $\mathcal{H}$- (or $\mathcal{S}$-) parallelotopes, [BoGKL90], [GrK93a]; see also [GrK94a].

In view of the results of Subsection 5.1, it is not surprising that the variants of MAXIMUM γ-PROJECTION and MINIMUM γ-PROJECTION that ask for the actual volumes of optimal projections are #$\mathbb{P}$-hard. However, it turns out that MAXIMUM γ-PROJECTION is hard for other reasons as well. In fact, even for $\gamma(n) = n-1$ (the case of projections onto hyperplanes), the problem MAXIMUM γ-PROJECTION is $\mathbb{NP}$-complete even for the class of all ($\mathcal{V}$- or $\mathcal{H}$-) simplices ([BuGK94a]), even though the $(n-1)$-dimensional volume of any projection of a (rational) simplex on a (rationally presented) hyperplane can be computed in polynomial time. On the other hand, minimizing projections of simplices on hyperplanes is easy, but MINIMUM γ-PROJECTION is $\mathbb{NP}$-hard for many classes of functionals γ and polytopes P (see [BuGK94a]).

Recalling from Subsection 4.4 that for any $z \in \mathbb{S}^{n-1}$,

$$nV([0,1]z, \overbrace{K, \ldots, K}^{n-1}) = V(\Pi_{\mathrm{lin}\{z\}^\perp}(K)),$$

these results imply that the problem of maximizing $V([0,1]z, \overbrace{K, \ldots, K}^{n-1})$ is already $\mathbb{NP}$-hard for K being a simplex, while the problem of minimizing $V([0,1]z, \overbrace{K, \ldots, K}^{n-1})$ is $\mathbb{NP}$-hard for arbitrary $\mathcal{H}$-polytopes (but easy for simplices). Extensions of these and other results can be found in [BuGK94a].

For some interesting theoretical results on projections see FILLIMAN [Fi88], [Fi90] and [Fi92]. The problem of estimating the intrinsic volume $V_i(K)$ of a body K from the intrinsic volumes $V_i(\Pi_{S_j} K)$ of K's projections onto certain j-dimensional subspaces $S_1, \ldots, S_m$ (with $1 \leq i \leq j \leq n-1$) has been studied by BETKE & MCMULLEN [BeM83].

Problems similar to those for projections can also be investigated for sections (with some additional constraints in the case of minimizing sections). In fact, for $\gamma \equiv 1$, the problem MAXIMUM γ-PROJECTION is the same as the (appropriately defined) problem MAXIMUM γ-SECTION, and the latter is hence again $\mathbb{NP}$-hard. Additional algorithmic results can be found in [BuGK94b].

The general problem of finding the maximum of the volumes of the j-dimensional sections of P (i.e., of the j-dimensional convex sets formed by intersecting P with a j-flat) is discussed by FILLIMAN [Fi92], who finds geometric conditions that must be satisfied by critical sections. For results related to extremal j-sections of simplices and cubes, see [Wa68], [Fi92].

Finally, we mention the survey article of MARTINI [Ma94], which discusses a variety of questions related to sections and projections.

8.2. EXPECTED VOLUMES

For a proper body K in $\mathbb{R}^n$ and an integer $m > n$, let $\varphi_m(K) = \mu_m(K)/V(K)$, where $\mu_m(K)$ is the expected volume of the convex hull of m points chosen independently and at random from the uniform distribution over K. For each m this is an affine invariant of K, because volume ratios are invariant under nonsingular affine transformations. The literature contains many results concerning the functions φ_m, and a good short survey with many references was given by CROFT, FALCONER & GUY [CrFG91, pp.54–57]. See BÁRÁNY & BUCHTA [BaB93] for later results and references.

Despite the extensive literature, we are not aware of any general algorithmic approach to the computation of $\varphi_m(P)$ when P is a given n-polytope. Indeed, even the numbers $s(n) = \varphi_{n+1}(S)$ for an n-simplex S have proved to be resistant. (These numbers are of interest for a comparison of the efficiency of two algorithms for the analysis of multicomponent phase diagrams; see [Kl65].) Although it is easy to see that $s(1) = 1/3$ and has long been known that $s(2) = 1/6$, for many years the best that could be done with $s(3)$ was to approximate it by means of Monte Carlo experiments (see [BuR92] and its references). Recently, however, BUCHTA & REITZINGER [BuR92] showed in a tour de force that

$$s(3) = \frac{13}{720} - \frac{\pi^2}{15015}.$$

For $n > 3$, $s(n)$ is still not known precisely.

It seems reasonable to conjecture that for each fixed n, $\varphi_{n+1}(K)$ is a maximum when K is a simplex, but the conjecture is open for all $n \geq 3$. GROEMER [Gr73] showed for all n that $\varphi_{n+1}(K)$ is a minimum when K is an ellipsoid, and the value in this case had been computed by KINGMAN [Ki69].

8.3. VOLUMES OF UNIONS AND INTERSECTIONS OF SPECIAL BODIES

The special bodies to be discussed in this subsection are boxes, balls, and simplices. For $a = (\alpha_1, \ldots, \alpha_n)^T, b = (\beta_1, \ldots, \beta_n)^T \in \mathbb{R}^n$ with $\alpha_i \leq \beta_i$ for all i, let $B(a,b)$ denote the box

$$B(a,b) = \{x = (\xi_1, \ldots, \xi_n)^T : \alpha_i \leq \xi_i \leq \beta_i \quad \text{for } 1 \leq i \leq n\}.$$

Now suppose that $B(a_1,b_1),\ldots,B(a_k,b_k)$ are boxes with $a_j=(\alpha_{j,1},\ldots,\alpha_{j,n})$ and $b_j=(\beta_{j,1},\ldots,\beta_{j,n})$ for $1\le j\le k$, and for $1\le i\le n$ let

$$\overline{\alpha}_i=\max\{\alpha_{1,i},\ldots,\alpha_{k,i}\},$$

and

$$\underline{\beta}_i=\min\{\beta_{1,i},\ldots,\beta_{k,i}\}.$$

Then the intersection $\bigcap_{i=1}^k B(a_i,b_i)$ is empty if $\overline{\alpha}_i>\underline{\beta}_i$ for some i, and otherwise the intersection is the box $B(\overline{a},\underline{b})$. This representation yields a fast algorithm for computing volumes of intersections of boxes.

Note that any algorithm for computing volumes of intersections of bodies of a special sort yields also an algorithm for computing volumes of unions of the same sort of bodies. That is true because the volume function is a valuation and hence

$$V\left(\bigcup_{i=1}^m K_i\right)=\sum_i V(K_i)-\sum_{i<j}V(K_i\cap K_j)+$$

$$+\sum_{i<j<k}V(K_i\cap K_j\cap K_k)-\ldots+(-1)^{m-1}V\left(\bigcap_{i=1}^m K_i\right).$$

However, this direct use of the principle of inclusion and exclusion is often not the best way to compute volumes of unions. For better ways to compute the volume of a union of k boxes in $\mathbb{R}^n$, see FREDMAN & WEIDE [FrW78] for an optimal $O(k\log k)$ algorithm when $n=1$; see VAN LEEUWEN & WOOD [VaW81] for an $O(k\log k)$ algorithm (due to J. L. Bentley) when $n=2$ and for an $O(k^{n-1})$ algorithm when $n\ge 3$.

Among the papers that contain algorithms for computing volumes of unions and intersections of balls, we mention [Au86], [AvBI88], [Sp85], and especially EDELSBRUNNER [Ed93] and EDELSBRUNNER & FU [EdF93]. See Subsection 9.11 for a suggested use of such algorithms in experimental computation on a famous unsolved problem.

From the viewpoint of computational complexity, the most interesting problem to be mentioned in this subsection is that of computing the volume of a union of n d-simplices in $\mathbb{R}^d$. (The change in notation – d rather than n for the dimension – is necessary in order to conform to the notation in the term *n^2-hard* below that is standard in the relevant part of the literature.) When $d=2$, this problem belongs to the class of so-called *n^2-hard* problems introduced by GAJENTAAN & OVERMARS [GaO93] with the aid of quadratic transformations from a "base problem" that is linearly equivalent to the following:

> Given three sets of integers A, B, and C with $|A|+|B|+|C|=n$, decide whether there exist $a\in A$, $b\in B$, and $c\in C$ such that $a+b=c$?

For the problem of computing the area of a union of triangles, as for other problems in the class, there are no known subquadratic algorithms. (In addition to the original paper of [GaO93], see [Or94], [EfLS93], [ErS93], and their references for more details. The present account is taken from [Or94].) From the viewpoint of computational convexity, it would be interesting to know what can be said for $d > 2$ about the complexity of computing the volume of the union of n simplices in $\mathbb{R}^d$, and what happens when the dimension d is part of the input.

8.4. More about dissections and triangulations

As we have seen, any sort of deterministic computation of the volume of a polytope P is apt to be time-consuming. However, since the volume of a simplex is so easy to compute, and since dissecting P into simplices is easy to understand and not hard to program (see Subsection 3.1), the use of such a dissection is the most convenient method in many practical cases. When faced with a polytope P whose volume is to be computed by means of dissection into simplices, it is natural to wonder what is the *minimum* number of simplices possible. That suggests the following decision problems.

8.4.1 Given a positive integer n, a proper $\mathcal{H}$-polytope (or $\mathcal{V}$-polytope) $P \subset \mathbb{R}^n$, a positive integer k.

(A) Can P be dissected into k or fewer n-simplices, each having all of its vertices among those of P?
(B) Can P be dissected into k or fewer n-simplices, each having all of its vertices at points of P that have rational coordinates?
(C) Can P be dissected into k or fewer n-simplices?

Note that in 8.4.1 (A), only the vertices of P can be used in forming the n-simplices of the dissection. It seems plausible that the minimum number of such simplices cannot be reduced by the use of additional vertices. However, we are not aware of any proof of this even for the case in which P is an n-cube. In 8.4.1 (B), additional vertices are permitted but are required to have all rational coordinates (as do the vertices of P), while in 8.4.1 (C) there is no restriction on the position of additional vertices. Hence it is conceivable that these three similar-sounding problems are of different computational complexity. It can be shown that 8.4.1 (A) belongs to the class $\mathbb{NP}$. However, for 8.4.1 (B) it does not seem obvious even that there exists a finite decision algorithm. (This is vaguely reminiscent of the fact that the problem of deciding whether a given polytope is combinatorially equivalent to one with exclusively rational vertices is algorithmically equivalent to the problem of deciding whether a diophantine equation is solvable in rationals (see Sturmfels [St87]) – and it is not known whether there is an algorithm for the latter problem (see [KlW91, p.95]).) For 8.4.1 (C), the existence of a finite decision algorithm follows from the decision theory of Tarski [Ta61] (see also [ChK73] and Renegar [Re92a], [Re92b], [Re92c]) because the existence of a dissection of the desired sort can be expressed in terms of the consistency (now over $\mathbb{R}^n$ rather than $\mathbb{Q}^n$) of a system of polynomial equalities and inequalities involving the vertex-coordinates.

When the above problems are posed for *triangulations* (as opposed to dissections), the above statements about computational complexity still apply to 8.4.1 (A) and 8.4.1 (C). However, in the case of triangulations, the same decision algorithm that works for 8.4.1 (C) applies also to 8.4.1 (B). That is a consequence of the following fact: If P is a polytope whose vertices are all at rational points, and $\mathcal{T}$ is a triangulation of P, then the nonrational vertices of $\mathcal{T}$ can be moved, one at a time, to nearby rational points so as to produce a triangulation $\mathcal{T}'$ of P whose vertices are all at rational points.

Though it has nothing to do with computing the volume of a polytope, we want to mention the following fact, simply for its intrinsic interest: An n-cube can be dissected into k n-simplices of equal volume if and only if k is a multiple of $n!$. This was first proved for $n = 2$ in [Mo70], then extended to arbitrary n in [Me79]. The proof depends in an essential way on valuation theory. For further results on dissecting polygons into triangles of equal area, see [KaS90] and [Mo90].

There is an extensive literature concerning the following questions.

8.4.2 What is the minimum number $T(n)$ $\langle S(n) \rangle$ of n-simplices into which an n-cube can be triangulated $\langle$dissected$\rangle$?

The number $S(n)$ has been fully determined only for $n \leq 4$, $T(n)$ only for $n \leq 7$ (see HUGHES [Hu93] and HUGHES & ANDERSON, [HuA94] and as $n \to \infty$ the best asymptotic lower and upper bounds are far apart. The best asymptotic lower bounds result from volume considerations (see [Ha91] for references), and the best asymptotic upper bounds come from the construction of specific triangulations in low dimensions together with a simple but elegant method of HAIMAN [Ha91] for extending these to higher dimensions. For $n \leq 8$, the best lower bounds for both $S(n)$ and $T(n)$ come from a linear programming approach proposed by SALLEE [Sa82] and developed further in [Hu93] and [HuA94].

Triangulations of n-cubes are of interest for their role in complementary pivoting algorithms used to find approximately fixed points of continuous mappings [To76]. In this connection, TODD [To76] proposes the number $(\mathrm{card}(\mathcal{T})/n!)^{1/n}$ as a measure of the efficiency of a triangulation $\mathcal{T}$ of the n-cube. The construction of [Ha91] shows that any value of this measure that is attained for some fixed n is also attainable asymptotically. However, this measure does not tell the whole story of efficiency, for it often happens that triangulations into fewer simplices require more complicated pivoting rules. See [ToT93] and [HuA94] for the details of some recent triangulations and for references to earlier work.

To end this subsection, we mention that ONG [On89], [On94], has analyzed a triangulation of the 3-cube that is notable for a number of geometric properties that make it especially convenient for use in the finite-element method for approximating solutions to partial differential equations. It would be worthwhile to produce and study higher-dimensional analogues of her triangulation.

9. Applications

This last section collects some of the more recent and probably less known applications of volume and mixed volume computation. The applications 9.3–9.5 and part of 9.9 were explored in more detail in [DyF91], the applications of 9.1, 9.2, 9.6–9.8 and part of 9.9 are dealt with in [DyGH94].

9.1. Counting integer points in lattice polytopes

As we have seen already in Subsection 3.5, there is a close connection between Volume Computation and counting lattice points. Here we show how a "mixed-volume-like" approach to lattice-point enumeration can be used to deduce some complexity results.

Let $\mathbb{L}$ denote an integer lattice of $\mathbb{R}^n$ whose vectors span $\mathbb{R}^n$, denote by $\mathcal{P}^n(\mathbb{L})$ the set of all polyopes in $\mathbb{R}^n$ whose vertex set belongs to $\mathbb{L}$, and let $G_{\mathbb{L}}$ denote the *lattice-point enumerator*, i.e. the functional $G_{\mathbb{L}} : \mathcal{P}^n(\mathbb{L}) \to \mathbb{N}_0$ defined by $G_{\mathbb{L}}(P) = \operatorname{card}(P \cap \mathbb{L})$. By Ehrhart [Eh67], [Eh68], [Eh69], [Eh77] there are functionals $G_{\mathbb{L},i} : \mathcal{P}^n(\mathbb{L}) \to \mathbb{N}_0$ such that for every $P \in \mathcal{P}^n(\mathbb{L})$ and $k \in \mathbb{N}$,

$$G_{\mathbb{L}}(kP) = \sum_{i=0}^{n} k^i G_{\mathbb{L},i}(P).$$

The polynomial on the right-hand side is often referred to as the *Ehrhart-polynomial*; see Stanley [St86b] for basic facts on this polynomial, and see [GrW93] for a survey on lattice-point problems. For simplicity we restrict the further considerations to the case where $\mathbb{L}$ is the standard integer lattice $\mathbb{Z}^n$ and omit the subscript $\mathbb{Z}^n$.

Note that $G_n(P)$ is just the volume of P; see 3.5. Suppose now, we could determine in polynomial time the number $G(P)$ of lattice points of a polytope $P \in \mathcal{P}(\mathbb{Z}^n)$. We could run this algorithm for the polytopes $1 \cdot P, \dots, n \cdot P$, and obtain $V(P) = G_n(P)$ by solving the system

$$Mx = \begin{pmatrix} 1 & 0 & \dots & 0 \\ 1 & 1^1 & \dots & 1^n \\ 1 & 2^1 & \dots & 2^n \\ \vdots & \vdots & & \vdots \\ 1 & n^1 & \dots & n^n \end{pmatrix} \begin{pmatrix} G_0(P) \\ G_1(P) \\ \vdots \\ G_n(P) \end{pmatrix} = \begin{pmatrix} G(0 \cdot P) \\ G(1 \cdot P) \\ \vdots \\ G(n \cdot P) \end{pmatrix} = b$$

of linear equations. Application of Theorems 5.1.4, 5.1.5 and 5.1.7 shows that the problem of evaluating $G(P)$ for integer $\mathcal{H}$-polytopes, integer $\mathcal{V}$-polytopes, or integer $\mathcal{S}$-zonotopes is #$\mathbb{P}$-complete.

Recall from Section 3.5 that a recent result of Barvinok [Ba93b] shows that in fixed dimension, $G(P)$ can be computed in polynomial time, see also [DyK93].

9.2. Zonotopes and Mixture-Management

The typical approach to standard problems in mixture-management models the problem as a linear program by assigning costs to each basic mixture. However, Girard & Valentin [GiV89] remark that for many applications in the petroleum industry the cost of these basic mixtures are essentially identical, and this accounts for the fact that the linear programming model may not always be the best one. They propose an approach that involves zonotopes.

Suppose that a seller has m containers, each of which contains a mixture of n basic chemicals. For $i \in \{1, \dots, m\}$ let $z_i = (\zeta_{i1}, \dots, \zeta_{in})^T$ represent the mixture in container i, where ζ_{ij} is the quantity of chemical j in container i. (The ζ_{ij} are assumed to be nonnegative rationals.)

Suppose, further, that a customer demands a certain mixture $b = (\beta_1, \dots, \beta_n)^T$ that consists of the quantities $\beta_1, \dots, \beta_n$ of chemical $1, \dots, n$, respectively.

In order to satisfy this demand the seller takes, for $i = 1, \dots, m$, a proportion λ_i of container i's content such that

$$\beta_j = \sum_{i=1}^{m} \lambda_i \zeta_{ij} \qquad \text{for } j \in \{1, \dots, n\}.$$

Hence, the zonotope

$$Z = \sum_{i=1}^{m} [0, 1] z_i$$

is the set of all possible demands that the seller can satisfy. In general, there is more than one way to satisfy the demand; thus the seller will have a choice of vectors

$$l \in \Lambda(b) = \left\{ l = (\lambda_1, \dots, \lambda_m)^T \in [0, 1]^m : \sum_{i=1}^{m} \lambda_i z_i = b \right\},$$

and the question is: What is a good strategy for the choice of l so as to be able to satisfy the widest possible variety of possible future demands. If the seller has no information on the distribution of these demands it might be reasonable to assume that they are uniformly distributed.

Then the objective for the seller is to maximize the volume of the zonotope $Z(l)$ that is the set of all mixtures that are still possible after the current demand b has been satisfied by the choice l. This maximization criterion was suggested in [GiV89]. Of course, the volume of $Z(l) = \sum_{i=1}^{m} [0, 1](1 - \lambda_i) z_i$ is a homogenous polynomial in the $(1 - \lambda_i)$'s, and its nth root is concave by the Brunn-Minkowski theorem 2.4.4. Hence the maximization problem is algorithmically tractable if the computation of function values is easy. However, Theorem 5.1.7 shows that the problem of computing $V(Z(l))$ is #$\mathbb{P}$-hard. Thus the algorithm suggested by [GiV89] is not efficient unless the number n of basic chemicals is small. Note, however, that the randomized algorithm of Theorem 7.1.1 could be used.

9.3. Integration over bodies

Suppose that K is a convex body of $\mathbb{R}^n$ and that the function $f : K \to \mathbb{R}$ is nonnegative and concave. Then

$$\int_K f(x)dx = V(K_f), \qquad \text{where} \quad K_f = \left\{ \begin{pmatrix} x \\ \xi_{n+1} \end{pmatrix} : x \in K, \, 0 \le \xi_{n+1} \le f(x) \right\}.$$

Since K_f is a convex body in $\mathbb{R}^{n+1}$, we can use the algorithm of Theorem 7.1.1 to approximate $\int_K f(x)dx$.

In order to bound the running time of the corresponding randomized algorithm, we need to make some assumptions about the *a priori* guarantees for K and f. Naturally, we will assume again that K is given by a centered well-bounded membership oracle with parameters r, R and b. Dyer & Frieze [DyF91] suggest, as measures for the size of f, the size L_1 of an upper bound of f on K and the size L_2 of a positive lower bound on f's average $\int_K f(x)dx/V(K)$ on K. Using 7.7.1, the integral can then be approximated in time that is polynomial in size(K) and L_1 and L_2.

DYER & FRIEZE [DyF91] further show that this approach can be extended to quasi-concave functions satisfying a Lipschitz-condition, and they derive a *pseudopolynomial* randomized algorithm for general integrable functions.

9.4. STOCHASTIC PROGRAMMING

As is pointed out by DYER & FRIEZE [DyF91], the randomized algorithm for computing the volume of convex bodies can, in certain cases, be used to approximate the expected value of certain stochastic programming problems. Examples discussed in [DyF91] include the problem of computing the expected value of the functional $\varphi(b)$ that is defined by

$$\varphi(b) = \max f(x)$$

$$g_i(x) \le \beta_i, \quad i = 1, \dots, m,$$

where f is a concave functional, $g_1, \dots, g_m$ are convex, and $b = (\beta_1, \dots, \beta_m)^T$ is chosen uniformly from a convex body $K \in \mathcal{K}^m$.

Another example of [DyF91] deals with a question that comes up in the sensitivity analysis for linear programs. When, in the linear program

$$\min\langle c, x\rangle$$

$$Ax = b$$

$$x \ge 0,$$

the parameters $(b^T, c^T)^T \in \mathbb{R}^m \times \mathbb{R}^n$ are chosen uniformly from a convex body K in $\mathbb{R}^{m+n}$, sensitivity analysis may ask for the probability that a specific nonsingular $(m \times m)$ submatrix B of A gives an optimal basic solution. This can be expressed in terms of volumes as follows. Since B is nonsingular, the condition $Ax = b$ is equivalent to $B^{-1}Ax = B^{-1}b$. Now, let $x_B = B^{-1}b$, let $\hat{x}_B$ denote the corresponding n-vector that is obtained from x_B by augmenting components 0 whenever the corresponding column of A does not belong to B, and let c_B denote the m-vector obtained from c by deleting all components that do not correspond to a column in B. Then it is well-known from duality theory of linear programming (and quite easy to derive, see e.g. [GoT89] or [GrK93b]) that $\hat{x}_B$ is an *optimal basic feasible solution* if and only if $\hat{x}_B$ is primal feasible and $\hat{y}_B = (c_B^T B^{-1})^T$ is dual feasible. Since, by definition, $A\hat{x}_B = b$, this is equivalent to

$$x_B \ge 0 \quad \text{and} \quad c_B^T B^{-1} A \le c^T.$$

Hence the choices $(b^T, c^T)^T$ of K for which B is an optimal basis are those which belong to the subset

$$K_B = K \cap \left\{ \begin{pmatrix} b \\ c \end{pmatrix} : B^{-1}b \ge 0,\, c_B^T B^{-1} A \le c^T \right\}$$

of K, and the probability of B being an optimal basis is given as the volume ratio

$$V(K_B)/V(K).$$

Again, under some reasonable assumptions, the randomized algorithm 7.1.1 for volume computation can be used.

9.5. Learning a halfspace

Another application for volume computation, due to Dyer, Frieze & Kannan (see [DyF91]), is related to certain questions in "learning theory." Suppose, an algorithm $\mathcal{A}$ wanted to "learn" an unknown inequality $\langle a, x\rangle \geq \alpha$ where $a \in [-1,1]^n$ and $\alpha \in [-1,1]$. Suppose, further, that there is a sequence $(x_i)_{i\in\mathbb{N}}$ of points provided, and at step i, a guess is made by $\mathcal{A}$ as to whether x_i satisfies the inequality. It is then revealed to $\mathcal{A}$ whether the guess is correct. The goal for $\mathcal{A}$ is to devise a strategy which minimizes the proportion of errors made.

Now, each query point x_i leads to two halfspaces

$$H_i^+ = \left\{ \begin{pmatrix} a \\ \alpha \end{pmatrix} : \langle a, x_i\rangle \geq \alpha \right\} \quad \text{and} \quad H_i^- = \left\{ \begin{pmatrix} a \\ \alpha \end{pmatrix} : \langle a, x_i\rangle \leq \alpha \right\},$$

and each "verification" if the guessed answer is correct or not rules out one of the halfspaces. Hence after step i, $\mathcal{A}$ knows a polytope P_i, and a good strategy for deciding, whether for point x_{i+1} the guess should be "yes" or "no" may be guided by the volumes of the two parts

$$P_i \cap H_{i+1}^+, \quad P_i \cap H_{i+1}^-.$$

An analysis of this approach and a comparison with a method of Maass & Turán [MaT89] can be found in Dyer & Frieze [DF91]. See Blum & Kannan [BlK93] for a polynomial time method for learning an intersection of a constant number of halfspaces over a uniform distribution of query points.

9.6. Permanents

For $i, j \in \{1, \dots, n\}$ let α_{ij} be a nonnegative integer, and set $Z_i = \sum_{j=1}^n [0, \alpha_{ij}] e_j$. As we have seen in Subsection 5.2,

$$n! V(Z_1, \dots, Z_n) = \mathrm{per}(A)$$

is the permanent of the matrix $A = (\alpha_{ij})_{i,j=1,\dots,n}$.

We have used the #$\mathbb{P}$-hardness of the problem of computing the permanent of a matrix to show in 5.2.2 that the problem of computing the mixed volume $V(Z_1, \dots, Z_n)$ of the rectangular parallelotopes $Z_1, \dots, Z_n$ is #$\mathbb{P}$-hard. However, the correspondence goes both ways, and any progress for mixed volume computation leads to new results on the "positive side" for permanent computation. It follows, for instance, from the results stated in Subsection 7.2 that one can approximate the permanent of matrices with positive integer entries of quasi-polynomial size, at least if they have the property that all but $n - o(\log n / \log\log n)$ of the rows are identical, see [DyGH94].

Note that the fastest deterministic algorithm known for computing the permanent of a square 0-1 matrix with n rows runs in time $n2^{n-1}$ (see Ryser [Ry63] and the improvement by Nijenhuis & Wilf [NiW78]), while the best known randomized algorithms for producing a relative approximation p with

$$\text{prob}\left\{\left|\frac{p}{\text{per}(A)}-1\right|\leq\epsilon\right\}\geq 1-\beta.$$

still use time of order $2^{O(\sqrt{n}\log^2(n))}\epsilon^{-2}\log(1/\beta)$, JERRUM & VAZIRANI [JeV91], see also KARMARKAR, KARP, LIPTON, LOVÁSZ & LUBY [KaKLLL93].

Let us point out that besides the well-known applications in mathematical programming and combinatorics, there is need for computing or approximating the permanents of certain matrices that arise in particle physics (see ZHU [Zh93]). The lack of efficient procedures for this task leads to difficulties in the study of the Bose-Einstein correlation between particles.

It seems appropriate to end this subsection by mentioning van der Waerden's 1926 conjecture that on the $(n-1)^2$-dimensional polytope formed by all $n \times n$ doubly stochastic matrices, the permanent attains its minimum at the matrix $\frac{1}{n}J_n$ whose entries are all equal to $1/n$. This was finally proved by EGORYCHEV [Eg81], who showed that the minimum is in fact attained *uniquely* at $\frac{1}{n}J_n$. An essential tool in his proof was the Aleksandrov-Fenchel inequality 2.4.3 for mixed volumes. KNUTH [Kn81] later gave a relatively elementary, self-contained proof of Egorychev's result, and it turned out that FALIKMAN [Fa81] had independently proved the conjecture (but not the uniqueness) by different methods. Nevertheless, we believe that mixed volumes will continue to be a useful tool in dealing with specific problems that may at first not appear to have any connection with mixed volumes. In the words of Egorychev [Eg81, p.299]: "The method of mixed volumes is ideally suited to solving extremal problems and problems of uniqueness, and obtaining deep new inequalities. It is reasonable to assume that in the future the method of mixed volumes will stand with that of generating functions as one of the basic analytical tools of combinatorial analysis."

9.7. POLYNOMIAL EQUATIONS

Mixed volumes play an important role in algebraic geometry. Let us here discuss the relation of mixed volumes and the number of solutions of a system of equations involving *Laurent polynomials.*

We use a notation similar to that introduced in Subsection 3.4: when $x = (\xi_1,\ldots,\xi_n)$, and $q = (\kappa_1,\ldots,\kappa_n) \in \mathbb{Z}^n$, then

$$x^q = \xi_1^{\kappa_1}\cdot\xi_2^{\kappa_2}\cdot\ldots\cdot\xi_n^{\kappa_n}.$$

Now, let $S_1, S_2, \ldots, S_n \subset \mathbb{Z}^n$, and let for $i = 1,\ldots,n$,

$$f_i(x) = \sum_{q\in S_i} c_q^{(i)} x^q,$$

where the coefficients $c_q^{(i)}$ are fixed complex numbers. Hence $f_i \in \mathbb{C}[x_1, x_1^{-1}, \ldots, x_n, x_n^{-1}]$, and we are interested in determining the number $L(F)$ of distinct common roots of the system $F = \{f_1,\ldots,f_n\}$. By a result of BERNSTHEIN [Be75] (see also [BuZ88, Chapter 27]) this number depends, if the coefficients $c_q^{(i)}$ $(q \in S_i)$ are chosen generically (i.e. in sufficiently general position), only on the *Newton polytopes*

$$P_i = \text{conv}(S_i)$$

of the polynomials, and, more precisely,

$$L(F) \ = \ n! \cdot V(P_1, P_2, \ldots, P_n).$$

HUBER & STURMFELS [HuS93] use this result in conjunction with an approach similar to those of Subsection 4.2 to devise a numerical continuation algorithm for computing all isolated common roots of a system of polynomial equations. See GELFAND, KAPRANOV & ZELEVINSKY [GeKZ90] and also [GrS93] for further information about Newton polytopes and related concepts, and VERSCHELDE & COOL [VeC92], PETERSEN & STURMFELS [PeS93], CANNY & ROJAS [CaR93], and VERSCHELDE, VERLINDEN & COOL [VeVC94] for further results along these lines.

9.8. BASES OF UNIMODULAR MATROIDS

Let $\mathcal{M}$ be a unimodular matroid of rank n with representation $v_1, v_2, \ldots, v_m$ over the reals. Let $S_1, \ldots, S_r$ be a partition of $\{1, 2, \ldots, m\}$ and let $i_1, \ldots, i_r$ be nonnegative integers such that $\sum_{j=1}^{r} i_j = n$. STANLEY [St81] shows that the number of bases of $\mathcal{M}$ with i_j elements in S_i for $i = 1, \ldots, r$ is

$$\binom{n}{i_1, \ldots, i_r} V(\overbrace{Z_1, \ldots, Z_1}^{i_1}, \ldots, \overbrace{Z_r, \ldots, Z_r}^{i_r}),$$

where Z_i is the zonotope

$$Z_i = \sum_{j \in S_i} [0,1] v_j.$$

Hence the results of Subsection 5.2 can be applied to yield #$\mathbb{P}$-hardness results for this counting problem. See BJÖRNER, LAS VERGNAS, STURMFELS, WHITE & ZIEGLER [BjLSWZ93] for a state-of-the-art account of oriented matroids.

9.9. PARTIAL ORDERS AND LINEAR EXTENSIONS

When dealing with the strong #$\mathbb{P}$-hardness of volume computation for $\mathcal{H}$-polytopes in Subsection 5.1, we showed that the number of linear extensions of a given ordering $\mathcal{O}$ is equal to the volume of the order polytope $P_{\mathcal{O}}$. Now we outline an extension of this result, also due to STANLEY [St81], that involves mixed volumes.

Let $N = \{p_1, \ldots, p_r, q_1, \ldots, q_{n-r}\}$ be a poset, let $N_p = \{p_1, \ldots, p_r\}$, $N_q = \{q_1, \ldots, q_{n-r}\}$, and suppose that $p_1 < p_2 < \cdots < p_r$. For $i_1, \ldots, i_r \in \{1, \ldots, n\}$ let $e(i_1, \ldots, i_r)$ be the number of linear extensions π of N such that $\pi(p_j) = i_j$ for $j = 1, 2, \ldots, r$.

Now, define for $j = 0, \ldots, r$ the order polytopes P_j of $\mathbb{R}^{n-r}$ as the sets of points $x = (\xi_1, \ldots, \xi_{n-r})^T$ that satisfy the following constraints:

$$\begin{array}{ll} 0 \le \xi_i \le 1 & \text{for } i = 1, \ldots, n-r; \\ \xi_i \le \xi_k & \text{if } q_i < q_k; i, k = 1, \ldots, n-r; \\ \xi_i = 0 & \text{if } j > 0 \text{ and } q_i < p_j; \ i = 1, \ldots, n-r; \\ \xi_i = 1 & \text{if } j < r \text{ and } q_i > p_{j+1}; \ i = 1, \ldots, n-r. \end{array}$$

Note that the polytopes P_j reflect the restriction of N to N_q that lies "between" p_j and p_{j+1}.

STANLEY [St81] showed that (with $i_0 = 0, i_{r+1} = n+1$),

$$e(i_1, i_2, \ldots, i_r) = (n-r)!V(\overbrace{P_0, \ldots, P_0}^{i_1 - i_0 - 1}, \ldots, \overbrace{P_r, \ldots, P_r}^{i_{r+1} - i_r - 1}).$$

As a side remark, observe that, when $r = 1$, the Aleksandrov-Fenchel inequalities 2.4.3 imply that for $i = 1, \ldots, n-1$

$$e(i)^2 \geq e(i-1)e(i+1),$$

and hence the sequence $e(1), \ldots, e(n)$ is unimodal.

9.10. EXPERIMENTAL COMPUTATION: POINTS ON SPHERES

There is a large collection of unsolved mathematical problems involving volumes or mixed volumes of polytopes or other bodies. In many cases, an important source of difficulty is a lack of intuition or computational experience that might guide one to a solution. That is especially true of extremum problems, and especially true in higher dimensions. We expect that improved methods of volume computation, in conjunction with heuristic optimization methods, will facilitate computational experiments to provide increased insight concerning these problems. Here we describe two problems that are not completely solved even in low dimensions and for which the range of ignorance increases rapidly as the dimension grows.

The sort of development that we have in mind is well illustrated by the history of the following problem:

9.10.1 How should k points be arranged on the unit sphere $\mathbb{S}^{n-1}$ in $\mathbb{R}^n$ so as to maximize the volume of their convex hull?

For $n = 2$, the solution is the obvious one: place the points at the vertices of a regular k-gon inscribed in the unit circle $\mathbb{S}^1$ [Fe53]. As far as we know, the only case of 9.10.1 that has been settled for all n is that in which $k = n+1$, where the regular arrangements (placing the points at the vertices of a regular n-simplex) are the only ones that maximize the volume (FEJES TÓTH [Fe64, p.313], SLEPIAN [Sl69]). FEJES TÓTH [Fe64] discusses the difficulties in the case $n = 4$, $k = 120$, saying that "It may be taken as certain that of the 4-dimensional polytopes with 120 vertices and unit circumradius $\{3, 3, 5\}$ (the regular one) has the greatest possible volume ... But so far we have no methods for proving these conjectures ...." Of course, computer experimentation cannot provide proofs of such conjectures. However, algorithms combining volume computation with optimization methods may provide useful clues in cases where there is no obvious candidate for the optimizing shape or when the "obvious" candidate turns out not to be the optimum. Consider, for example, the case $n = 3$, $k = 8$, where the 8 vertices of a cube do not yield even a relative maximum. The volume-maximizing arrangement of 8 points was first discovered (as a relative maximum) by computer experimentation (GRACE [Gr63]), and was later proved by BERMAN & HANES [BeH70] to be the maximum. We expect that, as algorithms for volume computation are improved, higher-dimensional analogues of this sort of compute-conjecture-prove development will occur.

The problem of the preceding paragraph is closely related to the problem of finding, for a given n and k, the n-polytopes of minimum volume among those that have k facets and are circumscribed about a given sphere in $\mathbb{R}^n$. Again, the regular solution is known to be optimum for all k when $n = 2$ [Fe53] and for all n when $k = n+1$. Beyond that, the solution is known when $n = 3$ for a few values of $k > 4$, but ignorance is almost total in the higher-dimensional cases. For the case $n = 3$, see GOLDBERG [Go35] for the history of this problem and SCHOEN [Sc86] for an algorithmic approach.

The following problem turns out to involve a specific mixed volume.

9.10.2 How should k points be arranged on the unit sphere $\mathbb{S}^{n-1}$ in $\mathbb{R}^n$ so as to maximize the mean width of their convex hull?

As pointed out at the end of Section 2, the mean width is just a multiple of the first intrinsic volume V_1. Let us add that for k-polytopes P in $\mathbb{R}^n$, there is a formula of SHEPHARD [Sh68]:

$$(1+(-1)^k)V_1(P) = \sum_{i=1}^{k-1} \sum_{F\in\mathcal{F}_i(P)} (-1)^{i+1} V_1(F).$$

The extent of ignorance concerning 9.10.2 is even greater than that for 9.10.1. For the important case in which $k = n+1$, several authors ([Gi52], [Ba63], [Ba65], [We68]) have assumed the existence of a proof that the regular arrangement maximizes the mean width. However, we are not aware of any such proof. (It is known, ALEXANDER [Al77], that the regular arrangement of $n+1$ points maximizes the *width* of the inscribed simplex.)

The problem 9.10.2 is of interest because of its connection with a problem in communication theory. Suppose that Z denotes the Gaussian distribution in $\mathbb{R}^n$ that has zero mean and whose covariance matrix is the $n \times n$ identity matrix. For a fixed $k > n$, a fixed $\lambda > 0$, and a fixed set U consisting of k points of $\mathbb{S}^{n-1}$, let Y be the vector random variable of the form $Y = \lambda U + Z$. Upon receiving Y, we are asked to decide which point of U has been transmitted, all points of U being equally likely a priori. The problem is to arrange the points of U so as to maximize the probability of this detection. The *simplex code conjecture* asserts that when $k = n+1$, the optimum arrangement is the regular one. A claimed proof [LaS66] of the conjecture was shown by FARBER [Fa68] and TANNER [Ta70] to be invalid, but the conjecture itself is still open. See [Fa68], [Ta74] for stronger forms of this conjecture, BALAKRISHNAN [Ba61], [Ba65], and TANNER [Ta70] for the relationship of the conjecture to mean widths of simplices, and CHAKERIAN & KLAMKIN [ChKl73] for other conjectures on mean widths of simplices.

9.11. EXPERIMENTAL COMPUTATION: PUSHING BALLS TOGETHER

For each point $c \in \mathbb{R}^n$, let $B(c) = \{x \in \mathbb{R}^n : \|x - c\|_2 \leq 1\}$, the Euclidean ball of unit radius centered at the point c. Now suppose that $p_1, \ldots, p_k$ and $q_1, \ldots, q_k$ are points of $\mathbb{R}^n$ such that for all i, j, $\|q_i - q_j\|_2 \leq \|p_i - p_j\|_2$ – in other words, the q_i are pairwise at least as close together as the p_i. It has been conjectured that under these circumstances, the greater extent of overlapping of the balls $B(q_i)$ insures that

$$V\left(\bigcap_{i=1}^{k} B(q_i)\right) \geq V\left(\bigcap_{i=1}^{k} B(p_i)\right)$$

and

$$V\left(\bigcup_{i=1}^{k} B(q_i)\right) \leq V\left(\bigcup_{i=1}^{k} B(p_i)\right).$$

(see [Th54] and [Kn55] for the latter conjecture). Both conjectures have been proved for the case in which $k \leq n+1$ and for unrestricted k when $n = 1$, but for unrestricted k both are open for all $n \geq 2$. Concerning the intersections, it follows from a theorem of Kirszbraun [Ki34] that if $V(\bigcap_{i=1}^{k} B(p_i)) > 0$ then $V(\bigcap_{i=1}^{k} B(q_i)) > 0$, and concerning the unions, it was proved by M. Kneser [Kn55] that

$$V\left(\bigcup_{i=1}^{k} B(q_i)\right) \leq 3^n V\left(\bigcup_{i=1}^{k} B(p_i)\right).$$

See [KlW91] for a detailed discussion of the above conjectures and some of their relatives, including a stronger conjecture of Kneser that implies the above conjectures. Kneser's interest in these questions arose from his study [Kn55] of a measure of surface area proposed by Minkowski.

Despite the plausibility of the above conjectures, it would not surprise us if they fail even when $n = 2$. It seems that for small n and for k not too much larger than $n+1$, it should be possible to design a computer experiment that would greatly improve the multiplier 3^n and would at the same time have a good chance of discovering a counterexample to the original conjecture (if one exists). Such an experiment would require fast algorithms for computing the volumes of intersections and unions of balls. See subsection 8.3 for references to such algorithms.

References

[Al37] Aleksandrov, A.D., *On the theory of mixed volumes of convex bodies, II. New inequalities between mixed volumes and their applications* (in Russian), Math. Sb. N.S. **2** (1937), pp. 1205–1238.

[Al38] Aleksandrov, A.D., *On the theory of mixed volumes of convex bodies, IV. Mixed discriminants and mixed volumes* (in Russian), Math. Sb. N.S. **3** (1938), pp. 227–251.

[Al77] Alexander, R., *The width and diameter of a simplex*, Geometriae Dedicata **6** (1977), pp. 87–94.

[AlS86] Allgower, E.L. and Schmidt, P.M., *Computing volumes of polyhedra*, Math. of Comput. **46** (1986), pp. 171–174.

[ApK90] Applegate, D. and Kannan, R., *Sampling and integration of near log-concave functions*, Proc. 23rd ACM Symp. Th. of Comput. (1990), pp. 156–163.

[Au88] Aurenhammer, F., *Improved algorithms for disks and balls using power diagrams*, J. Algorithms **9** (1988), pp. 151–161.

[AvBI88] Avis, D., Bhattacharya, B.K. and Imai, H., *Computing the volume of the union of spheres*, The Visual Computer **3** (1988), pp. 323–328.

[AvF91] Avis, D. and Fukuda, K., *A pivoting algorithm for convex hulls and vertex enumeration of arrangements of polyhedra*, Proc. 7th Ann. Symp. Comput. Geom., June, 1991, pp. 98–104; Discrete Comput. Geom. **8** (1992), pp. 295–313.

[Ba61] Balakrishnan, A.V., *A contribution to the sphere-packing problem of communication theory*, J. Math. Anal. Appl. **3** (1961), pp. 485–506.

[Ba63] Balakrishnan, A.V., *Research Problem No. 9: Geometry*, Bull. Amer. Math. Soc. **69** (1963), pp. 737–738.

[Ba65] Balakrishnan, A.V., *Signal selection for space communication channels*; In: *Advances in Communication Systems*, (ed. by A.V. Balakrishnan), Academic Press, New York, 1965, pp. 1–31.

[BaT24] Banach, S. and Tarski, A., *Sur la decomposition des ensembles de points en parties respectivement congruents*, Fund. Math. **6** (1924), pp. 244–277.

[BaB93] Bárány, I. and Buchta, C., *Random polytopes in a convex polytope, independence of shape and concentration of vertices*, Math. Ann. **297** (1993), pp. 467–497.

[BaF86] Bárány, I. and Füredi, Z., *Computing the volume is difficult*, Proc. ACM Symp. Th. Comp. **8** (1986), pp. 442–447; Discrete Comput. Geom. **2** (1987), pp. 319–326.

[BaS79] Barrow, D.L. and Smith, P.W., *Spline notation applied to a volume problem*, Amer. Math. Monthly **86** (1979), pp. 50–51.

[Ba91] Barvinok, A.I., *Calculation of exponential integrals* (in Russian), Zap. Nauchn. Sem. LOMI, Teoriya Slozhnosti Vychislenii **192** (1991), pp. 149–163.

[Ba93a] Barvinok, A.I., *Computing the volume, counting integral points, and exponential sums*, Discrete Comput. Geom. **10** (1993), pp. 123-141.

[Ba93b] Barvinok, A.I., *A polynomial time algorithm for counting integral points in polyhedra when the dimension is fixed*, Preprint (to appear).

[BaL93] Bayer, M. and Lee, C., *Combinatorial aspects of convex polytopes*; In: *Handbook of Convex Geometry*, (ed. by P. Gruber and J. Wills), North-Holland, Amsterdam , 1993, pp. 251–305.

[Be92] Ben–Israel, A., *A volume associated with m by n matrices*, Lin.Alg.Appl., **167** (1992), pp. 87–111.

[BeBG85] Bérad, P., Besson, G. and Gallot, A.S., *Sur une inégalité isopérimétrique qui géneralise celle de Paul Levy - Gromov*, Invent. Math. **80** (1985), pp. 295–308.

[BeZ65] Berezin, I.S. and Zhidkov, N.P., *Computing Methods, Vol.1*, Pergamon Press, Oxford, 1965.

[BeH70] Berman, J. and Hanes, K., *Volumes of polyhedra inscribed in the unit sphere in E^3*, Math. Ann. **188** (1970), pp. 78–84.

[Be75] Bernshtein,D.N., *The number of roots of a system of equations*, Funct. Anal. Appl. **9** (1975), pp. 183–185.

[Be'92] Betke, U., *Mixed volumes of polytopes*, Arch. Math. **58** (1992), pp. 388–391.

[BeH93] Betke, U. and Henk, M., *Approximating the volume of convex bodies*, Discrete Comput. Geom. **10** (1993), pp. 15–21.

[BeM83] Betke, U. and McMullen, P., *Estimating the size of convex bodies from projections*, J. London Math. Soc. (2) **27** (1983), pp. 525–538.

[BiN83] Bieri, H. and Nef, W., *A sweep-plane algorithm for computing the volume of polyhedra represented in boolean form*, Lin. Alg. Appl. **52/53** (1983), pp. 69–97.

[BjLSWZ93] Björner, A., Las Vergnas, M., Sturmfels, B., White, N. and Ziegler, G., *Oriented Matroids*, Cambridge University Press, Cambridge, 1993.

[BlK93] Blum, A.L. and Kannan, R., *Learning an intersection of k halfspaces over a uniform distribution*, Proc. 34th Symp. Found. Comput. Sci. (1993), pp. 312–320.

[Bl53] Blumenthal, L.M., *Distance Geometry*, Oxford Univ. Press, London, 1953.

[BlG43] Blumenthal, L.M. and Gillam, B.E., *Distribution of points in n-space*, Amer. Math. Monthly **50** (1943), pp. 181–185.

[BoGKL90] Bodlaender, H.L., Gritzmann, P., Klee, V. and Van Leeuwen, J., *The computational complexity of norm-maximization*, Combinatorica **10** (1990), pp. 203–225.

[BoHW72] Bokowski, J., Hadwiger, H. and Willis, J.M., *Eine Ungleichung zwischen Volumen, Oberfläche und Gitterpunktanzahl konvexer Körper im n-dimensionalen euklidischen Raum*, Math. Z. **127** (1972), pp. 363–364.

[Bo78] Boltyanskii, V.G., *Hilbert's Third Problem*, (trans. by R. Silverman), Winston, Washington, D.C., 1978.

[BoF34] Bonnesen, T. and Fenchel, W., *Theorie der konvexen Körper* Springer, Berlin, 1934; (reprinted: Chelsea, New York), 1948; *Theory of Convex Bodies*, (English edition), BCS Associates, Moscow, Idaho, U.S.A., 1987.

[BoH82] de Boor, C. and Höllig, K., *Recurrence relations for multivariate B-splines*, Proc. Amer. Math. Soc. **85** (1982), pp. 397–400.

[BrW91] Brightwell, G. and Winkler, P., *Counting linear extensions*, Order 1991, pp. 225–242.

[Br88] Brion, M., *Points entiers dans les polyèdres convexes*, Ann. Sci.Ècole Norm. Sup. (4) **21** (1988), pp. 653–663.

[Br83] Brøndsted, A., *An Introduction to Convex Polytopes*, Springer, New York, 1983.

[BuR92] Buchta, C. and Reitzinger, M., *What is the expected volume of a tetrahedron whose vertices are chosen at random from a given tetrahedron?*, Anz. Österr. Akad. Wiss., Math. Naturwiss. Kl. (1992), pp. 63–68.

[BuZ88] Burago, Y.D. and Zalgaller, V.A., *Geometric inequalities*, Springer, Berlin, 1988.

[BuGK94a] Burger, T., Gritzmann, P. and Klee, V., *Finding optimal shadows of polytopes*, in preparation.

[BuGK94b] Burger, T., Gritzmann, P. and Klee, V., *Optimizing sections of polytopes*, in preparation.

[CaR93] Canny, J. and Rojas, J.M., *An optmality condition for determining the exact number of roots of a polynomial system*, (to appear).

[ChK73] Chang, C.C. and Keisler, H.J., *Model Theory*, North-Holland, Amsterdam, 1973.

[ChKl73] Chakerian, D.G. and Klamkin, M.S., *Minimum triangles inscribed in a convex curve*, Math. Mag. **46** (1973), pp. 256–260.

[Ch93] Chazelle, B., *An optimal convex hull algorithm in any fixed dimension*, Discrete Comput. Geom. **10** (1993), pp. 377–409.

[ChHJ92] Chen, P.-C., Hansen, P. and Jaumard, B., *Partial pivoting in vertex enumeration*, RUTCOR Research Report #10 – 92 (1992).

[ChY77] Chung, K.C. and Yao, T.H., *On lattices admitting unique Lagrange interpolation*, SIAM J. Numer. Analysis **14** (1977), pp. 735–743.

[CoH79] Cohen, J. and Hickey, T., *Two algorithms for determining volumes of convex polyhedra*, J. Assoc. Comp. Mach. **26** (1979), pp. 401–414.

[CrFG91] Croft, H.T., Falconer, K.J. and Guy, R.K., *Unsolved Problems in Geometry*, Springer, New York, 1991.

[Da63] Dantzig, G.B., *Linear Programming and Extensions*, Princeton University Press, Princeton, 1963.

[DaR84] Davis, P.J. and Rabinovitz, P., *Methods of Numerical Integration*, (2nd ed.) Academic Press, Orlando, 1984.

[De00] Dehn, M., *Über raumgleiche Polyeder*, Nachr. Akad. Wiss. Göttingen, Math.-Phys. Kl. (1900), pp. 345–354.

[Dö65] Dörrie, H., *100 Great Problems of Elementary Mathematics*, Dover, New York, 1965.

[DuHK63] Dubins, L., Hirsch, M. and Karush, J., *Scissor congruence*, Israel J. Math. **1** (1963), pp. 239–247.

[Dy83] Dyer, M.E., *The complexity of vertex enumeration methods*, Math. Oper. Res. **8** (1983), pp. 381–402.

[Dy91] Dyer, M.E., *On counting lattice points in polyhedra*, SIAM J. Comput. **20** (1991), pp. 695–707.

[DyF88] Dyer, M.E. and Frieze, A.M., *The complexity of computing the volume of a polyhedron*, SIAM J. Comput. **17** (1988), pp. 967–974.

[DyF91] Dyer, M.E. and Frieze, A.M., *Computing the volume of convex bodies: a case where randomness provably helps* In: *Probabilistic Combinatorics and its Applications*, (ed. by Béla Bollobás), Proceedings of Symposia in Applied Mathematics Vol. 44, American Mathematical Society, 1991, pp. 123–169.

[DyFK89] Dyer, M.E., Frieze, A.M. and Kannan, R., *A random polynomial time algorithm for estimating volumes of convex bodies*, Proc. 21st Symp. Th. Comput. (1989), pp. 375–381.

[DyFK91] Dyer, M.E., Frieze, A.M. and Kannan, R., *A random polynomial time algorithm for approximating the volumes of convex bodies*, J. Assoc. Comp. Mach. (1989), pp. 1-17.

[DyGH94] Dyer, M.E., Gritzmann, P. and Hufnagel, A., *On the complexity of computing (mixed) volumes*, manuscript (1994).

[DyK93] Dyer, M.E. and Kannan, R., *On Barvinok's algorithm for counting lattice points in fixed dimension*, manuscript (1993).

[Ed87] Edelsbrunner, H., *Algorithms in Combinatorial Geometry*, Springer, New York etc., 1987.

[Ed93a] Edelsbrunner, H., *Computational geometry*; In: *Handbook of Convex Geometry A*, (ed. by P. Gruber and J. Wills), North-Holland, Amsterdam , 1993, pp. 699–735.

[Ed93b] Edelsbrunner, H., *The union of balls and its dual shape*, Proc. 9th Ann. Sympos. Comput. Geom. (1993), pp. ??-??.

[EdF93] Edelsbrunner, H. and Fu, P., *Measuring space filling diagrams*; (Rept. 1010) Nat. Center Supercomputer Appl., Univ. Illinois, Urbana, Illinois, 1993.

[EdM90] Edelsbrunner, H. and Mücke, P., *Simulation of simplicity: a technique to cope with degenerate cases in geometric algorithms*, ACM Trans. Graphics **9** (1990), pp. 66–104.

[EdOS86] Edelsbrunner, H., O'Rourke, J. and Seidel, R., *Constructing arrangements of lines and hyperplanes with applications*, SIAM J. Computing **15** (1986), pp. 341–363.

[EdSS91] Edelsbrunner, H., Seidel, R. and Sharir, M., *On the zone theorem for hyperplane arrangements*; In: *New Results and Trends in Computer Science*, (ed. by H. Maurer), Springer Lecture Notes in Computer Science 555, Berlin, 1991, pp. 108–123; SIAM J. Comput. **22** (1993), pp. 418–429.

[Ed70] Edmonds, J., *Submodular functions, matroids, and certain polyhedra*; In: *Combinatorial Structures and their Applications*, (ed. by R. Guy, H. Hanani, N. Sauer and J. Schönheim), Gordon and Breach, New York, 1970, pp. 69–87.

[EfLS93] Efrat, A., Lindenbaum, M. and Sharir, M., *On finding maximally consistent sets of halfspaces*; In: *Proc. 5th Canad. Conf. Comput. Geom.*, Univ. of Waterloo , Waterloo, Canada, 1993, pp. 432–436.

[Eg81] Egorychev, G.P., *The solution of van der Waerden's problem for permanents*, Advances in Math. **42** (1981), pp. 299–305.

[Eh67] Ehrhart, E., *Sur un probléme de géometrie diophantienne linéaire*, J. reine angew. Math. **226** (1967), pp. 1–29; **227** pp. 25–49.

[Eh68] Ehrhart, E., *Démonstration de la loi de réciprocité*, C.R. Acad. Sci. Paris **265** (1968), pp. 5–9, 91–94.

[Eh69] Ehrhart, E., *Démonstration de la loi de réciprocité*, C.R. Acad. Sci. Paris **266** (1969), pp. 696–697.

[Eh77] Ehrhart, E., *Polynômes arithmétiques et méthode des polyédres en combinatoire*, Birkhäuser, Basel, 1977.

[El86] Elekes, G., *A geometric inequality and the complexity of computing volume*, Discrete Comput. Geom. **1** (1986), pp. 289–292.

[ErS93] Erickson, J. and Seidel, R., *Better lower bounds on detecting affine and spherical degeneracies*, Proc. 34th Ann. IEEE Sympos. Found. Comput. Sci. (FOCS93), (1993), pp. 528–536.

[Fa81] Falikman, D.I., *A proof of the van der Waerden conjecture on the permanent of a doubly stochastic matrix*, Mat. Zametki **29** (1981), pp. 931–938; English translation: *Math. Notes*, Acad. Sci. USSR **29** (1981), pp. 475-479.

[Fa68] Farber, S.M., *On the signal selection problem for phase coherent and incoherent communication channels* , Tech. Report No. 4, Communications Theory Lab., Dept. of Electrical Engineering, Calif. Inst. Tech. (1968).

[Fe53] Fejes Tóth, L., *Lagerungen in der Ebene, auf der Kugel, und im Raum*, Springer, Berlin, 1953.

[Fe64] Fejes Tóth, L., *Regular Figures*, Pergamon, Oxford, 1964.

[Fe36] Fenchel, W., *Inégalités quadratique entre les volumes mixtes des corps convexes*, C.R. Acad. Sci. Paris **203** (1936), pp. 647–650.

[Fi88] Filliman, P., *Exterior algebra and projections of polytopes*, Discrete Comput. Geom. **5** (1990), pp. 305–322.

[Fi90] Filliman, P., *The extreme projections of the regular simplex*, Trans. Amer. Math. Soc. **317** (1990), pp. 611–629.

[Fi92] Filliman, P., *Volumes of duals and section of polytopes*, Mathematika **39** (1992), pp. 67–80.

[Fi76] Firey, W.J., *A functional characterization of certain mixed volumes*, Israel J. Math. **24** (1976), pp. 274–281.

[FrW78] Fredman, M.L. and Weide, B., *On the complexity of computing the measure of* $\bigcup[a_i, b_i]$, Comm. Assoc. Comp. Mach. **21** (1978) pp. 540–544.

[GaO93] Gajentaan, A. and Overmars, H., n^2*-hard problems in computational geometry*, RUU-CS-93-15 (1993), Dept. of Comp. Sci., Univ. of Utrecht, Utrecht, Netherlands.

[GaG94] Gardner, R.J. and Gritzmann, P., *Successive determination and verifications of polytopes by their X-rays*, J. London Math. Soc. (1994), (to appear).

[GaW89] Gardner, R.J. and Wagon, S., *At long last, the circle has been squared*, Notices Amer. Math. Soc. **36** (1989), pp. 1338–1343.

[GaJ79] Garey, M.R. and Johnson, D.S., *Computers and Intractability. A Guide to the Theory of NP-Completeness*, Freeman, San Francisco, 1979.

[GeKZ90] Gelfand, I.M., Kapranov, M.M. and Zelevinsky, A.V., *Newton Polytopes and the classical resultant and disciminant*, Advances in Math. **84** (1990), pp. 237–254.

[Ge1833] Gerwien, P., *Zerschneidung jeder beliebigen Anzahl von gleichen gradlinigen Figuren in dieselben Stücke*, J. reine angew. Math **10** (1833), pp. 228–234.

[Gi52] Gilbert, E.N., *A comparison of signaling alphabets*, Bell System Tech. J. **31** (1952), pp. 504–522.

[GiV89] Girard, D. and Valentin, P., *Zonotopes and mixture management*; In: *New Methods in Optimization and their Industrial Uses*, (ed. by J.P. Penot), ISNM87, Birkhäuser, Basel, 1989, pp. 57–71.

[Go35] Goldberg, M., *The isoperimetric problem for polyhedra*, Tôhoku Math. J. **40** (1935), pp. 226–236.

[GoT89] Goldfarb, D. and Todd, M.J., *Linear programming*; In: *Handbooks in Operations Research and Management Science, Vol. 1, Optimization*, (ed. by G.L. Nemhauser, A.G.H. Rinnooy Kan and M.J. Todd), North-Holland, Amsterdam, 1989, pp. 73–170.

[Gr63] Grace, D., *Search for largest polyhedra*, Math. Comput. **17** (1963), pp. 197–199.

[GrH94] Gritzmann, P. and Hufnagel, A., *An algorithmic version of Minkowski's reconstruction theorem*, in preparation.

[GrK89] Gritzmann, P. and Klee, V., *On the 0-1 maximization of positive definite quadratic forms*; In: *Operations Research Proceedings*, Springer, Berlin, 1989, pp. 222–227.

[GrK93a] Gritzmann, P. and Klee, V., *Computational complexity of inner and outer j-radii of polytopes in finite dimensional normed spaces*, Math. Programming **59** (1993), pp. 163–213.

[GrK93b] Gritzmann, P. and Klee, V., *Mathematical programming and convex geometry*; In: *Handbook of Convex Geometry, Vol. A*, (ed. by P. Gruber and J. Wills), North-Holland, Amsterdam, 1993, pp. 627–674.

[GrK94a] Gritzmann, P. and Klee, V., *On the complexity of some basic problems in computational convexity: I. Containment problems*, Discrete Math, (1994) (to appear); Reprinted in: *Trends in Discrete Mathematics*, (ed. by W. Deuber, H.-J. Prömel und B. Voigt), Topics in Discrete Mathematics North-Holland, Amsterdam, 1994, to appear.

[GrK94b] Gritzmann, P. and Klee, V., *On the complexity of some basic problems in computational convexity: III. Probing and reconstruction*, in preparation.

[GrK94c] Gritzmann, P. and Klee, V., *On the complexity of some basic problems in computational convexity: IV. Some algebraic applications*, in preparation.

[GrKL94] Gritzmann, P., Klee, V., and Larman, D.G., *Largest j-simplices in n-polytopes*, preprint (1994).

[GrS93] Gritzmann, P. and Sturmfels, B., *Minkowski addition of polytopes: computational complexity and applications to Gröbner bases*, SIAM J. Discrete Math. **6** (1993), pp. 246–269.

[GrW93] Gritzmann, P. and Wills, J.M., *Lattice points*; In: *Handbook of Convex Geometry B*, (ed. by P.M. Gruber and J.M. Wills), North-Holland, Amsterdam, 1993, pp. 765–798.

[Gr73] Groemer, H., *On some mean values associated with a randomly selected simplex in a convex set*, Pacific J. Math **45** (1973), pp. 525–533.

[GrLS81] Grötschel, M., Lovász, L., and Schrijver, A., *The ellipsoid method and its consequences in combinatorial optimization*, Combinatorica **1** (1981), pp. 169–197; Corr. **4** (1984), pp. 291–295.

[GrLS88] Grötschel, M., Lovász, L., and Schrijver, A., *Geometric Algorithms and Combinatorial Optimization*, Springer, Berlin, 1988.

[Gr67] Grünbaum, B., *Convex Polytopes*, Wiley-Interscience, London, 1967.

[Ha55] Hadwiger, H., *Eulers Charakteristik und kombinatorische Geometrie*, J. reine angew. Math. **194** (1955), pp. 101–110.

[Ha57] Hadwiger, H., *Vorlesungen über Inhalt, Oberfläche und Isoperimetrie*, Springer, Berlin, 1957.

[Ha70] Hadwiger, H., *Zentralaffine Kennzeichnung des Jordanschen Inhalts*, Elemente Math. **25** (1970), pp. 25–27.

[Ha75] Hadwiger, H., *Das Wills'sche Funktional*, Monatshefte Math. **79** (1975), pp. 213–221.

[HaG51] Hadwiger, H. and Glur, P., *Zerlegungsgleichheit ebener Polygone*, Elemente Math. **26** (1951), pp. 97–106.

[Ha91] Haiman, M., *A simple and relatively efficient triangulation of the n-cube*, Discrete Comput. Geom. **6** (1991), pp. 287–289.

[HaW68] Hardy, G.H. and Wright, E.M., *An Introduction to the Theory of Numbers*, Clarendon Press, 1968.

[Hi00] Hilbert, D., *Mathematische Probleme*, Nachr. Königl. Ges. Wiss. Göttingen, Math.-Phys. Kl. (1900), pp. 253–297; Bull. Amer. Math. Soc. **8** (1902), pp. 437–479.

[HuS94] Huber, B. and Sturmfels, B., *A polyhedral method for solving sparse polynomial systems*, Math. Comput., (1994) (to appear).

[Hu93] Hughes, R.B., *Minimum-cardinality triangulations of the d-cube for $d = 5$ and $d = 6$*, Discrete Math. **118** (1993), pp. 75–118.

[HuA94] Hughes, R.B. and Anderson, M.R., *Simplexity of the cube*, manuscript, 1994.

[JeV91] Jerrum, M.R. and Vazirani, U., *A mildly exponential approximation algorithm for the permanent*, manuscript, (1991).

[Je68] Jessen, B., *The algebra of polyhedra and the Dehn-Sydler theorem*, Math. Scand. **22** (1968), pp.241–256.

[Je72] Jessen, B., *Zur Algebra der Polytope*, Göttingen Nachr. Math. Phys. (1972), pp. 47–53.

[Jo42] John, F., *An inequality for convex bodies*, Univ. of Kentucky, Research Club Bull. **8** (1942), pp. 8–11.

[Jo48] John, F., *Extremum problems with inequalities as subsidiary conditions*; In: *Studies and Essays, Courant Anniversary Volume*, Interscience, New York, 1948, pp. 187–204.

[Ka94] Kaiser, M., *Deterministic algorithms that compute the volume of polytopes*, in preparation.

[KaLS94] Kannan, R., Lovász, L. and Simonovits, M., *Random walks, isotropic position, and volume algorithms*, in preparation.

[KaKLLL93] Karmarkar, N., Karp, R., Lipton, R., Lovász, L., Luby, M., *A Monte-Carlo algorithm for estimating the permanent*, SIAM J. Comput. **22** (1993), pp. 284–293.

[KaK90] Karzanov, A. and Khachiyan, L.G. *On the conductance of order Marcov chains*, Rutgers Univ. Tech. Report (1990).

[KaS90] Kasimatis, E.A. and Stein, S.K., *Equidissections of polygons*, Discrete Math. **85** (1990), pp. 281–294.

[Ke1615] Kepler, J., *Nova Stereometria doliorum vinariorum*, 1615; see: *Johannes Kepler Gesammelte Werke*, (ed. by M. Caspar), Beck, München, 1940.

[Kh88] Khachiyan, L.G., *On the complexity of computing the volume of a polytope*, Izvestia Akad. Nauk. SSSR, Engineering Cybernetics **3** (1988), pp. 216–217.

[Kh89] Khachiyan, L.G., *The problem of computing the volume of polytopes is #P-hard*, Uspekhi Mat. Nauk. **44** (1989), pp. 199–200.

[Kn93] Khachiyan, L.G., *Complexity of polytope volume computation*; In: *New Trends in Discrete and Computational Geometry*, (ed by J. Pach), Springer, Berlin, 1993, pp. 91–101.

[KhT93] Khachiyan, L.G. and Todd, M.J., *On the Complexity of approximating the maximal inscribed ellipsoid for a polytope*, Math. Prog. **61** (1993), pp. 137–159.

[Ki69] Kingman, J.F.C., *Random secants of a convex body*, J. Appl. Probability **6** (1969), pp. 660-672.

[Ki34] Kirszbraun, M., *Über die zusammenziehenden und Lipschitzen Transformationen*, Fund. Math. **22** (1934), pp. 77–108.

[KlSW89] Klebaner, F.C., Sudbury, A. and Watterson, G.A., *The volumes of simplices, or find the penguin*, J. Austral. Math. Soc. (ser. A) **47** (1989), pp. 263–268.

[Kl65] Klee, V., *Problem in barycentric coordinates*, J. Appl. Physics **36** (1965), pp. 1854–1856.

[KlW91] Klee, V. and Wagon, S., *Old and New Unsolved Problems in Plane Geometry and Number Theory*, Math. Assoc. Amer., Washington, D.C., 1991.

[Kn55] Kneser, M., *Einige Bemerkungen über das Minkowskische Flächenmass*, Archiv Math. **6** (1955), pp. 382–390.

[Kn81] Knuth, D., *A permanent inequality*, Amer. Math. Monthly (1981), pp. 731–740.

[KoK94] Koshleva, O. and Kreinovich, V. *Geombinatorics, computational complexity and saving environment: let's start*, Geombinatorics **3** (1994), pp. 90–99.

[Ko82] Kozlov, M.K., *An approximate method of calculating the volume of a convex polyhedron*, USSR Comp. Math. Math. Phys. **22** (1982), pp. 227–233.

[Ko86] Kozlov, M.K., *Algorithms for volume computation, based on Laplace transform*, Technical Report; (1986), Computing Center USSR Acad. Sci.

[La90] Laczkovich, M., *Equidecomposability and discrepancy: a solution of Tarski's circle-squaring problem*, J. Reine Angew. Math. **404** (1990), pp. 77–117.

[LaS66] Landau, H.J. and Slepian, D., *On the optimality of the regular simplex code*, Bell System Tech. J. **45** (1966), pp. 1247–1272.

[LA83] Lasserre, J.B., *An analytical expression and an algorithm for the volume of a convex polyhedron in $\mathbb{R}^n$*, J. Opt. Th. Appl. **39** (1983), pp. 363–377.

[La91] Lawrence, J., *Polytope volume computation*, Math. Comput. **57** (1991), pp. 259–271.

[LeR82a] Lee, Y.T. and Requicha, A.A.G., *Algorithms for computing the volume and other integral properties of solids. I. Known methods and open issues*, Comm. ACM **25** (1982), pp. 635–641.

[LeR82b] Lee, Y.T. and Requicha, A.A.G., *Algorithms for computing the volume and other integral properties of solids. II. A family of algorithms based on representation conversion and cellular approximation*, Comm. ACM **25** (1982), pp. 642–650.

[LL90] Lin, S.-Y. and Lin, Y.-F., *The n-dimensional Pythagorean theorem*, Lin. Multilin. Algebra, **26** (1990), pp. 9–13.

[Li86] Linial, N., *Hard enumeration problems in geometry and combinatorics*, SIAM J. Algeb. Discr. Meth. **7** (1986), pp. 331–335.

[Lo92] Lovász, L., *How to compute the volume?*, Jahresbericht Deutsche Math. Verein. (1992), pp. 138–151.

[Lo94] Lovász, L., *Random walks on graphs: A survey*, In: Combinatorics: *Paul Erdős is 80*, Vol II, (ed. by D. Miklos, V.T. Sós and T. Szőnyi), Bolyai Society, Budapest, 1994, to appear.

[LoS90] Lovász, L. and Simonovits, M., *The mixing rate of Markov chains, an isoperimetric inequality and computing the volume*, Proc. 31st Ann. Symp. Found. Comput. Sci. (1990), pp. 364–355.

[LoS93] Lovász, L. and Simonovits, M., *Random walks in a convex body and an improved volume algorithm*, Random Structures Alg. **4** (1993), pp. 359–412.

[MaT89] Maass, W. and Turán, G., *On the complexity of learning from counterexamples*, Proc. 30th Ann. IEEE Symp. Found. Comput. Sci. (1989), pp. 262–267.

[Ma85] Martini, H., *Some results and problems around zonotopes*, Coll. Math. Soc. J. Bolyai, **48**, Intuitive Geometry, Siófok (1985), pp. 383–418.

[Ma94] Martini, H., *Cross-sectional measures*; In: *Coll. Math. Soc. J. Bolyai, 63 (Intuitive Geometry, Szeged 1991*, North-Holland, Amsterdam, 1994, pp. ??–??.

[McS85] McKenna, M. and Seidel, R., *Finding the optimal shadow of a convex polytope*, Proc. Symp. Comp. Geom. (1985), pp. 24–28.

[Mc70] McMullen, P., *The maximum number of faces of a convex polytope*, Mathematika **17** (1970), pp. 179–184.

[Mc75] McMullen, P., *Non-linear angle-sum relations for polyhedra, cones and polytopes*, Math. Proc. Cambridge Philos. Soc. **78** (1975), pp. 247–261.

[Mc77] McMullen, P., *Valuations and Euler-type relations of certain classes of convex polytopes*, Proc. London Math. Soc. **35** (1977), pp. 113–135.

[Mc90] McMullen, P., *Monotone translation invariant valuations on convex bodies*, Arch. Math. **55** (1990), pp. 595-598.

[Mc93] McMullen, P., *Valuations and dissections*; In: *Handbook of Convex Geometry B*, (ed. by P.M. Gruber and J.M. Wills), North-Holland, Amsterdam, 1993, pp. 933–988.

[McS83] McMullen, P. and Schneider, R., *Valuations on convex bodies*; In: *Convexity and its Applications*, (ed. by P.M. Gruber and J.M. Wills), Birkhäuser, Basel, 1983, pp. 170–247.

[McS71] McMullen, P. and Shephard, G.C., *Convex polytopes and the upper bound conjecture*, Cambridge University Press, 1971.

[Me79] Mead, D.G., *Dissection of the hypercube into simplices*, Proc. Amer. Math. Soc. **76** (1979), pp. 302–304.

[MiM85] Miel, G. and Monney, R., *On the condition number of Lagrangian numerical differentiation*, Appl. Math. Comput. **16** (1985), pp. 241–252.

[Mil897] Minkowski, H., *Allgemeine Lehrsätze über konvexe Polyeder*, Nachr. Ges. Wiss. Göttingen (1897), pp. 198–219.

[Mi03] Minkowski, H., *Volumen und Oberfläche*, Math. Ann. **57** (1903), pp. 447–495.

[Mi11] Minkowski, H., *Theorie der konvexen Körper, insbesondere Begründung ihres Oberflächenbegriffs*; see: *Gesammelte Abhandlungen*, Vol. 2, Leipzig, Berlin, 1911.

[Mo70] Monsky, P., *On dividing a square into triangles*, Amer. Math. Monthly **77** (1970), pp. 161–164.

[Mo90] Monsky, P., *A conjecture of Stein on plane dissections*, Math. Z. **205** (1990), pp. 583–592.

[Mo89] Montgomery, H.L., *Computing the volume of a zonotope*, Amer. Math. Monthly **96** (1989), p. 431.

[Mo91] Moser, W.O.J., *Problems, problems, problems*, Discrete Appl. Math. **31** (1991), pp. 201–225.

[Mu94] Mulmuley, K., *Computational Geometry: An Introduction through Randomized Algorithms*, Prentice Hall, New York, 1994.

[NiW78] Nijenhuis A. and Wilf, H.S., *Combinatorial Algorithms*, Academic Press, New York, 1978.

[Ol86] Olmsted, C., *Two formulas for the general multivariate polynomial which interpolates a regular grid on a simplex*, Math. Comput. **47** (1986), pp. 275–284.

[On89] Ong, M.E.G., *Hierarchical basis preconditioners for second order elliptic problems in three dimensions*, Ph.D. dissertation, Appl. Math. Dept., Univ. Wash., Seattle (1989).

[On94] Ong, M.E.G., *Uniform Refinement of a tetrahedron*, SIAM J. Scientific Comput., to appear.

[Or94a] O'Rourke, J., *Computational Geometry in C*, Cambridge Univ. Press, 1994.

[Or94b] O'Rourke, J., *Computational geometry column 22*, SIGACT News **25** (1994), pp. 31-33.

[Pa92] Palmon, O., *The only convex body with extremal distance from the ball is the simplex*, Israel J. Math. **80** (1992), pp. 337–349.

[PaY90] Papadimitriou, C.H. and Yannakakis, M., *On recognizing integer polyhedra*, Combinatorica **10** (1990), pp. 107–109.

[PeS93] Pedersen, P. and Sturmfels, B., *Product formulas for resultants and Chow forms*, Math. Z., **214** (1993), pp. 377–396.

[Pi89] Pisier, G., *The volume of convex bodies and Banach space geometry*, Cambridge University Press, 1989.

[Po80] Podkorytov, A.N., *Summation of multiple Fourier series over polyhedra* (in Russian), Leningrad. Univ. Mat. Mekh. Astronom. **1** (1980), pp. 51–58.

[PrS85] Preparata, F.P. and Shamos, M.I., *Computational Geometry*, Springer, New York etc., 1985.

[Re92a] Renegar, J., *The computational complexity and geometry of the first order theory of the reals. I. Introduction. Preliminaries. The geometry of semi-algebraic sets. The decision problem for the existential theory of the reals*, J. Symbolic Comput. **13** (1992), pp. 255–299.

[Re92b] Renegar, J., *The computational complexity and geometry of the first order theory of the reals. II. The general decision problem. Preliminaries for quantifier elimination*, J. Symbolic Comput. **13** (1992), pp. 301–327.

[Re92c] Renegar, J., *The computational complexity and geometry of the first order theory of the reals. III. Quantifier elimination*, J. Symbolic Comput. **13** (1992), pp. 329–352.

[Ri75] Rivlin, T.J., *Optimally stable Lagrangian numerical differentiation*, SIAM J. numer. Anal. **12** (1975), pp. 712–725.

[Ri90] Rivlin, T.J., *Chebyshev polynomials. From approximation theory to algebra and number theory. 2nd ed.*, John Wiley, New York, 1990.

[Ry63] Ryser, H., *Combinatorial Mathematics*, The Carus Mathematical Monographs, 14, Math. Assoc. Amer., Washington, D.C., 1963.

[Sa79] Sah, C.-H., *Hilbert's Third Problem: Scissors Congruence*, Pitman, San Francisco, 1979.

[Sa74] Sahni, S., *Computationally related problems*, SIAM J. Comput. **3** (1974), pp. 262–279.

[Sa82] Sallee, F., *A note on minimal triangulations of the n-cube*, Discrete Appl. Math. **4** (1982), pp. 211-215.

[Sa'74] Salzer, H.E., *Some problems in optimally stable Lagrangian differentiation*, Math. Comput. **28** (1974), pp. 1105–1115.

[Sa93] Sangwine-Yager, J.R., *Mixed volumes*, In: *Handbook of Convex Geometry*, (ed. by P.M. Gruber and J.M. Wills), North-Holland, Amsterdam, 1993, pp. 43–72.

[Sc93] Schneider, R., *Convex Bodies: The Brunn-Minkowski Theory*, Encyclopedia of Mathematics and its Applications, Vol. 44, Cambridge University Press, 1993.

[Sc94] Schneider, R., *Polytopes and Brunn–Minkowski theory*, THIS VOLUME, pp. 273–299.

[Sc86] Schoen, A.H., *A defect-correction algorithm for minimizing the volume of a simple polyhedron which circumscribes a sphere*, Research Report, Computer Science Dept., Southern Illinois Univ., Carbondale, Ill. (1986).

[Se87] Seidel, R., *Output-Size Sensitive Algorithms for Constructive Problems in Computational Geometry*, Ph.D. Thesis, Department of Computer Science, Cornell University, Ithaca, New York, 1987.

[Se91] Seidel, R., *Small dimensional linear programming and convex hulls made easy*, Discrete Comput. Geom. **6** (1991), pp. 423–434.

[Sh68] Shephard, G.C., *The mean width of a convex polytope*, J. London Math. Soc. **43** (1968), pp. 207–209.

[Sh74] Shephard, G.C., *Combinatorial properties of associated zonotopes*, Canad. J. Math. **26** (1974), pp. 302–321.

[ShH54] Shoemaker, D.P. and Huang, T.C., *A systematic method for calculating the volumes of polyhedra corresponding to Brillouin zones*, Acta Cryst. **7** (1954), pp. 249–259.

[SiJ89] Sinclair, A. and Jerrum, M., *Approximate counting, uniform generation and rapidly mixing Markov chains*, Inform. Comput. **82** (1989), pp. 93–133.

[Sl69] Slepian, D., *The content of some extreme simplices*, Pacific J. Math. **31** (1969), pp. 795–808.

[So29] Sommerville, D.M.Y., *An Introduction to the Geometry of N Dimensions*, Methuen, London, 1929.

[Sp86] Speevak, T., *An efficient algorithm for obtaining the volume of a special kind of pyramid and application to convex polyhedra*, Math. Comput. **46** (1986), pp. 531–536.

[Sp85] Spirakis, P.G. , *The volume of the union of many spheres and point location problems*, Proc. Second Ann. Symp. Th. Comput. Sci.; *Lecture Notes in Computer Science*, No. 182, Springer, Berlin, 1985, pp. 328–338.

[St81] Stanley, R., *Two combinatorial applications of the Aleksandrov-Fenchel inequalities*, J. Comb. Th. A **17** (1981), pp. 56–65.

[St86a] Stanley, R., *Two order polytopes*, Discrete Comput. Geom. **1** (1986), pp. 9–23.

[St86b] Stanley, R., *Enumerative Combinatorics, Vol. 1*, Wadsworth-Brooks/Cole, Pacific Grove, California, 1986.

[St91] Stanley, R., *A zonotope associated with graphical degree sequences*; In: *Applied geometry and discrete mathematics: The Victor Klee Festschrift*, (ed. by P. Gritzmann, B. Sturmfels), Amer. Math. Soc. and Assoc. Comput. Mach., 1991, pp. 555–570.

[St1840] Steiner, J., *Über parallele Flächen*, Jahresber. Preuss. Akad. Wiss., (1840), pp. 114–118; see: *Gesammelte Werke, Vol. II*, Reimer, Berlin, 1882, pp. 173–176.

[St79] Stromberg, K., *The Banach-Tarski paradox*, Amer. Math. Monthly **86** (1979), pp. 151–161.

[St71] Stroud, A.H., *Approximate Calculation of Multiple Integrals*, Prentice Hall, Englewood Cliffs, 1971.

[St69] Struik, D.J., *A Source Book in Mathematics, 1200-1800*, Harvard Univ. Press, Cambridge, Mass., 1969.

[St87] Sturmfels, B., *On the decidability of Diophantine problems in combinatorial geometry*, Bull. Amer. Math. Soc. **17** (1987), pp. 121–124.

[Sy65] Sydler, J.-P., *Conditions nécessaire et suffisante pour l'equivalence des polyèdres de l'espace euclidien à troi dimensions*, Comm. Math. Helv. **40** (1965), pp. 43–80.

[Sw85] Swart, G., *Finding the convex hull facet by facet*, J. Algorithms **6** (1985), pp. 17–48.

[Ta70] Tanner, R.M., *Contributions to the simplex code conjecture*, Tech. Report No. 6151-8, Information Systems Lab., Stanford Univ., (1970).

[Ta74] Tanner, R.M., *Some content maximizing properties of the regular simplex*, Pacific J. Math. **52** (1974), pp. 611–616.

[TaKE88] Tarasov, S.P., Khachiyan, L.G. and Erlich, I.I., *The method of inscribed ellipsoids*, Soviet Math. Doklady **37** (1988), pp. 226–230.

[Ta61] Tarski, A., *A Decision Method for Elementary Algebra and Geometry*, Univ. of California Press, Berkeley, 1961.

[Tc57] Tchakaloff, V., *Formules de cubature mécaniques à coefficients non negatifs*, Bull. Sci. Math. **81** (1957), pp. 123–134.

[Th54] Thue Poulsen, E., *Problem 10*, Math. Scand. **2** (1954), p. 346.

[To76] Todd, M.J., *The Computation of Fixed Points and Applications*, Lecture Notes in Economic and Mathematical Systems, no. 124, Springer, Berlin, 1976.

[ToT93] Todd, M.J. and Tuncel, L., *A new triangulation for simplicial algorithms*, SIAM J. Discrete Math. **6** (1993), pp. 167–180.

[Va77] Valiant, L.G., *The complexity of computing the permanent*, Theor. Comput. Sci **8** (1979), pp. 189–201.

[VaW81] Van Leeuwen, J. and Wood, D., *The measure problem for rectangular ranges in d-space*, J. Algorithms **2** (1981), pp. 282–300.

[VeC92] Verschelde, J. and Cool, R., *Nonlinear reduction for solving deficient polynomial systems by continuation methods*, Numer. Math. **63** (1992), pp. 263–282.

[VeVC94] Verschelde, J., Verlinden, P. and Cool, R., *Homotopies exploiting Newton polytopes for solving sparse polynomial systems*, SIAM J. Numer. Anal. (1994), (to appear).

[Vo81] Von Hohenbalken, B., *Finding simplicial subdivisions of polytopes*, Math. Prog. **21** (1981), pp. 233–234.

[Wa85] Wagon, S., *The Banach-Tarski Paradox*, Cambridge Univ. Press, Cambridge, 1985.

[Wa68] Walkup, D.W., *A simplex with a large cross-section*, Amer. Math. Monthly **75** (1968), pp. 34–36.

[WaW69] Walkup, D.W. and Wets, R.J.B., *Lifting projections of convex polyhedra*, Pacific J. Math **28** (1969), pp. 465–475.

[We68] Weber, C.L., *Elements of Detection and Signal Design*, McGraw-Hill, New York, 1968.

[Ya90] Yao, F., *Computational geometry*; In: *Handbook of Theoretical Computer Science, A*, (ed. by J. van Leeuwen), Elsevier, Amsterdam, 1990, pp. 345–390.

[ZaC83] Zamanskii, L.Y. and Cherkasskii, V.L., *Determination of the number of integer points in polyhedra in R^3: polynomial algorithms*, Dokl. Acad. Nauk Ukrain. USSR, Ser. A **4** (1983), pp. 13–15.

[ZaC85] Zamanskii, L.Y. and Cherkasskii, V.L., *Generalization of the Jacobi-Perron algorithm for determining the number of integer points in polyhedra*, Dokl. Acad. Nauk USSR, Ser. A **10** (1985), pp. 10–13.

[Zh93] Zhu, X., *Progress in evaluating permanents of matrices for HBT Event simulation*, Manuscript, Nuclear Physics Lab, Univ. of Washington, Seattle (1993).

THE DIAMETER OF POLYTOPES AND RELATED APPLICATIONS

PETER KLEINSCHMIDT *

Universität Passau
Wirtschaftswissenschaftliche Fakultät
Lehrstuhl für Wirtschaftsinformatik
94030 Passau
Germany

Abstract. Upper bounds for the diameter of the edge-graph of polytopes deliver upper bounds for the worst possible behavior of best possible edge following algorithms for linear programming. We review some recent results for such bounds for general and integer polytopes. Results concerning the diameter of the polyhedra of dual transportation polyhedra are being used for efficient algorithms for assignment and transportation problems. An application to the classification of human chromosomes is presented.

1. Bounds for the diameter of polytopes

Let P be a (convex) d-dimensional polyhedron and $G(P)$ its edge-graph. Let $\delta(P)$ denote the diameter of $G(P)$, i.e. the number of edges of the longest path among all shortest paths joining any pair of vertices of P.
Let $\Delta(d,n) := \max\{\delta(P) | P \text{ is a } d\text{-polyhedron with } n \text{ facets }\}$.

The Hirsch-conjecture which was first formulated by W.M. Hirsch in 1957 states that $\Delta(d,n) \leq n-d$. It arose from an attempt to understand the computational complexity of edge-following algorithms for linear programming. The Hirsch-conjecture was proved to be false for unbounded polyhedra for $d \geq 4$ in [KW]. For polytopes it is still open for all $d \geq 4$. However, it is correct for several interesting classes of polyhedra occuring in combinatorial optimization. This includes the class of $0-1$ polytopes. There is quite a gap between the bound $n-d$ in the conjecture and the currently best known bounds. It is generally believed that the Hirsch-conjecture is false for polytopes in general. However, it would be of interest to have results concerning the behavior of the functional $\Delta(d,n)$. In particular, it is open whether $\Delta(d,n)$ is bounded by a polynomial in both d and n.
Finding a pivot selection rule for linear programming which guarantees a worst case behavior bounded by a polynomial in the size of the constraint matrix would yield

* This research was partly supported by the following grants:
Research grant I-84-095.6/88 of the German-Israeli-Foundation for Scientific Research and Development;
Research grants Kl 365/4-1 and Kl 365/4-2 of the German-Research-Society (DFG);
Research grant 03-KL7PAS - 6 of the German Federal Ministry of Research and Technology;
Research grant PL 92130 of the Biomed-1 Program of the European Union

T. Bisztriczky et al. (eds.), POLYTOPES: Abstract, Convex and Computational, 467–492.

a polynomial bound for $\Delta(d, n)$.
A comprehensive survey about results concerning the Hirsch-conjecture and its relatives up to 1987 is contained in [KK].

In this article, we will report on some more recent results. The best known bounds for $\Delta(d, n)$ have been proved in [L] and [KaKl]:

Theorem 1 $\Delta(d, n) \leq \min\{2^{d-3}n, n^{1+\log d}\}$.

The bound $n^{1+\log d}$ of [KaKl] was obtained by a "divide-and-conquer" method which uses very little of the geometry of polyhedra. Some work of Matoušek [Mat] indicates that for fairly general objects this bound is essentially best possible. For better results for polyhedra it seems necessary to use more of the geometry of the problem. Somewhat indirectly the diameter approach from [KaKl] lead Kalai to find a randomized pivot rule for linear programming which requires an expected number of arithmetic operations bounded from above by $n^{4\sqrt{d}}$ for every linear programming problem with d variables and n inequality constraints [Ka]. Similar results were obtained independently in [MSW]. Of particular interest in optimization are those polytopes whose vertices have only 0 and 1 as coordinates ($0-1$ polytopes). It was proved in [N] that a d-dimensional $0-1$ polytope P satisfies $\delta(P) \leq d$ which implies that the Hirsch conjecture is true for this class. For the $0-1$ polytope whose vertices are the incidence vectors of Hamiltonian circuits of the complete digraph on n nodes (the asymmetric travelling salesman polytope) it was proved in [PR] that its diameter is 2. The same holds for the assignment polytope (see section 2).

The following theorem from [KO] generalizes the result from [N] to integer polytopes. For its formulation we need some terminology.

For a polytope P in $\mathbb{R}^n$ and a linear functional w on $\mathbb{R}^n$ set
$k(P, w) := \text{card}\{< w, v > | v \in \text{vert}P\} - 1$, where *vert* P is the vertex-set of P.
The symmetrization of P is denoted by $P - P := \{x - y | x, y \in P\}$.

Theorem 2 Let P be a d-polytope in $\mathbb{R}^n$, let $\{w_1, \ldots, w_m\}$ be a subset of $\mathbb{R}^n$ whose linear span contains $P - P$, and assume $k(P, w_1) \geq \ldots \geq k(P, w_m)$.
Then $\delta(P) \leq \sum_{i=1}^{d} k(P, w_i)$.
Applying Theorem 2 to the standard basis $\{e_1, \ldots, e_n\}$ of $\mathbb{R}^n$, this yields the following bound on the diameter of an integer polytope.

Corollary 1 If $V \subset \{0, 1, \ldots, k\}^n$ then $\delta(\text{conv}V) \leq k \cdot \dim(\text{conv}V)$.

A polytope P in $\mathbb{R}^d$ is called *lattice free* provided $P \cap \mathbb{Z}^d = \text{vert}P$.
Among other interesting facts the following Theorems are proved in [DO].

Theorem 3 There exists a constant c such that every d-dimensional lattice-free polytope satisfies $\delta(P) \leq cd^3$.

Theorem 4 There exists a constant c for which the following holds: the diameter of any d-dimensional integer polytope P with no more than n facets and the property that for some $k \leq d$ no face of P of dimension larger than k contains an integer point in its relative interior satisfies $\delta(P) \leq cd^2(d-k) + (n - (d-k))2^k$.

2. Dual transportation polyhedra

This and the following sections are devoted to transportation problems and to efficient algorithms for solving them. Those algorithms are more or less directly related to diameter results of Balinski [B1].

We believe that further efficient algorithms for combinatorial optimization problems may be found by exploiting constructive diameter results for the underlying polytopes.

The transportation problem (TP) can be formulated as follows:

$$\min \sum_{i=1}^{m} \sum_{j=1}^{n} c_{ij} x_{ij} \quad \text{subject to}$$

$$\sum_{j=1}^{n} x_{ij} = a_i \quad , i = 1, \ldots, m \qquad (1)$$

$$\sum_{i=1}^{m} x_{ij} = b_j \quad , j = 1, \ldots, n \qquad (2)$$

$$x_{ij} \geq 0. \qquad (3)$$

The quantities a_i (the supplies) and b_j (the demands) are given positive integers for which it is assumed that $\sum_{i=1}^{m} a_i = \sum_{j=1}^{n} b_j$. The given quantities c_{ij} (the costs) are assumed to be arbitrary integers.

TP has the following interpretation:

There are m sources each having a supply a_i of some commodity and n sinks each having a demand b_j of that commodity. Shipping one unit of the commodity from source i to sink j has a cost of c_{ij}. The objective is to ship all the supplies from the sources to the sinks satisfying the demands at minimum total cost. The variable x_{ij} denotes the quantity that is shipped from source i to sink j.

For a given *TP* we denote by *PTP* (*primal transportation polytope*) the polytope of all feasible solutions of (1), (2) and (3).

It is well known that *PTP* is an $(n-1)(m-1)$-dimensional polytope with nm facets. It is unknown whether the Hirsch-conjecture is correct for *PTP*, i.e. whether $\delta(PTP) \leq n + m - 1$. So far it has been proved only for some rather unimportant classes of transportation problems [EKK].

Some special transportation problems occur in various applications:

If all a_i and b_j are equal to 1, then this *TP* is called the *assignment problem (AP)*. If just the a_i's are 1 we have the *semi-assignment problem (SAP)*.

In the case of *AP* the diameter of *PTP* is known to be 2 [P]. However, this does not seem to have any practical value, as this polytope is rather degenerate (not simple).

The dual of TP is as follows:

$$\max \sum_{i=1}^{m} a_i u_i + \sum_{j=1}^{n} b_j v_j \quad \text{subject to}$$

$$u_i + v_j \leq c_{ij} \quad , i = 1, \ldots, m, \; j = 1, \ldots, n.$$

As one of the constraints in *TP* is redundant, we may w.l.o.g. assume that $v_1 = 0$. For the vector $(u_1, \ldots, u_m, v_1, \ldots, v_n)$ we write (u, v) for short. The (unbounded) polyhedron of all feasible solutions of the dual of *TP* is denoted by *DTP*, i.e.

$$DTP := \{(u, v) | v_1 = 0 \text{ and } u_i + v_j \leq c_{ij}, i = 1, \ldots, m, j = 1, \ldots, n\}.$$

DTP is an $(n + m - 1)$-dimensional polyhedron with nm facets.

Balinski [B1] has proved that the Hirsch-conjecture is correct for *DTP*.

Theorem 5 $\delta(DTP) \leq (n-1)(m-1)$ and this bound is achieved for some special cost functions. As this result and its method of proof is of some importance for algorithmic results, we will sketch some ideas of its proof. For this we need some terminology and some basic facts about *DTP*.

Let $I := \{1, \ldots, m\}$ denote the set of sources and $J := \{1, \ldots, n\}$ the set of sinks, respectively. We will view the sets I and J as sets of nodes of a graph. The elements of I are called the *row nodes* and the elements of I are called the *column nodes.*

Let G be the complete bipartite graph on the node set $I \cup J$. An edge between $i \in I$ and $j \in J$ is denoted by (i, j). A *basic feasible solution*, i.e. a representation of a vertex of *DTP* as an intersection of $n + m - 1$ of its facets, can be represented by a *dual feasible tree* T. This is a spanning tree of G with the property that there is some $(u, v) \in DTP$ with $u_i + v_j = c_{ij}$ whenever $(i, j) \in T$. Observe that T has $n + m - 1$ edges and that its edges correspond to some set of facet inequalities whose intersection is a vertex. As $v_1 = 0$, the vector (u, v) is uniquely determined by T. We sometimes write $T(u, v)$ instead of T to indicate the connection between T and the dual variables (u, v).
It is well known that some dual feasible tree T corresponds to an optimal solution of *TP* and its dual if it carries some primal feasible flow, i.e. if there exists some $x = [x_{ij}] \in PTP$ such that $x_{ij} = 0$ for all pairs $(i, j) \notin T$. This fact follows from the complementary slackness Theorem which says that a pair $(x, (u, v))$ of some $x \in PTP$ and $(u, v) \in DTP$ is optimal for *TP* and its dual, respectively, provided $x_{ij}(c_{ij} - u_i - v_j) = 0$ for all $i = 1, \ldots, m$ and $j = 1, \ldots, n$.
If the dual feasible tree $T(u,' v')$ is obtained from $T(u, v)$ by the deletion of some edge of $T(u, v)$ and the insertion of some new edge we call this operation a *dual pivot.* If (u, v) and (u', v') are distinct then $T(u, v)$ and $T(u', v')$ determine an edge of *DTP*. Otherwise the pivot is called degenerate. Given some dual feasible tree $T(u, v)$ a dual pivot on some edge (k, l) of $T(u, v)$ can be performed as follows:

Remove the edge (k, l) from $T(u, v)$ and call T^k and T^l the components of the remaining graph which contain k and l, respectively.
Determine $\varepsilon := \min\{c_{ij} - u_i - v_j | i \in T^l \cap I, j \in T^k \cap J\}$ which is achieved for some edge (g, h). Observe that $\varepsilon = c_{gh} - u_g - v_h \geq 0$.

Update the dual variables as follows:

If the column node 1 is in T^l :	$u_i' := u_i - \varepsilon$, for all	$i \in T^k$; $u_i' = u_i$,	otherwise;
	$v_j' := v_j + \varepsilon$, for all	$j \in T^k$, $v_j' = v_j$,	otherwise.
If the column node 1 is in T_k :	$u_i' := u_i + \varepsilon$, for all	$i \in T^l$; $u_i' = u_i$,	otherwise;
	$v_j' := v_j - \varepsilon$, for all	$j \in T^l$, $v_j' = v_j$,	otherwise.

Add the arc (g, h) to the graph to obtain the new dual feasible tree $T(u', v')$. Observe that it takes $O(nm)$ time to determine the incoming edge (g, h).
For a dual feasible tree T we call the vector $\sigma(T) := (s_1, \ldots, s_n)$ where s_j is the valency of the column node j the *signature* of T.

Balinski proved in [B1]:

Lemma 1 If $\sigma(T(u, v)) = \sigma(T(u', v'))$ then $(u, v) = (u', v')$.

This key lemma shows that for proving Theorem 5 it suffices to show:

Lemma 2 There exists a sequence of at most $(m-1)(n-1)$ dual pivots to transform a given tree T with $\sigma(T) = (s_1, \ldots, s_n)$ into a tree with a given signature $\sigma' = (s'_1, \ldots, s'_n)$.

Proof A column node j of T is called a deficit (surplus) node if $s_j < s'_j (s_j > s'_j)$. The quantity $d(T) := \sum_{s_j < s'_j} (s'_j - s_j)$ is called the *net deficit.*
As every column node has valency at least 1 and at most the remaining $m-1$ valencies can contribute to $d(T)$ we have $d(T) \leq m - 1$.

The following algorithm transforms T into a tree with signature $(s'_1, \ldots, s'_n)$:

```
choose a surplus node s and a deficit node t in T;
while σ(T) ≠ σ' do begin
    compute the unique path from s to t in T;
    perform a dual pivot on the edge of the path which contains s;
    let (g, h) be the incoming edge;
    if h was not a deficit node before the pivot then
        s := h { case 1}
        else
        chose some surplus node s and some deficit node t {case 2};
end;
```

Within the while-block, case 1 can occur at most $n-2$ times in a row because eventually h must be t or some other deficit node. This means that after at most $n-1$ pivots the net deficit will drop by 1. As the initial net deficit is at most $m-1$, we have $d(T) = 0$ after at most $(n-1)(m-1)$ steps. In this case the signature of the tree is σ'. This bound is achieved if we set $c_{ij} := (n-j)(i-1)$.
As the most costly part in the algorithm is the computation of the incoming edge (which takes $O(mn)$ time), the total work of the algorithm is $O(n^2m^2)$. However, by the following argument this can be reduced to $O(nm^2)$ by a more refined bookkeeping:
During the stage of the algorithm where the net deficit does not change, the node t is always the same and the set of column nodes which are used for the computation of the incoming arcs is strictly shrinking while the set of row nodes involved in this computation is strictly growing. This means that only those quantities $c_{ij} - u_i - v_j$ have to be compared which concern those row nodes which have left the component which contains t. Hence, the total number of comparisons during a level is $0(nm)$. As there are at most $m-1$ levels the refined bound follows.

The following lemma (first observed in [B2] for the assignment problem) was proved in [KLS]. It is the main link between Balinski's Hirsch-conjecture result for *DTP* and the applications to algorithms for the *TP*.

Lemma 3 If T is a dual feasible tree for the semi-assignment problem and if $\sigma(T) = (b_1 + 1, \ldots, b_{j-1} + 1, b_j, b_j + 1, \ldots, b_n + 1)$ then T is an optimal tree, i.e. it carries an optimal flow for the semi-assignment problem.

Proof Let the column node j be the *root* of T. Set $x_{ik} := 1$ if $(i,k) \in T$ and i is a son of k with respect to the root j and set $x_{ij} := 0$, otherwise. As this is a primal feasible solution and as T is dual feasible, it follows from complementary slackness that x is optimal.

Corollary 2 The semi-assignment problem can be solved in $O(mn^2)$ time.

Starting from an arbitrary dual feasible tree the underlying algorithm for Corollary 2 is just an application of Lemma 2 to the signature of Lemma 3.
For the assignment problem Corollary 2 specializes to the bound $O(n^3)$.

In [KLS] the question was raised whether Lemma 3 can be generalized to the transportation problem:
For a given *TP* we call an integer vector $S = (s_1, \ldots, s_n)$ with $\sum_{j=1}^{n} s_j = n + m - 1$ and $s_j \geq 1, j = 1, \ldots, n$, an *optimal signature* provided that every tree (not necessarily only the dual feasible trees) with that signature carries a primal feasible flow for the *TP*. If an optimal signature exists (as it does for the semi-assignment problem by Lemma 3) then the *TP* can be solved in time $O(mn^2)$ provided the optimal signature is known beforehand.

The following Theorem from [KLS] shows that the class of *TP's* which have an optimal signature is rather restricted.

Theorem 6 An integer vector $(s_1, \ldots, s_n)$ with
$\sum_{j=1}^{n} s_j = n + m - 1$ and $s_j \geq 1, j = 1, \ldots, n$ is an optimal signature for a *TP* if and only if the following property holds:
For all $U \subset I$ and $V \subset J$ with $\sum_{i \in U} a_i < \sum_{j \in V} b_j$ the inequality $\sum_{j \in V} s_j \geq |U| + |V|$ holds.

This result shows the limitation of the signature method for the *TP* in general. The signature method is guided by the **excess in the number of arcs** in which the column nodes of the current tree lie in comparison to the number of arcs which an optimal tree should have.
We generalized this idea to develop algorithms for the *TP* which are guided by the **excess in incoming flow** which the column nodes have within a current tree. For semi-assignment problems our generalized algorithms coincide with the signature method. In this sense our algorithms of the next two sections are based on diameter results. A less detailed description of those algorithms and their complexity is contained in [KS].

3. A non-polynomial algorithm for the transportation problem

In the description of the *TP* in section 2 we assumed that the variables x_{ij} are defined for all arcs of the complete bipartite Graph G with node set $I \cup J$ and edge set $E = I \times J$. In most applications, the *TP* occurs as a "sparse" problem, i.e. the

set of edges which carry some flow is a subset of $I \times J$.
In what follows we assume that $E \subset I \times J$ and we set $k := |E|$. The constraints of TP and its dual have to be changed accordingly. With respect to some given flow we define the excess $\overline{a}_i$ of a row node and the excess $\overline{b}_j$ of a column node as follows:

$$\overline{a}_i := a_i - \sum_{(i,j)\in E} x_{ij} \quad \text{for any fixed} \quad i = 1, \ldots, m;$$

$$\overline{b}_j := \sum_{(i,j)\in E} x_{ij} - b_j \quad \text{for any fixed} \quad j = 1, \ldots, n.$$

Set $J_S := \{j \in J | \overline{b}_j > 0\}$ (the set of "surplus" nodes) and $J_D := \{j \in J | \overline{b}_j < 0\}$ (the set of "deficit" nodes).

The *reduced costs* with respect to some $(u, v) \in DTP$ are defined by $\overline{c}_{ij} := c_{ij} - u_i - v_j$.
The rough structure of our non-polynomial algorithm NTP for TP is described below. If applied to the assignment problem it is identical to an algorithm from [HR] which in turn was proved in [KLS] to be equivalent to the signature method.

```
algorithm NTP;
begin
    Initialize;
    while J_S ≠ ∅ do begin
        Maketree;
        Augment;
    end;
end;
```

We describe the procedures of NTP below.

```
procedure Initialize;
begin
    v_j := 0 for all j ∈ J;
    x_ij := 0 for all (i,j) ∈ E;
    ā_i := a_i for all i ∈ I;
    b̄_j := b_j for all j ∈ J;
    T := ∅ (T denotes the edge set of the future dual feasible tree);
    for all i ∈ I do begin
        compute c_ij* := min{c_ij | (i,j) ∈ E}   (j* ∈ J is a node for which
                                                   the minimum is achieved);
        u_i := c_ij*;
        T := T ∪ (i, j*);
        x_ij* := a_i;
        ā_i := ā_i − x_ij*;
        b̄_j* := b̄_j* + x_ij*;
        end;
    J_S := {j ∈ J | b̄_j > 0} and J_D := {j ∈ J | b̄_j < 0};
end;
```

At this stage it follows from the way the u_i are defined that we have a dual feasible (u,v). Of course, we may also detect here that the *TP* is infeasible if for some $i \in I$ the node j^* does not exist. For expository reasons we will always assume that the *TP* is feasible, i.e. that all arcs which occur in the various procedures do exist. The case of infeasibility can easily be handled by the use of artificial arcs with "infinite" costs.

After Initialize the flow x satisfies all the primal constraints for the row-nodes as all $\overline{a}_i$ are zero. Note that if J_S is empty it follows that J_D is empty as well and the column constraints are satisfied, too. This means that x is already an optimum for *TP*. Otherwise we continue with the while block of *NTP*. At the beginning of this block we will always have the following facts:

(1) $\overline{a}_i = 0$ for all $i \in I$;

(2) $x_{ij} \neq 0$ only if $(i,j) \in T$;

(3) $\overline{c}_{ij} \geq 0$ for all $(i,j) \in T$, i.e. (u,v) is dual feasible;

(4) $\overline{c}_{ij} = 0$ if $(i,j) \in T$.

If at the beginning of the while block T is not already the edge-set of a spanning tree then "Maketree" adds edges (with flow zero) to T and preserves dual feasibility. In "Augment" some "surplus" from J_S is shipped to J_D in order to reduce $\sum_{\overline{b}_j>0} \overline{b}_j$. The flow changes happen only within T. Arcs whose flow becomes zero are removed from T.

In detail this works as follows:

If T is already a spanning tree we directly call Augment. Otherwise T will be a collection of trees. In addition, there may be some isolated column nodes which do not lie in any edge of T. For $j \in J$ we call T_j the component of T which contains j or if j is isolated, T_j is j itself.

Then Maketree works as follows:

```
procedure Maketree;
begin
  choose some r ∈ J_S as the root;
  T̄ := T_r;
  while T is not a spanning tree do begin
    find ε := min{c̄_ij | i ∈ T̄ ∩ I, j ∈ J\T̄}(= c_i*j*);
    for all i, j ∈ T_j* do begin
      v_j := v_j + ε;
      u_i := u_j − ε;
    end;
    T̄ := T̄ ∪ T_j* ∪ (i*, j*);
  end;
  T := T̄;
end;
```

We will prove later (Lemma 5) that Maketree always produces a dual feasible tree. The following Lemma is important for the complexity of *NTP* and its strongly polynomial version of section 4.

Lemma 4 The procedure Maketree can be implemented in $O(k + n \log n)$ (amortized) time.

Proof The complexity bound of Lemma 4 can be obtained if we use a Fibonacci-heap (F-heap for short) as a data structure for the computation of ε. Fibonacci-heaps introduced in [FT] have meanwhile become a standard tool in this context and so we restrict ourselves only to some essential points here:

Initially, for each T_l (except T_r)as defined above we keep an *item* of our F-heap. The item of T_l is labelled by the arc (i^*, j^*) for which $\varepsilon_l := \min\{\overline{c}_{ij} | i \in \overline{T} \cap I, j \in T_l \cap J\}$ is achieved and ε_l is called the *key* of the item.

One could also keep an item for every node $j \in J \backslash \overline{T}$. This is usually done in other network flow algorithms. However, the special structure of our algorithms allows us to maintain the smaller F-heap. This may be useful for an implementation even though the theoretical bound for the size of both F-heaps is $O(n)$.
The F-heap allows us to find each ε in Maketree in constant (amortized) time.

When some T_j is added to $\overline{T}$ we can delete the corresponding item in $O(\log n)$ time. As this can happen at most $n-1$ times we spend $O(n \log n)$ time with deletions. When we compute the labels and keys of the items we have to access every arc only once. This is due to the fact that the sets $\overline{T} \cap I$ and $J \backslash \overline{T}$ in Maketree are growing and shrinking, respectively. So the total time spent on accesses is $O(k)$.

Recomputing the key of some item can result in a decrease-key operation caused by an arc that was not considered before. As every decrease-key operation takes constant amortized time with F-heaps this results in $O(k)$ work.
In "Augment" we pick some $t \in J_D$ (note that $J_D \neq \emptyset$ if $J_S \neq \emptyset$) and determine the unique path P from r to t in T. Taking the arc of P containing r as the first one, we call the arcs of P *odd* or *even*. Observe that the arcs with flow zero which were added in Maketree are by construction even and hence all odd arcs carry positive flow.

Then Augment is as follows:

procedure Augment;
begin
  choose some $t \in J_D$ and determine P from r to t as above;
  $\Delta_1 := \overline{b}_r; \Delta_2 := |\overline{b}_t|; \Delta_3 := \min\{x_{ij} | (i,j) \in P \text{ odd}\}$;
  (Note that all Δ_i are positive.)
  $\Delta := \min\{\Delta_1, \Delta_2, \Delta_3\}$;
  for all $(i,j) \in P$ do
    if (i,j) is odd then
      $x_{ij} := x_{ij} - \Delta$.
    else
      $x_{ij} := x_{ij} + \Delta$;
  if $x_{ij} = 0$ and $(i,j) \in T$ then $T := T \backslash (i,j)$;
  $\overline{b}_r := \overline{b}_r - \Delta$;
  $\overline{b}_t := \overline{b}_t + \Delta$;
  update J_S and J_D;
end;

Note that Augment preserves properties (1)-(4).

As Δ is positive, the quantity $\sum_{\bar{b}_j>0} \bar{b}_j$ is strictly decreasing after every call of Augment. Hence, the algorithm terminates. As $\Delta \leq \Delta_1$ and $\Delta \leq \Delta_2$, the sets J_S and J_D are not growing during Augment.
It is clear that the algorithm stops at optimality. Also it follows from Lemma 4 that its complexity is $O(k + n \log n)$ times the number of augmentations. The latter, however, may be in the order of magnitude of $\sum_{i=1}^{m} a_i$ in the case that all the Δ's are small. This means that the complexity of NTP may not be bounded by a polynomial in n and m.

In [S] several modifications for the algorithm are suggested which prove useful for an implementation. For example, one may perform several augmentations in Augment at a time as long as the paths P do not share an edge. Also one can identify all $j \in J_S$ as one collective root in Maketree.
Some preliminary experience with the implementation has given us the feeling that a "tuned" version might be quite efficient. As it stands, the test version compares roughly to the running time of the RELAX II code of Bertsekas and Tseng [BT].

For the correctness of NTP we still have to prove the following:

Lemma 5 Maketree preserves dual feasibility.

Proof Let $\varepsilon_1, \ldots, \varepsilon_l$ denote the various ε's in the order in which they are computed in Maketree and let $T_1, \ldots, T_l$ denote the components of T in the order in which they are added to T_r in Maketree.
At the beginning of Maketree we have dual feasibility and $\bar{c}_{ij} = 0$ on all arcs of $T_r \cup T_1 \cup \ldots \cup T_l$.
We claim $\varepsilon_1 \leq \varepsilon_2 \leq \ldots \leq \varepsilon_l$.
Let (u, v) be the dual variables before T_{p+1} is added to $T_r \cup T_1 \cup \ldots \cup T_p$, for some $p \leq l - 1$.
For $i \in (T_r \cup T_1 \cup \ldots \cup T_{p-1}) \cap I$ and $j \in T_{p+1} \cap J$ we have $\bar{c}_{ij} \geq \varepsilon_p$ by the definition of ε_p.
For $i \in T_p \cap I$ and $j \in T_{p+1} \cap J$ we have $\bar{c}_{ij} = c_{ij} - u_i - v_j = c_{ij} - (\overline{u}_i - \varepsilon_p) - v_j = c_{ij} - \overline{u}_i - v_j + \varepsilon_p$, where $\overline{u}_i$ is the value of the dual variable for i before T_p is added. As $c_{ij} - \overline{u}_i - v_j$ is the value of $\bar{c}_{ij}$ at the start of Maketree it is nonnegative. Hence, we have $\bar{c}_{ij} \geq \varepsilon_p$ again.
As ε_{p+1} is equal to $\min\{\bar{c}_{ij} \mid i \in (T_r \cup T_1 \cup \ldots T_p) \cap I, j \in T_{p+1} \cap J\}$, we have $\varepsilon_{p+1} \geq \varepsilon_p$, which proves the claim.
Using this fact, we will prove Lemma 5 by proving that (u, v) is dual feasible on $T_r \cup T_1 \cup \ldots \cup T_{p+1}$ after the addition of T_{p+1} provided (u, v) is dual feasible on $T_r \cup T_1 \cup \ldots \cup T_p$ before the addition of T_{p+1}.

Case 1: $i \in (T_r \cup T_1 \cup \ldots \cup T_p) \cap I, j \in T_{p+1} \cap J$.

Let $\bar{c}_{ij}$ denote the value of the reduced costs after T_{p+1} has been added. Then $\bar{c}_{ij} = c_{ij} - u_i - (v_j + \varepsilon_{p+1}) \geq 0$ by the definition of ε_{p+1}.

Case 2: $i \in T_{p+1} \cap I, j \in (T_1 \cup \ldots \cup T_p) \cap J$.

Let $\overline{v}_j$ be the value of the dual variable before the call of Maketree and let

$$\begin{aligned} j \in T_g, g \leq p. \text{ Then } \bar{c}_{ij} &= c_{ij} - (u_i - \varepsilon_{p+1}) - v_j \\ &= c_{ij} - (u_i - \varepsilon_{p+1}) - (\bar{v}_j + \varepsilon_g) \\ &= c_{ij} - u_i - \bar{v}_j + \varepsilon_{p+1} - \varepsilon_g. \end{aligned}$$

As $c_{ij} - u_i - \bar{v}_j$ is the value of $\bar{c}_{ij}$ at the beginning of Maketree it is nonnegative. The nonnegativity of $\bar{c}_{ij}$ then follows from the monotonicity of the ε's.

Case 3: $i \in T_{p+1} \cap I, j \in T_r \cap J$.

We have $\bar{c}_{ij} = c_{ij} - (u_i - \varepsilon_{p+1}) - v_j = c_{ij} - u_i - v_j + \varepsilon_{p+1}$.
The quantity $c_{ij} - u_i - v_j$ has the same (nonnegative) value as at the beginning of Maketree and, hence $\bar{c}_{ij} \geq 0$ by the nonnegativity of ε_{p+1}.

As in the case of the semi-assignment problem we can have at most $O(m)$ augmentations we have the following:

Corollary 3 *NTP* solves the semi-assignment problem in $O(m(k + n \log n))$ time. In the case of the assignment problem we have $m = n$ and, hence, it can be solved in $O(n(k + n \log n))$ time by *NTP*.

4. The strongly polynomial algorithm STP for the transportation problem

The fact that the quantities Δ which occur in the algorithm *NTP* of the previous section may be rather small prevented that algorithm from being *strongly polynomial.* This means that the number of arithmetic operations, accesses and comparisons is not bounded from above by a polynomial in n and m.

The algorithm *STP* in this section solves the *TP* in time $O(m \log m(k + n \log n))$ assuming w.l.o.g. that $m \geq n$. This goal is achieved by "excess-scaling", a procedure which makes sure that the quantities Δ occuring in *NTP* are "large".
The algorithm is closely related to an algorithm of Orlin [O] for the minimum-cost flow problem. When applied to *TP*, that algorithm has a running time of $O(m \log m(k + m \log m))$.

Our slight improvement over that bound comes from the fact that by a change in Augment we are able to keep the F-Heap only on the column nodes rather than on all nodes.
For the excess scaling we will introduce below the sets $J_S(\Delta)$ and $J_D(\Delta)$ which are parametrized analogues of the sets J_S amd J_D of section 3.

The structure of the algorithm *STP* is as follows:

```
algorithm STP;
    begin
    Δ := min{max_{i∈I} a_i, max_{j∈J} b_j};
    Initialize;
    while Δ ≥ 1 do begin
       Augmentrows
       Contract;
       Skipphase;
       while J_S(Δ) ≠ ∅ and J_D(Δ) ≠ ∅ do begin
           Maketree;
           Augment;
           end;
       Δ := Δ/2;
       end;
end;
```

Throughout we make the assumptions that the problem is feasible, that all considered arcs exist in E and that Δ is even at all times.

These assumptions are made for expository reasons only. The problem of feasibility and nonexisting arcs can be treated as in the non-polynomial version. Problems with the parity of Δ can be settled as in [O] by multiplications of the flows, excesses and capacities by 2.

At all times, it will be evident that the flows on all arcs are integer multiples of Δ. We will use this fact at various places without further mention.

We now describe the algorithm in more detail.

```
procedure Initialize;
begin
for all i ∈ I do ā_i := a_i;
for all j ∈ J do b̄_j := −b_j;
I(Δ) := {i ∈ I | ā_i ≥ Δ}; I' := I(Δ); J' := J;
for all j ∈ J do v_j := 0;
for all (i, j) ∈ E do x_ij := 0;
T := ∅;
end;
```

Observe that it follows from the definition of Δ that $I(\Delta)$ is nonempty. The variables u_i will be initialized in the procedure "Augmentrows".

In analogy to the non-polynomial version where the set $I^* := \{i \in I \mid \overline{a}_i > 0\}$ was empty immediately after the initialization, we will make $I(\Delta)$ empty in the procedure "Augmentrows", i.e. we will always maintain the property $0 \leq \overline{a}_i \leq a_i$ and $\overline{a}_i \leq \Delta$.

I' will be the "working set" of row nodes in T. A row node will enter T only if its excess exceeds Δ for the first time. J' is the set of column nodes initially. However, in the contraction process of the procedure Contract J' may change.

The main difference to the non-polynomial version is the Δ-scaling which essentially means that the sets J_D, J_S and I^* are parametrized by Δ. The additional ingredient which is essential for the complexity proof is edge contraction. Without contractions *STP* is almost identical to *NTP*. The "row augmentations" are performed as follows:

procedure Augmentrows;
begin
$I(\Delta) := \{i \in I \mid \overline{a}_i \geq \Delta\}$; $I' := I(\Delta) \cup I'$;
for all $i \in I'$ which are in an arc of T do begin
  choose an arc $(i,j) \in T$; $x_{ij} := x_{ij} + \Delta$;
  $\overline{a}_i = \overline{a}_i - \Delta$; $\overline{b}_j := \overline{b}_j + \Delta$;
  end;
for all $i \in I'$ which are not in an arc of T do begin
  determine $c_{ij^*} := \min\{c_{ij} \mid (i,j) \in E\}$;
  $u_i := c_{ij^*}$; $h := \max\{\alpha \in \mathbb{Z} \mid \alpha\Delta \leq \overline{a}_i\}$;
  $x_{ij^*} := h\Delta$; $\overline{a}_i := \overline{a}_i - x_{ij^*}$; $\overline{b}_{j^*} := \overline{b}_{j^*} + x_{ij^*}$;
  $T := T \cup (i, j^*)$;
  end;
end;

At this point $I(\Delta)$ is empty. We call the flow changes which occur in Augmentrows *row augmentations* in contrast to the augmentations which will be performed in Augment. Note that the way "Augmentrows" is described it appears as if I' can only grow. This is true if no contractions occur. A contraction may cause a node to be deleted from I'.

procedure Contract;
begin
for all $(i,j) \in T$ with $x_{ij} > 4(n+m)\Delta$ do begin
  contract the edge (i,j) into a new column node v;
  update I', J' and T accordingly;
  if $i \in I'$ then $b_v := b_j - a_i$ and $\overline{b}_v := \overline{a}_i + \overline{b}_j$;
  if $i \in J'$ then $b_v := b_i + b_j$ and $\overline{b}_v := \overline{b}_i + \overline{b}_j$;
  end;
end;

The contraction is an ordinary edge contraction on T. We keep the directions and the flows of the arcs in T. The bipartite structure of the column and row nodes will be destroyed because after a contraction there may be arcs between column nodes (we keep calling I' and J' row and column nodes, respectively and we still use a_v ($\overline{a}_v$) or b_v ($\overline{b}_v$) for the capacities (excesses) of row and column nodes). Observe that b_v may be negative for a contracted node $v \in J'$. For expository reasons we chose to preserve the "uncontracted" data of T in a separate data structure because this will allow us to simplify the explanation of dual updates and augmentations.

The following procedure guarantees that no "unnecessary" Δ-phases are considered.

procedure Skipphase;
begin
If $x_{ij} = 0$ for all $(i,j) \in T$ and $|\bar{b}_j| < \Delta$ for all $j \in J'$ then begin
set $\Delta := \max\{(\bar{a}_i,\ i \in I \text{ uncontracted}),\ (|\bar{b}_j|, j \in J')\}$;
start a new Δ-phase; (i.e. start the outer while-loop of STP with the new Δ)
end;
end;

Set $J_S(\Delta) := \{j \in J' \mid \bar{b}_j \geq \Delta\} \cup \{j \in J' \mid \bar{b}_j \geq \frac{\Delta}{2}, b_j > 0\}$
and $J_D(\Delta) := \{j \in J' \mid \bar{b}_j \leq -\Delta\} \cup \{j \in J' \mid \bar{b}_j \leq -\frac{\Delta}{2}, b_j < 0\}$.

These sets have to be updated at the beginning of each while-block.

If no contractions are performed the procedures "Maketree" and "Augment" are identical to the corresponding procedures in the non-polynomial version. Otherwise there are a few differences.

In "Maketree" we choose some $r \in J_S(\Delta)$ which may now be a contracted new node. The progress of building up the tree is then performed on the "blown up" tree exactly as in the non-polynomial version. The first starting $\overline{T}$ is the blown up component of T which contains r and "I" has to be replaced by "I'". The incoming arcs computed in "Maketree" are added both to the contracted tree T and its uncontracted counterpart. After the construction of T in "Maketree" let P be a path (ignoring directions) starting from r to any other column node in J'. Observe that the arcs which are added by Maketree and which lie on P are by construction directed "away from r".

procedure Augment;
begin
choose some $t \in J_D(\Delta)$;
compute the "augmenting" path P in T from r to t ignoring directions;
for all arcs $e \in P$ do
if e is directed away from r then $x(e) := x(e) + \Delta$
else $x(e) := x(e) - \Delta$; (such arcs have by construction a positive flow)
$\bar{b}_r := \bar{b}_r - \Delta$; $\bar{b}_t := \bar{b}_t + \Delta$;
delete all arcs with $x(e) = 0$ from T;
update $J_D(\Delta)$ and $J_S(\Delta)$;
end;

Observe that from the way the excesses are updated we always have

$$\bar{b}_j = \sum_{(i,j)\in T} x_{ij} - \sum_{(j,i)\in T} x_{ji} - b_j,$$

$$\bar{a}_i = a_i - \sum_{(i,j)\in T} x_{ij} \quad \text{and}$$

$$\sum_{\substack{i\in I,\\ i \text{ uncontracted}}} \bar{a}_i + \sum_{j\in J'} \bar{b}_j = 0.$$

This concludes the description of the algorithm.

By an *augmentation* we mean a call of "Augment".

5. Complexity and correctness of STP

The algorithm of section 4 is quite similar to Orlin's algorithm [O] for the transshipment problem. The main difference is that we maintain the different roles of the row nodes and column nodes throughout. This allows us to compute the incoming arcs with Fibonacci-heaps whose size is bounded by n rather than $n+m$. Another additional feature of our algorithm are the row augmentations which also imply some changes in the complexity proof.

In the sequel we will restrict ourselves to those aspects of the complexity proof which contain essential differences to Orlin's arguments. Also we will sometimes omit the discussion of all possible case distinctions as they can be treated very similarly.

Theorem 7 *The algorithm STP solves TP correctly in time* $O(m \log m(k + n \log n))$.

As "Maketree" runs in time $O(k + n \log n)$ for the same reasons as in the non-polynomial algorithm, the essential facts that have to be shown for the complexity bound are an $O(m \log m)$-bound both for the number of augmentations and for the number of Δ-phases. The other operations trivially are subsumed under those bounds.

A column node $j \in J'$ (an uncontracted row node $i \in I$) is called *activated* at the beginning of a Δ-phase if $j \notin J_S(\Delta') \cup J_D(\Delta')$ ($i \notin I(\Delta)$) at the end of the preceding Δ'-phase and $j \in J_S(\Delta) \cup J_D(\Delta)$ ($i \in I(\Delta)$) at the beginning of the Δ-phase.

Lemma 6. $|\bar{b}_j| \leq (n+m)\Delta$ for all $j \in J'$ at the end of a Δ-phase.

Proof. The Δ-phase may end because $J_S(\Delta) = \emptyset$ (the other cases are treated similarly). Then $\bar{b}_j < \Delta$ for all $j \in J'$. So the claim follows for all j with $\bar{b}_j \geq 0$. From $\sum_{i \in I} \bar{a}_i + \sum_{j \in J'} \bar{b}_j = 0$ it follows that

$$\sum_{\bar{b}_j \leq 0} |\bar{b}_j| = \sum_{\substack{i \in I \\ i \text{ uncontracted}}} \bar{a}_i + \sum_{\bar{b}_j > 0} \bar{b}_j < m\Delta + n\Delta$$

and the claim follows for negative $j \in J'$ as well.
The last inequality follows from the fact that $I(\Delta)$ is always empty after the row augmentations have taken place.

Lemma 7. If some $j \in J'$ becomes activated in the Δ-phase then $|b_j| \geq \Delta$.

Proof. Assuming that the preceding scaling parameter was 2Δ (the case of a skipped phase is trivial) four cases for the activation of j have to be considered:

(1) $\Delta/2 \leq \bar{b}_j < \Delta$, $b_j > 0$
(2) $-2\Delta < \bar{b}_j \leq -\Delta$, $b_j > 0$
(3) $\Delta \leq \bar{b}_j < 2\Delta$, $b_j < 0$
(4) $-\Delta < \bar{b}_j \leq -\Delta/2$, $b_j < 0$.

The cases (1) and (3) mean that $j \notin J_S(2\Delta)$ at the end of the 2Δ-phase but $j \in J_S(\Delta)$ at the beginning of the Δ-phase. The cases (2) and (4) are the corresponding cases for $J_D(2\Delta)$ and $J_D(\Delta)$. From the fact that $b_j+\bar{b}_j = \sum_{(i,j)\in T} x_{ij} - \sum_{(j,i)\in T} x_{ji}$ is an integer multiple of 2Δ the claim follows by inspection of the cases.

Lemma 8. If at the end of a Δ-phase $|b_j| \geq 4(n+m)^2\Delta$ for some $j \in J'$ then there is an arc incident to j in T such that the arc will be contracted in the next phase. The corresponding statement holds for some $i \in I'$ provided $a_i \geq 4(n+m)^2\Delta$.

Proof. We will prove only the first part of the Lemma and only for $b_j \geq 0$. The case $b_j < 0$ and the second part are proved similarly.

From the definition of $\bar{b}_j$, from $|b_j| \geq 4(n+m)^2\Delta$, from $\bar{b}_j| \leq (n+m)\Delta$ (Lemma 6) and from the nonnegativity of $\sum_{(j,i)\in T} x_{ji}$ it follows that:

$$\sum_{(i,j)\in T} x_{ij} = b_j + \bar{b}_j + \sum_{(j,i)\in T} x_{ji} \geq 4(n+m)^2\Delta - (n+m)\Delta. \tag{1}$$

Let $x := \max_{(i,j)\in T} x_{ij}$. Then it follows from (1) and the fact that T has at most $n+m-1$ arcs that

$$x \geq (4(n+m)^2\Delta - (n+m)\Delta) \;/\; n+m-1 \geq 4(n+m)\Delta.$$

Then the arc which carries the flow x will be contracted in the next phase.

For (1) we used the fact that $|\bar{b}_j| \leq (n+m)\Delta$ at the end of a Δ-phase.

For the corresponding argument in the second part of the Lemma we need the fact that $I(\Delta) = \emptyset$, i.e. $\bar{a}_i < \Delta$ at the end of the Δ-phase.

Lemma 9. A node is activated $O(\log m)$ times.

Proof. Let $i \in I$ be activated for the first time. Then $a_i = \bar{a}_i \geq \Delta$. From Lemma 8 it follows that after $\lceil \log(4(n+m)^2) \rceil = O(\log m)$ Δ-phases we have for the current scaling parameter Δ':

$$\Delta' \leq \frac{\Delta}{4(n+m)^2} \leq \frac{a_i}{4(n+m)^2}.$$

Hence Lemma 8 applies for i and Δ', i.e. i will be contracted in the next Δ-phase and can no longer be activated.

For $j \in J$ we use the same argument except that we need Lemma 7 for the fact that $|b_j| \geq \Delta$ after an activation.

Lemma 10. The number of augmentations in a Δ-phase is bounded by $A + 2C$ where A is the number of nodes activated at the beginning of the Δ-phase and C is the number of nodes contracted in this phase.

Proof. The proof of this Lemma contains the key facts of the complexity results. It is relatively complicated and has many parallels to the arguments in [O]. However, it is also quite different in parts, e.g. because some types of augmentations which are charged to contractions in [O] have to be charged to rowaugmentations in our case. To avoid too lengthy arguments we chose to restrict ourselves to the case that the last phase ended because $J_S(2\Delta)$ became empty. This case includes all the essential ideas.

For the beginning of the Δ-phase this implies the following facts which follow from $J_S(2\Delta) = \emptyset$ and the definition of activated nodes:

(1) $\bar{b}_j < 2\Delta$, $j \in J'$;
(2) if $b_j > 0$ then $\bar{b}_j < \Delta$;
(3) Some $j \in J'$ with $\bar{b}_j \geq 0$ is activated if and only if
   (a) $\Delta \leq \bar{b}_j < 2\Delta$, $b_j < 0$ or
   (b) $\Delta/2 \leq \bar{b}_j < \Delta$, $b_j > 0$;
(4) Let $F := \sum_{\bar{b}_j \geq 0} \lfloor \frac{\bar{b}_j}{\Delta} \rfloor$. Then at the beginning of the Δ-phase F is equal to $F_1 :=$ the number of activated column nodes of type (3a); (this uses (2));
(5) Let F_2 be the number of row nodes activated at the beginning of the Δ-phase (not counting those activated row nodes which send flow to some j with $\bar{b}_j < 0$ in Augmentrows). Then after Augmentrows $F \leq F_1 + F_2$;
(6) Let F_3 be the number of contractions in the Δ-phase. Then after all contractions have happened we have $F \leq F_1 + F_2 + F_3$ (a contraction contributes at most 1 to F);

This is the status of F before the augmentations in the while-block take place.

If an augmentation starts with some root r we either have $\bar{b}_r \geq \Delta$ ("type 1") or $\Delta/2 \leq \bar{b}_r < \Delta$, $b_r > 0$ ("type 2"). From (6) it follows that

(7) The number of augmentations of type 1 is bounded above by $F_1 + F_2 + F_3$.

Then Lemma 10 will follow from (7) and

(8) The number of augmentations of type 2 is bounded from above by $F_3 + F_4 + F_5$ where
   $F_4 :=$ number of activated column nodes of type (3b),
   $F_5 :=$ number of activated row nodes not counted in F_2 (i.e. row nodes which send flow to some j with $\bar{b}_j < 0$).

Clearly the activated nodes counted in F_1, F_2, F_4 and F_5 are all different which implies the correctness of Lemma 10.

For the proof of (8) we consider how a node $j \in J'$ could become the start of an augmentation, i.e. how the situation

$$(*)\ \Delta/2 \leq \bar{b}_j < \Delta \text{ and } b_j > 0$$

can arise. There are three exclusive cases to be considered:

(i) j was created by a contraction; we can charge this to F_3;
(ii) $\bar{b}_j < 0$ at the beginning of the Δ-phase and $(*)$ was generated by Augmentrows; we can charge this to F_5;
(iii) $(*)$ was already true at the beginning of the Δ-phase; this can be charged to F_4.

From (2) it follows that no other cases are possible.

The following is an easy consequence of Lemma 10.

Lemma 11. If the arc (i,j) is in T in the Δ-phase then x_{ij} will change by at most $4(n+m)\Delta$ over all subsequent phases.

Proof. The following changes are possible:

- Changes due to augmentrows (summed over all phases) $\leq \Delta + \Delta/2 + \cdots + 1 \leq 2\Delta$;
- changes due to augmentations counted in F_1, F_2, F_4 and F_5 of the proof of Lemma 10 $\leq (n+m)(\Delta + \Delta/2 + \cdots + 1) \leq 2(n+m)\Delta$;
- changes counted in F_3 in (7) and (8) of the proof of Lemma 10 $\leq 2 \cdot \Delta \cdot$ total number of contractions $\leq 2 \cdot \Delta \cdot (n+m-1)$.

These three quantities sum up to $4(n+m)\Delta$.

As there are at most $(n+m-1)$ contractions over all phases, the following Lemma follows directly from Lemma 9 and Lemma 10.

Lemma 12. There are $O(m \log m)$ augmentations and row-augmentations. Consequently there are $O(m \log m)$ Δ-phases in which an augmentation or a row augmentation takes place.

Lemma 13. There are $O(m \log m)$ Δ-phases.

Proof From Lemma 12 it follows that we only have to show that there are $O(m \log m)$ Δ-phases in which no flow changes and no contractions take place.

The following three cases are possible for the beginning of a Δ-phase without flow changes and contractions:

(1) There is some $j \in J'$ with $|\bar{b}_j| > \Delta/8(n+m)^2$ or some $i \in I$ with $\bar{a}_i > \Delta/8(n+m)^2$;
(2) (1) does not hold and all $x_{ij} \in T$ are zero;
(3) (1) does not hold and there is some nonzero $x_{ij} \in T$.

In all cases it is easy to see that there can be at most $O(\log m)$ Δ-phases without further flow changes and contractions until a contraction will take place. Hence there can be at most $O(m \log m)$ such phases.

As all other complexity issues are straightforward or very similar to arguments in [O] this completes our proof of the complexity bound in Theorem 7.

It remains to show the correctness of the algorithm. At the end of the algorithm all excesses are zero in the contracted graph. As in [O] one can successively blow up the contracted arcs one at a time in the reverse order in which they were contracted and one can show that at each step the flow of the blown up arc can be set to a nonnegative value such that the excesses of the nodes in the blown up graph are zero, too. The nonnegativity of these values follows from the limited changes of flow values as in Lemma 11. We omit the details here. As all excesses are zero in the final blown up tree we have a primal feasible solution on a tree which by construction is dual feasible. This completes the proof of the Theorem.

For having the possibility of studying the actual behavior of the strongly polynomial algorithm we wrote a code in C which is not "tuned" for fast performance but which has extensive protocol functions. Our hope is to get some insight from this in order to obtain some strongly polynomial algorithm which runs also fast in practice.

This source code can be obtained upon request from the author.

Jim Orlin pointed out to us that our algorithm applies without changes (also of the complexity bounds) to a generalization of the transportation problem in which we have "row nodes" with positive supplies and "column nodes" with positive or negative supplies. We may allow arcs between column nodes but row nodes are only contained in arcs which are directed to column nodes.

6. Two applications of the assignment and the semi-assignment problem

In this section we will describe an application of the semi-assignment problem to the classification of human chromosomes and an application of a constrained assignment problem to manpower planning.

Karyotyping is the process by which chromosomes in a dividing cell, suitably stained are identified and allocated to one of a number of groups. This is an important clinical process, since the identification of abnormalities in chromosomes of particular groups may be diagnostic of certain clinical syndromes.

Fig. 1. digital image of a metaphase cell

This process, if supported by a computer system is divided into the following steps:

1. Recording a microscopic image of a suitably prepared human cell in its metaphase.
2. Producing a digital image of the cell.
3. Obtaining a high quality image of the cell by using image processing methods (see Fig. 1).
4. Separating touching or overlapping chromosomes to obtain each chromosome as an individual image object.
5. Extracting "features" of every chromosome such as length, area, perimeter, bounding pattern information.
6. Assigning every chromosome to its karyotype (Fig.2)

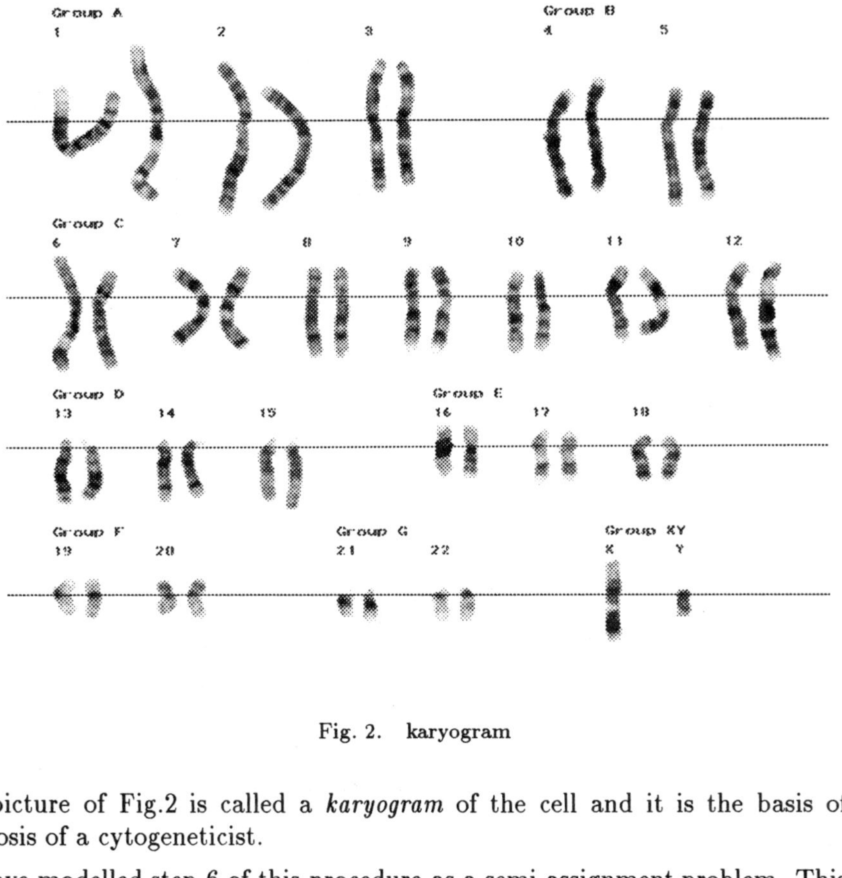

Fig. 2. karyogram

The picture of Fig.2 is called a *karyogram* of the cell and it is the basis of the diagnosis of a cytogeneticist.

We have modelled step 6 of this procedure as a semi-assignment problem. This will be described in the sequel.
In step 5, for every chromosome up to 30 characteristic features are computed.
These features can be divided into geometric information and information about the banding pattern which is generated in the staining process. The geometric features include area, length, size of the smallest enclosing rectangle, size of the enclosing rectangle of minimum width and the relative position of the centromere, an identation of the chromosome.
For the extraction of those features efficient algorithm from 2-dimensional compu-

tational geometry are being used [GP].
A sufficiently large set of cells serves as a *training set* for karyotyping. For the training set a correct classification is known by a careful analysis of a cytogeneticist. The features for the training set are computed.

Every human chromosome belongs to one of 24 types. Every class of a normal human cell contains two chromosomes for classes 1,...22. In a female cell there are two additional chromosomes in class 23 (the X-chromosomes) and in a male cell there is one chromosome in class 23 and one chromosome in class 24 (the Y-chromosome).

Based on the training data we compute for a given chromosome i and a given class j the *Mahalanobis-distance* d_{ij}. This quantity can be considered as the Euclidean distance between chromosome i and a "typical" chromosome of class j weighted by the inverse of the covariance matrix of the features of the training set.

In detail d_{ij} is defined as follows:

Let f_i be the feature vector of chromosome i and $\overline{f}_j$ the vector whose entries are the mean values of the features of the chromosomes of type j in the training set. The matrix V_j is an $(r \times r)$-matrix where r is the number of features. The diagonal entries of V_j are the variances and the off-diagonal entries are the covariances of the features of the training set.
Then the d_{ij} is defined by $d_{ij} := \sqrt{(f_i - \overline{f}_j)^T V_j^{-1}(f_i - \overline{f}_j)}$.

The aim of our model is to find an assignment of the chromosomes of a cell to their types such that the total sum of the Mahalanobis-distances is minimized.

If one knows beforehand that the cell is a normal female cell then this can be modelled as follows:

$$\min \sum_{i=1}^{46} \sum_{j=1}^{23} d_{ij} \text{ subject to}$$

$$\sum_{j=1}^{23} x_{ij} = 1 \quad , i = 1, \ldots, 46$$

$$\sum_{i=1}^{46} x_{ij} = 2 \quad , j = 1, \ldots, 23,$$

$$x_{ij} \in \{0, 1\}.$$

This is a semi-assignment problem which can be solved very efficiently.
Of course, in clinical practice the sex of the cell is not known in general. Also, the cell may be abnormal and its number of chromosomes could differ from 46.
For this purpose we have to set up a more complicated model. However, it will still be a semi-assignment problem.
Let $n \leq 49$ be the number of chromosomes in a cell.
We introduce an artificial chromosome $n + 1$ which we force to be assigned to class 23 or 24. If this chromosome is assigned to class 24, we have two free slots for class 23 (the case of the female cell) and if it is assigned to class 23, we have one free slot for class 24 and one for class 23 (the case of a male cell).
If $n > 46$, then only 46 chromosomes are assigned to classes $1, \ldots, 24$. The remaining chromosomes are assigned to an artificial type 25. In a subsequent stage (indepen-

dent of the semi-assignment problem) those extra chromosomes are assigned to the class from which they have minimum Mahalanobis distance.
The chromosomes $n+2,\ldots,50$ are artificial chromosomes which are assigned to arbitrary classes. They just serve the purpose of filling the free places to which no actual chromosomes are assigned.
Instead of using the Mahalanobis-distance d_{ij} itself as the cost function, it turned out that the best classification results are obtained by using its natural logarithm. We have no theoretical explanantion for this fact. However, by inspection of the data it turned out that one explanation may be that data errors which come from "bad" images are less punished by the logarithmic scaling of the d_{ij}.

Our semi-assignment model is as follows:

The costs are defined as:

$$c_{ij} := \ln d_{ij} \quad , i = 1, \ldots, n \quad j = 1, \ldots, 24;$$

$$c_{ij} := 0 \quad , i = n+1, \quad j = 23, 24;$$

$$c_{ij} := \inf \quad , i = n+1, \quad j = 1, \ldots, 22, j = 25;$$

$$c_{ij} := c \quad , i = n+2, \ldots, 50, j = 1, \ldots, 24;$$

$$c_{ij} := c \quad , i = 1, \ldots, 50 \ (i \neq n+1), j = 25.$$

Here inf is a very large number and c is a constant less than inf.
The objective function to be minimized is $\sum_{i=1}^{50} \sum_{j=1}^{25} c_{ij} x_{ij}$.

The constraints are:

$$\sum_{j=1}^{25} x_{ij} = 1 \quad , i = 1, \ldots, 50;$$

$$\sum_{i=1}^{50} x_{ij} = 2 \quad , j = 1, \ldots, 22;$$

$$\sum_{i=1}^{50} x_{ij} = 2 \quad , j = 23;$$

$$\sum_{i=1}^{50} x_{ij} = 1 \quad , j = 24;$$

$$\sum_{i=1}^{50} x_{ij} = 3 \quad , j = 25;$$

$$x_{ij} \in \{0, 1\}$$

This model yielded far better classification error rates than previous methods. It has been integrated into several karyotyping systems.

For a large test set of cells coming from clinical routine in the Rigshospitalet, Copenhagen, we obtained the following results:
From 2804 cells containing 128990 chromosomes we classified the chromosomes of 2041 cells without any errors. A total of 2599 chromosomes were misclassified in the remaining cells. This corresponds to an error rate of 2%. Per cell we make 0.9

errors on the average.
We have refined this method by using a second classifier with a different objective function. By automatically rejecting cells where the two classifiers disagree up to a certain degree, we obtain an error rate of 0.37% on the remaining set of cells [KMR]. This comes close to the error rate of 0.1% to 0.5% which has been reported for human operators.

Finally, we will describe some result from [H] which concerns the assignment problem with some further constraints. This result is useful for a system which supports the assignment of workers to jobs according to qualification and requirements. This task often occurs in the literature as a paradigm of an application of the assignment problem. However, there hardly exist any program systems which use this fact. One of the reasons is that in practice there are many additional facts that have to be respected. We have developed such a system [BK] in which we have incorporated algorithms for the assignment problem with additional constraints. This way, we can handle several problems occuring in manufacturing environments.
One of such problems is as follows: Assume some worker i can work at job $j(i)$ and another worker i' can work at job $j(i')$ different from $j(i)$. In certain cases it is required that either the workers i and i' are assigned to $j(i)$ and $j(i')$, respectively, or they are both not assigned to those jobs. This may be due to their capability of forming a team on those jobs.
In [Ab] and [AN] Aboudi and Nemhauser introduced a special case of this situation, which they called the couple constrained assignment problem (CCAP), and they proved that this problem is NP-hard. An alternative proof for the NP-completeness of CCAP which is a transformation from the 3-satisfiability problem is given in [H].

For NP-complete assignment problems with additional constraints (such as the asymmetric travelling salesman problem) it has proved useful to find facets of the polytope whose vertices represent the feasible (integer) solutions.
Even though a full knowledge of all facets is usually not available, a large subset of the set of facets can be useful for cutting plane approaches.
In [H] an exponential number of facets for CCAP was found. We will describe those facets in what follows.

CCAP is the linear assignment problem with the additional constraints

$$x_{2i-1,2i-1} - x_{2i,2i} = 0 \quad , \quad \forall i \in \{1, \ldots, m\} \tag{1}$$

where $m \leq \lfloor n/2 \rfloor$ is some positive integer. When formulated in the complete digraph model, CCAP is the problem of finding a maximum weight assignment in D_n, the complete digraph on n nodes, such that for every $i \in \{1, \ldots, m\}$ the nodes $2i-1$ and $2i$ either both have a loop or both have no loop.

In [AN] Aboudi and Nemhauser used a polyhedral approach to solve CCAP. They studied the CCAP-Polytope P, defined by

$$P = \text{conv}\{x \in \{0,1\}^{n^2} | x \text{ satisfies the assignment constraints and } (1)\},$$

and they identified a class of valid inequalities which they called SVI inequalities, defined by

$$\sum_{i\in K(I,J)\cup\tilde{K}(I,J)} x_{ij} + \sum_{i\in I}\sum_{j\in N\setminus J} x_{ij} \leq |I|$$

where $N = \{1,\ldots,n\}$, $\emptyset \neq I, J \subseteq N$, $|I| + |J| = n-1$,

$$\begin{aligned} K(I,J) &= \{2i-1\,|\{2i-1,2i\} \subseteq I\cap J, 1\leq i\leq m\}, \\ \tilde{K}(I,J) &= \{2i-1\,|\{2i-1,2i\} \subseteq (N\setminus I)\cap(N\setminus J), 1\leq i\leq m\} \end{aligned}$$

and $\tilde{K}(I,J) \neq \emptyset$ (in [AN] the SVI inequalities are written in a slightly different form which is equivalent to our representation.)

In [H] the SVI inequalities were strengthened to the following inequalities:

$$\sum_{i\in\overline{K}(I,J)} x_{ij} + \sum_{i\in I}\sum_{j\in J} x_{ij} \leq |I| \tag{2}$$

where $\emptyset \neq I, J \subseteq N, |J| = |I|+1, \{2i-1|\{2i-1,2i\} \subseteq J\setminus I, 1\leq i\leq m\} \neq \emptyset$ and $\bar{K}(I,J) = \{2i-1|\{2i-1,2i\} \subseteq I\oplus J, 1\leq i\leq m\}$.
(here $I\oplus J$ denotes the symmetric difference between I and J, i.e. $I\oplus J = (I\setminus J)\cup(J\setminus I)$.)

Furthermore, it is shown in [H] that the inequalities (2) define facets of P if $I\cap J = \emptyset$ or $I\cup J = N$.

Since $K(I,J)\cup\tilde{K}(I,J) \subseteq \bar{K}(I,N\setminus J)$ each SVI inequality given by I and J is dominated by the inequality (2) given by I and $N\setminus J$. If $K(I,J)\cup\tilde{K}(I,J)$ is even a proper subset of $\bar{K}(I,N\setminus J)$ then the SVI inequality is strictly dominated by the inequality (2). (Consider the example $n=5, m=2$ and $I=J=\{4,5\}$ that this can in fact happen).

References:

[A] Aboudi, R., *A constrained matching problem: a polyhedral approach*, PhD Thesis, School of Operations Research and Industrial Engineering, Cornell University, Ithaca, New York (1986).

[AN] Aboudi, R. and Nemhauser, G. L., *Some facets for an assignment problem with side constraints*, Oper. Res. 39 (1991), 244-250.

[B1] Balinski, M. L., *The Hirsch-conjecture for dual transportation polyhedra*, Math. of Oper. Res. 9 (1984), 629-633.

[B2] Balinski, M. L., *Signature methods for the assignment problem*, Oper. Res. 33 (1985), 527-536.

[BK] Bühner, R. and Kleinschmidt, P., *Rechnergestützter Personaleinsatz – dargestellt am Beispiel einer Fertigungsinsel-Organisation*, Die Betriebswirtschaft 49 (1989), 761-768.

[BT] Bertsekas, D. P. and Tseng, P., *The relax codes for linear minimum cost network flow problems*, Annals of Oper. Res. 13 (1988), 125-190.

[DO] Deza, M. and Onn, S., *On lattice-free polytopes*, (1993) submitted for publication.

[EKK] Emeličev, V. A., Kravcov, M. K. and Kračkowskii, A. P., *On some classes of transportation polytopes*, Soviet Math. Dokl. 19 (1978), 889-893.

[FT] Fredman, M. L. and Tarjan, R. E., *Fibonacci heaps and their uses in improved network optimization algorithms*, J. Assoc. Comput. Mach. 34 (1987), 596-615.

[GP] Granum, E. and Piper, J., *On fully automatic feature measurement for bounded chromosome classification*, Cytometry 10 (1989), 242-255.

[H] Hefner, A., *A polyhedral approach to constrained matching problems*, manuscript (1993).

[HR] Hung, M. S. and Rom, W. O., *Solving the assignment problem by relaxation*, Oper. Res. 28 (1980), 969-982.

[Ka] Kalai, G., *A subexponential randomized simplex algorithm*, in: Proc. 24th ACM Symposium on the Theory of Computing (STOC) (1992).

[KaKl] Kalai, G. and Kleitman, D. J., *A quasi-polynomial bound for the diameter of graphs and polyhedra*, Bull. Amer. Math. Soc. 26 (1992) 315-316.

[KK] Klee, V. L. and Kleinschmidt, P., *The d-step conjecture and its relatives*, Math. of Oper. Res. 12 (1987), 718-755.

[KLS] Kleinschmidt, P., Lee, C. W. and Schannath, H., *Transportation problems which can be solved by the use of Hirsch-paths for the dual problems*, Math. Progr. 37 (1987), 153-186.

[KMR] Kleinschmidt, P., Mitterrreiter, I. and Rank, C., *A hybrid method for automatic chromosome karyotyping*, Pattern Recognition Letters, to appear.

[KO] Kleinschmidt, P. and Onn, S., *On the diameter of convex polytopes*, Discrete Math. 102 (1992), 75-77.

[KS] Kleinschmidt, P. and Schannath, H., *A strongly polynomial algorithm for the transportation problem*, submitted to Math. Progr.

[KW] Klee, V. L. and Walkup, D. W., *The d-step conjecture for polyhedra of dimension* $d < 6$, Acta Math. 133 (1967), 53-78.

[L] Larman, D. G., *Paths on polytopes*, Proc. London Math. Soc. 20 (1970), 161-178.

[M] Matoušek, J., *Lower bounds for a subexponential optimization algorithm*, Preprint B 92-15, FU Berlin (1992).

[MSW] Matoušek, J., Sharir, M. and Welzl, E., *A subexponetial bound for linear programming*, in: Proc. Eight Annual ACM Symp. Computational Geometry (Berlin 1992), ACM Press 1992, 1-8.

[N] Naddef, D., *The Hirsch- conjecture is true for* $0-1$ *polytopes*, Math. Progr. 45 (1989), 109-110.

[O] Orlin, J. B., *A faster strongly polynomial minimum cost flow algorithm*, Proc. 20th ACM Symp. on the Theory of Comp. (1988), 377-387.

[PR] Padberg, M. W. and Rao, M. R., *The travelling salesman problem and a class of polyhedra of diameter two*, Math. Progr. 7 (1974), 32-45.

[S] Schannath, H., *Polynomiale und streng polynomiale Algorithmen für Netzwerkflußprobleme*, Ph.D. Thesis, Universität Passau (1991).

CONTRIBUTED PROBLEMS

JONATHAN SCHAER,
Department of Mathematics and Statistics
The University of Calgary
2500 University Dr. N.W.
Calgary, Alberta T2N 1N4, Canada

1. (Thomas Banchoff) What is the minimal number of vertices of a 4-polytope whose facets are combinatorial cubes, and that has more facets than vertices?

2. (Robert Dawson) Do there exist five simplices in $\mathbb{R}^3$ with mutually disjoint interiors such that each has, on every facet, face-to-face contact with another? This is related to the more general questions that J. Zaks asked (1978): Is it true that every non-empty neighborly family of convex polytopes contains a "free facet" (that is, a facet of one of its members, having interior which meets no other member)?

3. (Jonathan Fine) Given a matroid M with ground set $\{1, 2, ..., n\}$, we define a *matroid polytope* in $\mathbb{R}^n$ to be the convex hull of the points $\{e_{i_1} + e_{i_2} + ... + e_{i_r} \mid \{i_1, i_2, ..., i_r\}$ is a basis of $M\}$.
   What is the meaning, if any, of the h–vector of a matroid polytope?

4. (Jonathan Fine) There are formulas for the number of integer lattice points in a convex polytope (higher-dimensional analogues of Pick's theorem). The coefficients in Pick's theorem have a nice interpretation.
   Are there similar good interpretations of the coefficients in these higher dimensional formulas?
   (See J.E. Pommersheim: Toric varieties, lattice points, and Dedekind sums, *Math. Annalen*, **295** (1993), 1–24, for related work.)

The next three questions are related to the '*homology*' $H\triangle$ of a polytope $\triangle$. This homology theory, discussed in workshops at the ASI, is a reworking of ideas of Danilov (V.I. Danilov: The geometry of toric varieties, *Russ. Math. Survey*, **33**:2 (1978), 97–154). It is dual to the theory of stress, which is a cohomology theory. If $\triangle$ is a simple polytope, then this homology coincides with the homology of the associated toric variety, with real coefficients.

T. Bisztriczky et al. (eds.), POLYTOPES: Abstract, Convex and Computational, 493–497.

5. (Jonathan Fine) In Euclidean space, translations generate the relations which define $H\triangle$ for a convex polytope. This does not carry over to hyperbolic space which has no coherent concept of translation. However, stress on Euclidean polytopes gives a cohomology theory.
Is there an analogous theory of stress for hyperbolic polytopes, which is interesting? (What does "interesting" mean? For instance, according to Thurston many manifolds are covered by hyperbolic space, with a convex polytope as fundamental region. A good theory of stress may help study these polytopes.)

6. (Jonathan Fine) P. McMullen (On simple polytopes, *Inventiones Math.*, **113** (1993), 419–444) gives Riemann-Hodge-Minkowski inequalities for $H\triangle$ when $\triangle$ is a simple polytope. So far there is yet no definition for $H\triangle$ for a general convex polytope $\triangle$. This problem is really a warning that constructing $H\triangle$ in this general setting seems to require many such inequalities (which yet have to be defined or proved).

7. (Jonathan Fine) Write a computer program to calculate, and to allow calculations within, $H\triangle$, even for simple $\triangle$.

8. (Peter Gritzmann and Victor Klee) For a 3-dimensional convex body K, let $\alpha(K)$ denote the radius of the largest circular disk contained in K, let $\beta(K)$ denote the radius of the smallest circle through which K can be passed, and let $\gamma(K)$ denote the radius of the smallest circular cylinder along which K can be translated (that is, $\gamma(K)$ is the minimum of the circumradii of all orthogonal projections of K). Then $\alpha(K) \leq \beta(K) \leq \gamma(K)$, and there are examples of strict inequalities.
How can $\alpha(K)$, $\beta(K)$, and $\gamma(K)$ be computed efficiently when K is a convex polytope? (The answer is not clear even when K is a tetrahedron.)

9. (Peter Gruber) A lattice covering of E^d by simplices consists of the translates of a simplex by all vectors of a lattice, whose union covers E^d. It is known that for $d = 2$ the density of any lattice covering by triangles is at least $\frac{3}{2}$.
Is the density of a lattice covering of E^d by simplices always at least $\dfrac{d+1}{2}$?

10. (Wlodzimierz Kuperberg) Characterize all convex polyhedra P in $\mathbb{R}^d$ which permit a space tiling whose every member is a translate of P or of $-P$. (For $d = 2$ they are precisely the p-hexagons, that is, convex hexagons having a pair of opposite, parallel equal-length sides, including degenerate cases like triangles.)

11. (Wlodzimierz Kuperberg) Let C be a subset of $\mathbb{R}^3$. A convex body K is *trapped in* C if no point of C is in the interior of K, and it is impossible to move K by continuous rigid motion far away from its original position, so that during the entire motion the interior of K avoids C. A *tetrahedral cage* is the union of the edges of a tetrahedron.
Can every convex body in $\mathbb{R}^3$ be trapped in a tetrahedral cage?

12. (Marek Lassak) The thickness of a convex body $C \subseteq \mathbb{E}^d$ is the minimal distance between two parallel supporting hyperplanes of C. C is called *reduced* if no convex body properly contained in C has thickness equal to the thickness of C. Every regular n-gon in $\mathbb{E}^2$ is reduced if n is odd. On the other hand, the regular d-simplex is not reduced for $d \geq 3$.
Do there exist reduced d-polytopes for $d \geq 3$?

13. (Marek Lassak) Let $C \subseteq \mathbb{E}^d$ be a convex body, and $d(a,b)$ be the Euclidean distance of a and b. For $a \neq b$, define the *C-distance* as $\dfrac{2d(a,b)}{d(a',b')}$, where $\mathrm{conv}\{a',b'\}$ is the longest chord of C parallel to $\mathrm{conv}\{a,b\}$. Note that the C-distance is the usual norm associated to C, if C is centrally symmetric. A simplex is called *C-regular* if all pairs of vertices have the same C-distance.
What is the maximal C-distance between the pairs of vertices of a C-regular d-simplex contained in the convex body $C \subseteq \mathbb{E}^d$?
For what convex body C in $\mathbb{E}^d$ is this C–largest C–regular simplex smallest? For $d = 2$ it is conjectured to be the regular pentagon (or the regular octagon, if C is restricted to centrally symmetric C).

14. (Lauren Rose) Let P be a pure polyhedral complex of dimension d, that is, all maximal faces of P have dimension d. Assume further that each $(d-1)$-face lies either on one or on two d-faces, so that P is a *pseudomanifold* with or without boundary. The graph $G(P)$ is defined on the set of d-faces of P: Two d-faces form an edge if and only if they meet in a $(d-1)$-face.
Which graphs are $G(P)$ for some P?
Which graphs are $G(\Delta)$ for some simplicial complex Δ?
In particular, given a pseudomanifold P, does there exist a locally Cohen-Macaulay simplicial complex Δ with $G(\Delta) = G(P)$?

15. (Günther Rote, attributed to Bernard Charelle) Consider some arrangement of points in the unit square such that
(i) the distance of any two points is at least δ, and
(ii) the maximal area of intersection of a parallel strip and the unit square that does not contain any of the points is ϵ.
Determine the asymptotic dependence of the smallest possible ϵ on δ, as $\delta \to 0$. It is conjectured that $\dfrac{\epsilon}{\delta^2} \to \infty$.
Note that the number of points in an optimal arrangement for a given δ is $0\left(\frac{1}{\delta^2}\right)$, and, say, always $\epsilon > \frac{1}{4}\,\delta^2$. A selection of n random points results in

$\epsilon = 0\left(\frac{\log n}{n}\right)$ with high probability. One can probably also achieve $\delta \geq \frac{const}{\sqrt{n}}$ and still get $\epsilon = 0\left(\frac{\log n}{n}\right)$ by a random process.

16. (Günther Rote, attributed to Ken Clarkson) Given a real function $f(x, y)$ of two variables, and n points (x_i, y_i) in the plane. Consider all triangulations T of the convex hull C of the points, with the given points as vertices. For any such T, let $f^T(x, y)$ be the continuous approximation of f which agrees with f at the given points and is linear on every triangle of T.
For which triangulation T is the approximation error $\sup_{(x,y)\in C} \left|f^T(x, y) - f(x, y)\right|$ smallest?
Note: If f is convex, the optimal T can be obtained by taking the lower convex hull of the points $(x_i, y_i,\ f(x_i, y_i))$ and projecting it on the x-y-plane. In particular, for $f(x, y) = x^2 + y^2$, T is the Delone triangulation of the point set. But if $f(x) = x \cdot y$, for instance, can the optimal triangulation be computed in polynomial time? Are there any structural properties characterizing an optimal triangulation?

17. (Günther Rote) Let P be a 3-dimensional polytope each of whose facets is tangent to the unit sphere (a circumscribed polytope). For any edge e, consider the point x where one of the facets containing e touches the unit sphere, and denote by ω_e the angle with vertex x whose rays pass through the endpoints of e.
Prove that $\sum_e \omega_e \geq 2\pi$ where the sum is taken over all edges of a closed cycle of edges, and show that equality holds only if the cycle forms the boundary of a face.
Remark: It has been shown by Rivin, Hodgson and Smith that the existence of numbers $\omega_e \in (0, \pi)$ satisfying the above inequalities and equations is necessary and sufficient for a combinatorial type of polytope to be circumscribable. The above assertion would provide a direct proof of the necessity of the conditions.

18. (Rolf Schneider) Let P be a possibly non-convex polyhedron in $\mathbb{E}^3$. Denote by A the surface area of P, by l_i and α_i, respectively, the length of the ith edge e_i of P and the dihedral angle at e_i, and set $M = \sum l_i \left|\pi - \alpha_i\right|$. Is then $M^2 > 4\pi A$? According to Minkowski's classical theorem, the inequality holds if P is convex.

19. (Richard Stanley) A convex polytope P is *totally self-dual* (t.s.d.) if every section is self dual, that is, every interval of the face lattice $L(P)$ is a self-dual poset. For example a segment, any polygon, and various 3-polytopes are totally self-dual. Note that the free join (as complexes) of t.s.d. polytopes is also t.s.d.
The main question is whether there exist other types of t.s.d. polytopes of dimension more than 3. What is the dimension of the space spanned by the f-vectors of t.s.d. polytopes?
More generally, in the problems above, one may replace total self-duality with the property that every section has a symmetric f-vector $(f_0, f_1, ..., f_k)$, i.e. $f_i = f_{k-i}$. Similarly, what about Eulerian posets in this context?

20. (Walter Whiteley) A polyhedron (possibly non-convex and with self-intersection) in $\mathbb{R}^3$ is called *acoptic* by Grünbaum if it can be realized as an embedding in $\mathbb{R}^3$ with each face as a polygon (possibly non-convex) without self-intersection. The problem is to characterize acoptic polyhedra. In particular, characterize the 3–connected planar graphs with acoptic realization.

21. (Walter Whiteley, attributed to Stokker) Two polyhedra in $\mathbb{R}^3$ of the same combinatorial type are *isogonal* if they have the same dihedral angles on corresponding edges, and they have the same facial angles on corresponding face-vertex incidences. For instance, congruent or parallel polyhedra are isogonal.
Are there two convex polyhedra of the same combinatorial type with corresponding dihedral angles equal, but some pair of corresponding facial angles distinct (so that the polyhedra are not isogonal)?
There are polyhedra P and Q, P convex and Q non-convex, with the same dihedral angles but different facial angles.

22. (Walter Whiteley) A polyhedron P in $\mathbb{R}^3$ is *ϵ edge deformable* if for some $\epsilon > 0$ and for any small change in the edgelengths (the sum of the squares of the changes $< \epsilon$), there exists a polyhedron Q with the changed edgelengths, of the same combinatorial type as P.
Conjecture: *For every combinatorial type of polyhedron some (almost all) realizations are ϵ edge deformable for sufficiently small ϵ.*
The conjecture is true for all convex triangulated spheres (by a theorem of Dehn); it is also true for all combinatorial types with less than four non-triangular faces; it is false for the regular cube, but apparently true for all other combinatorial cubes.
This conjecture would imply a related conjecture (Whiteley and Connelly): *Every combinatorial type of convex polyhedron can be realized in $\mathbb{R}^3$ with rational edgelengths.*

I wish to thank Laura Anderson, Károly Bőröczky jr. and Peter Brass for collecting these problems during the Problem Sessions.

THREE PROBLEMS ABOUT 4-POLYTOPES

GÜNTER M. ZIEGLER
Konrad-Zuse-Zentrum für
Informationstechnik Berlin ZIB
Heilbronner Str. 10
D-10711 Berlin, Germany

To some extent, we can claim to "understand" 3-dimensional polytopes. In fact, Steinitz' Theorem

– "the combinatorial types of 3-polytopes are given by the simple, 3-connected, planar graphs"
  (Steinitz, see Steinitz & Rademacher [12])

reduces much of the geometry of 3-polytopes to entirely combinatorial questions. Its powerful extensions answer basic questions about representing combinatorial types by actual 3-dimensional polytopes:

– "every 3-polytope can be realized with rational vertex coordinates"
  (a trivial consequence of the inductive proof for Steinitz' theorem),
– "every combinatorial type of 3-polytopes can be realized with the shape of one facet (2-face) arbitrarily prescribed"
  (a theorem obtained by subtle adaption of the proof, by Barnette & Grünbaum [2]),
– "the space of all realizations of a convex 3-polytope, up to affine equivalence, is contractible, and thus in particular connected"
  (this is what Steinitz actually proved, see [12]).

For high-dimensional polytopes, there seems to be no simple analogue to Steinitz' theorem, and analogues of all the extensions mentioned above are known to fail. In this note we wish to point out, however, that satisfactory answers for 4-polytopes are not available, very basic problems are open. In fact, systematic statements are not available for polytopes in any fixed dimension d.

This is due to our methods to construct and analyze high-dimensional polytopes: affine Gale diagrams and the Lawrence construction. If they are employed to transfer results about planar configurations to polytopes, the polytopes produced will be very high-dimensional. (A method of Sturmfels [14] that would produce 6-polytopes is in error.)

In the following, we briefly sketch the current situation for the 4-dimensional analogues of the three extensions mentioned above. Via Schlegel diagrams, 4-polytopes can be visualized by 3-dimensional polytopal complexes. Thus there may be ways to attack the problems below by direct geometric methods; for example, explicit constructions of Schlegel diagrams in $\mathbb{R}^3$.

However, we have to carefully distinguish 3-diagrams (polytopal complexes that "look like Schlegel diagrams") from actual Schlegel diagrams that can be lifted to

T. Bisztriczky et al. (eds.), POLYTOPES: Abstract, Convex and Computational, 499–502.

4-polytopes. All the questions we mention may (and should) be posed independently for 3-diagrams, and for 2-dimensional polytopal complexes in $\mathbb{R}^3$.

For all the basic terminology, basic constructions, examples and further references, we refer to Grünbaum's fundamental volume [9], and also to [15]. In particular:

- [9, Chap. 13], [15, Chap. 4] for Steinitz' Theorem,
- [9, Sect. 3.3], [15, Chap. 5] for Schlegel diagrams and d-diagrams,
- [9, Chap. 5], [15, Chap. 6] for Gale diagrams,
- [3], [4, Sect. 9.3], [15, Sect. 6.6] for the Lawrence construction, and
- [11], [7, Sect. 6.3], [4, Sect. 8.6] for Mnëv's Universality Theorem.

1. Prescribing the shape of a 2-face

For 4-polytopes, the shape of a facet (that is, a 3-face) cannot be prescribed. An example with a minimal number of vertices (8 vertices, 14 facets) to that effect was constructed by Kleinschmidt [10]. Kleinschmidt's example has one octahedron facet, all other faces are simplices. An example with 8 vertices and only 9 facets (one octahedron, four square pyramids and four tetrahedra) will be given in [15, Ex. (6.12)], improving on an example with 11 facets in the preprint version.

Similarly, an example with a minimal number of facets (10 vertices, 7 facets) was independently constructed and analyzed by Barnette as a Schlegel diagram [1], and by Sturmfels via the affine Gale diagram [13]. They both studied the same example, namely, the prism over a square pyramid.

However, one is tempted to ask whether "prescribing a facet" is the right high-dimensional version of the Grünbaum-Barnette Theorem. Perhaps one can at least prescribe a 2-face?

It turns out that even this is impossible for $d \geq 5$. In fact, in [15, Sect. 6.5(c)], we construct (the affine Gale diagram of the polar of) a 5-polytope (12 vertices, 10 facets) with a hexagon 2-face that cannot be arbitrarily prescribed.

This leaves us with the question:

- for every 4-dimensional polytope, can one prescribe the shape of one 2-face?

It would be very surprising if the anwer was "yes". However, Gale diagrams seem to be the wrong tool to construct counterexamples. Perhaps one can construct explicit Schlegel diagrams, in the spirit of [1]?

2. Non-rational 4-polytopes

A polytope P is *rational* if there is a combinatorially equivalent polytope P' such that all the vertices of P' have rational coordinates. Klee had asked a long time ago [9, p. 92] whether every polytope is rational.

The first non-rational polytope was an 8-polytope with 12 vertices, constructed (in terms of its Gale diagram) by Perles, see [9, p. 94], [15, Sect. 6.5(a)]. This is still the smallest known example. In fact, d-polytopes with at most $d+3$ vertices are easily represented with rational coordinates (they have 1-dimensional affine Gale diagrams). Moreover, it can probably be shown that any non-rational d-polytope with $d+4$ vertices must have $d \geq 8$, using the fact (which follows from [8]) that every planar configuration of 8 points has a rational realization.

However, what about polytopes with more vertices and facets? It would be very surprising if one could show that every 4-polytope is rational. On the other hand, no non-rational polytope with $d < 8$ is known. We formulate the problem for $d = 4$, since this is the first open case, and since it might be within the reach of a direct geometric construction (via Schlegel diagrams):

– is every 4-dimensional polytope rational?

The same question is open for 3-diagrams, which are polytopal complexes in $\mathbb{R}^3$. It is even still open for 2-dimensional polyhedral complexes in $\mathbb{R}^3$ (cf. [9, p. 93]). This might be a good point to start geometric investigations.

3. Universality of realization spaces

Let P be a d-polytope P with vertex set $\{v_1, v_2, \ldots, v_n\}$, and for simplicity, assume that the vertices $v_1, v_2, \ldots, v_i$ affinely span a face of dimension $i-1$, for $i \leq d+1$.

The *realization space* $\mathcal{R}(P)$ — the set of coordinatizations of P, up to affine equivalence — can then be represented by a set of all $(d \times n)$-matrices $X = (\mathbf{x}_1, \ldots, \mathbf{x}_n) = (x_{ij})$ such that the first $d+1$ columns are fixed (for example, $x_{ij} = \delta_{ij}$ for $j \leq d+1$), and such that the convex hull of the columns of X is combinatorially isomorphic to P via $\mathbf{x}_j \longmapsto v_j$.

It is not hard to see that the realization space $\mathcal{R}(P)$ is a semi-algebraic variety: a subset of $\mathbb{R}^{d \times n}$ that can be defined by polynomial equations and strict inequalities. In particular, it is a topological space with interesting structure. The realization space is connected if and only if every coordinatization of P can be "deformed" into any other coordinatization in such a way that the combinatorial type is preserved during the deformation. The *isotopy problem* for polytopes asks whether this is always the case.

Steinitz' theorem (as above) states that the realization space is contractible, and hence connected, for every 3-polytope. In contrast to this, we know that this fails badly for high-dimensional polytopes. In fact, from Mnëv's Universality Theorem together with the Lawrence construction we know that the realization spaces of polytopes can have the homotopy types of arbitrary finite simplicial complexes. This is true even if we restrict to the case of simplicial polytopes [7, Sect. 6.2]. However, the polytopes which prove this (derived from planar point configurations via the Lawrence construction) are high-dimensional, by construction.

Still, we know one example of a simplicial 4-polytope with disconnected realization space, realizing Kleinschmidt's sphere from [5] with 10 vertices and 28 facets. It has a combinatorial symmetry that cannot be realized. Mnëv observed [11] that by general principles (Smith Theory), this implies that the realization space cannot be contractible. Bokowski & Guedes de Oliveira [6], [7, Sect. 6.2] found that, indeed, it is disconnected.

One can connect several copies of this polytope to get 4-polytopes whose realization space has 2^k connected components. However, it is not proved that the realization space can be arbitrarily complicated (for example, have the homotopy type of a circle). We certainly think that it can — but a proof method (possibly deriving this from Mnëv's Universality Theorem for point configurations) is not currently available. Thus, we ask whether 4-polytopes are universal:

– for every simplicial complex Δ, is there a 4-polytope whose realization space is homotopy equivalent to Δ?

More precisely, for every semi-algebraic variety one could ask for a polytope whose realization space is stably equivalent to it. A positive answer is not known for d-polytopes in any fixed dimension d.

References

1. Barnette, D.W.: *Two "simple" 3-spheres*, Discrete Math. **67** (1987), 97-99.
2. Barnette, D.W. and Grünbaum, B.: *Preassigning the shape of a face*, Pacific J. Math. **32** (1970), 299-302.
3. Billera, L.J. and Spellman Munson, B.: *Polarity and inner products in oriented matroids*, European J. Combinatorics **5** (1984), 293-308.
4. Björner, A., Las Vergnas, M., Sturmfels, B., White, N. and Ziegler, G.M.: *Oriented Matroids*, Encyclopedia of Mathematics **46**, Cambridge University Press, Cambridge 1993.
5. Bokowski, J., Ewald, G. and Kleinschmidt, P.: *On combinatorial and affine automorphisms of polytopes*, Israel J. Math. **47** (1984), 123-130.
6. Bokowski, J. and Guedes de Oliveira, A.: *Simplicial convex 4-polytopes do not have the isotopy property*, Portugaliae Mathematica **47** (1990), 309-318.
7. Bokowski, J. and Sturmfels, B.: *Computational Synthetic Geometry*, Lecture Notes in Mathematics **1355** (1989), Springer-Verlag Berlin Heidelberg.
8. Goodman, J.E. and Pollack, R.: *Proof of Grünbaum's conjecture on the stretchability of certain arrangements of pseudolines*, J. Combinatorial Theory Ser. A **29** (1980), 385-390.
9. Grünbaum, B.: *Convex Polytopes*, Interscience, London 1967; revised edition (V. Klee, P. Kleinschmidt, eds.), Springer-Verlag, in preparation.
10. Kleinschmidt, P.: *On facets with non-arbitrary shapes*, Pacific J. Math. **65** (1976), 97-101.
11. Mnëv, N.E.: *The universality theorems on the classification problem of configuration varieties and convex polytopes varieties*, in: "Topology and Geometry – Rohlin Seminar" (O. Ya. Viro, ed.), Lecture Notes in Mathematics **1346**, Springer-Verlag, Berlin Heidelberg 1988, 527-544.
12. Steinitz, E. and Rademacher, H.: *Vorlesungen über die Theorie der Polyeder*, Springer-Verlag, Berlin 1934; reprint, Springer-Verlag 1976.
13. Sturmfels, B.: *Boundary complexes of convex polytopes cannot be characterized locally*, J. London Math. Soc. **35** (1987), 314-326.
14. Sturmfels, B.: *Some applications of affine Gale diagrams to polytopes with few vertices*, SIAM J. Discrete Math. **1** (1988), 121-133.
15. Ziegler, G.M.: *Lectures on Polytopes*, Notes of a course at TU Berlin 1992/93; Technical Report TR 93-6, ZIB Berlin, June 1993, 196 pages; book in preparation, (Springer-Verlag, New York, 1994).

Index

R

S

T

U

V

W

Z